5th International Conference on Accelerated Carbonation for Environmental and Material Engineering 2015

New York, New York, USA
21-24 June 2015

ISBN: 978-1-5108-1549-0

Printed from e-media with permission by:

Curran Associates, Inc.
57 Morehouse Lane
Red Hook, NY 12571

Some format issues inherent in the e-media version may also appear in this print version.

Printed by Curran Associates, Inc. (2015)

For permission requests, please contact AIChE
at the address below.

AIChE
120 Wall Street, FL 23
New York, NY 10005-4020

Phone: (800) 242-4363
Fax: (203) 775-5177

www.aiche.org

Additional copies of this publication are available from:

Curran Associates, Inc.
57 Morehouse Lane
Red Hook, NY 12571 USA
Phone: 845-758-0400
Fax: 845-758-2634
Email: curran@proceedings.com
Web: www.proceedings.com

TABLE OF CONTENTS

Carbon Upcycling to Improve Ready-Mixed Concrete

S. Monkman[1*], M. MacDonald[1], P. Sandberg[1], D. Hooton[2]

[1] CarbonCure Technologies, Dartmouth, Canada
[2] Department of Civil Engineering, University of Toronto, Toronto, Canada
*Corresponding author: smonkman@carboncure.com

Abstract

Environmental awareness has prompted research into the use of carbon dioxide as a feedstock in the production of concrete. Industrial trials have examined the use of CO_2 in ready-mix concrete production. A CO_2 gas injection system delivered carbon dioxide into a ready-mix truck during batching and mixing of concrete. The carbon dioxide was thereby absorbed into the concrete and formed stable carbonate reaction products. The impacts of carbon dioxide dose, injection method and time of injection were examined. The concrete slump and temperature was measured. Compressive strength and resistivity were measured at ages from 24 hours to 58 days. Strength improvements of up to 14% at 24 hours and 26% at 28 days were observed. The resistivity was not affected by the carbonation treatment. The strength increase is attributable to the reaction of carbon dioxide with the cement very early after hydration has started. A combination of isothermal calorimetry and microstructural analysis suggests that nano-scale carbonates form and seed the hydration and contribute to the development of a stronger microstructure. The strength benefit suggests that carbonation is a possible tool to provide acceleration that could allow the greater use of supplementary cementitious materials in concrete mixtures. The use of waste carbon dioxide in the process offers a means to upcycle the CO_2 as a beneficial additive to concrete that could be widely employed throughout the built environment.

Introduction

Cement production is the source of about 5% of the world's annual CO_2 emissions[1] and 80 to 90% of the embodied CO_2 in concrete production results from the use of Portland cement. The industry has recognized a number of approaches to reduce the emissions intensity of the cement produced and used including improvements in thermal and electrical efficiency, increased use of alternative fuels, increased cement clinker substitution and carbon capture and storage[2].

It is clear, however, that practical limits on the impacts of these measures mean that it will be difficult[3] to attain the industry goal of reducing emissions to 50% below 2006 levels by 2050 that is outlined in the IEA & WBCSD roadmap. Innovative approaches are being sought and are likely to be a part of a portfolio strategy. Future emissions improvements will likely be incremental. Thus, a range of further approaches will also have to be pursued.

One potential method is to upcycle carbon dioxide into concrete products. The main cement phases, tricalcium silicate and dicalcium silicate, are known to react with carbon dioxide in the presence of water to form calcium carbonate and calcium silicate hydrate gel[4] as shown in equations 1 and 2:

$$3CaO{\cdot}SiO_2 + (3\text{-}x)CO_2 + yH_2O \rightarrow (3\text{-}x)CaCO_3 + xCaO{\cdot}SiO_3{\cdot}yH_2O \quad (1)$$

$$2CaO{\cdot}SiO_2 + (2\text{-}x)CO_2 + yH_2O \rightarrow (2\text{-}x)CaCO_3 + xCaO{\cdot}SiO_3{\cdot}yH_2O \quad (2)$$

Further any calcium hydroxide present in the hydrating cement system will react with carbon dioxide, as show in equation 3:

$$Ca(OH)_2 + CO_2 \rightarrow CaCO_3 + H_2O \quad (3)$$

The reaction proceeds in the aqueous state when Ca^{2+} ions from the cementitious phases meet CO_3^{2-} ions from the applied gas. The carbonation reactions are exothermic with heats of reaction of 347 kJ/mol for tricalcium silicate, 184 kJ/mol for β-dicalcium silicate[5] and 74 kJ/mol for calcium hydroxide[6]. The addition of gaseous CO_2 to freshly mixing concrete results in the formation of in situ, nano-scale and homogenously distributed carbonate reaction products.

Investigation of the use of carbon dioxide as a feedstock in industrial concrete production has been undertaken with the aim of improving the concrete properties.

Experimental

Industrial experiments were conducted where carbon dioxide was delivered to ready-mixed concrete during its production. The trials took place at Metrix Ready Mix in North York, Ontario. Liquid CO_2 was metered and injected into the mixing drum of a ready mix truck at a specified flow rate over a fixed injection interval whereupon it reacted with the hydrating cement. The concrete was then subjected to assessment and testing.

Methods

The fresh concrete was assessed in terms of slump and temperature. The concrete was further cast into 4" x 8" (100 x 200 mm) cylinders for compressive strength testing. The heat evolution was assessed via a Calmetrix iCal 8000 isothermal calorimeter set at 22 °C (71 °F)

Concrete performance was assessed through compressive strength testing at 1, 3, 7, 28 days with one trial concluding with a test at 59 days and the other concluding with 56 and 91 day tests. The 1 and 28 day measurements were performed in triplicate, while the measurements at other ages were performed in duplicate. The specimens tested beyond 1 day were subjected to standard moist curing conditions prior to testing. All concrete specimens were prepared and tested in accordance with CSA A23.2. Bulk resistivity was assessed using a Merlin Mark III by Germann Instruments.

Industrial trials

The trials were conducted on trucks loaded with 4 m^3 (half full) concrete using a 25 MPa (3625 psi) mix design as summarised in Table 1.

Table 1. Ready-mixed concrete mix design, per m^3

Component	Quantity	Per m^3	Per yd^3	Fraction
Concrete Sand	Fine aggregate	959 kg	1617 kg	42%
20mm Stone	Coarse aggregate	1080 kg	1821 kg	49%
Slag	Binder	53 kg	89 kg	2%
Cement	Binder	212 kg	357 kg	9%
Water Reducer	Admixture	190 ml	6.4 oz	
Mix Water	Water	155 kg	261 kg	

The requested load of concrete was batched into the truck and received final water adjustments by the truck operator. A sample of uncarbonated (control) concrete was removed, a slump test was performed, and test specimens were cast. The truck was then subjected to three sequential doses of carbon dioxide with assessment of slump and casting of the treated concrete between each dose. The sequential dosing of carbon dioxide was pursued to determine an optimum CO_2 dosage for that mixture.

The CO_2 delivery was controlled by a mobile gas injection system. The injection proceeded for 30-90 seconds with an additional 90 -180 seconds of high speed mixing after the injection was completed. Total CO_2 dosages were 0.1%, 0.3% and 0.6% by weight of cement (bwc).

The first trial served as a proof-of-concept effort that a low dose injection could be integrated into a conventional ready-mix operation. The second trial was to increase confidence in the results and to investigate an alternate CO_2 injection mode wherein the gas was added during the initial batching/mixing phase.

Results and Discussion

First Trial

A complete overview of the conditions produced during the first trial day is presented in Table 2. The weather was 19 °C and sunny.

Table 2. First trial production variables, CO_2 settings and fresh results

Code	Condition	Discharge Time (min)	Slump mm (inches)	Mix temperature °C (°F)
1401	Control	23	89 (3.5)	23.9 (75.0)
1402	0.1% CO_2	45	76 (3.0)	-
1403	0.3% CO_2	59	76 (3.0)	25.6 (78.1)
1404	0.6% CO_2	74	51 (2.0)	26.5 (79.7)

The control concrete was sampled 23 minutes after mixing started. It was observed that with sequential doses of carbon dioxide the temperature of the mix was increasing and the slump of the mixture was decreasing. While a temperature increase would be expected due to the exothermic nature of the carbonation reaction, it cannot be concluded whether the observed effects were attributable wholly to the carbonation process, or further represented normal changes due to the passing of time between measurements, mixing, progress of hydration, increasing ambient temperature, etc.

Isothermal Calorimetry The evolution of heat from the fresh concrete was measured from 0 to 20 hours after mixing. The amount of heat released during early hydration can be used as a proxy for the development of mechanical properties (such as setting or compressive strength) at very early ages[7]. A summary of the power data for the four batches is presented in Figure 1. The power data is normalized and presented in arbitrary units for ease of comparison. The plot excludes any data either acquired during the initial hydration acceleration and deceleration period. Data is plotted starting just after the induction period has ended at 4 hours of age.

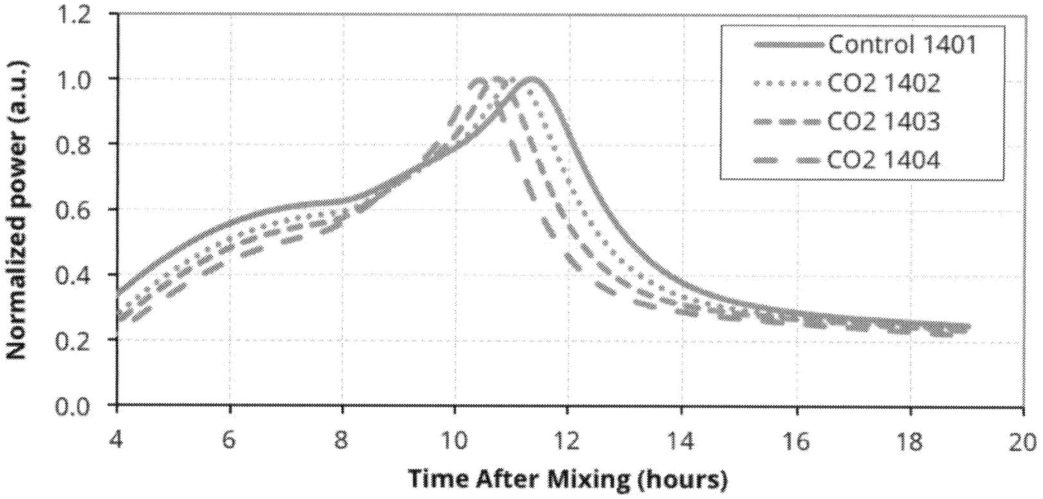

Figure 1 Normalized power during early hydration via isothermal calorimetry of batches produced in first ready-mix concrete trial

The acceleratory stage of hydration for each sample was underway by 4 hours. By 7 hours the heat evolution of the carbonated samples occurred at an increased rate (as noted in a shift to the left of the shape of the curves) where the effect was greater with the greater dose of carbon dioxide. Further, the heat release at the peak of the early hydration was found to increase in magnitude and be shifted to earlier times as the carbon dioxide dose increased.

The cumulative energy release relative to the control can be quantified at various ages and used as a metric of hydration progress. The energy released from each carbonated batch had matched that of the control after 9-11 hours. The low dose batch remained equivalent thereafter. The second dose of CO_2 improved to 12% better at 12 hours and finished at 6% more energy released through 20 hours. The highest dose was 13% higher at 12 hours and 4% higher through 20 hours.

Hardened properties The average compressive strength measured for each carbonated condition across the five test ages for the first trial is given in Figure 2. Each data point represents the average measured compressive strength in MPa with the relative comparison to the corresponding control value displayed as a percentage.

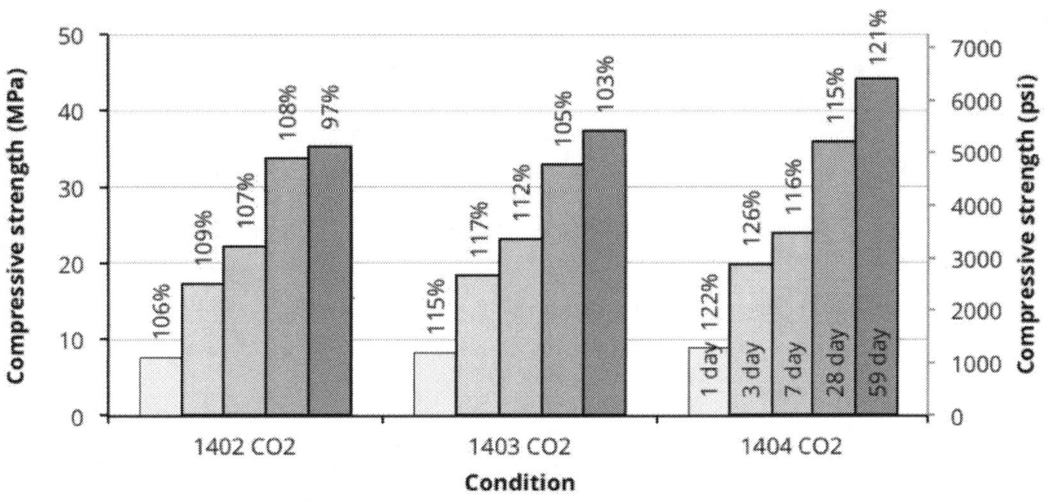

Figure 2 Average compressive strengths of all CO_2-treated specimens from trial 1 further expressed as a percentage of the control value

The compressive strength of the carbonated mixes increased for the higher two doses. The lowest dose of CO_2 had little impact on the compressive strength ranging between being 9% better than the control at 28 days but between 0 and 5% better at the other test ages. Batch 1403, with the middle dose of CO_2, was 11% stronger than the control at the two earlier ages, 10% at 7 days and 7% thereafter. The greatest benefit was seen in the final carbonated batch with 0.6% CO_2. A 19% increase in strength was observed at 1 day, 16% increase at 28 days and 26% increase at 59 days.

The bulk resistivity data is presented in Table 3. Higher resistivity values are indicative of lower permeability. The readings showed that the carbonated batches tended to be marginally lower than the control.

Table 3. Bulk Resistivity (Ω·m) for first trial test specimens at five different ages

		Test Age		
Code	**Condition**	**7 days**	**28 days**	**59 days**
1401	Control	33.6	78.7	106.5
1402	0.1% CO_2	25.5	74.1	112.0
1403	0.3% CO_2	24.7	67.1	97.7
1404	0.6% CO_2	24.1	70.2	101.8

Resistivity ranges that are relatable to the chloride ion penetrability levels described in ASTM C1202 confirm that the chloride penetrability risk for all samples was moderate at 28 days and low at 59 days. The resistivity data reassures that the carbon dioxide injected concrete is equivalent to the control concrete.

Second trial

An overview of the conditions produced during the second trial day is presented in. Table 4. The weather was again 19 °C (66 °F) and sunny. The trial repeated the sequential carbon dioxide dosing at the wash rack that had been explored in the first trial. A second truck was tested wherein the carbon dioxide was injected during the batching stage (batch 805).

Table 4. Ready-mixed concrete trial batches

Code	Condition	Discharge Time (min)	Slump mm (inches)	Mix temperature ° C (° F)
801	Control	21	89 (3.5)	22.4 (72.3)
802	0.1% CO_2	27	76 (3.0)	24.0 (75.2)
803	0.3% CO_2	43	64 (2.5)	24.7 (76.5)
804	0.6% CO_2	52	38 (1.5)	27.3 (81.1)
805	0.3% CO_2	0	76 (3.0)	23.3 (74.0)

Testing of the first truck showed, again, that the temperature of the mix increased and the slump of the mixture decreased with increasing doses of CO_2. Testing on the second truck showed that the slump was essentially equivalent to that of the control with a slightly increased temperature. The small temperature rise could be associated with the exothermic carbonation reaction but the second truck was batched/mixed 90 minutes after the first and thus the weather could have warmed the input materials.

Isothermal Calorimetry A summary of the power data for the five batches is presented in Figure 3.

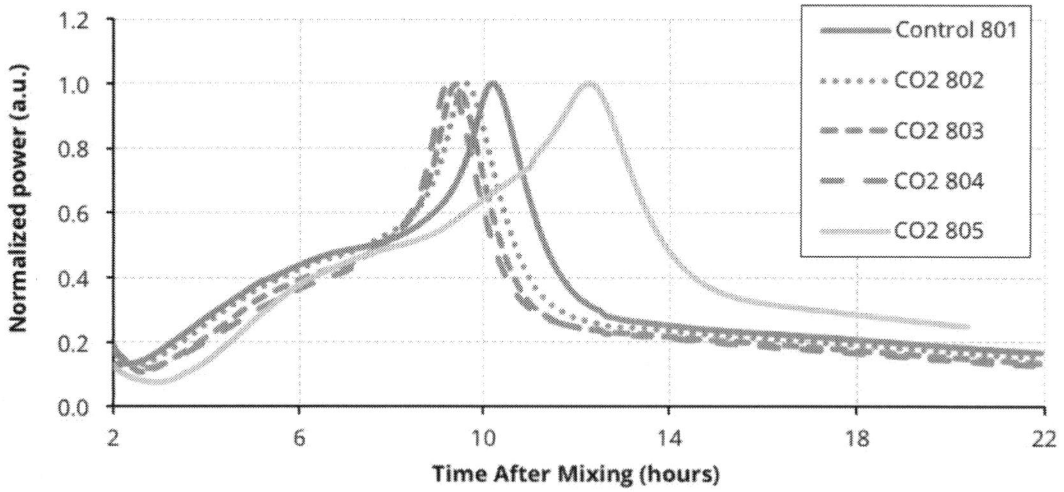

Figure 3 Normalized power during early hydration via isothermal calorimetry of batches from first truck in second ready-mix trial

The calorimetry of the concrete from first truck in the second trial was consistent with that of the first. The carbon dioxide was found to offer an accelerating effect. As observed in the first trial, the energy released from each carbonated batch had matched the control by 9-11 hours. The low dose of CO_2 improved to 13% better at 10 hours and finished at equivalent to the control. The second dose of CO_2 improved to

16% better at 10 hours and finished at equivalent to the control. The highest dose was 9% higher at 10 hours and 5% lower through 20 hours.

The batch that was dosed with CO_2 during batching (805) showed a calorimetry response that appeared to show some retardation relative to the control.

The extended induction period observed in batch 805 suggests that the carbonate reaction products contributed slowed dissolution of the silicates. Heat evolution lagged as low as 83% of the control at 11 hours. However, the hydration thereafter accelerated to release 5% more energy at 15 hours and 7% more at 20 hours. Once the initial product coating on the silicates has been overcome, the hydration accelerated due to increased product surface area providing nucleation sites.

Hardened Properties The average compressive strength measured for each carbonated condition across the six test ages is summarized for the September trial in Table 5. Both the average strength in MPa and the strength relative to the appropriate control are presented.

Table 5. Compressive strength (MPa, psi, relative to control) for concrete produced in second trial

		Test Age					
Code	Condition	1 days	3 days	7 days	28 days	58 days	91 days
0801	Control	8.0	14.8	19.2	30.8	32.8	36.8
0802	CO_2	8.0	15.8	20.4	32.0	27.9	37.8
0803	CO_2	8.6	16.6	21.4	33.9	37.9	38.9
0804	CO_2	8.7	16.1	22.1	32.8	36.4	39.6
0805	CO_2	9.2	18.6	23.2	35.5	38.7	42.4
0801	Control	1167	2142	2783	4465	4753	5343
0802	CO_2	1161	2290	2952	4643	4041	5488
0803	CO_2	1242	2401	3099	4916	5491	5644
0804	CO_2	1265	2333	3200	4759	5283	5739
0805	CO_2	1330	2694	3370	5155	5619	6146
0801	Control	100%	100%	100%	100%	100%	100%
0802	CO_2	99%	107%	106%	104%	85%	103%
0803	CO_2	106%	112%	111%	110%	116%	106%
0804	CO_2	108%	109%	115%	107%	111%	107%
0805	CO_2	114%	126%	121%	115%	118%	115%

As in the first trial, it was again seen that the smallest dose had the least effect. Likely the 0.10% bwc dose does not result in sufficient reaction product development to impact the concrete properties. The strengths ranged between 99% and 107% of the control across the test periods (excluding an anomalous reading at 56 days). The strength benefit from the 0.3% CO_2 bwc dose exceeded 6%, notably reaching 12% at 3 days and 16% at 56 days. The 0.6% dose in the first truck offered a strength benefit that ranged between 7% and 15%. The final batch showed the greatest promise given that the strength benefit exceeded 14% at all ages including a 26% benefit at 3 days and 15% at 91 days.

The bulk resistivity measurements were consistent with what was observed in the first trial insofar as the assessments of the carbonated batches were equivalent to the control. The chloride penetrability risk for all samples was assessed to be moderate at 28 days and low at 91 days.

Analysis

The formation of the carbonate reaction products is analogous to growing an in-situ nanoparticle $CaCO_3$ addition that would act as nucleation sites and impact later hydration product development. The calorimetric evidence parallels observations of ex-situ additions of nano-$CaCO_3$ to tricalcium silicate[8]. Model system work theorized that the nano-carbonates acted on the surface or in close proximity to calcium silicate particles. The in-situ carbonates could affect the stability of the Si-rich reaction product that covers the surface of the hydrating calcium silicates during the induction period and/or could act as heterogeneous nuclei for C-S-H growth after the induction period.

The batch 805 test showed that adding carbon dioxide during the batching resulted in retardation from 7 to 13 hours before causing an overall increase in early hydration. The CO_2 was delivered during the earliest exposure of cement to water and the carbonate reaction products likely formed on the cement particles. Slower initial hydration suggests that fewer active sites fuelling the growth of the hydration products but the later hydration increase may be due to the creation of nano-carbonate nuclei.

The carbon dioxide that was added to concrete at the wash rack was added 45-70 minutes after mixing started. After 45 min of hydration the aqueous phase of the concrete mix would be saturated in Ca^{2+} ions. Acceleration due to reaction product formation in solution at a distance from the cement particle surfaces is possible. Work with C-S-H additions to hydrating cement systems has suggested that the early hydration rate and total amount of early hydration could be increased though the creation of new nucleation sites away from the particle surfaces[9]. Such a mechanism is particularly relevant to the reactions at hand. Reflecting upon the chemical equations 1 and 2 previously presented, it is apparent that C-S-H gel formation is expected alongside the calcium carbonate development.

The amounts of carbon dioxide used were 0.6% by weight of cement or lower. This amount of CO_2 would correspond to $CaCO_3$ at 1.4% by weight. The small amount of CO_2 involved attests to the potency of the impact. In addition, in-situ production addresses one of the challenges of incorporating nano-$CaCO_3$ into cement systems, namely, achieving effective dispersion and fine distribution[10] which is conventionally addressed by nano-suspension through ultrasonication[11].

The two dosing approaches used represent three different potential industrial approaches. (1) Dosing the carbon dioxide during batching offers minimal interruption to the conventional batching and mixing process. Gas delivery equipment could be integrated into the standard ready mix production. (2) In cases where ready-mix concrete trucks pause after batching, such as at a wash rack, the carbon dioxide could be supplied concurrent with other normal production operations. (3) Finally, if an extended hydration time is found to have advantages, then the CO_2 could be dosed at the job site immediately prior to placing. It is possible that a small portable carbon dioxide injection system could be installed on trucks or at big projects.

One outcome of using a tool that can unlock a consistent strength benefit could be to sell a given concrete mix for at a price premium. A 15% strength improvement to a 35 MPa mix would allow for the production 40 MPa mix. Alternately, instead of maintaining the mix design and offering a strength-improved 40 MPa mix, the cement content could be reduced in order to deliver the same 35 MPa performance as before. The amounts of carbon dioxide used (and thereby absorbed) in these tests is small and possibly only offers a slight inherent environmental benefit but if carbon dioxide can be used as tool to reduce cement loadings there would be an direct and tangible reduction in the carbon footprint of the concrete so produced.

Conclusions

Industrial scale experiments assessed the viability of adding small doses of carbon dioxide to a ready-mix concrete production cycle in order develop performance benefits associated with the in-situ development of carbonate reaction products. Isothermal calorimetry indicated that hydration could be accelerated.

Compressive strength results show that a strength benefit of 20% at 3 days, 16% at 28 days and 26% at 58 days was achieved with an optimized dosage of carbon dioxide supplied at the wash rack. In practice this could potentially be a single optimized dose of carbon dioxide provided at the job site. Delivery of carbon dioxide into the concrete during batching resulted in retardation in the 7 to 13 hour timeframe but a small acceleration beyond 14 hours. The strength was 26% better than the control at 3 days, 15% better at 28 days and 18% at 56 days. Bulk resistivity, an indicator of permeability, was not changed by the carbonation. The difference in heat evolution between the dosing during batching vs the dosing at the wash rack hinted at possible differences in the reaction product formation.

The approach potentially can offer strength benefits or allow for cement reductions. Several avenues of investigation are being pursued to further characterize the mechanism of the reaction, understand variations between dosing approaches, and complete a broad durability assessment program.

Acknowledgements

The authors thank Metrix Ready Mix for supporting the work. Further, the assistance provided by University of Toronto graduate students Gita Charmchi and Soley Einarsdottir during the production and analysis phase was greatly appreciated. Research funding was received from Sustainable Development Technology Canada (SDTC) and the National Research Council's Industrial Research Assistance Program (IRAP).

References

1. Damtoft, J.S., Lukasik, J., Herfort, D., Sorrentino, D., Gartner, E.M. 2008. Sustainable Development and Climate Change Initiatives. *Cement and Concrete Research*, 38(2), 115–127.
2. IEA, 2009. *Cement Technology Roadmap: Carbon Emissions Reductions up to 2050.* OECD Publishing.
3. Barcelo, L., Kline, J., Walenta, G., Gartner, E. 2014. Cement and Carbon Emissions. *Materials and Structures*, 47(6), 1055–1065.
4. Berger, R.L., Young, J.F., Leung, K. 1972. Acceleration of Hydration of Calcium Silicates by Carbon-Dioxide Treatment. *Nature: Physical Science*, 240, 16–18.
5. Goodbrake, C.J., Young, J.F., Berger, R.L. 1979. Reaction of Beta-Dicalcium Silicate and Tricalcium Silicate with Carbon Dioxide and Water Vapor. *Journal of the American Ceramic Society*, 62(3-4), 168–171.
6. Moorehead, D.R., 1986. Cementation by the Carbonation of Hydrated Lime. *Cement and Concrete Research*, 16(5), 700–708.
7. Ge, Z.K., Wang, P.J., Sandberg, J., Ruiz, M. 2009. Characterization and Performance Prediction of Cement-Based Materials Using a Simple Isothermal Calorimeter. *Journal of Advanced Concrete Technology*, 7(3), 355–366.
8. Sato, T., Diallo, F. 2010. Seeding Effect of Nano-$CaCO_3$ on the Hydration of Tricalcium Silicate. *Transportation Research Record: Journal of the Transportation Research Board*, 2141(1), 61–67.
9. Thomas, J.J., Jennings, H.M., Chen, J.J. 2009. Influence of Nucleation Seeding on the Hydration Mechanisms of Tricalcium Silicate and Cement. *The Journal of Physical Chemistry C*, 113(11), 4327–34.
10. Kawashima, S., Seo, J.W.T., Corr, D., Hersam, M.C., Shah, S.P. 2014. Dispersion of $CaCO_3$ Nanoparticles by Sonication and Surfactant Treatment for Application in Fly Ash–cement Systems. *Materials and Structures*, 47(6), 1011–23.
11. Wang, X., Wang, K., Tanesi, J., Ardani, A. 2014. Effects of Nanomaterials on the Hydration Kinetics and Rheology of Portland Cement Pastes. *Advances in Civil Engineering Materials*, 3(2), 142–159.

Utilization of Carbonated BOF Slag as Partial Replacement of Aggregate in Cement Mortars

M. Bodor[1*], R.M. Santos[2], G. Cristea[1], M. Salman[3], Ö. Cizer[3], R.I. Iacobescu[4], Y.W. Chiang[5], K. Van Balen[3], M. Vlad[1] and T. Van Gerven[6]

[1] Department of Environmental Engineering and Metallurgical Technological Systems, Dunarea de Jos University of Galati, Galati, Romania
[2] School of Applied Chemical and Environmental Sciences, Sheridan Institute of Technology, Brampton, Ontario, Canada
[3] Department of Civil Engineering, KU Leuven, Leuven, Belgium
[4] Department of Materials Engineering, KU Leuven, Leuven, Belgium
[5] School of Engineering, University of Guelph, Guelph, Ontario, Canada
[6] Department of Chemical Engineering, KU Leuven, Leuven, Belgium
*Corresponding author: bodormarius@yahoo.com

Abstract
After direct mineral carbonation, a material rich in carbonates and with reduced quantities of free oxides is obtained. The aim of this work was to show that such materials can be used in the construction domain. Basic Oxygen Furnace (BOF) slag from the steelmaking process has been traditionally seen as unfit for bounded applications due to its propensity to swelling, resulting from hydration of its high free lime content. Here, BOF slag was crushed to suitable particle sizes, carbonated in an aqueous solution of carbonic acid, and utilized to replace 50% of natural sand aggregate in cement mortars. The mechanical and chemical properties of these mortars were compared to mortars containing non-carbonated slags, and a standard cement mortar as a reference. Tests were conducted to determine mortar flow and soundness, and cured mortar compressive strength and leaching tendencies. The results showed a satisfactory performance for all considered aspects (comparable with the reference) of the mortar sample containing 37.5 wt% (1.5 in 4 parts solids) carbonated BOF slag of <0.5 mm particle size.

Introduction

With the purpose of using steelmaking slags in the construction domain, and after much research conducted internationally, different standards have been implemented regarding the conditions that should be met by these industrial wastes before utilization. When intending to use slags as binders, the hydraulic activity of the material is very important.[1] However, with the use of BOF slag as a binder requiring intense grinding, and with the high metallic iron content of BOF slag, this processing step represents a high energy demand, and consequently a high cost. Lower processing costs can be achieved when using BOF as an aggregate, as the extent of particle size reduction (from monolith to millimeter-sized grains) can be reduced. For this application, the main parameters that must meet stringent specifications are the volume stability, referred to as soundness, the basicity, and the leaching of regulated toxic heavy metals and metalloids.[2] Wang et al.[3] developed a usability criterion based on the free lime content, the specific gravity and bulk relative gravity of the slag. Xie et al.[4] found that the failure mechanism in BOF dense-graded asphalt mixtures in moist environment is determined by the formation of portlandite ($Ca(OH)_2$) and calcium-silicate-hydrates (C-S-H), which leads to partial structural failure and facilitates water infiltration.

To obtain acceptable values of hydraulicity and soundness, one possible route is the reaction of the alkaline oxides in BOF slag with CO_2, a process called mineral carbonation that leads to the formation of geochemically stable carbonates (e.g. $CaCO_3$), thus preventing the formation of expansive hydroxides and implicitly obtaining a stable material.[5] An advantage of this approach is that both forms of free lime found in BOF slag (residual and precipitated) can be made to react with CO_2, thus substantially reducing basicity[2] and likely propensity for swelling, although the latter has not been reported in literature.

Mineral carbonation has also been reported to yield positive effects in terms of basicity and the leaching behavior of alkaline earth metals, heavy metals and metalloids from steelmaking slags.[6] However, Van Gerven et al.[7] observed that the leaching evolution of some metals increases at high carbonation degrees; this means that an optimal level of carbonation might exist that minimizes leaching.

The noble purpose of sequestering carbon dioxide, however, is insufficient for extending the process of mineral carbonation to an industrial scale if the economical balance is not a positive one. For this reason, the materials obtained after mineral carbonation must be valorized by utilizing them in large-scale commercial applications. The purpose of this work was to assess the mechanical and chemical properties of cement mortars containing non-carbonated or carbonated BOF slag as a fine aggregate. This work also aimed at evaluating the effect of slag particle size on the carbonation extent, and the effects of slag particle size and carbonation degree on the mortar properties.

Materials and Methods

BOF Slag Characterization

The BOF slag used in this study was acquisitioned from a steel plant in Belgium. Prior to analyses or experimentation, the material was first crushed using a jaw-crusher (Retsch BB100) to obtain particles smaller than 1.6 mm. The obtained material was further divided into three particle size classes using standard sieves: <0.08 mm, 0.08-0.5 mm, 0.5-1.6 mm.

The chemical composition of the slag was obtained by X-ray Fluorescence (XRF, Panalytical PW 2400) analysis. High concentration of calcium (35.0 wt%) was observed, which imparts this material its alkaline properties and high reactivity towards mineral carbonation. Also, BOF slag contains large quantities of iron (20.6 wt%), an undesirable loss of product from the steelmaking process, and a significant quantity of silicon (5.8 wt%), important for the formation of silicate minerals.

The mineralogical composition of the slag was determined by Quantitative X-Ray Diffraction (QXRD) on a Philips PW1830, mineral identification was done in Diffrac-Plus EVA (Bruker) and mineral quantification was performed by Rietveld refinement technique using Topas Academic v4.1 (Coelho Software). Predominant minerals detected consist of Ca–ferrite and Ca–silicates, with Srebrodolskite (32.3 wt%) and β–polymorph of dicalcium silicate (30.8 wt%) being present in the highest quantities. Also observed are high quantities of calcium (8.8 wt%) and magnesium (1.1 wt%) oxides, which are the minerals detrimental to soundness.

Pressurized Slurry Carbonation

Each particle size class of the slag was individually carbonated under pressure and elevated temperature in a 1.1 L stirred batch autoclave reactor (Buchi Ecoclave 300 type 3E). For all the experiments, 200 g of material was added to 800 ml ultrapure water (18.2 MΩ·cm); the slurry was stirred by a turbine impeller rotating at 1000 rpm. Carbonation was conducted at 90 °C and 20 bar,gauge CO_2 partial pressure, for 2 hours durations. The carbonated slurry was filtered and the recovered solids were dried for 24 hours at 105 °C prior to analyses and further utilization. Carbonation degree of BOF slag was determined by Thermogravimetric Analysis (TGA) using a NETZSCH STA 409 PC. Visualization of trapped carbon and of other elements within a carbonated BOF slag sample was realized through elemental mapping, performed based on Wavelength Dispersive Spectroscopy (WDS) by means of an Electron Probe Micro-Analyzer (EPMA) equipped with a Field Emission Gun (FEG) (JEOL JXA-8530F).

Preparation and testing of mortar samples

In order to test the possibility of using BOF slag as partial replacement of fine aggregate in building materials, mortar specimens were prepared according to the standard EN 196-1:1994, by using a binder/aggregate ratio of 1/3 and a binder/water mass ratio of 0.625. The natural aggregate used was standard quartz sand (CEN 196-1, at 0-2 mm particle diameter), and the binder used was cement CEM I 42,5 R HES (CCB Italcementi Group). The same mix ratios were also used when carbonated and non-carbonated BOF slag was used as replacement for half of the aggregate amount by weight. When one of the two finest slag classes (<0.08 mm and 0.08-0.5 mm) was used to replace the finer half of the natural aggregate, natural aggregate having particle size >0.5 mm (separated by sieving) was used as the other half. When the coarsest slag class was used (0.5-1.6 mm), an equal amount of natural aggregate having particle size <0.5 mm was used.

Consistency of the fresh mortar mixes was tested using a flow table (according to EN 1015-3:1998), and then cast in a mould that permitted four specimens (with dimensions 20x20x160 mm) of the same mortar sample to be obtained. After 60 cycles of shock compaction, the material was kept in the mould for 24 hours, and the demoulded specimens were placed in a moist room at ≥ 93% humidity and at constant temperature of 20 °C. Different curing durations (1 to 28 days) were used to assess its effect on the compressive strength. Also, some samples were subjected to accelerated curing in boiling water to assess the effect of this curing condition on compressive strength. The intention was to magnify the effect of unhydrated oxides on volumetric expansion and compressive strength by boiling in water for an extended period. After curing, specimens were dried for 48 hours at a temperature of 40 °C to remove excess pore water. After drying, each specimen was cut into four cubes (20x20x20 mm), and each cube was tested for compressive strength (according to EN 196-1) in a Schenk Trebel screw tester at a compression rate of 0.2 mm/s.

Each mortar mixture was also used for soundness testing (Le Chatelier test, according to EN 196-3). Samples mounted in the appropriate moulds and under load were kept in a water bath at 20 °C for 24 hours, and then boiled for 3 hours. Distances between indicator arms before and after boiling were measured by caliper and the difference (equivalent to sample expansion/swelling) were recorded.

Basicity and leaching behavior of the non-carbonated and carbonated slag powders and that of the mortar samples were tested using batch leaching test methodology (EN 12457-4). The crushed samples from compressive strength tests were first jaw-crushed and sieved to <1.6 mm. The solids were added to sealed bottles containing ultrapure water at a ratio of 10 g/100 ml and shaken for 24 hours at 160 rpm and 25 °C on a shaking table (Gerhardt Laboshake). Basicity was determined by pH measurement of the final slurry. The slurry was then centrifuged for 5 minutes at 4000 rpm (Heraeus Labofuge 400) to separate the supernatant, which was then diluted in 0.3 M nitric acid solution and analyzed in triplicate for metal and metalloid content by Inductively Coupled Plasma Mass Spectrometry (ICP-MS, Thermo Electron X Series).

Water absorption of non-carbonated and carbonated slag powders, and of the standard sand, was measured using an adapted methodology based on norms EN 1097–6 and BS:812 Part 2. Dry samples of the materials were mixed with DI water at a mass to volume ratio of 25 g to 100 ml in sealed bottles. The samples were allowed to hydrate statically for 24 hours. The resulting slurry was then vacuum-filtered on Whatman paper 5 to recover the moist solids. The difference in mass between the wet solids and the initially dry solids is taken as the total water absorption. The moist solids were then dried in an oven at 105 °C for 24 hours and re-weighed. The difference in mass between the moist and the dried solids is taken as physically absorbed water, while the difference in mass between the dried and the initially dry solids is taken as chemically absorbed water (i.e. water that is consumed by the formation of hydroxides and hydrates).

Results and Discussion

CO_2 Uptake

The CO_2 uptake of carbonated materials was determined, and thus the evolution of CO_2 uptake based on particle size distribution was obtained. An increase of CO_2 uptake is observed once the particle size is reduced (10.0 wt% uptake for BOF slag <1.6 mm, compared with 16.2 wt% uptake for BOF slag <0.08 mm). This trend is easily explainable by the material's exposed surface increase that occurs when particles size is diminished, thus promoting contact between carbonic acid from the solution with the reactive minerals on the particle surfaces. Once the carbonation extent increases, the formation of passivating layers prevents access of the carbonic acid to the unreacted particle core.[8] Greater mineral conversion is thus obtained for finer particles. Based on the calcium content of the fresh BOF slag, the maximum theoretical CO_2 uptake of the slag is 27.8 wt%. Accordingly, the Ca-conversion of the finest sample is 58.3 %, compared to 36.0 % for the coarsest sample.

EPMA analysis allowed insight on the distribution of the main constituent elements and mineral phases in the carbonated BOF slag <0.5 mm sample. Notably, no particle containing calcium but absent in other elements, which would correspond to free lime, is found. This signifies extensive free lime conversion. This result is corroborated by the elimination of chemical water absorption of the carbonated slags, according to Figure 1, which occurs in non-carbonated slag due to the hydration of free CaO into $Ca(OH)_2$.

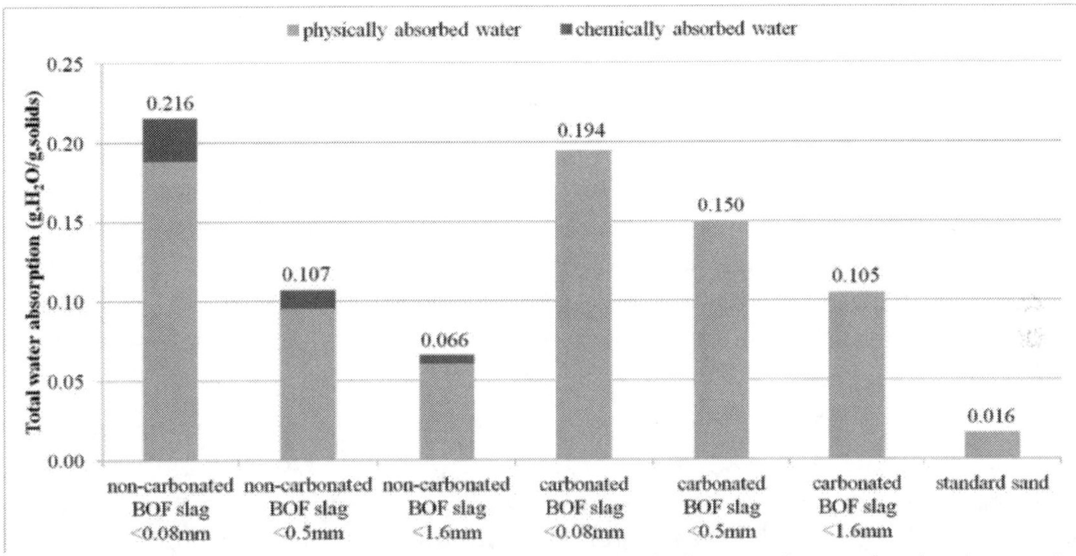

Figure 1. Water absorption (physical, chemical and total) of non-carbonated and carbonated BOF slags and of standard sand

Evidence was found by EPMA that lime may readily dissolve in the carbonation medium, and precipitates in the bulk solution as large crystals. In contrast, $CaCO_3$ originating from silicates precipitates immediately on the silicate surfaces, seeing that all dicalcium silicate grains were enveloped by $CaCO_3$ layers. Visual evidence was also found that dicalcium silicate reacts more extensively than srebrodolskite.

Mortar Paste Consistency

For all mortar samples, a constant water/cement weight ratio was used (0.625); hence consistency of mortar pastes was not controlled, but rather was a response variable. Flow table values were recorded to assess the impact of aggregate replacement (sand by slag) on the particle size distribution of the aggregate fraction, which has important implications on mortar consistency and strength properties. The presence of small sized aggregate particles in an excess amount leads to increased physical water absorption, as shown in Figure 1. This effect can lead to a decrease in mortar flow, and can further affect its density through formation of larger voids that reduce compressive strength.

Good agreement between paste consistency and mortar density was observed for samples containing non-carbonated or carbonated BOF slag <0.08 mm, and for that containing carbonated <1.6 mm BOF slag. For these samples, results from flow table (110 mm, which signifies no flow) and density (1.5-2.2 g/cm^3) measurements were the smallest. Small specimen densities are a consequence of higher voids, which is a direct result of poor flow behavior of the mortar. Flow table values obtained for mortar samples containing non-carbonated <1.6 mm BOF slag or non-carbonated/carbonated <0.5 mm BOF slag (123-136 mm, compared to 161 mm for the reference specimen) were reasonable for mortar applications. These specimens had densities of 2.4 g/cm^3, compared to 2.3 g/cm^3 for the reference. These results have important effects on a specimen's compressive strength.

Soundness of Mortars

For a material to be considered stable, the difference between the value read before boiling and the value read after boiling, during the Le Chatelier test, should not be greater than 5 mm.[9] In this study case, the reference mortar sample indicated a swelling expansion of only 0.9 mm. Other samples that remained under 5 mm variation were the mortar samples with 50% aggregate replaced by carbonated BOF slags of particle sizes <0.08 mm (-0.3 mm contraction) and <0.5 mm (1.1 mm swelling). All remaining mortar samples, including all which contained non-carbonated slag, registered swelling values substantially greater than 5 mm (11.4 to 91.7 mm swelling), placing them in the category of materials unsuited for use in the construction domain.

Notably, coarser slags had greater swelling. It had been rather expected that smaller particle size of the untreated BOF slag would conduce to greater swelling, since greater contact surface would be available to react with the aqueous solution. Instead, it appears that the finer slag becomes extensively hydrated, but not completely, already during paste preparation, and thus it suffers diminished delayed hydration (i.e. during curing). These results confirm the theory that free oxides, particularly lime, have a detrimental effect regarding volume stability.

A swelling value of 15.7 mm was obtained for the mortar sample containing carbonated <1.6 mm BOF slag. In this case, the reason for volume instability can be attributed to insufficient carbonation extent of slag. This means that a more intense process should be used for carbonation of materials with larger particle sizes, or further comminution should be done prior to, or during carbonation.

Compressive Strength of Mortars

The main objective of the compressive strength test was to assess the effect of aggregate replacement relative to the reference composition. The compressive strength data from each mortar mixture is presented in Figure 2. These are average values of the four specimens cured under different conditions (duration and standard or accelerated curing); no significant statistical difference was observed from the different curing conditions. On the other hand, significant differences were observed between mortars of different compositions.

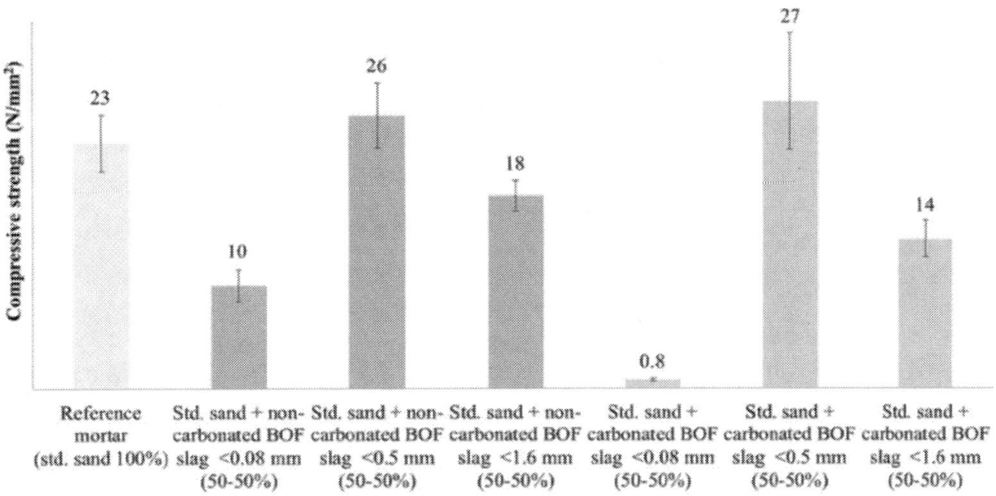

Figure 2. Average values of compressive strength for all mortar specimens (reference, non-carbonated BOF slag aggregate replacement, and carbonated BOF slag aggregate replacement); error bars are based on data from samples treated under different curing conditions (1 and 7 days followed by boiling, and 7 and 28 days without boiling) and quadruplicates of each measurement

From Figure 2 it is observed that the smallest values of compressive strength belong to the fairly dry mortar mixtures that showed insufficient flow, which included the two samples prepared with the finest BOF slag fractions (<0.08 mm). At the same time, the mortar samples prepared with the coarsest BOF slags (<1.6 mm) had compressive strengths significantly lower than the reference. This is in agreement with their volume instability being greater than that of samples prepared with the finer BOF slag fractions. Hence, the only mortar specimen that obtained similar flow value, soundness and compressive strength values as the reference mortar specimen, was the one with carbonated BOF slag of <0.5 mm particle size. For this mortar, the compressive strength value was averaged at 27.3 N/mm^2, even higher than the value obtained for the reference mortar (23.4 N/mm^2). This result seems to be due to the fact that mortars with carbonated slag as aggregate possess less pore water, as indicated with their flow table values, which in turn can be attributed to higher water absorption of the carbonated slag particles when compared to that of sand particles.

Results from compressive strength test, supported by the other tests (flow table and soundness), demonstrates the utility of the carbonation process as a treatment method for BOF slag towards its use as aggregate in the construction domain. Also, it can be concluded that if BOF slag is used as stated, this material should present a particle size distribution according to the standard EN 196-1:1994, and the free lime conversion after mineral carbonation should be high enough to ensure an acceptable value of soundness for the final building material.

Leaching from Mortars

Based on the abovementioned results, the only BOF slag fraction (carbonated and non-carbonated) used in leaching tests was that with particle size <0.5 mm, along with the crushed and ground mortar samples containing this type of slag, and the crushed and ground reference mortar sample used for comparison. The aim was to compare the leaching tendency of some regulated metalloid and heavy metals from the slag, and assess the effect played by both mineral carbonation and partial containment in a cementitious matrix on this tendency. At the same time, by comparing the leaching results from the reference mortar sample with those from the slag containing mortar samples, the differential influence on leaching tendency by including the BOF slag in the mortar mixture can be deduced.

It was found (Figure 3) that after mineral carbonation the leaching tendency of several elements drops, namely barium, nickel and lead, while it remains nearly unchanged for chromium and copper, and it increases substantially for vanadium. The leaching tendencies of cobalt and molybdenum remained below the detection limit (0.1 mg/kg). However, including the BOF slags in a cementitious mortar matrix led to full containment of the metalloid's leaching tendency. Leaching of lead and nickel similarly decreased from the slags contained in mortars.

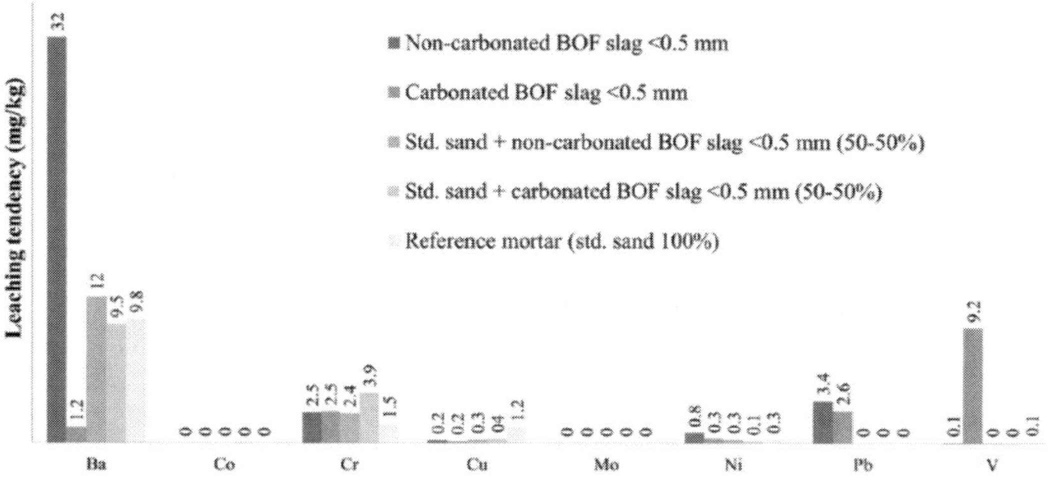

Figure 3. Leaching tendency of eight metalloids and heavy metals from <0.5 mm BOF slag before and after carbonation, from mortar samples containing <0.5 mm BOF slag (carbonated or non-carbonated) as partial aggregate replacement, and from reference mortar sample; value of '0' means <0.1 mg/kg

The only result of concern was that of chromium leaching from the carbonated BOF slag-containing mortar. Its Cr leaching tendency was the highest of all tested materials. Yet, the leaching extent from this material is still comparatively low; the Belgian-Walloon limit for Cr(VI) leaching from re-used waste materials is 20 mg/kg.[10] According to Fernández-Bertos et al.[11] and Cornelis et al.[12], leaching of chromium at high pH in alkaline materials is controlled by Cr substitution in calcium-silicate-hydrates (CSH), the main component of hydrated cement. Thus, over time, it is possible that chromium leached from BOF slag can become chemically bonded to the surrounding matrix, thus minimizing the leaching risk from the construction material for the duration of its life cycle, including its CDW (construction and demolition waste) phase.

Differing leaching tendencies from slag-containing mortar samples compared to leaching values of the same elements from the pure slags are linked to the basicity of the materials, and thus the pH values of the leaching solutions. A decrease of approximately one pH unit was registered for the carbonated BOF slag compared to the non-carbonated slag. The mortar samples, however, had elevated pH ranging from 12.4 for the reference sample to 12.7 for the mortar containing non-carbonated BOF slag. The high pH values of the mortars represent the reason for which vanadium leaching was kept low, since V is known to mobilize in response to pH reduction.[13]

The results regarding leaching from mortar samples containing carbonated BOF slag as partial replacement of natural sand aggregate provides further confirmation that mineral carbonation process can be considered an effective treatment method for metallurgical slags prior to utilization of this material in the construction domain. A further research direction that can be investigated is the inclusion of sorbent materials as partial aggregate replacement to further attenuate the leaching tendency of elements of concern, such as chromium in the present case.

Conclusions

The research presented in this paper aimed to demonstrate that mineral carbonation can be considered a route to treat steelmaking converter slags in order to further use this residual material as a partial aggregate replacement in the domain of civil constructions. Mineral carbonation of BOF slags leads to the sequestration of meaningful amounts of CO_2, 0.11–0.19 kg,CO_2/kg,slag at the conversions achieved herein. Taking into account a 50% replacement ratio of natural aggregate, and an aggregate to binder ratio of 3, the prepared mortars contained up to 74 grams of CO_2 per kilo; this categorizes it as a carbon sink.

The mineral carbonation process also represents a means of treatment by forming stable carbonates through reaction of calcium and magnesium oxides with CO_2. This is especially important for the stabilization of free lime, which was shown to have detrimental effects on the hydration-driven volume expansion of mortars prepared with non-carbonated BOF slag. In contrast, for mortar samples containing carbonated slag, soundness increased in proportion to the carbonation conversion, reaching values similar to the soundness of reference cement mortar.

Particles size had an important influence on flow value of the fresh mortar and, implicitly, on compressive strength of the hardened mortar specimens. From this point of view, it is concluded that maintaining the particle size distribution of the aggregate mixture according to the standard norms is critical for ensuring suitable consistency and mechanical strength.

The best result obtained in terms of the combination of tested parameters (consistency via flow table test, soundness, compressive strength and leaching of toxic elements) corresponded to the mortar mixture prepared with carbonated BOF slag of <0.5 mm particle size. In light of these results, it is possible to conclude that BOF slag utilization as partial aggregate replacement in construction domain can be a realistic method to valorize and recycle this kind of industrial residue.

Acknowledgements

This work was supported by the KU Leuven Industrial Research Fund, through the Knowledge Platform on Sustainable Materialization of Residues from Thermal Processes into Products (SMaRT-Pro²), by the European Union, the Romanian Government and the "Dunarea de Jos" University of Galati, through the project POSDRU/107/1.5/S/76822, and by the Natural Sciences and Engineering Research Council of Canada (NSERC), through a PGS-D scholarship for Rafael Santos.

References

1. Belhadj, E., Diliberto, C., Lecomte, A. 2012. Characterization and activation of basic oxygen furnace slag. *Cement & Concrete Composites*, 34, 34-40.
2. Santos, R.M., Ling, D., Sarvaramini, A., Guo, M., Elsen, J., Larachi, F., Beaudoin, G., Blanpain, B., Van Gerven, T. 2012. Stabilization of basic oxygen furnace slag by hot-stage carbonation treatment. *Chemical Engineering Journal*, 203, 239-250.
3. Wang, G., Wang, Y., Gao, Z. 2010. Use of steel slag as a granular material: Volume expansion prediction and usability criteria. *Journal of Hazardous Materials*, 184, 555-560.
4. Xie, J., Wu, S., Lin, J., Cai, J., Chen, Z., Wei, W. 2012. Recycling of basic oxygen furnace slag in asphalt mixture: Material characterization & moisture damage investigation. *Construction and Building Materials*, 36, 467-474.
5. Mikhail, S.A., Turcotte, A.M. 1995. Thermal behaviour of basic oxyen furnace waste slag. *Thermochimica Acta*, 263, 87-94.
6. Bodor, M., Santos, R.M., Van Gerven, T., Vlad, M. 2013. Recent developments and perspectives on the treatment of industrial wastes by mineral carbonation - A review. *Central European Journal of Engineering*, 3(4), 566-584.
7. Van Gerven, T., Cornelis, G., Vandoren, E., Vandecasteele, C., Garrabrants, A.C., Sanchez, F., Kosson, D.S. 2006. Effects of progresive carbonation on heavy metal leaching from cement-bound waste. *AIChE Journal*, 52(2), 826-837.
8. Santos, R.M., François, D., Mertens, G., Elsen, J., Van Gerven, T. 2013. Ultrasound-Intensified Mineral Carbonation. *Applied Thermal Engineering*, 57(1-2), 154-163.
9. Li, Z., Zhao, S., Zhao, X., He, T. 2013. Cementitious property modification of basic oxygen furnace steel slag. *Construction and Building Materials*, 48, 575-579.
10. Ministère de la Région Wallonne. 2001. Arrêté du Gouvernement wallon favorisant la valorisation de certains déchets. *Moniteur Belge*, 2, 23859-23883.
11. Fernández-Bertos, M., Simons, S.J.R., Hills, C.D., Carey, P.J. 2004. A review of accelerated carbonation technology in the treatment of cement-based materials and sequestration of CO_2. *Journal of Hazardous Materials*, B112, 193-205.
12. Cornelis, G., Johnson, A., Van Gerven, T., Vandecasteele, C. 2008. Leaching mechanisms of oxyanionic metalloid and metal species in alkaline solid wastes: A review. *Applied Geochemistry*, 23, 955-976.
13. Chaurand, P., Rose, J., Domas, J., Bottero, J.-Y. 2006. Speciation of Cr and V within BOF steel slag reused in road constructions. *Journal of Geochemical Exploration*, 88(1-3), 10-14.

Recent Advances in the Development of a New pH Swing Method Based on a Regenerable Precipitant-Solvent System for Metals Recovery and Greenhouse Gas Control

K. M. Steel* and R. D. Balucan

School of Chemical Engineering, The University of Queensland, St Lucia 4072, Australia. *Corresponding author: karen.steel@uq.edu.au

Abstract

The challenge that exists for making a CO_2 mineralisation process viable is a means to recover both acid and base. Acid is needed to dissolve cations such as Mg and Ca from oxide ores which are then converted to carbonate minerals such as $MgCO_3$. Base is needed to raise solution pH to the level needed for rapid carbonate precipitation. And the process steps, including acid and base recovery, need to have low energy requirements. Triethylamine is proposed to be an extremely promising reagent for enabling the pH swing needed for a CO_2 mineralisation process. We also propose that a CO_2 mineralisation process could align itself with Ni recovery as the current hydrometallurgical processes for Ni production requires the processing of large quantities of ore as the Ni is present at low levels (1-2 wt%), and therefore the process also generate vast quantities of Mg-sulphate tailings. In this study we have examined the interaction of triethylamine with sulphuric acid and attempted to recover both reagents back. Triethylamine rapidly complexes protons to generate a pH > 10 which is more than adequate for CO_2 mineralisation. We have found that is it possible to recovery the triethylamine using fractional distillation methods, whereby co-evaporation of water is avoided and the sulphuric acid is co-generated. So far we have recovered a sulphuric acid solution with a pH < 2, which is a level capable of dissolving Mg from serpentinite. Future work is needed to further study the selectivity of the separation and the ability for both reagents to be continuously recycled and not lost through side reactions. Examination of the energy requirements is also needed.

Introduction

The major barrier to CO_2 sequestration is the lack of economic incentives for investors. A carbon tax or emissions trading scheme provides some incentive, however, the money to be saved does not match the capital investment required to implement new technologies, particularly when new technologies carry a high level of risk. For underground storage of CO_2, there are many studies, in particular those where CO_2 has been used for enhanced oil recovery, indicating that the risk of CO_2 leakage is low. However, there is not yet an overwhelming agreement that zero leakage can be maintained over the long-term.

This research sets out to develop a process that sequesters CO_2 with the following major benefits:

1. Does not rely solely on a carbon tax because the process provides its own economic benefits.

2. The CO_2 is in a permanently locked mineral carbonate form.

This research relates to another mineral carbonation process; the carbonation of red mud waste[1], which is a commercial success, capturing 70k tonnes of CO_2 p.a. as well as reducing the alkalinity and improving the solid-like characteristics of the waste.

Details of the process examined here is shown in Figure 1. It does not take into account the water balance.

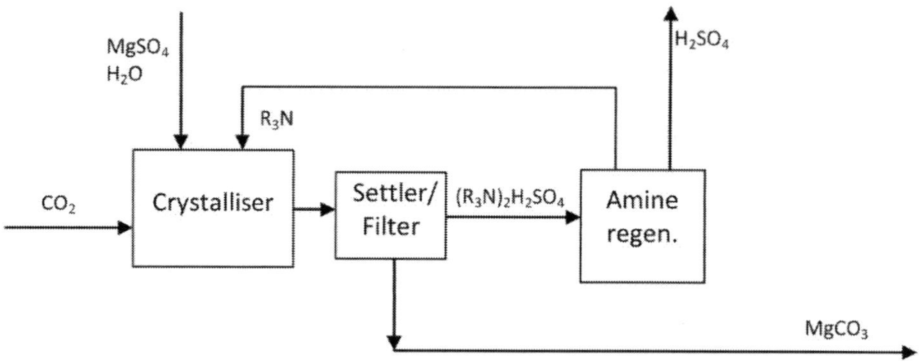

Figure 1: Proposed acid recovery block flow diagram

An aqueous stream of $MgSO_4$ and a tertiary amine are added to a continuously stirred tank crystalliser in which CO_2 gas is continually sparged through the base. The CO_2 reacts with water to form protons, bicarbonate and carbonate ions. The amine acts as a buffer, complexing protons and maintaining the pH > 8 which provides a carbonate ion concentration high enough to bind with Mg^{2+} ions and precipitate $MgCO_3$. The reaction occurring in the crystalliser is as follows:

$$MgSO_4 + H_2SO_4 + CO_2 + H_2O + 4\,R_3N \leftrightarrows 2\,(R_3N)_2H_2SO_4 + MgCO_3 \quad ...1$$

The contents of the reactor are continuously withdrawn. The solid $MgCO_3$ is separated from the liquid phase by filtration. The acid bound amine is heated to dissociate the amine and separate it via evaporation.

It is proposed that a process for converting $MgSO_4$ into $MgCO_3$ with the concomitant recovery of H_2SO_4 could be highly suited to the High Pressure Acid Leaching (HPAL) process used to extract nickel from laterite ores, for which vast quantities of $MgSO_4$ tailings are generated because the amount of Ni in the ore is only about 1-2 wt%. Figure 2a shows the current method and Figure 2b shows how the acid recovery process shown in Figure 1 could be implemented to reduce CO_2 emissions and recover acid.

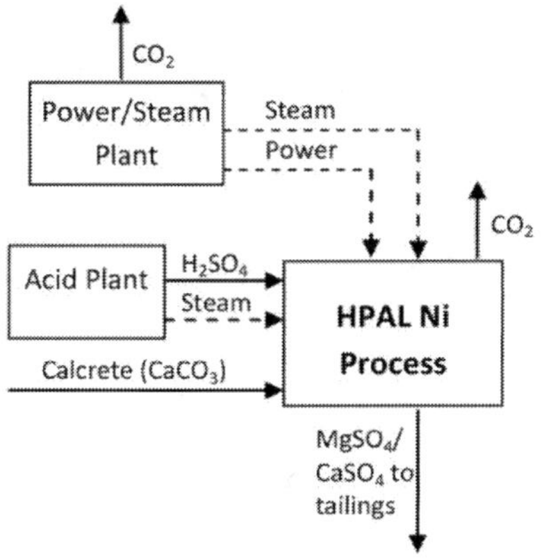

Figure 2a: Current method for tailings management in Ni laterite processing

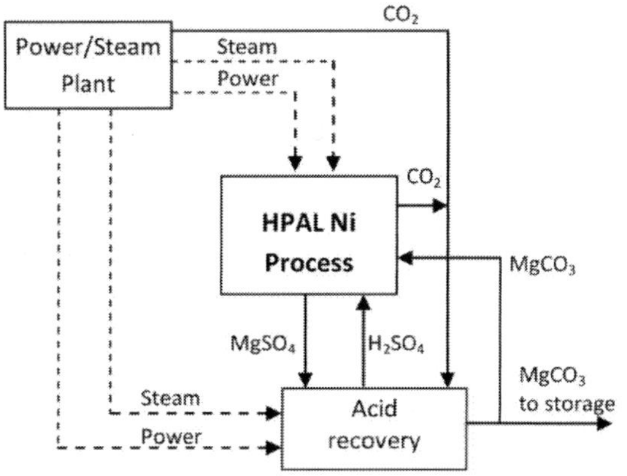

Figure 2b: Proposed new process for tailings management in Ni laterite processing

Part of the $MgCO_3$ produced is recycled back into the HPAL Ni process to be used for neutralisation while the remainder is sent for storage. The acid is recycled back to the HPAL Ni process and the amine is cooled and recycled back to the crystalliser.

While the current process consumes H_2SO_4 and $CaCO_3$, and produces CO_2, the new process has closed-loop streams for both acid and base, and consumes CO_2. While the current process avoids Mg-rich deposits (as it means a loss of acid), the new process is suited to these and, consequently, is capable of prolonging the life of the mine, allowing the deeper saprolite reserves, which are often higher in Ni, to be processed. As the Mg content increases, a point is reached whereby the plant can take in extra CO_2 from neighbouring industrial operations. It follows that Ni plants could play a role in helping industries to achieve emission reduction targets.

The motivation for this work is in improving the economics for Ni production and reducing the environmental impact from CO_2 emissions. There is a second environmental issue driving this work. While tailings disposal in the desert regions of Australia is not problematic, as evaporation assists with solids production and hence disposal, in areas of heavy rainfall land disposal of tailings is prohibitive. For many sites the only option is deep sea disposal which receives considerable objections from local communities over the potential effects to fragile reef systems and to fishing. At Goro in New Caledonia an expensive reverse osmosis (RO) unit was installed to ensure the levels of dissolved elements were close to that of the surrounding ocean water. Given that the processing of tailings is becoming necessary for many plants it seems logical to consider re-designing the downstream part of the process to reduce waste.

The innovation in this research is in the use of tertiary amines for converting $MgSO_4$ and H_2CO_3 to $MgCO_3$ and H_2SO_4. It has been recognised that a two-stage process with strong acid to dissolve Mg from serpentinite and strong base to precipitate $MgCO_3$ are needed to enable mineral carbonation within reasonable time scales, as discussed by

Park and Fan, who called this need "pH swing"[2]. Unfortunately, the only base that occurs naturally in large quantities is carbonates, which obviously cannot be used. Other bases (such as NaOH) come from high energy processes, and are therefore prohibitive. The need for pH swing and lack of an available technology is the main reason why mineral carbonation is stalling as a viable technology for CO_2 sequestration.

Attention has turned to regenerable buffers where heat is used for regeneration. Ammonia has been suggested[3], however, evaporation of water, in addition to high temperatures (>300°C) to break the ammonium sulphate bond, are needed. The use of tertiary amines has been suggested[4,5,6]. Tertiary amines are already used for CO_2 capture in conjunction with primary amines. While the primary amine bonds strongly with CO_2, enabling a high capture efficiency, tertiary amines cannot bond with CO_2, however, they do bond with protons, enabling capture of CO_2 as bicarbonate and carbonate ions. Although the capture efficiency is lower for tertiary amines, the energy needed for regeneration is lower (1 MJ/kg CO_2 for tertiary amines as opposed to 3-5 MJ/kg CO_2 for primary amines), and so the combination of primary and tertiary amines provides a good compromise for traditional CO_2 capture with amines[7].

This research is focused on examining whether tertiary amines can provide a low energy means to convert $MgSO_4$ and H_2CO_3 to $MgCO_3$ and H_2SO_4. Unlike the traditional use of tertiary amines for CO_2 capture, in the system here carbonate ions are continually withdrawn from solution as solid carbonate, and this is considered to provide a strong driving force for CO_2 absorption, and raise the capture efficiency to a level higher than usually found with the use of tertiary amines for CO_2 capture.

This paper presents our attempt to regenerate acid from the proton-bound tertiary amine. Our previous work has shown that an amine capable of generating a pH > 8 is needed to produce $MgCO_3$ and that triethylamine is capable of generating a pH > 10[5]. Therefore, triethylamine was chosen. Sulphuric acid was chosen as it is already used in laterite processing.

Experimental

The triethylamine (R_3N) used in this work had a purity ≥ 99 % and the sulphuric acid was AR-grade.

The recovery of amine was studied using a 250 mL Quickfit three-necked, round-bottomed flask equipped with a 3-sectioned glass Snyder fractionating column in one of the necks. Floating glass ball valves prevented the backflow of volatised components back into the distilling flask. A high temperature KNO_3 gel-based pH probe and thermometer to allow compensation for temperature effects were fitted in the other two necks. Solutions were continuously mixed via magnetic stirring at 500 rpm whilst an oil bath maintained the solution temperatures at the desired value within ± 5 °C. A Quickfit receiver assembly was placed at the top of the column to allow the vapour at the top to be condensed and collected.

Results and Discussion

Tertiary amine and proton bonding

Equation 1 shows the general reaction that occurs when serpentinite reacts with acid to release Mg^{2+} into solution.

$$Mg_3Si_2O_5(OH)_4 + 3\ H_2SO_4 \rightarrow 3\ MgSO_4 + 2\ SiO_2 + 5\ H_2O$$
...2

While extraction of Mg is undesirable in the traditional HPAL process where Mg-rich laterites are avoided as it represents loss of acid, it is desired here.

Excess acid is used to obtain high extraction efficiencies. Therefore, amine is needed to bind with protons from the excess acid, protons generated in solution when CO_2 is injected into the solution, and needed to bind with the small amount of protons which drive the pH up to 10, as shown by reactions 3-5, respectively.

$$2\ R_3N + H_2SO_4 \leftrightarrows 2\ R_3NH^+ + SO_4^{2-} \quad ...3$$

$$2\ R_3N + CO_2 + H_2O \leftrightarrows 2\ R_3NH^+ + CO_3^{2-} \quad ...4$$

$$R_3N + H_2O \leftrightarrows R_3NH^+ + OH^- \quad ...5$$

Tertiary amine regeneration

Our previous work has shown that increasing temperature reduces the strength of the amine-proton bond slightly, thereby regenerating some of the amine involved in binding with the small amount of protons which drives the pH from 7-10[5]. Solvent extraction of amine using kerosene was also found to be capable of regenerating these protons. However, neither heating or solvent extraction was capable of regenerating the amine involved in bonding excess protons or protons generated from CO_2 injection.

To examine regeneration of amine involved in bonding excess protons, triethylamine was mixed with H_2SO_4 such that the triethylamine was in slight excess and reaction 3 had taken place. This solution underwent multiple stage distillation to recover the amine and the pH of the solution was measured whereby a decrease in pH would indicated that amine was being separated. The experiments involved heating 30 mL aqueous solution of fully neutralised H_2SO_4 (20 mmol; 20 mL 0.5 M H_2SO_4) with Et_3N (71 mmol; 10 mL 99 % purity R_3N) from 25°C to the boiling temperature (i.e., 105 °C).

Figure 3 shows the pH of the solution. The pH decreased to approximately 7 as the temperature increased before decreasing significantly to pH 2.7 as the temperature remained constant during amine volatilisation. The residual solution volume was 65 % of the original. Given that the triethylamine comprised approximately 33 % of the system's initial volume, the selectivity achieved with multiple stage distillation is excellent.

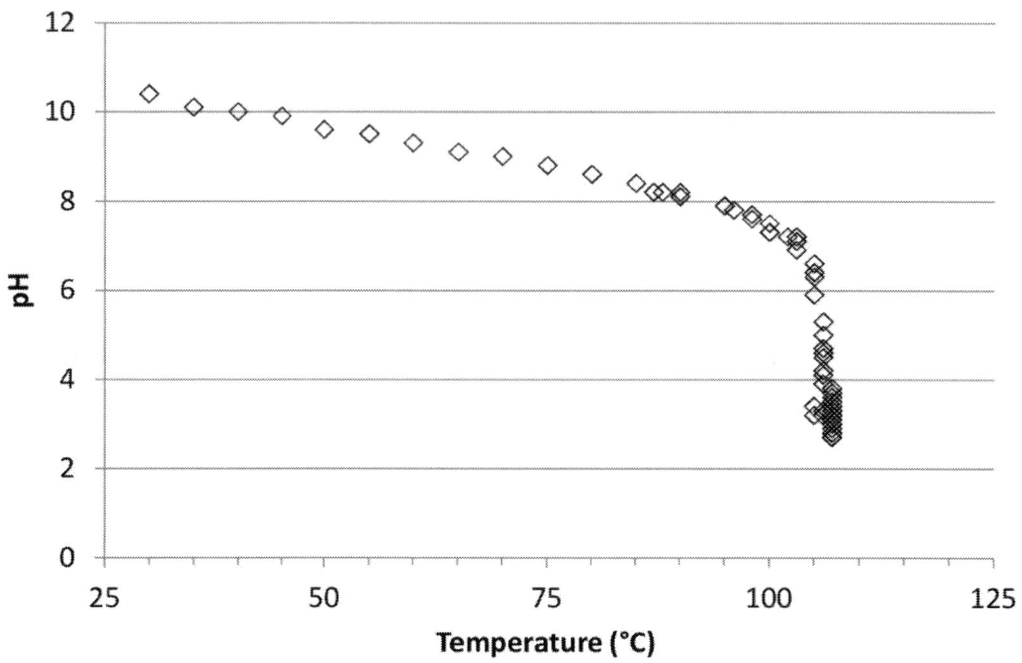

Figure 3. Aqueous pH trends of Et_3N–H_2SO_4–H_2O system with respect to temperature in multi-stage (fractional) distillation arrangements. The pH of the cooled residues, residual volumes, % V_{res}, and total time, t_{total}, min, are included.

The acid being generated during regeneration from the Et_3N–H_2SO_4–H_2O system is expected to be predominantly HSO_4^- and H^+ according to reaction 6 below.

$$2\ R_3NH^+_{(aq)} + SO_4^{2-}{}_{(aq)} \rightarrow 2\ R_3N_{(g)} + HSO_4^-{}_{(aq)} + H^+_{(aq)} \quad ...6$$

The regeneration of HSO_4^- and H^+ at pH 2.7 is consistent with predictions from equilibrium constants. H_2SO_4 is expected to become appreciable only at lower pH levels. When the solution in the distillation flask was cooled to ambient temperature the pH decreased further to 1.9 which is expected based on the temperature dependence of the equilibrium constants for the sulphuric acid system.

The next phase of work will involve further examining the recovery of triethylamine using fractional distillation methods, focusing on selectivity of the separation and energy requirements. Further work will also involve examination of the continued use of the regenerated amine and whether any unwanted reactions occur leading to the loss of amine.

Conclusion

This study has examined the use of triethylamine as a means to complex protons in solution to enable solution pH to rise to high levels needed for rapid formation of $MgCO_3$. We have found that triethylamine will complex protons readily and achieve an equilibrium solution pH in excess of 10. We reacted triethylamine with sulphuric acid and then found that it was possible to recover the amine by fractional distillation methods, whereby co-evaporation of water is avoided. This recovery also generates an aqueous sulphuric acid solution with a pH < 2. At this current stage in research, the use of a triethylamine and sulphuric acid are considered to be a promising means to provide the pH swing capability needed for CO_2 mineralisation. The sulphuric acid is used to dissolve Mg from serpentinites and the triethylamine is used to raise the pH up 10 to produce $MgCO_3$. The process could be operated symbiotically with the current High Pressure Acid Leaching of laterite ores to extract Ni.

References

1. Davidson, S., **2005**. A red mud remedy. ECOS magazine, Vol 127, page 35.

2. Park A.-H.A.; Fan, L.-S. **2004.** CO_2 mineral sequestration: physically activated dissolution of serpentine and pH swing process. *Chem. Eng. Sci.*, 59, pp 5241–5247.

3. Wang, X., Maroto-Valer, M., **2011**. Integration of CO_2 capture and mineral carbonation by using recyclable ammonium salts. *Chemsuschem*, 4, pp 1291-1300.

4. Steel K. M. **2013**. Sequestration of carbon dioxide. International patent WO 2013/010226 A1.

5. Steel K. M., Alizadehhesari, K., Balucan, R. D., Bašić B., **2013** 'Conversion of CO_2 into mineral carbonates using a regenerable buffer to control solution pH' *Fuel*, 111, pp 40-47.

6. Wang, W., Liu, X., Wang, P., Zheng, Y., Wang, M., Enhancement of CO_2 mineralization in Ca^{2+}-/Mg^{2+}-rich aqueous solutions using insoluble amine, **2013**, *Ind. Eng. Chem. Res.*, 52, pp 8028-8033.

7. Aroonwilas, A., Veawab, A., **2007**. Integration of CO_2 capture unit using single- and blended-amines into supercritical coal-fired power plants: Implications for emission and energy management. *International Journal of Greenhouse Gas Control*, 1, pp 143-150.

8. Jaques, A. L., Huleatt, M. B., Ratajkoski, M., Towner, R. R., **2005**. Exploration and discovery of Australia's copper, nickel, lead and zinc resources 1976-2005. Geoscience Australia.

Mineral Carbonation of CO_2 with Steel Slags and Waste Concrete

A. Ben Ghacham[1*], L.C Pasquier[1], E. Cecchi[1], J.F Blais[1] and G. Mercier[1]

[1] INRS-ETE, Quebec, Canada

*alia.ben_ghachem@ete.inrs.ca

Abstract

In this study, samples of waste concrete and steel slags were tested for CO_2 sequestration using direct aqueous mineral carbonation in milder conditions. Tests were conducted at ambient temperature, with a simulated flue gas (18.2 vol.%CO_2), at maximum pressure of 10.68 barg. The effect of the aggregates separation from the concrete matrix on reaction efficiency was studied. Results show that the resulting concrete powder was more reactive. Conversion of 0.055 g CO_2/g sample was observed compared with 0.036 g CO_2/g sample obtained with raw waste concrete. Tests realised with aggregates reveals that they are little reactive. Tests were also realised by treating successive batches of gas with renewing the liquid phase after each batch. The effect of solid material reconditioning by grinding was monitored. It revealed that these methodologies have slightly enhanced the material sequestration potential. In addition, the use of concrete powder presents two main advantages; the enhancement of the CO_2 conversion and the possibility to sell the resulting aggregates which will have beneficial effect on the process economy. Concerning steel slags, results showed the possibility to obtain relatively good conversion of 0.062 g CO_2/g sample in milder conditions.

Introduction

The atmospheric CO_2 concentration has increased by 30% since the industrial revolution[1]. This increase is directly linked with the intensive use of fossil fuels. Therefore, it is important to adapt new technologies within industries in order to offset their emissions. Carbon Capture and Storage (CCS) become an important research topic for CO_2 reduction. Among the various routes proposed, mineral carbonation (MC) represents a thermodynamically favourable and safe alternative to the CO_2 underground storage. Natural silicates were first studied as a MC feedstock due to their interesting carbonation potential. In fact, it is reported that that forsterite (olivine), lizardite (serpentine) and wollastonite presents a sequestration capacity of 0.55, 0.4 and 0.36 g CO_2/g sample respectively[2]. However, high pressure and temperature conditions, the necessity of material mining and their transportation to the mineralization plant represent major disadvantages toward an industrial application[2-4]. Nevertheless, other alkaline materials could be used for MC such as wastes streams containing calcium or magnesium. Most alkaline residues sources studied are; iron and steelmaking slags[3, 5], cement kiln dust (CKD)[6] and alkaline combustion ashes coal[7, 8]. In this study, we proposed to use waste concrete and steel slag as MC feedstock. The choice is based on the abundance of these materials and their large availability around industrial CO_2 emitters. Construction and demolition wastes (C&D) are heterogeneous materials produced during construction and demolition operations. A crude estimation of global C&D waste production per year is 1.4-5.9 Gt (21% from construction and 79% demolition)[9, 10]. In the United States and Canada, approximately 42% of the total quantity of C&D waste is typically reused or recycled and the rest is managed in landfills[11]. Concrete represents more than 50% of construction and demolition wastes[12]. Therefore, the valorisation of waste concrete in MC could represent a good manner to ally the reduction of landfilled concrete quantities and reduce industrial CO_2 emissions. MC is a common phenomenon naturally affecting concrete. The concrete carbonation mechanism is based on diffusion of atmospheric CO_2 which dissolve into carbonic acid inducing a pH decrease resulting in calcium ions leaching from the solid phase[13]. The products of carbonation (mainly $CaCO_3$) precipitate thereafter in the pore space[13]. MC of concrete implies essentially calcium-bearing phases obtained from cement hydration. The principal reactions are transformation of Calcium Silicate Hydrate (CSH gel) ($xCaO,SiO_2,nH_2O$) and portlandite ($Ca(OH)_2$) (Eq. 1 and Eq. 2), ettringite ($Ca_6Al_2O_6(SO_4)_3,32H_2O$), and AFm ($Ca_4Al_2O_6(SO_4),12H_2O$) into calcium carbonate, gypsum, and an assemblage of amorphous aluminum and silica oxyhydroxides[14].

$$\mathrm{x\,CaO, SiO_2, nH_2O + CO_2(g) \rightarrow xCaCO_3 + SiO_2 + nH_2O} \quad (1)$$

$$\mathrm{Ca(OH)_2 + CO_2(g) \rightarrow CaCO_3 + H_2O} \quad (2)$$

In the literature, the MC using cementitious materials as feedstock is essentially related to the concrete cure or the use of fresh cement as waste cement[15-18]. Few studies have examined the use of the waste concrete properly. Concerning steel slags, they are abundant co-products of steel industry. The annual world production of slags from iron and steel industries reaches almost 50 million tons[19]. It contains typically around 38-46% CaO and 6-12% MgO[10, 19, 20]. This characteristic imparts their alkaline properties for MC. It is estimated that the present production of slag could sequester 44 to 59 Mt Carbon per year[10]. In addition, it is reported by the Intergovernmental Panel on Climate Change that the steel industry accounts for 3% - 4% of total world greenhouse gas emissions[21]. Therefore, it presents the advantage to have a carbonation feedstock directly at the CO_2 producing locations. This avoids cost and energy consumption associated with material transportation. Anterior studies using steel slags for MC present interesting results[3, 22, 23]. However, few studies used low gas pressure, temperature, reaction time and non-concentrated CO_2 stream. Thus, the objective of this study was to evaluate waste concrete and steel slag sequestration potential in milder conditions when treating a flue gas like composed with 18.2 vol.% CO_2.

Mineral Carbonation of waste concrete

Waste concrete was collected from an ecocenter of Quebec city. The sample was first grinded to reach a particle size of 21.38 µm. MC tests were performed at ambient temperature, with a gas containing 18.2 vol.% CO_2, under a pressure of 9.92 barg (pCO_2 = 1.80 barg), with a ratio L/S=10 wt.% during 10 minutes. The best CO_2 uptake obtained with raw waste concrete was 54.16 % corresponding to a conversion capacity (into carbonates) of 0.036 g CO_2/g sample.

Concrete samples contain a large quantity of coarse aggregates in their structure. Such materials are not composed of cementitious paste making them inert for carbonation. In order to enhance the concrete potential reactivity and to limit the amount of inert materials, aggregates were separated from the cement matrix by crushing and sieving the samples prior to carbonation. The powder which present a particle size <500 µm have been retained, grinded in order to obtain particles with an average diameter of 33.31 µm and tested in MC. A schematic drawing of the process is presented in the figure 1. The table 1 present the chemical composition of the raw concrete, concrete powder and aggregates.

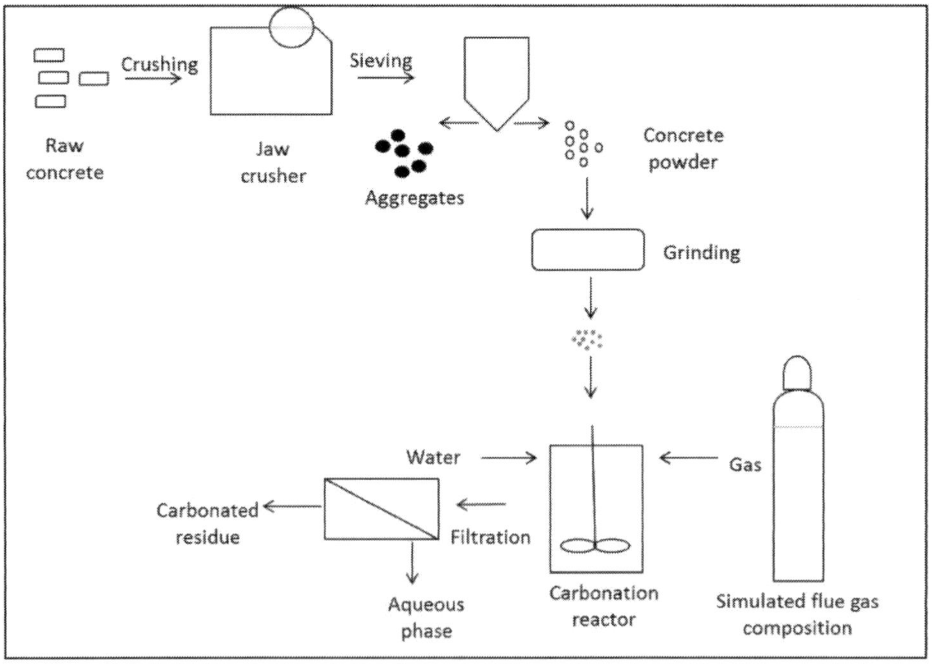

Figure 1. Schematic drawing of the concrete separation and carbonation process

Table 1. Chemical composition of the raw concrete, concrete powder and aggregates (%)

Sample	Al	Ca	Fe	Mg	Si	inorg.C
Raw concrete	3.36	20.88	1.31	1.27	17.74	4.75
Concrete powder	4.39	22.04	1.62	1.22	18.95	3.04
Aggregates	2.33	28.86	1.21	1.35	15.97	7.8

The figure 2 shows the results obtained. The CO_2 removal using concrete powder is 74.78 % which correspond to 0.055 g CO_2/g sample. The results obtained with concrete powder were more important. This confirmed that the aggregates contained in the concrete matrix are little or non-reactive. To confirm this hypothesis, a sample only composed with aggregates was grinded and tested in similar experimental conditions. Results showed that the CO_2 removal is very close to the results obtained when contacting only distilled water with the gas (figure 2). Thus, only the dissolution of CO_2 in the aqueous phase is responsible for CO_2 depletion in the gas phase.

Statistical tests were used to confirm these observations. Comparisons between the CO_2 removal percentages were realized for assays with raw concrete, concrete powder, aggregates and water only (blank reaction). The test HSD of Tukey-Kramer with α =0.05 was used. The p-value obtained when comparing concrete powder and raw concrete gave a value of 0.0019. It confirms the significant difference between the samples reactivity. However p-value obtained after comparison between blank reaction and aggregates gave 0.0755 which reflects that the difference between their reactivity is no significant.

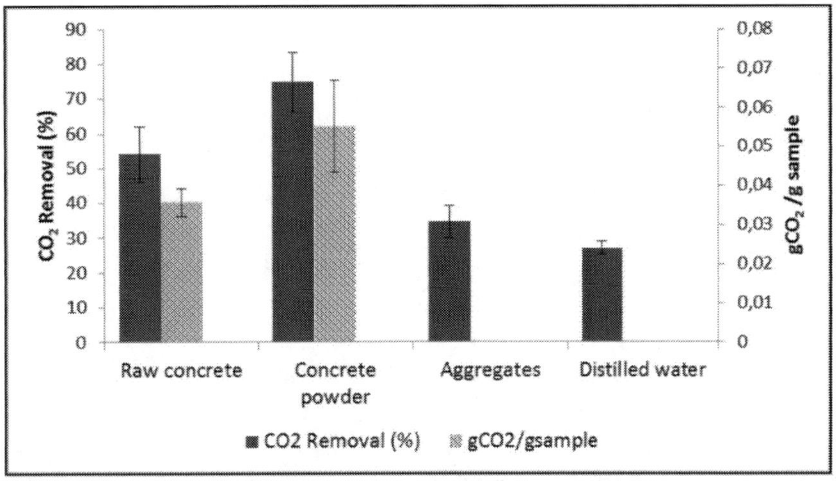

Figure 2. % CO_2 Removal and CO_2 conversion (g CO_2 /g sample) with raw concrete, concrete powder, aggregates and distilled water

In a second part, successive batches of gas were contacted with the same concrete powder sample in order to evaluate its maximal reactivity. A total of three successive batches of gas using the same liquid and solid phases were conducted. In addition, the effect of renewing the liquid phase and reconditioning the solid phase (grinding) after each batch of gas was also monitored. It was observed that the CO_2 uptake capacity decreased along the batches of gas (fig. 3). The better conversion rate (0.061 g CO_2/g sample) was obtained when the liquid phase was renewed and solid phase grinded. However, the improvement of conversion rate wasn't significant between these three routes tested. It could be noted that higher difference is shown in the case of the 3rd batch between direct successive batches and methodologies including renewing water. This essentially reflects the decreasing capacity of the CO_2 to dissolve in the aqueous phase as it probably reaches saturation in the case of direct successive batches methodology.

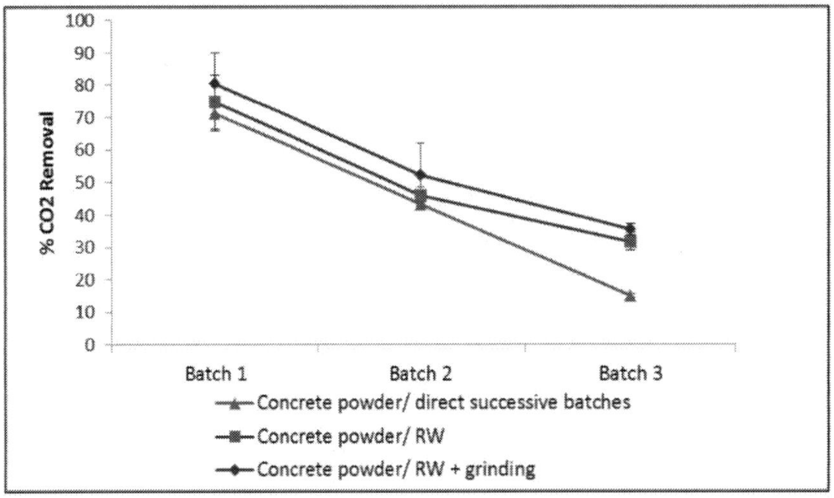

Figure 3. Evolution of % CO_2 Removal during gas successive batches for different methodologies (RW: Renewing Water)

Mineral carbonation of steel slags

Tests were realized using electric furnace slags (EAFS). Their Chemical composition is presented in the table 2.

Table 2. Chemical composition of steel slag (%)

Sample	Al	Ca	Fe	Mg	Si	inorg.C
EAFS	3.56	23.72	27.01	7.31	7.81	0.046

Steel slags are highly rich in calcium (23.72 wt.%) and therefore have a significant carbonation potential (0.26 gCO_2/g sample). After grinding, a mean particles size of 24.04 µm was obtained. Tests were conducted at ambient temperature, under a pressure of 10.68 barg (pCO_2 = 1.94 barg) with a ratio L/S = 10 wt.%. Tests were made for different reaction time from 2 to 30 minutes. The results presented in figure 4 show that an efficiency of 0.062 g CO_2 /g sample mass could be reach after 30 minutes which correspond to 68.25 % of CO_2 removal. This represents carbonation efficiency of 23.85%.

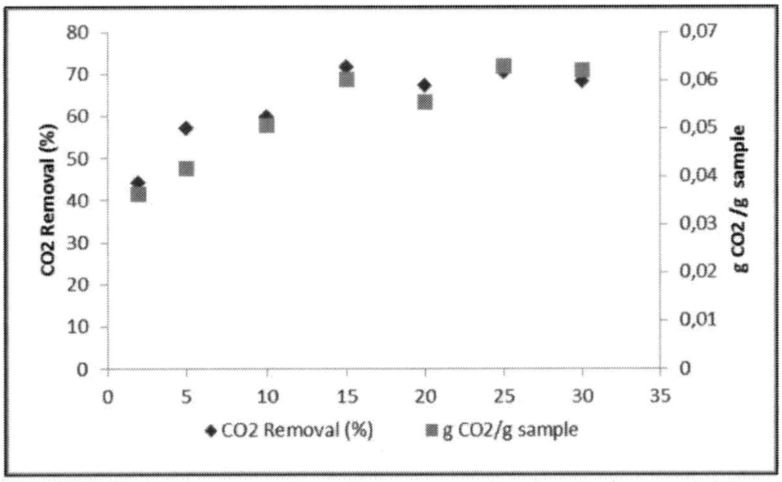

Figure 4. Evolution of % CO_2 Removal and CO_2 conversion (g CO_2 /g sample) for steel slag as function of time

It should be noted that after 10 minutes of reaction, we obtained 0.051 g CO_2/g sample corresponding to 59.79% CO_2 removal. It corresponds to 19.23% of the theoretical potential sequestration of steel slag. In almost similar conditions, Bonenfant[24] reported an efficiency of 0.0174 g CO_2/g sample after 24h reaction. In this study, the conversion rate after 10 minutes is almost 3 times higher. However, Baciocchi[22] reported 0.28 g CO_2/g sample after 5 hours of reaction for experiments conducted at 100 °C, 100 barg with pure CO_2. Even if these results are much higher than those presented in this work, the relatively long reaction time, the high temperature and pressure conditions are avoided in our case which may allow to diminishing the process cost.

Conclusion

In the present manuscript, the results obtained with waste concrete and EAF steel slag for MC in aqueous phase at ambient temperature and moderate pressure were presented. It was demonstrated that powder concrete obtained after separation of aggregates contained in raw concrete was more reactive than raw concrete himself. This observation was related to the fact that aggregates removed were nor or little reactive. It could be assumed that aggregate would be mostly limestone due to their higher inorganic carbon content, and de facto, not reactive. Another type of aggregate could lead to different results. Successive batches gas allowed to slightly enhance the CO_2 conversion rate: from 0.055 g CO_2/g sample to 0.061 g CO_2 /g sample for one batch of gas treated and 3 successive batches of treated respectively. The best results involved the renewing of the liquid phase and the solid phase reconditioning between each batch of gas. The tests conducted with steel slags at moderate conditions showed that we can obtain 19 % carbonation efficiency within 10 minutes. These promising results encourages pursuing the development of what could end to be a realistic process at higher scale. Further research to evaluate the energy requirements and the economic feasibility of such process is in progress.

References

1. Olajire, A.A., *A review of mineral carbonation technology in sequestration of CO2.* Journal of Petroleum Science and Engineering, 2013. **109**(0): p. 364-392.

2. O'connor, W., et al., *Aqueous mineral carbonation.* Final Report, DOE/ARC-TR-04-002 (March 15, 2005), 2005.

3. Huijgen, W.J.J. and R.N.J. Comans, *Mineral CO2 Sequestration by Steel Slag Carbonation.* Environmental Science & Technology, 2005. **39**(24): p. 9676-9682.

4. Sanna, A., et al., *A review of mineral carbonation technologies to sequester CO 2.* Chemical Society Reviews, 2014. **43**(23): p. 8049-8080.

5. Chang, E.-E., et al., *Performance evaluation for carbonation of steel-making slags in a slurry reactor.* Journal of hazardous materials, 2011. **186**(1): p. 558-564.

6. Huntzinger, D.N., et al., *Carbon dioxide sequestration in cement kiln dust through mineral carbonation.* Environmental Science & Technology, 2009. **43**(6): p. 1986-1992.

7. Montes-Hernandez, G., et al., *Mineral sequestration of CO2 by aqueous carbonation of coal combustion fly-ash.* Journal of Hazardous Materials, 2009. **161**(2–3): p. 1347-1354.

8. Nyambura, M.G., et al., *Carbonation of brine impacted fractionated coal fly ash: Implications for CO< sub> 2</sub> sequestration.* Journal of environmental management, 2011. **92**(3): p. 655-664.

9. Lawson, N., et al., *Recycling construction and demolition wastes-a UK perspective.* Environmental Management and Health, 2001. **12**(2): p. 146-157.

10. Renforth, P., et al., *Silicate Production and Availability for Mineral Carbonation.* Environmental Science & Technology, 2011. **45**(6): p. 2035-2041.

11. Venta, G.J. and M. Nisbet, *Waste Streams from Building Construction and Demolition with a Specific Focus on Concrete*, S.p.f.P.W.G.S. Canada, Editor. 2001: Ottawa.

12. Nisbet, M., G. Venta, and S. Foo. *Demolition and deconstruction: Review of the current status of reuse and recycling of building materials.* 2012 April 3, 2012]; Available from: ftp://ftp.tech-env.com/pub/Retrofit/AWMA%20paper_wm1b.pdf.

13. Bertos, M.F., et al., *A review of accelerated carbonation technology in the treatment of cement-based materials and sequestration of CO2.* Journal of Hazardous Materials, 2004. **112**(3): p. 193-205.

14. Venhuis, M.A. and E.J. Reardon, *Vacuum Method for Carbonation of Cementitious Wasteforms.* Environmental Science & Technology, 2001. **35**(20): p. 4120-4125.

15. Uliasz-Bocheńczyk, A. and R. Pomykała, *Mineral sequestration of CO 2 with the use of cement waste.* Energy Procedia, 2011. **4**: p. 2855-2860.

16. Kashef-Haghighi, S. and S. Ghoshal, *CO2 sequestration in concrete through accelerated carbonation curing in a flow-through reactor.* Industrial & Engineering Chemistry Research, 2009. **49**(3): p. 1143-1149.

17. Shao, Y., et al., *Accelerated Carbonation of Portland Limestone Cement.* Journal of Materials in Civil Engineering, 2014. **26**(1): p. 117-124.

18. Short, N., P. Purnell, and C. Page, *Preliminary investigations into the supercritical carbonation of cement pastes*. Journal of materials science, 2001. **36**(1): p. 35-41.

19. Proctor, D.M., et al., *Physical and Chemical Characterisctics of Blast Furance, Basic Oxygem Furnace, and Electric Arc Furnace Steel Industry Slags*. Environ. Sci. Technol., 2000. **34**: p. 1576-1582.

20. Poh, H., G.S. Ghataora, and N. Ghazireh, *Soil stabilization using basic oxygen steel slag fines*. Journal of materials in Civil Engineering, 2006. **18**(2): p. 229-240.

21. IPCC, *Les éléments scientifiques* in *IPCC Fourth Assessment Report : Climate change 2007*. 2007.

22. Baciocchi, R., et al., *Thin-film versus slurry-phase carbonation of steel slag: CO 2 uptake and effects on mineralogy*. Journal of hazardous materials, 2015. **283**: p. 302-313.

23. Said, A., et al., *Production of precipitated calcium carbonate (PCC) from steelmaking slag for fixation of CO 2*. Applied Energy, 2013. **112**: p. 765-771.

24. Bonenfant, D., et al., *CO2 sequestration potential of steel slags at ambient pressure and temperature*. Industrial & Engineering Chemistry Research, 2008. **47**(20): p. 7610-7616.

Elucidating the CO_2 chemisorption process on sodium cobaltate ($Na_xCoO_{2\pm\delta}$)

E. Vera* and H. Pfeiffer

Instituto de Investigaciones en Materiales, Universidad Nacional Autónoma de México, México D.F., México.

*Corresponding author: lizvebe@gmail.com

Abstract

Sodium cobaltate was synthesized, characterized and tested as a possible CO_2 captor. The structural and microstructural characterization, by XRD and elemental analysis, indicated a chemical composition of $Na_{0.69}CoO_{1.85}$ for the synthesized material. The CO_2 capture was studied dynamically and isothermally, showing that the $Na_{0.69}CoO_{1.85}$ is able to chemisorb CO_2 in a wide temperature range. The major capture was evidenced between 600 and 725°C, with a maximum capture efficiency of 65.3% at 700 °C. The characterization (structural and microstructural) of the analyzed products showed the formation of Na_2CO_3 and different cobalt oxides. In addition, a cobalt reduction dependent on the temperature was evidenced. Finally, a first order kinetic analysis determined a $\Delta H^{\neq}$ value of 82 kJ/mol for the CO_2 chemisorption process.

Introduction

Carbon dioxide (CO_2) is a major greenhouse gas that is located naturally in the atmosphere. However, many human activities have considerably raised the concentration levels of this gas, approximately 40 ppm in the last ten years. [1] In order to reduce the emissions of CO_2 into the atmosphere and to storage it, many ceramic materials, in which a reversible chemisorption occurs, have been developed. [2] Among these materials, lithium ceramics have shown high capacities to capture CO_2 at high temperatures. [3] The replacement of lithium by sodium in some of these materials has notoriously improved its capture capacity. [4] The CO_2 capture process in these materials consists of two stages, first there is a superficial reaction of the material with the CO_2, where a fine carbonate surface is formed. Once this reaction is completed, the sodium atoms diffusion to the surface occurs and they continue to react with the CO_2. Sodium cobaltate ($NaCoO_2$) has a lamellar structure where the sodium atoms are located between $(CoO)^{+1}$ layers,[5] therefore the sodium atoms may easily diffuse through them when chemisorbing CO_2. The aim of this work is to test and elucidate the CO_2 capture process in a $NaCoO_2$ phase.

Results and discussion

$NaCoO_2$ was synthesized via solid state reaction using sodium carbonate (Na_2CO_3, Aldrich) and cobalt carbonate hydrate ($CoCO_3{\cdot}xH_2O$, Aldrich). The powders were mechanically mixed and calcined using two thermal treatments, first at 750 °C and then 850 °C, both for 12 hours. A 27% excess of Na_2CO_3 was used due to sodium's tendency to sublimate. The synthesized material was later characterized structural and microstructurally by X-ray diffraction (XRD), elemental analysis (atomic absorption), nitrogen adsorption-desorption and scanning electron microscopy (SEM).

XRD pattern fitted to the $Na_{0.74}CoO_2$ phase, and in order to accurately measure the concentrations of sodium (Na) and cobalt (Co) in the sample, an elemental analysis was performed. Therefore, the real chemical sample composition was $Na_{0.69}CoO_{1.85}$. The N_2 adsorption-desorption isotherm corresponded to a type II isotherm, associated to a non-porous material and the BET model provided a small area of 0.6 m^2/g. This result is in good agreement with synthesis method, solid-state reaction. Secondary electron images showed particles with sizes that vary from 2 to 10 μm. Most of the particles are sintered and form agglomerates with sizes from 20 to 100 μm.

The synthesized material was thermogravimetrically analyzed in a Q500HR thermobalance (TA Instruments). The dynamic analysis was performed from 30 to 900 °C at a heating rate of 5 °C/min under a CO_2 flow of 60 mL/min (Praxair, grade 3.0). The dynamic TG curve is presented in figure 1. It presents two temperatures ranges where weight increments were produced: 1) between 300 and 550 °C, the CO_2 is superficially captured, and 2) after 550 °C the CO_2 is volumetrically chemisorbed. Finally, at 750 °C the desorption process begins. From this experiment different temperatures were selected to perform isothermal analysis, varying from 400 to 765 °C. The final weight increment increased with the temperature from 2.4 to 9.5% for 400 and 700 °C, respectively. According to these results, the $Na_{0.69}CoO_{1.84}$ presents the highest CO_2 capture between 600 and 725 °C. At 700 °C the material reaches its highest capture capacity, achieving to capture up to 2.15 mmol of CO_2 per gram of ceramic, corresponding to a 65.3% efficiency of capture. This result is comparable to the efficiencies obtained in other sodium and lithium ceramics with higher alkali/metal molar ratios. [6-10]

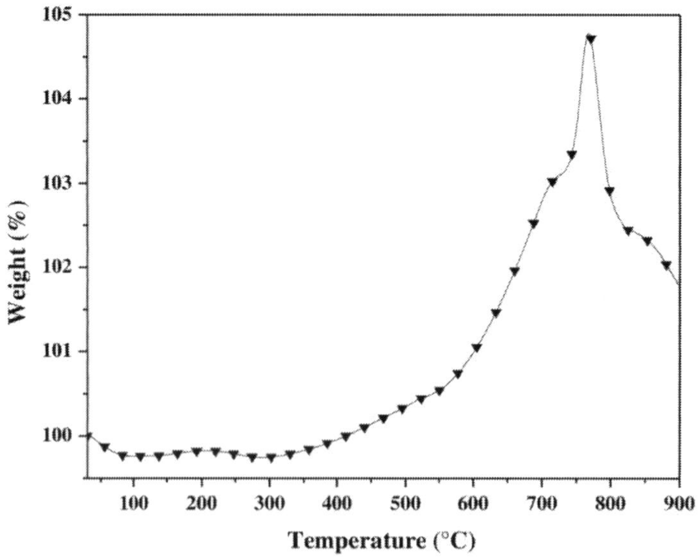

Figure 1. $Na_{0.69}CoO_{1.85}$ dynamic thermogram in a CO_2 flow.

After the isothermal treatments, some of the products were analyzed by SEM and XRD. The diffractograms evidenced the formation of Na_2CO_3 and different cobalt oxides (Co_3O_4 and CoO), where the cobalt oxide formation depends on temperature. Backscattered electron images are presented in figure 2. The size particle and morphology did not present significant changes. Nevertheless, the particles are more defined at higher temperatures, though the agglomerates are denser than in the pristine sample. The formation of different phases was evidenced through the contrast difference, corresponding to Na_2CO_3 (dark phase) and cobalt oxides (light phase).

Finally a kinetic analysis of CO_2 chemisorption was performed. Considering that isothermal experiments were performed in a CO_2 saturated atmosphere it could be assumed that the capture process corresponds to a first-order reaction with respect to the $Na_{0.69}CoO_{1.84}$. Hence, the first seconds of the isothermal TG curves, before the diffusion process begins, were fitted to first-order equations. From this analysis, the reaction rate constants from the superficial capture process were obtained. As it would be expected the kinetic constant values increased as a function of temperature. Then, these values were fitted to the Eyring's model. In this case, only the kinetics constants between 450 and 765 °C fitted, linearly, to the model. Thus, in this temperature range the activation enthalpy determined for the CO_2 chemisorption process was $\Delta H^{\ddagger}$ =82.01 kJ/mol. This value shows that the CO_2 chemisorption is more dependent on temperature than Na_2ZrO_3, but significantly less dependent than Na_2TiO_3, who have $\Delta H^{\ddagger}$ values of 33 kJ/mol and 140.9 kJ/mol, respectively. [9,10]

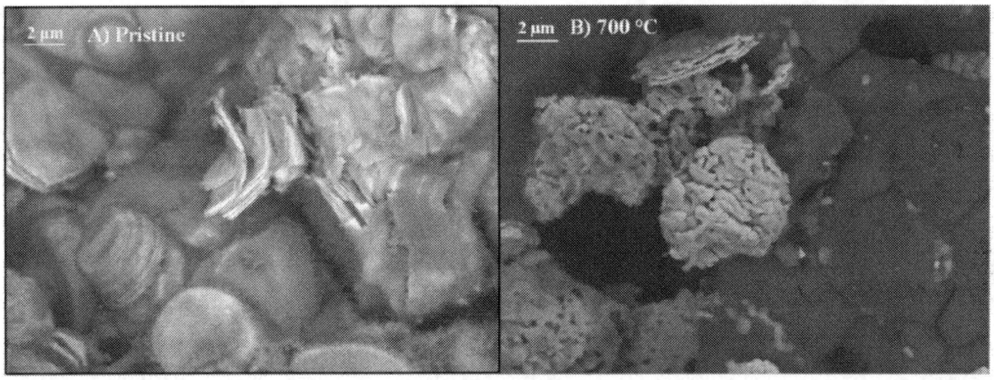

Figure 2. SEM images of the $Na_{0.69}CoO_{1.85}$ pristine sample (A) and of the isothermal products at 700 °C (B).

Conclusions

$Na_{0.69}Co_{1.85}$ was synthesized via solid state reaction. The sample structure and microstructure were analyzed by different techniques. The CO_2 chemisorption process was studied dynamically and isothermally by thermogravimetric analyses. A re-characterization of the products showed the formation of Na_2CO_3 and different cobalt oxides, depending on the temperature. These results show that $Na_{0.69}CoO_{1.85}$ is able to trap CO_2 chemically in a wide range of temperatures. While the superficial processes are important, the CO_2 chemisorption is mainly associated to diffusion processes.

Although the $Na_{0.69}CoO_{1.84}$ shows a low capture capacity, the change in the cobalt valence indicates a complex capture process in which not only superficial and diffusion are present but also electronic exchanges are taking place. The kinetic analysis also showed a capture process which presents a high activation enthalpy.

Acknowledgements

This work was financially supported by SENER-CONACYT 150358 and PAPIIT-UNAM IN-102313 projects. E. Vera wishes to thank CONACYT for financial support. The authors wish to thank Adriana Tejeda and Omar Novelo for technical assistance.

References

1. Monthly Mean Concentrations at Mauna Loa Observatory (PPM), National Oceanic & Atmospheric Administration, Earth System Research Laboratory, Trends in Carbon Dioxide.
2. Markewitz P., Kuckshinrichs W., Leitner W., Linssen J., Zapp P., Bongartz R., Schreiber A., Müller T.E., Worlwide Innovations in the Development of Carbon Technologies and the Utilization of CO_2. *Energy & Environmental Science*, 2012, 5, 7281-7305.
3. Pfeiffer H. Advances in CO_2 Conversion and Utilization; *ACS Symposium Series*; Hu, Y. H., Ed.; American Chemical Society: Washington, DC, 2010; Vol. 1056, pp 233-253.
4. Li Z., Yang J., Hou J. G., Zhu Q., First-principles lattice dynamics of NaCoO2, *Physical Review B*, 2004, 70, 144518.
5. Palacios-Romero L.M., Lima E., Pfeiffer H., Structural and CO_2 chemisorption study on nonstechiometric lithium cuprates ($Li_{2+x}CuO_{2+x/2}$), *Journal of Physical Chemistry A*, 2009, 113, 193-198.
6. Ávalos-Rendón T., Casa-Madrid J., Pfeiffer H., Thermochemical capture of carbon dioxide on lithium aluminates ($LiAlO_2$ and Li_5AlO_4): a new option for the CO_2 absorption, *Journal of Physical Chemistry A*, 2009, 113, 6919–6923.
7. Alcántar-Vázquez B., Diaz C., Romero-Ibarra I. C., Lima E., Pfeiffer H., Structural and CO_2 chemisorption analyses on $Na_2(Zr_{1-x}Al_x)O_3$ solid solutions, *Journal of Physical Chemistry C*, 2013, 117, 16483–16491.
8. Alcérreca-Corte I., Fregoso-Israel E., Pfeiffer H., CO_2 absorption on Na_2ZrO_3: a kinetic analysis of the chemisorption and diffusion processes, *Journal of Physical Chemistry C*, 2008, 112, 6520–6525.
9. Sánchez-Camacho P., Romero-Ibarra I. C., Duan Y., Pfeiffer H., Thermodynamic and kinetic analyses of the CO_2 chemisorption mechanism on Na_2TiO_3: experimental and theoretical evidences, *Journal of Physical Chemistry C*, 2014, 118, 19822–19832.

CO_2 Chemisorption and Structural Analyses of the $Na_2(Zr_{1-x}Al_x)O_{3-x/2}$ Solid Solution

B. Alcántar-Vázquez [1,2*], J.F. Gómez-García [3], G. Tavizon [3], C. Diaz[2], E. Lima[1] and H. Pfeiffer[1]

[1]Instituto de Investigaciones en Materiales, Universidad Nacional Autónoma de México, México DF, Mexico.
[2]Facultad de Ciencias Químicas e Ingeniería, Universidad Autónoma de Baja California, Tijuana B.C. Mexico.
[3]Facultad de Química, Universidad Nacional Autónoma de México, México DF, Mexico.
*Corresponding author: bcecilia_19@iim.unam.mx

Abstract

To determine the influence of the aluminum content in the sodium zirconate during the CO_2 chemisorption process, specific compounds of the $Na_2(Zr_{1-x}Al_x)O_{3-x/2}$ solid solution were synthesized via a solid-state reaction. Samples were characterized using X-ray diffraction and ^{27}Al solid-state nuclear magnetic resonance. Then, samples were tested as CO_2 captors using dynamic and isothermal analyses. Additionally, ionic conduction experiments were performed in different sodium containing phases to determine the effect of these phases in the CO_2 chemisorption process. Characterization results show that aluminum ions can occupy zirconium or sodium sites in the Na_2ZrO_3 structure. Thus, the aluminum dissolution is compensated by different structural defects. The CO_2 capture evaluation shows that the aluminum presence into the Na_2ZrO_3 structure improves the CO_2 chemisorption within certain aluminum content under specific thermal conditions. The CO_2 chemisorption differences were attributed to different sodium secondary phases produced in each case and their corresponding diffusion properties. Finally, cyclic performance tests indicate that $Na_2(Zr_{1-x}Al_x)O_{3-x/2}$ ceramics exhibited high and stable CO_2 capture behaviors.

Introduction

Since the industrial revolution, fossil fuels such as coal, petroleum and natural gas were and still are the most important energy sources. As a consequence, the atmospheric CO_2 concentration has gradually increased. In order to mitigate global warming, it is important to reduce the emissions of CO_2, which is a major greenhouse gas. For this purpose, innovative technologies for carbon capture and storage have been developed in the last years.

Different alkaline ceramics, mainly lithium or sodium, show excellent CO_2 capture properties.[1-3] Among them, zirconates have been reported as good candidates as CO_2 solid sorbents. In 2004 it was reported that Na_2ZrO_3 is able to absorb CO_2 at 600 °C. Since then, several works have reported that Na_2ZrO_3 presents much better characteristics as CO_2 captor in comparison with Li_2ZrO_3.[4-6] Na_2ZrO_3 has a lamellar crystalline structure, where sodium atoms are located among the $(ZrO_3)^{2-}$ layers, which favors sodium diffusion porcesses.[4]

In order to enhance the CO_2 chemisorption on these ceramics, different structural and/or chemical modifications have been carried out. For example, different solid solutions such as $Li_{2-x}Na_xZrO_3$, $Li_{3.7}Al_{0.1}SiO_4$, $Li_{3.7}Fe_{0.1}SiO_4$ and $Li_{4-x}Na_xSiO_4$ have been evaluated as possible CO_2 captors.[5,7-9] In general, all these solid solutions have shown an improvement on different properties of the CO_2 chemisorption reaction, in comparison to their respective pure alkaline ceramics. The improvements observed on the different solid solutions have been attributed to the point defects formation into the lattice or to the secondary phases formation, and these improve diffusion processes, which are the limiting steps of the CO_2 capture. Therefore, the aim of the present work was to analyze both structural and microstructural characteristics of the compounds of the $Na_2(Zr_{1-x}Al_x)O_{3-x/2}$ solid solution and the CO_2 chemisorption properties. [10,11]

Results and discussion

To determine the influence of the aluminum content in the sodium zirconate during the CO_2 chemisorption process, various compounds of the $Na_2(Zr_{1-x}Al_x)O_{3-x/2}$ solid solution were synthesized via a solid-state reaction. The compounds were obtained by mixing mechanically the corresponding amounts of ZrO_2, Na_2CO_3 and $Al(NO_3)_3{\cdot}9H_2O$, with a Na/M molar ratio of 2.1:1 (M = Zr + Al) and x values between 0.05 and 0.3. Powder mixtures were subsequently calcined at 800 and 900 °C for 6 h, respectively. The Na_2ZrO_3 was prepared and analyzed for comparison purposes, following the same synthesis procedure and the corresponding stoichiometric amounts. The samples were structurally characterized using powder X-ray diffraction (XRD) and solid-state nuclear magnetic resonance (MAS NMR).

XRD results show that Na_2ZrO_3 and the compounds of $Na_2(Zr_{1-x}Al_x)O_{3-x/2}$ solid solution correspond to the monoclinic Na_2ZrO_3 crystalline phase (JCPDS file 35-0770), indicating a complete aluminum dissolution up to $x \leq 0.3$. In these cases, the XRD patterns did not show any evident peak shift or intensity changes. After the crystalline verification of the samples, some microstructural properties were determined by N_2 adsorption-desorption. The samples showed N_2 adsorption-desorption type II isotherms, with a very narrow H3-type hysteresis loop, according to the IUPAC classification.[12] This behavior corresponds to non-porous materials frequently obtained by solid state synthesis. Then, the surface areas were measured using the BET method, obtaining very similar values, $\leq 1 m^2/g$, in all cases.

Besides, the ^{27}Al solid-state NMR results showed two resonance peaks centered at 70–78 and 7–9 ppm (Figure 1), which means that the compounds of $Na_2(Zr_{1-x}Al_x)O_{3-x/2}$ structures contain two types of aluminum, tetra- and hexa-coordinated, respectively. Therefore, the aluminum atoms are located in tetrahedral and octahedral positions, corresponding to zirconium and sodium interlayer atoms, respectively. In any case, the presence of aluminum atoms (Al^{3+}) in the tetrahedral (substituting Zr^{4+}) or the octahedral positions (substituting Na^{1+}) induces vacancy generation due to the cation charge variations. Thus, the aluminum dissolution is compensated by different structural defects.

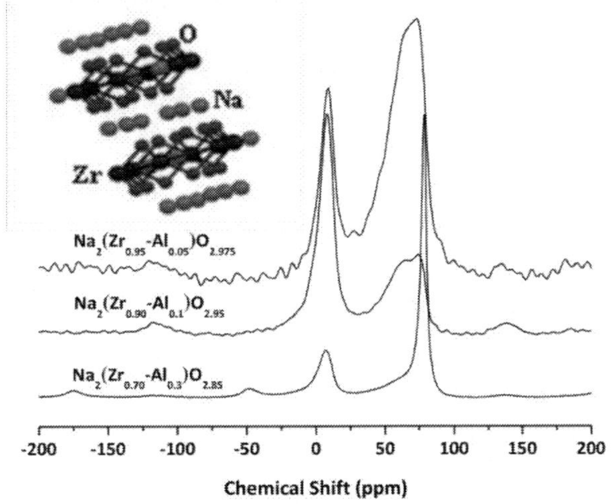

Figure 1. ^{27}Al MAS NMR of the compounds of the $Na_2(Zr_{1-x}Al_x)O_{3-x/2}$ solid solution. The inset shows the scheme of the Na_2ZrO_3 crystalline structure.

Then, samples were tested as CO_2 captors using dynamic and isothermal analyses with Q500HR equipment, from TA Instruments (Figure 2). The samples were dynamically heated from room temperature to 800 at 5 °C/min. These analyses were carried out under a saturated CO_2 atmosphere. For the isothermal analysis, the samples were initially heated to 850 °C using a N_2 flow. This initial thermal step was performed to eliminate any previous sample carbonation. Then, each sample was cooled to its respective isothermal temperature (between 300 and 700 °C) to perform independent CO_2 chemisorption processes. As the sample reached the corresponding temperature, the gas flow was switched from N_2 to CO_2. Figure 2 presents the dynamic thermograms into a CO_2 flux. In the compounds of the $Na_2(Zr_{1-x}Al_x)O_{3-x/2}$ solid solution, the aluminum addition evidently enhances the superficial CO_2 chemisorption process. The weight increases from 2.8 to 9.3 wt % for Na_2ZrO_3 and $Na_2(Zr_{0.70}Al_{0.3})O_{2.85}$, respectively. This tendency corroborates that the particle surface basicity is increased by the aluminum addition. Then, between 300 and 400 °C all samples lost around 0.3 wt %, attributed to the CO_2 chemisorption–desorption superficial equilibrium. Therefore, the superficial desorption process does not seem to be modified by the aluminum presence. After that, the CO_2 capture controlled by diffusion processes is activated at temperatures higher than 400 °C, as in the Na_2ZrO_3 sample. However, qualitatively, the aluminum addition does not seem to improve the final amount of CO_2 captured in this temperature range.

These results were corroborated kinetically ($\Delta H^{\ddagger}$ were determined). In fact, isothermal experiments at 550 °C (see inset in Figure 2), where the diffusion processes had been activated, shows that all the samples reaches the plateau during the first minutes. The

Na_2ZrO_3 is the sample that chemisorbed CO_2 at a slower rate in comparison with the compounds of the $Na_2(Zr_{1-x}Al_x)O_{3-x/2}$ solid solution. After only 8 min, the samples captured between 10 and 14 wt % of CO_2. The final weight gained in $Na_2(Zr_{0.95}Al_{0.05})O_{2.975}$ and $Na_2(Zr_{0.70}Al_{0.3})O_{2.85}$ is the same (17 wt %). Isotherms were fit to double or triple exponential models: (1) the CO_2 chemisorption over the surface of the $Na_2(Zr_{1-x}Al_x)O_{3-x/2}$ particles, (2) the CO_2 chemisorption kinetically controlled by diffusion processes, and (3) the CO_2 desorption process. The last process was detected only between 200 and 400 °C. The kinetic constant values indicated that the CO_2 chemisorption kinetically controlled by diffusion processes is the rate-limiting step for the whole process. Additionally, $\Delta H^{\ddagger}$ values tended to increase as a function of the aluminum content.

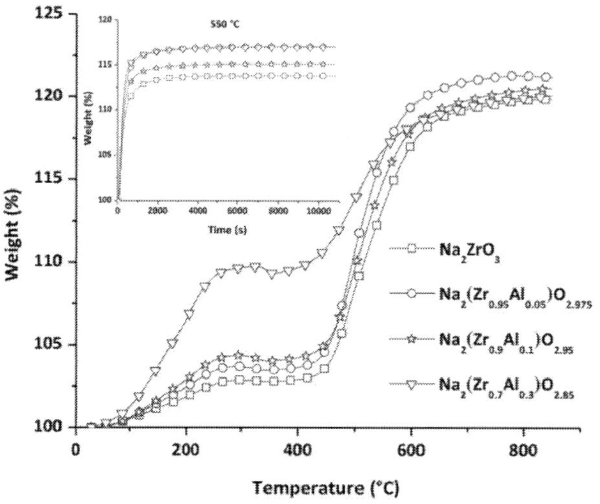

Figure 2. Thermogravimetric dynamic and isothermal analysis at 550 °C (inset) of the $Na_2(Zr_{1-x}Al_x)O_{3-x/2}$ solid solution in a flux of CO_2.

To explain the CO_2 capture increases observed in the Al containing samples, we must take into account the following factors: i) the formation of vacancies during synthesis should favor some intercrystalline diffusion processes (at high temperatures, 600-700 °C); ii) the presence of aluminum modifies the external shell composition, like Na_2CO_3–ZrO_2–$NaAlO_2$–Al_2O_3, where the $NaAlO_2$ and Al_2O_3 contents increase as a function of the aluminum content; and iii) at moderate temperatures (400-550 °C) the external shell mesoporosity decreased with the aluminum, limiting the CO_2 diffusion.

Additionally, to determine the effect of these phases in the CO_2 chemisorption process, the ionic conductivity of the following sodium phases was analyzed: Na_2ZrO_3, $Na_2(Zr_{0.9}Al_{0.1})O_{2.95}$, Na_2CO_3, and $NaAlO_2$. For these experiments, fine grain powders of the samples were pressed in the form of pellets of 1 cm in diameter and about 0.1cm in thickness. Then, each pellet was heated at the sintering temperature of each compound, and the top and bottom pellet surfaces were coated with gold (as oxygen-ion blocking electrode), using a sputtering technique. The measurements were performed in a quartz cell coupled to a vacuum system (800-700 mTorr) and heated between 400 and 800 °C. Both sample sets were analyzed using the 2-point DC technique, and platinum wires as electrodes. If the electronic contribution to conductivity is neglected, and the blocking electrode is taken into account, ionic conductivity will be proportional to the sodium diffusion. Figure 3 shows the temperature dependence of the ionic conductivity for the samples. As it could be expected, all the phases presented a higher ionic conduction as a function of temperature, except for the Na_2CO_3, in which the ionic conductivity decreases between 520 and 595 °C. This is associated with a Na_2CO_3 crystalline phase transition, from monoclinic (at low temperature) to hexagonal (at high temperature).[13] From this figure, it can be seen that the ionic conductivity decreases as follows in the sodium zirconate and aluminate phases: $Na_2ZrO_3 > Na_2(Zr_{0.9}Al_{0.1})O_{2.95} \gg NaAlO_2$. Therefore, the sodium ionic conduction must be favored in the pristine zirconate phase without aluminum. Additionally, the carbonate external shell is composed of $NaAlO_2$ and Na_2CO_3. In this case, if the sodium ionic conduction in the $NaAlO_2$ phase is compared with that of the Na_2CO_3 phase, it is clearly evident that sodium ionic conduction is considerably lower in the $NaAlO_2$ phase.

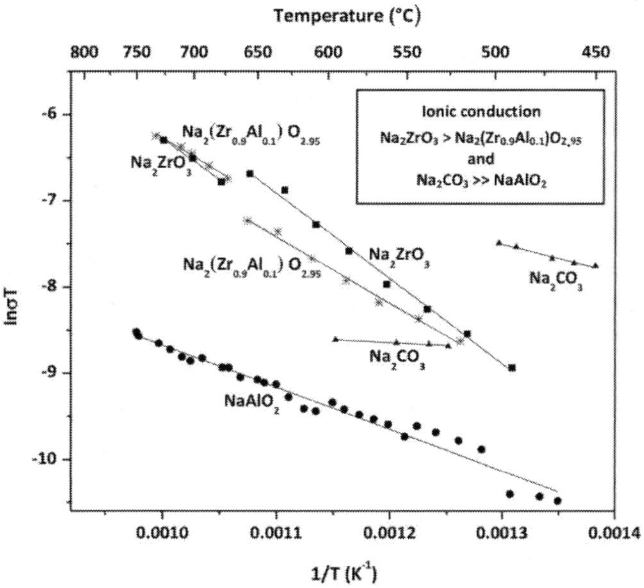

Figure 3. Temperature dependence of the ionic conductivity for the Na_2ZrO_3, $Na_2(Zr_{0.9}Al_{0.1})O_{2.95}$, $NaAlO_2$, and Na_2CO_3 samples.

Moreover, to evaluate the regeneration properties and the thermal stability after several cycles of CO_2 chemisorption/desorption the compounds $Na_2(Zr_{0.95}Al_{0.05})O_{2.975}$ and $Na_2(Zr_{0.70}Al_{0.3})O_{2.85}$, were tested using a multi cycle method. The cyclic behavior was analyzed at 550 and 600 °C for $Na_2(Zr_{0.95}Al_{0.05})O_{2.975}$ and $Na_2(Zr_{0.70}Al_{0.3})O_{2.85}$, respectively, because at these temperatures the highest CO_2 chemisorptions were obtained. Results presented in Figure 4 show that the CO_2 chemisorption capacity after 20 cycles stabilizes to 19.1 wt % for $Na_2(Zr_{0.95}Al_{0.05})O_{2.975}$ and 16.3 wt % for

$Na_2(Zr_{0.70}Al_{0.3})O_{2.85}$. Namely, the cyclic experiments indicate that $Na_2(Zr_{1-x}Al_x)O_{3-x/2}$ ceramics exhibited high and stable CO_2 capture behaviors.

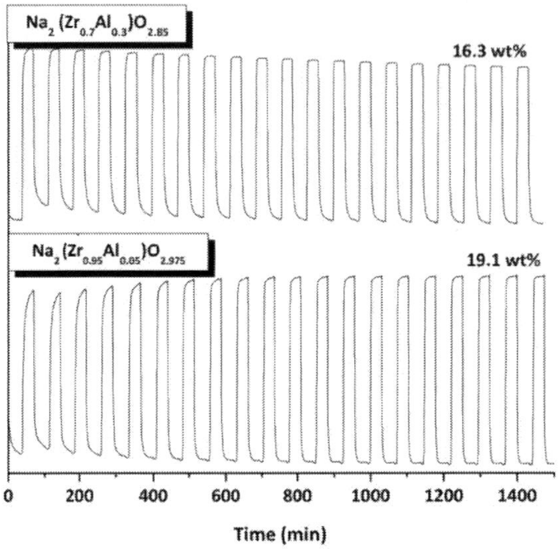

Figure 4. Multicycle performance of CO_2 chemisorption/desorption on $Na_2(Zr_{0.95}Al_{0.05})O_{2.975}$ and $Na_2(Zr_{0.7}Al_{0.3})O_{2.85}$, where the CO_2 absorptions were performed at 550 and 600 °C for 30 min, respectively. Desorption process was performed at 800 °C under a N_2 flow.

Conclusions

The results presented here confirm that aluminum atoms are incorporated into the Na_2ZrO_3 structure, occupying zirconium and sodium atom sites. Thus, the aluminum dissolution is compensated by different structural defects. The compounds of the $Na_2(Zr_{1-x}Al_x)O_{3-x/2}$ solid solution, with aluminum atoms localized at octahedral sites, were able to chemisorb CO_2 in a wide temperature range (200–700 °C), exhibiting higher CO_2 chemisorptions than Na_2ZrO_3. The CO_2 chemisorption differences were attributed to several factors such as the presence or absence of different structural vacancies and the presence of different secondary phases in the carbonated external shell. In this regard, the ionic conductivity processes in the external shell are limited by the $NaAlO_2$ presence. Finally, the cyclic experiments indicate that the $Na_2(Zr_{0.95}Al_{0.05})O_{2.975}$ compound presents high and stable behaviors.

Acknowledgements

This work was financially supported by the projects SENER-CONACYT 150358 and PAPIIT-UNAM IN-102313. B.A.-V. thanks CONACYT for financial support. We thank Adriana Tejeda and G. Zedillo for technical help.

References

1. Kumar, S., Saxena, S.K. 2014. A Comparative Study of CO_2 Sorption Properties for Different Oxides. *Materials for Renewable and Sustainable Energy*, 3, 1–15.
2. Wang, J., Huang, L., Yang, R., Zhang, Z., Wu, J., Gao, Y., Wang, Q., O'Hare, D., Zhong, Z. 2014. Recent Advances in Solid Sorbents for CO_2 Capture and New Development Trends. *Energy & Environmental Science,* DOI: 10.1039/C4EE01647E
3. Ortiz-Landeros, J., Ávalos-Rendón, T., Gómez-Yáñez, C., Pfeiffer, H. 2012. Analysis and Perspectives Concerning CO_2 Chemisorption on Lithium Ceramics Using Thermal Analysis. *Journal of Thermal Analysis and Calorimetry,* 108, 647–655.
4. Alcerreca-Corte, I., Fregoso-Israel, E., Pfeiffer, H. 2008. CO_2 Absorption on Na_2ZrO_3: A Kinetic Analysis of the Chemisorption and Diffusion Processes. *Journal of Physical Chemistry C,* 112, 6520–6525.
5. Pfeiffer, H., Vazquez, C., Lara, V.H., Bosch, P. 2007. Thermal Behavior and CO_2 Absorption of $Li_{2-x}Na_xZrO_3$ Solid Solutions. *Chemistry of Materials,* 19, 922–926.
6. Zhao, T., Ochoa-Fernández, E., Rønning, M., Chen, D. 2007. Preparation and High-Temperature CO_2 Capture Properties of Nanocrystalline Na_2ZrO_3. *Chemistry of Materials,* 19, 3294–3301.
7. Mejia-Trejo, V., Fregoso-Israel, E., Pfeiffer, H. 2008. Textural, Structural, and CO_2 Chemisorption Effects Produced on the Lithium Orthosilicate by its Doping with Sodium ($Li_{4-x}Na_xSiO_4$). *Chemistry of Materials,* 20, 7171–7176.
8. Gauer, C., Heschel, W. 2006. Doped Lithium Orthosilicate for Absorption of Carbon Dioxide. *Journal of Materials Science,* 41, 2405–2409.
9. Pfeiffer, H., Lima, E., Bosch, P. 2006. Lithium-Sodium Metazirconate Solid Solutions, $Li_{2-x}Na_xZrO_3$ ($0 \le X \le 2$), a Hierarchical Architecture. *Chemistry of Materials,* 18, 2642-2647.
10. Alcántar-Vázquez, B., Diaz, C., Romero-Ibarra, I.C., Lima, E., Pfeiffer, H. 2013. Structural and CO_2 Chemisorption Analyses on $Na_2(Zr_{1-x}Al_x)O_3$ Solid Solutions. *Journal of Physical Chemistry C,* 117, 16483–16491.
11. Alcántar-Vázquez, B., Gómez-García, J.F., Tavizon, G., Ibarra, I.A., Diaz, C., Lima, E., Pfeiffer, H. 2014. Structural and Ionic Conduction Analyses of the $Na_2(Zr_{1-x}Al_x)O_{3-x/2}$ Solid solution, During the CO_2 Chemisorption Process. *Journal of Physical Chemistry C,* 118, 26212–26218.
12. Lowell, S., Shields, J.E., Thomas, M.A. 2004. Characterization of Porous Solids and Powders: Surface Area, Pore Size and Density. Particle Technology Series, Kluwer Academic Publishers, London.
13. Swainsonf, I. P., Dovet, M.T., Harrisfill, M.J. 1995. Neutron powder diffraction study of the ferroelastic phase transition and lattice melting in sodium carbonate, Na_2CO_3. *Journal of Physics: Condensed Matter*, 7, 4395–4417.

CO_2 capture in lithium cuprate (Li_2CuO_2) at low temperatures; Effect of the water vapour addition

Hugo A. Lara-García[1] and Heriberto Pfeiffer[1]

[1]Instituto de Investigaciones en Materiales, Universidad Nacional Autónoma de México, Circuito Exterior s/n Cd. Universitaria, Del. Coyoacán, CP 04510, México DF, Mexico.

hugolfc28@gmail.com

Abstract

Li_2CuO_2 was synthesized using a solid-state reaction, and then different water sorption experiments were performed using N_2 and CO_2 as carrier gases. When the N_2–H_2O flow gas was used, Li_2CuO_2 showed a water vapor sorption because of two different processes: superficial hydration and hydroxylation. However, if CO_2 was used as the water vapor carrier gas, the Li_2CuO_2 carbonation process was observed at low temperatures (30–80 °C), the carbonation process was induced by the lithium ceramic superficial hydroxylation. The Li_2CuO_2 captured 6.7 mmol of CO_2 per gram of ceramic (efficiency of 73%), a considerably high CO_2 amount among different materials.

Introduction

Carbon dioxide (CO_2) emissions are one of the most threatening pollution problems in the world. One approach to solve such environmental problem is the CO_2 capture and storage. Therefore, CO_2 capture in solid materials have been considered to be the most practical option as an efficient and cheap alternative to reduce greenhouse gases. In this way, different materials have been proposed as CO_2 captors such as zeolites, porous active carbons, metal–organic frameworks (MOFs), alkali metal-promoted alumina and carbonates, alkaline and alkaline-earth ceramics, and layered double hydroxides.[1]

In addition, water vapour is commonly found to coexist with CO_2, during post-combustion. In a typical powder plant, the final flue gases contain approximately N_2 (72%), CO_2 (8-12%), H_2O (8-10%) and smaller concentrations of the other polluting species. Consequently, it is very important to understand the CO_2-H_2O system.[2]

Lithium cuprate (Li_2CuO_2) has been studied as CO_2 chemisorbent at high temperature (>200ºC) presenting interesting results. The process of CO_2 capture, using Li_2CuO_2 as a solid absorbent, takes place according to the next reaction:

$$Li_2CuO_{2(s)} + CO_{2(g)} \leftrightarrow Li_2CO_{3(s)} + CuO_{(s)}$$

For this reaction, the theoretical capacity corresponds to 9.1mmol/gr, and the maximal capture experimental reported is about 9.05 mmol/gr.[3,4,5,6]

On the other hand, in some lithium ceramic it has been reported that the presence of water vapour couses the capture of CO_2 at low temperature. Therefore the objective of this work is the study of the lithium cuprate under these conditions.

Experimental Section

Lithium cuprate (Li_2CuO_2) was synthesized by solid-state method. Initially, lithium oxide (Li_2O, Aldrich) and copper oxide (CuO, Acros Organics) were mixed mechanically, in order to get a good homogeneity of the reagents, mixtures were prepared using a lithium excess of 10%, based on the stoichiometric lithium content on Li_2CuO_2. Then, the powders were calcined at 800 °C for 6 h in air.

A diffractometer (Siemens D-5000) coupled to a cobalt anode X-ray tube was used to identify the phases obtained. Additionally the microstructural characteristics of the Li_2CuO_2 sample was determined via N_2 adsorption-desorption and scanning electron microscopy (SEM). N_2 adsorption (Bel-Japan Minisorp II) was used to determine the sample surface area using the BET model. Then, the sample morphology was analyzed by SEM, which was performed on a JEOL JMS-7600F.

Dynamic and isothermal experiments were carried out on a humidity-controlled thermobalance (TA Instruments, model Q5000SA) at different temperatures and relative humidity (RH). The experiments were performed using distilled water and two different gases: nitrogen (N_2, Praxair grade 4.8) and carbon dioxide (CO_2, Praxair grade 3.0). The total flow gas used in all the experiments was 100 mL/min and the RH percentages were controlled automatically by the Q5000SA equipment. Dynamic water vapor sorption/desorption experiments were generated at different temperatures (between 30 and 80 °C), varying the RH from 0 to 80% and then from 80 to 0%. Additionally, different isothermal experiments were performed at specific temperatures and RH for 180 min, using N_2 or CO_2 as carrier gases.

Afterward, the products were characterized to identify the hydration products. The samples were analyzed using infrared spectroscopy (FTIR), and thermogravimetric analysis (TGA). For FTIR espectroscopy samples were analyzed in an Alpha-Platinum spectrometer from Bruker. The TGA measurements were performed under a nitrogen atmosphere using a TA Instruments model Q500HR thermobalance from 30 to 930 °C.

Results and Discusion

The Li_2CuO_2 XRD pattern of the prepared sample, it was fitted to JCPDS file 00-084-1971 and secondary phases were not detected. After, the sample microstructure was analyzed by N_2 adsorption and SEM. The N_2 adsorption-desorption isotherm was type II according to the IUPAC classification.[7] This kind of N_2 adsorption-desorption isotherm is associated to macroporous or nonporous materials. Then, surface area was determined using the BET model (0.2 m^2/g). This result is in good agreement with the synthesis method used, solid-state reaction.

Experimentally, Li_2CuO_2-N_2-H_2O and Li_2CuO_2-CO_2-H_2O systems were evaluated at different temperatures (30-80 °C). Initially, the water sorption–desorption curves were obtained using N_2 as a carrier gas. In these cases, the sorption isotherms corresponded to type III, according to the IUPAC classification.[7] The water sorption varied as a function of the temperature. The same isothermal trends were observed when CO_2 was used as carrier gas. Nevertheless, the weight increments were much higher in the CO_2 cases. Figure 1 shows the Li_2CuO_2-N_2-H_2O and Li_2CuO_2-CO_2-H_2O sorption-desorption curves, it is clear that the final weight gained is twice larger in the CO_2-H_2O system (at 80 ºC) in comparison to the N_2-H_2O case. Therefore, the CO_2-H_2O system produces different reactions than N_2-H_2O, where hydration, hydroxylation, and carbonation processes must be taking place.

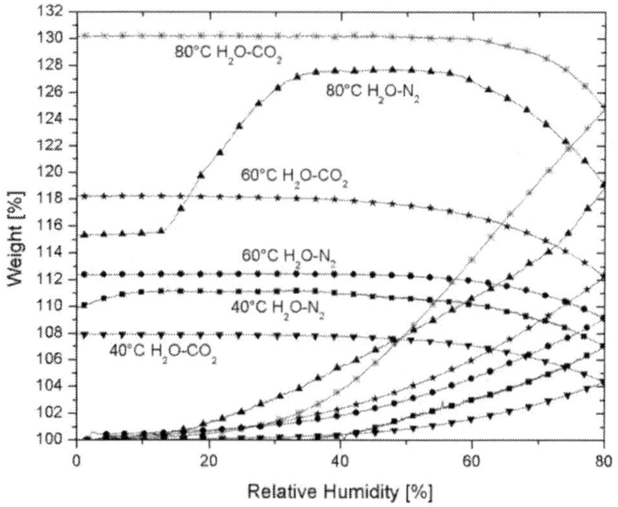

Figure 1. Li_2CuO_2-H_2O-N_2 and Li_2CuO_2-H_2O-CO_2 sorption-desorption curves performed between 40 and 80 °C.

To further understand the influence of RH and temperature in the Li_2CuO_2–CO_2–H_2O system, different isothermal experiments were performed at different temperatures and RH. For each temperature, the weight increment rates increased as a function of the RH, as it could be expected. The final weight increments increased as a function of temperature and RH, it can be well represented if the isotherms with 80 % of RH are compared at different temperatures. The final weights in these cases were 21.5, 24.5 and 37.5 wt% at 40, 60 and 80 ºC, respectively (Figure 2).

To confirm and quantify that CO_2 was chemisorbed on Li_2CuO_2, the products were analyzed using infrared spectroscopy (FTIR) and termogravimetric analysis. The IR spectrums presented the metal–oxygen vibration bands (Li–O and Cu–O) between 400 and 750 cm^{-1}, and presented all vibration bands of lithium carbonate (1499, 1434, 1086, 862, 738 and 498 cm^{-1}). The thermograms show three different decomposition processes. Initially, between room temperature and 120 °C, the samples lost small quantities of weight (between 1 and 2.5 wt%), which could be attributed to dehydration process. The second weight decrement was observed between 350 and 470 °C, and it can be attributed to the dehydroxylation process and the decarbonation process was produced in two steps between 610 and 760 ºC and between 760 and 925 °C. These two processes can be described as superficial and bulk decarbonation processes.

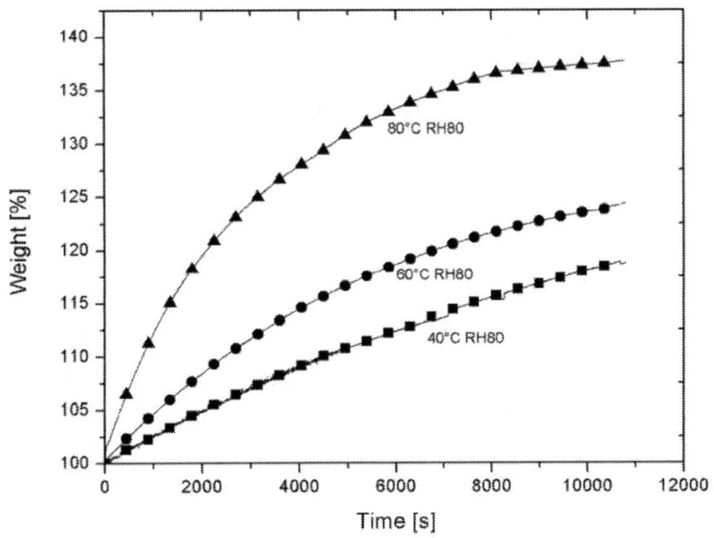

Figure 2. Li_2CuO_2-CO_2-H_2O kinetic isotherms performed at different temperatures (40, 60 and 80 °C) at a RH 80 %.

The decomposition results confirm that a higher temperature and relative humidity the lithium cuprate capture more CO_2. The maximum capture is at 80 °C with a 80 % of humidity relative the Li_2CuO_2 capture about 6.68 mmol/gr to CO_2, while at 80°C and 60% of RH capture 5.27 mmol/gr and, at 60°C and 80% of RH 4.39 mmol/gr to CO_2.

Conclusions

The CO_2–H_2O sorption process of Li_2CuO_2 was evaluated at low temperatures (30–80 °C). Lithium cuprate have shown good CO_2 capture properties at these temperatures in the presence of water vapor, in comparison to dry conditions. The explanation given for this effect has been associated to the ceramic hydroxylation process, which promotes the CO_2 reactivity.

Reference

[1] Choi, S.; Dress, J. H.; Jones, C. W. ChemSusChem. 2009, *Adsorbent materials for carbon dioxide capture from large anthropogenic point sources,* 2 (9), 796–854.

[2] Avalos-Rendon, T.; Pfeiffer, H. Energy & Fuels. 2012, *Thermochemical capture of carbon dioxide on lithium aluminates ($LiAlO_2$ and Li_5AlO_4): a new option for the CO_2 absorption,* 26 (5), 3110-3114

[3] Palacios-Romero, L. M.; Pfeiffer, H. Chem. Lett. 2008, *Lithium cuprate (Li_2CuO_2): A new. possible ceramic material for CO_2 chemisorption*, 37, 862–863.

[4] Palacios-Romero, L. M.; Lima, E.; Pfeiffer, H., J. Phys. Chem. A. 2009, *Structural Analysis and CO2 Chemisorption Study on Non-stoichiometric Lithium Cuprates ($Li_{2+x}CuO_{2+x/2}$),* 113 (1), 193–198

[5] Yusuke Matsukura, Takeshi Okumura, Ryota Kobayashi, Katsuyoshi Oh-ishi; Chem. Lett. 2010, *Synthesis and CO_2 Absorption Properties of Single-phase Li_2CuO_2 as a CO_2 Absorbent,* 39 (9), 966-967

[6] Katsuyoshi Oh-ishi, Yusuke Matsukura, Takeshi Okumura, Yuuki Matsunaga, Ryota Kobayashi J. Solid State Chem. *2014, Fundamental research on gas–solid reaction between CO_2 and Li_2CuO_2 linking application for solid CO_2 absorbent,* 211 ,162-169.

[7] S. Lowell, J. E. Shields, M. A. Thomas; 2004, *Particle Technology Series,* Kluwer Academic Publishers, London.

Dissolution of Thermally Activated Serpentine at Flue Gas Conditions – Near Equilibrium Dynamics

S. Hariharan, M. Werner and M. Mazzotti*

Institute of Process Engineering, ETH Zurich, Zurich, Switzerland

*Corresponding author: marco.mazzotti@ipe.mavt.ethz.ch

Abstract

This article briefly communicates our recent work on the near-equilibrium dissolution dynamics of thermally activated serpentine (dehydroxylated lizardite) under conditions relevant for CO_2 mineralization. A parametric study on the non-steady state dissolution of dehydroxylated lizardite particles was performed over temperature ranges 30°C ≤ T ≤ 90°C and CO_2 partial pressures of 0.025 ≤ pCO_2 ≤ 1.0 atm in batch operation, for durations up to 3 hours. Non-stoichiometric release of Mg(aq) and SiO_2(aq) were observed. The far-from-equilibrium dissolution model was unable to predict the dissolution profiles under the new operating conditions of high pH, high aqueous species activities and the approach of dissolution equilibrium.

Introduction

The rate limiting step for a CO_2 mineralization process is in ensuring a continuous supply of free Mg^{2+}/Ca^{2+} cations at high concentrations for carbonate precipitation. The metal cations are usually obtained by dissolving silicate rocks. The great abundance of the serpentine group makes it an attractive feedstock for mineral carbonation. However, it suffers from slow dissolution kinetics when compared to several other less abundant natural minerals. Thermal activation through dehydroxylation has shown to accelerate the dissolution kinetics for serpentine[1]. However, heat activation is an energy intensive process and adds substantially to the costs. The use of direct flue-gas mineralization in aqueous medium avoids the costs associated with the capture of CO_2.

Dissolving activated serpentine offers several advantages over a process that uses pH-tuning additives to dissolve natural serpentine. These include the simplicity of the aqueous chemistry, the non-necessity for additive recovery and the dissolution at lean operating conditions. Earlier, we had studied the far-from-equilibrium, non-steady state dissolution kinetics for 75% dehydroxylated lizardite particles under temperatures (*T*) and partial pressures of CO_2 (pCO_2) not exceeding 120°C and 2 bar, respectively. The dissolution was performed in a liquid and gas flow-through set-up operated at high flow rates to keep the Mg^{2+} concentrations low. Up to 80% of the magnesium could be extracted within 100 min [2]. We had also developed the first kinetic model that describes the dissolution of dehydroxylated lizardite particles at far-from-equilibrium conditions [3]. In a real CO_2 mineralization process, the dissolution of these particles will be subject to influences arising from thermodynamic constraints and secondary precipitation phenomena. Nevertheless, the far-from-equilibrium kinetic study experimentally demonstrated the rapid dissolution kinetics for dehydroxylated lizardite. It also allowed for the study of the intrinsic dissolution kinetics, independent of thermodynamic effects. This is important since when one develops an optimal dissolution strategy, solutions aimed at tackling kinetic and thermodynamic limitations will not be identical. Having understood the far-from-equilibrium kinetics, we have followed it up by studying the dissolution dynamics under environments representative of a real process condition, i.e. near-equilibrium conditions. The preliminary result of this study is presented in this communication.

Experimental and Results

A parametric study on the non-steady state dissolution of dehydroxylated lizardite particles was performed over temperature ranges 30°C ≤ T ≤ 90°C and CO_2 partial pressures of 0.025 ≤ pCO_2 ≤ 1.0 atm in batch operation inside a 100 mL Teflon reactor. A conductivity probe (InLab®731, Mettler Toledo) was calibrated that enabled continuous online measurement of the solution conductivity during dissolution. Experiments were performed for durations up to 3 hours. Aqueous samples were also taken and analyzed for dissolved silica concentrations. Different amounts of activated serpentine were dissolved in order to achieve high values for the activity ratio term $aMg^{2+}/(aH^{+})^{2}$. The upper limit values for this activity ratio term were determined by the solubility of the kinetically relevant magnesium carbonate at the operating conditions applied. Two different particle size fractions were also tested.

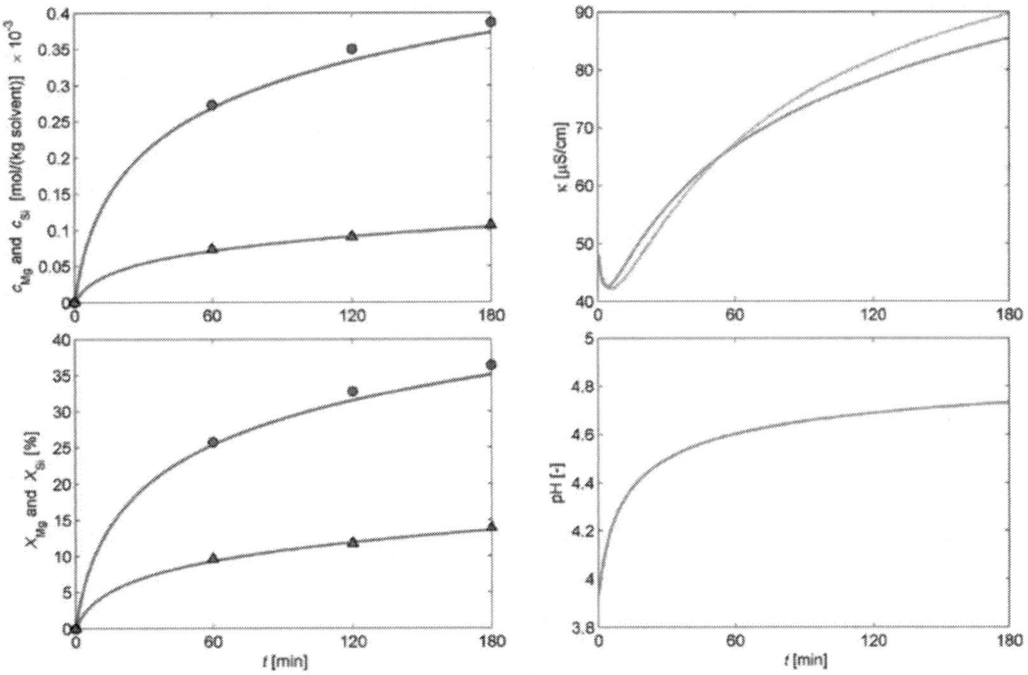

Figure 1: Experimental and simulated profiles for the batch dissolution of 10 mg of dehydroxylated lizardite at 30°C and 1 atm CO_2 conditions. The dissolved magnesium profiles are shown in red and the dissolved silica profiles are shown in blue. Experimental and simulated solution conductivity profiles (κ) are shown as black and green lines respectively. The simulated pH of the solution is also shown.

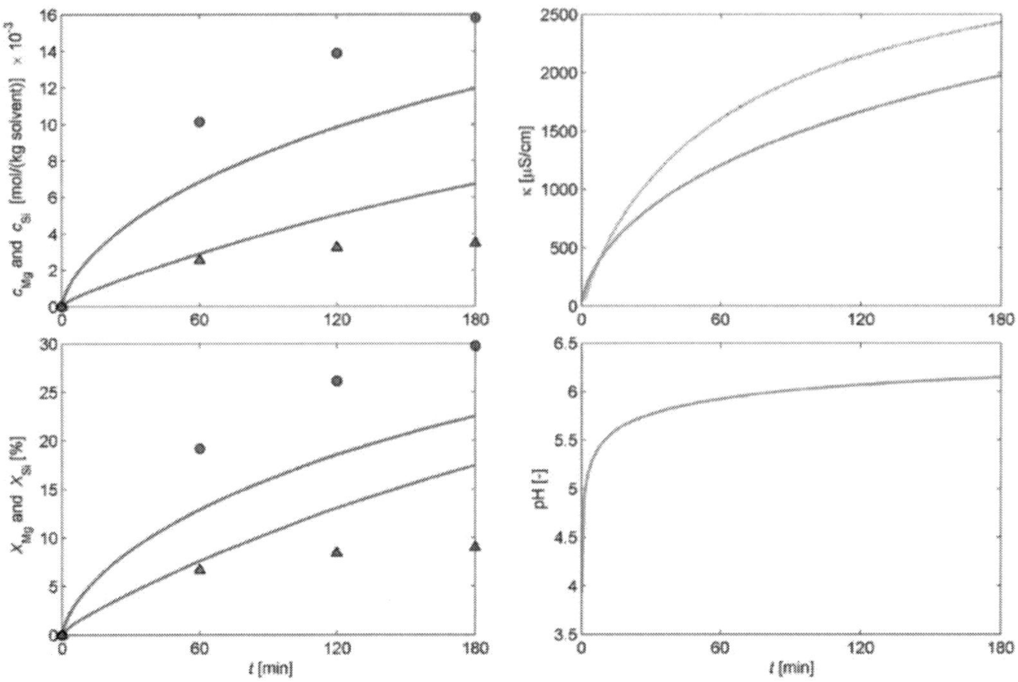

Figure 2: Experimental and simulated profiles for the batch dissolution of 500 mg of dehydroxylated lizardite at 30°C and 1 atm CO_2 conditions. The dissolved magnesium profiles are shown in red and the dissolved silica profiles are shown in blue. Experimental and simulated solution conductivity profiles (κ) are shown as black and green lines respectively. The simulated pH of the solution is also shown.

Figure1 shows the dissolution profiles for 10 mg of dehydroxylated lizardite particles at 30°C and under 1 atm CO_2. The dissolution conditions observed in this experiment were still representative of far-from-equilibrium operation. The far-from-equilibrium model was successfully able to predict the dissolution profiles accurately, thus validating the new operating procedures of the batch experimental set-up. However, upon increasing the amount of particles loaded to 500 mg, the prediction of the model is poorer (Figure2). Interestingly, the model under-estimates the magnesium dissolution rate, while it over-estimates the silica dissolution rate. It has to be kept in mind that the obtained magnesium and silica concentrations are at least two orders of magnitude higher than those in our far-from-equilibrium experiments. Therefore, there is a need to update the dissolution rate expressions to accommodate the new effects of higher pH, high aqueous species activities and the approach of dissolution equilibrium.

References

1. O'connor, W. K., Dahlin, D. C., Rush, G. E., Gerdemann, S. J., Penner, L. R., & Nilsen, D. N. 2005. Aqueous mineral carbonation: Mineral availability, pretreatment, reaction parametrics, and process studies. *A report produced by the Albany Research Centre for the US Department of Energy*.
2. Werner, M., Hariharan, S., Zingaretti, D., Baciocchi, R., & Mazzotti, M. 2014. Dissolution of dehydroxylated lizardite at flue gas conditions: I. Experimental study. *Chemical Engineering Journal*, 241, 301-313.

3. Hariharan, S., Werner, M., Hänchen, M., & Mazzotti, M. 2014. Dissolution of dehydroxylated lizardite at flue gas conditions: II. Kinetic modeling. *Chemical Engineering Journal*, 241, 314-326.

Critical role of water during CO_2 sequestration in nickel mining residues under ambient conditions

A. Entezari-Zarandi,[1] F. Larachi,[*,1] G. Beaudoin,[2] B. Plante[3]

[1]Department of Chemical Engineering, [2]Department of Geology and Geological Engineering, Université Laval, Québec, QC, Canada G1V 0A6; [2]Université du Québec en Abitibi-Témiscamingue, Rouyn-Noranda, Québec, Canada J9X 5E4

*Corresponding e-mail address: Faical.Larachi@gch.ulaval.ca

Abstract

Mineral carbonation is among the most promising options being researched for permanent carbon capture and storage owing to the known reactivity of magnesium silicates and brucite with carbon dioxide to form stable and environmentally benign carbonates. Milling process residues from the Dumont Nickel Project (DNP, Québec) will contain considerable amounts of chrysotile, lizardite and brucite. A differential diffusion carbonation cell was built to monitor the kinetics of mineral carbonation under ambient conditions. The kinetic measurements revealed the complex role of water both as reacting medium and moiety in the carbonation pathway. Mg^{2+} and OH^- ions, produced by brucite dissolution, are essential components in regulating medium alkalinity and formation of carbonated products. Instead of forming anhydrous $MgCO_3$, the highly hydrated nature of Mg^{2+} favored the formation of metastable hydrous carbonates. Nesquehonite ($Mg(HCO_3)(OH)\cdot 2H_2O$) was identified by XRD analysis as the neoformed carbonate mineral. In this study emphasis was placed on the formation and consumption of water during ambient mineral carbonation and possible carbonation reaction pathways. The study's results shed light on the carbonation mechanisms occurring within mining residues in contact with ambient air.

Introduction

Developing economically feasible strategies for long-term storage of carbon dioxide has become over the past few years a major stake in response to the concerns over global warming. CO_2 sequestration in geological formations and mineral carbonation rank high among the most promising strategies. While the former emphasizes trapping gaseous, liquid, or supercritical carbon dioxide deep in geological formations , which will require monitoring of potential leakages into groundwater or back into the atmosphere,[1] the latter is implemented with minimal aboveground infrastructure to produce environmentally and chemically benign carbonates with no need of post-monitoring.[2, 3]

Availability of large amounts of magnesium-rich mining wastes, which are regarded to be virtually free, typically fine grained and highly reactive material, could be considered as an economically attractive carbon capture and storage (CCS) option to capture CO_2 from the atmosphere.[4] Efforts are being made to introduce and optimize techniques on an industrial and global scale to permanently sequester atmospheric

CO_2 in the form of stable mineral carbonates starting with mining and industrial residues.[5-14] Alkaline wastes such as steel slag,[15] cement kiln dust,[11] coal fly ash,[16] saline waste water[17] and municipal solid wastes[18] have been recently evaluated for their usefulness in carbonation.

In ambient conditions, secondary carbonates such as nesquehonite ($Mg(HCO_3)(OH)\cdot 2H_2O$) and hydromagnesite ($4MgCO_3\cdot Mg(OH)_2\cdot 4H_2O$) are reported to normally form[13, 19, 20] on the surface of waste piles with the additional advantage of cementing the particles together to prevent some environmental issues related to dispersion of materials such as mineral fibers. Normally, under ambient conditions, the metastable hydrated carbonate phases interfere with the production of magnesite because of a slower magnesite nucleation and crystal growth.[21] The formation of nesquehonite at ambient temperature and atmospheric CO_2 pressure whether in aqueous media[20, 22] or with brucite-rich mining residues[8, 20, 23] is well documented in the literature. Lansfordite, $MgCO_3\cdot 5H_2O$,[24] may form during low-temperature carbonation but transforms into nesquehonite at temperatures over 10 °C. Swanson et al. reviewed the different magnesium carbonate phases depending on the Mg^{2+} source, e.g., magnesium salt or hydroxide.[21] They showed that seeding the carbonating medium with magnesite favored the formation of such anhydrous carbonate at the expense of species as nesquehonite and hydromagnesite.[21] Table 1 lists the common magnesium carbonates that may form in near ambient conditions:

Table 1. Common magnesium carbonates forming near ambient conditions.

Carbonate mineral	*Formula*	*reference*
Artinite	$Mg_2(CO_3)(OH)_2\cdot 3H_2O$	25
Nesquehonite	$MgCO_3\cdot 3H_2O$	6-8, 20, 22, 26
Hydromagnesite	$Mg_5(CO_3)_4(OH)_2\cdot 4H_2O$	27, 28
Dypingite	$Mg_5(CO_3)_4(OH)_2\cdot 5H_2O$	7, 28
Magnesite	$MgCO_3$	29

While ultramafic wastes are promising sinks for CO_2 sequestration, their alkaline nature upon contact with rain can pose serious concerns on alkaline rock drainage. Although high alkalinity of the medium is essential for formation of carbonated species, high pH run-off waters can be harmful to the environment either because compounds which are sparingly soluble in neutral water may be become soluble in alkaline media, or, on contrary, being already dissolved in neutral pH may precipitate in the alkaline environment.[30-32] Also, Canada's metal mining effluent regulations (MMER) stipulates that discharge of alkaline waters with pH > 9.5 is prohibited. Mineral carbonation includes consumption of metal and hydroxide ions for formation of (hydrated) metal carbonates and water. To the best of these authors' knowledge, the dynamics of water formation/consumption during the evolution of hydrated carbonates has hitherto been scarcely covered.

Research conducted in our group to date showed that carbonation is very sensitive to water saturation and watering frequency of mineral layers. Similar effects of water saturation on the carbonation of NiMR beds were also reported Awoh et al.[33] While there is virtually no carbonation of dry material in contact with dry gas (i.e., moisture free), the degree carbonation of partially wetted materials exposed to dry/humid gas was noticeable. Also, little carbonation was found in the case of dry material traversed for several days by humid gas streams.[29]

Monitoring of CO_2 partial pressure decline through carbonation in a modified eudiometer enabled us to measure CO_2 consumption continuously in near ambient conditions. These experiments already confirmed that pore-water saturation and partial pressure of CO_2 are the most influential factors during material carbonation. Nesquehonite and dypingite were detected as the dominant carbonate species, respectively, at 21°C and 33°C.[7] To assess the impact of residue mineralogy and grain size, gas composition, water saturation and watering frequencies on carbonation, a fixed-bed reactor was designed and instrumented for in-situ monitoring of the headspace composition while solid samples were also taken for ex-situ analyses. Post-mortem analysis of the reacted materials showed that chrysotile carbonation was controlled by fiber length. It was observed that carbonation has a cementing action keeping the particles together and preventing the fibers from being blown away by convective circulation of air in the packed-bed cell.[34] Parallel to the convective carbonation tests in the packed bed cell, the carbonation behavior was also monitored in a diffusion cell where dynamic carbonation tests were run in specially designed dual-compartment differential fixed-bed diffusion–carbonation reactor. Low carbonation rates under both dry and fully saturated conditions were likewise observed. While water was found to be a crucial medium for supply of leached magnesium and dissolved CO_2, blockage of all pores by water was found to be the main reason for the slow carbonation kinetics of a completely saturated medium.[4] Also, a sharp decline in water saturation was detected while carbonating brucite-rich nickel mining residues which was ascribed to consumption of water in the formation of nesquehonite.

The intent of this study is to gain knowledge on the carbonation behavior in different conditions of temperature and water availability mimicking the realistic environment of northern Quebec, with both rainy and warm summers as well as the cold and freezing winters. Emphasis will be placed on water formation and consumption during ambient mineral carbonation and different mineralogy of the carbonated material. Possible carbonation reaction pathways will be discussed. The results of this study can bring insights on the carbonation mechanisms occurring within mining residue piles in contact with ambient air. Evidence of water formation and consumption during carbonation reactions is scrutinized specifically in this work to further our understanding of the mineral carbonation pathways under ambient conditions.

Methods and materials

Sample characterization

Ultramafic zones of serpentinite, peridotite and dunite, and mafic zones of clinopyroxenite and gabbro are the major rocks of the Dumont sill in the Abitibi region of Québec, Canada.[35] Nickel mining residue, NiMR, used in this study originated from Dumont Nickel Project. Mineral identification was performed on as-received (after sieving) and carbonated NiMR powders by means of TG/DTA, PXRD and IR techniques. The carbonation tests concerned the sieved sample fraction with D_{80} = 150 µm for which the BET (TRISTAR 3000) surface area and bulk density were 11 m^2/g and 1.61 g/cm^3, respectively. X-ray powder diffraction (PXRD) data were collected on a Siemens D5000 diffractometer using Cu K_α radiation (λ = 1.54059 Å) generated at 40 kV and 30 mA emission current, at Δθ scanning rate of 1°/min over the 5–65° 2θ region (step size of 0.04° and 1.2 s dwell time). Chrysotile and lizardite set aside, minor amounts of crystalline brucite, magnetite and wustite were also detected (figure 1). However, no crystalline carbonate phases were detected in the as-received NiMR. Free brucite was quantified by means of a thermogravimetric method (PerkinElmer Pyris Diamond TG/DTA) described in our earlier work.[36] An Advance Optima continuous gas analyzer (AO2000 Series, Uras 14) CO_2 infrared analyzer coupled with an induction furnace at 810 °C was used for solid calcination. Accordingly, possible amorphous carbonates and organic carbon were evaluated to contribute for less than 0.2 wt. %.

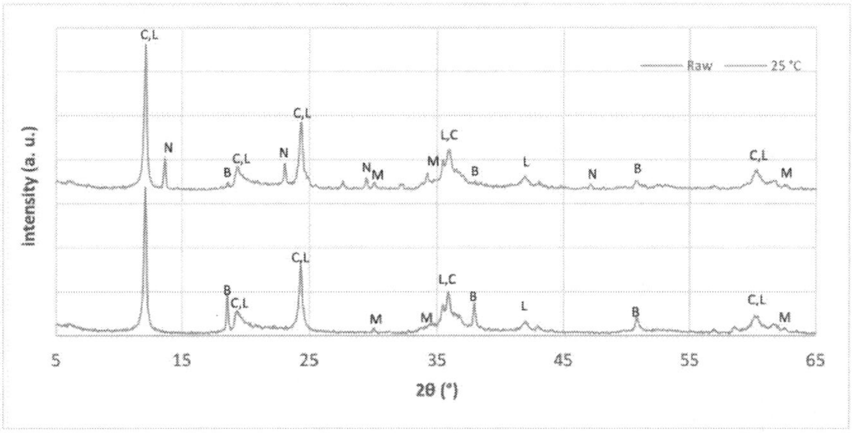

Figure 1. a) X-ray powder diffraction patterns of as-received and carbonated NiMR sample; C: chrysotile, L: lizardite, N: nesquehonite, B: brucite and M: magnetite;

Experimental set-up

The experimental set-up used in this study is a modified version of a differential diffusion fixed-bed reactor originally devised to monitor CO_2 diffusion and sequestration (Figure 2).[4] It consists of three main parts: a gas chamber (lower compartment) where CO_2 is loaded, a sample holder which contains the NiMR powder and a diffusion chamber (upper compartment). The specified mixture of gas (10% CO_2 balanced with N_2) charged in the gas chamber diffuses through the solid layer to the diffusion chamber and by this means CO_2 reacts with the whole length of material. The new double wall reactor is suitable for performing temperature dependent tests, while there will be no pressure drop due to CO_2 consumption as it is equipped with a fixed volume gas bag. General description of the reactor described elsewhere.[4, 37]

The setup has two sets of instrumentations in gas and solid phases. Gas and diffusion chambers are equipped with CO_2 sensors (Vaisala GMT221), integrated hygrometer–thermocouples (Vaisala HMT333) and low velocity fans, while the NiMR layer was instrumented with a time domain reflectometry (TDR) probe (Delta-T WET-2) for online monitoring of water saturation, bulk ionic conductivity and layer temperature. A pH probe (HI 1053B, Hanna Instruments) was also used to monitor the instantaneous pore-water pH variations.

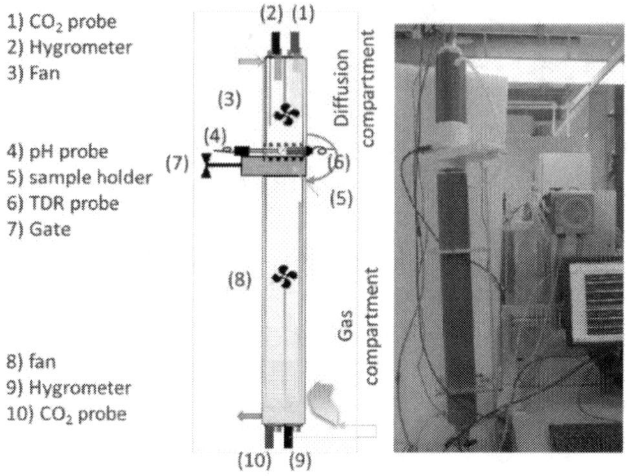

Figure 2. Schematic representation of the experimental setup.

Carbonation methodology

For each experiment run, *ca.* 40 g of NiMR, dried overnight in 110°C, was placed in a sample holder, pressed lightly by tapping to eliminate air cavities and flatten the surface. The liquid saturation, β, of the residue layer was obtained by stepwise addition of calibrated amounts of water untill reaching the desired level (40, 50, 60 or 80%). Initially, the gas chamber (~ 1.5 L) was isolated from both the diffusion chamber (~ 0.5 L) and the sample to be carbonated by closing a separation gate (Figure 2).

The initial CO_2 mole fraction in the gas chamber was yCO_2 = 7 and 10 vol. % CO_2 (balanced with N_2). During cell preparation, the diffusion chamber was also preventively purged with pure N_2 to get rid of oxygen and CO_2. Opening the gate coincided with the reaction starting time and tests continued until CO_2 consumption was stopped. To assess material maximum capacity for CO_2 storage, as well as to objectify various carbonation regimes, several shots of refilling of the gas chamber were carried out without sample removal. PXRD analysis was performed, in addition to raw NiMR, on the carbonated material to identify and follow the evolution of the neo-formed carbonate phases.

Results and discussion

Each carbonation experiment consisted in monitoring simultaneously seven different variables. For the headspace in both gas and diffusion chambers: CO_2 composition, temperature and humidity, and for the residue layer: temperature, pH, water saturation and ionic conductivity. Reactions R1 – R4 below depict the general steps for brucite carbonation and will be used to discuss the general behavior of NiMR in terms of possible mineral carbonation pathways. Serpentinite dissolution, instead of R1, is also a source for soluble magnesium cations.

$Mg(OH)_2 + H_2O \rightarrow Mg^{2+} + 2OH^-$ (R1)

$CO_2 + H_2O \rightarrow HCO_3^- + H^+$ (R2)

$HCO_3^- + H^+ \rightarrow CO_3^{2-} + H^+$ (R3)

$Mg^{2+} + HCO_3^- + 2OH^- + H^+ + H_2O \rightarrow Mg(HCO_3)(OH)\cdot 2H_2O$ (R4)

There is still debate in the literature regarding the correct chemical formula of nesquehonite which is commonly written either as $Mg(HCO_3)(OH)\cdot 2H_2O$ or $MgCO_3\cdot 3H_2O$.[38, 39] Recent work by Frost and Palmer[38] deciphering OH vibrations from infrared emission spectroscopy, both in stretching and deformation modes, support the idea that nesquehonite's stoichiometry would be better defined as $Mg(HCO_3)(OH)\cdot 2H_2O$.

Effect of particle size

Assima et al.[4] demonstrated the sensitivity of carbonation to water saturation by bringing into contact dry, half-saturated and fully saturated residue layers with gaseous CO_2. Low carbonation rates under dry and fully saturated conditions were observed, while half-saturated NiMR layers showed relatively fast carbonation rate. As-received NiMR was sieved into three classes of particle sizes identified as fine (< 106 μm), medium (106 < d < 180 μm) and coarse (180 < d < 212 μm) in order to test the influence of particle size on the carbonation potential in terms of possible change in mineralogy due to liberation and concentration of specific minerals in each size class. Figure 3 shows the CO_2 consumption/diffusion profiles of fine, medium and coarse NiMR layers subjected to 40, 60 and 80 % water saturation conditions.

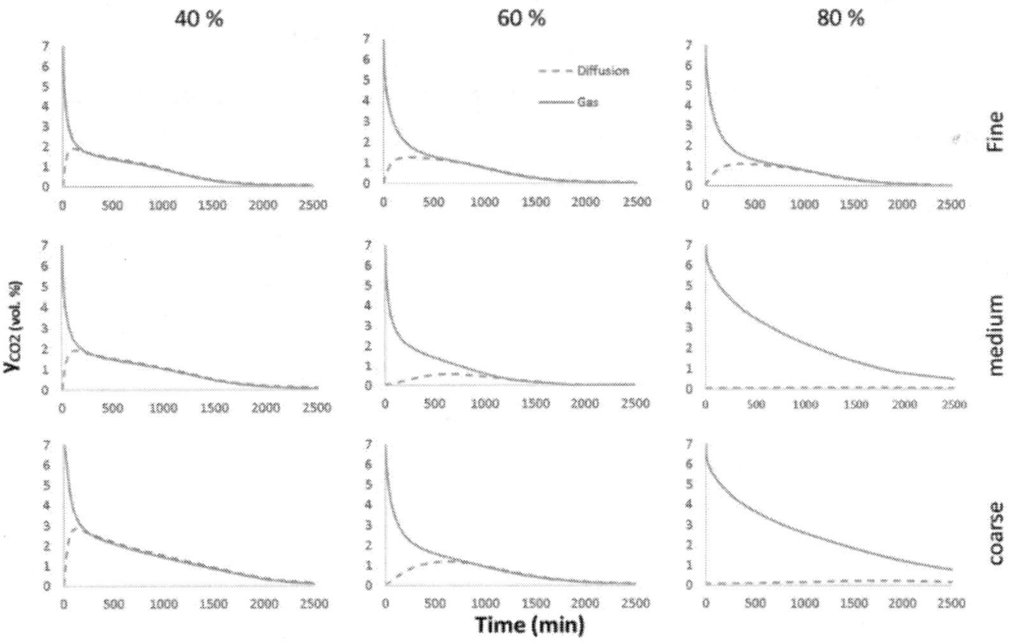

Figure 3. In-situ monitoring of CO_2 consumption/diffusion across NiMR layer in terms of evolution of CO_2 fraction (in percent) in the diffusion (dotted line) and gas (solid line) compartments; 45 g NiMR, 7 % CO_2 (balanced with N_2).

As expected, increased water saturation decreased gas diffusion mainly due to blockage of the pores with water. However, in the case of fine particles at 80 % water saturation, unlike for the medium and coarse samples, CO_2 diffusion was observed to some extent although maintaining the same decreasing trend. Having inspected the reactor, it was noticed that tiny cracks were formed at the surface of the solid layer which showed a clay-like appearance with the fine particles (Figure 4). The extent of cracks and their size was negligible at 60 % saturation, whereas no cracks were identified in the other cases. Lower water saturation was found to promote gas diffusion through the layer, however, the reaction kinetics, as stated before, is highly dependent on the availability of water. Thus, one may state that in the saturation range of 80 – 100 %, blockage of the pores led to severely reduced gaseous diffusion while water saturations between 40% and 60 % were optimal for enabling relatively rapid ambient mineral carbonation.

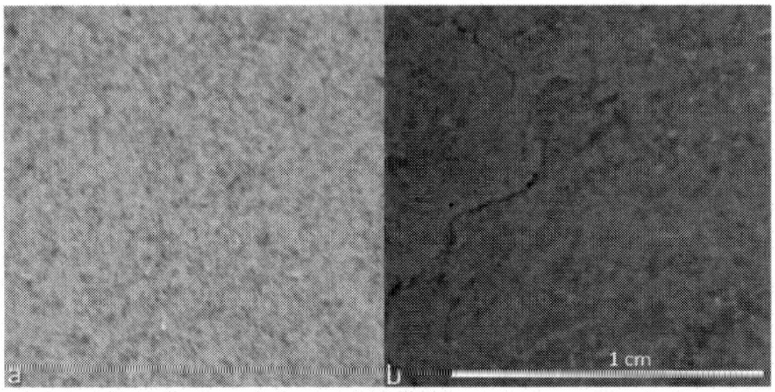

Figure 4. Fine particle NiMR layer, a) dry and b) 50% water saturated, note the hairline cracks formed on the surface of the wetted sample.

Figure 5 depicts the PXRD spectra for the different particle sizes. Brucite content of each size class is also provided in Table 2 (details on brucite quantification will be discussed later). Slightly higher brucite content of fines in comparison to coarse particle size can be justified by liberation and concentration due to excessive grinding. Thus, higher available specific surface area of fine particles in combination with more abundant brucite is responsible for the faster carbonation kinetics. Details on brucite quantification will be discussed in the subsequent sections.

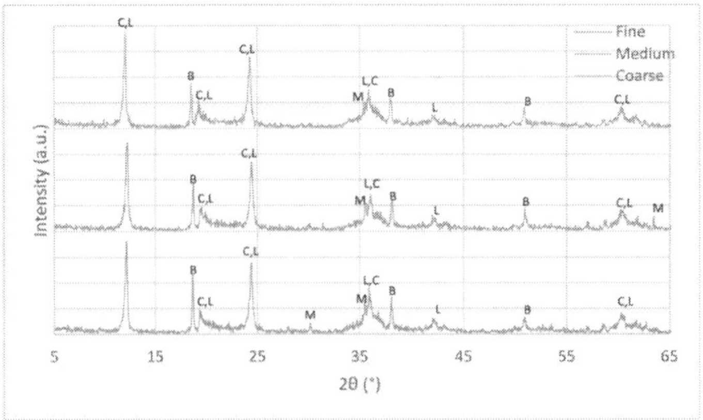

Figure 5. PXRD patterns of different particle sizes for sieved as-received NiMR.

Table 2. Weight percent brucite content of different size classes of NiMR

	Fine	*Medium*	*Coarse*
Brucite content, %	12	11.5	11.2

Kinetics of carbonation

The as-received NiMR wetted up to 50% water saturation was contacted with CO_2 (initial yCO_2 = 10 vol. % CO_2 balanced with pure nitrogen) to study accelerated mineral carbonation at ambient temperature. CO_2 instantaneous composition in gas and diffusion chambers is displayed in Figure 6a. Owing to CO_2 absorption from gas chamber, poral water acidification in the sample prompts magnesium leaching from the material. Simultaneously, while traveling across the NiMR layer, part of CO_2 diffuses in the diffusion chamber and its mole fraction starts to increase, up to a point where compositional gradients between the two chambers are smoothed out. Afterwards, both chambers exhibit a decreasing trend in CO_2 composition during the reaction course where CO_2 consumption follows almost a monotonous trend. The total volume of residual CO_2 (V_c) is calculated as:

$$V_r(t) = y_{CO2}^{G}(t)V^{G} + y_{CO2}^{D}(t)V^{D} \quad \text{(Eq.1)}$$

where y_{CO2} is the CO_2 mole fraction and V is the compartment volume, and superscripts G and D refer to gas and diffusion chambers, respectively.

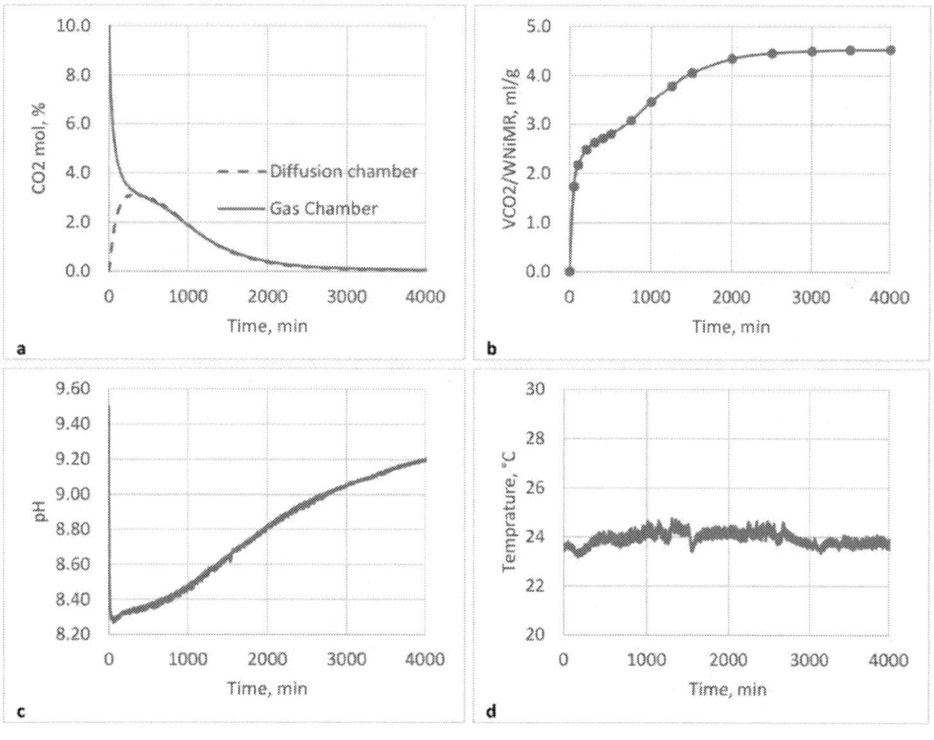

Figure 6. In-situ monitoring of a) CO_2 consumption/diffusion, b) CO_2 sequestered per unit mass of NiMR, c) pH variations and d) temperature changes over reaction time.

One can echo the sharp pH drop occurring in the early minutes of reaction with the fast consumption of CO_2 by comparing their corresponding dynamics in Figures 6a,c. At t = 0, the initial pH value (~9.5) of the poral water resulted mainly from brucite-like dissolution (R1) and to a lesser extent to chrysotile and lizardite dissolutions. We made sure that equilibrium conditions whereby poral water was saturated with brucite dissociation species were achieved before opening the gate at t = 0. Afterwards by opening the gate, the partially wetted material and CO_2 are brought into contact triggering CO_2 hydration/dissociation (R2, R3).

It could be inferred from Figures 6 and 7 that as CO_2 content approaches zero, ionic conductivity, pH and water saturation tend to retrieve their initial values. Owing to high pH values, bicarbonate, carbonate and water formation is expected. The dynamics featured by the water saturation profile is in support of the speculation of water formation/consumption during mineral carbonation (Figure 7b). This means that water formation is the direct result of acid-base neutralization which has also been reflected in Figure 7b by an increase in water saturation in the early stage of the reaction. Time-dependent X-ray diffraction patterns of brucite slurry carbonation in ambient temperature and pCO_2 = 15 atm, published by Zhao et al.,[20] showed that carbonation begins nearly after the first hour of the experiments. Using carbon abundance analysis, Harisson et al.,[8] have also reported that nesquehonite formation occurs within 6 – 8 h of brucite slurry contact with 10% CO_2 and within 2 h of reaction with 50% and 100% CO_2. Interestingly, this time span of 6-8 h for nesquehonite formation is in agreement with our experimental data.

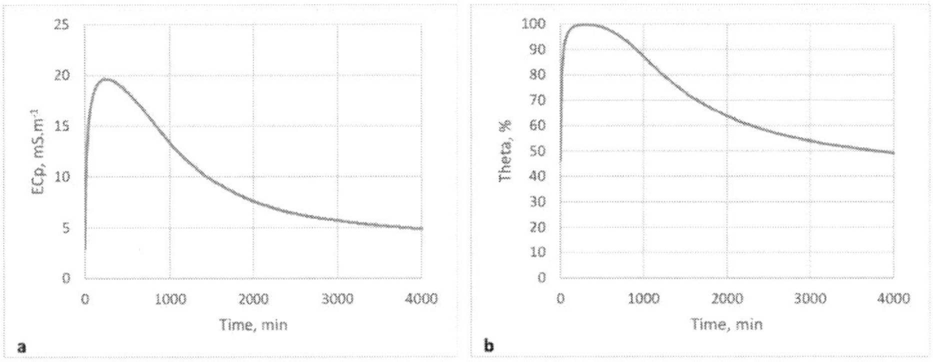

Figure 7. In-situ monitoring of a) ionic conductivity and b) water saturation variation over time.

In Figure 7b, three different evolution trends are noticed in terms of water behavior in the course of carbonation. Water production continues until all the pores are filled by water causing fully saturated NiMR to act as a barrier to CO_2 diffusion. This phenomenon has been observed in Figure 6a where a change in the profile slope indicates a decline in CO_2 consumption rate. Derivatives with respect to time of water saturation (Figure 7b) and CO_2 composition (Figure 6a) are plotted in Figure 8. As can be observed, the local maximum in the CO_2 consumption curve is coincident with water formation rate equal zero, at $t \approx 340 - 360$ min for β = 100%. This also represents the point where the two compartments have reached equal CO_2 concentrations. At this stage, based on the observations of Harrison et al.,[8] nesquehonite formation starts with water incorporation in its lattice. Thus, decreasing the water saturation tends to gradually accelerate the reaction again and finally lower availability of CO_2 in the reactor will slow down the reaction.

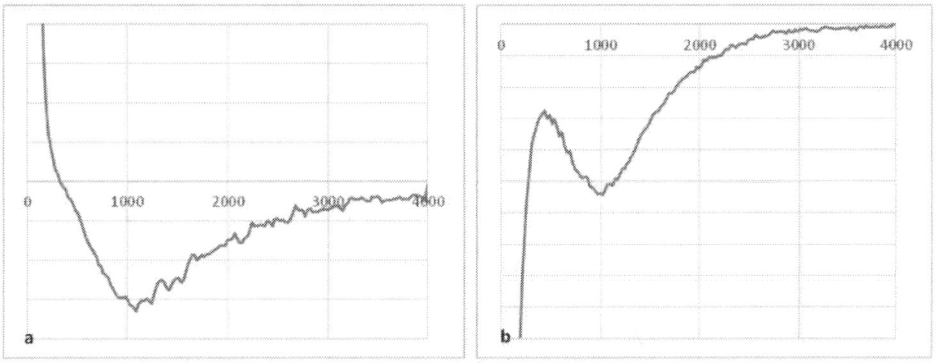

Figure 8. Time derivative plots of a) water saturation (figure 7b) and b) CO_2 depletion from gas chamber over time (figure 6a).

Water consumption in Figure 7b is supported by PXRD spectra of the carbonated material (Figure 1). At room temperature, nesquehonite readily precipitates from magnesium bicarbonate solutions.[40] Each nesquehonite molecule diverts three water molecules from solution to be incorporated in its crystal structure; the emergence of its characteristic peaks in XRD pattern confirms water consumption in the course of reaction as suggested by the tail of Figure 7b.

Effect of gas chamber refills

Figure 9 illustrates the instantaneous total CO_2 consumption profiles as well as pH variations upon successive refills of the gas chamber. The aim is to study the behavior of partially carbonated material in carbonation conditions while tracking the maximum capacity of carbon sequestration. The first contact between NiMR and CO_2 results in nesquehonite formation. However, as Figure 9a suggests, carbon uptake gets slower at each refilling step.

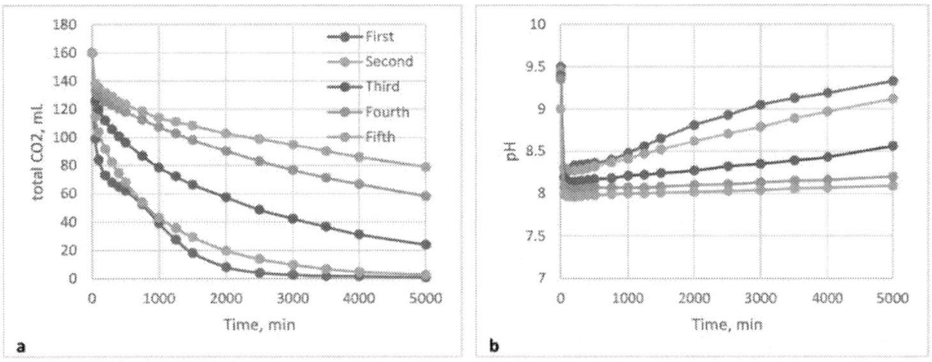

Figure 9. Monitoring of NiMR behavior upon successive chamber replenishments with CO_2 a) CO_2 consumption and b) pH variations

Kinetics of carbonation is highly dependent on the available brucite content of the NiMR. As the reaction proceeds, it is the brucitic part that is primarily involved and therefore its slowed down leaching in further fillings of the gas chamber led to slowed down carbonation of brucite-depleting and partially carbonated NiMR. Thus, one may conclude from Figure 9a that from the third refill onward, lack of Mg^{2+} and OH^- in pore water, due to decreasing amounts of brucite as well as possible passivation of particles, translated in lower CO_2 consumptions. Also, Vermilyea reported "aged" samples referring to already leached brucite showing lower dissolution rates in a further contact due to elimination of "active" regions.[41] In this way, the NiMR carbonation sites can be characterized as: a) fast carbonation sites mainly the brucitic species present in NiMR and b) slow carbonation site of brucite-poor and partially carbonated NiMR. Although NiMR tends to render poral water more alkaline, the progress of carbonation reaction has an effect on the solution pH by counterbalancing the rates of dissolution of brucite and hydration/dissociation CO_2 in the poral water. The pH trend and its dynamics are also influenced by the concurring passivation of particle surfaces by developing layers of carbonates, silica gel and/or iron hydroxide contaminants.[29] In accordance with the CO_2 consumption, pH restoration slows down after every refill of the carbonation cell (Figure 9b). This may also be indicative of the slower bulk leaching of NiMR into pore water due to the lower solubility of chrysotile/ lizardite in comparison with brucite.

The total carbon dioxide sequestered per unit mass of NiMR (C_{tot}) can be calculated as:

$$C_{tot} = \sum_{i=1}^{n} (V_1^i - (V_2^i + V_1^{*i})) \Big/ W_{NiMR} \qquad \text{(Eq.2)}$$

Where V_1^i is the volume of CO_2 in the gas chamber in t = 0 and V_2^i and V_1^{*i} are the volumes of residual CO_2 in the diffusion and gas chambers after reaction. W_{NiMR} is the mass of NiMR sample (Figure 10).

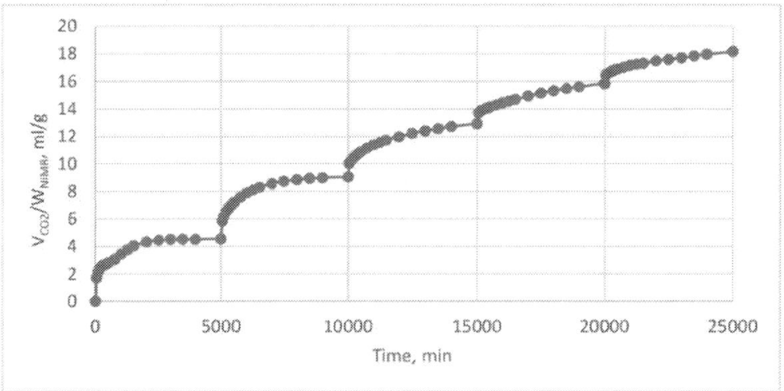

Figure 10. Monitoring of total CO_2 sequestration per unit mass of NiMR over successive refills.

TGA analysis of carbonation products

Thermal decompositions of approximately 30 mg of the as-received and partially carbonated NiMR samples were carried out under nitrogen flow (Figure 11). Samples were first placed in an open ceramic crucible and heated at 10°C min^{-1} up to 110°C and thence for one hour to get rid of unbound water. Afterwards, heat up at 1.0°C min^{-1} up to 1000°C was resumed. Table 3 gathers the registered values of mass loss of Figure 11 over four characteristic temperature segments.

The first segment of TGA diagram is related to the elimination of weakly bound (physical) water up to 110 °C. The second segment for carbonated NiMR corresponds to a mass loss of 1.4 % at about 115 – 125 °C which is attributed to the dehydration of nesquehonite into a stable amorphous magnesium carbonate most likely hydromagnesite.[42] The next mass loss in the temperature range 300 – 450 °C is related to brucite dehydroxylation. In the case of the carbonated sample, it appears that there are unidentified components that contribute also to the mass loss interfering thus with brucite dehydroxylation. While nesquehonite was the only crystalline carbonate phase that was objectified via XRPD in the carbonated material, it is not excluded that there may be other carbonates with low abundance and/or with amorphous structures that contributed to the mass loss in this temperature region. The mass loss corresponding to lizardite/chrysotile dehydroxylation occurred in the temperature range of 500 – 850 °C. For the carbonated material, there is a small mass loss of *ca.* ~ 1.2 % in the range 680 – 750 °C which could suggest that some thermally-driven changes occurred in the lizardite/chrysotile structure. The difference in mass loss, *ca.* ~ 0.1%, between the as-received and carbonated NiMR was low indicating no more activity than dehydroxylation of the magnesium silicate polymorphs.

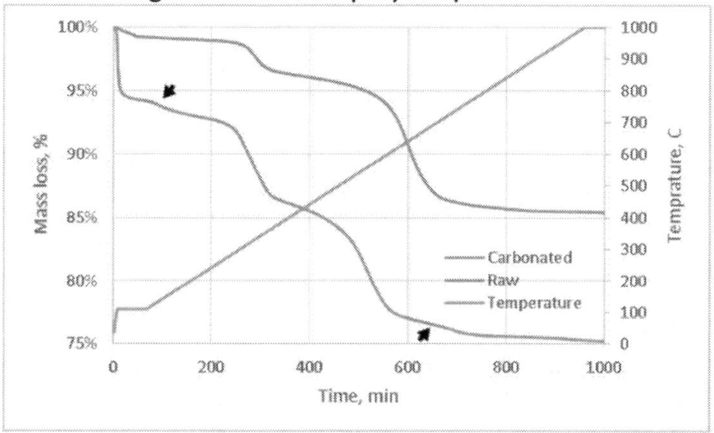

Figure 11. TGA data monitoring of as-received and carbonated NiMR thermal decomposition.

Table 3. Calculated mass loss percentage of as-received and carbonated NiMR samples of Figure 11.

	Mass loss, °C			
trend	25-110	110-150	300-450	500-850
Raw NiMR	0.8	0.2	3.3	10.2
Carbonated	5.8	1.4	7.4	10.1

Conclusion

Mineral carbonation of an ultramafic mining waste was studied under ambient conditions. Ambient mineral carbonation was found to dependent on the water poral saturation. It was observed that water will be formed in the early stages of the reaction and will be consumed later to form hydrated carbonate species. As a result, NiMR carbonation proceeded as two main processes: a) a fast carbonation step mainly due to brucite consumption and, b) slow carbonation step of brucite-depleted NiMR. Extending carbonation of partially carbonated materials after brucite pre-carbonation at various degrees showed indeed slower subsequent carbonation kinetics. This was interpreted as a result of surface hindrance by carbonate product layer on NiMR particles as well as passivation by silica gel and/or iron hydroxide. Pore water pH was shown to be greatly influence by carbonation.

Acknowledgements

The authors are grateful for financial support from the Natural Sciences and Engineering Research Council and from Royal Nickel Corp.

References

1. E. J. Wilson, T. L. Johnson and D. W. Keith, *Environ Sci Technol*, 2003, 37, 3476-3483.
2. L.-C. Pasquier, G. Mercier, J.-F. Blais, E. Cecchi and S. Kentish, *Applied Geochemistry*, 2014, 50, 66-73.
3. A. Hemmati, J. Shayegan, J. Bu, T. Y. Yeo and P. Sharratt, *International Journal of Mineral Processing*, 2014, 130, 20-27.
4. G. P. Assima, F. Larachi, J. Molson and G. Beaudoin, *Minerals Engineering*, 2014, 59, 39-44.
5. A.-H. A. Park and L.-S. Fan, *Chemical Engineering Science*, 2004, 59, 5241-5247.
6. G. P. Assima, F. Larachi, J. Molson and G. Beaudoin, *Chemical Engineering Journal*, 2014, 240, 394-403.
7. J. Pronost, G. Beaudoin, J. Tremblay, F. Larachi, J. Duchesne, R. Hébert and M. Constantin, *Environmental Science & Technology*, 2011, 45, 9413-9420.
8. A. L. Harrison, I. M. Power and G. M. Dipple, *Environmental Science & Technology*, 2013, 47, 126-134.
9. C. Julcour, F. Bourgeois, B. Bonfils, I. Benhamed, F. Guyot, F. Bodénan, C. Petiot and É. C. Gaucher, *Chemical Engineering Journal*, 2015, 262, 716-726.
10. N. A. Meyer, J. U. Vögeli, M. Becker, J. L. Broadhurst, D. L. Reid and J. P. Franzidis, *Minerals Engineering*, 2014, 59, 45-51.
11. D. N. Huntzinger, J. S. Gierke, L. L. Sutter, S. K. Kawatra and T. C. Eisele, *Journal of Hazardous Materials*, 2009, 168, 31-37.
12. F. Bodénan, F. Bourgeois, C. Petiot, T. Augé, B. Bonfils, C. Julcour-Lebigue, F. Guyot, A. Boukary, J. Tremosa, A. Lassin, E. C. Gaucher and P. Chiquet, *Minerals Engineering*, 2014, 59, 52-63.
13. H. C. Oskierski, B. Z. Dlugogorski and G. Jacobsen, *Chemical Geology*, 2013, 358, 156-169.
14. M. Bodor, R. Santos, T. Van Gerven and M. Vlad, *cent.eur.j.eng*, 2013, 3, 566-584.
15. A. Kirchofer, A. Becker, A. Brandt and J. Wilcox, *Environmental Science & Technology*, 2013, 47, 7548-7554.
16. G. Montes-Hernandez, R. Pérez-López, F. Renard, J. M. Nieto and L. Charlet, *Journal of Hazardous Materials*, 2009, 161, 1347-1354.
17. R. M. Dilmore, B. H. Howard, Y. Soong, C. Griffith, S. W. Hedges, A. D. DeGalbo, B. Morreale, J. P. Baltrus, D. E. Allen and J. K. Fu, *Environmental Engineering Science*, 2009, 26, 1325-1333.
18. N. Um, S.-Y. Nam and J.-W. Ahn, *MATERIALS TRANSACTIONS*, 2013, 54, 1510-1516.
19. A. L. Harrison, G. M. Dipple, I. M. Power and K. U. Mayer, *Geochimica et Cosmochimica Acta*, 2015, 148, 477-495.
20. L. Zhao, L. Sang, J. Chen, J. Ji and H. H. Teng, *Environmental Science & Technology*, 2010, 44, 406-411.

21. E. J. Swanson, K. J. Fricker, M. Sun and A.-H. A. Park, *Physical Chemistry Chemical Physics*, 2014, 16, 23440-23450.
22. M. Hänchen, V. Prigiobbe, R. Baciocchi and M. Mazzotti, *Chemical Engineering Science*, 2008, 63, 1012-1028.
23. L. N. Lammers, G. E. Brown Jr, D. K. Bird, R. B. Thomas, N. C. Johnson, R. J. Rosenbauer and K. Maher, *Geochimica et Cosmochimica Acta*, 2015, 155, 30-46.
24. D. W. Ming and W. T. Franklin, *Soil Sci. Soc. Am. J.*, 1985, 49, 1303-1308.
25. M. F.A. and T. C.S., Berkeley, 1967.
26. G. P. Assima, F. Larachi, J. Molson and G. Beaudoin, *The Canadian Journal of Chemical Engineering*, 2014, 92, 2029-2038.
27. Y. Xiong and A. S. Lord, *Applied Geochemistry*, 2008, 23, 1634-1659.
28. S. A. Wilson, G. M. Dipple, I. M. Power, J. M. Thom, R. G. Anderson, M. Raudsepp, J. E. Gabites and G. Southam, *Economic Geology*, 2009, 104, 95-112.
29. G. P. Assima, F. Larachi, G. Beaudoin and J. Molson, *Industrial & Engineering Chemistry Research*, 2012, 51, 8726-8734.
30. C. Cravotta, III, presented in part at the 7th International Conference on Acid Rock Drainage (ICARD), St. Louis, 2006.
31. D. Banks, V. P. Parnachev, B. Frengstad, W. Holden, A. A. Vedernikov and O. V. Karnachuk, *Geological Society, London, Special Publications*, 2002, 198, 287-296.
32. C. Cravotta, III, *Mine Water Environ*, 2007, 26, 128-149.
33. A. S. Awoh, B. Plante, B. Bussière and M. Mbonimpa, GeoMontreal conference, Canada, September 29 to Thursday October 3, 2013.
34. G. P. Assima, F. Larachi, G. Beaudoin and J. Molson, *International Journal of Greenhouse Gas Control*, 2013, 12, 124-135.
35. M. Sciortino, J. E. Mungall and J. Muinonen, *Economic Geology*, 2015, 110, 733-761.
36. G. P. Assima, F. Larachi, J. Molson and G. Beaudoin, *Thermochimica Acta*, 2013, 566, 281-291.
37. G. P. Assima, F. Larachi, J. Molson and G. Beaudoin, *Chemical Engineering Journal*, 2014, 245, 56-64.
38. E. E. Coleyshaw, G. Crump and W. P. Griffith, *Spectrochimica Acta Part A: Molecular and Biomolecular Spectroscopy*, 2003, 59, 2231-2239.
39. R. L. Frost and S. J. Palmer, *Spectrochimica Acta Part A: Molecular and Biomolecular Spectroscopy*, 2011, 78, 1255-1260.
40. L. Hopkinson, K. Rutt and C. Gordon, *The Journal of Geology*, 2008, 116, 387-400.
41. D. A. Vermilyea, *Journal of The Electrochemical Society*, 1969, 116, 1179-1183.
42. J. Lanas and J. I. Alvarez, *Thermochimica Acta*, 2004, 421, 123-132.

Carbon Sequestration and Metals Stabilization Characteristics of a Magnesium Oxide-Based Binder

K.-Y. Hwang[1], H. Q. H. Phan[2], and I. Hwang[1*]

[1] School of Chemical/Biomolecular and Environmental Engineering, Pusan National University, 30 Jangjeon-Dong Geumjeong-Gu, Busan 609-735, Korea
[2] Department of Chemical and Environmental Engineering, I405, Lac Hong University, 10 Huynh Van Nghe, Buu Long Ward, Dong Nai Province, Vietnam
*Corresponding author: ihwang@pusan.ac.kr

Abstract

This study aims to evaluate an MgO-based binder in sequestrating CO_2 under the accelerated carbonation conditions and in simultaneously stabilizing heavy metals in contaminated sediments. The contaminated sediment treated by the MgO-based binder was exposed to CO_2 at concentrations of 0.034%, 50%, and 100%, respectively. The treated sediment was then evaluated with respect to the kinetics and extent of carbonation and to the stabilization of heavy metals. The rate and extent of carbonation increased as the CO_2 concentration increased. The capacity of carbonation was estimated to be 118 kg CO_2 per tonne of the treated sediment. Magnesite and nesquehonite were identified as major products of carbonation. Sequential extraction analyses on the treated sediment showed that the MgO-based binder produced very persistent solid phases that could efficiently stabilize heavy metals.

Introduction

A low carbon MgO-based binder with a formula of $MgO_{0.5}$-($L_{0.1}$-($FA_{0.4}BFS_{0.6})_{0.9})_{0.5}$ was previously developed by mixing MgO, CaO (L), fly ash (FA), and blast furnace slag (BFS).[1] Utilization of MgO as a supplementary cementitious material is timely because it is produced at much lower kiln temperatures (<750°C) than at the temperature where Portland cement is produced. MgO production is less energy intensive than cement production. MgO is also known to sequestrate substantial amounts of CO_2 at the ambient temperature and pressure.[2]

The current study focused on the application of the MgO-based binder to treating the dredged sediment contaminated with heavy metals under accelerated carbonation conditions. Initially, kinetics and mechanisms of carbonation by the MgO-based binder were evaluated. Then the metal stabilization characteristics were further investigated.

Materials and Methods

Materials

The materials used for the binder were as follows: MgO (light, assay (ignited) min. 98.0%, Yufeng Chemical Co., China), CaO (at least 98%, extra pure reagent, Junsei, Japan), fly ash (SamchenpoCoal-fired Power Plant, Republic of Korea), and blast furnace slag (POSCO, Republic of Korea).The components of fly ash and blast furnace slag are presented in Table 1. The binders were mixed in the proportion of 50%, 5%, 18%, 27 wt% with MgO, lime, fly ash, blast furnace slag, respectively.

Sediment samples (20kg, 5-10cm depth) were collected from Noksan area (35°7′21.78″N and 128°54′2.32″E) in the lower reaches of Seonakdong River, Busan, Korea. The water content, TOC, and pH values of the sediment samples were 47.6%, 5.3%, and 8.29, respectively. Dredged sediment samples were dried and crushed by a ball mill and were sieved to yield sizes below 75μm.

Table 1. Chemical composition of fly ash and blast furnace slag

Components	Content (%)	
	Fly ash	**Blast furnace slag**
SiO_2	51.3	34.8
Al_2O_3	20.8	15.4
Fe_2O_3	8.95	0.30
CaO	9.3	44.3
MgO	3.78	3.46
etc	3.17	1.51

Experiment procedures

The sediment samples were homogenized and dried in an oven at 105°C for 2 hours prior to carbonation experiments. Dried-samples were then placed on 90-mm petri dishes and were mixed with the MgO-based binder with a binder to soil ratio of 0.3. The mixtures were then spread to yield a circular plate specimen with a thickness of around 2 to 3 mm and a surface area ranging from 19.6 to 20.3 cm^2.

The specimens were incubated in a chamber containing CO_2 (Figure 1). Humidity was controlled at three different regions, which were low (17±2%), intermediate (61±3), and high relative humidity (98±1%) conditions. To obtain the target humidities, a saturated solution of NaCl and silica gel was used. The treated sediment samples were carbonated under different CO_2 concentrations of 0.034%, 50%, and 100%, respectively. Nitrogen as a balance gas was mixed with CO_2 gas to achieve target CO_2 concentrations.

At each sample time, samples were retrieved from the chamber and were dried at 105°C for 2 hours to determine the extent of carbonation by quantifying the weight changes. The samples were prepared in duplicate.

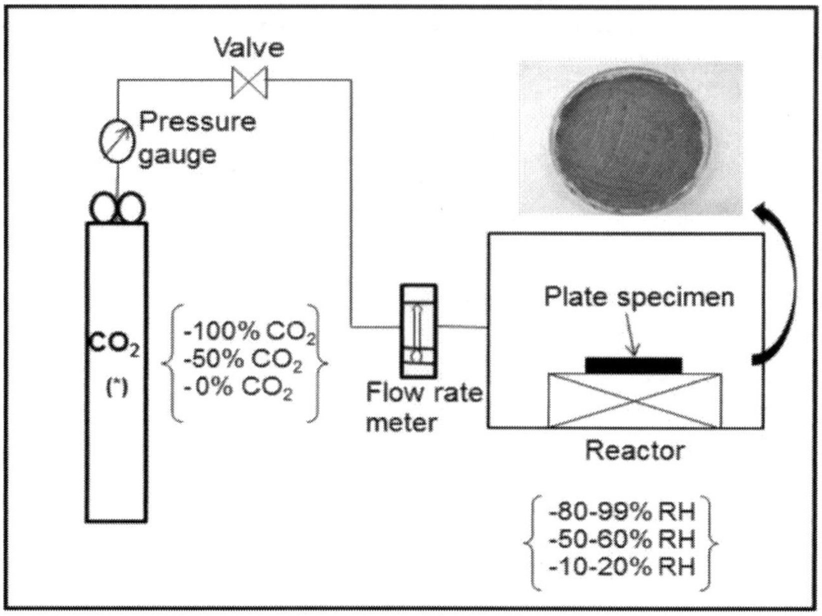

Figure 1. Experimental set-up for carbonation experiments.

To evaluate the performances of the MgO-based binder in stabilizing heavy metals, the artificially contaminated sediments were treated by the binder. The contaminated sediments were prepared by spiking the heavy metals (Pb, Ni, Cu, Cd, and Zn) to the dredged sediments to yield a concentration of 2,000mg/kg each. The heavy-metal-spiked sediments were then aged in the atmosphere for 7 weeks. The dosage of the binder was 30% of the sediment on a weight basis and the water to solid ratio was 0.6. The specimens were manually prepared to have spherical shapes with a diameter of 4-5 mm. To evaluate the speciation and binding of heavy metals in the treated sediment, five-step extractions were carried out using a modified sequential extraction procedure.[3]

Results and Discussion

Characteristics of carbonation by the MgO-based binder

Figure 1 shows that the extent of CO_2 sequestration is increased, as the CO_2 concentration is increased. Within 12 hours, the degree of carbonation higher than 80% was achieved. The maximum value greater than 90% was obtained after 1 day at 100% CO_2. pH values of the samples dropped from around 11.5 to 9.4-9.5 when CO_2 concentrations were 50% 100% CO_2. Especially, during the first 3 hours of carbonation, pH values decreased by 1 unit to 10.5. On the other hand, pH values were stable around 11.5 at the ambient condition.

The results imply that the rate of CO_2 diffusion through the pores of solids was enhanced as the concentration of CO_2 was increased. This is consistent with the observation in the direct gas-solid carbonation in an air pollution control system experiment.[4]

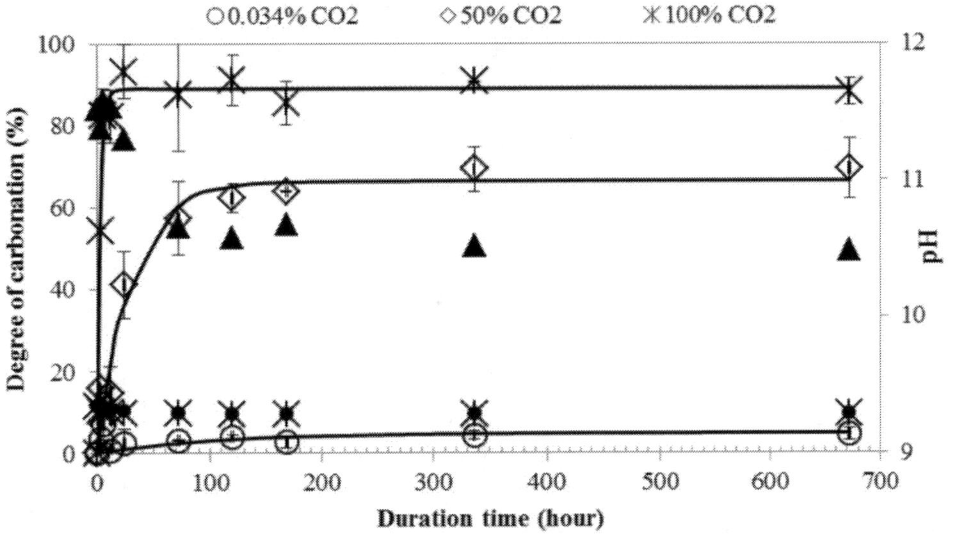

Figure 2. Kinetics of carbonation by the sediments treated with the MgO-based binder at different CO_2 concentrations (0.034%, 50%, and 100% CO_2). (•) represents pH changes at the 50% CO_2 condition. (▲) represents pH changes at the 0.034% CO_2 condition.

Figure 3 illustrates X-Ray diffraction patterns for different samples after 7 days of carbonation. In the 0% CO_2 sample, brucite ($Mg(OH)_2$) formation was observed that is a hydration product of periclase. In the presence of CO_2 at 0.034%, 50%, and 100%, brucite was transformed to carbonated crystals such as magnesite ($MgCO_3$) and nesquehonite ($MgCO_3.3H_2O$). Products were observed at different CO_2 conditions were similar. This is consistent with the results from the carbonation experiments that were conducted at the ambient temperature.[5,6]

The formation of hydrated carbonates of magnesium, i.e., nesquehonite, is reported to be thermodynamically favorable at a partial pressure of CO_2 of 1 bar in MgO-CO_2-H_2O system. The amount of water in the system also is a key factor for the formation of nesquehonite.[7] The initial water to solid ratio of 0.7 in this study was favorable for the formation of nesquehonite in the MgO-CO_2-H_2O system. Kaolinite, $Al_2Si_2O_5(OH)_4$, which was supposed to be a clay mineral of the dredged sediment, was also detected. The presence of $CaCO_3$ are attributable to the carbonation of calcium in the binders and also to the presence of the impurities. However, our previous study showed that the contribution of $CaCO_3$ to the carbonation was not significant in the MgO-based binder system.[1]

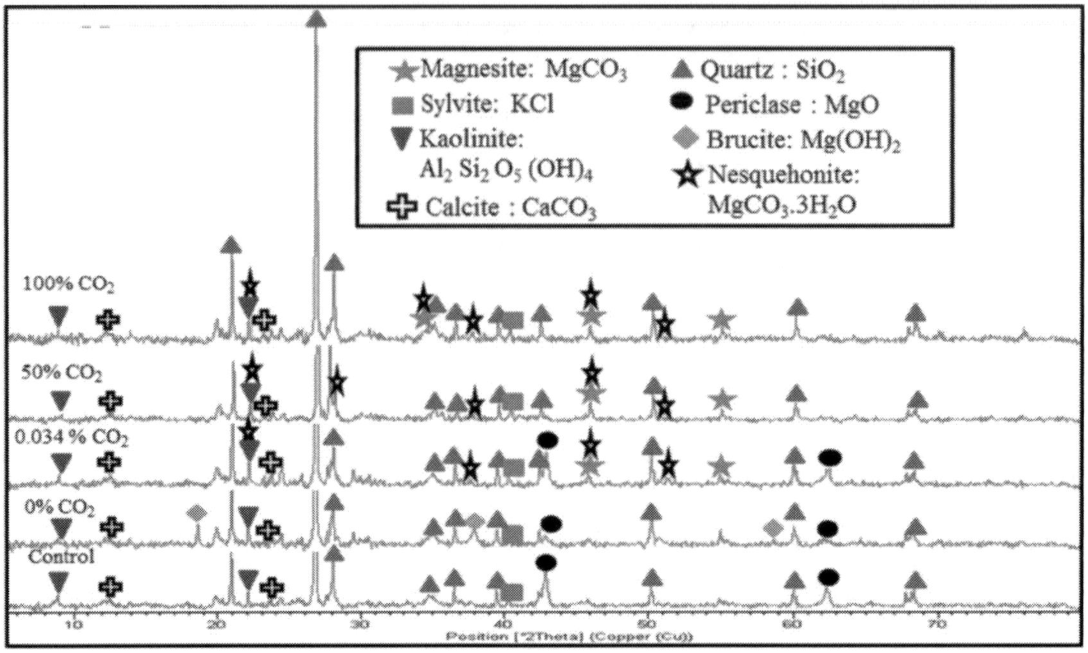

Figure 3. The XRD patterns for the sediments treated with the MgO-based binder at different CO_2 concentrations (0.034%, 50%, and 100% CO_2).

Metal stabilization in the sediment treated by the MgO-based binder

Despite the ambiguity and the lack of selectivity in the sequential extraction, this extraction technique has been widely used for obtaining an understanding on the heavy metals speciation in solid samples. Although sequential extraction was originally developed for sediments and its application in the solidification/stabilization treated wastes needs further development, results of this test can help to understand mechanisms of metal binding in the treated wastes. The recoveries of the heavy metals from the sequential extraction of treated samples were determined as the ratio of sum of the five different fractions to the sum of concentrations of the spiked heavy metals. The results are presented in Table 3.

It is interesting to observe that total recoveries of the heavy metals were substantially lower in the systems containing MgO, although the compressive strength of the cement system was similar to the other systems (results not shown). The average recoveries for the heavy metals followed the order: Cement/MgO < MgO-based binder < MgO < Cement. The presence of MgO decreased the recoveries. When the samples that were treated by the binders with MgO were extracted to the final step of sequential extraction, substantial portions of the samples were not digested. Nonrecovered heavy metals were believed to have existed in the non-digested phases. The results indicate the MgO-based binder has very effective metal stabilization mechanisms.

Table 2. Total recoveries of heavy metals from treated samples using sequential extraction.

Target	SEP recovery			
	Cement	MgO	Cement/MgO	MgO-based binder
Pb	106.12	76.35	42.18	62.29
Ni	78.39	53.38	31.95	47.95
Cu	88.51	55.81	32.67	50.74
Cd	56.45	121.1	25.99	39.83
Zn	74.99	47.79	30.27	45.87
Average	80.89	70.89	32.61	49.34

Conclusions

This study demonstrates that the MgO-based binder is able to sequestrate substantial amount of CO_2 under accelerated carbonation conditions and that the binder can stabilize heavy metals very effectively when it is applied to treating contaminated sediments. Magnesite and nesquehonite were identified as major products of mineral carbonation. The capacity of CO_2 sequestration was estimated to be 118 kg CO_2 per tonne of the treated sediment. Sequential extraction analyses on the treated sediment showed that the MgO-based binder produced very persistent solid phases that could efficiently stabilize heavy metals.

Acknowledgement

This work (Grants No. 00045220) was supported by Business for Cooperative R&D between Industry, Academy, and Research Institutes funded by the Korea Small and Medium Business Administration in 2011.

References

[1] Hwang, K.-Y., Seo, J.-Y., Phan, H. Q. H., Ahn, J.-Y., Hwang, I. 2014. MgO-Based Binder for Treating Contaminated Sediments: Characteristics of Metal Stabilization and Mineral Carbonation, *Clean-Soil, Air, Water*, 42(3), 355-363.

[2] Churakov, A. V., Iannuzzi, M., Parrinello, M., 2004. Ab Initio Study of Dehydroxylation-Carbonation Reaction on Brucite Surface, *Journal of Chemical Physics*, *108*, 11567-11574.

[3] Tessier, A., Campbell, P.G.C., Bisson, B. 1979. Sequential extraction procedure for particulate trace metals, Anal Chem, 51, 844–851

[4] Tian, S., Jiang, J., 2012. Sequestration of Flue Gas CO_2 by Direct Gas–Solid Carbonation of Air Pollution Control System Residues, *Environmental Science & Technology, 46*, 13545–13551.

[5] Mo, L., D., Panesar, D. K. 2012, Effects of accelerated carbonation on the microstructure of Portland cement pastes containing reactive MgO, *Cement and Concrete Research, 42*, 769–777.

[6] Hänchen, M., Prigiobbe, V., Baciocchi, R., Mazzotti, M. 2008, Precipitation in the Mg-carbonate system-effects of temperature and CO_2 pressure, *Chemical Engineering Science, 63*, 1012 – 1028.

[7] Silva, P. D., Bucea, L., Moorehead, D. R., Sirivivatnanon, V. 2009, Chemical, microstructural and strength development of calcium and magnesium carbonate binders, *Cement and Concrete Research, 39*, 460–465.

Purification of Slag-Derived Leachate and Selective Carbonation for the Production of High-Quality Precipitated Calcium Carbonate

K. De Crom[1], Y.W. Chiang[2], T. Van Gerven[1] and R.M. Santos[3*]

[1] Department of Chemical Engineering, KU Leuven, Leuven, Belgium
[2] School of Engineering, University of Guelph, Guelph, Ontario, Canada
[3] School of Applied Chemical and Environmental Sciences, Sheridan Institute of Technology, Brampton, Ontario, Canada
*Corresponding author: rafael.santos@sheridancollege.ca

Abstract

Much research has been performed to synthesize precipitated calcium carbonate (PCC) from calcium extracted from blast furnace (BF) slag. A significant challenge has been obtaining a chemically pure sample with a homogeneous crystal structure and narrow particle size distribution, which can have profitable marketability (e.g. in the papermaking industry). The aim of this study was to investigate the influence of alternative and combinatory processing strategies on the chemical, morphological and mineralogical properties of BF slag-derived PCC. In a first phase, the physicochemical removal of impurities from the leachate, prior to the carbonation, was investigated, by means of temperature reduction and pH elevation; this led to selective precipitation of Al, Mg and Si. In a second phase, the operational conditions of the carbonation step were optimized, by lessening the severity of the operational conditions (temperature and CO_2 pressure) and by reducing the amount of acid-neutralizing base (sodium hydroxide/bicarbonate) added during the carbonation process; this led to more selective calcium carbonation. Using these strategies, it was possible to synthesize chemically pure PCC (>98% Ca) with uniform morphology (scalenohedral), homogeneous mineralogy (>88% calcite) and narrow (1.09 uniformity), small (D_{50} = 1.1 μm) particle size distribution.

Introduction

In recent years, research has focused on employing waste residues from the iron and steel industry, rather than natural minerals, as source materials for mineral carbonation, since these waste residues are inexpensive (or even have a negative price), are readily available, and are often produced near large CO_2 sources.[1] Special attention has been given to waste materials with a high calcium content, mainly blast furnace (BF) slag and basic oxygen furnace (BOF) slag.[1-5] The choice for calcium-rich materials stems from the fact that precipitated calcium carbonate (PCC) has a larger market demand than magnesium carbonates.[2]

Two main types of mineral carbonation exist: direct and indirect carbonation. In the former, the source material is directly carbonated, while in the latter the calcium or magnesium is first extracted (e.g. using an organic lixiviant such as acetic acid or ammonium acetate) and subsequently carbonated. This results in post-carbonation precipitates with a higher purity and thus increasing the economic value, compared to direct carbonation.[1]

Kakizawa et al.[6] proposed a reaction scheme (see equation (1) and (2)) to indirectly synthesize PCC from calcium present in wollastonite ($CaSiO_3$). Teir et al.[2] stated that, in this reaction scheme, wollastonite may be replaced by BF slag, which consists of mainly calcium silicate minerals. Subsequently, Eloneva et al.[1] experimentally synthesized PCC starting from BF slag, based on the proposed reaction scheme. In a first step, the calcium was leached out from the BF slag with an acetic acid solution. Next, the leachate was separated from the post-extraction BF slag and subsequently carbonated. During the carbonation step, an acid neutralizing agent was added to induce precipitation of calcium carbonate crystals; sodium hydroxide was selected.

$$CaSiO_{3(s)} + 2CH_3COOH_{(aq)} \rightarrow Ca^{2+}{}_{(aq)} + 2CH_3COO^-{}_{(aq)} + SiO_{2(s)} + H_2O_{(l)} \quad (1)$$

$$Ca^{2+}{}_{(aq)} + 2CH_3COO^-{}_{(aq)} + CO_{2(aq)} + H_2O_{(l)} \rightarrow CaCO_{3(s)} + 2CH_3COOH_{(aq)} \quad (2)$$

The PCC synthesized by Eloneva et al.[1], however, was not commercial grade; the calcium content of the post-carbonation precipitate was only between 63–89 wt%, as small amounts of aluminum, magnesium and silicon present in the initial BF slag also ended up in the post-carbonation precipitate. Furthermore, calcite, aragonite and calcite-magnesian phases were detected in the PCC sample, making for a mineralogically and morphologically heterogeneous product.

Chiang et al.[5] attempted to optimize the extraction step by decreasing the acetic acid concentration, which led to a more selective leaching of calcium from the BF slag. However, concurrently, the overall calcium extraction extent became smaller. To balance the extent and selectivity of the calcium extraction, Chiang et al.[5] proposed a two-step leaching process. This by-passed the negative effect of high acidity on the selectivity while still maintaining a sufficient calcium extraction yield. Still, after carbonation of the leachates, the calcium concentration of the post-carbonation precipitates remained in the range of 69 – 90 wt%, and different mineralogical phases were also present. Based on these past studies, it appears that improvement of PCC purity may only be reached by removing the main impurities prior to the carbonation step, for example by selective precipitation.[7]

The goal of the present work was to synthesize marketable PCC intended for use as filler in paper products. For this application, the carbonate has to be chemically pure (~98 wt% calcium), should have a homogeneous mineralogical structure, a small average particle size and a narrow size distribution.[8] The approach was to investigate the influence of alternative and combinatory processing strategies on the chemical, morphological and mineralogical properties of BF slag-derived PCC, in view of generating a high-quality product. The research focus was divided in three stages:

(a) the separation of the leachate and the post-extraction BF slag, targeting the removal of (nearly) all suspended particles (mainly silica) and residual BF slag prior to the carbonation;

(b) investigation towards the physico-chemical removal of the most prevalent impurities, i.e. silicon, aluminum and magnesium, prior to the carbonation; and

(c) optimization of the carbonation conditions to precipitate PCC with a sufficiently high calcium content, homogeneous crystal structure, small average particle size and narrow size distribution.

Materials and Methods

Materials

Ground granulated BF slag was obtained from a Belgian industrial steelworks. For the different experiments, the BF slag was manually milled with mortar and pestle. The resulting material had a D_{50} value of 2.1 mm and a uniformity coefficient (U) of 12 ($U = D_{60}/D_{10}$); these values were determined via classic sieving classification. Wavelength dispersive X-ray fluorescence (WDXRF, Panalytical PW2400) was used to determine the chemical composition. The slag was composed of following oxides: CaO (41.2 wt%), SiO_2 (35.3 wt%), Al_2O_3 (10.1 wt%), MgO (7.0 wt%); metallic components were Ti (0.8 wt%), Fe (0.4 wt%) and Mn (0.3 wt%). X-ray Diffraction (XRD, Philips PW1830) showed the slag to be largely amorphous; amorphicity is caused by the granulation process whereby the slag is rapidly cooled, preventing crystallization.

The lixiviant used was analytical grade acetic acid (CH_3COOH), the pH adjusting base used was analytical grade sodium hydroxide (NaOH), the carbon dioxide used was industrial grade (99.5 %), and water used was ultrapure (>18.1 MΩ·cm).

Experimental Methodology

The experimental procedure commenced by selectively extracting calcium from the BF slag in a two-step leaching process with acetic acid. The two-step method was used to avoid the detrimental effect of high acidity on the extraction selectivity while still attaining a sufficient extraction yield. In each step, 731 ml of a 2.0 mol/l acetic acid solution was mixed with the 100 g BF slag solids in a Büchi Ecoclave reactor (internal volume = 1.1 liter, electrically heated, cooled with water in a cooling mantle, mixed with a turbine impeller, baffled). The slurry was mixed at 30 °C for 60 min at 1000 rpm. After the second extraction step, both leachates were blended together. These leaching conditions were based on the work of Chiang et al.[5].

The next stage in the experimental process was to separate the leachate solution from the post-extraction residual solids (including the bulk of the suspended silica particles). Separation consisted of centrifugation and/or filtration (depending on the experimental goal). Centrifugation was performed using a Heraeus Labofuge 400 centrifuge, operated at 4000 RCF (relative centrifugal force) for 10 min; the leachate was collected as the supernatant. Filtration was performed using Whatman No. 2 filter paper (pore size = 8.0 μm) or Millipore MF membrane filters (pore size = 0.45 μm).

In a final processing stage, the leachate solution was carbonated. Carbonation of the leachate was performed at different reaction intensities (30 – 90 °C and 2 – 6 bar CO_2) for 60 min at 1000 rpm in the Büchi Ecoclave reactor. To promote precipitation of calcium carbonate and neutralize the regenerated acetic acid, sodium hydroxide was added, in amounts varying from 1.4 to 2.0 mol/l; the latter is equimolar with respect to the acetic acid used in the extraction step. The carbonation precipitates were collected via filtration using Whatman, No. 2 filter paper, thoroughly washed with distilled water and dried for 24 h at 105 °C.

In specific experiments to improve purity, an extra step was added to the experimental procedure, focusing on the removal of soluble impurities. This extra step was fitted between the solid/liquid separation stage and the carbonation stage. First, the temperature of the leachate was lowered (to between 0 and 20 °C) by placing the leachate in a refrigerator. Once cooled, the leachate was filtered with Millipore MF membrane filters to remove precipitates. In additional experiments, the pH of the leachate was increased by sodium hydroxide addition (0.50 – 1.25 mol/l), followed by filtration with Millipore MF membrane filters to remove precipitates. After these steps, the purified leachates from were carbonated as previously described.

Analytical Methods

The mineralogical composition of the different precipitates was determined by Rietveld refinement of XRD data using TOPAS Academic 4.1 (Coelho Software). To determine the chemical composition of the different precipitates, the solids were dissolved in nitric acid (65 wt%). Subsequently, the Al, Ca, Fe, Mg, Mn, Si and Ti contents were determined via Inductively Coupled Plasma Mass Spectroscopy (ICP-MS, Thermo Electron X Series). The concentrations of Fe, Mn and Ti were neglected due to their limited occurrence, and the sum of Al, Ca, Mg and Si contents was then normalized to 100% for ease of purity comparison. The aqueous concentrations of Al, Ca, Mg and Si in the leachates and in the post-carbonation liquid medium were also determined via ICP-MS measurements. The morphology of the solid materials was visualized by Scanning Electron Microscopy (SEM, Philips XL30 FEG). The average particle size diameter and size distribution of the solids were measured via wet Laser Diffraction Analysis (LDA, Malvern Mastersizer 3000).

Results and Discussion

This research was divided in three phases. In the first phase, the post-extraction solid/liquid separation was optimized. In the second phase, soluble impurities were removed from the leachate. In the third phase, the carbonation process was optimized. For brevity, these results are omitted from this text and will be published in full elsewhere.[9] In this section, the overall optimized process is discussed and the optimized product presented. In order to illustrate the attained improvements of the present work, a reference PCC sample was also synthesized using the same conditions of Chiang et al.[5], and its properties are first presented.

Reference Sample

The XRD diagram of the reference PCC sample is shown in Figure 1. The quantitative results of the XRD Rietveld refinement are shown in Table 1. The PCC sample consisted predominantly of the calcite mineral (67.2 wt%). However, different undesired mineral phases were detected, influencing the morphology of the PCC sample. Furthermore, the chemical purity of the post-carbonation precipitate was low: 77.5 wt%. As a result of the combination of low chemical and mineralogical purity, a very limited amount of homogeneous calcite crystals was formed, as visualized by SEM; the images are presented in Figure 2.

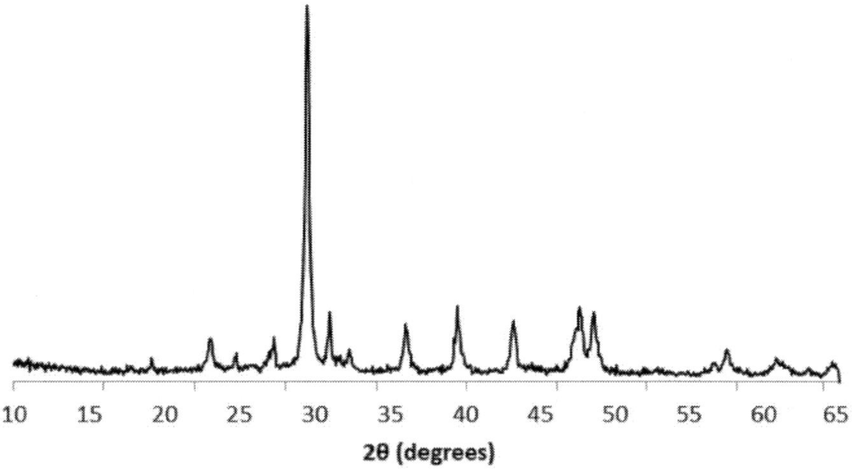

Figure 1. XRD diagram of the reference PCC sample

Table 1. Mineralogical composition (wt% of crystalline fraction) of the reference PCC sample, determined by XRD Rietveld refinement; < indicates that the presence of the specific mineral is less than the estimated quantification accuracy of ±2.0 wt%

Mineral phase	Mineral formula	wt(%)
Aragonite	$CaCO_3$	2.0
Calcite	$CaCO_3$	67.2
Magnesian calcite	$Ca_{1-0.85}Mg_{0-0.15}CO_3$	4.8
Dolomite	$CaMg(CO_3)_2$	4.1
Huntite	$Mg_3Ca(CO_3)_4$	2.1
Hydromagnesite	$Mg_5(CO_3)_4(OH)_2{\cdot}4H_2O$	3.6
Magnesite	$MgCO_3$	2.3
Monohydrocalcite	$CaCO_3{\cdot}H_2O$	<
Nesquehonite	$Mg(HCO_3)(OH){\cdot}2(H_2O)$	5.4
Vaterite	$CaCO_3$	5.8

Figure 2. SEM picture of the reference PCC sample

Optimized Process Experiments

Using optimal separation, purification and carbonation methods, the different improvements were combined in a novel slag-to-PCC synthesis process set-up, illustrated in Figure 3.

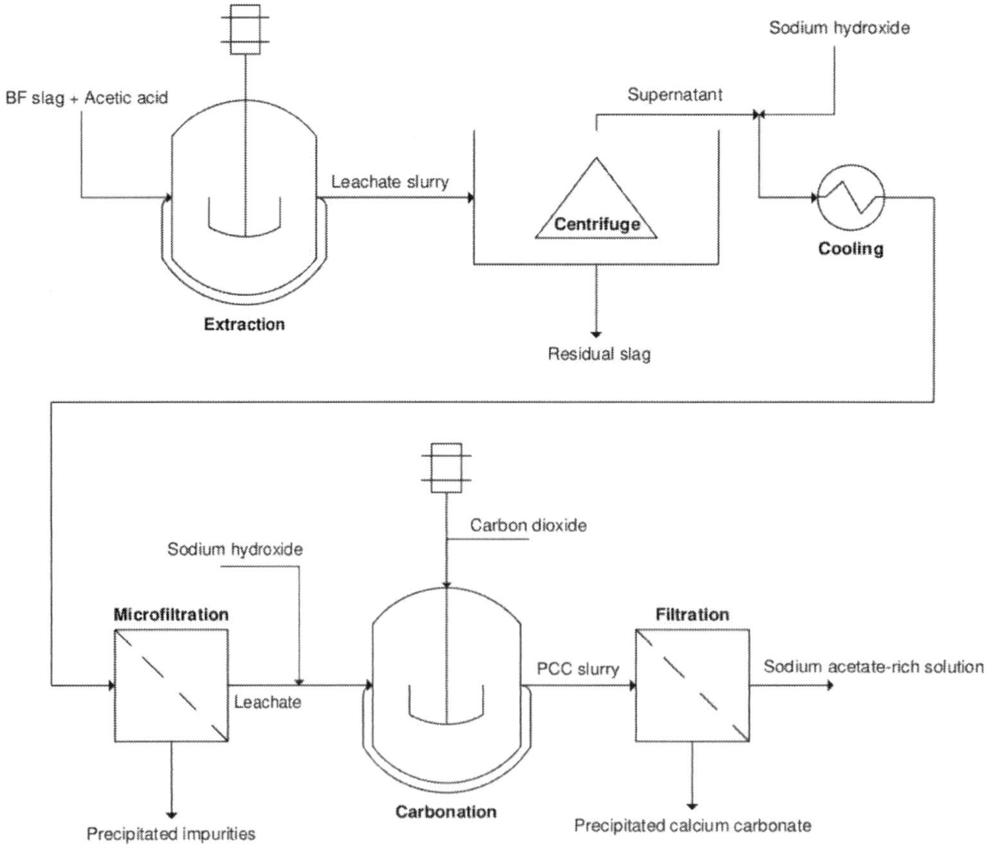

Figure 3. Envisioned slag-to-PCC synthesis process

In a first step, the calcium is selectively extracted from the BF slag with acetic acid. Subsequently, the post-extraction solid residue is separated from the calcium acetate-containing leachate. The optimal solid/liquid separation step comprises of centrifugation followed by membrane microfiltration; the former aims at the removal of the suspended silica, to eliminate filter clogging, and the bulk of the residual slag, while the latter consists of a polishing action. In the optimized process, the centrifuged supernatant is cooled down to 1 °C prior to microfiltration, in order to decrease the solubility of silica and precipitate it out of solution.

To subsequently synthesize PCC, additional sodium hydroxide must to be added during carbonation, to neutralize the reconstituted acetic acid. However, in contrast to previous practices, a sub-equimolar amount (1.7 mol,NaOH/l) of sodium hydroxide, with respect to the acetic acid employed in the leaching step (2.0 mol,CH_3COOH/l), is added. Furthermore, sodium hydroxide is added prior to the cooling step. In this manner, the main soluble impurities, i.e. aluminum and magnesium, are transformed from soluble acetates to poorly soluble hydroxides; the latter precipitate out of solution and are subsequently removed by microfiltration. The voluminous hydroxide flocs that form entrain much of the suspended small silica particles that remain after centrifugation.

As a consequence of this train of purification steps, a leachate with a very high purity is obtained, minimizing the co-precipitation of impurities together with the PCC in the next and final step: carbonation. This approach was successful in decreasing the co-precipitation of magnesium with the PCC, thus increasing the product purity. Furthermore, this carbonation was performed at a lower severity (30°C and 2 bar,CO_2), favoring the formation of calcite and preventing the formation of aragonite; this results in a post-carbonation precipitate with a homogeneous mineralogical content and morphology.

The calcium content of the hereby obtained post-carbonation precipitate was 98.1 wt%; this is sufficiently high to be applicable as filler material in paper,[8] and can truly be referred to as PCC. The main impurity present in this precipitate was magnesium (1.5 wt%), which from a toxicity perspective is passive and therefore not worrisome. The aluminum concentration was close to zero (0.1 wt%), while no silicon was detected (<0.1 wt%). Even higher purity, if required, may be obtained by more precisely optimizing the amount of sodium hydroxide added prior to the separation of the leachate.

The mineralogical composition of the optimized precipitate was determined via quantitative XRD measurement; the results are given in Table 2, while the XRD diagram is presented in Figure 4. The calcite fraction of the post-carbonation precipitate was large, 88 wt%. Furthermore, small amounts of nesquehonite and magnesian calcite were present, both around 3 wt%. Mineral phases present below 2 wt%, which summed to 5.8 wt%, are at the quantification limit of the technique, so their presence is uncertain and may be overestimated. It can thus be concluded that a product with a very uniform crystalline mineral composition was synthesized.

Table 2. Mineralogical composition (wt% of crystalline fraction) of the carbonation precipitates obtained with the optimized process set-up, determined by XRD Rietveld refinement; < indicates that the presence of the specific mineral is less than the estimated quantification accuracy of ±2.0 wt%

Mineral phase	wt%
Aragonite	<
Calcite	88.2
Magnesian Calcite	2.8
Dolomite	<
Huntite	<
Hydromagnesite	<
Magnesite	<
Monohydrocalcite	<
Nesquehonite	3.2
Vaterite	<

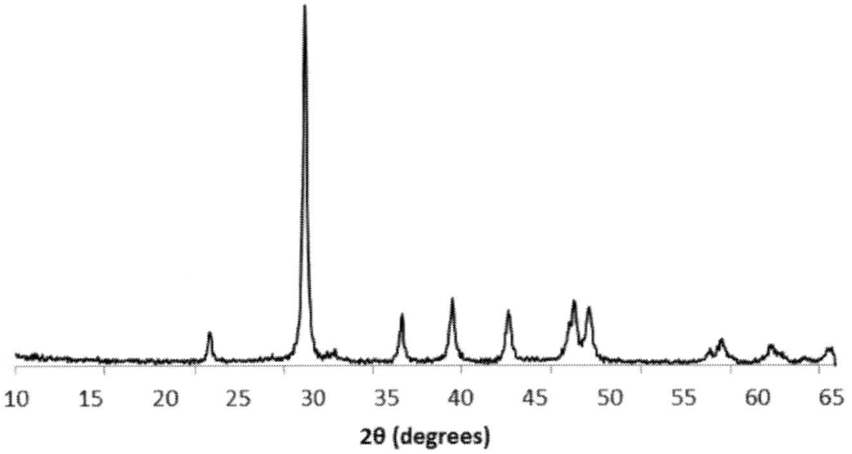

Figure 4. XRD diagram of the post-carbonation precipitate synthesized with the optimized process

A graphical representation of the particle size distribution of the obtained precipitate is shown in Figure 5. The precipitate met the desired qualities of a small average particle size and a narrow size distribution, as stated by Tavakkoli et al.[10] and Gill[8]. Further optimization of the particle size and size distribution may be attained by adjusting the carbonation conditions further.

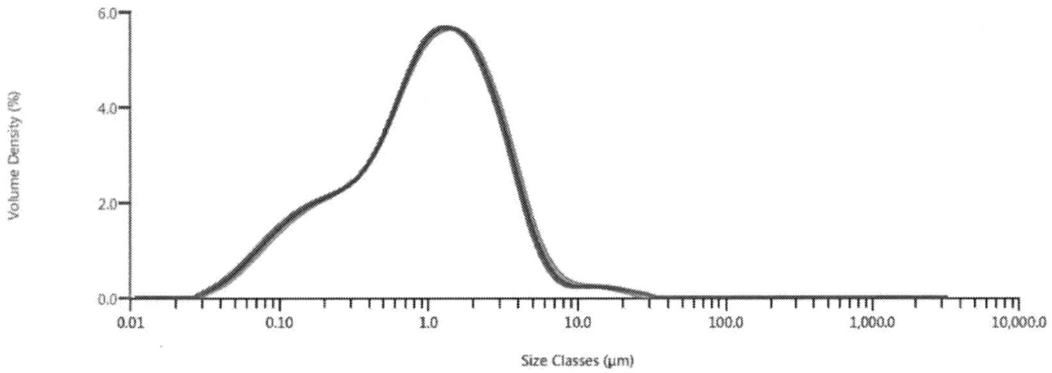

Figure 5. Particle size distribution of the optimized carbonation precipitate, determined by wet laser diffraction analysis; overlapping lines represent measurement replicates

Conclusions

The present work has investigated the possibility to produce precipitated calcium carbonate (PCC) with a high chemical purity, homogeneous mineralogical structure and narrow particle size distribution by reacting carbon dioxide with calcium extracted from blast furnace (BF) slag. To achieve this goal, a previously proposed three-step process was optimized and extended with additional novel purification steps, focusing on the removal of fine colloidal particles and soluble impurities. For the first time, it was possible to synthesize papermaking grade PCC from calcium present in BF slag.

The envisioned process opens a new pathway for PCC preparation, whereby an industrial residue is used in place of natural resources, which is attractive from a sustainability perspective due to the waste valorization being combined with non-renewable resource preservation. Still, further research and process development is needed to maximize the economic potential of the proposed process. The main hurdle with the current set-up is the high amount of liquid waste after the carbonation; more research towards lixiviant regeneration is warranted. The main opportunity that was not addressed in this work is the upgrading of the post-extraction slag residue, and of other produced solid residues, to valuable products.

Acknowledgements

Tom Van Gerven and Rafael Santos acknowledge funding from the KU Leuven Industrial Research Fund (IOF) for the SMaRT-Pro^2 Knowledge Platform. Rafael Santos is also grateful for postdoctoral financial support (PDM-kort) from the KU Leuven Special Research Fund (BOF). The KU Leuven Department of Earth and Environmental Sciences and the Department of Metallurgy and Materials Engineering are acknowledged for the use of XRD, LDA, XRF, and SEM equipment.

References

1. Eloneva, S., Teir, S., Salminen, J., Fogelholm, C.-J., Zevenhoven, R. 2008. Fixation of CO2 by carbonating calcium derived from blast furnace slag. *Energy*, 33(1), 1461-1467.
2. Teir, S., Eloneva, S., Fogelholm, C.-J., Zevenhoven, R. 2007. Dissolution of steelmaking slags in acetic acid for precipitated calcium carbonate production. *Energy*, 32(1), 528-539.
3. Bao, W., Li, H. 2012. Synthesis of aragonite superstructure from steelmaking slag via indirect CO_2 mineral sequestration. *World Academy of Science, Engineering and Technology*, 65, 1036-1042.
4. Said, A., Mattila, H.-P., Järvinen, M., Zevenhoven, R. 2013. Production of precipitated calcium carbonate (PCC) from steelmaking slag for fixation of CO_2. Applied Energy, 112, 765-771.
5. Chiang, Y.W., Santos, R.M., Elsen, J., Meesschaert, B., Martens, J.A., Van Gerven, T. 2014. Towards zero-waste mineral carbon sequestration via two-way valorization of blast furnace slag. *Chemical Engineering Journal*, 249(1), 260-269.
6. Kakizawa, M., Yamasaki, A., Yanagisawa, Y. 2001. A new CO_2 disposal process via artificial weathering of calcium silicate accelerated by acetic acid. *Energy*, 26(1), 341-354.
7. Bao, W., Li, H., Zhang, Y. 2014. Experimental investigation of enhanced carbonation by solvent extraction for indirect CO_2 mineral sequestration. *Greenhouse Gases: Science & Technology*, 4, 785-799.
8. Gill, R.A. 2005. Fillers for papermaking. *In: TAPPI Introduction to Wet End Chemistry Short Course.* Panama City Beach, Florida, 9-11 May 2005.
9. De Crom, K., Chiang, Y.W., Van Gerven, T., Santos, R.M. Purification of slag-derived leachate and selective carbonation for high-quality precipitated calcium carbonate synthesis. *Advances in Materials Science and Engineering*, under review.
10. Tavakkoli, B., Söteman, J., Pohl, M., Schmölzer, T. 2012. Production of high purity precipitated calcium carbonate. *European Patent Application*, EP2447213 A1.

Feasibility of carbonation of alkaline waste oil shale ash in the rocking-type autoclave reactor

O. Velts[1*], M. Uibu[1], K. Rijnsburger[2], P. Knops[2] and R. Kuusik[1]

[1]Laboratory of Inorganic Materials, Tallinn University of Technology, Ehitajate tee 5, 19086 Tallinn, Estonia

[2]Innovation Concepts B.V., NL4201 BH Gorinchem, Netherlands

*Corresponding author: olga.velts@ttu.ee

Abstract

This paper elucidates the method for carbonation of oil shale waste ash in the rocking-type autoclave reactor under various conditions. Current study elaborates the feasibility of the process and gives a detailed characterization of the treated ash samples. The effect of the operating parameters (additives, pressure and temperature, ash type) on the CO_2-binding efficiency is examined. The overall carbonation efficiency calculated on the bases of all CO_2-binding components in ash varied in the range of 57-72% dependent on the applied conditions and the type of ash. Optimal results were obtained at 90°C and 100 bar/180°C and 50 bar without additives. Investigated technological approach allows Ca-silicates to be incorporated into the CO_2 binding process in addition to free lime.

Introduction

The increasing carbon dioxide emissions from fossil fuel combustion, have led to concerns about climate change. Mineral carbonation, which includes the reaction of Ca or Mg containing minerals with CO_2, has been proposed as one of the options for reducing carbon dioxide emissions, especially in case the suitable waste materials and industrial by-products are available.

In Estonia, local low-grade carbonaceous fuel, kukersite oil shale, is used as a primary energy source. The mineral matter of oil shale is dominated by calcite and dolomite, which also contribute to its high specific carbon emission factor (27.85 tC/TJ)[1]. Estonia is one of the highest per-capita CO_2 emitters (13.4 mt/year in 2013) among EU Member States; its power sector is the main CO_2 emitter and also a producer of large amounts of calcium rich (up to 30% of free CaO and ~30% Ca-Mg silicates) waste ash (~7 Mt/year)[2]. Due to the presence of the free CaO the residual material is very alkaline. The oil shale based power sector can benefit from mineral carbonation through the stabilization and neutralization of waste ashes as well as CO_2 capture and storage.

In this study, the feasibility of an innovative carbonation approach for CO_2 mineralization on the basis of Estonian oil shale waste ashes was studied in a rocking-type autoclave reactor. The influence of process parameters (temperature, CO_2 partial pressure, addition of additives, ash type) on the ash carbonation efficiency was investigated, seeking the optimal conditions for maximizing the utilization of ash as a carbon sink. The rocking-type autoclave was used as a representative process in a future full scale Gravity Pressure Vessel[3].

Materials and methods

The initial samples were characterized by chemical analysis, quantitative X-ray diffraction (XRD), BET nitrogen dynamic desorption analysis (BET) and scanning electron microscopy (SEM) methods (Figure 1). The oil shale cyclone ash (CaO_{free}=20%, CaO_t=51%, TC=0.344%, BET SSA=0.61 m^2/g) used in carbonation experiment was collected from pulverized firing boiler at Estonian power plants (PF). Another type of ash from circulated fluidized bed combustion boiler (8% CaO_{free}, CaO_t=29.5%, TC= 1.25%, BET SSA=8.0 m^2/g) was used for comparison (Table 1, sample 9).

Figure 1. SEM micrographs of (a) pulverized firing and (b) circulated fluidized bed ash

The carbonation of ash suspension (12.5% S/L ratio) was performed in a rocking-type autoclave reactor (Figure 2) by injecting CO_2 gas (90% CO_2, 10% N_2) at the 100 bar and 180°C (the reference test). The effect of additives ($NaHCO_3$, organic acids (OA)), temperature and pressure on the carbonation efficiency were investigated by varying the operating conditions (Table 1). The duration of the carbonation stage was 60 minutes.

The lab-scale rocking autoclave (length 1000mm, internal diameter 47mm) is constructed to achieve the similar flow as a full scale Gravity Pressure Vessel installation[4,5].

The solids are falling down, as the gas bubbles are moving upwards through the liquid. As soon as the solids hit the bottom of the reactor, the tube rotates and solids and gas switch their positions, after which the process repeats. The CO_2 is injected during the carbonation process through a high pressure tube and a special rotating joint. The heating takes place by 3 heat elements, the cooling is achieved by a water cooled jacket.

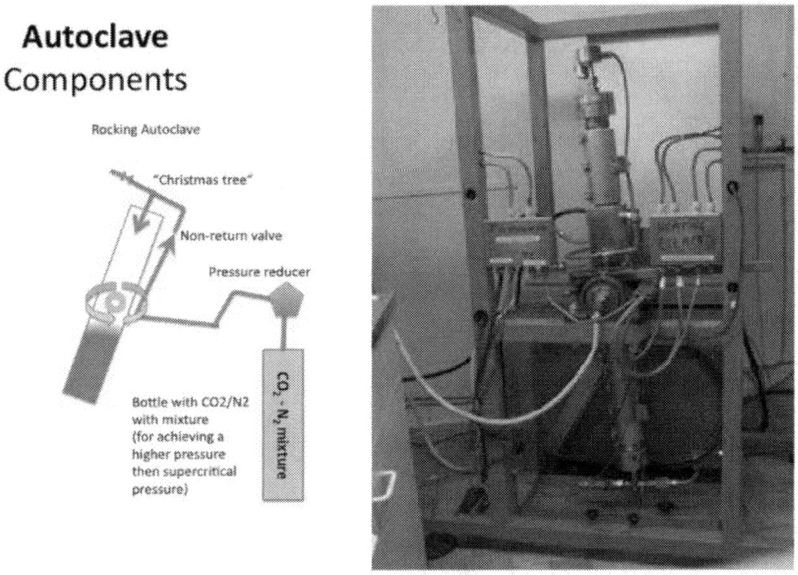

Figure 2. Scheme (a) and photo (b) of the rocking-type autoclave

After carbonation, the suspension was dewatered by settling of the solids and dried at 110°C. The characteristics (composition, morphology, surface area, particle size, etc.) of the initial and treated samples as well as the liquid phase were determined using various analytical techniques: total CaO (slightly modified EVS-EN 196-2:1997); free CaO[6]; total carbon (TC) analysis for estimation of the carbonate content (ELTRA CS-580 Carbon Determinator), total S and its bonding forms (EVS 644:1995), scanning electron microscopy (SEM) analysis for surface morphology characterization (Zeiss EVO15), particle size distribution (Horiba LA-950V2 laser scattering analyser), specific surface area (SSA) by BET nitrogen dynamic desorption analysis method (Sorptometer KELVIN 1042, Costech Microanalytical Ltd.); pH (Mettler Toledo GWB SG2). The mineralogical analysis was performed by the means of powder X-ray diffraction (XRD), using a Bruker 8D diffractometer with Ni-filtered CuKa radiation (scanning step of 0.02°2θ from 2 to 75°2θ, counting time of 0.5 s). A quantitative mineral content of powdered and unoriented whole-rock preparations was measured using Rietveld-based code, Siroquant-3[7,8].

The efficiency of the carbonation process was described by the degree of CO_2-binding BD_{CO2} (Eq. 1). This parameter indicates the fraction of the theoretical maximum ash-binding capacity, $\%ThCO_2$ (Eq. 2)[9] actually utilized:

$$BD_{CO_2},\% = \frac{\frac{CO_2 \times CaO^i_{tot}}{CaO_{tot}} - CO^i_2}{\%ThCO_2} * 100, \qquad (1)$$

where CO_2 and CO_2^i are the CO_2 contents analytically determined in the treated and initial samples; CaO_{tot} and CaO^i_{tot} are the total CaO contents analytically determined in the treated and initial samples, respectively.

$$\%ThCO_2 = 0.785(\%CaO - 0.56 \times \%CaCO_3 - 0.7 \times \%SO_3) + 1.091 \times \%MgO + 0.71 \times \%Na_2O + 0.468(\%K_2O - 0.632 \times \%KCl) \quad (2)$$

Results and discussion

The carbonation tests in rocking-type autoclave reactor were performed with PF oil shale ash at various operating parameters (with and without the additives) (Table 1, samples 1-8). For comparative investigations, carbonation (without the additives) was undertaken also with ash from circulated fluidized bed combustion boiler (CFB ash) at P=100 bar, T=180°C (Table 1, sample 9).

The Total Carbon (TC) and Total Inorganic Carbon (TIC) analyses showed that the carbonated samples contained 44-55% of $CaCO_3$ and residual lime and portlandite in fractional quantity at the applied conditions.

Table 1. Operating conditions and main characteristics of the carbonation products

Sample	Operating parameters				Solid product characteristics				
	T, C	P, bar	$NaHCO_3$, M	OA***, %	CO_2^{TC}, %	CO_2^{TIC}, %	BD_{CO2}, %	SSA, m^2/g	CaO_{free}, %
1	180	100	-	-	23.6		69.5	13.0	0.13
2	180	100	0.64	-	23.8		70.1	3.7	0.58
3	180	100	-	0.05	25.7	20.9	61.2*	13.3	0.10
4	180	100	0.64	0.05	24.2	23.2	68.3*	6.0	0.38
5	180	50	-	-	24.0		70.7		0.42
6	90	100	0.64	-	19.5		56.9		1.20
7	90	100	-	-	24.4		72.0		0.42
8	180	50	0.64	-	20.9		61.2		1.57
9**	180	100	-	-	19.5		69.4	16.7	0.10

*Calculated on the basis of total inorganic carbon
** Ash from CFB boiler used for carbonation
***OA - Organic acid

The morphology of the precipitated particles was examined using SEM and the XRD analysis. According to XRD analysis, in addition to amorphous phase (*e.g.* glass, hydrated Al-Si phases), the main new phases in the treated (at P=100 bar, T=180°C, with and without the additives) samples were pure calcite, aragonite and magnesian calcite. The presence of anhydrite, melilite, wollastonite etc was also confirmed (Table 2). It should be pointed out, that no dolomite phase was found. Pure calcite was the main component in the carbonated FBC ash (Table 2, sample 9). Interestingly, carbonation with addition of organic acids (contains oxalate) provoked the formation of whewellite (Table 2, sample 3).

Table 2. Phase composition of the original ashes and carbonation products

	Sample						
Component, %	**PF ash**	**CFB ash**	**1**	**2**	**3**	**4**	**9**
Amorp. content	32.4	17.7	39.6	34.3	32.8	28.5	29.9
Quartz	4.7	18.4	2.3	2.7	2.2	2.7	10.2
Calcite	3.5	8.7	28	19.7	19.8	20.6	35
Calcite, magnesian				31.7	31.6	38.4	
Aragonite			22.4				5.1
K-feldspar	2.3	12.4	1.8				8.1
Anhydrite	7.3	10.5	1.6		1.7		4.2
Merwinite	4.5	1.0		1.8	0.9	1.5	
Melilite	1.6	1.0	1.9	2.9	2.6	2.7	
Periclase	3.9	2.4		1.4	0.5	1	
C_2S, beta	16.8	6.4		4.0	1.6	2.9	
Wollastonite				1.4	0.4	1.7	
Illite/mica		3.8	0.3				2.8
Bassanite			1.8				3.5
Hematite		2.1	0.3				1.1
Lime	18.1	3.3					
Portlandite	5.0	12.4					
Whewellite					6.0		

Both XRD and SEM analyses confirmed the presence of significant amount of aragonite in the sample 1 (Figure 3, a, b). A distinguish morphology of needle-like aragonite crystals is also observed in the carbonated CFB ash (Figure 3, f). However, in the samples 2-4, carbonate is present in the form of mainly magnesian calcite and pure rombohedral calcite crystals. The results suggest that the addition of additives triggered formation of magnesian calcite. SEM micrographs also show the precipitation of $CaCO_3$ on the input particles (surface coating). Similar process rate controlling phenomena has been noticed in case of continuous mode aqueous carbonation of low-porosity PF ashes, as the recirculating aqueous phase was saturated with carbonate ions[10]. Furthermore, SSA increased from 0.6 to 13 m^2/g for the samples 1 and 3, but only to 4-6 m^2/g for samples 2 and 4. This indicates that the addition of $NaHCO_3$ could have influenced this parameter. In the case of CFB ash, about 2x increase of the SSA took place (Table 1, sample 9). SEM analysis also detected about 2-3% of S in some of the samples, which indicated partial substitution of CO_3 ions in calcite structure by SO_4.

According to particle size distribution analysis, the mean particle size of the initial ash samples was in the range of 33-45 μm and depending on the carbonation conditions increased to 38-50 μm.

The liquid phase of the carbonated suspension contained mainly Mg (0.7-1.2 g/L), Ca (0.12-0.3 g/L) and Si (0.07-0.14 g/L) ions. The pH of the effluent was in the range of 7-8.0.

Figure 3. SEM micrographs of the treated ash samples (Table 1): (a, b) sample 1; (c) sample 2; (d) sample 3; (e) sample 4; (f) sample 9

In addition to lime, portlantide, anhydrate, components such as Ca-silicates ($CaSiO_3$, and Ca_2SiO_4) and MgO present in oil-shale ash also take part in CO_2-binding reactions. The theoretical maximum ash-binding capacity (*Th*CO_2%) calculated according to Eq. 2 is 42.9% for PF ash and 23.5% for CFB ash used in this work. When all CO_2-binding components are taken into account, the BD_{CO2} reached 57-72% for PF ash, depending on the process conditions, and ~69% for CFB ash.

Due to the lower content of free CaO, the CO_2-binding capacity of CFB ash was lower. The amount of CO_2 bound by 1 t of ash, reached 244-309 kg and 163 kg for PF and CFB ashes, respectively. To compare, our developments involving ash-water suspension carbonation method demonstrated that oil shale ashes are able to bind up to 290 kg CO_2 per ton of ash[11,12].

Although the experiments with olivine[13,14] showed intensification of the carbonation through additives, there was no beneficial effect in case of oil shale ash at the reference conditions (T=180°C and P=100 bar). Carbonation efficiency of ~70% was achieved with or without the addition of $NaHCO_3$. Using organic acids actually diminishes the carbonation efficiency as the carbon is bond by the organics: ~1.7% of C for sample 3, which could explain the formation of the whewelite (Table 2).

At lower T or P (T=90°C; P=50-100bar), similar results were observed for samples 5 and 7: the CO_2-binding degree of 71-72% was achieved (Table 1). However, the addition of $NaHCO_3$ at these conditions decreased noticeably the CO_2 binding degree (BD_{CO2} ~57-61%, Table 1, sample 6 and 8).The content of residual lime (1.2-1.6%) was higher as well in these conditions.

The results show, that the K and Na compounds do not take part in the carbonation process, whereas Mg is only partially involved.

Conclusions

The prospects of oil shale waste ash carbonation in the rocking-type autoclave reactor were demonstrated. It can be concluded, that the investigated ashes displayed good CO_2-binding efficiency and the potential of the free lime in oil shale ash was utilized in full amount under studied conditions. The results showed, that in addition to free lime, this method involve also Ca-silicates into the CO_2 mineralization process. The overall carbonation efficiency calculated on the bases of all CO_2-binding components in ash varied in the range of 57-72%, depending on the applied conditions and the type of ash.

The anticipated intensification of the CO_2-binding process through addition of additives did not take place. At the same time, reduction of the reference temperature or pressure allowed deeper carbonation: highest CO_2-binding degree (up to 72%) achieved at 90°C and 100 bar or 180°C and 50 bar without the addition of additives. The impact of operating conditions will be investigated further for process optimization.

Acknowledgements

Authors express their gratitude to Dr. P. Somelar (University of Tartu) for quantitative XRD analysis and SEM measurements as well as to Dr. R. Santos (KU Leuven) for the help with the analyses. This work was supported by institutional research funding IUT (IUT33-19) of the Estonian Ministry of Education and Research, by the Estonian

Ministry of Education and Research (3.2.0501.10.002) and the Estonian Science Foundation (Grant No 9334).

References

1. Method for determining the amount of carbon dioxide discharged into the atmosphere. Regulation of the Minister of the Environment No. 94 [In Estonian]
2. Eesti Energia Annual Report 2013, https://www.energia.ee/-/doc/10187/pdf/concern/annual_report_2013_eng.pdf, 19.03.2015
3. Santos, R.M., Van Gerven, T., Knops, P., Rijnsburger K. Opportunities and developments on the intensification of chemical and geochemical processes by gravity pressure vessel technology. *Presentation at the Fourth European Process Intensification Conference:* https://lirias.kuleuven.be/bitstream/123456789/399157/2/Santos_EPIC2013.pdf, 19.03.2015
4. Kaufmann, L.A., Peterscheck H. 1986. Modeling Vertech's mile long multiphase reaction vessel. Chemical Engineering Science, 41(4), 685-692.
5. Innovation Concepts B.V.: http://www.innovationconcepts.eu/ 19.03.2015
6. Reispere, H.J. 1966. Determination of free CaO content in oil shale ash. *Transact. Tallinn Polytechnical Institute,* 245, 73–76.
7. Taylor, J.C. 1991. Computer programs for standardless quantitative analysis of minerals using the full powdered diffraction profile. *Powder Diffraction,* 6, 2-9.
8. Colin, R.W., Taylor, C.J., Cohen, D.R. 1999. Quantitative mineralogy of sandstones by X-Ray diffractometry and normative analysis. *Journal of Sediment Research,* 69, 1050-1062.
9. Huntzinger, D.N., Gierke, J.S., Kawatra, K., Eisele, T.C., Sutter, L.L. 2009. Carbon dioxide sequestration in cement kiln dust through mineral carbonation. *Environmental Science & Technology*, 43, 1986-1992.
10. Uibu, M., Kuusik, R. 2014. Main physicochemical factors affecting the aqueous carbonation of oil shale ash. *Minerals Engineering*, 59, 64-70.
11. Uibu, M., Kuusik, R., Andreas, L., Kirsimäe. K. 2011. The CO_2-binding by Ca-Mg-silicates in direct aqueous carbonation of oil shale ash and steel slag. *Energy Procedia*, 4, 925-932.
12. Uibu, M., Velts, O., Kuusik, R. 2011. Aqueous carbonation of oil shale wastes from Estonian power production for CO_2 fixation and PCC production. *Proceedings of the Conference of Young Scientist on Energy Issues* (Kaunas, Lithuania, May 26-27, 2011), 415-424.
13. Sanna, A., Uibu, M., Caramanna, G., Kuusik, R., Maroto-Valer, M.M. 2014. A review of mineral carbonation technologies to sequester CO_2. Chemical Society Reviews, 43, 8049-8080.
14. O'Connor, W.K., Dahlin, D.C., Rush, G.E., Gerdemann, S.J., Penner, L.R., Nilsen, D.N. 2005. Aqueous mineral carbonation: Mineral availability, pretreatment, reaction parametrics, and process studies. *DOE/ARC-TR-04-002*; Albany Research Center, US.

Case Study for Utilisation of CO_2 from Flue Gases in PCC Production from Steelmaking Slag

S. Teir [1*], T. Kotiranta[2], T. Parviainen[3] and H.-P. Mattila[4]

[1] VTT Technical Research Centre of Finland Ltd, Espoo, Finland
[2] Outotec Research Center, Pori, Finland
[3] Outokumpu Stainless Oy, Tornio, Finland
[4] Åbo Akademi University, Turku, Finland
*Corresponding author: sebastian.teir@vtt.fi

Abstract
This article presents a techno-economic assessment for producing precipitated calcium carbonate (PCC) from steelmaking slag. The studied case consists of a stainless steel plant generating slag and a paper plant consuming PCC situated at a relatively close distance. The process uses ammonium salts such as ammonium chloride as solvent for extracting calcium from the slag and subsequently precipitating calcium carbonate using lime kiln flue gas. The process was modelled and the investment costs and operational costs estimated. In addition, the environmental footprint of the process was assessed. The results from this work indicates that the process could in its current stage of development be economic for producing calcium carbonate of similar grade as ground calcium carbonate (GCC) used by paper mills. At the same time, the process would avoid 0.3 t CO_2 emissions per tonne PCC produced. Although the operational costs for the process were found to be similar to that of a conventional PCC process, the investment is more than twice the cost of a conventional PCC process. The main reason for the high investment costs is that the studied process uses an ammonium salt solution needs to be recovered for reuse and removed from the PCC product, while a conventional PCC process uses only water as a process solution.

Introduction

Synthetic calcium carbonate, commonly known as precipitated calcium carbonate (PCC), is used as a filler and coating pigment in papers and plastics. In conventional production, PCC is produced by carbonation of hydrated lime slurry (calcium hydroxide) in batches. Although the production of PCC binds CO_2 as carbonate, more CO_2 is emitted when the limestone that provides lime as raw material for the process is calcined. A recent PCC production concept enables extraction of calcium from steelmaking slags using ammonium salt solutions, which omits the need for fresh limestone and its calcination. This alternative concept has the potential to reduce CO_2 emissions and simultaneously utilize steelmaking slag.

In this work, an economic assessment for producing PCC from steelmaking slag was made. The studied case consisted of a stainless steel plant generating slag and a paper plant consuming both PCC and GCC (ground calcium carbonate) situated at a relatively close distance. The process was modelled using Outotec HSC Chemistry software. The environmental footprint of the process was calculated with GaBi software. Additional experiments, such as calcium extraction from steelmaking slag and filterability of PCC, were performed for acquiring a better estimate of certain process parameters.

Process concept and case study

In the process concept studied here, also known as the Slag2PCC process, calcium is extracted from steelmaking slag using ammonium salt solutions in the extractor:

$$CaO(s) + 2NH_4^+(aq) \rightarrow Ca^{2+}(aq) + H_2O(l) + 2NH_3(aq) \quad (1)$$

$$Ca(OH)_2(s) + 2NH_4^+(aq) \rightarrow Ca^{2+}(aq) + 2H_2O(l) + 2NH_3(aq) \quad (2)$$

$$Ca_2SiO_4(s) + 2NH_4^+(aq) \rightarrow Ca^{2+}(aq) + CaSiO_3(s) + H_2O(l) + 2NH_3(aq) \quad (3)$$

PCC is subsequently precipitated in the carbonator by bubbling CO_2-containing flue gas through the solution[1]:

$$2NH_3(aq) + Ca^{2+}(aq) + H_2O(l) + CO_2(g) \rightarrow CaCO_3(s) + 2NH_4^+(aq) \quad (4)$$

The PCC is separated by filtration and the solvent is returned to the extractor, as the ammonium salt is regenerated in the carbonator (Figure 1). Both the PCC and the unreacted part of the slag from the extractor (slag residue) are washed in order to remove solvent remnants (ammonium salt). The ammonium salt in the washing water is recovered by a reverse osmosis unit to minimize the need for solvent make-up and to clean up the water spent in the process.

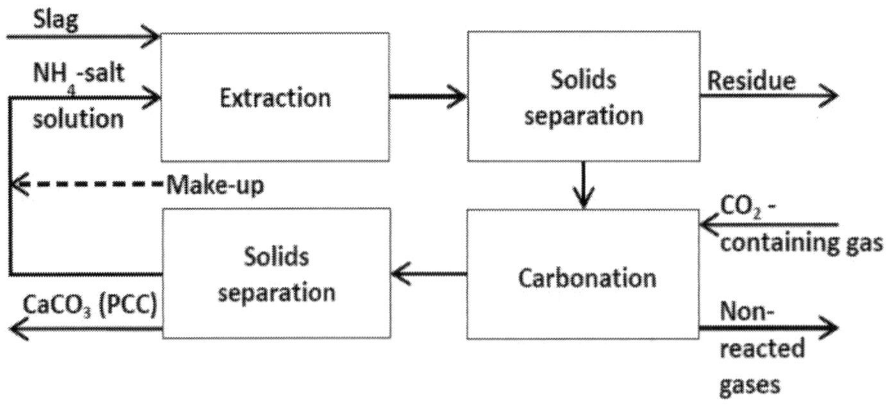

Figure 1. The Slag2PCC concept.

The purpose of this study was to determine the cost and environmental impact of a new PCC plant that would produce PCC from steelmaking slag for use in a paper mill. The paper mill of this case study requires annually 60,000 t fine-ground (60% < 1 μm) GCC for coating purposes and 80,000 t scalenohedral PCC (1-2 μm median particle size, narrow distribution) as filler material in the production of fine paper. The stainless steel plant, producing steelmaking slag, is located 40 km away from the paper mill. The steel plant produces ~280,000 t per year of air-cooled Argon Oxygen Decarburization (AOD) slag and ladle slag that would be readily available for use by the Slag2PCC process. Since the distance between the two plants is relatively short, previous calculations have shown that the cheapest alternative considering logistics is to build the Slag2PCC process next to the steel mill and transport the produced PCC by trucks to the paper mill.[2] A lime kiln is currently available at the steel mill site, which provides a suitable CO_2 source for the Slag2PCC process (22% vol-CO_2, 80 – 120°C, 100,000 - 140,000 t CO_2 per year).

Reactivity of AOD slag

The reactivity of various AOD slags from the steel mill of the case study was tested for compatibility with the process. Based on XRF analysis, the slag samples consisted mostly of Ca (41-43 wt-%), Si (12-14 wt-%) and Mg (4-5 wt-%)[3]. Mineral phases identified by XRD were bredigite (12-18%), larnite (11-17%), periclase (3-4%) and fluorite (1-2%), but the majority was amorphous (60-70%). Ground, water-cooled AOD slag, filler slag (the fine portion of AOD and ladle slag), and air-cooled AOD slag were leached in 4 M ammonium acetate (solid concentration 100 g/l) for 120 min at ambient temperature and pressure. Calcium was selectively extracted from all samples, with the air-cooled AOD slag showing the highest extraction of calcium (60-63% of Ca in slag extracted) and water-cooled AOD slag the lowest extraction (34-39%)[3]. Currently, only air-cooled AOD slag is being produced at the steel mill of the case study, indicating that a 50 % calcium extraction should be a realistic target.

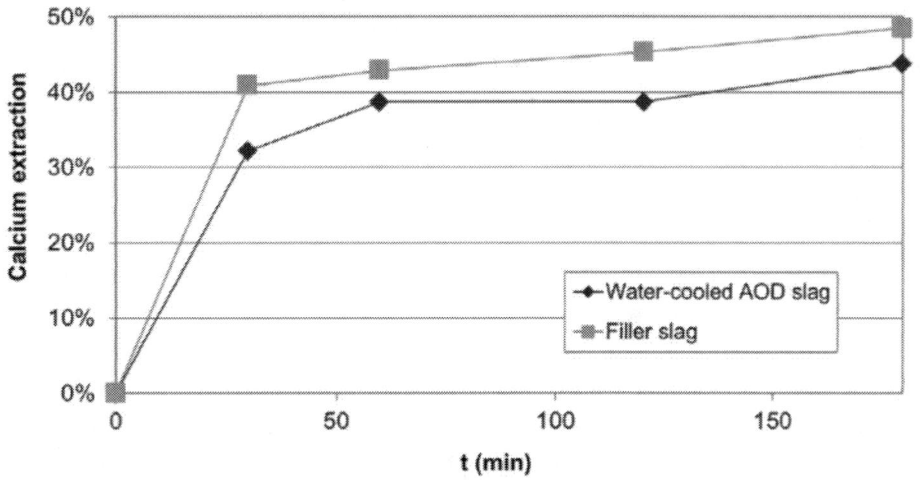

Figure 2. Extraction of calcium from water-cooled AOD slag and filler slag in 4 M ammonium acetate.

The rate of extraction was also tested with two separate experiments: one for water-cooled AOD slag and one for filler slag. The experiments showed that a significant part of calcium was extracted already during the first half hour of the experiments (Figure 2). Only trace amounts of other elements were extracted[3], which made it possible to precipitate, by adding CO_2 to the resulting solution, a very pure $CaCO_3$ product with an ISO brightness of 97. While the experiments were performed with ammonium acetate, previous work has shown that there is hardly any difference in extraction yield when leaching steelmaking slag in ammonium chloride or ammonium nitrate in comparison to ammonium acetate[4,5]. Thus, ammonium chloride was used in the calculations for the techno-economic assessment, because it is the cheapest option of these three salts.[6]

Technology restrictions

Åbo Akademi University and Aalto University are currently developing the Slag2PCC process further within the Finnish Carbon Capture and Storage R&D program (CCSP). A laboratory pilot has been constructed at Aalto University. It has been shown that by adjusting process parameters calcium carbonate particles of different sizes can be synthesized – from 20-50 µm particle clusters down to sub-micron (<1 µm) particles. Similarly, the particle shape can be adjusted – both rhombohedral and scalenohedral calcite can be produced, as well as aragonite[6,7]. The PCC produced from slag is of similar quality to commercial PCC, with preliminary experiments from ongoing research indicating that it is suitable for paper making applications.

One of the challenges for the Slag2PCC process is the requirements for washing of the PCC. In order to minimize the solvent remnants on fine (1-2 µm diameter) PCC particles, a three-stage washing and filtration equipment setup would be needed. This complex filtration setup would make up a large part of the investment costs for this facility, making a Slag2PCC process 3-4 times more expensive than a similar sized conventional PCC plant[2]. Instead, the Slag2PCC process parameters could be adjusted to produce larger (20-50 µm) particles, to be subsequently ground down to a finer size suitable for paper making purposes[8]. Although this lowers the investment costs significantly, grinding makes the particle size distribution less narrow than that needed for the filler purposes. At the current development stage it seems more economically feasible to aim for a product that can compete with fine-ground aragonite (GCC) instead of fine, scalenohedral PCC.

Process modelling

The new PCC production process was modelled using Outotec HSC Chemistry software. In this case, the process is adjusted to produce aragonite, which is ground very fine in order to meet the requirements of the GCC used for coating (60% < 1 µm) by the paper mill. The flow sheet of the process model with process streams are shown in Figure 3. The process streams were calculated by specifying the yield for each component (Table 1, 2 and 3). The slag composition is approximated as 57% CaO (41% Ca). The slag is ground down before fed to the extractor (80% <300 µm). The extraction reaction was

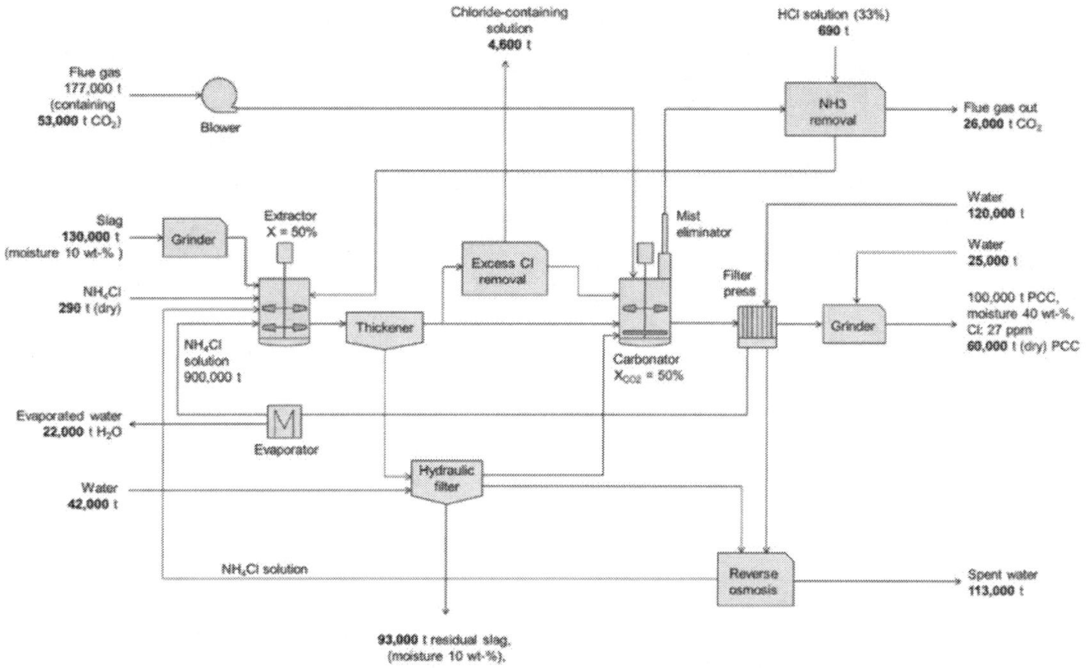

Figure 3. Flow sheet of the studied process. Mass flow presented on an annual basis.

Table 1. Input data for extractor, thickener and hydraulic filter.

Parameter	Value	Comment
NH_4^+ concentration	36 g/l in extractor input	2 M NH_4Cl
NH_4^+ concentration	14 g/l in extractor output	Based on stoichiometry, 0.7 M Ca in solution
Temperature	60 °C	Based on experiments[6]
Extraction yield	50%	Estimate, based on experimental data
Solid concentration	81 g/l	Used in experiments
Thickener underflow solids	1000 g/l	Assumption
Slag filter cake moisture	10%	Assumption
Slag filter washing efficiency	90%	Assumption
Slag filter washing water amount	0.5 m^3/t	Assumption

simplified by assuming that all calcium is available as CaO. The extraction was then calculated according to Eq. 1, using 50% conversion of calcium. In the carbonator, 99.5% of the ammonia participates in the precipitation reaction in Eq. 4 while 0.5% is assumed to evaporate[9]. Ammonia in the carbonator off-gas needs therefore to be washed with an acidic solution as ammonia gas has a relatively high partial pressure and therefore cannot be captured with water washing. The incoming flue gas to the

Table 2. Input data for carbonator and PCC filter.

Parameter	Value	Comment
Temperature	60°C	Based on experiments[6,7]
Feed gas: CO_2	22 vol-%	Case-specific feature
Feed gas: moisture	6 vol-%	Case-specific feature
CO_2 consumption	50% of feed	Assumption
Ammonia evaporation	0.5%	Based on experiments at room temperature[9]
PCC production	9.6 t/h	Case-specific requirement
PCC filter cake moisture	20%	Ongoing test work
Filter washing efficiency	99.9%	Needed for the defined Cl content (dry solids) in the PCC product
Filter washing water amount	2 m^3/t	Ongoing test work

Table 3. Input data for reverse osmosis unit and evaporator.

Parameter	Value	Comment
Total salt concentration in the recovered solution	60 g/L	Assumed 100% recovery
Evaporated water	3.5 t/h	Needed to keep extractor input at 2 M NH_4Cl
Steam requirements	1.2 t/h	Three-stage evaporator

carbonator contains 6 vol-% water and the carbonator is operated at 60°C temperature. The water partial pressure in the carbonator is assumed to be the same as for pure water, the dew point showing that some water is evaporated in the carbonator. The amount is not large enough to evaporate the water fed into the process in washing of solids and therefore, an evaporator is needed for removing excess water from the process.

The current steel mill plant layout was investigated and a suitable location found for the new PCC plant next to the current slag processing facility. The flue gas from the lime kiln would be transported 1200 m by pipeline. Hence, the pipeline could be designed to bring the flue gas temperature below that of the carbonator to cool the heat produced by the exothermic reaction in the carbonator. As aragonite can be produced by keeping the carbonator temperature at 60°C the whole process cycle is kept at 60°C, as the extractor is much less sensitive to temperature than the carbonator[10]. The plant is operated 6257 h/a (assuming three shifts with weekend interruption).

Environmental footprint

The environmental footprint was calculated with GaBi software based on the HSC Chemistry model and estimation on the process equipment electricity consumption. In Figure 4 the global warming potential (GWP) per ton of produced PCC is presented. The calculation takes into account CO_2 capture, electricity consumption, process steam consumption and its production, slag residue disposal, waste water treatment, transportation of PCC, fresh water usage and HCl consumption.

The main emission source is the electricity production. The electricity consumption in the process is 209 kWh/t PCC. It includes PCC grinding 107 kWh/t PCC, reverse osmosis (RO) electricity consumption 44 kWh/t PCC, general electricity use (stirrers, pumps, etc.) 42 kWh/t PCC and slag grinding 16 kWh/t PCC. The second biggest emission source is the steam consumed in the evaporation unit. It is assumed to be produced with a heavy fuel oil fired unit. The net GWP value was -0.3 tonne CO_2 equivalent per tonne of produced PCC.

In the current calculations, steam is produced using heavy fuel oil, which gives high CO_2 emissions. However, since the steel plant is located at the same site as the PCC plant there might be possibilities to perform the evaporation in more environmentally friendly ways, for instance by a cooling tower-type evaporator that use waste heat from the steel mill. The footprint calculation does not include the ammonium chloride since that was not found from the software database. However, the impact of ammonium chloride production on the result is expected to be very small.[6]

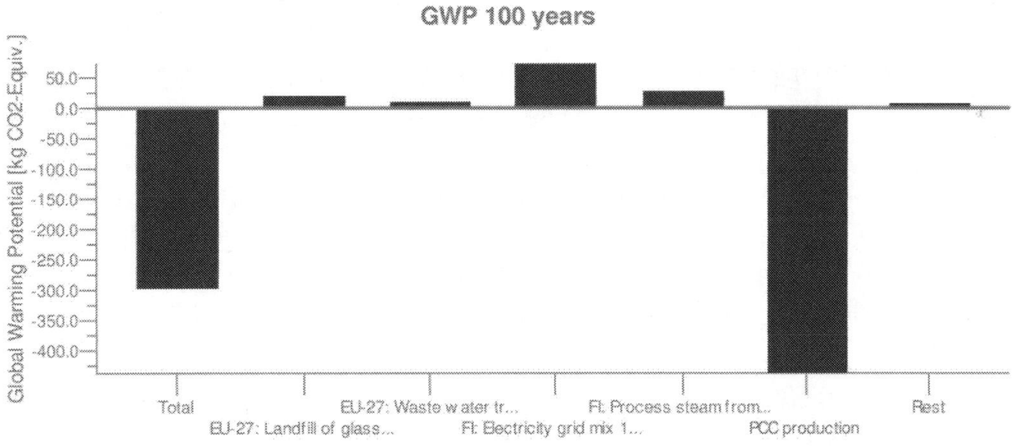

Figure 4. Process environmental footprint per tonne of produced PCC.

Summary of cost calculations

The investment costs and commodity costs were evaluated on a coarse level, using the data from the process calculations. The results give an approximate of the level of cost for the process. The operational costs consist of commodity cost (NH_4Cl make-up, HCl, slag, reverse osmosis unit use, heat for evaporation and electricity for grinder, stirrers, pumps and filters), transportation costs of PCC, maintenance and personnel costs. An electricity cost of 60 €/MWh was used, as well as 30 €/MWh for steam (needed for evaporation). The process calculations show that the largest operational cost is expected to be the cost for the slag residue handling (estimated coarsely at 15 €/t), representing 34% of the total operational costs. The largest heat requirements comes from the evaporation (7.3 GWh/a) and the largest power requirements come from PCC grinding (6.4 GWh/a). The total operational costs were 4.1 M€/a.

The total investment for the new PCC process was estimated to 26 M€, of which 38% was for the PCC filtration equipment alone. The investment costs were estimated only for the main equipment of the process using in-house process component cost data and experience. Costs for intermediate storage and handling of raw materials and products are excluded. The main equipment prices include the share of construction, piping, automation, etc. The share of main equipment prices is 34% of the total investment price. The filtration investments were estimated on engineering, procurement and construction (EPC) basis by Outotec. For comparison, the investment cost for a conventional PCC plant of the same PCC production capacity is about 10 M€ and the operational costs about 4.4 M€/year.

The investment payback time was calculated for a loan based on constant payments and a constant interest rate of 5% using Excel's PMT function, which calculates the annual constant payment according to this function:

$$PMT = \left(PV + \frac{PV}{(1+r)^{n}-1}\right) \times r \qquad (5)$$

where *PMT* = annual payment, *PV* = present value (i.e. investment cost), *r* = the (annual) interest rate, *n* = number of periods (i.e. investment payback time in years).

The economic feasibility of the process is highly dependent on the value of the produced calcium carbonate. If the produced PCC was sold to a price of 160 €/t (the price for paper grade GCC of a similar particle size distribution) the annual revenue would be 5.5 M€ with a turnover of 9.6 M€. Assuming that all the revenue is used for payments of the loan, the investment payback time would be 6 years.

Since the steel mill is part of the EU emissions trading system, it is possible that an economic benefit could be had from the CO_2 emissions avoided thanks to the Slag2PCC process. Currently, emission allowance prices (EUA) have been very low and was 7 €/t CO_2 at the time of writing. At this price, the benefit from emission allowances would only be 0.13 M€ per year.

Discussion and conclusions

The case study performed indicates that the Slag2PCC process could in its current stage of development be economic for producing calcium carbonate to replace the GCC used by paper mills. At the same time, the Slag2PCC process would sequester 0.3 tonne CO_2 per tonne PCC produced. Although the operational costs for the Slag2PCC process were found to be similar to that of a conventional PCC process, the investment is more than twice as expensive. The main reason is that the Slag2PCC process uses an ammonium salt solution that needs to be recovered for reuse and removed from the PCC product, while a conventional PCC process uses only water as a process solution.

However, several aspects of the process have not been studied in detail and need more work, including CO_2 dissolution rate and ammonia evaporation in the carbonator, filterability of the produced PCC and use of the produced PCC in paper making applications. The ammonia lost to the gases in the carbonation is now assumed to be 0.5%, based on the room temperature experiments and needs to be verified at current process conditions. Tests on the filterability of the produced PCC are currently ongoing. In addition, extensive tests are needed to compare the use of PCC produced from slag to commercial materials (PCC and GCC) in papermaking applications.

NH_3 removal from the used flue gas can alternatively be performed with sulphuric acid. In that case the produced ammonium sulphate would be discarded and ammonium chloride feed to the process would increase, while the excess chloride removal process would not be needed.

The solution after the thickener might contain small particles originating from the slag, which would end up to the PCC product. Thus, the solution would need an additional filter for clarification after the thickener.

The CO_2 absorption in the carbonator needs more research. The gas feed to the process stage is larger than what is usual for conventional reactors. Although the conventional PCC production is operated with similar gas flows, the difference is that the pH in the studied process is lower, possibly affecting the CO_2 absorption. Therefore this process might need another kind of reactor design for the carbonation stage.

Acknowledgements

This work was carried out in the Carbon Capture and Storage Program (CCSP) research program coordinated by CLEEN Ltd. with funding and support from Stora Enso Oyj, Tapojärvi Oy, Outotec Oyj and the Finnish Funding Agency for Technology and Innovation, Tekes. The initial work was carried out in a separate project with funding and support from Stora Enso Oyj and Outokumpu Stainless Oy. Jouko Pakarinen assisted with process details and with the cost assessment. Prof. Ron Zevenhoven is acknowledged for his comments.

References

1. Teir, S., Eloneva, S., Revitzer, H., Zevenhoven, R., Salminen, J., Fogelholm, C.-J., Pöyliö, E. 2009. Method of producing calcium carbonate from waste and by-products. International application published under the patent cooperation treaty (PCT), WO 2009/144382, PCT/FI2009/050455.
2. Teir, S. PCC from steelmaking slag – case study. Research report VTT-R-07647-13, VTT Technical Research Centre of Finland. CCSP Deliverable D517. Restricted report.

3. Teir, S. Kronqvist, P., Kaartinen, T., Parviainen, T. 2015. Leaching of Residual Steel Slag from a Mineral Carbonation Process. *In: Proceedings of the 5th International Conference on Accelerated Carbonation for Environmental and Material Engineering (ACEME)*, New York, U.S.A, June 21-24, 2015.
4. Eloneva, S., Teir, S., Revitzer, H., Salminen, J., Said, A., Fogelholm, C.-J., Zevenhoven, R., 2009. Reduction of CO_2 emissions from steel plants by using steelmaking slags for production of marketable calcium carbonate. Steel Research International, 80(6), 415-421.
5. Rinne, J. 2008. CO_2 sequestration potential of industrial by-products and waste materials in Finland. Master Thesis, Department of Energy Engineering, Helsinki University of Technology.
6. Mattila, H.-P. 2014. Utilization of steelmaking waste materials for production of calcium carbonate (CaCO3). Doctoral dissertation. Åbo Akademi University, Turku, Finland. http://urn.fi/URN:NBN:fi-fe2014092444878
7. Mattila, H.-P., Slotte, M., Said, A., Zevenhoven, R. 2013. BOF slag carbonation in aqueous NH4Cl-solutions: steering product quality in a continuous process. *In: Proceedings of the 4th International Conference on Accelerated Carbonation for Environmental and Materials Engineering (ACEME)*, Leuven, Belgium. pp. 547-550.
8. Kotiranta, T. 2014. PCC production from steel mill slag, update. Outotec Research Center, Pori. Report number 14175-ORC-T. Restricted report.
9. Eloneva, S., Mannisto, P., Said, A., Fogelholm, C.-J. 2011. Ammonium salt-based steelmaking slag carbonation: Precipitation of $CaCO_3$ and ammonia losses assessment. *Greenhouse Gases: Science and Technology*, 1, 305-311.
10. Mattila, H.-P., Grigaliūnaitė, I., Zevenhoven, R. 2012. Chemical kinetics modelling and process parameter sensitivity for precipitated calcium carbonate production from steelmaking slags. Chem. Eng. J., 192, pp. 77–89. DOI: 10.1016/j.cej.2012.03.068

Demonstration pilot plant for the production of precipitated calcium carbonate (PCC) from steelmaking slag and carbon dioxide

A. SAID[1,*] and M. JÄRVINEN[1]

[1] Department of Energy Technology, School of Engineering,
Aalto University, Espoo, Finland
*Corresponding author: arshe.said@aalto.fi, +358-50-5718886

Abstract

The production of PCC from steel slag has been proposed as a potential method of simultaneously reducing the CO_2 emissions from the steel industry and turning a waste stream into a valuable product. To avoid mining activities and associated CO_2 emissions, the conventional PCC production based on limestone can be replaced by the PCC produced via Slag2PCC. In 2014, researchers at Aalto University, Department of Energy Technology have designed, constructed and tested the world's first Slag2PCC pilot plant test facility that converts steel slag and CO_2 into PCC. The researchers have scaled-up the Slag2PCC process for 400 times, from 500 mL to 200 L. In the calcium extraction process, the pilot plant can handle up to 190 L of liquid solvent and 20 kg of solid steel slag. Based on these amounts, the pilot plant can produce 5-10 kg of PCC depending on the Ca extraction efficiency. The feasibility of the Slag2PCC process route has been successfully demonstrated at the pilot scale. Although the quality of the produced PCC is not yet of commercial standard, but we have successfully produced aragonite and rhombohedral calcite PCC crystals. This paper will present results achieved from the pilot plant.

Introduction

Carbon dioxide sequestration by mineral carbonation of Ca/Mg silicate minerals has been considered as a potential technology to mitigate global warming[1]. In addition to the natural sources of these silicate materials, a number of industrial wastes and by-products such as steelmaking slag is an attractive feedstock for CO_2 sequestration. Steel manufacturing which is one of the biggest industrial sources of CO_2 emissions could benefit from this option by utilizing its own by-products, steelmaking slags to combine with CO_2. Slags have a high calcium content and thus, if the calcium is selectively extracted prior the carbonation, the end product will be pure precipitated calcium carbonate (PCC).

The production of PCC from steel slag (Slag2PCC) has been proposed as a potential method of simultaneously reducing the CO_2 emissions from the steel industry and turning a waste stream into a valuable product[2-9]. In the Slag2PCC process route, an ammonium salt solvent (e.g. NH_4Cl, NH_4NO_3 or CH_3COONH_4) is used to extract calcium from BOF (Basic Oxygen Furnace) slag. The resulting Ca-rich solution is reacted with CO_2 gas to form PCC. In order to reduce mining activities and associated CO_2 emissions, the conventional PCC production methods based on limestone can be replaced by the PCC produced via Slag2PCC process route.

This PCC can be replaced by the conventional PCC production methods based on limestone to avoid mining activities and associated CO_2 emissions.

PCC is mainly consumed by the paper industry as a filler and coating pigment of paper and paperboard[10]. In Finland, where both the iron and steel as well as pulp and paper industries have a strong presence, the research on production of PCC from steelmaking slag has been intensified.

Researchers at Åbo Akademi University have built a sub-demo test facility for Slag2PCC process route[2]. In 2014, researchers at Aalto University, Department of Energy Technology have designed, constructed and tested the world's first Slag2PCC pilot plant test facility that converts steel slag and CO_2 into PCC. The researchers have scaled-up the Slag2PCC process for 400 times, from 500 mL to 200 L. In the calcium extraction process, the pilot plant can handle up to 190 L of liquid solvent and 20 kg of solid steel slag. Based on these amounts, the pilot plant can produce 5-10 kg of PCC depending on the Ca extraction efficiency. The feasibility of the Slag2PCC process route has been successfully demonstrated at the pilot scale, although the quality of the PCC produced from the process is not yet of commercial standard.

This paper will present results achieved from the Slag2PCC pilot plant test facility as well as the process design and the challenges during the scale-up work

Scale-up of Slag-to-PCC process

Figure 1 shows the process flow diagram of the Slag2PCC pilot plant at Aalto University. The pilot plant consists of three 200 L reactors: an extraction reactor (Reactor 1) and two carbonation reactors (Reactors 2 and 3). In the Reactor 1, calcium is extracted from steel slag, after which this Ca-rich solution is filtered through one of the extraction filters (Extraction filter 1 or 2) equipped with 1 μm filter bags. To remove particles smaller than one micron, the liquid is pumped via a secondary filtration stage, consisting of two 1 μm and one 0.45 μm filter in series. After filtration, the Ca-rich solution can be stored in a reserve tank (Tank 1) or carbonated directly in Reactor 2.

In fact, Reactor 2 is the main carbonation reactor where the Ca-rich solution reacts with CO_2 to precipitate PCC, while Reactor 3 is for solvent purification prior to recycling it back to the extraction reactor. Therefore, most of the Ca-rich solution (~ 85 %) is pumped to the Reactor 2, pure CO_2 is bubbled from the bottom of the reactor via a gas sparger of 470 holes with diameter 1 mm. After carbonation is complete, the liquid from the reactor is filtered to collect the PCC using an identical filtration system as the extraction filters (PCC filter 1 or 2) and the solvent is stored in a solvent tank (Solvent tank 1). The filtrate is then mixed with the remaining Ca-rich solution (~ 15 %) in the Reactor 3 to precipitate any unreacted bicarbonate (HCO_3^-) and carbonate (CO_3

$^{2-}$) ions preventing them from precipitating as PCC in the extraction reactor when the solvent is recycled.

The pilot plant also includes two plate and frame heat exchangers (Heat exchanger 1 and 2) which can be used to heat or cool the solvent going into the reactors. Each reactor is equipped with a temperature and a pH indicator, a level controller and a three-level 2-bladed variable speed drive (VSD) agitator with a maximum speed of 202 rpm for the Reactor 1 and 170 rpm for Reactors 2 and 3.

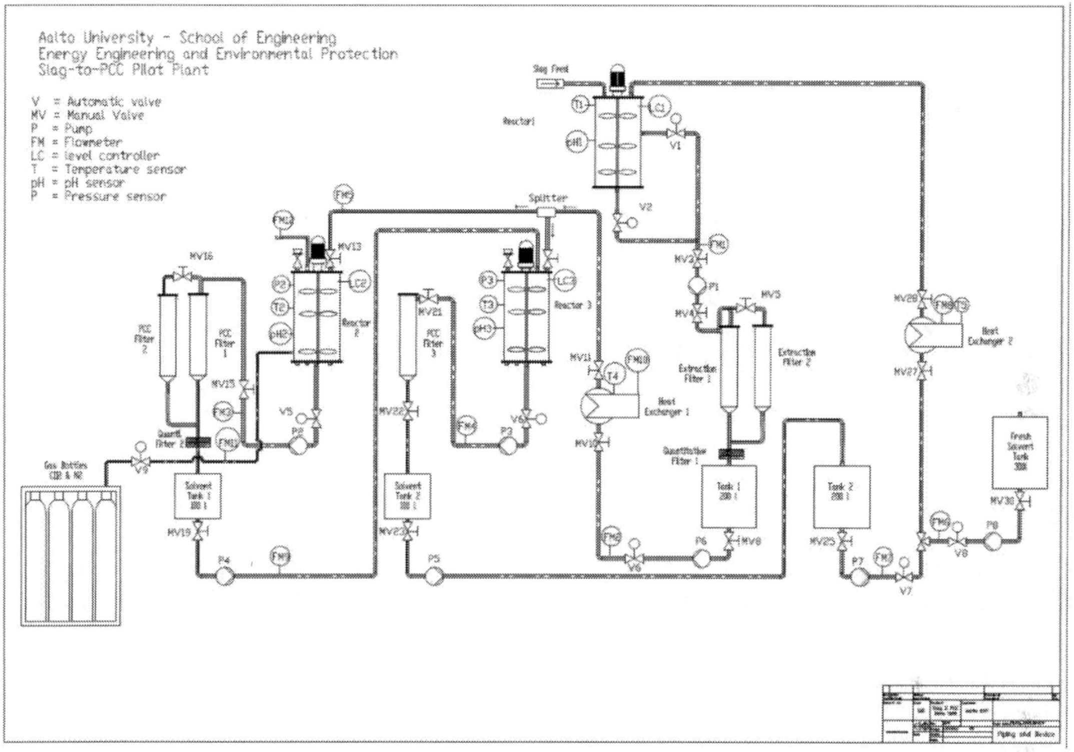

Figure 1. Process Design of Slag2PCC pilot plant test facility at Aalto University, Finland

Figure 2. Slag2PCC pilot plant test facility at Aalto University, Finland

Experimental Procedure

Steel converter slag (<250 µm) supplied by Ruukki Metals Oy (currently SSAB Europe Oy), Raahe, Finland, was mixed with 1M $NH_4Cl(aq)$ solvent to selectively dissolve calcium from the slag at room temperature (~20°C). According to the XRF analysis, the slag contains 51 wt-% of calcium. For each test, the solid to liquid ratio was kept constant 100 g/L. 150 L of solvent was pumped from the Fresh Solvent Tank into the extraction reactor (Reactor 1) and mixed with 15 kg of slag. The mixture was agitated at 202 rpm for one hour. Then, the mixture was pumped to the filtration unit (Extraction filtration) where the Ca-rich solution was separated from the residual slag. The Ca-rich solution was pumped to the carbonation reactor (Reactor 2). The CO_2 gas flow was set using a rotameter and carbonation time began when the flow of CO_2 started. Pure CO_2 was used in all tests. The pH and temperature of the extraction and carbonation processes solution were logged automatically at 10 second intervals. The Ca-rich solution from the extraction stage was heated up through in Heat exchanger 1 up to the desired initial carbonation temperature (Table 1). The precipitated PCC was separated from the regenerated $NH_4Cl(aq)$ solvent in a filter (PCC filter 1 and 2). The PCC samples were taken for SEM imaging.

Table 1. Experimental conditions for the pilot plant tests

	Extraction				Carbonation			
Test	Solvent (L)	SLR* (g/L)	$[NH_4Cl]$ (M)	Slag (kg)	T (°C)	Solution (L)	Time (min)	CO_2 (L/min)
Test 1	150	100	1	15	20	135	60	14
Test 2	150	100	1	15	45	135	41	14
Test 3	150	100	1	15	58	135	39	14

*Solid-to-liquid ratio

Results

Extraction process

The extraction process aimed to selectively dissolve calcium from the slag with ammonium chloride in an intensive stirred reactor. As a result, after an hour of extraction, 78 % of the calcium in the slag was dissolved. Figures 3 and 4 show the recorded pH and temperature data, respectively. Immediately after the slag had been added into the solution both the pH and temperature rose (from 8.2 to 9.0 pH units and from 23.08°C to 25.32°C, respectively).

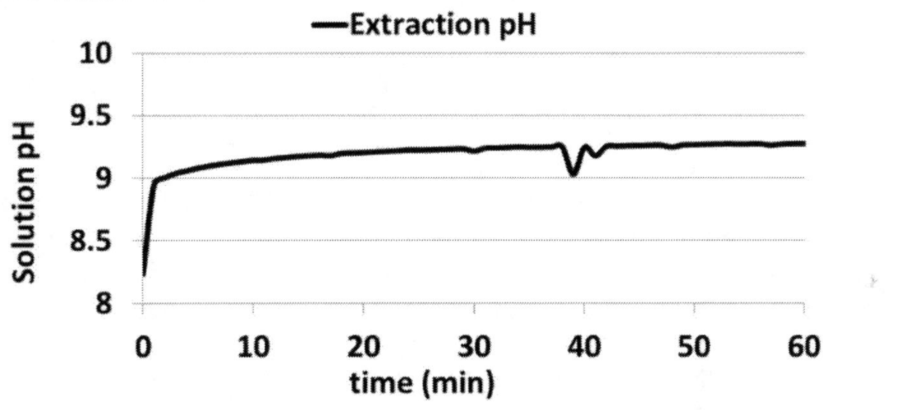

Figure 3. Online solution pH recordings of the calcium extraction process

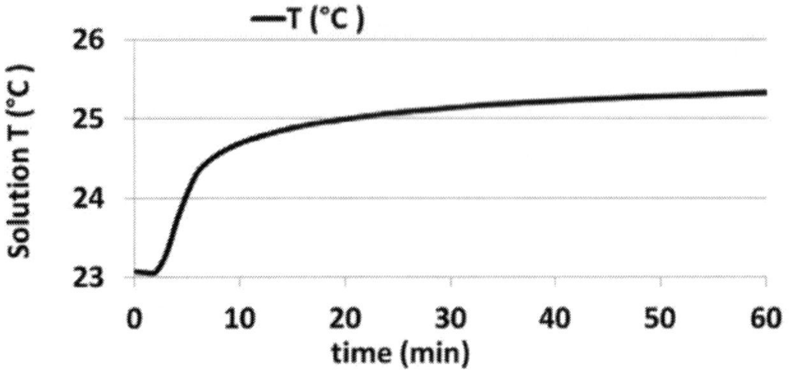

Figure 4. Online solution temperature recordings of the calcium extraction process

Carbonation process

The Ca-rich solution from the extraction process reacted with pure CO_2 in the carbonation reactor (Reactor 2) producing precipitated calcium carbonate (PCC). Generally, PCC can precipitate in multiple crystalline phases (calcite, aragonite, and vaterite). Thus, a precise control over precipitation conditions is important for ensuring a high quality of the PCC product[11]. In these tests, we varied only the initial temperature of the carbonation solution, the aim being to produce three different PCC morphologies, rhombohedral calcite at 20°C, scalenohedral calcite at 45°C and aragonite at 58°C.

The temperature and pH during the carbonation tests was recorded (Fig. 5 and 6). Each test was stopped once the pH had passed the neutral equivalence point and entered the slightly acidic region (pH range of 6.4-7). The higher was the solution temperature, the faster the equivalence point was reached, leading to shorter carbonation times.

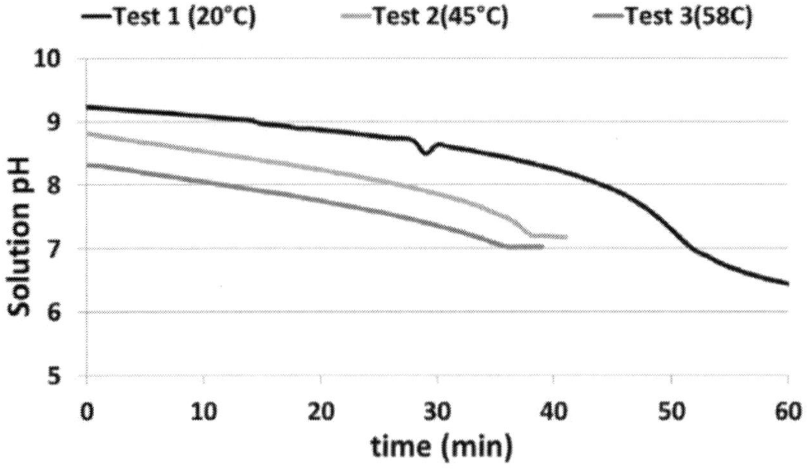

Figure 5. Online solution pH recordings of the carbonation process

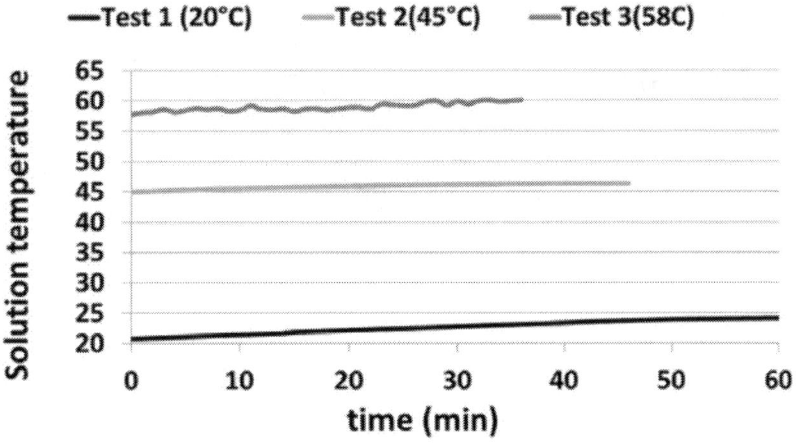

Figure 6. Online solution temperature recordings of the carbonation stage

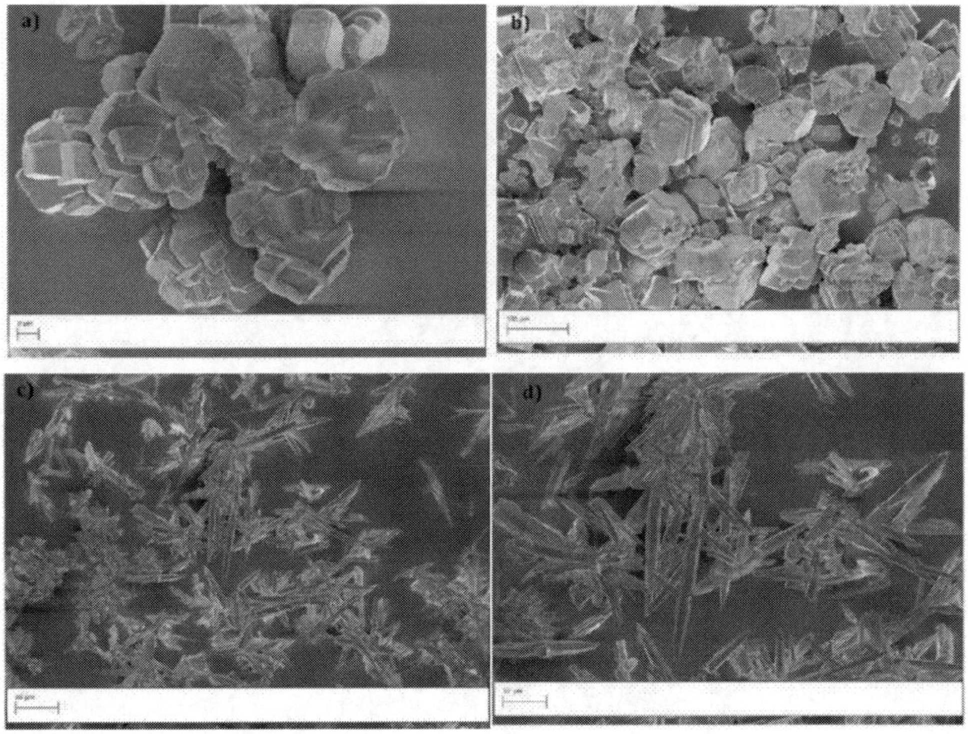

Figure 7. SEM micrographs of PCC produced with the Slag2PCC pilot plant: a) Test 1 at 20°C, rhombohedral calcite. A scale bar of 2 µm. b) Test 2 at 45°C, rhombohedral calcite. A scale bar of 100µm. c) and d) Test 3 at 58°C, aragonite. A scale bar of 20 µm and 10 µm, respectively

Fig. 7 shows the SEM images taken of PCC samples from each test. At temperatures of 20°C and 45°C agglomerated rhombohedral calcite was precipitated (Figs. 7 a and b, respectively), while essentially 100% aragonite needles of various sizes, ranging from

2-50 μm in length and maximum width of around 3 μm precipitated at 58°C (Fig. 7 c-d). Scalenohedral calcite was supposed to precipitate at 45°C (Test 2) but due to some reasons it could not precipitate at these process conditions. The solvent concentration can be an obstacle to the formation of scalenohedral crystals. Exactly how the solvent concentration affects the PCC morphology has not yet been studied. However, Mattila et al.[4] have, based on lab-scale experiments, proposed that low NH_4Cl concentration (e.g. 0.01 M) during the extraction process will improve the PCC morphology and minimize the PCC washing required to remove the NH_4Cl impurities.

Conclusions

This work has demonstrated that the scale-up work of the Slag2PCC process route was successful and PCC has been produced from steelmaking slag and CO_2 at a pilot scale, although the quality of the PCC produced from the Slag2PCC process route is not yet of commercial standard. The attempt to produce scalenohedral PCC at pilot scale has not yet been successful, but it should be possible to produce it using a different process condition. As future work, we will investigate process parameters that control the PCC morphology, agglomeration and particle size distribution.

Acknowledgements

This work was carried out in the Carbon Capture and Storage research program(CCSP) coordinated by CLEEN Ltd. with funding from the Finnish Funding Agency for Technology and Innovation, Tekes and the Academy of Finland (Projects 258319 and 26495). Special thanks to Professor Ron Zevenhoven, Dr. Hannu-Petteri Mattila at Åbo Akademi University and Mr. Tuukka Kotiranta at Outotec Research Centre, Pori, Finland for their valuable contributions and comments.

References

1. Eloneva S. Reduction of CO_2 emissions by mineral carbonation: Steelmaking Slags as Raw Material with a Pure Calcium Carbonate End Product PhD thesis, Aalto University School of Science and Technology, Espoo Finland 2010.
2. Mattila H.-P. 2014. Utilization of steelmaking waste materials for production of calcium carbonate ($CaCO_3$). Doctoral Thesis at Thermal and Flow Engineering Laboratory, Department of Chemical Engineering, Åbo Akademi University.
3. Eloneva S., Said A., Fogelholm C.-J., Zevenhoven R. Preliminary assessment of a method utilizing carbon dioxide and steelmaking slags to produce precipitated calcium carbonate. Appl. Energy 2012(90), 329-34.
4. Mattila H.-P. Hudd H. Zevenhoven R. 2014. Cradle-to-gate life cycle assessment of precipitated calcium carbonate production from steel converter slag. Journal of Cleaner Production. 2014(84), 611-618
5. Mattila H.-P., Grigaliunaite I., Said A., Fogelholm C.-J., Zevenhoven R. Production of papermaking grade calcium carbonate from steelmaking slag – product quality and development of a larger scale process. Proc. of SCANMET IV 2012(2), 19-28.

6. Mattila H.-P., Grigaliūnaitė I., Zevenhoven R. Chemical kinetics modeling and process parameter sensitivity for precipitated calcium carbonate production from steelmaking slags. Chem. Eng. J. 2012(192), 77-89.
7. Said A., Mattila H.-P., Fogelholm C.-J., Zevenhoven R. Effect of steelmaking slag properties to the extraction of calcium by using ammonium salts. International Conference on Applied Energy ICAE 2012, Paper ID: ICAE2012-A10464. Jul 5-8, 2012, Suzhou, China 2012
8. Said A., Mattila H.-P., Järvinen M., Zevenhoven R. Production of precipitated calcium carbonate (PCC) from steelmaking slag for fixation of CO_2. Appl. Energy 2013(112), 765-71.
9. Said A., Mattila O., Eloneva S., Järvinen M. Enhancement of calcium dissolution from steel slag by ultrasound. Chemical Engineering and Processing: Process Intensification 2015(89), 1-8
10. Specialty Minerals Inc. 2012. Specialty Minerals Precipitated Calcium Carbonates In Pharmaceuticals. www.specialtyminerals.com/specialty-applications/specialty-markets-for-minerals/pharmaceuticals/ Accessed 20.3.2014.
11. Flaten E.M., Seiersten M., Andreassen J. Polymorphism and morphology of calcium carbonate precipitated in mixed solvents of ethylene glycol and water. J. Cryst. Growth 2009(311), 3533-8

Unified Kinetics of Serpentine Dehydroxylation.

Jason P. Mann

Brian S. Haynes

THE UNIVERSITY OF SYDNEY

ACEME 2015

THE UNIVERSITY OF SYDNEY

Dehydroxylation Kinetics

Current Status

- $Mg_3Si_2O_5(OH)_4 \rightarrow 3MgO.2SiO_2.(2\text{-}x)H_2O + xH_2O$
- First kinetic study: 1965,
  - Water on the moon.
- Kinetic compensation effect,

$$k(T) = A \exp\left(-\frac{E}{RT}\right)$$

  - Unreliable design parameters outside experimental T range,
  - Why does it exist?

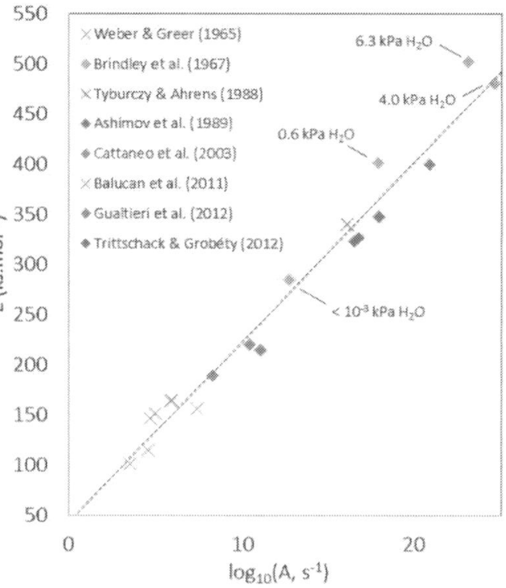

Dehydroxylation Kinetics
Solid State Kinetics

› Generalised isothermal rate equation

$$\frac{d\alpha}{dt} = k(T)f(\alpha) \xrightarrow{\text{Integrate}} \int_0^{\alpha} \frac{d\alpha}{f(\alpha)} = g(\alpha) = k(T)t$$

› $f(\alpha)$ & $g(\alpha)$ are typically structure dependent, e.g.

- Spherical, retreating interface (R3),

$$g(\alpha) = 1 - (1-\alpha)^{1/3}$$

- Spherical, diffusion controlled (D4),

$$g(\alpha) = 1 - \frac{2}{3}\alpha - (1-\alpha)^{2/3}$$

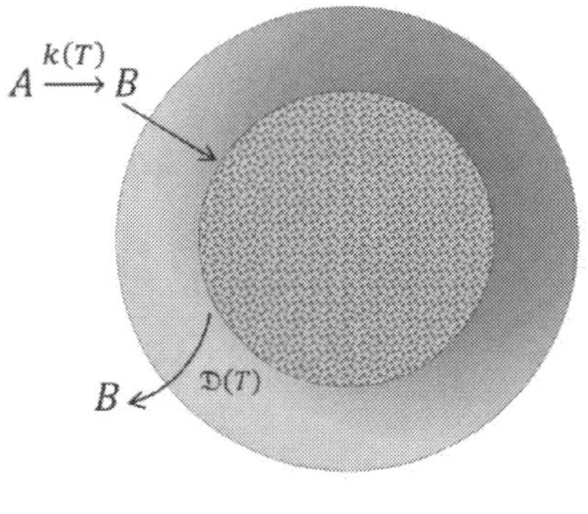

Dehydroxylation Kinetics
Previous Kinetic Data

› Isothermal data, reduced time (τ).

- All data falls between R3 & D4.
- Generalised model,
  - Mixed reaction and diffusion control,
  - Coupled by Damköhler number,

$$\mathrm{Da}_2 = \frac{k'}{K}\frac{d_p}{\mathfrak{D}}$$

$$\left[\frac{\mathrm{Da}_2}{4}\left(1 - \frac{2}{3}\alpha - (1-\alpha)^{2/3}\right) + \left(1 - (1-\alpha)^{1/3}\right)\right] = \frac{2k'}{d_p}t$$

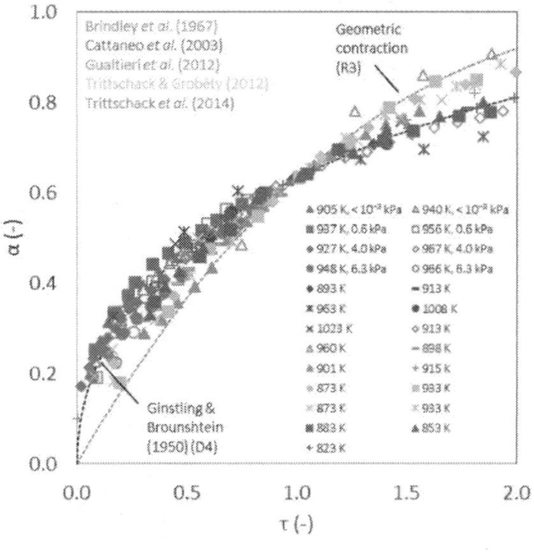

Dehydroxylation Kinetics
Methodology

› Non-isothermal thermogravimetric analysis.

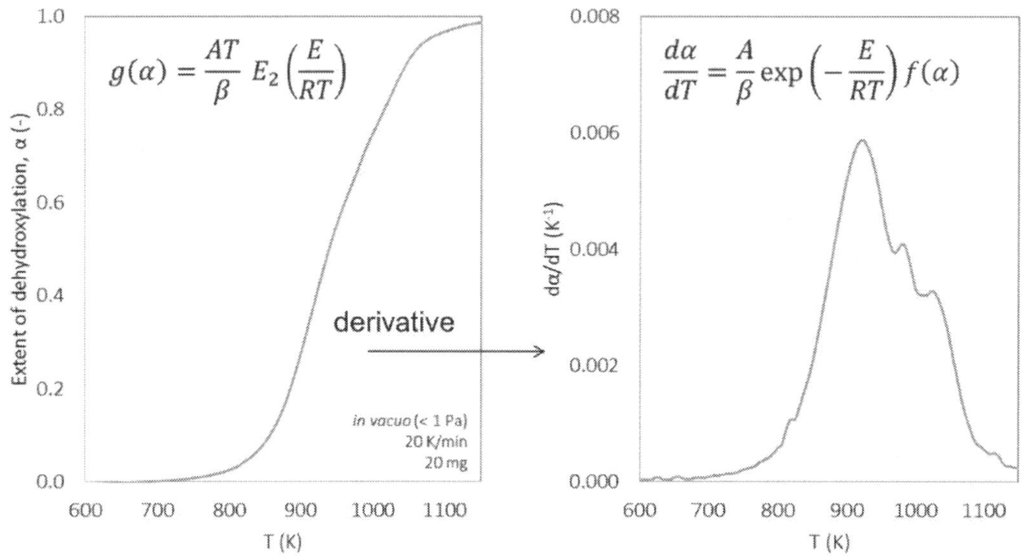

5

Dehydroxylation Kinetics
Findings

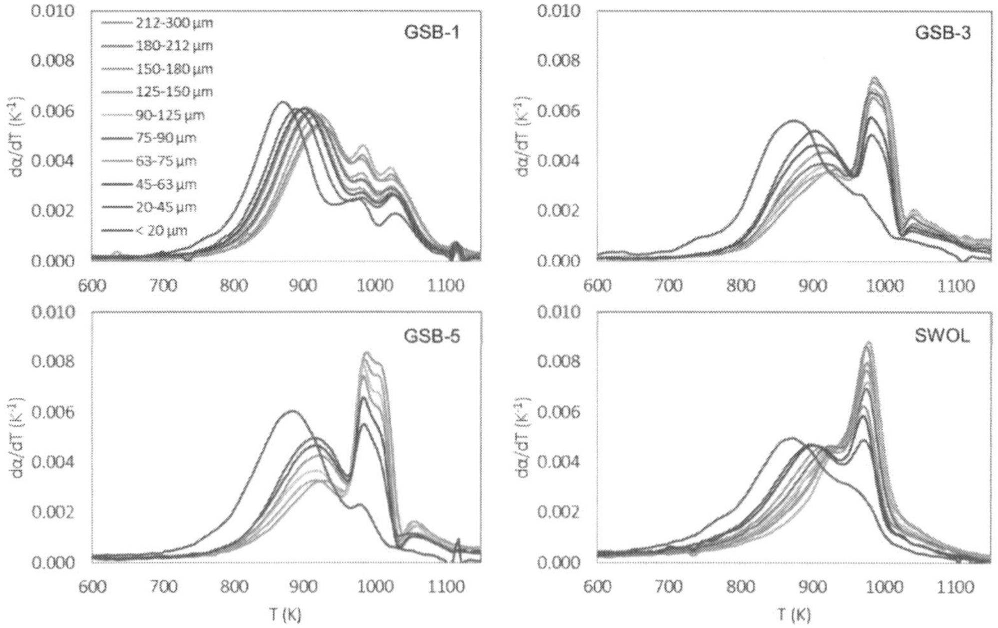

6

THE UNIVERSITY OF SYDNEY

Dehydroxylation Kinetics

Modelling: Variable rate feature

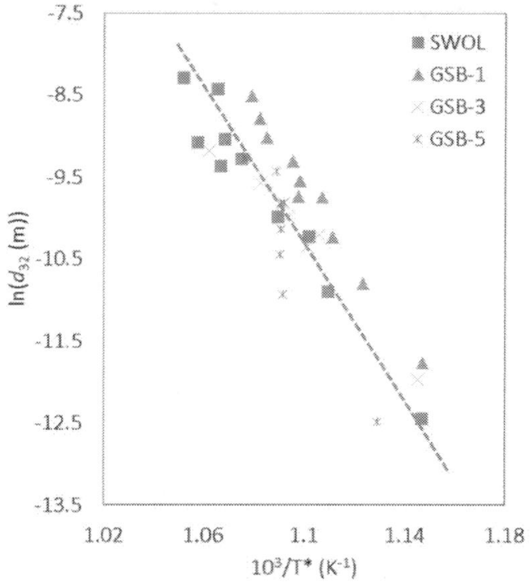

› Spherical, retreating interface (R3),

- $$\frac{d\alpha}{dT} = 2\frac{1}{\beta}\frac{A'}{d_p}\exp\left(-\frac{E}{RT}\right)3(1-\alpha)^{2/3}$$

› Maximum rate occurs at T*,

- $$\ln(d_{32}) = b' - \frac{E}{RT^*}$$

- $E = 320\ \mathrm{kJ.mol^{-1}}$
- $\log_{10}(A', \mathrm{m.s^{-1}}) = 11.4$

7

THE UNIVERSITY OF SYDNEY

Dehydroxylation Kinetics

Modelling: Variable rate feature

› Agreement with previous data?
- Yes, both experiment and theory.
- A-factor related through,

$$A = \frac{2A'}{d_p} .$$

› Another point in the line?
- No.

› Intrinsic to the mineral.

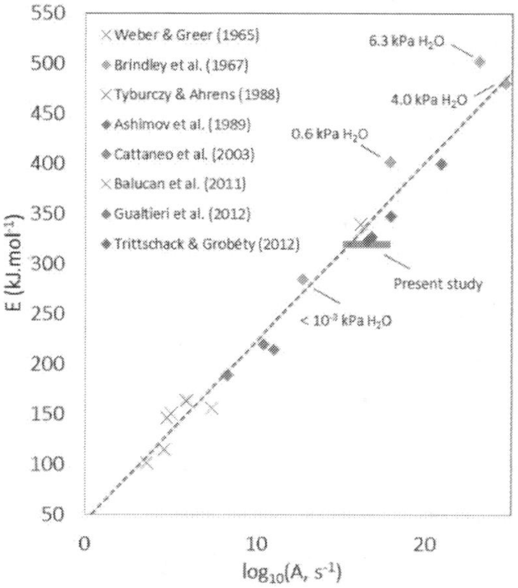

8

THE UNIVERSITY OF SYDNEY

Dehydroxylation Kinetics
Subsidiary Peaks

› Typically one to four peaks.

› Independent of particle size,
- Suggests homogeneous reaction,
- Unique location of T*.

› Final 30-40 % dehydroxylation.

› Significance?
- Higher calcination temperatures,
  - Recrystallisation.

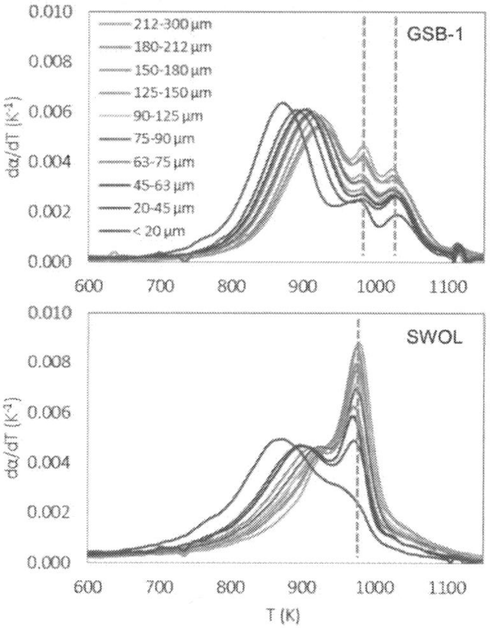

THE UNIVERSITY OF SYDNEY

Conclusions
Dehydroxylation kinetics

› Kinetic compensation effect,
- Particle size, kinetic interpretation, etc.,
- Can lead to grossly unreliable predictions.

› Dehydroxylation proceeds via two parallel pathways,
1. Spherical, retreating interface (R3),
   - Intrinsic to the mineral,
   - Particle size dependent,
   - Accounts for ~60 % dehydroxylation.
2. Apparent homogenous mechanism,
   - Particle size independent,
   - Last ~40 % dehydroxylation.
     - Higher activation energy.

THE UNIVERSITY OF SYDNEY

Thank you for listening
Acknowledgements & Questions

- Orica Mining Services,

ORICA

- AMMRF.

ammrf
Australian
Microscopy & Microanalysis
Research Facility

- Raw isothermal data,
  - Dr. Roy Trittschack,
  - Prof. Alessandro Gualtieri.

11

Utilization of oil shale ash as a concrete constituent and carbon sink

M. Uibu[1*], P. Somelar[2], L-M. Raado[3], N. Irha[4], A. Koroljova[5] and R. Kuusik[1]

[1] Laboratory of Inorganic Materials, Tallinn University of Technology, Ehitajate tee 5, 19086 Tallinn, Estonia
[2] Department of Geology, University of Tartu, Ravila 14A, 50411 Tartu, Estonia
[3] Laboratory of Building Materials, Tallinn University of Technology, Ehitajate tee 5, 19086 Tallinn, Estonia
[4] National Institute of Chemical Physics and Biophysics, Akadeemia tee 23, 12618 Tallinn, Estonia
[5] Eesti Energia AS, Laki tn. 24, 12915 Tallinn
*Corresponding author: mai.uibu@ttu.ee

Abstract
The study was focused on elaborating the strength development, mineralogy and leaching characteristics of different types of oil shale ash (OSA)-based concretes; optimizing the composite blends in order to utilize OSA flows from PF and CFB boilers as well as the solid residues from OSA accelerated carbonation process for mine backfilling. The OSA-based concretes achieved compressive strength of 1–5 MPa after 7 days of curing (maximum after 28 days >25 MPa). The strength development was mainly executed through the formation of ettringite and calcium silicate hydrates (CSH). Substantial amount of CO_2 was bound in the process (up to 30% of calcite in leached samples). The leachate pH and electrical conductivity (EC) were initially over the limits, but decreased considerably after 28–91 days of pre-curing (pH<11.5; EC<2500µs/cm). Carbonated OSA could be used as a neutral binder constituent in combination with PF ashes to stabilize pH levels at leaching conditions. In order to utilize the major ash flows from PF and CFB boilers as concrete constituents, the composite blends must be optimized focusing on the synthesis of stable calcium CSH phases.

Introduction

The Estonian kukersite oil shale (OS) is the largest industrially exploitable OS resource in the world (annual mining output 14-16 Mt).[1,2] The combustion of OS in electric power plants, based on circulating fluidized bed combustion (CFB) and pulverized firing (PF), is accompanied by the CO_2 emissions (ca 13 Mt annually) and generation of huge amounts of waste ash (45–48% of the OS dry mass[3]), which is predominantly deposited in ash fields near the power plants (5-7 Mt annually).[2] However, OSA could be used as a readily available and cheap Ca-source for CO_2 sequestration.[4]

Industrial wastes as well as CO_2 emissions could be reduced by increasing the use of supplementary cementitious materials (fly ashes, silica fume and steel slag) in concrete. However, the optimal replacement percentage is typically 20-50% and these concretes tend to have slower strength development since the hydraulic Portland cement reaction is much faster than the pozzolanic one.[5] These supplementary cementitious materials could also be used as controlled low-strength material (<8.3 MPa) for structural filling applications such as pipeline bedding, tilt-up construction, foundation backfill, filling abandoned mines, under slab voids, and sequential excavations in contaminated soils.[6]

Studies[2,7,8] have shown that the hydrous transformation of free lime (CaO), anhydrite ($CaSO_4$), secondary Ca(Mg)-silicate minerals and amorphous Al–Si glass phases control the cementation of OSA, mainly by the formation of secondary Ca-rich hydrate phases and by the carbonation of portlandite (Eqs 1-6).

$$CaO + H_2O \rightarrow Ca(OH)_2 \quad (1)$$

$$Ca(OH)_2 + CO_2 \rightarrow CaCO_3 + H_2O \quad (2)$$

$$2Ca_2SiO_4 + 4H_2O \rightarrow 3CaO \cdot 2SiO_2 \cdot 3H_2O + Ca(OH)_2 \quad (3)$$

$$SiO_2 + Ca(OH)_2 + H_2O \rightarrow CaO \cdot SiO_2 \cdot 2H_2O \quad (4)$$

$$CaSO_4 + H_2O \rightarrow CaSO_4 \cdot 2H_2O \quad (5)$$

$$3Ca(OH)_2 + 2Al(OH)_3 + CaSO_4 + 26H_2O \rightarrow Ca_6Al_2(SO_4)_3(OH)_{12} \cdot 26H_2O \quad (6)$$

The properties of high temperature PF OSA from electrostatic precipitators as a Portland cement constituent are listed in the standard EVS 636:2002.[9,10] However, the PF technology-based boilers will be partly shut down, and partly replaced with CFBC boilers by 2018.[2] The remaining ash flows from PF and CFB boilers differ noticeably by their binding characteristics.[10] Accelerated carbonation is expected to decrease the binder properties of OSA, therefore it could be used as binder constituent only in combination with initial ashes.

The study was focused on elaborating the strength development, mineralogy and leaching characteristics of different types of OSA-based concretes; optimizing the composite blends in order to utilize the OSA flows from PF and CFB boilers as well as the solid residues from accelerated carbonation process for mine backfilling.

Materials and Methods

The OS-fired power plant ashes were used as binders and OS mining residue (crushed limestone 4/16 mm) was used as the aggregate for concrete mixes. The PF ash samples (Table 1), including ash from the 1st field of the electrostatic precipitator (PFEF1), cyclone ash (PFCA) and ground cyclone ash (gPFCA), as well as CFB ash samples from the 1st field of the electrostatic precipitator (CFBEF1), were collected from the Estonian Power Plant (Estonia). The carbonated OSA samples (cPFA and cCFBA) were taken from continuous accelerated carbonation process which is described by Uibu et al.[4]

Table 1. Main characteristics of oil shale ashes

	PFEF1	PFCA/gPFCA	CFBEF1	cCFBA	cPFA
Phase composition					
Amorphous content, %	59.7	32.2	23.6		
Quartz SiO_2, %	3.8	4.9	17.4		
Calcite $CaCO_3$, %	6.1	11.9	9.4		
Anhydrite $CaSO_4$ %	6.9	7.6	8.7		
Lime CaO, %	11.3	21.6	8.5		
Hematite Fe_2O_3, %	1.3	1.7	4.9		
Adularia $KAlSi_3O_8$, %	1.2	1.7	13.9		
Illite $K_{0.65}Al_2[Al_{0.65}Si_{3.35}O_{10}](OH)_2$, %	0	0	3.5		
Periclase MgO, %	1.6	3.0	2.8		
C2S Ca_2SiO_4, %	4.1	7.7	2.9		
C4AF4CaO*Al_2O_3*Fe_2O_3, %	1.3	2.1	0		
Merwinite $Ca_3Mg(SiO_4)_2$, %	0.9	2.9	1.6		
Melilite$(Ca,Na)_2(Mg,Al)(Si,Al)_3O_7$, %	1.3	1.9	2.5		
Σ, %	99.5	99.2	99.7		
Chemical composition					
CaO^{free}%	17.9	22.5	9.13	0.93	4.47
CO_2, %	4.33	5.3	3.42	19.9	18.07
SO_3^{total}, %	9.03	5.45	5.83		
Physical parameters					
BET SSA m^2/g	1.13	0.91/1.5	3.69	19.38	11.28
Particle size d_{mean}, μm	16.96	55.78/40.30	32.48	32.78	50.16

All concrete mixes contained 50% OSA binder and 50% by mass of pre-dried aggregate. Individual ashes types (PFEF1, PFCA and CFBEF1) as well as ash mixtures Mix (25% CFBEF1+25% PFCA), Mixg (25% CFBEF1+25% gPFCA), Carb1 (12.5% PFEF1+ 25% PFCA+12.5% cCFBA), Carb 2(12.5% PFCA + 25% CFBEF1 + 12.5% cCFBA) and Carb 3 (12.5% PFEF1 + 12.5% CFBEF1 + 25% cPFA) were used as OSA binders. Concrete constituents were mixed in a laboratory concrete mixer. Water was added to gain equal consistency, with a slump of 10-12 cm according to EVS EN 12350-2:2009.[11] After 48 hours of hardening, concrete cubes were (10x10x10 cm) demoulded and further hardening was carried out for 7, 28 and 91 days at 95±5%RH and 20±2°C.

The strength development of OSA-based concretes hardened for 7, 28 and 91 days was tested by EVS-EN 12350-2:2009.[11] The water resistance was tested according to EVS-EN 12390-8[12] by determining the maximal depth of water penetration under the pressure of 200 kPa. The water penetration tests were carried out with 15x15x15 cm cubes hardened for 28 days at 95±5%RH and 20±2°C.

The whole-rock mineralogical analysis was performed by the means of powder X-ray diffraction (XRD), using a Bruker 8D diffractometer with Ni-filtered CuKa radiation (scanning step of 0.02°2θ from 2 to 75°2θ, counting time of 0.5 s). A quantitative mineral content of powdered and unoriented whole-rock preparations was measured using Rietveld-based code, Siroquant-3.[13] The proportions of the amorphous (glass-like) phases were estimated from its mass-adsorption effect on the patterns with an added spike of a known phase (ZnO). The Scanning Electron Microscopy (SEM) studies were carried out using a Jeol JSM-8404A microscope.

A batch leaching test according to European standard CEN/TS 15862:2012[14] was carried out to study the leaching characteristics of OSA-based monolithic composites. The sample bodies were contained in deionized water (Milli Q) at a liquid to surface area ratio of 12 cm^3/cm^2±2% for 24 hours at 20±2°C. Selected composites (pre-cured for 91 days) were leached for 126 days, changing water 8 times. At the end of extraction, the leachate was filtered through a 0.45 µm filter (HIMFIL, Estonia). The leachate pH (H_2O) and electrical conductivity (EC, µS/cm) were measured using a BENCH PC 510 pH/Conductivity Meter (Eutech Instruments Pte Ltd, Singapore/ Oakland Instruments, Vernon Hills, IL, USA). Two replicate samples were used and the test results were averaged. The concentration of SO_4^{2-} was determined with a Spectro Direct Lovibond No 400228 spectrophotometer and the concentrations of Ca^{2+}, K^+, Na^+ and Mg^{2+} were determined with a SpectrAA 55B Flame Atomic Absorption Spectrometer. The sample bodies were dried to a constant mass and analyzed further after leaching to determine the changes in compression strength and phase composition.

Results and Discussion

Strength development of OSA based concretes

Compressive strength as a key property for structural concrete is mainly attributed by the cement paste which acts as the glue for aggregates. Hydration kinetics of the cement is controlled by dissolution, diffusion of reactants to the site of chemical reaction, nucleation of the first product and the product growth.[15] The ash properties (Table 1) reflect in combustion technology and boiler temperature regimes (about 1400°C in PF and 800°C in CFB boilers).[10,16] Usage of various types of OSA as binders depends on the course of hydration which determines the strength development and water-durability of ash stone.

Test results indicated that due to weaker hydraulic properties of OSA concretes, longer pre-curing period in humid conditions are needed before water exposure. The compressive strength of 1–5 MPa was achieved after 7 days of hardening (Fig. 1). In most cases, the strength development of 7–91 day OSA concretes was promoted by leaching conditions, because the pozzolanic reactions, which were inhibited due to a lack of reactive water, could continue. Hydraulic PFEF1-based concretes were characterized by the fastest strength development, which also continued in leaching conditions (up to 26 MPa). The strength development of pozzolanic CFBEF1 concretes proceeds more slowly and the compressive strength remains lower (6–14MPa) due to lack of reactive species, but its water-durability was above average (Fig. 2A). The PFCA as a partly dead burnt air binder cracks due to volume expansion and has no durability in water (Figs. 1 and 2A). Grinding improves its properties as a binder constituent. Using it as a lime source in Mixg gave promising results with the maximum compressive strength <25 MPa (Fig. 1). Lime acts here as a cementitious agent by producing highly alkaline conditions, in which silica and alumina go into solution and react with the available calcium to form pozzolanic products.[17]

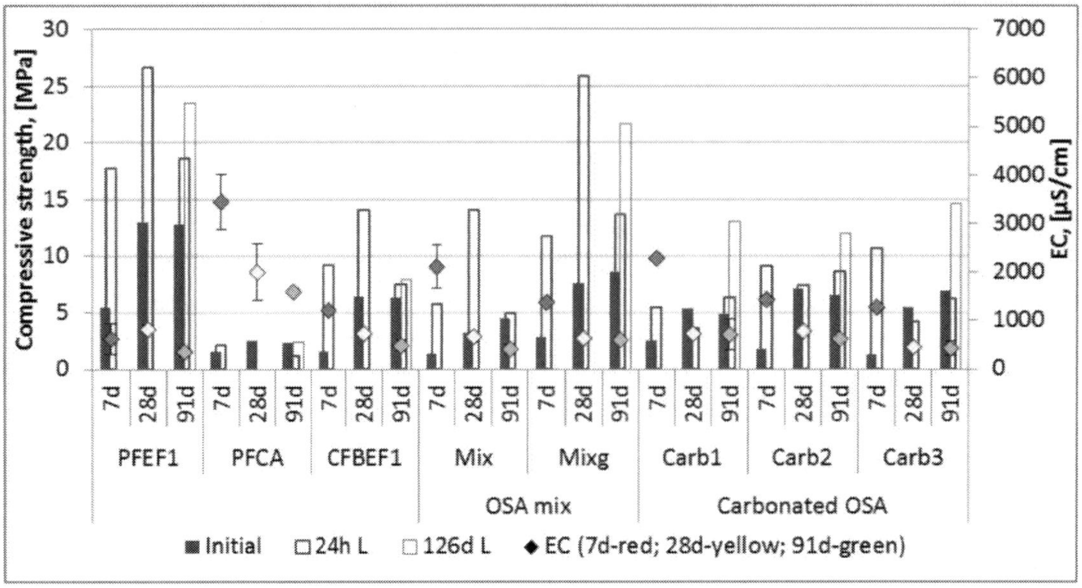

Figure 1. The compressive strength of OSA based composites after 7, 28 and 91 days of curing, before and after leaching for 24 hours and for 126 days.

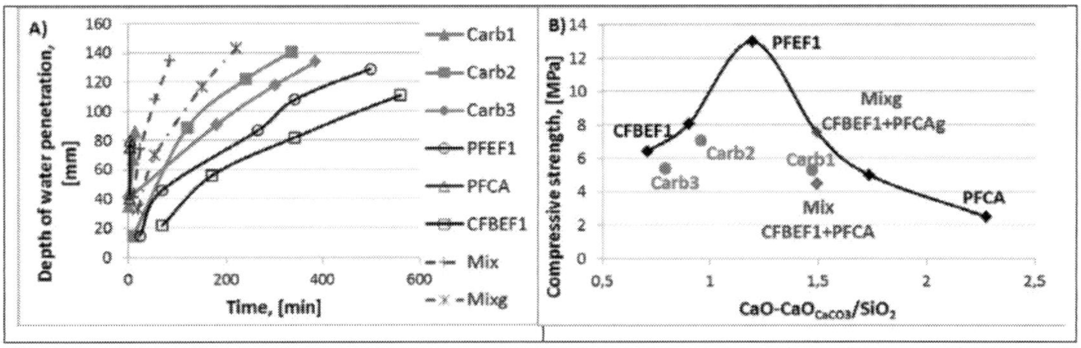

Figure 2. A)The depth of water penetration vs time under pressure of 200 kPa and B) Effect of $CaO\text{-}CaO_{CaCO3}/SiO_2$ ratio on compressive strength (the ratio is calculated according to the chemical composition of respective ash types and mixtures[16]) for 28-day OSA-concretes.

Carbonated OSA could be used as a neutral binder constituent in combination with PF ashes (Fig. 1) to stabilize the leachate pH. The Carb2 and Carb3 were characterized by average water durability CFBEF1 (Fig. 2A). As expected, increasing the share of coarse PFCA in the mixture has a decreasing effect on water durability (Fig. 2A, Carb1).

In order to synthesize the strength-giving phases (CSH, etc.) for different applications, the ratio of CaO/SiO_2 in the starting binder materials must be considered.[5,18] Fig. 2B indicates that the compressive strength values were linked to the $CaO\text{-}CaO_{CaCO3}/SiO_2$ (inactive $CaCO_3$ was discarded) ratio in the starting materials. Ratios in the range of 0.9–1.5 resulted in higher compressive strength values. The influence of particle size revealed through composite Mix, as the compressive strength remains lower due to the decreased hydration reactivity of coarse PFCA (Fig. 2B). The addition of carbonated OSA lowered the compressive strength values at similar $CaO\text{-}CaO_{CaCO3}/SiO_2$ ratios (Carb1, Carb2, Carb3), because aqueous carbonation changes the phase composition of OSA and significantly lowers its binding ability.

Phase transformations

The X-ray diffraction method does not allow for the direct determination of amorphous phases such as glass and the main cement hydration product calcium silicate hydrate (CSH) gels.[7] The estimated value of amorphous phase content varied from 3.4 to 53.6%, being highest in case of composite Carb3 (Fig. 4). The hardening of OSA stones starts with the formation of portlandite and gypsum as a result of hydration (Eqs. 1, 4), following the formation of ettringite (5–23%, Fig. 4) as the initial strength-giving phase (Eq. 5). OSA-based concretes also bind considerable amount of CO_2 during hardening and leaching. The calcite content was 9-23% after 91 days of curing, noted that Carb3 contained 12.5% cPFA. Carbonation was promoted by 24 h leaching at the expense of portlandite and ettringite (Fig. 4).

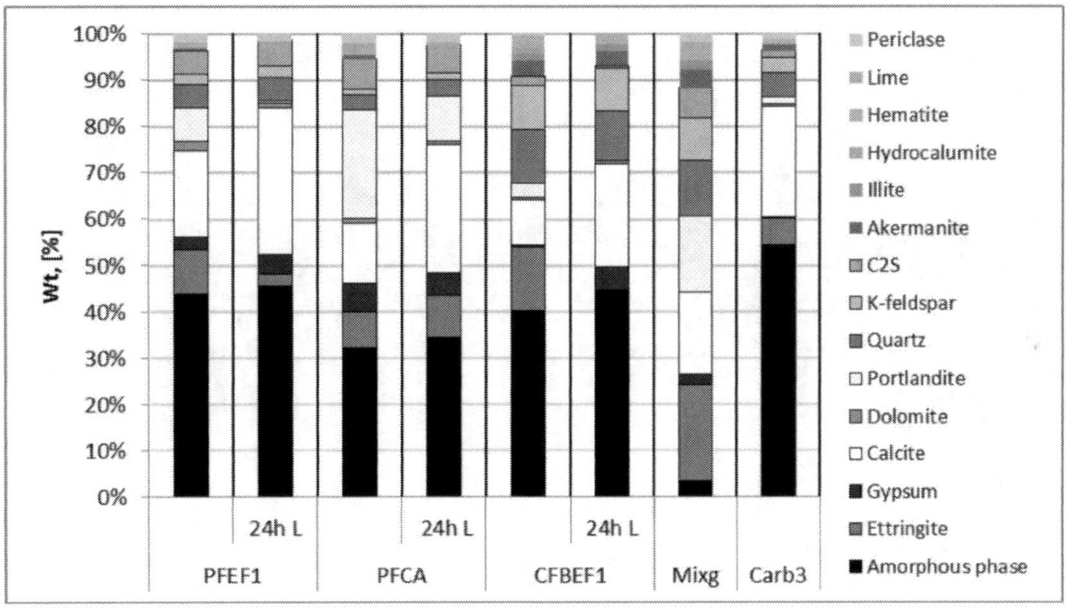

Figure 4. Mineral phase composition of OSA based concretes after 91 days of curing; before and after leaching for 24 hours.

Ettringite is a significant strength agent at the early stage of application, as its needle-like structure leads to crystal interlocking[17], but this type of cementation is quite unstable in leaching environments.[7,8] Ettringite is a stable mineral above a pH of 10.7, which dissolves into gypsum, Al-hydroxide and Ca-aluminate type phases under lower pH values.[19] Carbon dioxide decomposes ettringite, not only due to the pH decrease, but also through direct carbonation reactions.[17] The phase composition of OSA monoliths after leaching also indicated that the ettringite phase was replaced by gypsum and calcite (Fig. 4), especially when the leachate pH decreased below 10.7 (Fig. 5). Hydraulic reactions of calcium silicates and water (Eq. 3), predominantly in the case of PF OSA finer fractions, generate a much more stable type of cementation. Forming the hydration product - CSH - is the main factor responsible for strength development and the hydraulic behavior of cements.[10,20]

The content of potential cementing phases (CaO, secondary Ca-silicates and amorphous silicate glass) in CFB ashes is considerably lower. Its strength development is executed through the formation of ettringite (Fig. 4) and gel-like tobermorites ($Ca_5Si_6O_{16}(OH)_2{\cdot}4(H_2O)$), which fill the spaces between the particles and densify the matrix. Also, the lack of portlandite in concrete mixture could inhibit the pozzolanic reactions, as seen in the case of composite CFBEF1 (Fig 1 and 4).

Using coarse, PFCA as a lime additive did not result in a satisfactory concrete binder (Mix, Fig. 1) due to its decreased hydration reactivity – dead burned lime blocked with melted phase.[21] Much better results were obtained by mixing ground PFCA with CFBEF1 (Mixg, Fig. 1). The combination of the high-pH solution produced by the excess lime, soluble Al and SO_4^{2-} supplied by anhydrite and gypsum leads to the formation of ettringite[18,22] as in seen in Fig 4.

Carb2: 91d curing	CFBEF1: 91d curing + 126d leaching

Figure 5. Scanning electron microscopy images of the oil shale based concretes.

The SEM morphological analyses of broken surfaces (Fig. 5) indicated that the samples were covered with the fine layer of hydrated Ca- and Ca-Mg-silicates. The needle-like ettringite crystals were also evident, especially in leached samples (Fig. 5). The shell pieces of hydrated Ca-Mg silicates and ettringite loosened from the glass spheres are clearly evident in samples that contained carbonated ash (Fig. 5, Carb2).

Leaching dynamics of oil shale ash based concretes

Leaching characteristics of OSA based concretes was studied by containing the pre-hardened samples in deionized water for 24 hours. The pH values of the leachates were in the range of 11.5–12.2 after 7 days of curing, but declined over time close to the limits set for general wastes ($pH<11.5$[23]) (Fig. 6). The pH values were lowest ($pH=9.3–10.1$) in case of CFBEF1, Carb2 and Carb3 (Fig. 6). The electrical conductivity values crossed or were close to the EU limits (<2500µS/cm[24,25]) in most of the 7-day concretes, but decreased to 300–950 µS/cm after 91 days of curing in most cases (Fig. 1). As expected, the leachates of PFCA-based concretes, characterized by very low water resistance (Fig. 2) and considerable portlandite content (Fig. 4), remained deeply alkaline (pH=11.3–12.1) with the EC values well over the limits throughout the hardening process (Figs 1 and 6). Using ground PFCA in combination with CFB ash (Mixg) decreases leaching, which can also be explained by the increasing compressive strength (Fig. 1) and water resistance (Fig. 2).

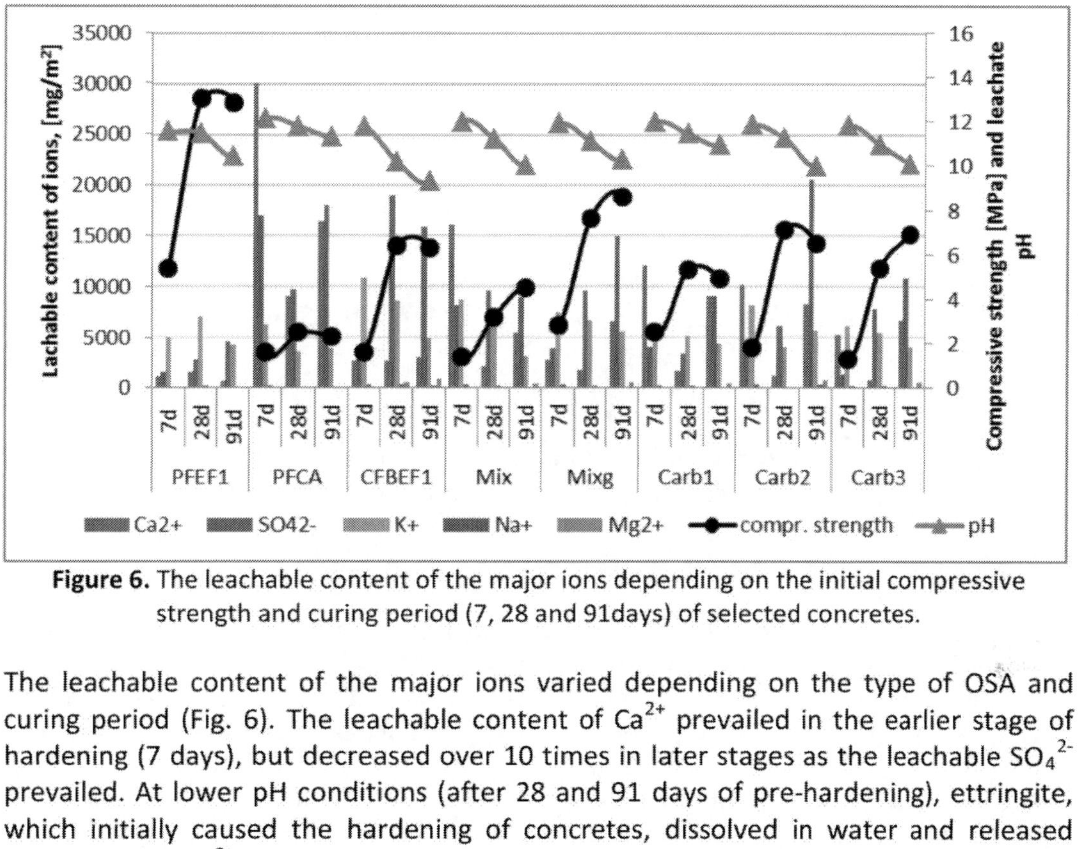

Figure 6. The leachable content of the major ions depending on the initial compressive strength and curing period (7, 28 and 91days) of selected concretes.

The leachable content of the major ions varied depending on the type of OSA and curing period (Fig. 6). The leachable content of Ca^{2+} prevailed in the earlier stage of hardening (7 days), but decreased over 10 times in later stages as the leachable SO_4^{2-} prevailed. At lower pH conditions (after 28 and 91 days of pre-hardening), ettringite, which initially caused the hardening of concretes, dissolved in water and released sulfates. Free Ca^{2+} ions were bound into forming Ca-hydrosilicates, calcite and in a limited amount (CFBEF1) back to gypsum (Fig. 4).

Conclusions

The study indicated that the major ash flows from PF and CFB boilers could be utilized as controlled low-strength materials for mine backfilling. Carbonated OSA could be used as a pH decreasing binder constituent in combination with PF ashes. Hydration, setting and hardening processes of the OSA-based binders have specific characteristics influenced by the firing temperature and the type of dust collector. The OSA-based concretes achieved compressive strength of 1–5 MPa after 7 days (maximum after 28 days >25 MPa). The strength development was mainly executed through the formation of ettringite and CSH. Also, OSA-based concretes react readily with air CO_2 in curing and leaching conditions to form calcite at the expense of portlandite and ettringite (up to 30% in leached samples).

As the formation of concrete structure takes time, the OSA-based concretes have initially very low water resistance. Nevertheless, in later stages of curing (>7 days), the pozzolanic reactions that were inhibited due to lack of reactive water, continued in leaching conditions. The PFCA as the air binder had no durability in leaching conditions, but it could be used after grinding as a lime activator in mixed binders. The pH and EC of the leachates crossed or were close to the limits set for general wastes in the earlier stage of curing but decreased considerably after 28–91 days (pH<11.5; EC<2500µs/cm). Concretes that contained carbonated OSA as a binder constituent were characterized by average strength development and water durability.

In order to utilize the major ash flows from PF and CFB boilers as concrete constituents, the composite blends must be optimized focusing on the synthesis of stable calcium silicate hydrate phases.

Acknowledgements

The financial support of the Archimedes foundation (project 3.2.0501. 10.0002), Estonian Ministry of Education and Research (SF0690001s09, SF0140082s08, IUT33-19) and the Estonian Science Foundation (Grant No 9334) are gratefully acknowledged. Participation is made possible by funding from the Baltic-American Freedom Foundation (BAFF). For more information about BAFF scholarships and speaker support, visit www. balticamericanfreedomfoundation.org.

References

1. Teedumäe, A. & Raukas, A. 2006. The possibility of integrating sustainability into legal framework for use of oil shale reserves. *Oil Shale* 23, 119-124.
2. Kuusik, R., Uibu, M., Kirsimäe, K., Mõtlep, R., Meriste, T. 2012. Open-air deposition of Estonian oil shale ash: formation, state of art, problems and prospects for the abatement of environmental impact. *Oil Shale* 29, 376-403.
3. Bauert, H., Kattai, V. 1997. Kukersite oil shale, Geology and Mineral Resources of Estonia, Estonian Academy Publishers, Tallinn.
4. Uibu, M., Velts, O., Kuusik, R. Developments in CO_2 mineral carbonation of oil shale ash. 2010. *Journal of Hazardous Materials* 174, 209-214.
5. Kuder, K., Lehman, D., Berman, J., Hannesson, G., Shogren, R. 2012. Mechanical properties of self consolidating concrete blended with high volumes of fly ash and slag. *Construction and Building Materials* 34, 285-295.
6. Naganathan, S., Hashim Abdul Razak, Siti Nadzriah Abdul Hamid. 2010. Effect of kaolin addition on the performance of controlled low-strength material using industrial waste incineration bottom ash. *Waste Management & Research* 28, 848-860.
7. Pihu, T., Arro, H., Prikk, A., Rootamm, R., Konist, A., Kirsimäe, K., Liira, M., Mõtlep, R. 2012. Oil shale CFBC ash cementation properties in ash fields. 2012. *Fuel* 93, 172-180.
8. Liira, M., Kirsimäe, K., Kuusik, R., Mõtlep, R. 2009. Transformation of calcareous oil-shale circulating fluidized-bed combustionboiler ashes under wet conditions. *Fuel* 88, 712-718.
9. EVS 636:2002 Burnt oil shale for production Portland burnt oil shale cement, Portland composite and masonry cement.

10. Raado, L.-M., Kuusik, R., Hain, T., Uibu, M., Somelar, P. 2014. Oil shale based stone formation - hydration, hardening dynamics and phase transformations. *Oil Shale* 31, 91-101.
11. EVS EN 12350-2:2009 Testing fresh concrete - Part 2: Slump test.
12. EVS-EN 12390-8:2009, Testing hardened concrete - Part 8: Depth of penetration of water under pressure.
13. Taylor, J.C. 1991. Computer programs for standardless quantitative analysis of minerals using the full powdered diffraction profile. *Powder Diffraction* 6, 2-9.
14. CEN/TS 15862:2012 Characterization of waste - Compliance leaching test – One stage batch leaching test for monoliths at fixed liquid to surface area ratio (L/A) for test portions with fixed minimum dimensions.
15. Bullard, J.W., Jennings, H.M., Livingston, R.A., Nonat, A., Scherer, G.W., Schweitzer, J.S., Scrivener, K.L., Thomas, J.J. 2011. Mechanisms of cement hydration. *Cement and Concrete Research* 41, 1208-1223.
16. Irha, N., Uibu, M., Jefimova, J., Raado, L.-M., Hain, T., Kuusik, R. 2014. Leaching behaviour of Estonian oil shale-based construction mortars. *Oil Shale* 31, 394–411.
17. Chrysochoou, M., Dermatas, D. 2006. Evaluation of ettringite and hydrocalumite formation for heavy metal immobilization: Literature review and experimental study. *Journal of Hazardous Materials* 136, 20-33.
18. Illikainen, M., Tanskanen, P., Kinnunen, P., Körkkö, M., Peltosaari, O., Wigren, V., Österbacka, J., Talling, B., Niinimäki, J. 2014. Reactivity and self-hardening of fly ash from the fluidized bed combustion of wood and peat. 2014. *Fuel* 135, 69-75.
19. Myneni, S.C.B., Traina, S.J., Logan, T.L. 1998. Ettringite solubility and geochemistry of the $Ca(OH)_2$–Al2$(SO_4)_3$–H_2O system at 1 atm pressure and 298 K. *Chemical Geology* 148, 1-19.
20. Matschei, T., Lothenbach, B., Glasser, F.P. 2007. Thermodynamic properties of Portland cement hydrates in the system CaO–Al_2O3–SiO_2–$CaSO_4$–$CaCO_3$–H_2O. *Cement and Concrete Research* 37, 1379-1410.
21. Kuusik, R., Uibu, M., Kirsimäe, K. 2005. Characterization of oil shale ashes formed at industrial scale boilers. *Oil Shale* 22, 407-420.
22. Anthony, E.J., Bulewicz, E.M., Dudek, K., Kozak, A. 2002. The long term behaviour of CFBC ash–water systems. *Waste Management* 22, 99-111.
23. RT I 2010, 46, 273, Keskkonnaministri 21. aprilli 2004. a määruse nr 21 «Teatud liiki ja teatud koguses tavajäätmete, mille vastava käitlemise korral pole jäätmeloa omamine kohustuslik, taaskasutamise või tekkekohas kõrvaldamise nõuded» lisa 2. Saasteainete leostuvuse piirväärtused.
24. RTL 2001, 100, 1369, Joogivee kvaliteedi- ja kontrollinõuded ning analüüsimeetodid.
25. Council of the European Union, Council directive 98/83/EC of 3 November 1998 on the quality of water intended for human consumption, Off. J. Eur. Commun. L330 (1998) 32–54.

"Asbestos destruction with CO_2"

Dr. Rafael Santos[1]

Gino Smeulders MSc.[2]

Pol Knops MSc.[3]

[1] Sheridan College, Canada

[2] de Biogeoloog, the Netherlands

[3] Innovation Concepts, the Netherlands

Sheridan | Get Creative

de Biogeoloog

Innovation Concepts

Chrysotile (fibrous asbestos)

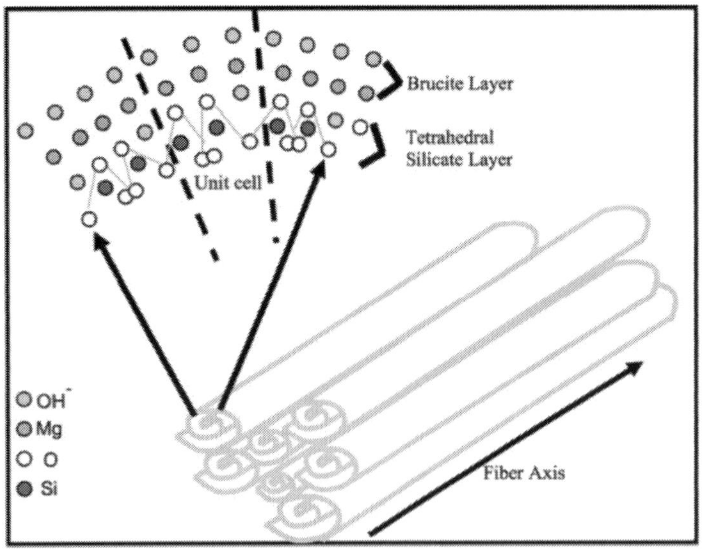

Fenton rxn's: $Fe^{2+} + H_2O_2 \rightarrow Fe^{3+} + HO^- + HO^\bullet$ (1)

$Fe^{2+} + ROOH \rightarrow Fe^{3+} + RO^\bullet + HO^-$ (2)

Naturally occurring serpentine mineral.

$(Mg)_3Si_2O_5(OH)_4$

Hazardous if inhaled due to acicular shape and iron content (2-4 wt%).

Objective: conversion to harmless minerals.

Brucite layer ($Mg(OH_2)$) readily dissolves and exposes next layer as the mineral sheet unrolls.

Weathering process analogous to parent mineral olivine (Mg_2SiO_4).

More surface area than olivine.

Weathering reactions

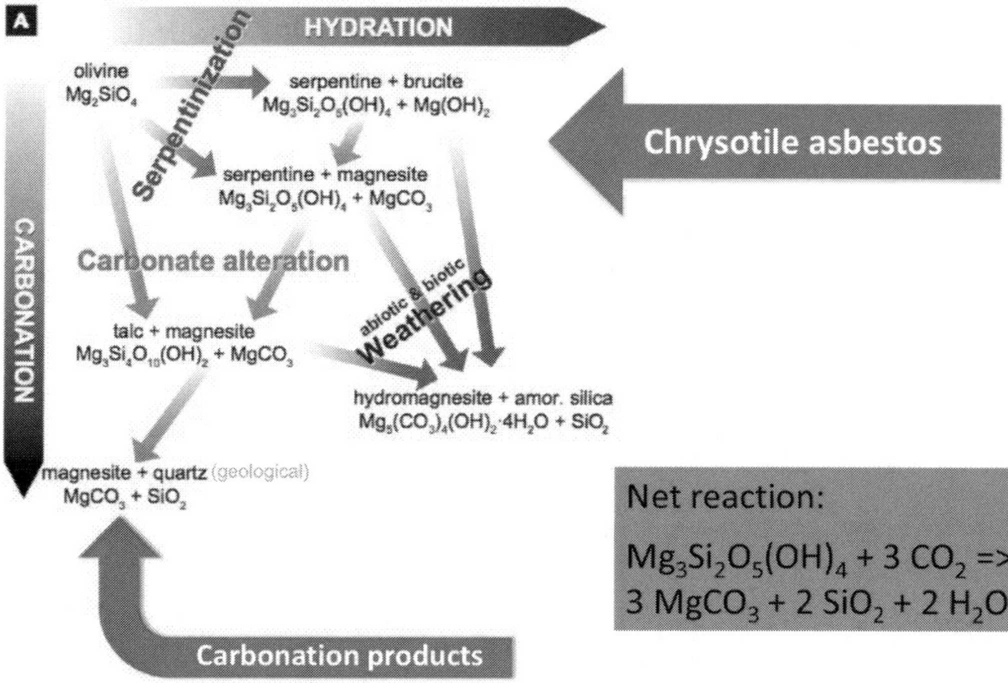

Net reaction:

$Mg_3Si_2O_5(OH)_4 + 3\ CO_2 \Rightarrow$
$3\ MgCO_3 + 2\ SiO_2 + 2\ H_2O$

Experimental protocol

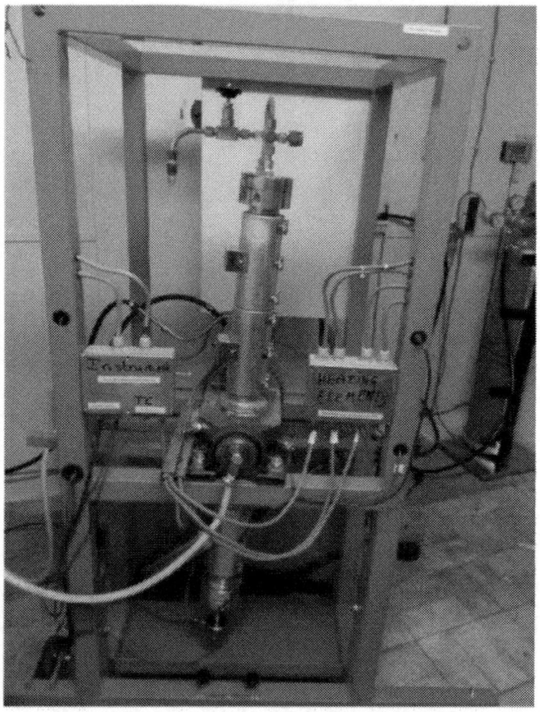

Rocking Autoclave

Elevated temperature and CO_2 pressure.

No stirrer = no mechanical breakdown due to abrasion.

Rocking mechanisms continuously keeps chrysotile fibers in suspension.

Prototype for upscaling of Gravity Pressure Vessel (GPV).

Proven concept for olivine and oil shale ash carbonation.

Experimental protocol

Analytical techniques

Optical and polarized light microscopy – visual assessment of particle morphology

X-Ray Diffraction + Rietveld refinement – quantitative mineralogical analysis

Electron microscopy + EDX – visual assessment of microstructure
– total elemental analysis

Advantages of mineral carbonation/GPV process

- Complete treatment in liquid state.
- Required milling can be done wet.
- CO_2 sequestration.
- Limited energy consumption.
- Can be upscaled to continuous process.
- Stable and safe end-products???

Performed tests

Sample type	Temp. (°C)	Process time (min.)	Additive	Pre-heating (°C)
Chrysotile fibers	*Reference material (not treated)*			
Chrysotile fibers	220	60	–	–
Chrysotile fibers	220	300	–	–
Chrysotile fibers	220	300	NaCl + $NaHCO_3$	700
Chrysotile fibers	220	300	NaCl + $NaHCO_3$	–
Asbestos cement	*Reference material (not treated)*			
Asbestos cement	220	60	–	–
Asbestos cement	220	300	–	–
Asbestos pulp	*Reference material (not treated)*			
Asbestos pulp	220	60	–	–
Asbestos pulp	220	300	–	–

Results – chrysotile, no additive/pre-heating

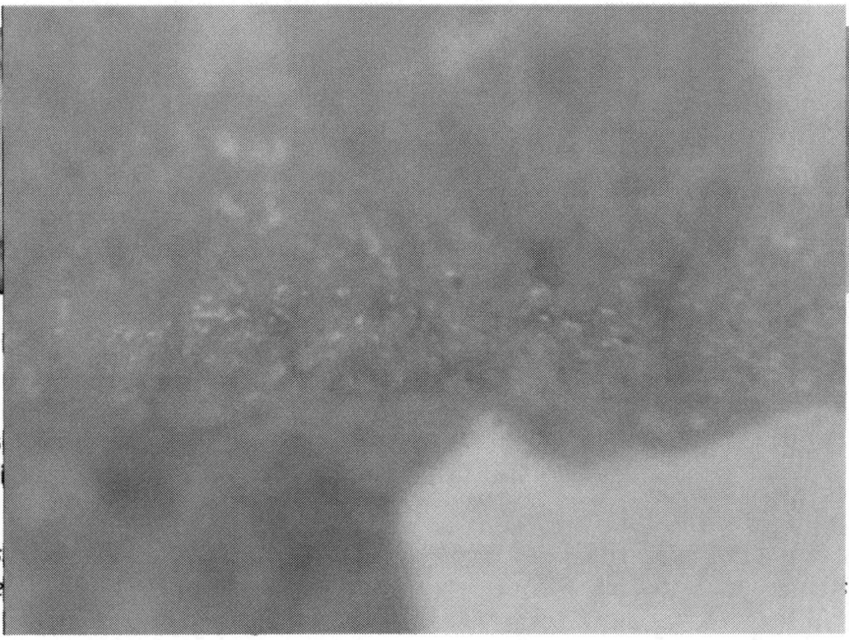

Reference

XRD:
90% chrys
10% chlori

- *Progres*
- *Weathe* ... *usually formed*

Results – chrysotile + additives/pre-heating

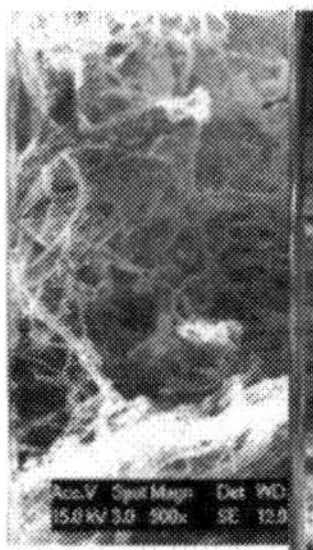

No additive/ … ating (5h)

XRD:
56% chryso
25% carbon
13% pyroxe

Chemistry (I … 966):
100% altere

Results – Morphology (size and habit)

- Asbestos fibers are considered hazardous if <u>ALL</u> of the following conditions are met:
  - i. it consists of chrysotile or one of the six amphibole minerals;
  - ii. the mineral has an acicular habit;
  - iii. the length is > 5 µm;
  - iv. the diameter is < 3 µm;
  - v. ratio of length:diameter is at least 3.
- False negatives are possible with the current ISO standards.

Chrysotile + Additives (no pre-heating)

Top: fibers are fused together as if molten, and have etched surfaces.
Bottom: fiber tips are fused together and curled, thus rendering them harmless.

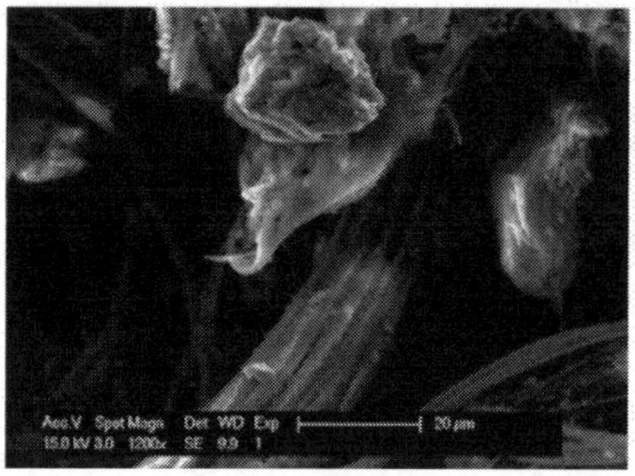

Chrysotile + Additives + Pre-Heating

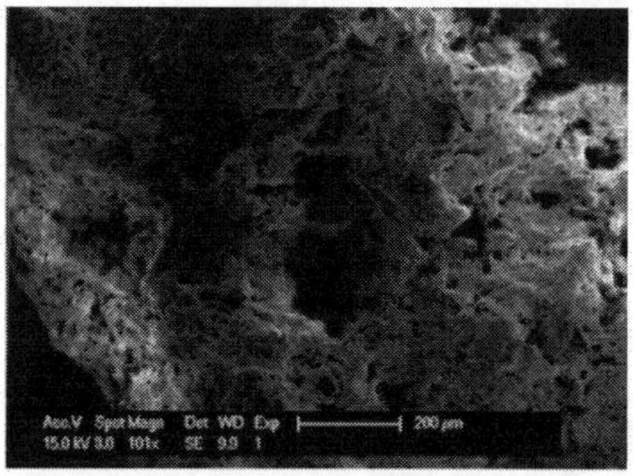
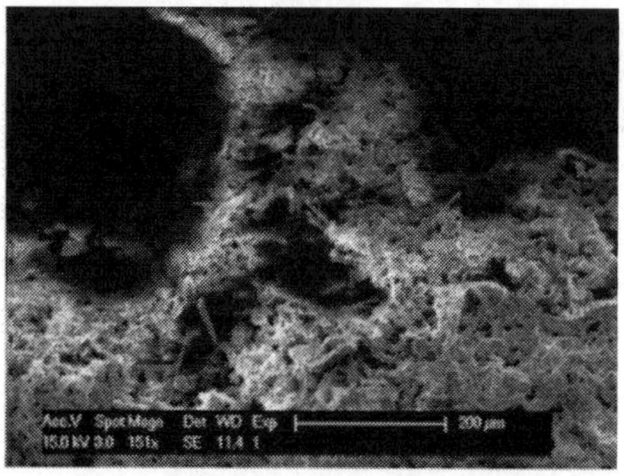

Bottom: Visible remnants of fibers very rare, only a few found after extensive searching. Those remnants were fused together and had curled amorphous tips.

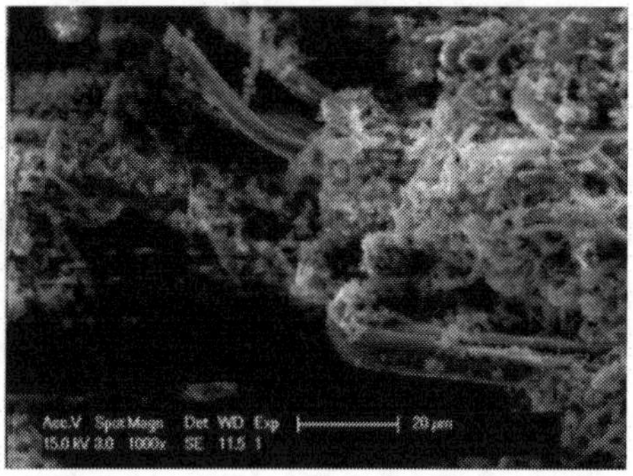

Results – Morphology (size and habit)

- The converted material does not contain fibers with acicular habit.
- The remains of the fibers do not meet the size criteria due to formation of aggregates.
- These characteristics presumably render the material "safe".
  - Risks are eliminated because fused minerals are too heavy to become airborne and without a pointy tip, minerals will not penetrate lung tissue.
- Still, more analyses (particle size distributions) and Fenton reactivity test should be conducted to conclude its safety.
- Discussions are ongoing with governmental authorities to clarify properties.

Results – Magnesium leaching from fibers

- In naturally occurring asbestos, Mg is between 38-42 wt%
- This is also documented in ISO 14966, which describes how asbestos can be identified by SEM-EDX.

- Most treated samples showed significantly reduced amounts of magnesium, which is removed from the chrysotile by carbonation and hydrolysis.

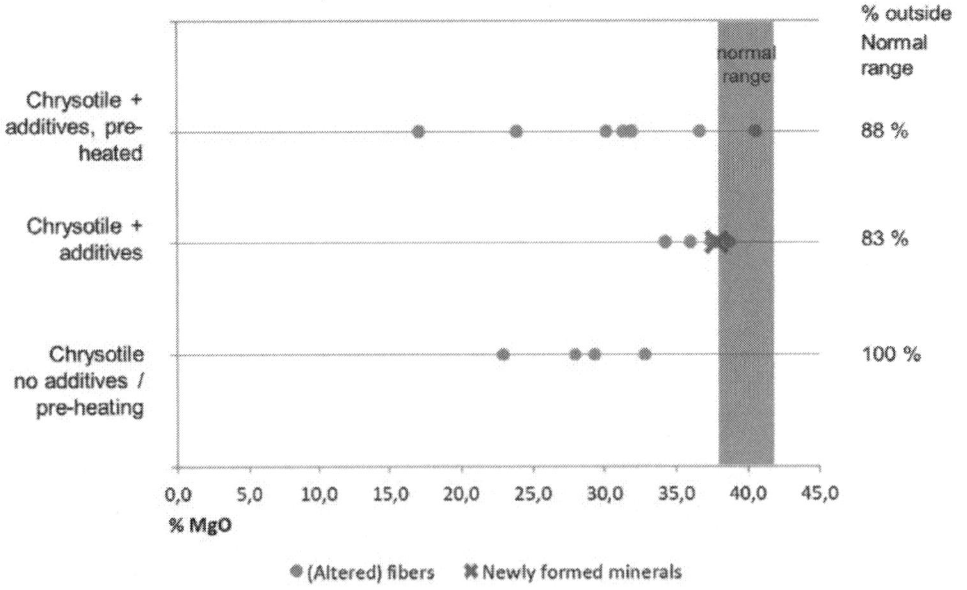

Summarized results

Method	XRD	SEM – Mg leaching	SEM – optical
Not treated	90% chrysotile		Acicular fibers
1 hour	78% chrysotile	Not investigated	Acicular fibers
5 hours	56% chrysotile	100% outside normal range	Acicular fibers covered with secondary minerals
5 hours + additives	58% chrysotile	83% outside normal range	Morphed and fused minerals with etched surfaces Curved amorphous fibre tips
5 hours + additives + pre-heating	34% chrysotile	88% outside normal range	Spongy texture Very few fibrous minerals that are all morphed, fused and have curved tips

Additional information from experiments with asbestos cement and pulp:

- Matrix dissolution is too slow – inhibits fiber breakdown. Higher loading and/or abrasives are necessary to open cement matrix.
- Amphibole asbestos reacts more slowly, but pre-heating not tested yet.

Conclusions

- Treatment without additives and pre-heating show slow but progressive breakdown.
- Additives did not speed up process, but are conducive to morphing acicular fiber tips and fusing minerals, thus eliminating risks.
- Pre-heating to expel OH-groups show nearly complete conversion; fibrous remnants not harmful.
- Complete carbonation may be possible if we were to rapidly cool the pre-heated chrysotile in water, which would completely open the mineral structure.

Thank you

#ACEME2015
#AcceleratedCarbonation

Results – Mineralogy (conversion)

- Why is carbonation not complete despite the very high surface area?
- Because the nature of the silicates is of importance.
  - Olivine the fastest weathering is an neso-silicate.
  - Quartz, almost completely inert is tecto silicate.
  - Phyllosilcates (sheet silicates) like chrysotile are more stable / weather slower than the olivine, which consists of isolated silica tetrahedra.
- Therefore the reactivity of silicates follows a certain binding order:

  Neso -> Soro -> Cyclo -> Ino -> Phyllo -> Tecto
- Complete carbonation may be possible if we were to rapidly cool the pre-heated chrysotile in water, which would completely open the mineral structure.

The Essential Mineralogy of Raw Materials Used for Mineral Carbonation

S. Sjöblom[1*] and O. Eklund[1]

[1] Åbo Akademi University, Faculty of Science and Engineering, Geology and Mineralogy, 20500 Turku, Finland

*Corresponding author: sonsjobl@abo.fi

Abstract

Geological and mineralogical characterization of rocks plays an important role in the success of mineral carbonation. Therefore, a more detailed mineralogical characterization has been the focus of the studies[1,2] aiming to develop "The ÅA Route" carbonation method. By specifying the essential mineralogy of the raw material, the mineralization process can be fully utilized. In "The ÅA Route" carbonation process $Mg(OH)_2$ is produced from the ground material and reacts with CO_2 in an exothermic reaction that produces magnesite, $MgCO_3$.[3,4] In general, ultramafic rocks are stated as promising raw material for mineral carbonation.[3,5] However, mineralogical characteristics like the crystal water- and the Mg-content in the minerals, the crystal structure as well as the composition of the parental rock and its metamorphism grade have a great influence to the outcome. Results state that a suitable raw material should be an ultramafic rock with sufficient amounts of Mg (> 17 %) as well as crystalline H2O (> 12.5 %). Ultramafic phyllosilicates produce the best results but their parent rocks influence their suitability.

Introduction

Within the last decade, the Thermal and Flow Engineering Laboratory at Åbo Akademi University has been developing a method for Carbon Capture and Mineralization (CCM). This method is based on extraction and precipitation of magnesium in the form of $Mg(OH)_2$ and a subsequent reaction with CO_2, resulting in a natural, stable carbonate, $MgCO_3$.[3,4,5] The results have been promising.[5,1] The method does not only create a storage option for anthropogenic CO_2-emissions but it exploits mine waste that may be suitable as raw material for this purpose. CCM would therefore combine two waste materials, mine waste and CO_2, to produce a profitable material to be used i.e. in the construction industry.[6]

Mineralogical knowledge of the raw material is essential for the development of the methods that use naturally occurring materials. Therefore, a geological and mineralogical point of view was introduced to "The ÅA Route" in 2012[1]. The original plan of studying the CCM-suitability of mine waste and minerals, the influence of their origin and what should be taken into account while selecting material for CCM, was presented in 2013 at the 4th ACEME conference[7]. This study presents the results gained from the studies that were executed with the geological aspect, aiming to improve of "The ÅA Route" –method.

Material

24 samples (rocks and minerals) were collected and studied. The samples represent all possible naturally occurring Mg-bearing silicate mineral groups, collected from different parts of Europe. Most of the samples were mine waste rocks. Sampled rocks and minerals that possessed the necessary characteristics to pass all the studied steps were pyroxene (inosilicate)[1], volcanic tuff[1], olivine (neso-/orthosilicate)[8], cordierite (cyclosilicate)[8], talc (phyllosilicate)[1,8], chlorite (phyllosilicate)[1], phlogopite (phyllosilicate)[8], vermiculite (phyllosilicate)[8] and different types of serpentinites (phyllosilicate) [1,2].

Methodology

Experiments were conducted with a specific carbonation method developed at Åbo Akademi University. In the conventional "The ÅA Route" -method, Mg is extracted from the raw material using ammonium sulphate at 400 °C, precipitated as $Mg(OH)_2$ in water and fused with CO_2 as an exothermic reaction. The produced end product is magnesium carbonate, $MgCO_3$.[3,4] The fate of magnesium is monitored with several methods in order to calculate reaction quantities and conversion levels in the different parts of the process (Figure 1).

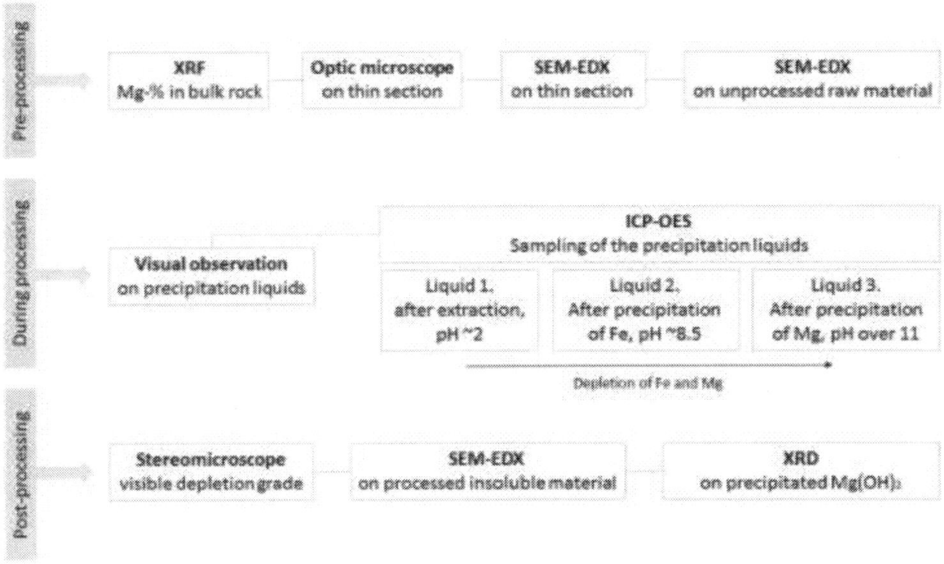

Figure 1. Method chart for Mg-monitoring in each step of the process.

Studies executed before processing include measuring the chemical composition of the bulk material with X-ray Fluorescence (XRF) and with Scanning Electron Microscope and Energy-dispersive X-ray spectroscopy (SEM-EDX). SEM-EDX is used for analyzing the sample surfaces and for measuring the relative Mg-amount of selected areas. With these methods it is possible to tell the overall Mg-%, main locations of the Mg (i.e. the Mg mineral phases) and how random the distribution of Mg in the bulk sample is.

During processing, observing the chemical reactions is often very indicative and informative. Solution samples are collected after each precipitation step and analyzed with an ICP-OES. The first aqueous solution tells the amount of Mg that was extracted from the bulk sample. The second solution shows if any Mg was precipitated during the precipitation of Fe. The third solution shows whether there was Mg left in the "theoretically empty" solution. Some Mg usually precipitates already during the precipitation of Fe while some is present in the final solution. The overlapping amounts need to be monitored in forthcoming studies for economic reasons.

After processing, the insoluble material is compared to the ground unprocessed raw material. This is done with a stereomicroscope and SEM-EDX. The difference in the Mg-% between the unprocessed and processed material gives estimation on how successful the extraction of Mg was. As it is possible that Mg precipitates as MgO (that is much less reactive than $Mg(OH)_2$[4]) the ratio of $Mg(OH)_2/MgO$ needs to be confirmed. This ratio is studied with X-ray Diffraction (XRD).

All the results gained with these methods are used in order to produce exact mass balance calculations and reliable data, which is used to draw the conclusions.

Results

Based on the experiments, it can be stated that a suitable raw material should be an ultramafic rock with sufficient amounts of Mg (> 17 %) as well as crystalline H_2O (> 12.5 %)[1]. However, depending on the rocks silicate group its structure may prevent its suitability for CCM[8]. Our results were processed into a flow chart that creates an easy way to see whether a rock may be suitable for mineral carbonation or not (Figure 2).

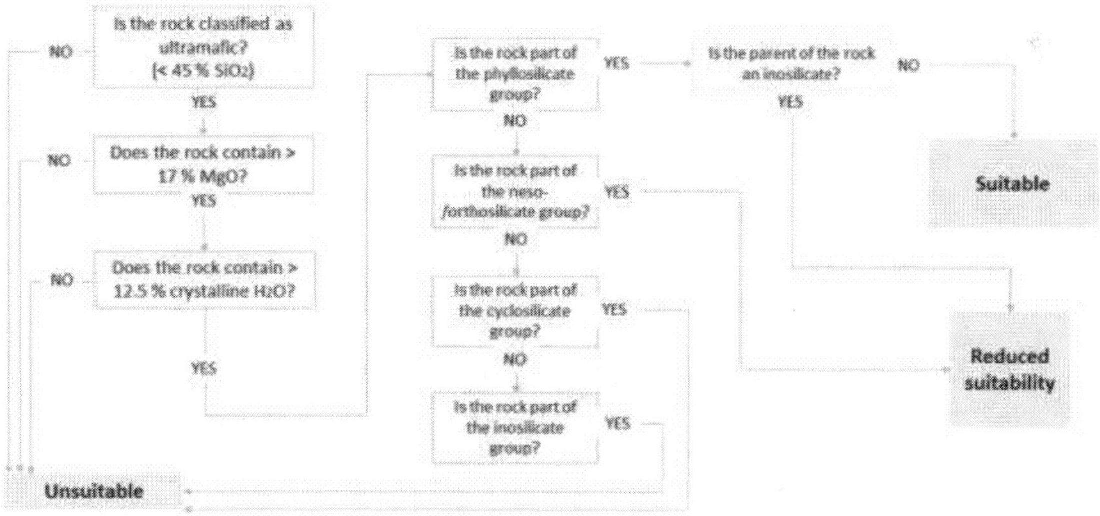

Figure 2. Suitability flow chart for the material selected for mineral carbonation.

Phyllosilicates consistently produced good results while neso-/orthosilicates produced poor results[8]. Mg-extraction from cyclo- and inosilicates was not successful at all[8].

Results from phyllosilicates also revealed that even though a rock itself may theoretically have the suitable characteristics, the suitability depends on if it originates from a parent rock with unfavorable characteristics[2]. This was noted as the serpentinite samples descending from phyllosilicates (Hitura antigorite seam, Elba lizardite, Piteå serpentinite) reacted very well, while serpentinites descending from inosilicates (Modum antigorite dominating serpentinite, Modum lizardite dominating serpentinite) had lower reactivity[2]. Metamorphosis or other deformation may not be able to destroy all structures of the parent rock and therefore their relicts reflect their characteristics. One significant notice was that the Mg/Fe-ratio in the samples was not a factor as long as Fe was carefully precipitated before precipitating the Mg.

Acknowledgements

The authors acknowledge the financial support received from K. H. Renlund Foundation and Ekokem. Ron Zevenhoven and his co-workers at the laboratory for Thermal and Flow Engineering are acknowledged for comments and support with the experimental work. Sten Lindholm is acknowledged for ICP-OES analyses.

References

1. Sjöblom, S., Eklund, O. 2014. Suitability of Finnish mine waste (rocks and tailings) for Mineral Carbonation. PROCEEDINGS OF ECOS 2014 - 27th International Conference on Efficiency, Cost, Optimization, Simulation and Environmental Impact of Energy Systems. June 15-19, 2014, Turku, Finland, paper 244.
2. Sjöblom, S., Eklund, O. 2015. The Significance of the Serpentinite Characteristics in Mineral Carbonation by "the ÅA Route". (Submitted to American Mineralogist)
3. Nduagu, E., Björklöf, T., Fagerlund, J., Wärnå, J., Geerlings, H., Zevenhoven, R. 2012a. Production of reactive magnesium from magnesium silicate for the purpose of CO_2 mineralization. Part 1. Application to Finnish serpentinite. Minerals Engineering, 2012; 30:75-86.
4. Fagerlund, J., Nduagu, E.,Romão, I., Zevenhoven, R. 2010. A stepwise process for carbon dioxide sequestration using magnesium silicates. Front. Chem. Eng. China 2010, 4(2): 133 – 141.
5. Nduagu, E., Björklöf, T., Fagerlund, J., Mäkelä, E., Salonen, J., Geerlings, H., Zevenhoven, R. 2012b. Production of reactive magnesium from magnesium silicate for the purpose of CO_2 mineralization. Part 2. Mg extraction modeling and application to different Mg silicate rocks. Minerals Engineering, 2012; 30:87-94.
6. Gunning, P. J., Hills, C.D., Carey, P.J. 2009. Production of lightweight aggregate from industrial waste and carbon dioxide. Waste Management, 29 (10). pp. 2722-2728.
7. Sjöblom, S.,Eklund, O. 2012. Geological and mineralogical aspects on mineral carbonation of rocks and mine tailings. In: R. Nasser, R. Santos, Ö. Cizer, T. Van Gerven (eds.) Proceedings of ACEME13, Leuven, Belgium, April 10-12, 2013, s. 497 - 500.
8. Sjöblom, S., Eklund O. (2015). Capability of different Silicate Groups in Extraction of Mg with "The ÅA Route"-method. (Manuscript, in prep.)

Leaching thermodynamics of oil shale waste key components

K. Tamm[1*], P. Kallaste[1], M. Uibu[1], J. Kallas[1] and R. Kuusik[1]

[1] Laboratory of Inorganic Materials, Tallinn University of Technology, Tallinn, Estonia
*Corresponding author: kadriann.tamm@ttu.ee

Abstract

Estonia is strongly dependent on locally mined oil shale (OS) which is the main fuel for power and oil production. Latter is accompanied with huge amount of solid waste, currently wet-deposited in open-air fields. The leaching behavior of the main water-soluble Ca-compounds ($Ca(OH)_2$, $CaSO_4$*$2H_2O$ and CaS) of three types of OS mineral waste (from pulverized firing - PF, circulating fluidized bed combustion - CFB and solid heat carrier - SHC) was investigated. For comparison, the binary ($Ca(OH)_2$-$CaSO_4$*$2H_2O$-H_2O) and ternary ($Ca(OH)_2$-$CaSO_4$*$2H_2O$-CaS-H_2O) model systems were prepared according to the composition of OS processing ashes. The equilibrium constants of the dissolution reactions for ash-water systems were evaluated from experimental data as well as thermodynamic calculations. Comparing the models with the real OSA water systems the results coincide in satisfactory extent.

Introduction

About 60% of the Estonia`s fuel balance is covered by local low-grade fuel, oil shale (OS), which is utilized mainly in heat and power as well as in shale oil production. Eesti Energia, the major power utility company, deposited 8.1Mt of oil shale ash (OSA) and emitted 13.4Mt of CO_2 in 2013[1]. Safe disposing of mineral waste and control of atmospheric emissions are among the most challenging problems caused by the extensive use of OS. Today, about 90% of the ash is wet deposited in the ash fields near the power plants.[2] The OSA is classified as a "hazardous waste" due to its high free lime content which dominates the leaching (pH values 12 - 13). On the other hand, this makes OSA (or OSA leachates) an attractive source for CO_2 mineralization[3] or producing precipitated calcium carbonate (PCC).[4] Formation of PCC from Ca-rich OSA is a multi-stage process. Better understanding of the OSA leaching would allow us to implement more effective methods for waste valorization and environmental pollution control.

The aim of the study was to elaborate the leaching equilibriums of the main water-soluble Ca-compounds of OSAs, as a basis for mathematical models, which simulate the processes from leaching to aqueous carbonation.

Materials and methods

Characterization of materials

The CFB ash, PF ash and SHC ash were used as Ca-rich raw materials. The CFB (a mixture of ashes taken from common ash silo, whereby the ash is collected from different equipment before depositing on the ash fields) and PF (from cyclone separator) ashes were collected from the electricity production boilers at Estonian power plants at 2011. The SHC ash from 2012 is the final residue from the shale oil production process at the Viru Keemia Group Ltd. retorting plant. The ash samples were analyzed using X-ray fluorescence spectroscopy (XRF, Rigakus Primus II) as well as quantitative X-ray diffraction (XRD, Bruker D8 Advanced) methods. The contents of free lime (CaO_{free}) (ethylene glycol method)[5], different forms of sulfur (sulfate and sulfide)[6], total carbon (TC) and inorganic carbon (TIC) (Electra CS - 580 Carbon/ Sulfur Determinator) have been determined. BET specific surface area (SSA) was estimated using Kelvin 1042 sorptometer. The chemical and mineral compositions in weight percentages as well as the main physical property of the ashes are given in Table 1.

Table 1. The composition of OS ashes in wt% and physical characteristic

Chemical composition	**CFB**	**PF**	**SHC**
CaO_{free}	12.31	22.37	1.89
$CaSO_4$	11.20	7.43	2.58
CaS	1.44	0.14	2.74
TC	3.03	0.30	7.82
TIC	3.03	0.30	6.45
TOC	0	0	1.37
Mineral composition	**CFB**	**PF**	**SHC**
Quartz	11.4	4.4	14.1
Calcite	27.7	3.3	29.4
Dolomite	1.4	0	10.0
Orthoclase	10.2	3.9	12.7
Lime	9.3	28.3	3
Anhydrite	10.1	8.6	0.7
Oldhamite	-	-	4.6
Periclase	2.7	5.1	8.7
Physical characteristic	**CFB**	**PF**	**SHC**
SSA, m^2/g	6.30	0.40	7.44

PF ash is characterized by the highest free lime concentration. The high concentration of sulfates in CFB ash is presented mainly as anhydrite ($CaSO_4$), which in water systems crystallize into gypsum ($CaSO_4*2H_2O$). In both CFB and PF ashes the content of sulfides (mainly present as CaS) is relatively low. SHC ash is characterized by high sulfide content. The organic matter is also present (COD 99±10 mg/L)[7] in SHC ash due to incomplete pyrolysis at lower temperatures. The values of specific surface area of CFB and SHC ashes are over ten times higher than that of the PF ash (Table 1). In ambient environment over series of chemical reactions, a part of the sulfides can be emitted into the atmosphere as gaseous H_2S, generating atmospheric pollution. Sulfides, which remain in the aqueous phase, may also give toxic effects depending on the pH value of the suspension.[8]

Experimental procedure

The experiments with PF, CFB and SHC water systems as well as corresponding binary ($Ca(OH)_2$-$CaSO_4$*$2H_2O$-H_2O) and ternary ($Ca(OH)_2$-$CaSO_4$*$2H_2O$-CaS-H_2O) model systems were performed at room temperature and atmospheric pressure. The selected ash/water ratios ranged from 0.004 - 0.200, based on previous studies[9] and currently used settings of hydro transport of OSA. Pure $Ca(OH)_2$ (BDH, 95% purity) (as Ca^{2+}-ion source), $CaSO_4$*$2H_2O$ (lach:ner, 99% purity) (as SO_4^{2-}-ion source) and CaS (Alfa Aeser, 99.9% purity) (as S^{2-}-ion source) were used for binary and ternary model systems. Model systems were prepared in accordance to the components percentages in OSAs for a comparable study. The studied systems were kept, in closed centrifuge tubes, for 3 h in an overhead shaker (GFL 3025) at 45 rpm throughout the experiment. The suspensions were prepared in 50 mL tubes sealed with airtight caps beforehand. The pH (MT SevenGo pH) and electrical conductivity (MT SevenGo Duo Pro) were measured immediately after the 3 h experiment. The obtained suspensions were then filtered using a vacuum filter (filter paper: Munktell, 100 g/m^2). The contents of calcium (Ca^{2+}-ions), total reduced sulfur and alkalinity (OH^--, CO_3^{2-}- and HCO_3^--ions) were determined with titration methods (ISO – 6058:1984[10], iodometrically[11] and ISO - 9963-1: 1994(E)[12], respectively). The contents of total sulfide (sum of dissolved S^{2-}, HS^- and H_2S) and SO_4^{2-}-ion were determined with a spectrometer (Lovibond Spectro Direct, methods: DPD/Catalyst and Bariumsulfate - Turbidity, respectively).

Eq. 1 – 4 represent the dissociation equilibriums for the ternary ($Ca(OH)_2$-$CaSO_4$*$2H_2O$-CaS-H_2O) model systems:

$$Ca(OH)_2 \leftrightarrow Ca^{2+} + 2OH^- \qquad K_{eq1} \qquad (1)$$

$$CaSO_4 \cdot 2H_2O \leftrightarrow Ca^{2+} + SO_4^{2-} + 2H_2O \qquad K_{eq2} \qquad (2)$$

$$CaS \leftrightarrow Ca^{2+} + S^{2-} \qquad K_{eq3} \qquad (3)$$

$$S^{2-} + H_2O \leftrightarrow HS^- + OH^- \qquad K_{eq4} \qquad (4)$$

The $Ca(OH)_2$ dissolution mechanism is presented as a one-step reaction system for a more universal approach.[13] Also, the formation of H_2S is not included in the proposed model. As a more complex approach, the two-step reaction system for lime dissolution and further hydrolysis of sulfide, producing H_2S, is included in a Aspen Plus model.

Results and discussion

The following section will discuss the thermodynamic calculations for ternary model systems and the solubility results of the experimentally measured binary and ternary model as well as OSA water systems.

Simulation approach

The equilibrium calculations were carried out using thermodynamic program based on ELECNRTL model with Aspen Plus V8.6 default databases. The Aspen Plus RGibbs reactor model which uses the Gibbs free energy minimization technique to determine the composition of each phase was employed for predicting the equilibrium composition of the ternary systems according to the composition of different ashes (CFB, PF, SHC). The thermodynamic model of a $Ca(OH)_2$-$CaSO_4$-CaS-H_2O system is shown on Figure 1.

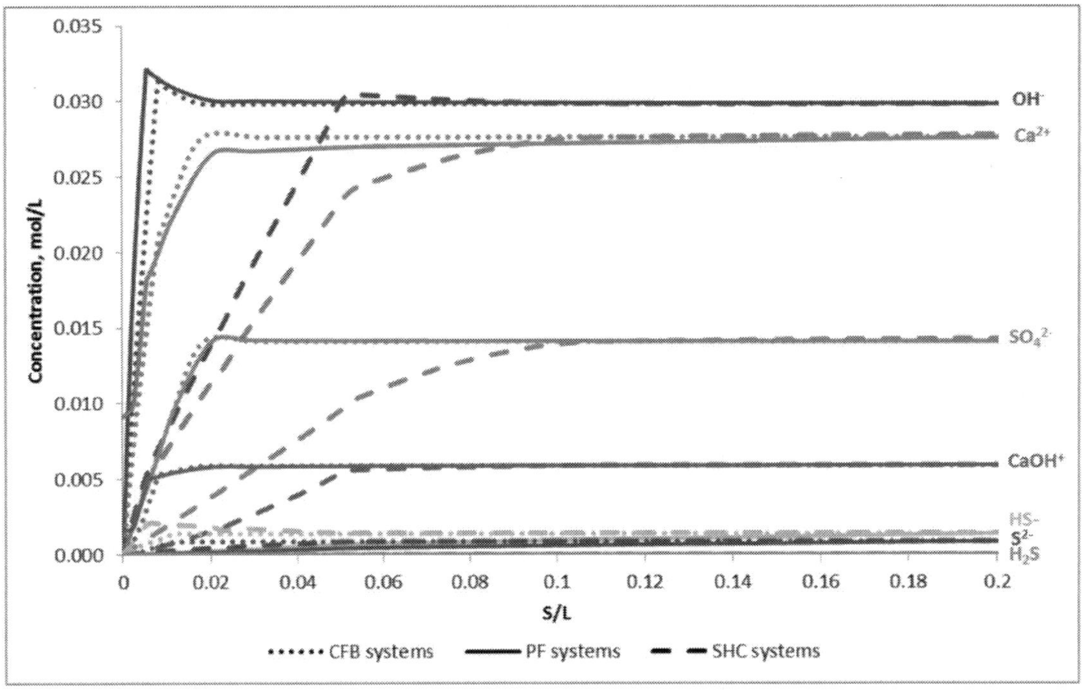

Figure 1. Equilibrium composition of ternary model systems (Aspen Plus V8.6 simulation)

It can be seen from the simulation results that the equilibrium state of ion distribution is reached for two of the most oversaturated systems (S/L ratio 0.1 - 0.2).

The equilibrium constant K_{eq4} for Eq. 4, calculated from the simulated equilibrium composition (Figure 1) was used to determine the sulfide forms in experimental model and real OSAs water systems (Figure 2). In case of PF ash, the sulfides content is two orders of magnitude lower compared to other ashes and are therefore excluded.

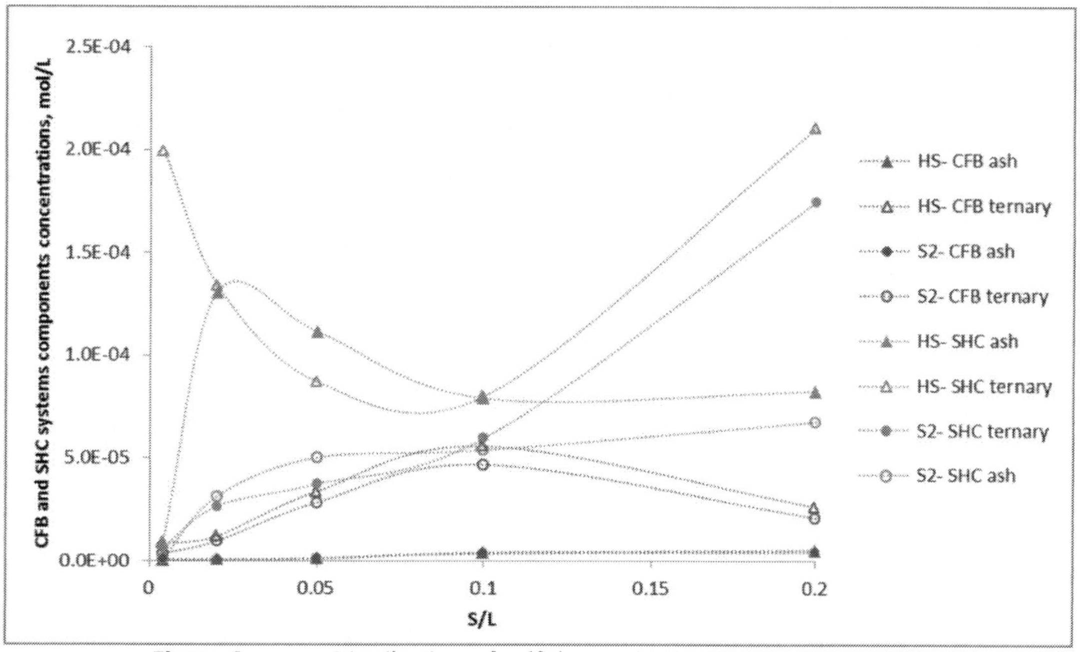

Figure 2. Distribution of sulfide species in OSA - water systems

For comparison with the experimental results, the totals of Ca^{2+}- ions (Ca^{2+} + $CaOH^{+}$), OH^{-}- ions (OH^{-} + $CaOH^{+}$) and S^{2-}- ions (S^{2-} + HS^{-} + H_2S) were used to determine the equilibrium constants (K_{eq1}, K_{eq2}, K_{eq3}) for Eq. 1 - 3.

Solubility results

By varying the concentrations of model components of OSAs according to the ash/water ratios from 0.004 to 0.200 and allowing the batch experiments to achieve equilibriums, system pH, conductivity and main ions (Ca^{2+}, SO_4^{2-}, S^{2-}) concentrations in liquid phase were obtained. The leaching equilibrium of the three types of ashes - CFB, PF and SHC - and the corresponding ternary model systems are demonstrated in Figure 3 - 5, respectively. The average pH and conductivity (E) are based on experimentally measured values of equilibrium state model and ash systems.

The pH values (12.31 - 12.66) of saturated binary and ternary systems coincided with the real OSA based systems. The highest pH values occur in PF ash based systems - this is due to the initial free lime (CaO) content which affects pH the most.

From the conductivity results, at saturated concentrations of CFB and SHC model systems, the values are higher (8.91 - 9.97 mS/cm) while the PF model systems stabilize at lower values (8.90 - 9.58 mS/cm). This reflects the fact that higher initial CaS concentration has a stronger influence on the conductivity (the ions remain in aqueous phase) than other components mentioned. In the real OSA systems, the lowest conductivity values occur in SHC ash systems (10.35 mS/cm) whereas the highest values are found in PF ash systems (up to 12.25 mS/cm).

Ca^{2+}- and $SO4^{2-}$- ions at saturated concentrations present similar values in all model systems (~0.033 - 0.035 and ~0.014 - 0.015 mol/L, respectively) being slightly higher in ternary model systems due to the dissolution (Eq. 3 - 4) of CaS and the oxidation of sulfide ions to intermediate forms (inc. $2HS^- + 2O_2 \rightarrow S_2O_3^{2-} + H_2O$) and sulfate ($HS^- + 2O_2 \rightarrow SO_4^{2-} + H^+$). The small variations of ~0.001 mol/L can be neglected; therefore the results of binary model systems are not expressed on Figures 3 - 5. The concentrations of Ca^{2+}- and SO_4^{2-}- ions (~0.034 and 0.0085 mol/L, accordingly) in SHC ash systems are lower compared to CFB and PF systems (accordingly ~0.04 and ~0.016 mol/L). Latter is related to the original composition of the SHC ash.

The SHC ash water systems of the highest CaS initial concentration are also clearly distinguished by the leaching of sulfides. Concentration of total sulfide in SHC ternary model system was up to 5.6E-4 mol/L, which is much higher compared to CFB and PF ternary model systems (7.9E-5 and 2.1E-5 mol/L, accordingly). While the S^{2-}- ion concentrations in the leachates of CFB and PF ash systems remain at 1.5E-5 mol/L, it is up to 20 times higher in SHC ash system, steadying at 2.2E-4 mol/L (the equilibrium concentration was not reached in our conditions). Even though the initial CaS concentration difference in the compared ashes is not twenty-fold, the leaching of CaS is favored in case of the SHC ash systems.

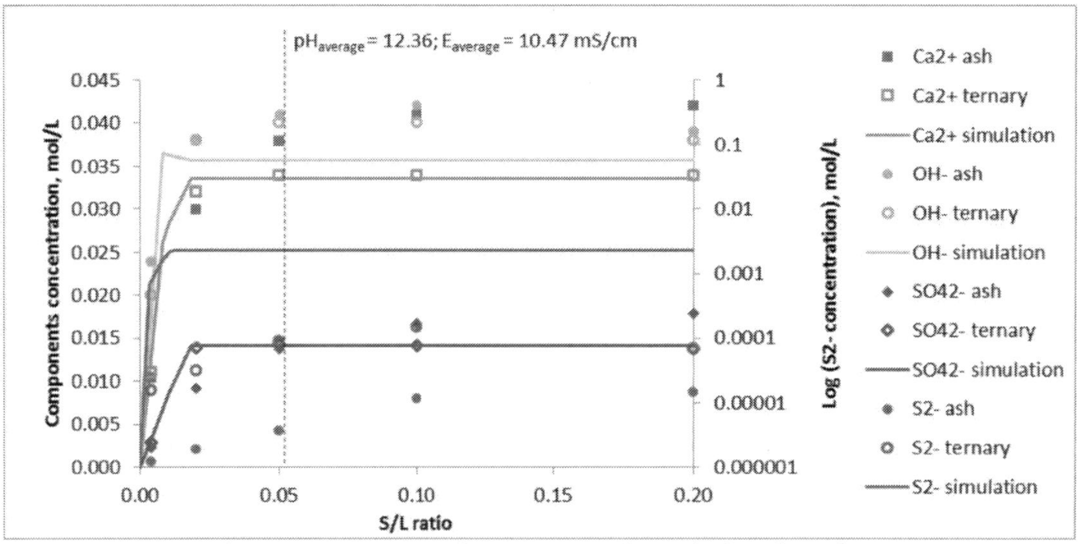

Figure 3. CFB ash experimental vs calculated results

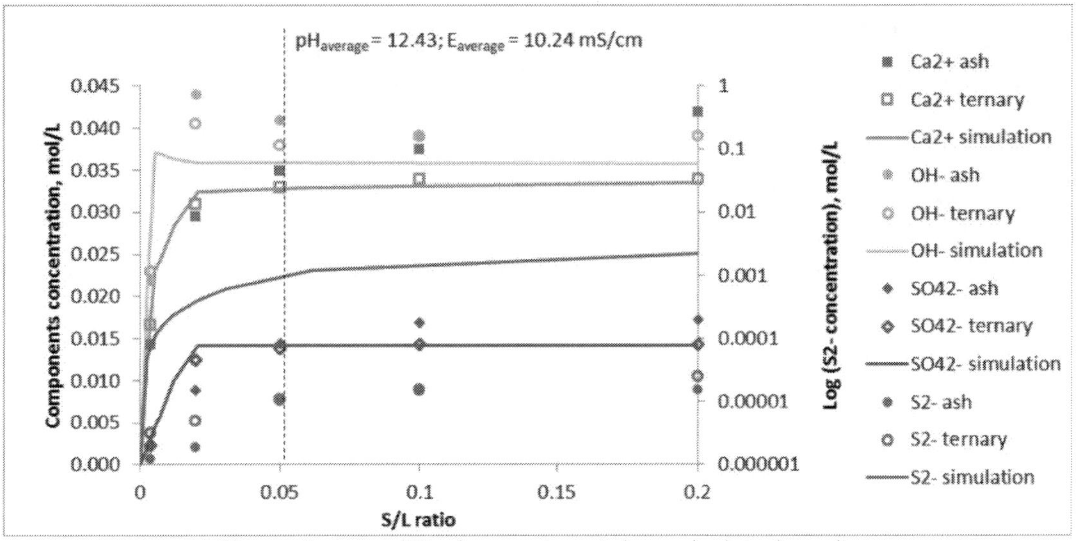

Figure 4. PF ash experimental vs calculated results

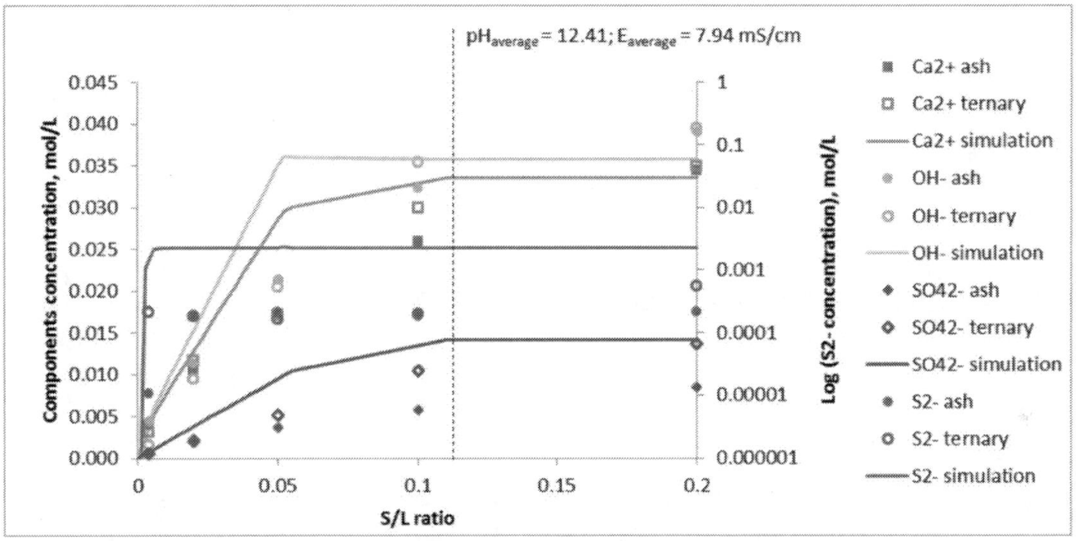

Figure 5. SHC ash experimental vs calculated results

The results of ternary model and real OSA water systems are comparable. The small differences are caused by the other water-soluble OSA components (Table 1). Consequently, the OSA leaching scheme can be modelled using ternary model system.

The predicted results (ELECNRTL, Aspen Plus V8.6) for OSAs dissociation (Eq. 1 - 4) were compared with the experimental data (Figure 3 - 5). The relatively small deviations between measured and predicted data confirm the ability of the proposed model to quite accurately describe the OSA dissolution process. The sulfide concentrations vary mostly due to the experimentally determined solubility of pure CaS (0.002292 mol/L) inserted into the Aspen Plus calculation platform.

The results showed the clear dependence between Ca^{2+}- ion concentration and conductivity (E) of the solution for all of the OSAs ternary ($Ca(OH)_2$-$CaSO_4$*$2H_2O$-CaS) model systems (R^2 = 0.986) and for the OSAs (CFB, PF and SHC) water systems (Figure 6).

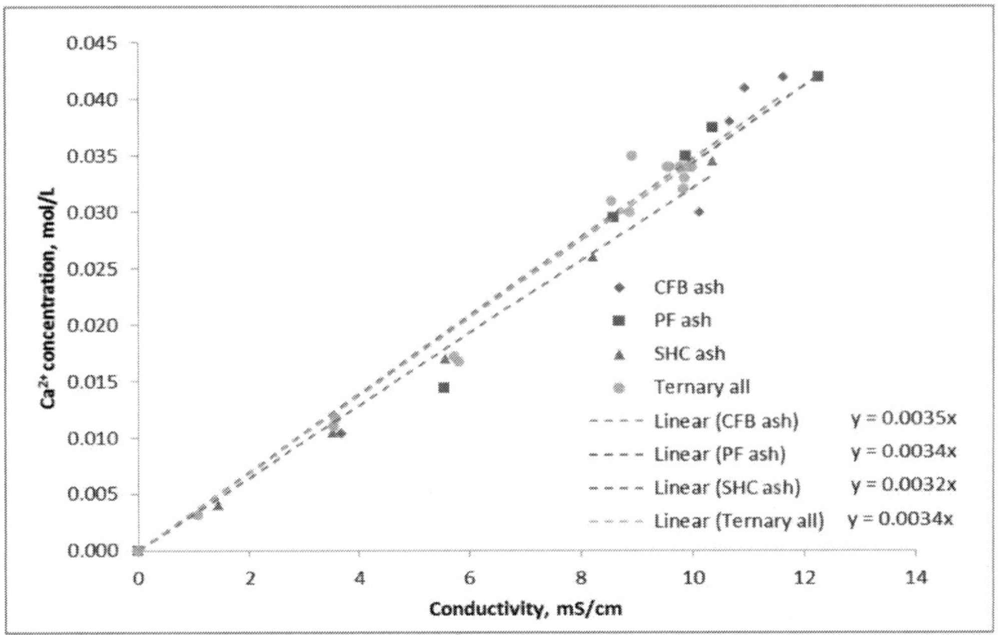

Figure 6. Relationship between Ca^{2+}- ion concentration and conductivity in ternary model and OSA based water systems

The correlation of Ca^{2+}- ion and E coincides in all systems to a satisfactory extent; therefore, it can be expressed with the following averaged equation (R^2 = 0.982):

$$[Ca^{2+}] = 0.003375 * E \tag{5}$$

Based on the latter, the Ca^{2+}- ion values for OSA water systems can be used to compare the kinetic model predicted values, which may play significant role in optimizing an industrial PCC route on the basis of OSAs.

The equilibrium constants at the equilibrium states were calculated, according to the experimental data and thermodynamic model, and presented as average values in Table 2.

The results indicated that the experimental and predicted values for K_{eq1} and K_{eq2} ($K_{eq1average}$ = 5.24E-5, $K_{eq2average}$ = 4.76E-5 and K_{eq1} = 4.28E-05, K_{eq2} = 4.73E-4, respectively) coincide quite well for the ternary model systems. Comparing the results of ternary model (experimental and simulated) and OSA water systems, the K_{eq1} and K_{eq2} values for real OSA systems are higher due to the additional ash components that could provide Ca^{2+} - and SO_4^{2-} - ions into solution. The value of K_{eq1} (5.24E-5) furthest agrees with SHC model system result (5.45E-5), which indicates that, in the SHC ash, the balance between calcium and hydroxide affects the main three components. The received results of K_{eq2} from ternary model and ash systems match better in case of

CFB and PF systems because the SO_4^{2-}- ion concentration does not reach the equilibrium in real SHC ash systems. The values of K_{eq3} are also similar for the OSA and ternary model systems (furthest agrees with systems of PF, low sulfide content), but differ more in comparison with the simulation, because currently the model considers the solubility of CaS in pure water. Consequently, the $Ca(OH)_2$-$CaSO_4$-CaS-H_2O model system should be adjusted according to the new experimental data.

Table 2. Equilibrium constants of the studied systems

CFB

Equilibrium constant, K_{eq}	Experimental			Simulation
	mol/L			
	Ash	**Ternary**	**Binary**	**Ternary**
K_{eq1}	6.66E-05	5.25E-05	4.52E-05	4.28E-05
K_{eq2}	6.52E-04	4.77E-04	4.77E-04	4.73E-04
K_{eq3}	4.05E-07	3.49E-06	-	7.68E-05
K_{eq4}	-	-	-	4.76E-02

PF

Equilibrium constant, K_{eq}	Experimental			Simulation
	mol/L			
	Ash	**Ternary**	**Binary**	**Ternary**
K_{eq1}	7.21E-05	5.03E-05	5.06E-05	4.28E-05
K_{eq2}	6.18E-04	4.74E-04	4.74E-04	4.73E-04
K_{eq3}	5.36E-07	5.78E-07	-	7.68E-05
K_{eq4}	-	-	-	4.76E-02

SHC

Equilibrium constant, K_{eq}	Experimental			Simulation
	mol/L			
	Ash	**Ternary**	**Binary**	**Ternary**
K_{eq1}	5.24E-05	5.45E-05	4.30E-05	4.28E-05
K_{eq2}	2.93E-04	4.76E-04	4.35E-04	4.73E-04
K_{eq3}	7.52E-06	1.96E-05	-	7.68E-05
K_{eq4}	-	-	-	4.76E-02

On the basis of the three types of OSAs differing by chemical and physical structure (Table 1) it has been shown that the proposed model can be implemented for describing the leaching behavior of the key Ca-compounds of $Ca(OH)_2$, $CaSO_4*2H_2O$ and CaS.

The equilibrium constants (Table 2) and Eq. 5 can be used for more complex OSA water system layout for compilation of OSA utilization reactor schemes and also, for aqueous systems based on analogical alkaline waste materials.

Conclusion

The leaching behavior of the key Ca - compounds in OSAs, as sorbents for CO_2 mineralization and PCC production, was studied in order to provide better understanding for handling OSA suspensions and leachates.

The thermodynamics on the basis of the comparative equilibrium, 3 h, batch experiments with the aqueous systems of three different types of OSAs (CFB, PF and SHC) as well as corresponding binary ($Ca(OH)_2$-$CaSO_4*2H_2O$-H_2O) and ternary ($Ca(OH)_2$-$CaSO_4*2H_2O$-CaS-H_2O) model systems, were demonstrated.

The results indicate that the OSA leaching behavior could be described by the ternary model system which agrees quite well with the Aspen Plus thermodynamic model. The pH, conductivity and Ca^{2+}- ion concentrations of OSAs water systems were slightly higher, comparing to ternary model systems, caused by the other water-soluble components in OSA raw materials.

The thermodynamic equilibrium constants (K_{eq1} - K_{eq4}) for $Ca(OH)_2$, $CaSO_4*2H_2O$ and CaS dissolution reactions in OSA based water systems were evaluated. Also, an equation was proposed for calculating Ca^{2+}- ion concentrations from the conductivity data at the equilibrium state.

On the basis of the three types of OSAs differing by chemical and physical structure it has been shown that the proposed model could be implemented for describing the leaching behavior of the key Ca-compounds in OSA.

The equilibrium constants could be used, firstly, to estimate the kinetics of dissolution and compose a complete OSA leaching model and, secondly, to design a pilot-scale OSA carbonation reactor.

Acknowledgements

This research was supported by European Social Fund´s Doctoral Studies and Internationalization Program Kristjan Jaak, Estonian Ministry of Education and Research (SF0140082s08), Estonian Science Foundation (Grant No 9334) and institutional research funding IUT (IUT33-19), which are gratefully acknowledged.

References

1. Eesti Energia Annual Report 2013 (https://www.energia.ee/-/doc/10187/pdf/concern/annual_report_2013_eng.pdf)
2. Kuusik, R., Uibu, M., Kirsimäe, K., Mõtlep, R., Meriste, T. 2012. Open-air deposition of Estonian oil shale ash: formation, state of art, problems and prospects for the abatement of environmental impact. *Oil Shale*, 29(4), 376 - 403.
3. Uibu, M., Uus, M., Kuusik, R. 2009. CO_2 mineral sequestration in oil-shale wastes from Estonian power production. *Journal of Environmental Management*, 90(2), 1253-1260.
4. Velts, O., Uibu, M., Kallas, J., Kuusik, R. 2011. Waste oil shale ash as a novel source of calcium for precipitated calcium carbonate: Carbonation mechanism, modeling, and product characterization. *Journal of Hazardous Materials*, 195, 139-146.

5. Riispere, H.J., Determination of free CaO content in oil shale ash. Transact. Tallinn Polytechnical Institute, series A. 1966. Nr. 245. P. 73-76 (in Estonian).
6. EVS 664:1995 Solid Fuels. Sulphur Content – Determination of Total Sulphur and its Bonding Forms.
7. APHA – American Public Health Association, American Water Works Association, Water Environment Federation. 2005. *Standard methods for examination of water and wastewater*. 21st ed., pg. 1368, Washington DC, USA.
8. Tamm, K., Kuusik, R., Uibu, M., Kallas, J. 2013. Transformations of sulfides during aqueous carbonation of oil shale ash. *Energy Procedia*, 37, 5905 – 5912.
9. Velts, O., Uibu, M., Rudjak, I., Kallas, J., Kuusik, R. 2009. Utilization of oil shale ash to prepare PCC: leachibility dynamics and equilibrium in the ash-water system. *Energy Procedia*, 1(1) 4843-4850.
10. ISO – 6058:1984, ISO 6058:1984, Water quality - Determination of calcium content - EDTA titrimetric method.
11. EKUK Virumaa affiliate Kohta-Järve laboratory of chemistry. 2002. The quality procedure guide TJ 17 – Determination of sulfide in water.
12. ISO – 9963-1: 1994(E)), Water quality – Determination of alkalinity – Part 1: Determination of total and composite alkalinity.
13. Ritchie, I.M., Bing-An, X. 1990. The Kinetics of Lime Slaking. *Hydrometallurgy*, 23, 377 – 396.

Negative About Carbon: A Positive Approach

P.Gunning[1*] and C.D.Hills[2]

[1] Carbon8 Systems Ltd, UK
[2] University of Greenwich, UK
*Corresponding author: petergunning@c8a.co.uk

Abstract

Carbon8 Systems Ltd. was formed in 2006 as spinout-company of the University of Greenwich. By using the Carbon8 process, industrial thermal residues are solidified and stabilised in a hardened pellet form that can be a direct substitute for natural aggregate e.g. for the production of concrete construction blocks. The aggregates produced were rigorously tested and given 'end-of-waste' designation by the UK Environment Agency. In early 2012, a bespoke commercial plant was commissioned at Brandon in Suffolk, UK, operated by Carbon8 Aggregates Ltd. In 2014, a second production line was added to the Brandon facility, increasing its capacity to 60,000 tonnes per year. The ACT-produced aggregate is carbon negative as it contains more combined carbon than is generated by its production. Consequently, the concrete construction blocks produced are also carbon negative, and are marketed under the name: 'Carbon Buster'. Plans are at an advanced stage for the construction of a second production facility in the UK. This is scheduled to be operational by late 2015. The present work discusses the more recent developments of the commercial application of accelerated carbonation technology.

Introduction

The UK currently incinerates 10 percent of municipal waste, compared to the European average of 17.3 percent.[1] Diminishing landfill space and rising disposal costs are promoting the adoption of energy recovery as a preferred management option.

Flue gases generated by municipal incineration contain pollutants, e.g. NO_x, heavy metals, which are removed via air pollution control (APC) systems. The injection of lime and other reagents generates air pollution control residues (APCr), containing these pollutants.[2] APCr generation is typically in the range of 2-6% of the total throughput of an incinerator.[3]

The majority of UK APCr is sent to either specialized hazardous landfill facilities or placed in underground storage. The properties of APCr often exceed the acceptance criteria for hazardous landfill, and can only be disposed at sites that have received a derogation permit. Methods of treating APCr are under development, including washing the dusts,[4] chemical treatments, and vitrification.[5] However, these are not operating commercially.

Accelerated carbonation is employed by Carbon8 to solidify and stabilise APCr. The process enables the fine-grained dust to be formed into aggregates with properties that make them suitable for use as construction aggregates.

Carbon8 Systems and its sister company, Carbon8 Aggregates, have been operating commercially since 2012. The present work discusses the development of the process, and the future potential of the technology.

A Decade of Development

The potential to rapidly harden pelleted materials was observed during pilot scale trials to assess the use of carbonation to promote the rate and extent of setting of stabilisation/solidification (s/s) treated soils.[6] The production of carbonation-hardened agglomerates from a number of industrial residues was progressed at the University of Greenwich.[7,8]

In January 2010, a pilot scale trial at a hazardous waste landfill receiving APCr was undertaken to demonstrate the larger-scale application of the technology. This led to the construction of a full-scale demonstration plant in November 2010.[9] The plant was used to manufacture in excess of 200 tonnes of aggregate, for use in validation trials as a gravel-replacement in concrete construction blocks.[10]

European waste legislation and its implementation in the UK has presented a number of challenges to the development of the carbonation process. At each stage of the scaling-up of the process, the support of the UK Environment Agency was required. Accredited laboratory testing of the aggregate and block to European Standard validated these materials as products. By working closely with the UK Environment Agency, 'End of Waste' status for the aggregate was approved.

Using the experience and know-how developed, a production facility capable of producing 18,000 tonnes of aggregate per year was built in 2012.

Commissioning the Plant

The commissioning of the plant established it as first commercial plant of its type in the world. Initially, a single production line was installed. However, demand for the aggregate necessitated the construction of a second line in early 2014, increasing production to 36,000 tonnes per year (see Figure 1).

Figure 1. APCr processing plant (second phase)

The full-scale process for the production of APCr aggregates by the Carbon8 process is shown in Figure 2.

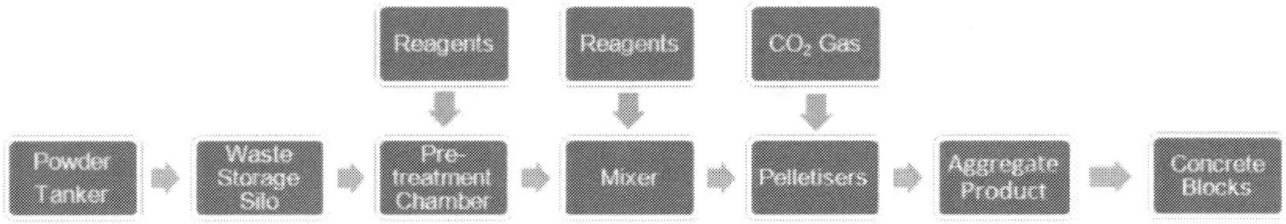

Figure 2. Flowchart of the full-scale process

APCr arrives at the site in 25 tonne powder tanker trucks and is transferred to two $85m^3$ silos. A refrigerated tank stores liquid CO_2 for use in the carbonation process. This CO_2 has been captured from gaseous waste emissions of industrial processes and thus has been diverted from atmospheric release.The APCr passes through a pre-treatment chamber, where initial carbonation of the APCr takes place, into a batch mixer where reagents are introduced as necessary, to produce a mixture that is then conveyed to pelletising units to be aggregated. Further addition of CO_2 induces chemical stabilisation and solidification of the aggregate. The aggregated product is then conveyed to storage bays, and screened to produce the required grading.

Full-Scale Production

Carbon8 operates under a strict quality control system in compliance with BS EN ISO14001, BS OHSAS18001 and BS EN ISO9001.[11,12,13] Daily checks on the quality of the incoming APCr (including bulk density and carbon dioxide reactivity) and outgoing aggregate product (compressive strength, particle size) are carried out to ensure that it meets the required specification set out in the 'End of Waste' documentation.

Aggregate Applications

The aggregate product (C8A), is a rounded pelleted material, which can be supplied in particle gradings between 50mm and sub-1mm (see Figure 3). The product is independently verified as carbon negative, imbibing more carbon dioxide than is generated in its manufacture. C8A typically has a carbon footprint of -46kg/tonne.

The aggregate has been trialled and is suitable for use in a range of products, including concrete construction blocks, concrete, and floor screeds (see Figure 3).

Figure 3. C8A applications; top left – C8A aggregate, top right – concrete construction blocks, bottom left – fragment of readymix concrete, bottom right – section from a screed floor

Concrete construction blocks. Blocks made with C8A can fully comply with European Standard BS EN 771-3. Lignacite Ltd is amongst several manufacturers utilising C8A in concrete block production, and is the UK's largest independent producer. Their 'Carbon Buster' block is certified as carbon negative up to -14kg/tonne, and contains up to 50% recycled content. Quality control testing has been carried out by an independent third party UKAS ISO 17025:2005 accredited laboratory. The most recent results are shown in Table 1. For comparison, a set of blocks manufactured using the traditional formulation was tested alongside the Carbon Buster blocks. The results show comparable properties between the two formulations in terms of strength (compressive and transverse failure) and density.

Table 1. Physical properties of APCr aggregate and control concrete blocks

Property	Without APCr Aggregate	With APCr Aggregate
Density (kg/m^3)	1660	1730
Compressive Strength (MPa)	11.8	11.2
Transverse Failure Load (kN)	7.75	7.78

Ready-mixed concrete. C8A is suitable for use as partial or full replacement of natural aggregates in many readymix concrete. Figure 4 shows the compressive strength of concrete with 100% of the coarse aggregate component composed of C8A.

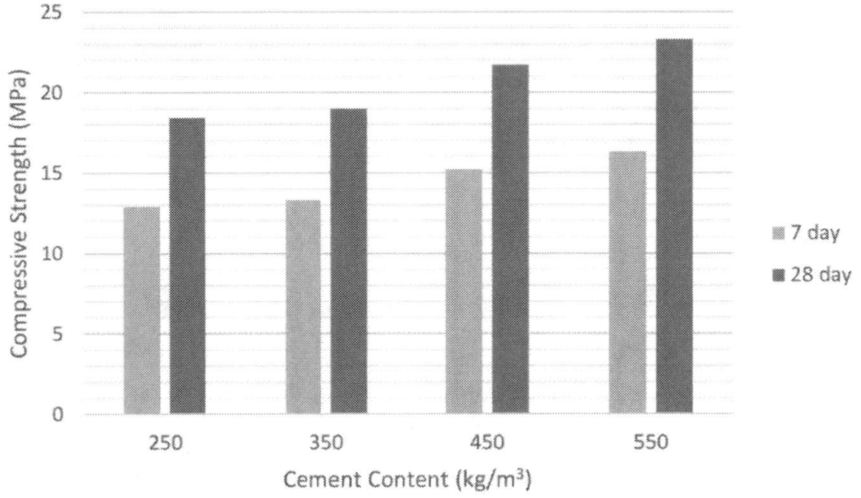

Figure 4. Strength of concrete containing C8A

Screed. Carbon8 Aggregate is ideal for no-fines base course in a lightweight screed system. Lightweight screed omits the fine aggregate normally used in concrete to produce a highly-porous and low density structure which has enhanced thermal and acoustic properties. C8A is shown in Figure 3 as part of a screed system comprising a 75mm layer of no-fines concrete with a 25mm mortar cap. Strength data is presented in Figure 5, with accompanying density data in Figure 6.

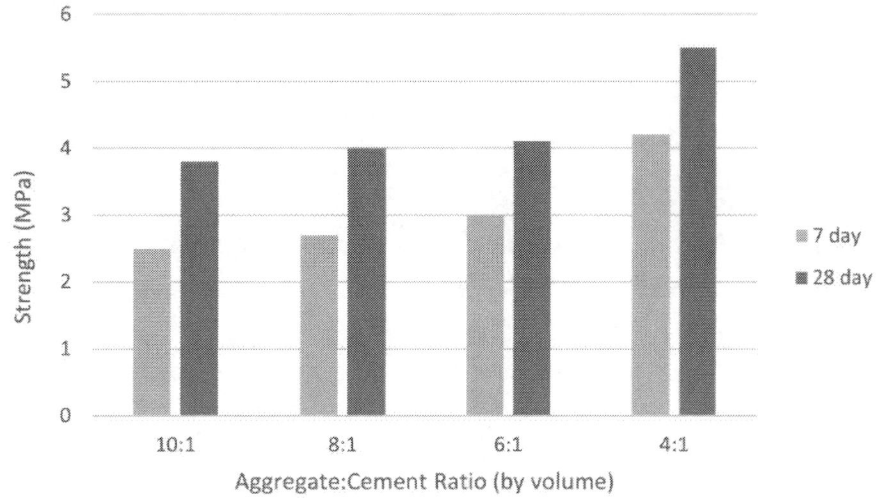

Figure 5. Strength of no-fines concrete containing C8A

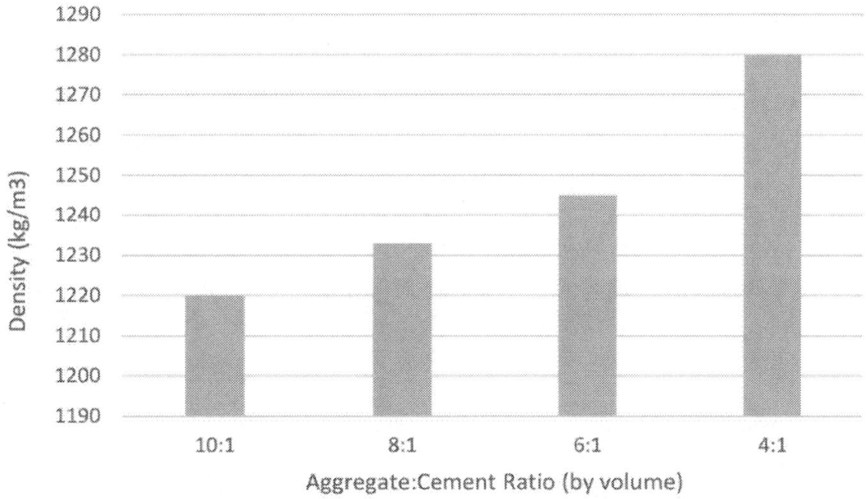

Figure 6. Density of no-fines concrete containing C8A

Plant No. 2,3,4....

A second plant at Avonmouth in South-West England is under construction, and is scheduled to begin production in late 2015. A third plant in the North-West is in the advanced design stage, and is due to begin construction in 2016. Plans for a fourth plant in the South-East and a fifth in the North-East are already being formulated

Summary

Carbon8 Ltd has developed and operates a commercial process for the recycling of municipal incinerator air pollution control residue as construction aggregate:

- The aggregate product has been rigorously tested and received 'end-of-waste' classification by the UK Environment Agency
- The world's first commercial carbonated aggregate facility was installed in Brandon, UK in early 2012
- C8A is direct replacement for natural gravel, and is certified carbon negative
- Product manufacture is according to a quality system in compliance with ISO14001, OHSAS18001 and ISO9001
- Continued monitoring of the APCr and aggregate product has identified trends in their physical and chemical properties and ensured compliance to the End of Waste License.
- A second plant is under construction, with a third plant in the advanced stages of planning, with a further two plants due for construction over the next two years.

Acknowledgements

The authors would like to thank the other members of the Carbon8 team; Dr. Paula Carey, Mark Orsbourn, Steve Greig, Andrew Doel, and Tom Bissett of Carbon8 Aggregates Ltd. Their invaluable contribution to this endeavour is gratefully acknowledged. Also the immense support of Grundon Waste Management Ltd and Lignacite Ltd, in particular Allan Eastwood, has been invaluable. Many thanks also to Shane Pearson, manager of the Brandon processing facility, and his team. Thank you also to Prof. Alan Maries for his assistance in preparing this manuscript.

References

1. CIWM http://www.ciwm.co.uk/CIWM/InformationCentre/AtoZ/IPages/Incineration.aspx. Accessed 04.02.2015.
2. DEFRA. 2013. Incineration of Municipal Solid Waste. Accessed 04/02/2015.
3. DEFRA. 2013. http://www.defra.gov.uk/statistics/environment/waste. Accessed 04.02.2015.
4. Chimenos, J.M, Fernandez, A.I., Cervantes, A., Miralles, L., Fernandez, M.A., Espiell, F. 2005. Optimizing the APC residue washing process to minimize the release of chloride and heavy metals. *Waste Management*, 25, 686–693.
5. Amutha Rani, D., Gomez, E., Boccaccini, A.R., Hao, L., Deegan, D., Cheeseman, C.R. 2008. Plasma treatment of air pollution control residues. *Waste Management*, 28, 1254–1262.
6. Antemir, A. 2010. Performance assessment of stabilised/solidified waste-forms. Ph.D. Thesis, University of Greenwich.
7. Gunning, P.J., Antemir, A., Hills, C.D., Carey, P.J. 2011. Secondary aggregate from waste treated with carbon dioxide. *Proceedings of the Institution of Civil Engineers: Construction Materials*, 164, 231-239.
8. Padfield, A.M., Carey, P.J., Hills, C.D., Poole, A.B. 2004. Reuse of quarry fines in production of secondary aggregates. *Engineering Sustainability*, 157, 149–154.

9. Gunning, P.J., Hills, C.D., Carey, P.J., Antemir, A. 2011. Novel approaches to the valorisation of ashes using aggregation by carbonation. *In: Proceedings of the 2nd International Slag Valorisation Symposium, Leuven, Belgium.*

10. Gunning, P.J., Hills, C.D., Carey, P.J., Grieg, S., Schilling, M. 2012. Carbonation of MSWI APCr for Re-Use as a Lightweight Aggregate. *Proceeding of the Ash Utilisation Conference 2012, Stockholm, Sweden.*

11. BS EN ISO 14001:2004. Environmental management systems — Requirements with guidance for use. European Committee for Standard.

12. BS OHSAS18001:2007. Occupational health and safety management systems – requirements. British Standard Institute.

13. BS EN ISO9001:2008. Quality management systems – requirements. European Committee for Standard.

Single and Multi-Step Processes for Flue Gas CO_2 Mineralization Using Thermally Activated Serpentine

S. Hariharan, M. Werner and M. Mazzotti*

Institute of Process Engineering, ETH Zurich, Zurich, Switzerland

*Corresponding author: marco.mazzotti@ipe.mavt.ethz.ch

Abstract

This contribution briefly communicates our work on the development of mineralization processes that integrates CO_2 capture and uses thermally activated serpentine (dehydroxylated lizardite) as the precursor material. A series of carefully monitored single-step batch experiments helped to explore the carbonation potential of activated serpentine, as well as to understand fate of the solutes and the precipitation regime in the $MgO\text{-}CO_2\text{-}H_2O$ system. The low carbonation efficiency obtained in this operating mode was attributed to insufficient magnesium (Mg) extraction efficiency, which could be greatly enhanced by separating the precursor dissolution step from the carbonate precipitation step. Our two-step process foresees a temperature swing ($T_{diss.} < T_{precip.}$) that can be aided by a low-level pCO_2 swing ($pCO2_{diss.} < pCO2_{precip.}$). The combined swing was tested successfully through preliminary experiments and simulations. The results point toward further optimization potential when using more than one dissolution reactor operated at increasing temperature, i.e. toward a multi-step process design.

Introduction

Originally, aqueous CO_2 mineralization was studied at high temperature with the acidifying help of high CO_2 partial pressures to extract alkaline earth metals from a solid feed, which then form carbonates by precipitation. Recent developments in mineralization research have seen process concepts that accept dilute CO_2 streams as input, thus avoiding the cost associated to a preliminary capture step. Usually, these concepts employ pH-tuning agents, which require full recovery, and/or a rather complex multi-step design to deal with the low reactivity of the solid feed at flue gas conditions. More simply, the CO_2 absorption, feed dissolution, and carbonate precipitation could take place simultaneously in a single reactor without additives. This, however, requires special feed materials, such as alkaline industrial residues (fly ashes, cement kiln dust, etc.) or the magnesium (Mg) silicate serpentine in its activated form. Alkaline residues are highly reactive, but suffer from limited availability. Natural serpentine is worldwide abundant, but needs to be thermally pre-treated to maximize its reactivity[1]. In contrast to older findings[2], an optimal thermal activation strategy that uses natural gas as the source of heat was shown to be relatively inexpensive[3].

Methods

The feasibility of single-step flue gas CO_2 mineralization was investigated in a first series of experiments. Ground (sub 125 μm) and heat-treated (75% dehydroxylated or

25% residual hydroxyls) serpentine was mixed with ultrapure water, and exposed to CO_2 at ambient pressure, in a 100 mL Teflon reactor equipped with a gas dip tube, reflux condenser, sampling port, and either a pH or Raman probe. Experiments were carried out at T = 30, 50, 60, and 90°C, and slurry densities S/L (solids to liquid ratio) of 5, 10, 15, and 20% by weight. Slurry samples were taken at regular intervals of t = 0.5, 1, 1.5, 2, 3, 4, 6, 8 h and filtered immediately. The filtrate was diluted and analyzed for solute concentration of Mg^{2+} and silica. The solids were dried under vacuum overnight and analyzed using TGA, XRD, and SEM image analysis.

In order to tackle the equilibrium inhibitions obtained in the single step experiments, a double step strategy was proposed[6]. Here, dissolution of activated serpentine and precipitation of carbonates occur in two different reactors that are operated at different reaction temperatures (see Fig. 1). A temperature swing in combination with a pCO_2 swing was used that exploits the fact that the solubility of the relevant Mg-carbonates decreases with increasing T and decreasing pCO_2. The solution that is transported from the dissolution to the precipitation side is rich in Mg and consequently in CO_2, so as excess CO_2 will degas inside the high-T, low-pCO_2 precipitation reactor. That CO_2 will increase the pCO_2 of the flue gas when using the latter as a stripping gas in said reactor.

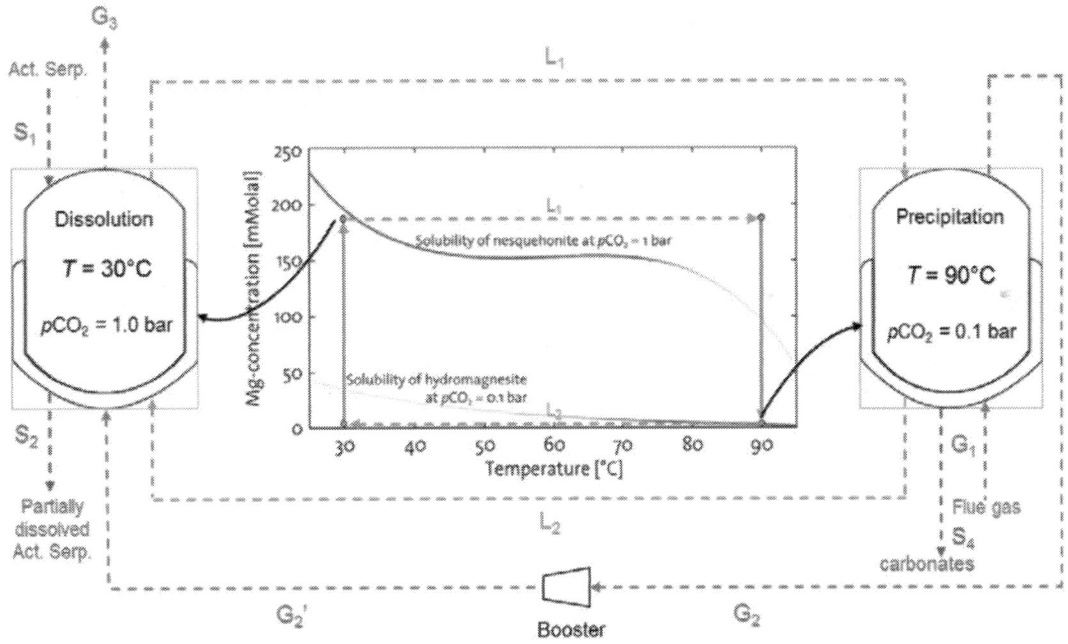

Figure 1. Schematic flow sheet for a double-step CO_2 mineralization process. Dissolution of activated serpentine is done at low temperatures (like 30°C) and moderate pCO_2 (like 1.0 bar or less), while carbonate precipitation is done at moderate temperatures (90°C) and low pCO_2 (0.1 bar, corr. to the pCO2 of the inlet flue gas). The liquid stream is transported between the reactors and carries the Mg^{2+} solute for carbonate precipitation.

A series of proof-of-concept experiments were performed that aimed to validate our proposed double-step strategy. For this purpose, two Teflon reactors were connected via a double-head peristaltic pump, which circulates aqueous Mg^{2+} between the two reactors. In contrast to the flow sheet as foreseen in Fig. 1, the gas phase (synthetic mixtures of CO_2 and N_2) was fed independently, i.e. without connecting the dissolution to the precipitation reactor, for the sake of simplicity.

Additionally, sensitivity analyses were performed for the flow sheet at steady state operating conditions[6], and suitable operating conditions were identified based on thermodynamics.

Results: Experiments

As expected from literature[4,5], the hydrated Mg-carbonate nesquehonite precipitated during low-T experiments (30-50°C), while a basic-hydrated Mg-carbonate (dypingite or hydromagnesite) was formed at 90°C. Transformation of the former into the latter was observed at 60°C. The onset of precipitation, t_{onset}, decreased with slurry density S/L and with temperature T. However, the conversion of the Mg in the feed to carbonates, R_x, barely exceed 20% at 30°C (cf. Fig. 2) and 50°C, and even lower values were measured at 60°C and 90°C ($R_{Mg} \leq 15\%$)[6].

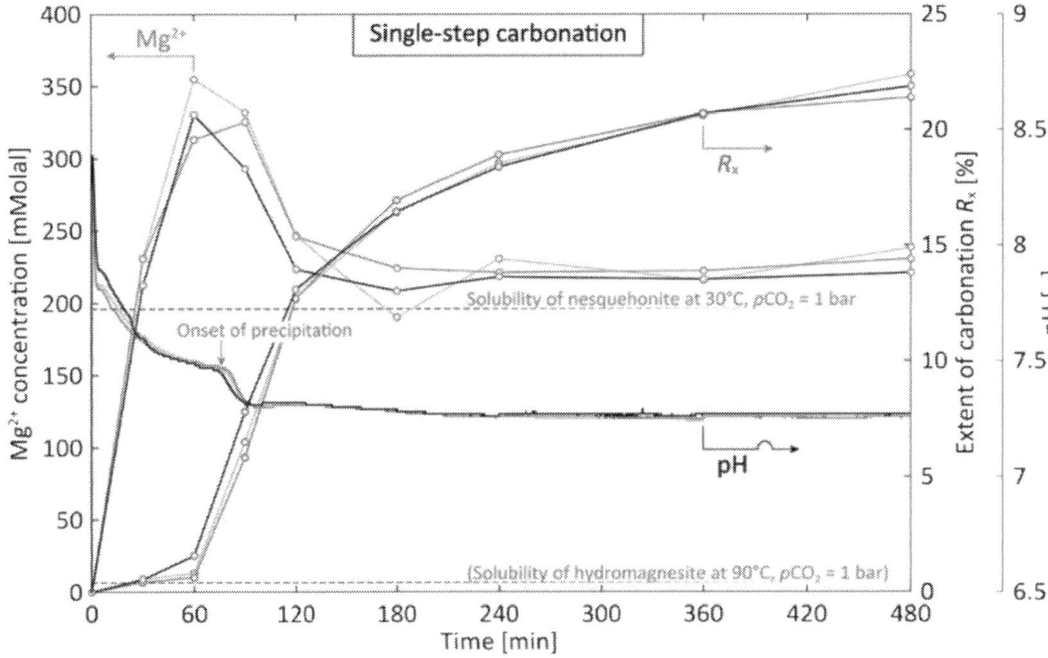

Figure 2. Three replicates of a single-step carbonation experiment at T = 30°C, pCO_2 = 1 bar, S/L = 20% wt.: shown is the evolution of the Mg^{2+} concentration (green circles), extent of carbonation R_x (red circles), and in-situ measured pH profiles.

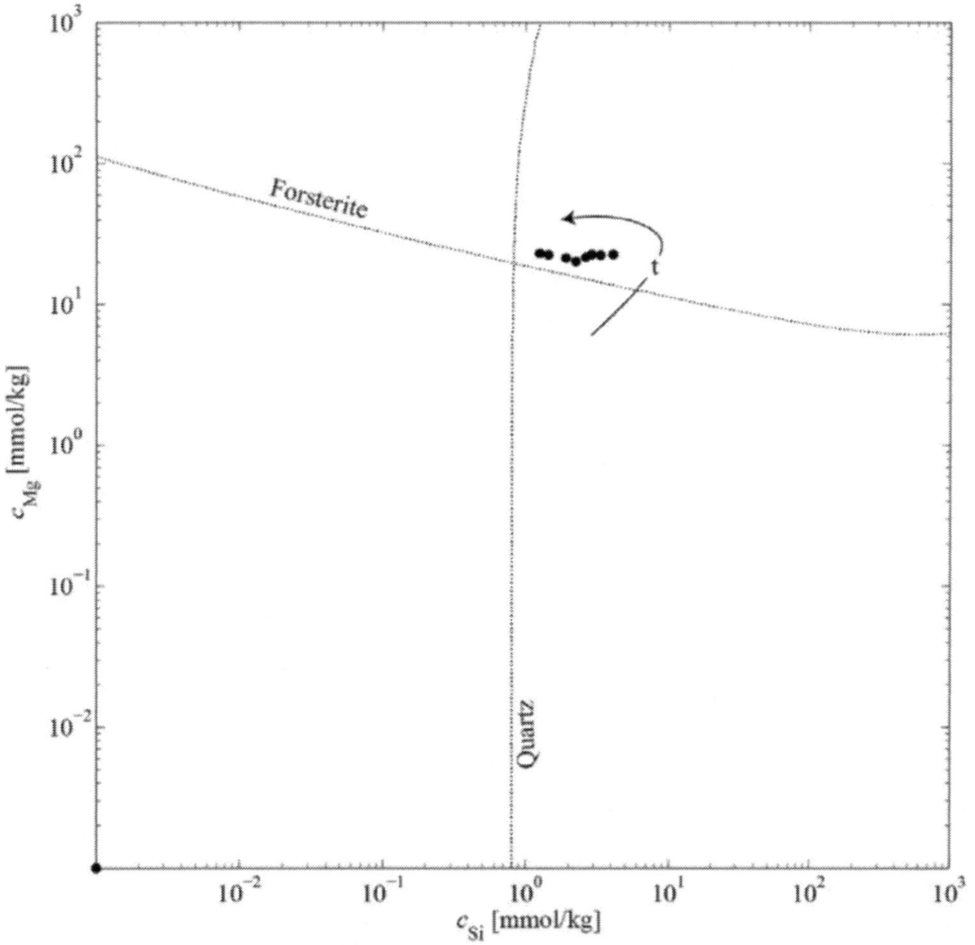

Figure 3. A single-step carbonation experiment at T = 90°C, pCO_2 = 1 bar, S/L = 5% wt. The experimental data points shown by the markers correspond to Mg^{2+} and aqueous silica measured at different dissolution time t during the experiment. The dotted lines correspond to the solubility of Forsterite and Quartz as a function of Mg^{2+} and aqueous silica concentrations at 90°C and pCO_2 = 1 bar.

A low value of R_x at higher temperatures is particularly surprising, as the far-from-dissolution kinetics is known to increase with T [7,8]. However upon analysing the experimental data, in particular the data at 90°C for equilibrium effects, we see that the dissolving system of activated serpentine is in equilibrium with forsterite (see Fig. 3). This prevents further dissolution of activated serpentine resulting in low values of R_x at higher temperatures under low pCO_2 values. Therefore, although activated serpentine dissolution kinetics are favorable at higher temperatures, equilibrium effects result in poor dissolution efficiency.

Regarding the double-step carbonation experiments, our preliminary results are promising, in the sense that we obtained a higher magnesium extraction efficiency, X_{Mg}, and carbonation efficiency, R_x, than from any of the single-step experiments (see Fig. 4). A further improvement in the process scheme was proposed by having multiple dissolution steps that leads to a multi-step mineralization process. In this regard, a linear temperature ramp between 30 and 90°C was introduced experimentally in the dissolution reactor (cf. Fig. 4). This once again performed better than our double step-process. An optimal dissolution strategy for activated serpentine is currently being devised that would enable us to obtain the maximum carbonation extent, R_x.

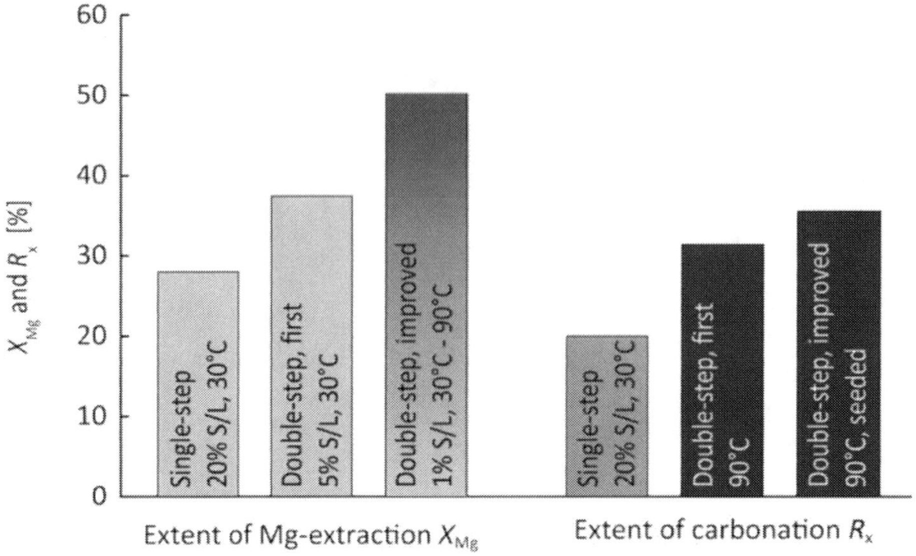

Figure 4. Comparison of the extent of Mg-extraction and carbonation between the best-case single-step experiment, our first semi-continuous double-step experiment, and another double-step experiment with improved operating conditions (mimicking a multi-step process with several dissolution reactors operated at increasing temperature).

The fact that only the solutes but not the unreacted activated serpentine are carried over to the precipitation reactor allows to obtain a pure Mg-carbonate product, as shown in Fig. 5 here as confirmed by our solids analysis (see Fig. 5). Given the relatively lean operating conditions applied in our experiments, that product consisted of a basic-hydrated Mg-carbonate with a degree of hydration in between that of hydromagnesite ($(MgCO_3)_4Mg(OH)_2{\cdot}4H_2O$) and that of dypingite ($(MgCO_3)_4Mg(OH)_2{\cdot}5H_2O$).

Concluding remarks

In summary, thermally activated serpentine can be carbonated under very lean operating conditions of less than 100°C and CO_2 pressures not higher than 1 bar, which serves the purpose of direct flue gas CO_2 mineralization. By exploiting basic thermodynamics, it is possible to separate and optimize the dissolution and the precipitation step without utilizing additional chemicals for pH control. Thereby, we have shown that such separation yields a pure basic-hydrated Mg-carbonate product. The logical next step is to realize the process cycle in the lab in fully continuous mode. This work is subject to our ongoing research efforts.

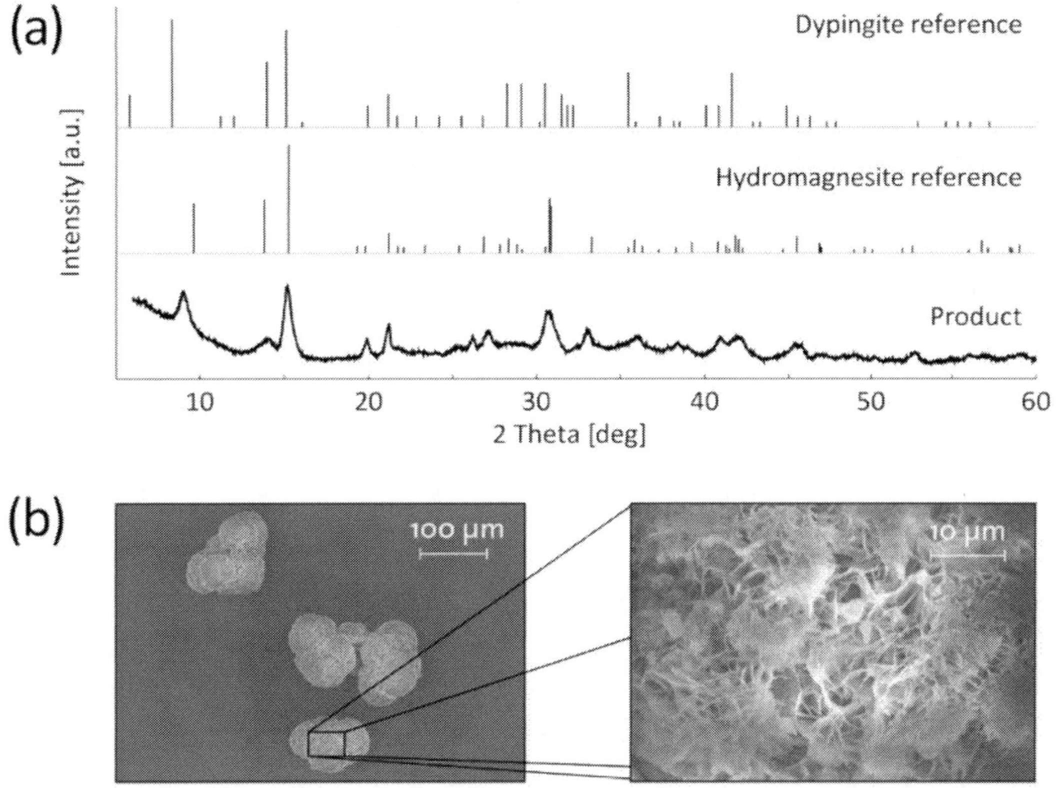

Figure 5. (a) XRD spectrum (including the reference spectra of the best matching phases) and (b) SEM images of the final product in the precipitator collected after a double-step experiment.

References

1. McKelvy, M.J., Chizmeshya, A.V.G., Diefenbacher, J., Béarat, H., Wolf, G., 2007. Exploration of the role of heat activation in enhancing serpentine carbon sequestration reactions. *Environmental Science & Technology*, 38(24), 6897-6903.
2. O'Connor, W.K., Dahlin, D.C, Rush, G.E., Gerdemann, S.J., Penner, L.R., Nilsen, D.N., 2005. Aqueous mineral carbonation: Mineral availability, pre-treatment, reaction parametrics, and process studies. *Albany Research Center Technical Report DOE/ARC-TR-04-002*, Final Project Report.
3. Balucan, R.D., Dlugogorski, B.Z., Kennedy, E.M., Belova, I.V., Murch, G.E., 2013. Energy cost of heat activating serpentinites for CO2 storage by mineralisation. *International Journal of Greenhouse Gas Control*, 17, 225-239.
4. Hänchen, M., Prigiobbe, V., Baciocchi, R., Mazzotti, M., 2008. Precipitation in the Mg-carbonate system - effects of temperature and CO_2 pressure. *Chemical Engineering Science,* 63(4), 1012-1028.

5. Prigiobbe, V., Mazzotti, M., 2013. Precipitation of Mg-carbonates at elevated temperature and partial pressure of CO_2. *Chemical Engineering Journal*, 223, 755-63.
6. Werner, M., Hariharan, S., Mazzotti, M., 2014. Flue gas CO_2 mineralization using thermally activated serpentine: From single- to double-step carbonation. *Physical Chemistry Chemical Physics*, 16, 24978-24993.
7. Werner, M., Hariharan, S., Zingaretti, D., Baciocchi, R., Mazzotti, M. 2014. Dissolution of dehydroxylated lizardite at flue gas conditions: I. Experimental study. *Chemical Engineering Journal*, 241, 301-313.
8. Hariharan, S., Werner, M., Hänchen, M., Mazzotti, M. 2014. Dissolution of dehydroxylated lizardite at flue gas conditions: II. Kinetic modeling. *Chemical Engineering Journal*, 241, 314-326.

Carbon Dioxide Sequestration in Wastes: The SAPICO2 Project

P.Gunning[1], C.D.Hills[1*], R.Gunning[1], A.Maries[2]

[1] University of Greenwich, UK

[2] AMSTaR Consultancy, London, UK

*Corresponding author: c.d.hills@gre.ac.uk

Abstract

The SAPICO2 project is an INTERREG IVA cross channel collaboration established to develop new technologies utilising accelerated carbonation for the creation of eco-construction materials from solid wastes. Over three phases, the project has identified carbon dioxide reactivity of solid wastes, produced prototype-construction materials (from these wastes), and manufactured bulk samples of carbonated products. The project aims to introduce carbonated building products into continental Europe, by demonstrating the potential of this emerging technology to meet the EU sustainability agenda. By collaborating with the major bodies responsible for environmental matters, it is hoped that new policy and mechanisms can be developed that facilitate the full commercial valorisation of waste using gaseous CO_2.

Introduction

The SAPICO2 (Sustainable Aggregate Product with Imbibed CO_2) project is a collaboration between the University of Greenwich (UK), the University of Picardie Jules Verne (France), and Carbon8 Systems (a spinout company from the University of Greenwich).[1] SAPICO2 is a European Regional Development Fund project, supported by the INTERREG's Manche (Channel) programme.[2] The 2-year project began in August 2013, and will soon be coming to a successful conclusion.

The SAPICO2 project aims to develop new technology for combining gaseous CO_2 with solid waste to manufacture carbonate-cemented eco-construction materials (including aggregates) and introduce this to the French Channel region.

The first phase of the project involved identifying available waste streams in France and in the UK for their suitability with accelerated carbonation treatment. By gathering samples of these wastes, a laboratory assessment of their chemical and physical properties, along with CO_2 reactivity, was made. In the second phase, those materials reactive with were developed into prototype aggregates for preliminary testing. The prototype aggregates meeting the agreed specification were developed further through a series of pilot scale trials to produce bulk quantities for rigorous end-use testing.

In the UK, consultations with the Environment Agency (EA), along with the Building Research Establishment, the Quarry Products Association, British Cement Association and the UK Quality Ash Association have taken place. In France, UPJV have had dialogues with the French Environment and Energy Management Agency (ADEME), through their Waste and Soils Management Office and the Picardie Regional and Departmental Councils, as well as with private waste management/processing organisations.

The project was recently showcased at the Large-Volume CO_2 Utilization International Conference in Lyon, France.[3] The world's first carbon negative 'welcome' bench was also unveiled to the public during the conference, and is now being used by the students of the University of Lyon.

Now in the third and final phase of the project, bulk samples have been manufactured using paper ash and municipal waste incinerator residues from the UK, and using biomass ashes and quarry fines from France. Plans are in place to conduct a proof-of-concept demonstration to manufacture 1-2 tonnes of biomass ash aggregate for end-use trials in conjunction with a construction block manufacturer in France. This will be a key step in obtaining validation of the aggregate as a product, by demonstrating that the end-of-waste criteria has been fulfilled.

The pressure to decrease the amount of waste sent to landfill through recycling has forced the waste classification system to be altered. Recently, the UK Environment Agency signalled changes to the de-classification of certain wastes into by-products, if they meet the following criteria:

i) there is an established use for the material;

ii) no further processing is required for this use; and

iii) the material is used as an integral part of the continuing production process.

Currently, a comparable 'end-of-waste' process does not exist in France. However, it is hoped that this project (and others like it) will motivate ADEME towards a change of approach and that a route to facilitate waste de-classification is possible in France.

Methodology

Materials

The European Waste Catalogue has been used to classify wastes in this project. The two relevant chapters are 10 "Inorganic wastes from thermal processes" and 19 "Wastes from treatment facilities, off-site waste water treatment plants and the water industry".[4]

More than 100 waste samples were obtained across the UK and France, and were grouped into 13 categories (Table 1).

Characterisation and Carbon Reactivity

The procedure for evaluating the carbon reactivity of the wastes is outlined below.

Sample Preparation. The materials were homogenised by riffling the sample down to a usable size (200g for fine powders, 500g for coarser materials). The materials were then dried to constant weight at 105°C, and reduced to less than 2mm using a jaw crusher, and ground to less than 125μm using a disc mill (where necessary).

Sample Characterisation. Each sample was analysed by X-Ray Fluorescence (XRF) (Bruker D8) to determine bulk chemistry, X-Ray Diffraction (XRD) (Bruker S4) to identify mineral phase composition, and total carbon (TC) content by thermal analysis (Hach Lange TOC-550TN).

Carbonation Technique. Complete carbonation of the materials was achieved by repeated cycles of exposure to CO_2, oven drying, crushing, re-wetting, and further exposure to CO_2. The ground waste was mixed with 10% by weight of water, and placed in a CO_2 chamber pressurised at 2 bar. Initial carbonation cycle lengths were one hour. However, as sample reactivity decreased with repeated carbonation (as verified by TC analysis), the cycle lengths were increased progressively up to 7 days. Total carbonation was verified through analysis of the wastes by XRD.

Table 1. Residues examined

No.	Waste Group	Source	No. of Samples
1	Biomass Ash - Bottom Ash (BBA)	Heavy material remaining in the grate after combustion	7
2	Biomass Ash - Fly Ash (BFA)	Lighter material carried from the combustion zone in the flue gasses before removal	6
3	Cement Kiln Dust (CKD)	Cement kiln dusts (CKDs), including bypass dusts (CBDs) consist of alkali-rich particulate matter removed from cement kiln exhaust gasses	5
4	Coal Combustion – Furnace Bottom Ash (FBA)	Grate residue from coal combustion	4
5	Coal Combustion – Pulverised Fuel Ash (PFA)	Flue gas particulate matter from coal combustion	7
6	Metallurgical (MTL)	Residues from metal production/processing	5
7	Municipal Solid Waste Incineration (MSWI) – APC	Fine grained hazardous dust produced by flue gas treatment systems	21
8	Municipal Solid Waste Incineration (MSWI) - Bottom Ash (BA)	Coarse grained residue remaining in the incineration zone after combustion	12
9	Municipal Solid Waste Incineration (MSWI) - Fly Ash (FA)	Airborne particulate matter carried in the combustion gas and collected in the emissions filtration system	10
10	Municipal Solid Waste Incineration (MSWI) - Boiler Ash (BLA)	Other residues arising from municipal incineration	3
11	Paper Sludge Incineration Ash (PSIA)	Incineration residue from paper making wastewater sludge	3
12	Sewage Sludge Ash (SSA)	Ash from incineration of wastewater treatment sludge	5
13	Steel Slags (SS)	Metallic waste from steel manufacture	13

Aggregate Development

Aggregate development was carried out in two phases. Firstly, aggregates were manufactured according to a standard set of formulations. These were subjected to basic testing (strength, leaching). The results were used to identify the aggregates conforming to the designed specification. Secondly, selected compliant aggregates were manufactured in larger quantities, and subjected to more rigorous testing.

Manufacture. Products were manufactured using a combined carbonation-induced forming technique described previously.[5]

Product Testing. The tests applied are summarised in Table 2.

Table 2. Tests applied

Test	Standard
Apparent Density[6]	EN 1097-6
Bulk Density[7]	EN 1097-3
Water Absorption[6]	EN 1097-6
Crushing Resistance[8]	EN 13055-1
Concrete test specimen manufacture[9]	EN 12390-2
Concrete Flexural Strength[10]	EN 12390-5
Concrete Compressive Strength[11]	EN 12390-3
Concrete Thermal Properties[12]	ASTM C177-13

Results

Waste Characterisation

The results yielded a valuable insight into the variability of the chemistry of wastes. The bulk chemistries of the wastes are shown in Table 3. If attention is focused upon the primary constituent oxides (aluminium, calcium, potassium, sulphur, silicon), the variability within each group can be appreciated. This reflects variations in the raw materials used, operating conditions, plant age, and waste treatment methods.

Figure 1 illustrates the CO_2 capacity of the wastes. The total CO_2 values include the carbon present in the as-received material and that chemically bound as a result of carbonation. The net CO_2 values include only that chemically bound when the wastes are subjected to carbonation. The disparity between the total and net values should be noted. Thus, a waste that might be initially expected to be suitable for carbonation, based upon a preliminary assessment of its chemical properties, may in fact be non-reactive. As reflected in the oxide compositions, the CO_2 capacities both within and between the waste groups were also highly variable (see Figure 2).

The relationship between oxide composition and CO_2 capacity is being evaluated statistically to improve the prediction of CO_2 reactivity of the wastes. Early results indicate that an equation model developed from the dataset is more accurate previous models.[13]

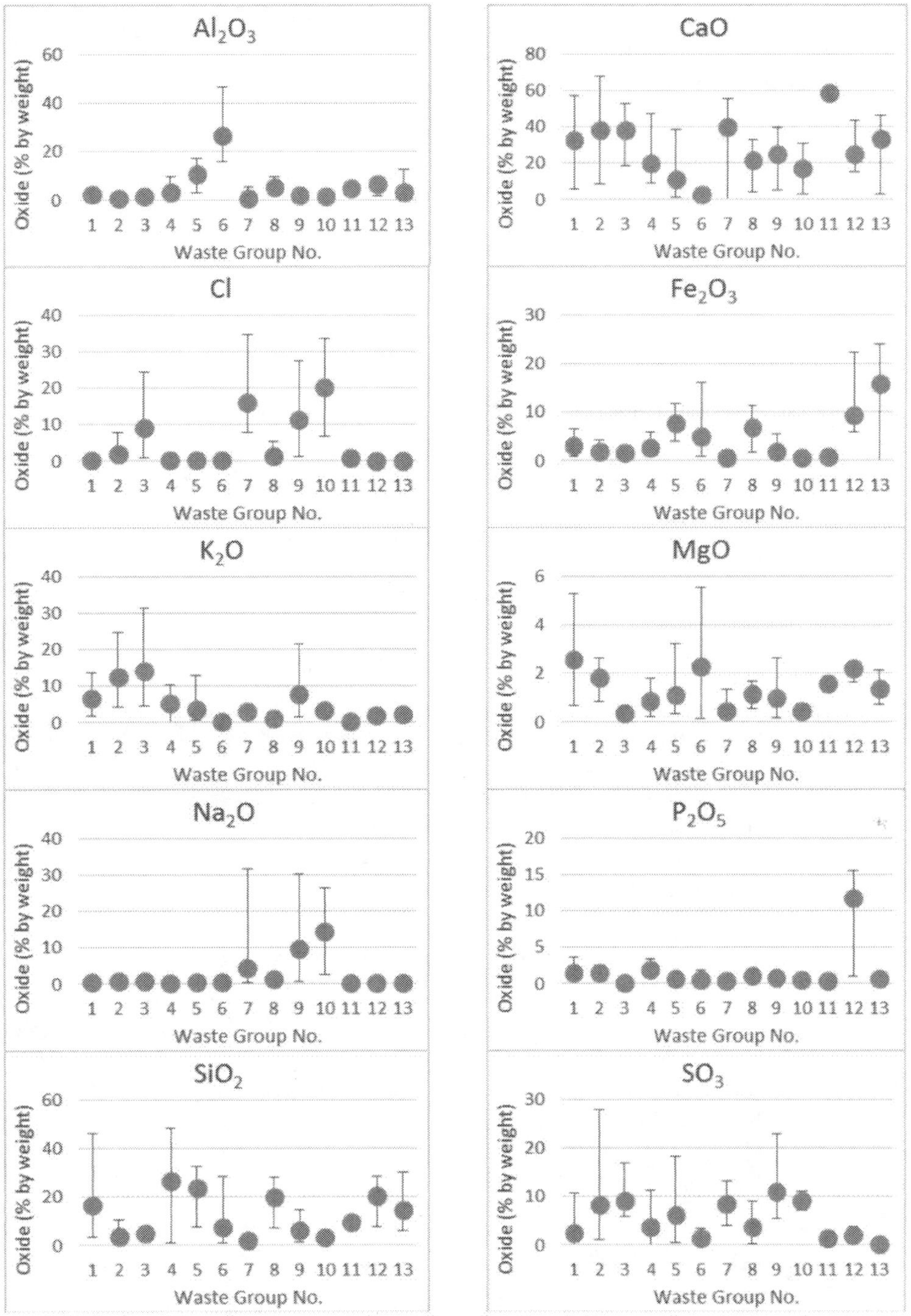

Figure 1: Bulk chemistry of waste groups (error bars denote minimum and maximum values in the group)

The next stage of analysis will employ more elaborate statistical analysis to explore further issues such as the fit for individual waste groups and possible interferences between variables (i.e. oxides).

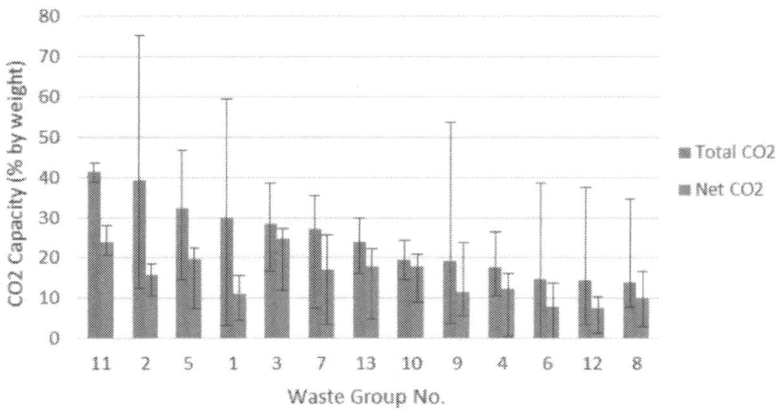

Figure 2: Carbon dioxide capacity of wastes (whiskers illustrate minimum and maximum values)

Aggregate Product Development

Aggregates produced from the wastes were tested against the product specifications developed by Carbon8. Aggregates derived from municipal solid waste incineration air pollution control residues (MSWI-APCr), paper sludge incineration ash (PSIA), and biomass bottom/fly ash (BBA/BFA), were selected for rigorous testing. Alongside the three prototypes, a control primary aggregate (flint gravel) and secondary aggregate (expanded clay) were also tested (see Table 3).

Table 3. Aggregate/concrete testing results

	Test	MSWI-APCr	PSIA	BBA/BFA	1° Aggregate	2° Aggregate
Aggregate	Apparent Density	2.6	2.5	2.5	2.7	2.0
	Bulk Density (kg/m^3)	1113	855	1050	1489	700
	Water Absorption (%)	16.1	24.1	17.8	<0.5	19.9
	Crushing Resistance (N/mm^2)	8.3	5.2	6.4	20.3	6.6
	Thermal Conductivity (W/m.K)	0.2	0.15	•	0.15	0.24
Concrete	Bulk Density (kg/m^3)	1956	1702	1898	2121	1558
	Flexural Strength (N/mm^2)	2.7	2.9	2.8	2.9	3.7
	Compressive Strength (N/mm^2)	12.5	14.3	12.2	28.4	11.1
	Thermal Conductivity (W/m.K)	0.7	0.6	•	0.9	0.6
	Specific Heat (J/kg.K)	755	828	•	815	739
	Thermal Effusivity (J/m^2.K.s1/^{2})	1031	943	•	1285	792

• *test results pending*

The prototypes exhibited bulk densities akin to the secondary aggregate, and were within BS EN 13055-1 requirements for lightweight aggregate. Crushing resistance was comparable to the secondary aggregate, but considerably less than that of the primary aggregate control. The density of the concretes produced from the prototypes fell between that of the secondary aggregate (1558 kg/m^3) and that of the primary aggregate (2121 kg/m^3). Compressive strength of the prototype aggregate concretes was commensurate to the secondary aggregate, but did not achieve the same flexural strength. The thermal properties of the concrete made from the prototypes was comparable to the secondary aggregate, thus was superior comparable to a standard concrete. These results indicate that the prototype aggregates have potential applications in the lightweight/medium-weight aggregate market.

A full end-use evaluation has been made for the MSWI-APCr aggregate, by validating its use in concrete construction block manufacture (see Table 4). The results show comparable properties between the standard product and MSWI-APCr aggregate substituted one in terms of strength (compressive and transverse failure), density, and geometry (flatness and parallelism). The MSWI-APCr aggregate has been granted end-of-waste status in the UK, and is now in full commercial production (currently 60,000 tonnes per year).

Table 4. Physical properties of MSWI-APCr aggregate and control concrete blocks

Property	Standard Aggregate	With APCr Aggregate*
Density (kg/m^3)	1660	1730
Compressive Strength (MPa)	11.8	11.2
Transverse Failure Load (kN)	7.75	7.78
Flatness (mm)	0.0	0.0
Parallelism (mm)	1.2	1.0
Drying Shrinkage (mm)	0.38	0.34
Moisture Expansion (mm)	0.13	0.11

**partial substitution*

Following a proof-of-concept trial planned to manufacture a bulk sample the BBA/BFA aggregate, an identical end-use evaluation for its use in concrete construction blocks will be undertaken. It is expected that the results of this work will be a major step forward in achieving end-of-waste status for this material.

Acknowledgements

The authors gratefully acknowledge the financial support of European Regional Development Fund through INTERREG IVA Manche (Channel) programme.

The invaluable work of the SAPICO2 academic partners at the Université de Picardie Jules Verne is acknowledged, namely: Professor Thierry Langlet, Dr Geoffrey Promis, Dr Haroun Sebei and Dr Alexandra Bourdot.

The patient support of Dr Paula Carey of Carbon8 during the course of this project has been much appreciated.

References

1. Project SAPICO2. https://www.u-picardie.fr/lti/theme-mee/mee_projets/sapico. Accessed 20.04.2015.
2. Interreg IV A France (Channel) - England programme. http://www.interreg4a-manche.eu/. Accessed 20.04.2015.
3. The CO_2 Forum. http://co2forum.cpe.fr/. Accessed 20.04.2015.
4. Waste Framework Directive http://ec.europa.eu/environment/waste/framework/list.htm. Accessed 20.04.2015.
5. Gunning, P.J., Hills, C.D., Carey, P.J. 2009. Production of lightweight aggregate from industrial waste and carbon dioxide. *Waste Management,* 29, 2722–2728.
6. European Committee for Standardisation. EN 1097-6. Tests for the mechanical and physical properties of aggregates - Part 6: Determination of particle density and water absorption. http://standards.cen.eu/index.html
7. European Committee for Standardisation. EN 1097-3. Tests for the mechanical and physical properties of aggregates - Part 3: Loose bulk density and voids. http://standards.cen.eu/index.html
8. European Committee for Standardisation. EN 13055-1. Lightweight aggregates - Part 1: Lightweight aggregates for concrete, mortar and grout. http://standards.cen.eu/index.html
9. European Committee for Standardisation. EN 12390-2. Testingn hardened concrete – Part 2: Making and curing test speciments. http://standards.cen.eu/index.html
10. European Committee for Standardisation. EN 12390-5. Testing hardened concrete – Part 5: Flexural strength of test speciments. http://standards.cen.eu/index.html
11. European Committee for Standardisation. EN 12390-3. Testing hardened concrete – Part 3: Compressive strength of test speciments. http://standards.cen.eu/index.html
12. American Society for Testing and Materials. ASTM C177-13. Standard test method for steady-state heat flux measurements and thermal transmission properties by means of the guarded-hot-plate apparatus. http://www.astm.org/
13. Steinour, H.H. (1959). Some effects of carbon dioxide on mortars and concrete – discussion. *Journal of the American Concrete Institute*, 55, 905–907.

Leaching of Residual Steel Slag from a Mineral Carbonation Process

S. Teir[1*], P. Kronqvist[1], T. Kaartinen[1] and T. Parviainen[2]

[1] VTT Technical Research Centre of Finland Ltd, Espoo, Finland
[2] Outokumpu Stainless Oy, Tornio, Finland
*Corresponding author: sebastian.teir@vtt.fi

Abstract
In an alternative concept for producing precipitated calcium carbonate (PCC), calcium is extracted from steelmaking slag using ammonium salt solutions, from which PCC can subsequently be precipitated using CO_2. A by-product of the process is a calcium-depleted slag residue. The objective of this study was to compare the composition and solubility of residual slag generated by the process to that of raw, untreated slag. Three different stainless steel slag products were tested. Samples representing residual slag were prepared by leaching each slag in 4 M ammonium acetate solutions. The solubility in water of each raw slag and residual slag was tested using two-stage and single-stage batch leaching tests. The leaching tests with residual slag showed a high concentration of calcium and dissolved organic carbon (DOC), which was most likely a remnant of the spent ammonium acetate solvent used in preparation of the residual slag. The processing does not appear to significantly affect the solubility of other elements. The results implies that the PCC process would have to filter and wash the residual slag to remove the remnants of organic carbon, which adds some costs to the process and increases the use of water.

Introduction

In an alternative concept for producing precipitated calcium carbonate (PCC), called Slag2PCC, calcium is extracted from steelmaking slag using ammonium salt solutions, from which PCC can subsequently be precipitated using CO_2 (Figure 1)[1]. As the PCC precipitates the solvent is regenerated, and can be returned to the extraction reactor for reuse in the process. While PCC is a desired product of use in polymers and paper, a less desired product of the process is the slag residue, or "residual slag", from which roughly half the original calcium content has been leached. In addition, the residual slag contains remnants of the ammonium salt solvent. The usability of the residual slag from the process can have a large effect on the economic feasibility of such a PCC process. As roughly half of the calcium contained in the raw slag is selectively removed in the process, the concentration of trace elements in the slag will be higher. Also, treatment of the slag with the ammonium salt solution might make certain elements of the slag more prone to leaching in water. This could be an interesting opportunity for recovering valuable elements, but it would also raise the cost of the PCC process due to possible requirements for further treatment of the residual slag. Therefore, a study was carried out to compare the composition of residual slag generated by such a process to that of the raw, untreated slag and their solubility in water.

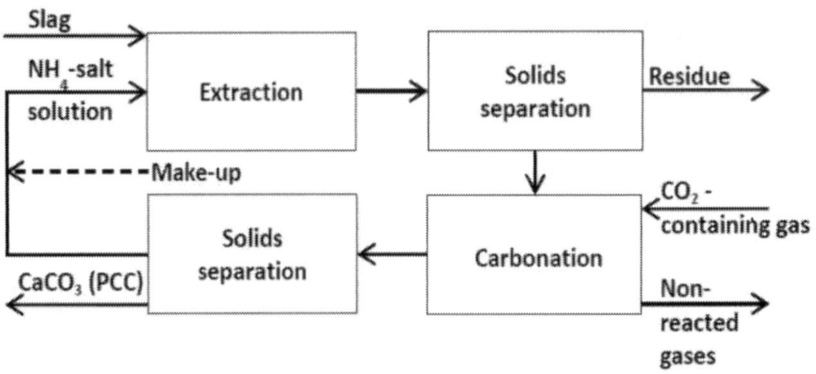

Figure 1. The Slag2PCC concept[2].

Methods

Three different stainless steel slags from the Outokumpu Tornio Works were used: ground, water-cooled Argon Oxygen Decarburization (AOD) slag, filler slag (the fine portion of AOD and ladle slag), and air-cooled AOD slag. In order to prepare solids representing residual slags from the extraction stage of the PCC process (Figure 1), each slag type was leached separately in 4 M ammonium acetate solutions for 120 minutes, after which the solids were separated by filtration (1 μm filter). The procedure was performed three times for each slag, producing three similar residual slag samples of each slag type. The slags were characterised by X-ray diffraction (XRD) and X-ray fluorescence (XRF; Table 1). Mineral phases identified by XRD were bredigite (12-18%), larnite (11-17%), periclase (3-4%) and fluorite (1-2%), but the majority was amorphous (60-70%). The concentration of calcium in solution after 120 min was measured using inductively coupled plasma atomic emission spectroscopy (ICP-AES), verifying that a significant portion of calcium was extracted from all samples: 60-63% of Ca in air-cooled AOD slag, 50-52% of Ca in filler slag and 34-39% of Ca in water-cooled AOD slag. In the residual slag samples the portion of amorphous material was higher (75-82%) and the share of crystalline phases lower.

Table 1. XRF analysis of slags used in the leaching tests (WC = water-cooled, AC = air-cooled).

	Fresh slag samples			Residual slag[a]		
Items	WC-AOD	Filler	AC-AOD	WC-AOD	Filler	AC-AOD
Ni	<0.1	<0.1	<0.1	<0.1	0.2	<0.1
MgO	7.2	7.4	7.8	7.4	8.7	8.9
SiO_2	30.2	26.5	28.6	32.4	31.0	31.4
Al_2O_3	1.5	2.6	1.6	1.7	3.0	1.8
TiO_2	0.4	0.6	0.2	0.4	0.7	0.3
MnO	0.3	0.4	<0.1	0.3	0.5	<0.1
CaO	58.2	57.4	59.6	42.9	35.4	29.5
Cr_2O_3	0.5	1.1	0.4	0.6	1.3	0.4
Fe_2O_3	0.1	0.6	<0.1	0.2	0.8	<0.1

[a]The average composition of three samples are presented for each residual slag.

The solubility in water of water-cooled AOD slag, filler slag, and their corresponding residual slags was tested using a standard, accredited, two-stage batch leaching test (CEN EN 12457-3)[3] with a cumulative liquid-solid ratio of 10. The concentrations of dissolved elements were analysed using ICP-MS (Mass Spectrometry) for As, Ba, Cd, Co, Cr, Cu, Mo, Ni, Pb, Sb, Se, V, and Zn, ICP-AES for Ca, CVAFS (Cold Vapour Atomic Fluorescence Spectroscopy) for Hg, ion selective electrode for F^-, ion chromatography for Cl^- and SO_4^{2-}, and infra-red analysis for organic carbon. The solubility of dry-cooled AOD slag and the corresponding residual slags was tested separately using a single-stage batch leaching test with a liquid-solid ratio of 10. For these solutions, concentrations of dissolved elements were measured only for a few selected elements.

Results

The results from the leaching tests of steelmaking slags and residual slags are shown in Table 2. All the residual slag leachates had a lower pH content than those of the raw slags. As the residual slags were not dried before the leaching tests, only filtered, the

Table 2. Dissolved elements (unit: mg/kg dry solids; WC = water-cooled; AC = air-cooled).

	Fresh slag samples			Residual slag[a]		
Items	WC-AOD	Filler	AC-AOD	WC-AOD	Filler	AC-AOD
As	<0.01	<0.01		<0.01	<0.01	
Ba	0.2	14		4.5	7.9	
Cd	0.02	<0.01		<0.01	<0.01	
Co	0.1	0.01		0.02	0.04	
Cr	0.05	1.3	<0.1	0.13	1.6	0.12
Cr^{6+}			<0.1			0.16
Cu	<0.01	0.01		0.01	0.02	
Hg	0.0009	0.0008		0.0004	0.0003	
Mo	11	1.3	<0.1	2.47	1.3	0.91
Ni	0.01	0.02		0.33	0.55	
Pb	<0.01	<0.01		<0.01	<0.01	
Sb	<0.01	<0.01		<0.01	<0.01	
Se	0.05	0.04		0.13	0.04	
V	0.02	<0.01		0.33	0.44	
Zn	0.09	0.12		0.18	0.2	
Ca	3000	8500		26000	29000	
Cl^-	<20	12		<200	<200	
F^-	63	110	8.9	114	99	39.9
SO_4^{2-}	12	<2.7		130	180	
DOC	9.8	15		73000	73000	
pH	12.0[b]	12.6[b]	11.8	9.9[b]	9.7[b]	
Conduct. (mS/m)	212[b]	805[b]	222	480[b]	684[b]	1519
Weight, wet (g)	175	176		282	287	
Weight, dry (g)	175	175		175	175	

[a]The average solubility from three tests are presented for each residual slag.
[b]Measured from the stage containing a liquid-solid ratio of 8.

residual slag samples contained a lot of solvent: 38-39 wt-%. Due to the use of ammonium acetate solvent the residual slag leachates contained a higher amount of calcium and dissolved organic carbon (DOC). The AOD slag leachates contained some molybdenum and fluoride, while the residual slag leachates had lower concentrations of molybdenum but higher concentrations of fluoride. No significant differences in the leaching behaviour of other elements were found.

Discussions and conclusions

Since the residual slags contained a significant amount of spent ammonium acetate solvent, it is most likely the source of calcium and dissolved organic carbon (i.e. calcium acetate formed when extracting calcium from raw slag using ammonium acetate). It is also likely that most of the trace elements in the residual slag leaching tests were extracted already with the ammonium acetate solution during the preparation of the residual slag and were transported to the leaching tests readily dissolved in the wet residual slag. However, the similar concentrations of other elements when comparing raw slags and residual slags indicate that ammonium acetate should not make the other elements of the slag more prone to leaching in water. Additionally, the results show that among the elements analysed calcium is very selectively extracted using ammonium acetate. Also, the experiments helped in identifying and demonstrating two requirements for a large-scale Slag2PCC process: 1) residual slag would need to be filtered in order to recover as much calcium and ammonium salt as possible, after which 2) the residual slag needs to be washed to remove remnants of the ammonium salt for further treatment. This adds costs to the process and increases the use of water in the Slag2PCC process.

Acknowledgements

The work presented here was carried out in 2011 with funding and support from Stora Enso Oyj and Outokumpu Stainless Oy. Jouni Ylipekkala is acknowledged for providing slags as well as XRD & XRF analyses of the raw slags and residual slags. Jouni also supervised the leaching tests of dry-cooled AOD slag and its residual slag.

References

1. Teir, S., Eloneva, S., Revitzer, H., Zevenhoven, R., Salminen, J., Fogelholm, C.-J., Pöyliö, E. 2009. Method of producing calcium carbonate from waste and by-products. International application published under the patent cooperation treaty (PCT), WO 2009/144382, PCT/FI2009/050455.
2. Teir, S., Kotiranta, T., Parviainen, T., Mattila, H.-P. 2015. Case study for utilisation of CO_2 from flue gases in PCC production from steelmaking slag. *In: Proceedings of the 5th International Conference on Accelerated Carbonation for Environmental and Material Engineering (ACEME)*, New York, U.S.A, June 21-24, 2015.
3. European Committee for Standardization (CEN). 2002. EN 12457-3 CEN, Characterization of Waste – Leaching – Compliance Test for Leaching of Granular Materials and Sludge.

Innovation Concepts

CO2 Energy Reactor

Pol Knops
Innovation Concepts B.V.
knops@innovationconcepts.eu
Tel: +31 6 5130 4842

6/29/15 2015 ACEME / Innovation Concepts 1

Mineralization challenges

- **Process**
  - 180°C, > 40 bar, 1 hour residence time
  + Gravity Pressure Vessel

- **Economic**
  - CO2 price
  + Reaction products

6/29/15 2015 ACEME / Innovation Concepts 2

Process challenges

180°C, 40 bar, 1 hour

Olivine:

$Mg_2SiO_4 + 2CO_2$ ➔ $\mathbf{2MgCO_3 + SiO_2}$

[ΔH = **-239.2 kJ/mol**]

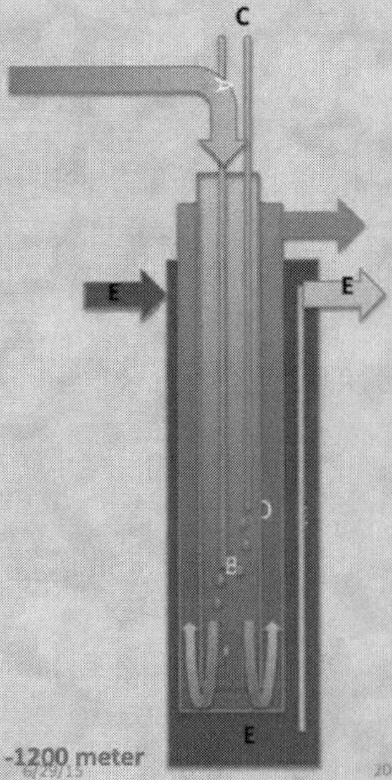

Reactor

A: Incoming material
B: DownComer tube
C: CO2 injection
D: UpComer tube
E: Surrounding cooling tube

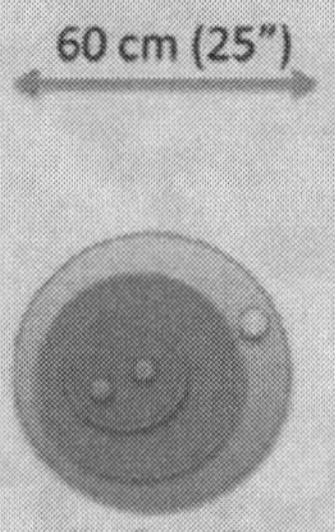

Gravity Pressure Vessel

- Pressure — Depth — 100 bar
- Residence time — Size — 1 hour
- Temperature — Exothermic — 180°C
- Heat exchanger
- Plugflow

Well head

Test autoclave

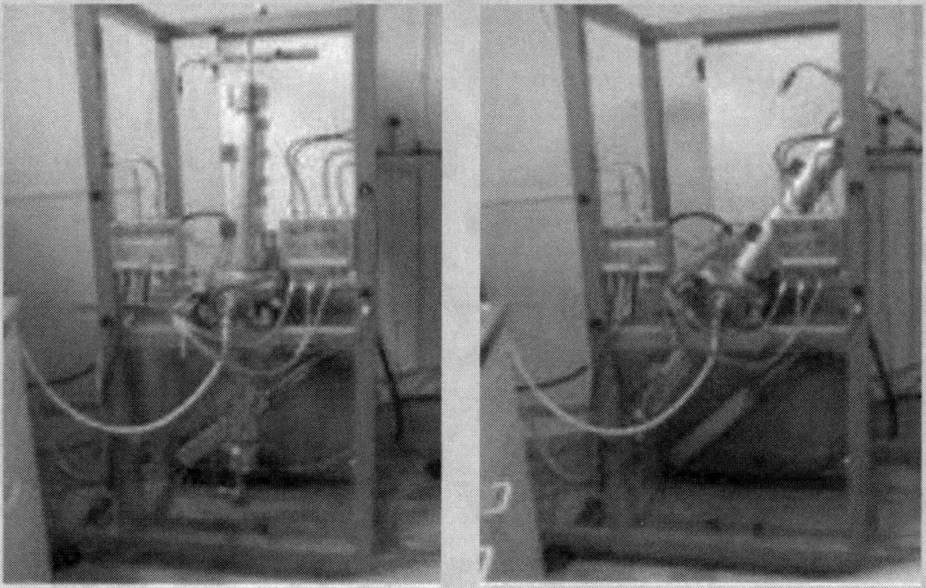

Economic challenges

180°C, 40 bar, 1 hour

Olivine:

$Mg_2SiO_4 + 2CO_2 \rightarrow 2MgCO_3 + SiO_2$

[ΔH = **-239.2 kJ/mol**]

Reaction products

9

Reaction products

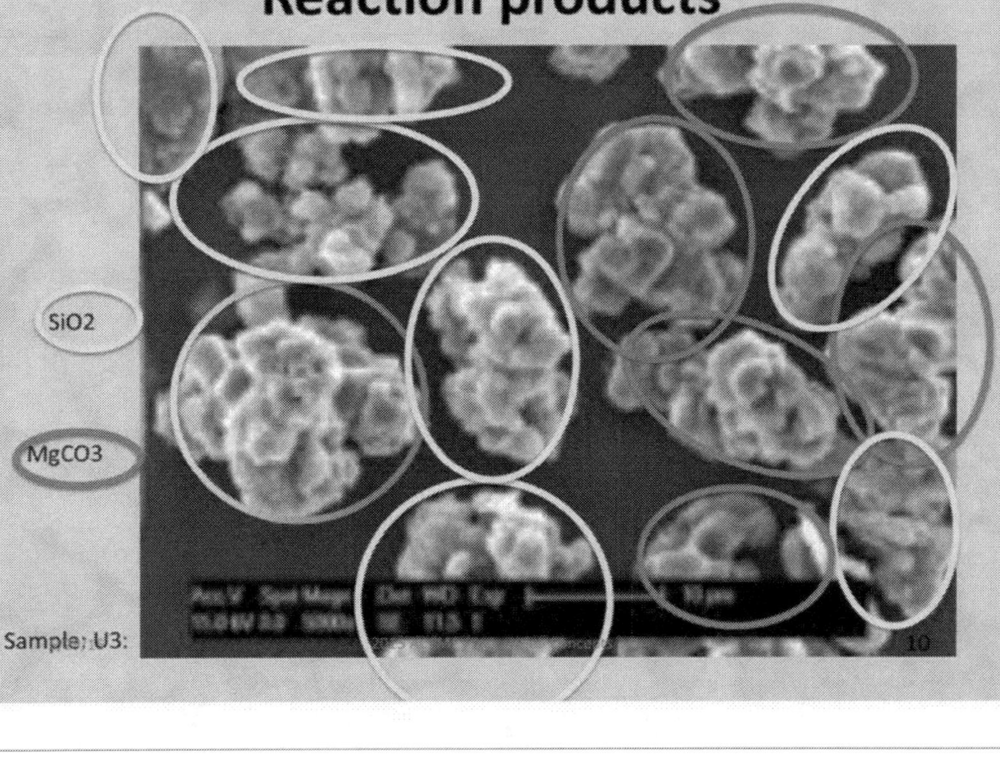

10

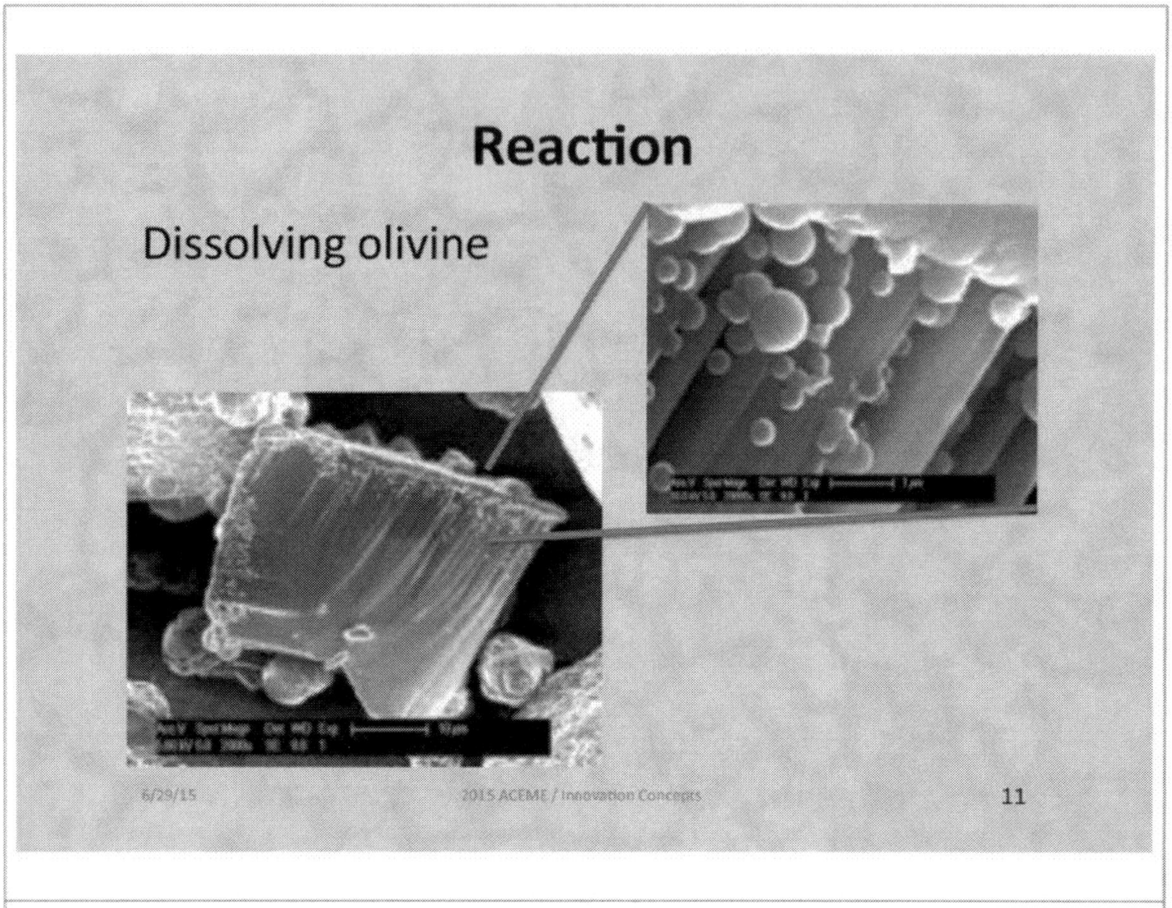
Reaction
Dissolving olivine
6/29/15
2015 ACEME / Innovation Concepts
11

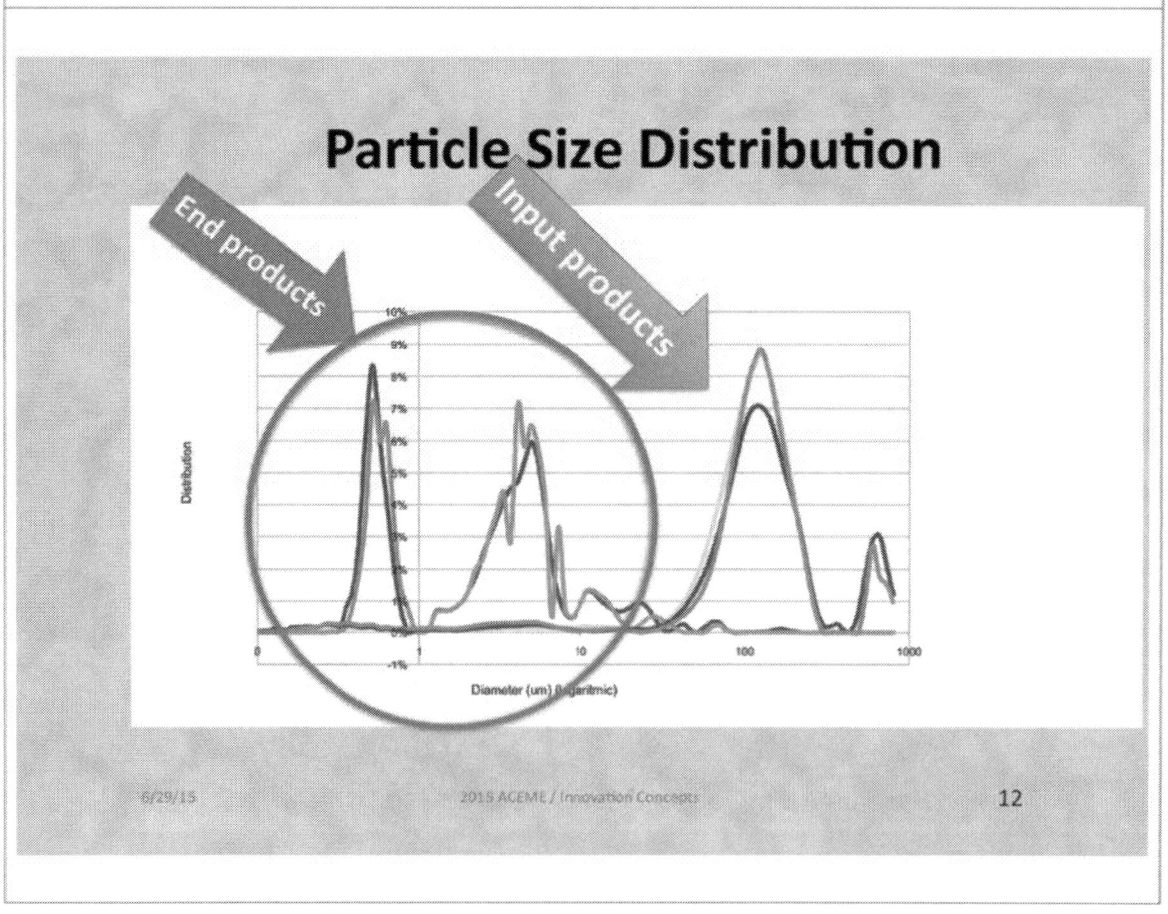
Particle Size Distribution
End products
Input products
Distribution
Diameter (um) (logaritmic)
6/29/15
2015 ACEME / Innovation Concepts
12

Comparing Rocking Autoclave / CSTR

CSTR

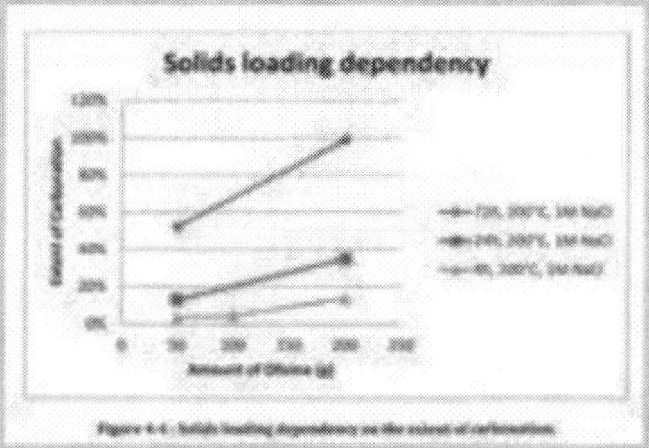

Rocking Autoclave

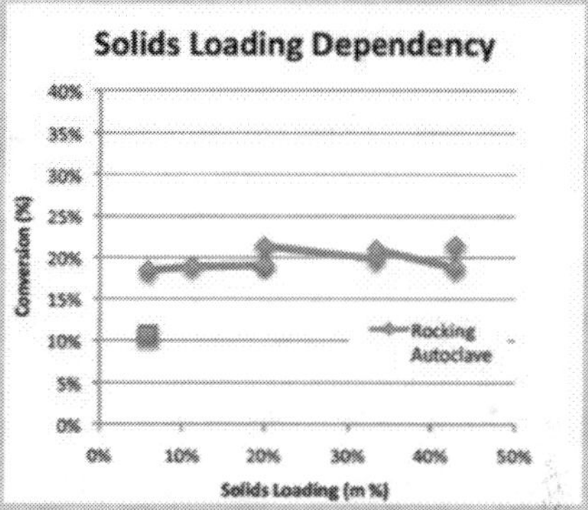

Comparing Rocking Autoclave / CSTR

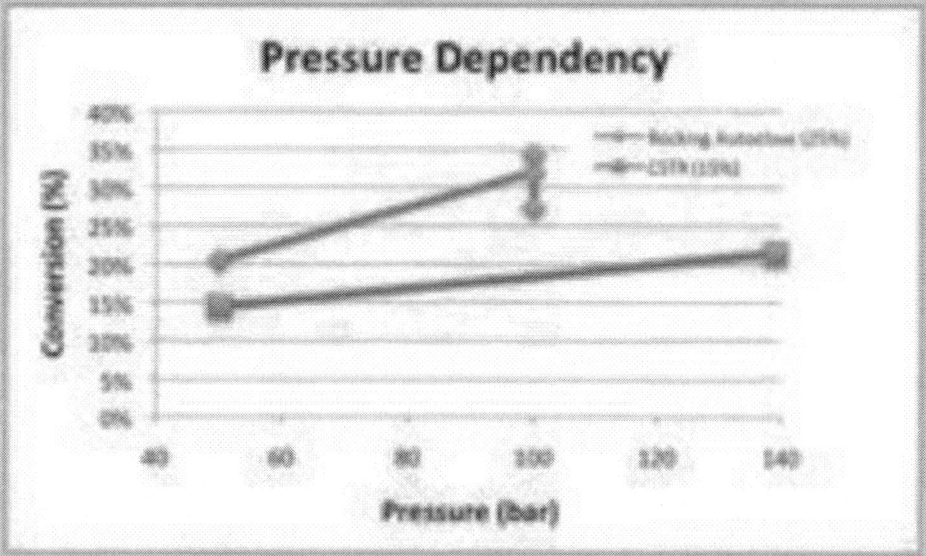

Applications

- Concrete — **Silica** (pozzolanic activity) ?
  Fine dispersed particles
- Coatings / paints — "functional filler"
  (particle size, color)
- Paper — Silica + valued filler
  (particle size, color)
- Heat — Integration Carbon Capture
- **CO2 captured**

6/29/15 2015 ACEME / Innovation Concepts 15

Process challenges

180°C, 40 bar, 1 hour

Olivine:

$$Mg_2SiO_4 + 2CO_2 \rightarrow 2MgCO_3 + SiO_2$$

[ΔH = -239.2 kJ/mol]

6/29/15 2015 ACEME / Innovation Concepts 16

Acknowledgment

- **Acknowledgments:**
  - **ISPT, A Major Steel company, Sibelco**

Questions ?

Pol Knops
Innovation Concepts B.V.
knops@innovationconcepts.eu
Tel: +31 6 5130 4842

Advances on the carbon dioxide (CO_2) reactivity on lithium and sodium ceramics and its catalytic conversion to added value products

H. Pfeiffer*

Instituto de Investigaciones en Materiales, Universidad Nacional Autónoma de México, Circuito exterior s/n, Cd. Universitaria, Del. Coyoacán, México DF, CP 04510, México.
*Corresponding author: pfeiffer@iim.unam.mx

Abstract

Lithium and sodium ceramics are among several materials that have potential applications in CO_2 removal. These ceramics are able to chemisorb CO_2 in a wide temperature range, presenting several interesting properties. On the other hand, the CO_2 catalytic conversion to added value products is a second issue that must be further analyzed. Then, the aim of this work is to analyze different experiments related to the CO_2 chemisorption on lithium and sodium ceramics to propose how different factors control this process and analyse different evidences about the CO2 high temperature conversion to syngas.

The anthropogenic amounts of carbon dioxide (CO_2) in the atmosphere have been raised dramatically, mainly due to the combustion of different carbonaceous materials used in energy production, transport and other important industries such as cement production, iron and steel making. To solve or mitigate, at least, this environmental problem, several alternatives have been proposed. A promising alternative for reducing the CO_2 emissions is to capture and concentrate the gas and its subsequent chemical transformation.

Different materials have been proposed as potential CO_2 captors. Among these materials, different lithium and sodium ceramics have been proposed as possible high temperature CO_2 captors. Some of the alkaline ceramics most studied are the lithium and sodium silicates (Li_4SiO_4, Li_8SiO_6, Li_2SiO_3 and Na_2SiO_3), lithium aluminate (Li_5AlO_4) lithium and sodium zirconates (Li_2ZrO_3 and Na_2ZrO_3) and lithium cuprate (Li_2CuO_2), among others. All these ceramics have been tested, producing interesting results at different temperature ranges between 30 and 700 °C.

In this ceramics CO_2 capture is produced throughout a chemisorption process. Initially, the alkaline ceramic particles react with CO_2 at the surface. The superficial reaction implies the formation of an external shell of Li_2CO_3 or Na_2CO_3 and secondary phases. Once the external shell is produced, the reaction mechanism is controlled by different diffusion processes. Therefore, the correct understanding of all these phenomena is a very important step during the selection of the ideal CO_2 capture conditions. In other words, the selection of materials, temperature range, gas and/or vapour concentrations, flows, etc. Then, the aim of this work is to present different experiments related to the CO_2 reactivity, chemisorption, on lithium and sodium ceramics and to propose how different factors control this process. The work will be focused on diffusion controlled CO_2 chemisorption, which has been shown to be the limiting step of the CO_2 chemisorption process. Diffusion controlled CO_2 chemisorption appears to be mainly influenced by the chemical composition of a product's external shell.

About the CO_2 chemisorption process, it has been observed that the crystalline structure of a material modifies the CO_2 capture properties of that material. An example is the CO_2 chemisorptions produced on the monoclinic and tetragonal Li_2ZrO_3 phases, as the tetragonal phase possesses a faster kinetic CO_2 chemisorption than the monoclinic one. Additionally, when lithium or sodium ceramics are partially doped with K, Li or Na, the CO_2 chemisorption process is significantly improved. This has been explained as the result of CO_2 diffusion through molten eutectic carbonate phases, which is a liquid at lower temperatures than the un-doped material. As this eutectic phase is liquid at temperatures where CO_2 chemisorption occurs, diffusion processes become more efficient. The external shell produced on the surface of lithium ceramic particles plays an important role in the CO_2 chemisorption process. However, if lithium or sodium carbonate control the second part of the CO_2 chemisorption process, all the lithium or sodium ceramics should behave identically once the external carbonate shell is produced. This similarity in chemisorption is not observed. Different papers in the literature have presented kinetic analyses showing that diffusion controlled CO_2 chemisorption is the limiting step of the reaction process.

Additionally, these ceramics have been studied as CO_2 captors at low temperatures (30–80 °C), in the presence of water steam producing lithium carbonate (Li_2CO_3) or sodium acid carbonate ($NaHCO_3$) in the sodium ceramics instead of sodium carbonate (Na_2CO_3). It means that sodium ceramics can trap 1 or 2 mol of CO_2 per mole of ceramic, depending on the dry or humid conditions at low temperatures. In general, it seems that water stem act as catalytic active specie which must modified the activation energy of the whole process, allowing the carbonation process in this temperature range.

Conversely, different catalytic reactions, where carbon dioxide (CO_2) and/or methane (CH_4) are used, have attracted the research attention because of environmental applications. Some of these reactions are the methane reforming (called dry reforming), the water splitting and the CO_2 reduction. All these reactions produce syngas ($CO + H_2$). The syngas produced by these reactions can be used in different energetic and organic applications. Additionally, these reactions are also environmentally important, as CO_2 and CH_4 (perhaps the two most important greenhouse gases) are catalytically converted into added value products, such as hydrocarbons and oxygenated compounds.

Therefore, this presentation will show some new advances in the CO_2 capture on alkaline ceramics and how the CO_2 chemically trapped can be catalyzed to added value products through the syngas production at different temperature ranges.

Acknowledgements

This work was financially supported by the projects SENER-CONACYT (150358) and PAPIIT-UNAM (IN-102313).

References

1. Ortiz-Landeros, J.; Ávalos-Rendón, T., Gómez-Yáñez, C., Pfeiffer, H. 2012. Analysis and perspectives concerning CO_2 chemisorption on lithium ceramics using thermal analysis. *J. Therm. Anal. Calorim.*, 108, 647.
2. Isahak, W. N. R. W.; Ramli, Z. A. C.; Hisham, M. W. M.; Yarmo, M. A. 2015. The formation of a series of carbonates from carbon dioxide: Capturing and utilization. *Renew. Sust. Ener. Rev.*, 47, 93.
3.
4. Choi, S., Drese, J. H., Jones, C. W. 2009. Adsorbent materials for carbon dioxide capture from large anthropogenic point sources. *ChemSusChem*, 2, 796.
5. Pfeiffer, H. 2010. Advances in CO_2 Conversion and Utilization, Chapter 15, ACS ed., 1056, 233.
6. Wang, J.; Huang, L.; Yang, R.; Zhang, Z.; Wu, J.; Gao, Y.; Wang, Q.; O'Hare, D.; Zhong, Z. 2014. Recent advances in solid sorbents for CO_2 capture and new development trends, *Ener. Environ. Sci.*, 7, 3478.
7. Mosqueda-Altamirano, H. A., Vázquez, C. Bosch, P. Pfeiffer, H. 2006. Chemical sorption of carbón dioxide (CO_2) on lithium oxide (Li_2O). *Chem. Mater.*, 18, 2307.
8. Nakagawa, K., Ohashi, T. 1998. A novel method of CO_2 capture from high temperature gases. *J. Electrochem. Soc.,* 145, 1344.
9. Nair, B. N., Burwood, R. P., Goh, V. J., Nakagawa, K., Yamaguchi, T. 2009. Lithium based ceramic materials and membranes for high temperature CO_2 separation. *Prog. Mater. Sci.*, 54, 511.
10. Vernon, P. D. F., Green, M. L. H., Cheetham, A. K., Ashcrof, A. T., 1992. Partial oxidation of methane to synthesis gas, and carbon dioxide as an oxidising agent for methane conversion. *Catal. Today,* 13, 417–426.
11. Sun, L., Tan, Y., Zhang, Q., Xie, H., Han, Y., 2011. Combined air partial oxidation and CO_2 reforming of coal bed methane to synthesis gas over co-precipitated Ni-Mg-ZrO_2 catalyst. *Inter. J. Hydrogen Ener.* 36, 12259–12267.

12. Subramani, V., Gangwal, S. K., 2008. A review or recent literature to search for an efficient catalyst process for the conversion of syngas to ethanol. *Ener. Fuels* 22, 814–839.

13. Alcántar-Vázquez, B., Diaz, C., Romero-Ibarra, I. C., Lima, E., Pfeiffer, H., 2013. Structural and CO_2 chemisorption analyses on $Na_2(Zr_{1-x}Al_x)O_3$ solid solutions. *J. Phys. Chem. C*, 117, 16483-16491.

14. Rodríguez-Mosqueda, R., Pfeiffer, H., 2013. High CO_2 capture in sodium metasilicate (Na_2SiO_3) at low temperatures (30-60 °C) through the CO_2-H_2O chemisorption process. *J. Phys. Chem. C*, 117, 13452-13461.

15. Alcantar-Vazquez, B.; Gomez-Garcia, J. F.; Tavizon, G.; Ibarra, I. A.; Diaz, C.; Lima, e.; Pfeiffer, H. 2014. Structural and ionic conduction analyses of the $Na_2(Zr_{1-x}Al_x)O_{3-x/2}$ solid solution, during the CO_2 chemisorption process. *J. Phys. Chem. C*, 118, 26212.

16. Romero-Ibarra, I. C., Ortiz-Landeros, J., Pfeiffer, H., 2013. Microstructural and CO_2 chemisorption analyses of Li4SiO4: Effect of surface modification by the ball milling process. *Thermochim. Acta* 567, 118-124.

17. Vera, E.; Alcantar-Vazquez, B.; Pfeiffer, H. 2015. CO_2 chemisorption and evidences of the CO oxidation-chemisorption mechanisms on sodium cobaltate. *Chem. Eng. J.*, 271, 106.

CO2 capture and storage by mineral carbonation
Humidity, Temperature and Exposure to Amorphous Silica Control the Stability of CO2 Storage in Hydrated Mg-Carbonate Minerals

Wilson, S. A., Monash University
Morgan, B., CSIRO Mineral Resources
Power, I. M., The University of British Columbia
Williams, T.,
Originally presented on: 6/23/2015 9:00:00 - 9:15:00

Increasing attention is being paid to technologies that employ hydrated Mg-carbonate minerals as traps for CO_2 pollution. However, relatively little is known about the environmental behaviour and long-term stability of these minerals under conditions relevant to CO_2storage. Thus, we have undertaken a long-term experiment to monitor and map the relative stability of four hydrated Mg-carbonate minerals over the range of atmospheric conditions experienced at Earth's surface.

Aliquots of synthetic lansfordite ($MgCO_3 \cdot 5H_2O$), nesquehonite ($MgCO_3 \cdot 3H_2O$), dypingite [$Mg_5(CO_3)_4(OH)_2 \cdot {\sim}5H_2O$] and hydromagnesite [$Mg_5(CO_3)_4(OH)_2 \cdot 4H_2O$] were maintained under set conditions of temperature (-25 °C $\leq$ T $\leq$ 75 °C) and relative humidity (2% $\leq$ RH $\leq$ 100%) for a period of approximately 20 months. Changes in mineralogy were assessed using powder X-ray diffraction data and scanning electron microscopy after approximately 2, 7 and 20 months.

Our experiments demonstrate irreversible decomposition of more hydrous Mg-carbonate minerals to less hydrous phases with increasing atmospheric relative humidity (RH) and/or temperature. Highly hydrated minerals such as lansfordite and nesquehonite decompose to relatively stable hydromagnesite at high RH. However, these phases and dypingite persist to unexpectedly high temperatures at low RH – even after 20 months. Therefore, transformation to more stable, less hydrous Mg-carbonate minerals appears to require a supply of H_2O to mediate reactions. Because Mg-carbonate minerals are hygroscopic this provides a mechanism by which atmospheric H_2O vapour may become sorbed at grain surfaces to facilitate reactions. Scanning electron microscope imaging reveals (1) pristine, unreacted crystals at low RH and (2) abundant evidence of decomposition reactions at grain surfaces in high RH experiments. These observations support the conclusion that Mg-carbonate phase transitions occur via vapour-mediated dissolution–reprecipitation reactions.

Phase transitions amongst hydrated Mg-carbonate minerals also appear to be dependant on reaction path. For instance, formation of hydromagnesite, the most stable of the hydrated Mg-carbonates, is more sluggish via decomposition of structurally and compositionally related dypingite than via decomposition of the more hydrated phases, lansfordite and nesquehonite. Thus, energy barriers to ensuring more stable storage of CO_2 within hydromagnesite appear to be lower for decomposition of lansfordite and nesquehonite to this phase.

Within four months of storage at high RH (~100%) and T $\geq$ 23 °C, all hydrated Mg-carbonate minerals begin to decompose to an X-ray amorphous phase that exhibits short range order on the nanoscale. Our scanning transmission electron microscopy and energy dispersive X-ray spectroscopy results indicate that this phase is Mg- and Si-bearing with a 15-Å layered structure. This likely indicates formation of a smectite-like phase, formed by reaction of the hydrated Mg-carbonates with the silica glass vessels in which they were stored. Because partial dissolution of Mg-carbonates produces high-pH solutions, this would lead to dissolution of the amorphous silica vessels. Indeed, the glass vessels used in several of the 100% RH, 75 °C experiments showed evidence of dissolution.

Our experimental results suggest that decomposition of hydrated Mg-carbonate minerals toward hydromagnesite, and ultimately magnesite, is likely mediated by dissolution–reprecipitation reactions. These may be driven by formation of thin films of H_2O on mineral surfaces at or below the deliquescence RH (~100%) of these hygroscopic minerals. Thus, atmospheric conditions will be a major control on decomposition of hydrated Mg-carbonate minerals to more stable phases in both industrial carbonation processes and geoengineered landscapes. Another important implication of this work is that long-term storage of hydrated Mg-carbonate minerals with amorphous silica may pose a risk for CO_2 release in environments that are subject to high humidity and precipitation. Both of these phases result from mineral carbonation reactions; thus, it may become important to separate these products prior to storage or to ensure storage under appropriate climatic conditions.

KINETICS OF MAGNESIUM EXTRACTION FROM ACTIVATED SERPENTINE BY CARBONIC ACID

ALI ZAFARANLOO

JASON P. MANN

BRIAN S. HAYNES

School of Chemical and Biomolecular Engineering

THE UNIVERSITY OF SYDNEY

ACEME
Columbia University, NY
21st – 24th June , 2015

BACKGROUND

- NETL Albany Process for Mineral Carbonation based on Serpentinite
  - Heat treatment to break down crystal structure and activate MgO
  - Mg is leached by protons associated with carbonic acid equilibria,

$$H_2O\ (aq) + CO_2\ (g) \rightleftharpoons H_2CO_3\ (aq) \rightleftharpoons H^+(aq) + HCO_3^-(aq) \rightleftharpoons 2H^+(aq) + CO_3^{2-}(aq)$$

$$MgO\ (s) + 2H^+(aq) \rightleftharpoons Mg^{2+}(aq) + H_2O$$

  - *In situ* precipitation of $MgCO_3$

$$Mg^{2+}(aq) + CO_3^{2-}(aq) \rightleftharpoons MgCO_3\ (s)$$

  - Addition of $NaHCO_3$ leads to better yield of $MgCO_3$ (s)

- What determines the overall rate?

$NaHCO_3$ / MgO / CO_2 / H_2O System

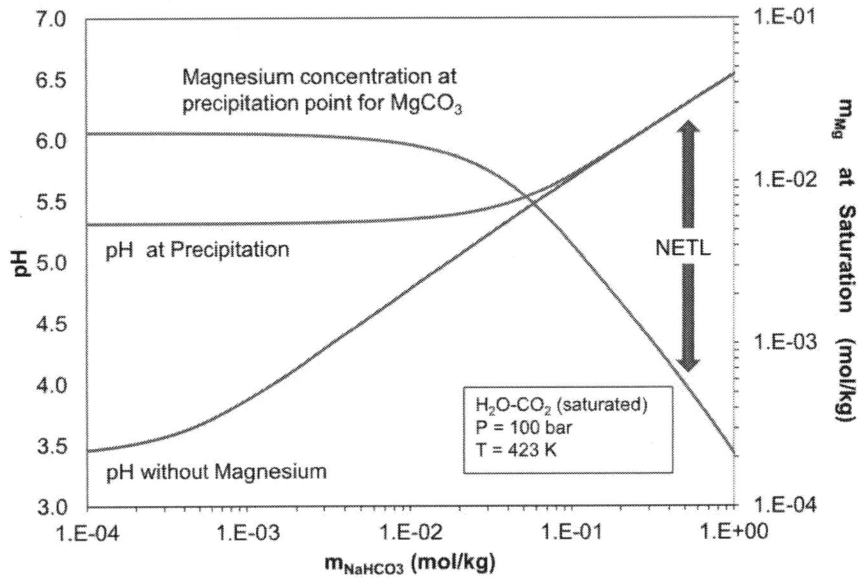

3

Mg Extraction

- Addition of $NaHCO_3$
  - suppresses $MgCO_3$ solubility BUT
  - also raises pH: acid extraction of Mg is suppressed
  - requires expensive additive (but it is recyclable)
- Extraction of Mg into solution
  - Raises pH and inhibits further acid extraction of Mg (self-inhibition)
- Two questions
  - Can (carbonic) acid extraction and carbonate precipitation be separated?
  - What are the kinetics of Mg extraction?

4

THE UNIVERSITY OF SYDNEY

Experimental

- High P-T semi-batch fluidised bed reactor,
- stainless steel 3/8 " bulkhead union
  - Length = 2.87 cm
  - Volume = 1.14 cm^3

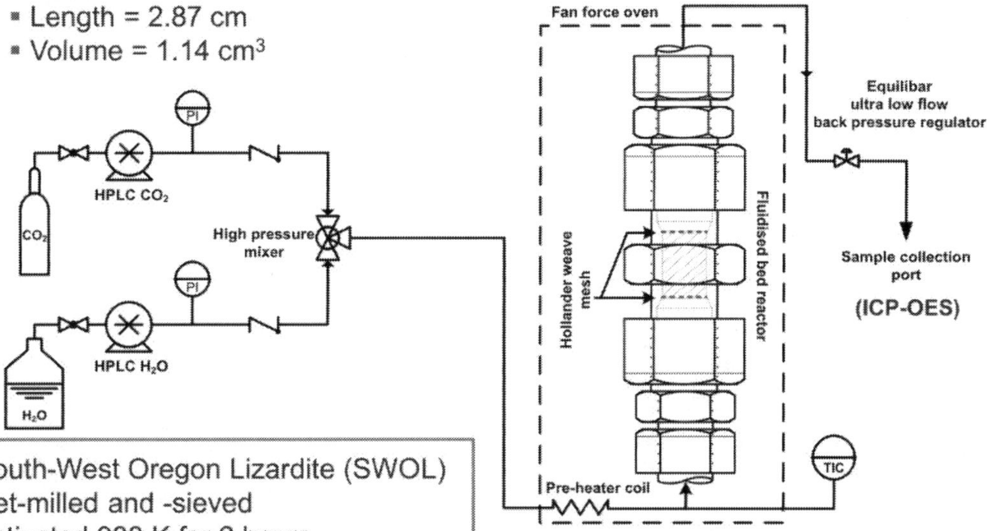

- South-West Oregon Lizardite (SWOL)
- wet-milled and -sieved
- Activated 900 K for 2 hours

3

THE UNIVERSITY OF SYDNEY

Stratification in a Packed Bed

- 250 mg activated SWOL, P_{CO2} = 100 bar, T = 423K
- Particle size = 45-63 μm
- 30 minute reaction time,

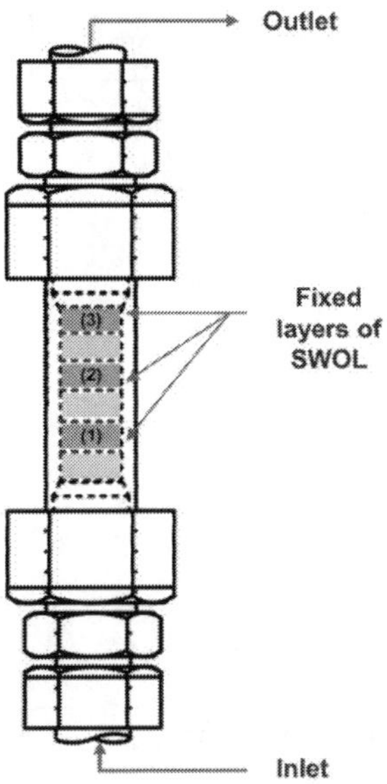

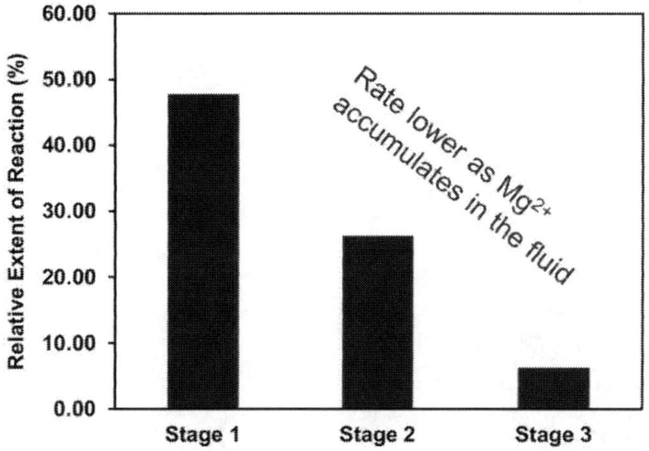

4

THE UNIVERSITY OF SYDNEY

FLUIDISED BED – TYPICAL RESULTS

- P_{CO2} = 100 bar, T = 423K
- Particle size = 45-63 μm
- Space time = 13 s

- Data analysis

$$C_{Mg} = \frac{-dm_{Mg}}{dt} = F y_{Mg}$$

- F = mass flow rate of aqueous effluent
  y_{Mg} = mass fraction Mg in the effluent
  m_{Mg} = mass Mg in the reactor

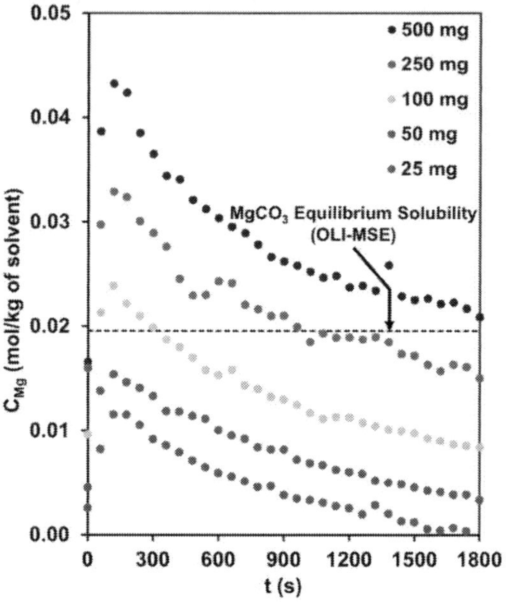

7

THE UNIVERSITY OF SYDNEY

CONVERSION AND pH PROFILES

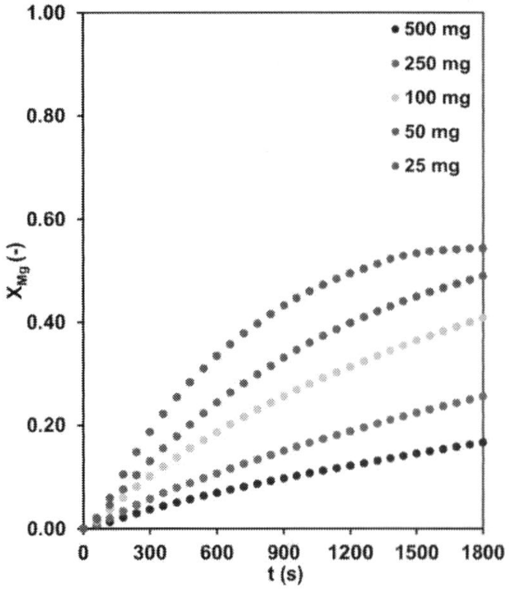

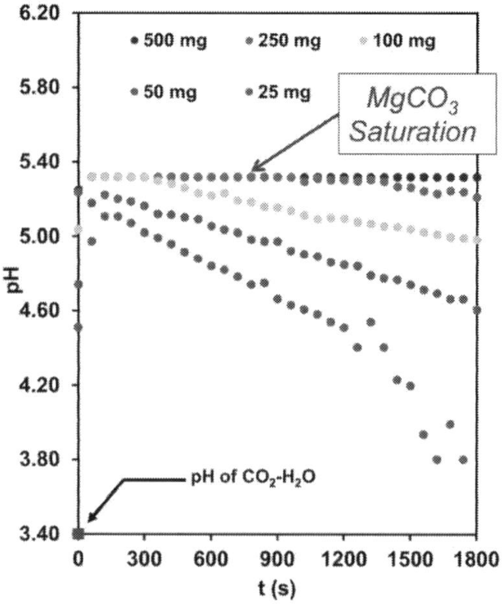

8

THE UNIVERSITY OF SYDNEY

Rate - Dependence on Conversion

$$\frac{dX_{Mg}}{dt} = \frac{d\left(1 - \frac{m_{Mg}}{m_{Mg}|_0}\right)}{dt} = \frac{F}{C_{Mg}|_0 w_0} y_{Mg}$$

- $C_{Mg}|_0$ = 24.83 wt% (initial Mg content)
- w_0 = initial sample mass

$$X_{Mg} = \int_0^t \frac{dX_{Mg}}{dt} dt$$

- After 4 hours with 25 mg

$$X_{Mg} \rightarrow 0.75$$

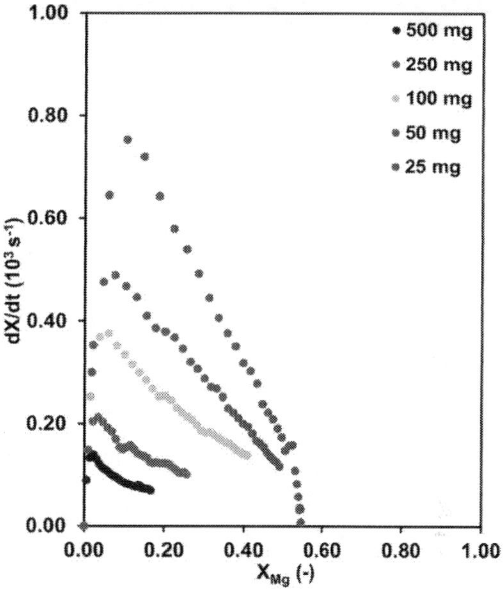

9

THE UNIVERSITY OF SYDNEY

Rate - Dependence on Conversion

- Expect

$$\frac{dX_{Mg}}{dt} = f(X) \times A_0 exp\left(-E/_{RT}\right) \times [H^+]^n$$

- Assume $n \sim 1$
- $f(X)$ ~ constant, except at high loadings

  i.e. when [Mg] ~ $[Mg]_{sat}$

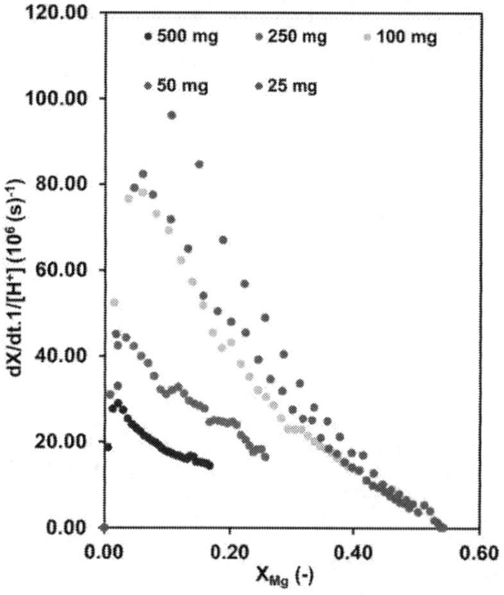

10

Rate – Effect of Temperature

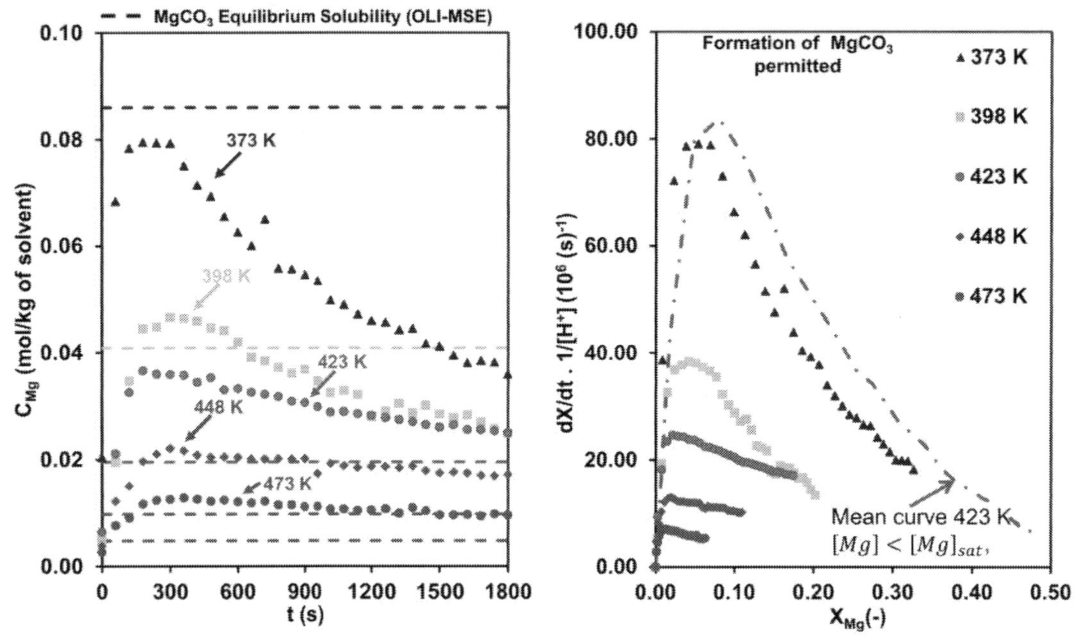

Relationship to Saturation

- Extraction rate and shape of rate profile appears to change when

$$[Mg] \rightarrow [Mg]_{sat}$$

- $[Mg] < [Mg]_{sat}$
  - ~invariant rate profile $f(X)$
  - rate and rate profile independent of temperature

- $[Mg] \sim [Mg]_{sat}$
  - Rate becomes much slower
  - No evidence of homogeneous precipitation of $MgCO_3$
    - No visible particles
    - Any $MgCO_3$ formed at high T would analyse as Mg^{2+}
    - Mass balance closure for Mg within 1%

What is Happening to the Particles?

Material		Surface area ($m^2 g^{-1}$)	Total Volume ($cm^3 g^{-1}$)	Average Pore Size (Å)
Unreacted SWOL		12	0.02	69.45
423 K	$[Mg] < [Mg]_{sat}$	128	0.11	34.30
423 K	$[Mg] > [Mg]_{sat}$	91	0.07	32.15
473 K	$[Mg] > [Mg]_{sat}$	44	0.03	31.39

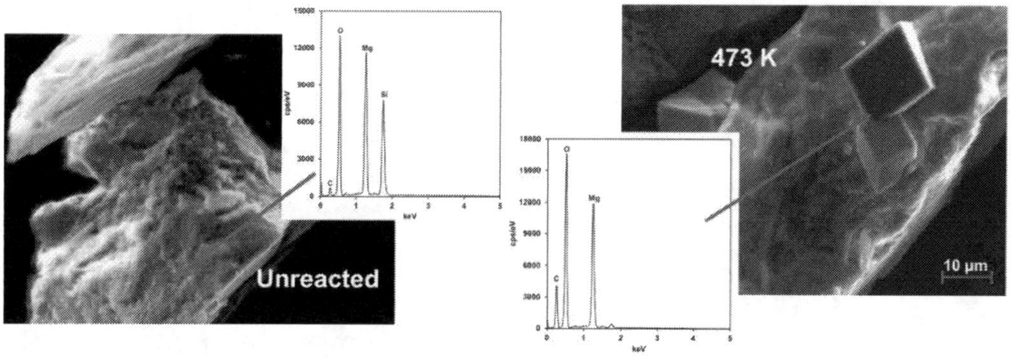

13

Silica

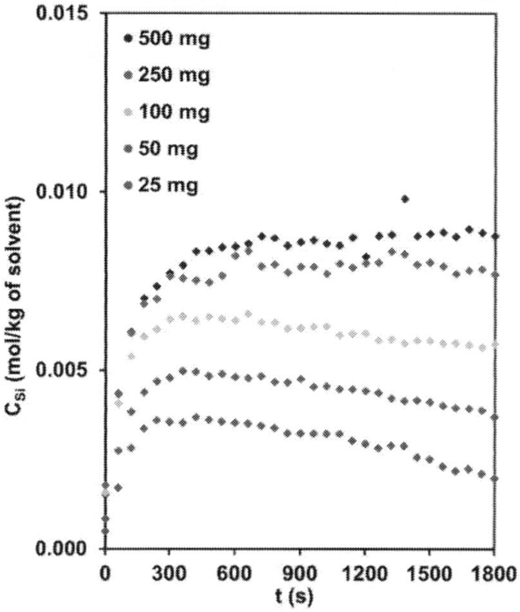

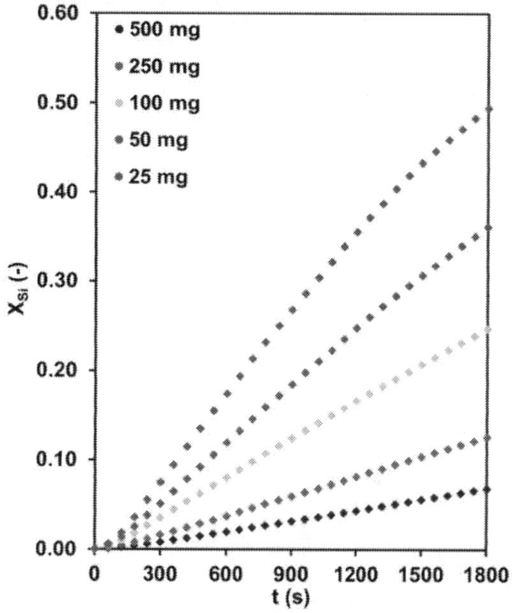

14

CONCLUSIONS

- Mg extraction far from $MgCO_3$ saturation condition depends
  - Directly on $[H^+]^n$: $n \sim 1$
  - Weakly on temperature: $E \sim 0$
  - Strongly on X

Evidence of a common rate function.

- Around $MgCO_3$ saturation
  - Rate falls dramatically
  - Solids formation only on surface of particles

Surface morphology changes – silica?
HCO_3^- ligand effects on rate?

15

ACKNOWLEDGEMENTS

- Orica Australia
- Mci
- AZ hold the HB & FM Scholarship

16

Utilization of Magnesium Hydroxy-Carbonate Hydrate (MHCH) materials formed by precipitation of CO_2 in alkaline aqueous solutions

G. Jauffret[1*], J. Morrison[1], F.P. Glasser[1], J.-L. Galvez-Martos[2], S. Yoon[2], M.S. Imbabi[2]

[1] Department of Chemistry, University of Aberdeen, AB24 3UE, UK
[2] School of Engineering, University of Aberdeen, AB24 3UE, UK
[*] Corresponding author: g.jauffret@abdn.ac.uk

Abstract

CO_2 capture from mixed gaseous emissions and subsequent utilisation has proven to be possible in many ways and at different points in the cycle. This paper describes a relatively simple route for the capture of CO_2 and its conversion to useful solid construction products. CO_2 from a combustion gas waste stream (*e.g.* from a cement plant) containing N_2 and water vapour is selectively dissolved into an alkaline aqueous solution at ~ 25 °C and 1 atm. The carbonate and bicarbonate ions thus formed react with Mg ions (*e.g.* from a seawater desalination brine) and precipitate Magnesium Hydroxy-Carbonate Hydrate (MHCH) phases; the presence of NaCl does not interfere significantly with the precipitation process. Nesquehonite, $MgCO_3 \cdot 3H_2O$, is the preferred product obtained at ~ 25 °C and 1 atm and typically occurs as micrometric well crystallized prismatic needles. After its thermal activation, the powder is mixed with water and the mix can be shaped into panels or blocks. The hardened materials contain ~ 30 wt% CO_2 and their properties depend on the applied parameters; they are non-toxic, non-combustible and can be recycled at end of life.

Introduction

The concentration of CO_2 in the atmosphere has increased steadily in the course of time, mostly due to an increased combustion of fossil fuels, and is now above 400 ppm[1]. In order to prevent excessive release of CO_2 into the atmosphere, considerable effort has focused on carbon capture. The conventional approach (carbon capture and storage, CCS) involves the separation of CO_2 from other gases, its liquefaction, transportation and storage underground (*e.g.* in suitable geological reservoirs); the technology is well known and several large scale projects have been reported worldwide[2]. An alternative approach (carbon capture and utilization, CCU) seeks to make use of the CO_2 in various industrial applications[3].

An interesting way to store CO_2 is as stable carbonates[4-6], either by carbonation of Mg-rich minerals (*e.g.* olivine or serpentine) or by use of solutions containing Mg (*e.g.* seawater or brines). Although magnesite ($MgCO_3$) is the most thermodynamically stable Mg carbonate phase, its precipitation is inhibited under usual conditions by the existence of a "dehydration barrier" which needs to be overcome[7]. Several phases in the MgO-CO_2-H_2O system that are encountered in nature have been synthesized[8-13];

these will be referred to collectively henceforth as Magnesium Hydroxy-Carbonate Hydrates (MHCH) phases, even though not all contain OH groups.

As carbonates obtained from the precipitation of CO_2 could be used rather than stored (*e.g.* as cement replacement or aggregates[14,15]), the utilization of the MHCH phases has been considered. This paper presents work by the authors on the production of materials based on MHCH phases, more particularly nesquehonite ($MgCO_3{\cdot}3H_2O$), and some of the measured properties of these materials.

Materials and methods

The syntheses involved the direct use of Na_2CO_3 and/or $NaHCO_3$ but the results would be similar if CO_2 was dissolved in a NaOH solution. The OH/CO_2 molar ratio was used to define the composition of the solution (and so in an industrial process the extent of CO_2 dissolution): when the OH/CO_2 ratio is equal to 2, the solution is Na_2CO_3; when it is equal to 1, the solution is $NaHCO_3$; any value between 1 and 2 implies a solution containing a mixture of carbonate and bicarbonate. Combining CO_2 at an early stage ensures a better control over its capture. In a first step towards using more complex real solutions, $MgCl_2{\cdot}6H_2O$ was used as the source of magnesium.

The samples were characterized by XRD with a Panalytical X'Pert Powder diffractometer (Cu $K_{\alpha1-\alpha2}$ radiation, 45 mA and 40 kV operating parameters, 0.013 °2θ step size, static air atmosphere) and by SEM with a Hitachi S-520 Scanning Electron Microscope (20 kV acceleration voltage, samples coated with gold under flowing Ar). The compressive strength was measured on a Hounsfield Universal press using a loading speed of 1 $mm.min^{-1}$ and a 10 kN sensor.

Experimental procedures

When mixing the Mg solution and the carbonate/bicarbonate solution at ~ 25 °C and 1 atm, the MHCH phase formed was found to be nesquehonite ($MgCO_3{\cdot}3H_2O$). A synthesis would typically last for 4 h with constant stirring. The slurry was filtered and the cake washed with water before drying; the cake was first disaggregated into smaller particles to facilitate drying then, once dry, ground manually into a powder.

Thermodynamic modelling using GEMS (Gibbs Energy Minimization Software[16]) was performed to predict which phase(s) would form and what the yield of the reaction would be for different experimental conditions. Because no thermodynamic data were available for dypingite, only the data for nesquehonite and hydromagnesite were used. However, the value for hydromagnesite had to be slightly modified for nesquehonite to be predicted under the experimental conditions used; the Gibbs energy of formation values used were -1724[17] and -5840 $kJ.mol^{-1}$ for nesquehonite and hydromagnesite respectively. The theoretical and experimental results were found to be in good agreement thus limiting the number of experiments required; for example, the Mg yield (*i.e.* the amount of Mg which has been incorporated in the solid) after 4 h at 25 °C for a reaction between 0.1 mol of $MgCl_2{\cdot}6H_2O$ and 0.1 mol of Na_2CO_3 was found to be 85 +/- 5 % experimentally against 88 % predicted by the model.

In order to achieve cementitious properties, nesquehonite samples were thermally activated in air in an oven for 24 h in the range 100-250 °C. Shapes (cubes and cylinders) were prepared by mixing the activated powder samples with water and moulding the mix with different methods, mainly casting with or without vibrating but also pressing; after 1 day of curing in a humid atmosphere, the shapes were demoulded and allowed to dry before measuring their compressive strength.

Results and discussion

Synthesis of the nesquehonite

Nesquehonite typically occurs as well crystallized prismatic needle-shaped crystals of different dimensions as can be seen in figure 1.

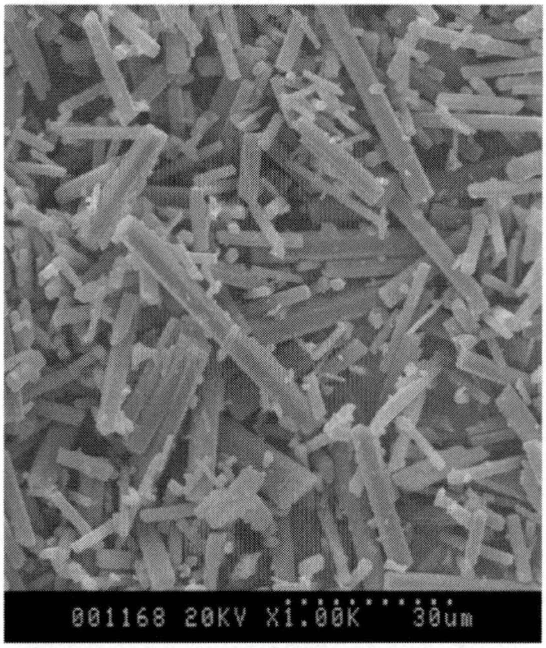

Figure 1. SEM photo of nesquehonite crystals

Nesquehonite is the main MHCH phase precipitated at ~ 25 °C and 1 atm but other MHCH phases (table 1) could be obtained as primary synthesis products depending on the experimental conditions; nesquehonite can also progressively convert to hydromagnesite and/or dypingite under some experimental conditions (*e.g.* temperature)[10,12,18-24].

Table 1. Main MHCH phases encountered during experiments

Phase	Formula	%MgO	%CO_2	%H_2O
Nesquehonite	$MgCO_3 \cdot 3H_2O$	29.1	31.8	39.1
Dypingite	$Mg_5(CO_3)_4(OH)_2 \cdot 5H_2O$	41.5	36.2	22.3
Hydromagnesite	$Mg_5(CO_3)_4(OH)_2 \cdot 4H_2O$	43.1	37.6	19.3

The OH/CO_2 ratio can be used to assess the consumption of alkalis, which is an important factor when considering the economy of the whole process. An ideal use of alkalis would be to form a solution of pure bicarbonate (*i.e.* with a OH/CO_2 ratio of 1) but the precipitation of nesquehonite was limited (*i.e.* the kinetics of the precipitation reaction were slow and the yield of solids was low); a ratio of 1.5-2 has been found to be more favourable. The advantage of using a dissolution stage for the capture of CO_2 is that, in the case of a mixture of gases (*e.g.* N_2, O_2 and CO_2), CO_2 is much more soluble and so automatically separates from the other, less soluble, gases.

As nesquehonite is not reported to contain OH groups in its structure[25,26] (unlike other MHCH phases like hydromagnesite[27,28]), the sole purpose of using NaOH is to raise the pH so that nesquehonite can precipitate (the hydroxide groups are not incorporated in the nesquehonite structure). However, because NaOH is diluted by the aqueous solution supplying Mg and is practically irrecoverable, other sources of alkalinity are being evaluated to compensate for this problem.

The presence of Na and Cl in the Mg-solution (*e.g.* a desalination reject brine) has been investigated and was not found to affect significantly the precipitation process: washing the product at the filtration stage removes residual Na and Cl ions present in the mother liquor and avoids contamination of the product by NaCl upon drying.

Cementitious properties of the materials

The thermal decomposition of nesquehonite takes place in two stages as water is first lost then CO_2; this implies that nesquehonite could be dehydrated in such a way that no CO_2 is lost, *i.e.* if the temperature is kept below 300-350 °C[19,21,29-33]. Upon dehydration, nesquehonite was found to form first a lower hydrate[21,23,29,31,33] (which we term "phase X") then an XRD amorphous phase[19,25,33] (figure 2).

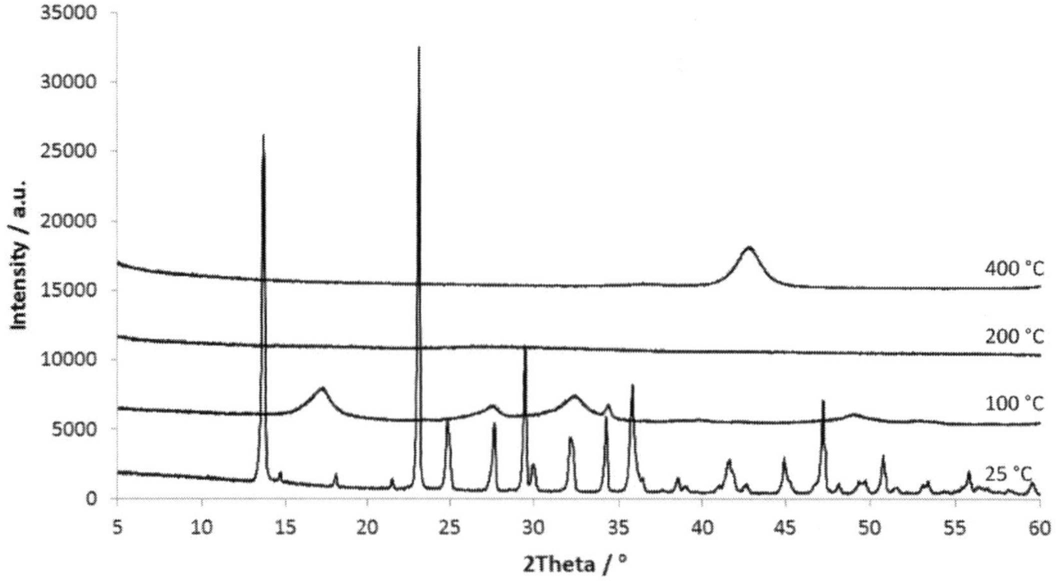

Figure 2. XRD patterns of nesquehonite, "phase X", amorphous phase and periclase, MgO (from bottom to top) along with the temperature at which they are observed

The thermal decomposition of nesquehonite is a complex process that is influenced by the experimental conditions used. Despite some changes, the morphology of the crystals was found to be overall preserved[21,33,34]. As long as decarbonation does not take place, nesquehonite can successfully be regenerated upon rehydration with water in excess[33,34]; this indicates that the structure of nesquehonite is at least partially preserved during dehydration.

The compressive strength of the nesquehonite-based shapes was found to be influenced by several factors including the activation temperature, the water/solid ratio and the method of preparation. In general, the compressive strength seems to reach a maximum for activation temperatures between 150 and 200 °C and to increase with decreasing water/solid ratio. A correlation between the density of the products and their compressive strength was observed: vibrating the moulded mixes or pressing the mixes into the moulds leads to less porous and stronger products. For example, cubes which were prepared from nesquehonite thermally activated at 180 °C, rehydrated with a water/solid ratio of 0.8 and dried in air at ~ 25 °C for 14 days (at which point their mass stabilized) developed a compressive strength of ~ 12 MPa but additional compressing process during casting increased the compressive strength to ~ 20 MPa (by comparison, a value of 7 MPa is typically required for the compressive strength of concrete blocks). Depending on the preparation and curing conditions, some conversion of nesquehonite into dypingite and/or hydromagnesite was observed. Some test samples have been recycled, *i.e.* successfully submitted to another cycle of dehydration and rehydration after crushing, but more work is still needed to follow the evolution of the properties with further recycling.

The study of the properties of the products is still ongoing but it appears to be possible to optimize the preparation conditions so that the properties would correspond to the desired application for the product (*e.g.* as a panel, figure 3).

Figure 3. Prototype of a 35 mm thick, A4-size panel prepared with activated nesquehonite (the bulk density of this product is ~ 785 $kg.m^{-3}$ and its thermal conductivity ~ 0.26 $W.m^{-1}.K^{-1}$)

Environmental performance and process economics

A preliminary assessment of the life cycle greenhouse gases emissions, in a comparatively low carbon intensity European context, indicates that the process and the product have a net value of greenhouse gases emissions (*i.e.* a positive carbon footprint) of 250 kg equivalent CO_2 (CO_2e) per tonne of material. This indicates that the emissions generated by the process and embodied in the raw materials are higher than the amount of CO_2e captured as carbonate. The carbon footprint is mostly associated with the production of alkalis which, when they are produced by electrolysis, have an average carbon footprint of 1.12 kg of CO_2e per kg of sodium hydroxide[35]. However, the process itself is flexible enough to use alkaline waste, with a null-carbon footprint by definition, to substitute part of the commercial alkali required.

Nevertheless, the environmental credit of the product cannot be neglected: while the carbon footprint is positive, it is still less than the carbon footprint of substituted materials such as plasterboard (around 300 kg CO_2e/tonne) or Portland cement (~ 700-800 kg CO_2e/tonne)[36]. In addition, the low density will enhance the substitution potential and increase emissions savings.

Trade-off impacts analysis shows a relatively high water footprint, around 10-20 L water per kg of material (compared to 2 L/kg for plaster), and embodied energy, 14 MJ per kg (compared to 6 MJ/kg for plaster). This preliminary assessment also shows that process economics are heavily affected by the consumption of alkalis and the scale of the process: production costs could be reduced to ~ 1 USD/kg (the foreseen threshold for the marketability of such product) but, to achieve that level, over 1000 m^3 of desalination brine (typically with 0.1 M Mg^{2+}) have to be carbonated per day (*i.e.* more than 2500 kg of Mg^{2+} per day). These calculations do not include savings arising from process optimization, which has yet to be done: for example, the use of alkaline waste or optimized use of water for the washing of the product could help reduce significantly the water and carbon footprints as well as the embodied energy.

Conclusions

This paper shows that cementitious materials based on nesquehonite can be produced from the precipitation reaction between a source of CO_2 and a source of Mg. These products would contain at least 30 wt% CO_2 and, like other carbonates, they are non-toxic and non-combustible; they can also be recycled at end of life. NaCl can be avoided in the product if the precipitation solid is washed and so the product could be used in conjunction with metallic elements (*e.g.* steel nails or staples).

The properties of the products depend on their preparation conditions and there is still scope for optimization. At present, the outlook appears favourable for use of these materials as combined interior finishing of walls and ceilings. Study of the durability of these products is also still ongoing.

The possibility of producing a useful construction material from wastes is interesting but, for a full scale process to be truly viable, ecological (carbon footprint, impact of rejects, etc...) and economical (source of alkalis and raw materials, transport costs, etc...) aspects need to be assessed thoroughly.

Acknowledgements

The authors would like to express their sincere gratitude to the Gulf Organization for Research and Development (GORD) in Qatar for funding this research.

References

1. Earth System Research Laboratory, Global Monitoring Division, Mauna Loa Observatory, Trends in Atmospheric Carbon Dioxide. *http://www.esrl.noaa.gov/gmd/ccgg/trends* (accessed on 28-03-15).
2. Global CCS Institute, Large Scale CCS Projects (Overview, Map & Database). *http://www.globalccsinstitute.com/projects/large-scale-ccs-projects* (accessed on 28-03-15).
3. Styring, P., Jansen, D., 2011. Carbon Capture and Utilization in the green economy. *Report no. 501*. The Centre for Low Carbon Futures 2011 and CO_2Chem Publishing 2012, Sheffield.
4. Lackner, K.S., 2002. Carbonate chemistry for sequestering fossil carbon. *Annual Reviews of Energy and Environment*, 27, 193-232.
5. Mazzotti, M., *et al.*, 2005. Mineral carbonation and industrial uses of carbon dioxide. *In: IPCC Special Report on Carbon dioxide Capture and Storage*, 319-338.
6. Haywood, H.M., Eyre, J.M., Scholes, H., 2001. Carbon dioxide sequestration as stable carbonate minerals – Environmental barriers. *Environmental Geology*, 41, 11-16.
7. Lippmann, F., 1973. *Sedimentary Carbonate Materials*. Springer-Verlag, Berlin.

8. De Vito, C., Ferrini, V., Mignardi, S., Cagnetti. M., Leccese, F., 2012. Progress in carbon dioxide sequestration via carbonation of aqueous saline waste. *Periodico di Mineralogia*, 81, 333-344.

9. Ma, J., Yoon, R.H., 2013. Use of reactive species in water for CO_2 mineralisation. *Energy and Fuels*, 27, 4190-4198.

10. Case, D.H., Wang, F., Giammar, D.E., 2011. Precipitation of magnesium carbonates as a function of temperature, solution composition, and presence of a silicate mineral substrate. *Environmental Engineering Science*, 28, 881-889.

11. Ferrini, V., De Vito, C., Mignardi, S., 2009. Synthesis of nesquehonite by reaction of gaseous CO_2 with Mg chloride solution: its potential role in the sequestration of carbon dioxide. *Journal of Hazardous Materials*, 168, 832-837.

12. Haenchen, M., Prigiobbe, V., Baciocchi, R., Mazzotti, M., 2008. Precipitation in the Mg-carbonate system—effects of temperature and CO_2 pressure. *Chemical Engineering Science*, 63, 1012-1028.

13. Kloprogge, J.T., Martens, W.N., Nothdurft, L., Duong, L.V., Webb, G.E., 2003. Low temperature synthesis and characterization of nesquehonite. *Journal of Materials Science Letters*, 22, 825-829.

14. Constanz, B., Youngs, A., Holland, T., 2009. CO_2 sequestering formed building materials. *Patent US20090519135*, 30-09-2009.

15. Camire, C., Lu, P., Kendall, T., Devenney, M., 2012. Methods and products utilising magnesium oxide for carbon dioxide sequestration. *Patent US20120291675*, 22-11-2012.

16. Paul Scherrer Institut, Gibbs Energy Minimization Software, *http://gems.web.psi.ch*.

17. Langmuir, D., 1965. Stability of carbonates in the system $MgO\text{-}CO_2\text{-}H_2O$. *The Journal of Geology*, 73, 730-754.

18. Ballirano, P., De Vito, C., Mignardi, S. and Ferrini, V., 2013. Phase transitions in the Mg-CO_2-H_2O system and the thermal decomposition of dypingite, $Mg_5(CO_3)_4(OH)_2{\cdot}5H_2O$: Implications for geosequestration of carbon dioxide. *Chemical Geology*, 340, 59-67.

19. Davies, P.J., Bubela, B., 1973. The transformation of nesquehonite into hydromagnesite. *Chemical Geology*, 12, 289-300.

20. Hopkinson, L., Kristova, P., Rutt, K., Cressey, G., 2012. Phase transitions in the system $MgO\text{-}CO_2\text{-}H_2O$ during CO_2 degassing of Mg-bearing solutions. *Geochimica et Cosmochimica Acta*, 76, 1-13.

21. Dell, R.M., Weller, S.W., 1959. The thermal decomposition of nesquehonite $MgCO_3{\cdot}3H_2O$ and magnesium ammonium carbonate $MgCO_3{\cdot}(NH_4)_2CO_3{\cdot}4H_2O$. *Transactions of the Faraday Society*, 55, 2203-2220.

22. Hopkinson, L., Rutt, K., Cressey, G., 2008. The transformation of nesquehonite to hydromagnesite in the system $CaO\text{-}MgO\text{-}H_2O\text{-}CO_2$: an experimental spectroscopic study. *The Journal of Geology*, 116, 387-400.

23. Bender, M., Sprague, R.S., 1965. The preparation and structure of magnesium carbonate trihydrate. *Journal of Inorganic and Nuclear Chemistry*, 27, 1872-1873.

24. Canterford, J.H., Tsambourakis, G., Lambert, B., 1984. Some observations on the properties of dypingite, $Mg_5(CO_3)_4(OH)_2{\cdot}5H_2O$, and related minerals. *Mineralogical Magazine*, 48, 437-442.

25. Stephan, G.W., MacGillavry, C.H., 1972. The crystal structure of nesquehonite, $MgCO_3{\cdot}3H_2O$. *Acta Crystallographica B*, 1972, 28, 1031-1033.

26. Giester, G., Lengauer, C.L., Rieck, B., 2000. The crystal structure of nesquehonite, $MgCO_3 \cdot 3H_2O$, from Lavrion, Greece. *Mineralogy and Petrology*, 70, 153-163.

27. Akao, M., Marumo, F., Iwai, S., 1974. The crystal structure of hydromagnesite. *Acta Crystallographica B*, 30, 2670-2672.

28. Akao, M., Iwai, S., 1977. The hydrogen bonding of hydromagnesite. *Acta Crystallographica B*, 33, 1273-1275.

29. Ballirano, P., De Vito, C., Ferrini, V., Mignardi, S., 2010. The thermal behaviour and structural stability of nesquehonite, $MgCO_3 \cdot 3H_2O$, evaluated by *in situ* laboratory parallel-beam X-ray powder diffraction: new constraints on CO_2 sequestration within minerals. *Journal of Hazardous Materials*, 178, 522-528.

30. Hales, M.C., Frost, R.L., Martens, W.N., 2008. Thermo-Raman spectroscopy of synthetic nesquehonite - Implication for the geosequestration of greenhouse gases. *Journal of Raman Spectroscopy*, 39, 1141-1149.

31. Lanas, J., Alvarez, J.I., 2004. Dolomitic lime: thermal decomposition of nesquehonite. *Thermochimica Acta*, 421, 123-132.

32. Vagvolgyi, V., Hales, M., Frost, R.L., Locke, A., Kristof, J., Horvath, E., 2008. Conventional and controlled rate thermal analysis of nesquehonite $Mg(HCO_3)(OH) \cdot 2(H_2O)$. *Journal of Thermal Analysis and Calorimetry*, 94, 523-528.

33. Jauffret, G., Morrison, J., Glasser, F.P., 201X. On the thermal decomposition of nesquehonite. *Journal of Thermal Analysis and Calorimetry*, under review (submitted on 17/12/14).

34. Morrison, J., Jauffret, G., Glasser, F.P., Galvez-Martos, J.L., Imbabi, M.S., 2014. Towards carbon negative cements. *In: Proceedings of the 34th Annual Cement and Concrete Science Conference and Workshop on Waste Cementation*, Sheffield, UK, 14-17 September 2014, paper no. 222.

35. Althaus, H.J., Chudacoff, M., Hischier, R., Jungbluth, N., Osses, M., Primas, A., 2007. Life Cycle Inventories of Chemicals. *Final report Ecoinvent data v2.0 No. 8*. Swiss Centre for Life Cycle Inventories, Dübendorf, CH.

36. Kellenberger, D., Althaus, H.J., Jungbluth, N., Künniger, T., 2007. Life Cycle Inventories of Building Products. *Final report Ecoinvent data v2.0 No. 7*. Swiss Centre for Life Cycle Inventories, Dübendorf, CH.

Alkali-based extraction of $Mg(OH)_2$ from Mg silicate minerals via solid state reaction with NaOH

S. Madeddu[1*], M. Priestnall[2], E. Godoy[3], V. Kumar[4], M. Evans[5], A. R. Badrol[1], R. Wang[1], and H. Kinoshita[1]

[1] Department of Materials Science and Engineering, The University of Sheffield, Sheffield, United Kingdom.
[2] Innovate UK, Swindon, United Kingdom (previously, Cambridge Carbon Capture Ltd.)
[3] Polarcus DMCC, Dubai, U.A.E.
[4] Department of Materials Science and Metallurgy, University of Cambridge, Cambridge, United Kingdom
[5] Cambridge Carbon Capture Ltd., Cambridge, United Kingdom.
*Corresponding author: mtq11sm@sheffield.ac.uk

Abstract
$Mg(OH)_2$ can capture CO_2 by forming Mg carbonate minerals, but its limited availability in natural deposits has hindered this application. In our previous study, we demonstrated that $Mg(OH)_2$ can be obtained from dunite, an ultramafic rock composed of Mg silicate minerals in highly concentrated NaOH aqueous systems at 180 °C. However, the large quantity of NaOH consumed could be an obstacle for the implementation of the technology.
The present study shows that the reduction of NaOH consumption is achievable via the solid state reaction of Mg silicate minerals with NaOH for the extraction of $Mg(OH)_2$. The results demonstrated that NaOH consumption was reduced by 97 % with respect to the NaOH aqueous system, maintaining a comparable yield of $Mg(OH)_2$ extraction, i. e. 55 – 62 %. The addition of H_2O in small amount was also found beneficial and improved the extraction of $Mg(OH)_2$.

Introduction

$Mg(OH)_2$ is a suitable feedstock material for CO_2 sequestration as it captures CO_2 by forming stable or metastable Mg carbonate minerals which can safely store the CO_2 over a long timescale [1, 2].
$Mg(OH)_2$ can carbonate under the exposure to gaseous or supercritical CO_2 or via dissolution in H_2O with purged CO_2, i. e. aqueous carbonation[3, 4, 5, 6]. $Mg(OH)_2$ can also be used to simultaneously separate and capture CO_2 from flue gases by forming $Mg(HCO_3)_2$[7]. However, the application of $Mg(OH)_2$ for CO_2 storage has been hindered by the scarce availability of mineral $Mg(OH)_2$, brucite, in natural deposits[1,8].
Mg silicate minerals, in particular serpentine and forsterite, are among the most abundant minerals in nature and are commonly used as feedstock materials for mineral carbonation[9,10]. The direct carbonation of these minerals is limited by their slow kinetic of reaction, and the intermediate chemical treatment to extract Mg is necessary to improve the carbonation rate[1,9].

The extraction of $Mg(OH)_2$ from Mg silicate minerals via chemical processing benefits from the large availability of the minerals, and at the same time improves the overall efficiency of the mineral carbonation process since $Mg(OH)_2$ has a faster kinetic of carbonation than Mg silicate minerals[5]. The extraction of $Mg(OH)_2$ from Mg silicate minerals could also be a profitable method to recycle asbestos or mine wastes[11, 12].

In a previous study we investigated the alkaline digestion of Mg silicate minerals into $Mg(OH)_2$[13]. Dunite, an ultramafic rock composed of Mg silicate minerals, reacted in highly concentrated NaOH aqueous systems at 180 °C and was almost fully digested into $Mg(OH)_2$ within 6 hours[13]. An issue for the implementation of the technology was the large consumption of NaOH, i. e. ~ 20.3 g NaOH per 1 g dunite[13].

The research has focused on the development of a new technology based on the solid state reaction of dunite with NaOH aiming at the reduction of the alkaline reactant required to extract $Mg(OH)_2$. The effect of NaOH was studied by mixing powdered dunite with different amount of NaOH. H_2O was added to the selected samples and the rate of $Mg(OH)_2$ extraction in *dry* and *wet* dunite-NaOH mixtures was compared. The results were also compared with those from the alkaline digestion of dunite with NaOH aqueous system previously studied.

Materials and Methods

Dunite

The powdered feedstock material was provided by Sibelco Ltd. and it was mined from dunite formation in Åheim, Norway. The dunite was sieved to obtain a homogeneous powder with particle size < 63 µm. The analysis via Laser Diffraction with dry dispersion showed an average distribution of the particles size around 25 µm, with 90 % of the particles < 58 µm.

X-ray Diffraction (XRD) analysis showed that the material was composed mainly of forsterite, Mg_2SiO_4, with minor Mg-bearing minerals: clinochlore, serpentine, enstatite, talc, hornblende and spinel. The Rietveld Refinement Quantitative Phase Analysis (QPA) conducted on the XRD pattern estimated the amount of forsterite ~ 73 ± 2 wt%. The TGA detected the presence of 0.42 wt% $Mg(OH)_2$.

Extraction of $Mg(OH)_2$ via solid state reaction

NaOH pearl reagent grade was provided by Fisher Chemical and it was mixed with powdered dunite using mortar and pestle. When the dunite-NaOH mixture was well blended it was compacted with a hand press at 3 tonnes for 1 minute to form a pellet. The dunite and NaOH were mixed with mole ratio of 1:2, 1:1.5, 1:1, and 1:0.5. The moles of dunite were calculated assuming that the material was composed of 100 % forsterite, Mg_2SiO_4.

Two series of experiments were conducted in *dry* and *wet* conditions. For the *wet* experimental series, droplets of distilled H_2O were added to the pellets to have $NaOH:H_2O$ mole ratio of 4:1. The compositions of the samples is summarised in Table 1. The pellets were placed into steel vessels lined with Teflon which were sealed and heated at 180 °C in an electric furnace for 6 hours. Subsequently, the pellets were removed from the vessel, crushed and ground to a fine powder which was washed with distilled water. The solid product was separated from the liquid phase via vacuum filtration and dried in an oven at 90 °C for 1 hour. The reaction products were analysed via X-ray diffraction and Thermogravimetric analysis.

Table 1: samples composition

Dry	*Wet*
Dunite:NaOH (mole/mole)	**Dunite:NaOH:H_2O** (mole/mole/mole)
1:2	1:2:0.5
1:1.5	1:1.5:0.375
1:1	1:1:0.25
1:0.5	1:0.5:0.125

Phase analysis

X-ray diffraction (XRD) was conducted on the reaction products with the Siemens D5000 with Cu Kα X-ray source, λ = 1.54 Å. The powdered samples were placed in a plastic sample holder and the scans were run from 10° to 70° 2θ with step size 0.05° 2θ and scan rate of 2.5°/minute.

Thermogravimetric analysis (TGA) was conducted using the PerkinElmer Pyris 1. The samples were placed in an alumina crucible and heated at a rate of 10 °C/minute from room temperature to 1000 °C, under N_2 flow. The first derivative of the TG curve, DTG, was also determined to assist the calculation of the weight loss occurred during the heating process. The amount of $Mg(OH)_2$ was calculated based on the dehydroxylation (Eq. 1) taking place between 350 – 450 °C[14].

$$Mg(OH)_{2(s)} \rightarrow MgO_{(s)} + H_2O_{(g)} \qquad (1)$$

Results and discussion

Mechanism of reaction

The XRD analysis was conducted on the reaction products from *dry* experimental series before and after they were washed with distilled H_2O. The analysis on the unwashed samples revealed the presence of $Mg(OH)_2$, Na_2SiO_3, Na_2CO_3 and unreacted mineral components of dunite. These data suggest that two fundamental reactions occurred in the dunite-NaOH mixtures, i. e. the reaction of Mg silicate minerals to form $Mg(OH)_2$ and Na_2SiO_3 (Eq. 2) and the carbonation of some of NaOH (Eq. 3).

$$Mg_2SiO_4 + 2NaOH + H_2O \rightarrow 2Mg(OH)_2 + Na_2SiO_3 \qquad (2)$$

$$2NaOH + CO_2 \rightarrow Na_2CO_3 + H_2O \qquad (3)$$

The study of the Gibbs free energy of reaction (ΔG) revealed that Eq. 2 is energetically favoured ($\Delta G < 0$) only in presence of H_2O, suggesting that H_2O is required to form Na_2SiO_3 as a by-product when Mg_2SiO_4 and NaOH react. H_2O can be present in air or provided from the carbonation of NaOH (Eq.3).
The overall reaction occurred in the system can be expressed by Eq. 4, as the sum of Eq. 2 and 3. The negative Gibbs energy change ($\Delta G < 0$) of this reaction suggests that, in presence of CO_2, $Mg(OH)_2$ and Na_2SiO_3 can form without externally adding H_2O to the system.

$$Mg_2SiO_4 + 4NaOH + CO_2 \rightarrow 2Mg(OH)_2 + Na_2SiO_3 + Na_2CO_3 \quad (4)$$

Effect of NaOH and H_2O

The XRD patterns of the raw dunite and reaction products from the *dry* experimental series are reported in Figure 1.
The two main reflection peaks for brucite, are found at 18.6 °2ϑ and 38 °2ϑ, and overlap with the reflection peaks of clinochlore and forsterite, respectively. The XRD patterns show that increasing the amount of NaOH mixed with dunite, the intensity of brucite reflection peaks progressively increased whereas the intensity of the reflection peaks from dunite mineral components decreased, suggesting that more brucite was extracted from dunite.

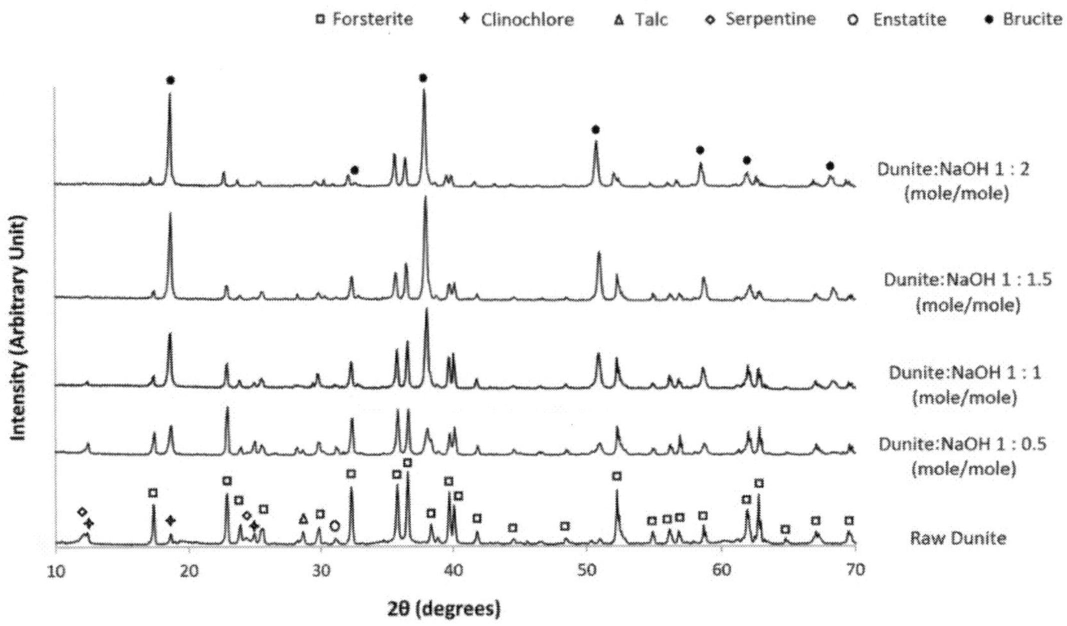

Figure 1: XRD patterns of raw dunite and reaction products from dunite-NaOH solid state reaction at 180 °C for 6 hours, *dry* experimental series.

The wt% of $Mg(OH)_2$ in the reaction products from *dry* and *wet* experimental series was estimated via TGA, and the results are compared in Figure 2. The concentration of $Mg(OH)_2$ increased linearly with the increase of NaOH and reached 46.7 and 54. 1 wt% at dunite:NaOH mole ratio of 1:2 for *dry* and *wet* experimental series, respectively. The results show that reducing the amount of NaOH significantly lowered the extraction of $Mg(OH)_2$.

The benefit of additional H_2O became significant for the samples with dunite:NaOH mole ratio of 1:2 and 1:1.5, although it was not obvious in the samples with dunite:NaOH mole ratio of 1:1 and 1:0.5.

In the *wet* experimental series, the samples were prepared with the same NaOH:H_2O mole ratio of 4:1, and thus, samples with larger amount of NaOH also contained more H_2O respect to dunite and the positive effect of H_2O became apparent.

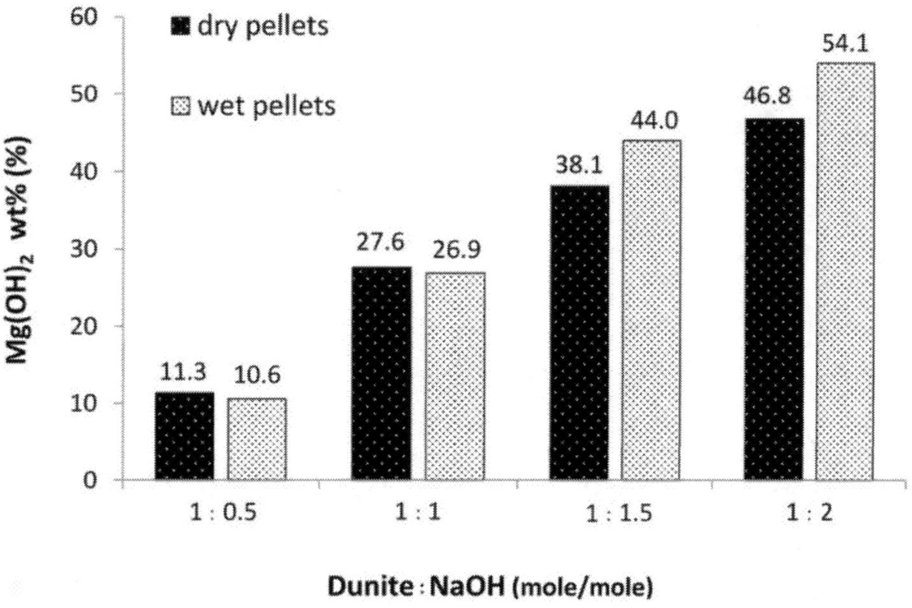

Figure 2: Comparison of $Mg(OH)_2$ concentration estimated via TGA in reaction products reacted at 180 ° C for 6 hours from *dry* and *wet* experimental series.

Comparison of solid state reaction and NaOH aqueous system

The yield of $Mg(OH)_2$ via solid state reaction was compared with that from the alkaline digestion of dunite with NaOH aqueous system. The yield was calculated according to the procedure described by Madeddu et al.[13]

The reaction products from the solid state reaction gave the highest yield of reaction with dunite:NaOH mole ratio of 1:2, i. e. 55.5 % and 62.6 % for the *dry* and *wet* experimental series, respectively. These results are comparable with the yield of 58 – 66 % achieved in the alkaline digestion of dunite with 50 mol/Kg NaOH aqueous system at 180 °C for 6 hours[13].

The alkaline digestion with 50 mol/Kg NaOH aqueous system required approximately 20.3 g of NaOH per 1 g of dunite processed whereas the solid state reaction used only 0.57 g of NaOH per 1 g of dunite, which corresponds to ~ 97 % reduction of the NaOH consumption. The consumption of H_2O was avoided in the *dry* experimental series or reduced by 99 % in the *wet* experimental series.

Conclusions

The solid state reaction of dunite and NaOH compacted into pellets and heated at 180 °C for 6 hours, resulted in reaction products with 46.7 wt% $Mg(OH)_2$ when dunite:NaOH mole ratio of 1:2 was used.

The addition of H_2O improved the extraction of $Mg(OH)_2$ in the samples with dunite:NaOH mole ratio of 1:1.5 and 1:2. The reaction product from the *wet* experimental series with dunite:NaOH mole ratio of 1:2 resulted in the highest concentration of $Mg(OH)_2$ achieved in this study, i. e. 54 wt%.

The NaOH required to extract $Mg(OH)_2$ from dunite via solid state reaction with dunite:NaOH mole ratio of 1:2, was 97 % lower than that used for the alkaline digestion of dunite with NaOH aqueous system to obtain a comparable yield of $Mg(OH)_2$, i. e. 55 – 62 %. The consumption of H_2O was also avoided or reduced by 99 %. The present study successfully demonstrated that it is possible to extract $Mg(OH)_2$ from Mg silicate minerals via solid state reaction with NaOH and that the consumption of the alkaline reactant can be reduced, achieving significant yield of reaction.

Acknowledgements

The authors thank Regione Autonoma della Sardegna, Agenzia Regionale per il Lavoro, Italy, and the EU-European Social Fund for their contribution to the work presented which is part of the PhD project funded by the Master and Back Programme 2011, Percorsi di Alta Formazione, P.O.R. FSE 2007–2013. The authors also thank The Armourers & Brasiers' Gauntlet Trust, Cambridge, UK, and the Learned Society of the Department of Materials Science and Engineering of The University of Sheffield, Sheffield, UK, for the financial support.

References

1. Metz, B., Davidson, O., De Coninck, H. C., Loos, M. and Meyer, L. A., (eds.) 2005. *IPCC: IPCC Special Report on Carbon Dioxide Capture and Storage*. Prepared by Working Group III of the Intergovernmental Panel on Climate Change, Cambridge University Press, Cambridge, United Kingdom and New York, NY, USA.
2. Wilson, S. A. and Maynard-Casely, H. E. 2013. Influence of relative humidity on the hydration state of magnesium carbonate minerals: Outlook for optimizing stability of CO_2 storage in minerals. *In: Proceedings of the 4th International Conference on Accelerated Carbonation for Environmental and Materials Engineering, ACEME 2013*. Leuven, Belgium, 9 - 12 April 2013.
3. Zhao, L., Sang, L., Chen, J., Ji, J., and Teng, H. H. 2010. Aqueous carbonation of natural brucite: relevance to CO_2 sequestration. *Environmental Science and Technology*, 44, 406 - 411.
4. Schaef, H. T., Windisch C. F., McGrail, P. F., and Rosso, K. M. 2011. Brucite [$Mg(OH)_2$] carbonation in wet supercritical CO_2: An in situ high pressure X-ray diffraction study. *Geochimica et Cosmochimica Acta*, 75, 7458-7471.
5. Fagerlund, J., Nduagu, E., Romão, I., and Zevenhoven, R. 2012. CO_2 fixation using magnesium silicate minerals part 1: Process description and performance. *Energy*, 41, (1), 184 – 191.
6. Loring, J. S., Thompson, C. J., Zhang, C. Z., Wang, Z., Schaef, H. T., and Rosso, K. M. 2012. In situ infrared spectroscopic study of brucite carbonation in dry to water-saturated supercritical carbon dioxide. *The Journal of Physical Chemistry A*, 116, 4768 - 4777.
7. Jung, K. S., Keener T. C., Khang, S.-J., and Lee, S.-K. 2004. A technical and economic evaluation of CO_2 separation from power plant flue gases with reclaimed $Mg(OH)_2$. *Clean Technologies and Environmental Policies*, 6, 201 - 212.
8. Simandl, G. J., Paradis, S. and Irvine M. 2007. Brucite - Industrial mineral with a future. *Geoscience Canada*, 34, (2), 57-64.
9. Lackner, K. S., Wendt, C. H., Butt, D. P., Joyce, E. L. and Sharp, D. H. 1995. Carbon dioxide disposal in carbonate minerals. *Energy*, 20, (11), 1153-1170.

10. Best, M. G. 2003, *Igneous and metamorphic petrology*, 2nd Ed., Blackwell Science Ltd. Oxford, UK.
11. Bobicki, E. R., Liu, Q., Xu, Z., and Zeng, H. 2012. Carbon capture and storage using alkaline industrial wastes. *Progress in Energy and Combustion Science*, 38, 302 – 320.
12. Sergeeva, I., and Kerestedjian, T. 2012. Experimental dissolution of asbestos, aimed on passivation of asbestos waste dumps. *In: Acta Mineralogica-Petrographica, Abstract Series, volume 7, Joint 5th Mineral Sciences in the Carpathians Conference and 3rd Central-European Mineralogical Conference*, Miskolc, Hungary, 20 – 21 April 2012.
13. Madeddu, S., Priestnall, M., Kinoshita, H. and Godoy, E. 2014. Alkaline digestion of dunite for $Mg(OH)_2$ production: An investigation for indirect CO_2 sequestration. *Minerals Engineering*, 59, 31-38.
14. Földvári, M. 2011. Handbook of thermogravimetric system of minerals and its use in geological practice. *Occasional Papers of the Geological Institute of Hungary*, volume 213, Geological Institute of Hungary, Budapest.

Waters Associated with the Carbonation of Ultramafic Mine Tailings, Woodsreef Asbestos Mine, Australia

Oskierski, H. C., Murdoch University
Dlugogorski, B. Z., Murdoch University
Jacobsen, G.,
Originally presented on: 6/22/2015 13:15:00 - 13:30:00

The Woodsreef Asbestos Deposit, New South Wales, Australia, is a chrysotile mineralisation hosted in the ultramafic rocks of the Great Serpentinite Belt, predominantly consisting of schistose and massive serpentinite, as well as partially serpentinised harzburgite. Chrysotile has been extracted from the deposit intermittently between 1906 and 1983, producing 24.2 Mt of ultramafic tailings. The tailings result from dry-grinding of chrysotile ore and are stored above ground on an area covering about 0.5 km^2.

Extensive carbonate crusts have formed on the tailings pile since the closure of the mine. Natural weathering dissolves Mg-silicate minerals present in the tailings and precipitates Mg-carbonate minerals in the form of crusts and cements. Isotopic signatures of the carbonate minerals ($\delta^{13}C$, $\delta^{18}O$, $F^{14}C$) indicate that carbonate crusts consisting of hydromagnesite predominantly incorporate CO_2 of atmospheric origin. Estimation of the carbonate content has shown that large amounts of CO_2 have been sequestered in the tailings at Woodsreef at rates significantly elevated above the background CO_2 uptake rate by chemical weathering of coherent silicate rocks. There is potential to further enhance the rates of CO_2 sequestration by optimizing the tailings storage for this purpose. Natural weathering of ultramafic tailings thus represents a viable option for low-energy, low-cost sequestration of CO_2, directly from the atmosphere.

Since the carbonation of mine tailings during weathering occurs in the aqueous phase additional information on the process can be unlocked by investigating the chemistry and isotopic composition of waters interacting with the tailings. The isotopic composition of these waters also represents an intermediate step in the pathway of the sequestered carbon and can thus serves to better constrain isotopic fractionation during formation of hydrated Mg-carbonates in these settings.

In this contribution we consider the chemistry and isotopic signatures of natural waters that are associated with the carbonation of the tailings of the Woodsreef Asbestos Mine. Measurements of pH, T, conductivity, cation content, δ^2H, $\delta^{13}C_{DIC}$, $\delta^{18}O$ and $F^{14}C$ of water samples are presented and used to discuss the interaction of these waters with the tailings material.

Insights into the reactivity of serpentine minerals and serpentinites under conditions relevant for *ex situ* and *in situ* CO_2 sequestration through mineral carbonation.

A. Lacinska[1,2], M.T. Styles[1] and P.D. Brown[2]

[1] British Geological Survey, Nottingham, United Kingdom
[2] Department of Mechanical, Materials and Manufacturing Engineering, Nottingham University, Nottingham, United Kingdom
*Corresponding author: alci@bgs.ac.uk

Abstract
Serpentine minerals and serpentinites are important Mg-rich materials suggested as a feedstock and host rock for *ex situ* and *in situ* carbon capture and storage by mineralisation, CCSM. Understanding of their functional properties under conditions relevant for both CCSM processes is essential for technology efficiency improvements and feasibility evaluation. Here we present the results of experimental investigation of these minerals and rocks. The results of short term, accelerated acid-leaching of a variety of serpentine minerals show that the process efficiency is dependent on mineral structure. Mg^{2+} leaching efficiency was determined and showed to decrease from chrysotile *ca.* 85% > lizardite 2H1 *ca.* 80% > lizardite 2H1 *ca.* 65% > polygonal serpentine *ca.* 45% > antigorite 30% > lizardite 1T *ca.* 20%. The results of long term carbonation experiments under conditions relevant for *in situ* CCSM show that the reactivity of serpentinite was much greater than that of harzburgite and dunite. Mineral carbonation did occur and the CO_2 was crystallographically captured in magnesite, however the extent of carbonation was low and appeared to be controlled by preferential dissolution of serpentine - filled veins rather than bulk rock serpentinite. Magnesite showed evidence for localised dissolution and thus the fate of CO_2 injected into ultramafic rock reservoir should receive more scientific and experimental attention.

Introduction

Serpentine minerals and serpentinites have long been recognised as a candidate material for CO_2 sequestration through mineral carbonation (CCSM). They are Mg silicates that provide Mg^{2+} cations for the reaction with CO_2 to form Mg carbonates, creating a permanent sink for the greenhouse gas. In both, *ex situ* and *in situ* CCSM scenarios, the Mg cations are released from the mineral structure by a combination of pre-treatment processes, all of which ultimately lead to accelerated disintegration of the serpentine. The degree of disintegration and hence the amount of Mg released puts constraints on the subsequent extent of carbonation and the efficiency of the overall technology. For CCSM technology to be implemented on an industrial scale,

this process step has to be efficient and relatively fast, providing a high % of cations at an acceptable cost. Studies to accelerate the reaction kinetics of *ex situ* Mg extraction include heat treatment [1, 2], reduction of particle size, the use of various organic or inorganic solvents [3-5] and bacterial leaching [6]. The *in situ* scenarios involve increase of the temperature above that naturally encountered at the injection and reaction depth for a general enhancement of reaction kinetics [7, 8].
This paper focuses on the reactivity of carefully selected, purest possible serpentine minerals and a serpentinite rock under conditions relevant for *ex situ* and *in situ* CCSM. Accordingly, two sets of experiments were undertaken to appraise the compositional, structural and chemical modification of the starting materials and the results were used to gain insights into the mineral/rock reactivity and determine the Mg^{2+} leaching efficiency LE, an essential parameter used in *ex situ* CCSM process viability evaluation.

Experimental section

Materials

Six, structurally and chemically diverse serpentine minerals were investigated under conditions relevant for *ex situ* CCSM; and a serpentinite rock, partially serpentinised harzburgite and partially serpentinised dunite were studied under conditions applicable for *in situ* CCSM (Table 1). The bulk chemical composition (XRF) of all samples presented here is shown in Table 2, whilst a detailed mineralogical investigation of the samples is presented elsewhere (Lacinska, 2015 PhD Thesis).

Table 1 Summary of samples investigated in this study.

Mineral/rock type	Sample no	Origin	Experimental application
Antigorite	Ant1	Pian del Re Monviso Massif, Italy	*ex situ*
Lizardite 1T*	Liz1	Monte Fico, Elba Island, Italy	*ex situ*
Lizardite 2H1*	Liz2	Gwendreath Quarry, Lizard Peninsula	*ex situ*
Lizardite 2H1	Liz3	United Kingdom	*ex situ*
Polygonal Serpentine	PLS1	Montecastelli, Tuscany, Italy	*ex situ*
Chrysotile	Chr2	New Amianthus Asbestos Mine, Kaapsche Hoop, Transvaal region, South Africa	*ex situ*
Serpentinite	Srp	Oman-UAE Ophiolite, Wadi Ham Masafi, UAE	*in situ*
Serpentinised Harzburgite	Hzb	Oman-UAE Ophiolite, Ismah, UAE	*in situ*
Serpentinised Dunite	Dun	Oman-UAE Ophiolite, UAE	*in situ*

Lizardite 1T – one layer trigonal polytype; Lizardite 2H1 – two layer hexagonal polytype

Table 2 XRF compositional and loss of ignition data. The major components are in bold. The chemistry reflects the mineralogical heterogeneity of the samples. The total iron is reported as Fe_2O_3

Sample name	*Ant1*	*Liz1*	*Liz2*	*Liz3*	*PLS1*	*Chr1*	*Srp*	*Hzb*	*Dun*
Predominant mineral phase	**Antigorite**	**Lizardite 1T (+chrysotile)**	**Lizardite $2H_1$**	**Lizardite $2H_1$**	**Polygonal Serpentine**	**Chrysotile**	**Serpentinite**	**Harzburgite**	**Dunite**
SiO_2	**41.58**	**39.99**	**42.39**	**42.65**	**40.06**	**37.77**	**34.1**	**40.8**	**35.1**
Al_2O_3	**1.54**	**2.93**	**0.89**	**0.2**	**3.17**	**0.12**	**0.3**	**2.1**	**0.2**
Fe_2O_3 total	**6.63**	**4.28**	**1.85**	**1.04**	**3.29**	**1.82**	**8.3**	**8.4**	**7.5**
Mn_3O_4	0.09	0.04	0.03	0.03	0.04	0.26	0.1	0.1	0.1
MgO	**39.14**	**40.34**	**42.25**	**42.08**	**40.97**	**34.64**	**39.6**	**38.2**	**44.4**
CaO	<0.01	0.14	0.01	0.02	0.02	7.96*	0.2	2.0	0.1
Cr_2O_3	0.3	0.01	<0.01	0.01	<0.01	<0.01	0.5	0.5	0.9
NiO	0.22	0.04	0.2	0.21	<0.01	<0.01	0.2	0.3	0.3
Total	**89.51**	**87.77**	**87.62**	**86.24**	**87.55**	**82.58**	**83.3**	**92.4**	**88.7**
LOI	12.1	13.2	13.94	14.46	13.6	18.16	17.9	8.2	12.2*

**the high CaO value in Chr1 corresponds to the presence of calcite in the starting material; whilst the high LOI value in dunite probably results from sample heterogeneity*

Experimental Procedure

Two sets of controlled laboratory experiments were performed:

(i) ***ex situ*** **CCSM** – hand-picked serpentine mineral grains were milled to pass through 250 μm sieve and the powders obtained were leached with a solution of 1.4 M ammonium bisulphate for a maximum of three hours at a constant temperature of 70°C and ambient pressure. Each serpentine powder was split into 10 g sets of four sub-samples, poured into separate HDEP bottles and filled with 200 ml of the acid pre-heated to 70°C. The bottles were placed on a rotation mixer (6 rpm). For each sample set, the leaching experiments were stopped after 30 (a), 60 (b), 120 (c) and 180 (d) minutes. Solid and liquid reaction products were collected, after each specified time interval, and analysed using a combination of complementary analytical techniques to appraise the time-dependant compositional and textural transformations;

(ii) ***in situ*** **CCSM** – cores (~27 mm long and 17 mm in diameter) of serpentinite, harzburgite and dunite were reacted under conditions relevant for the injection of CO_2 at ~ 1 km depth, *i.e.* 70°C and 100 bar (10 MPa) CO_2 pressure. The experiments were run continuously for eight months (246 days) under supercritical (sc) CO_2 and in CO_2 – saturated brine (synthetic sea water), as well as control experiments under an N_2 - atmosphere. The temperature selected was a compromise between that expected at the relevant injection depth (*ca.* 40-50°C) and the probability of a reaction to occur within a time-constrained experiment. Two cores per each experimental vessel were used one suspended in the head space and exposed to wet supercritical CO_2 and other immersed in CO_2-saturated brine at the bottom of the vessel.

Characterization methods

The characterisation of samples before and after processing was accomplished using a combination of complementary analytical techniques, including: X-ray diffractometry (XRD) to determine crystal structure, scanning electron microscopy (SEM) to appraise the morphology and micro-composition of particles and X-ray fluorescence spectroscopy (XRF) to analyse the bulk composition of the starting materials and reaction products.

Results

The experimental results for *ex situ* and *in situ* experiments are presented separately.

Ex situ study

Figure 1 illustrates the Mg^{2+} leaching efficiency (LE) from serpentine minerals calculated from the amount of magnesium remaining in solid reaction residuum (normalised XRF data). It shows that the maximum LE was obtained after 3 hours of leaching, and *ca.* 85% Mg was extracted from chrysotile followed by lizardite 2H1 (Liz3) *ca.* 80% > lizardite 2H1 (Liz2) *ca.* 65% > polygonal serpentine (PLS1) *ca.* 45% > antigorite (Ant1) *ca.* 30% > lizardite 1T (Liz1) *ca.* 20%.

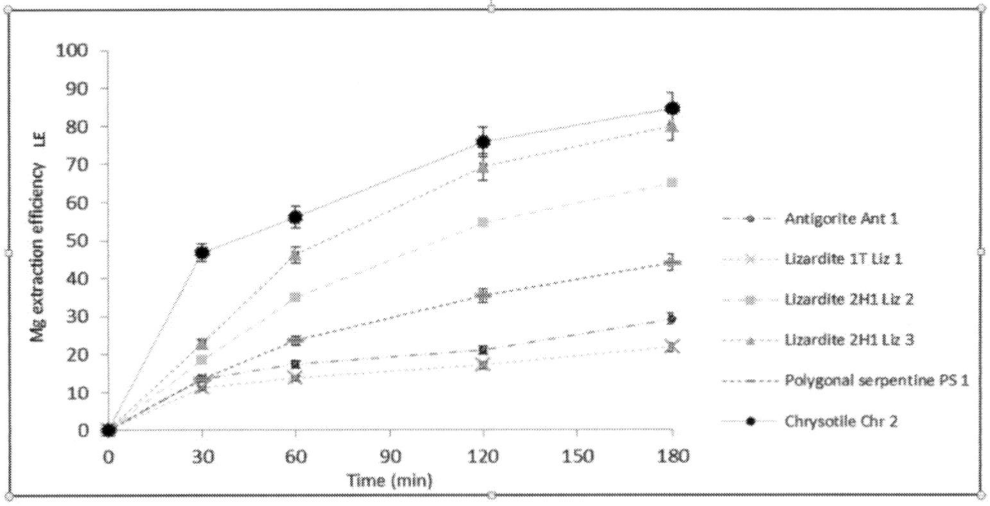

Figure 1 Accelerated acid-induced Mg^{2+} leaching efficiency (LE) from serpentine minerals.

The accelerated acid-leaching resulted in a number of compositional, structural and textural transformations, the extent of which varied from sample to sample. Primary compositional changes included (i) transformation of serpentine, *i.e.* Mg silicate to siliceous, largely amorphous residuum, containing varying amounts of residual Mg and Fe, and (ii) incorporation of the leached Mg or Ca (in Chr2 sample) into boussingaultite $(NH_4)_2Mg(SO_4)_2{\cdot}6H_2O$ or gypsum $CaSO4{\cdot}2H_2O$, respectively, *i.e.* by-products of the reaction between starting minerals and NH_4HSO_4, used as a leaching solvent. The boussingaultite, present in all samples but Liz1, acted as a sink of Mg^{2+} thereby potentially lowering the apparent leaching efficiency measured on the solutions.

The reacted powder particles exhibited a wide diversity of experiment-induced textural changes. Figure 2 illustrates features observed in lizardite $2H_1$ (Liz3) and chrysotile (Chr2), both being representative of the features observed in the remaining samples. Heterogeneous leaching patterns (Figure 2B) and a leached layer texture were distinguished with the later only observed in a few samples (Figure 2 A,C). The heterogeneous patterns were characterised by the coexistence of a localised preferentially leached, usually fractured, siliceous residuum and intact serpentine zones. In contrast, the leached layer formed a roughly concentric rim and appeared to have formed only in particles with an intrinsic textural homogeneity.

Chrysotile was leached extensively and the dominantly siliceous residuum is an extreme case with no intrinsic textures of the original particles. The original chrysotile sample contained *ca.* 10% of calcite. Upon contact with the reactive acid, Ca^{2+} was formed a hydrated calcium sulphate - gypsum, euhedral crystals of which were seen scattered within the dense siliceous residuum (Figure 2E).

Figure 2 BSEM images showing the time-dependent textural and compositional particle transformations in lizardite $2H_1$ (Liz3) and chrysotile (Chr2): A) low magnification image of 60 minutes-leached particles showing a roughly even distribution of leached and fractured shells; B) high magnification image showing texture-controlled preferential leaching; C) high magnification image of concentrically leached particle, with element distribution shown on the associated X-ray element distribution maps D) fine reaction residuum of 180 minutes-leached chrysotile and e) high magnification image showing a mass of siliceous residuum from deeply leached chrysotile, engulfing euhedral crystals of gypsum.

In situ study

The fluid-rock interaction involving serpentinite, dunite and harzburgite over an eight month period resulted in a number of compositional and textural transformations. The transformations were most evident in the reacted serpentinite and were manifested by the formation of white carbonate crust on the surface, and to a lesser extent within, the reacted core. The reactivity of dunite and harzburgite was significantly lower and the extent of carbonation after eight months was negligible.

Figure 3 illustrates the transformations of serpentinite cores reacted with supercritical wet CO_2 (top core – head space of the experimental vessel) and CO_2-saturated brine (bottom core). Apart from a clear evidence for compositional modification, the top core in particular, exhibited textural disintegration resulting in a substantial, *ca.* 30% core loss.

Figure 3 Transformational features of the serpentinite cores during fluid-rock interaction in the CO_2 – pressurised vessel. The white crust in both cores is dominated by Mg-carbonate, predominantly magnesite and minor amounts of calcite (Cal).

Microanalysis of the reacted serpentinite samples revealed that the dissolution of Mg-silicates progressed along original, serpentine-filled veins and did not affect the host rock. In addition, dissolution patterns were also observed along the original as well as authigenic Mg-carbonates. Recycling of original carbonates was particularly well demonstrated within serpentinite from a control experiment, *i.e.* pressurised with N_2 not CO_2. Figure 4 illustrates elongate crystals of aragonite and platy crystals of magnesite, both that formed as a result of dissolution and re-precipitation of original carbonates present in the serpentinite.

Figure 4 Authigenic carbonate mineralisation in serpentinite reacted with saline solution in an N_2 pressurised vessel: a) optical microscopy crossed polarised light image of aragonite found on the outside surface of the serpentinite; b) acicular aragonite crystals; c,d) secondary electron (SE) images showing the morphology of aragonite crystal; e,f) details of Mg-carbonate precipitated on the surface of the serpentinite core (SE).

Discussion and conclusions

This study focused on the reactivity of serpentine minerals and a serpentinite rock under conditions relevant for *ex situ* and *in situ* CCSM.

The results of accelerated acid-leaching of serpentine minerals show significant differences, that are consistent across the "polymorph group". Mg^{2+} LE , of the various types shows the following order: chrysotile *ca.* 85% > lizardite 2H1 (Liz3) *ca.* 80% > lizardite 2H1 (Liz2) *ca.* 65% > polygonal serpentine (PLS1) *ca.* 45% > antigorite (Ant1) *ca.* 30% > lizardite 1T (Liz1) *ca.* 20%.

This reactivity variation can be attributed to several factors amongst which crystal structure and chemistry play an important role. The tubular structure of chrysotile results in a higher inter-crystalline strain and a defective lattice, encouraging faster bond breaking and dehydroxylation [9] and thus producing high LE in this study. Further, the inner surface of the nanotubes provided additional reactive area, thus enhancing the amount of possible reactive pathways. In contrast, the wave-like structural arrangement in antigorite generally leads to a complex three-dimensional network of chemical bonds extending throughout the entire crystal [10 and references therein]. As a consequence a so called "antigorite chemical cement" forms that, as expected, proved acid-resistant in this study and led to overall low Mg^{2+} LE. The number of layers and their stacking order in lizardite results in formation of structural polytypes that as shown in this study, exhibit different reactivity, *i.e.* lizardite $2H_1$ up to 80% LE and lizardite 1T *ca.* 20%.

It is recognised that the reactivity is also controlled by the crystal chemistry of a mineral, in particular the effect of Al on the stabilisation of crystal structure is emphasized, together with the effect of particle size distribution and specific surface area as further discussed in Lacinska *et al* (in review).
In the context of the reactivity of serpentine minerals in the *ex situ* CCSM system, it is concluded that:

1. Any natural serpentine system is rarely mineralogically and crystallographically pure, and this heterogeneity greatly affects the sample reactivity and the extent of cation extraction – essential for technology efficiency. Consequently, the efficiency will be maximised by using a specific type of serpentine.
2. Chrysotile, predominantly due to its tubular morphology and strain-rich structure, is suggested as a best feed material for *ex situ* CCSM.
3. Identifying and quantifying the most reactive types of lizardite in natural rocks will be difficult and require detailed mineralogical studies.

The results of 8-month fluid-rock interaction experiments under conditions relevant for *in situ* CCSM and including three proposed types of host rocks, *i.e.* serpentinite, dunite and harzburgite, showed that the mineral composition and rock texture of the starting, samples, varied greatly. Serpentinite exhibited the greatest degree of dissolution and textural modification; both resulted in the formation of Mg-carbonate crust, thereby fixing the CO_2 in the crystal structure. Although, visually the extent of carbonation seemed significant (Figure 3), detailed microanalysis revealed that most of the carbonate formed on the outside of the core, leaving the inner zones largely intact. Further, it was shown that the original Mg (±Ca) - carbonate and serpentine - filled fractures were the sole source of Mg cations for further reaction with CO_2. No dissolution of the bulk rock was observed. A detailed discussion on the reactivity of serpentinite, dunite and harzburgite and rate of carbonation in comparison with natural analogues will be presented in a future publication. Based on the evidence presented here, it can be concluded that:

1. Amongst the three host rocks typically suggested for *in situ* CCSM, serpentinite proved most reactive in an experimental system used in this study. Dunite and harzburgite were virtually un-reactive.
2. Although CO_2 can be fixed efficiently in magnesite, the extent of this reaction seemed low and was largely controlled by the texture and reactivity of original Mg-rich mineral phases present in the host rock, *i.e.* favouring vein serpentine over the bulk rock serpentinite.
3. The significant textural disintegration of the serpentinite core from the head space of the experimental vessel suggests high reactivity of wet supercritical CO_2 fluid, promoting enhanced geochemical reactions.

4. This study showed that a certain amount of carbonate recycling did take place and attention should be paid to the possibility of carbonate recycling in the injected CO_2-host rock system. Dissolution of original carbonate will inevitably affect the geochemistry of solution. How this would affect the ultramafic reservoir is not known and the commonly quoted statement that CCSM does not require long term monitoring may not be accurate. Thus, the fate of CO_2 injected into ultramafic/serpentinite host rock should be further investigated experimentally and then after closely monitored if the process is implemented on a large scale.

References

1. Maroto-Valer, M.M., Kuchta, M.E., Zhang, Y.Z., Andresen, J.M., Fauth, D.J. 2004. Comparison of physical and chemical activation of serpentine for enhanced CO_2 sequestration. *Abstracts of Papers of the American Chemical Society*, 227, U1095-U1095.
2. Hariharan, S.B., Werner, M., Zingaretti, D., Baciocchi, R., Mazzotti, M. 2013. Dissolution of activated serpentine for direct flue-gas mineralization. *Ghgt-11*, 37, 5938-5944.
3. Park, A.H.A. and L.S. Fan. 2004. CO_2 mineral sequestration: physically activated dissolution of serpentine and pH swing process. *Chemical Engineering Science.* 59 (22-23) 5241-5247.
4. Wang, X. and M.M. Maroto-Valer. 2011. Dissolution of serpentine using recyclable ammonium salts for CO2 mineral carbonation. *Fuel*. 90 (3) 1229-1237.
5. Pundsack, F.L. and N.J. Somerville. 1967. Recovery of silica, iron oxide and magnesium carbonate from the treatment of serpentine with ammonium bisulfate. United States Patent Office, 1967.
6. Power, I.M., G.M. Dipple, and G. Southam, Bioleaching of Ultramafic Tailings by Acidithiobacillus spp. for CO_2 Sequestration. 2010. *Environmental Science & Technology.* 44(1) 456-462.
7. Hovelmann, J., Austrheim, H., Beinlich, A., Munz, I.A. Experimental study of the carbonation of partially serpentinized and weathered peridotites. 2011. *Geochimica Et Cosmochimica Acta.* 75(22) 6760-6779.
8. Van Noort, R., Spiers, C.J., Drury, M.R., Kandianis, M.T.Peridotite dissolution and carbonation rates at fracture surfaces under conditions relevant for in situ mineralization of CO_2. 2013. *Geochimica Et Cosmochimica Acta*, 106 1-24.
9. Gualtieri, A.F., C. Giacobbe, and C. Viti. 2012. The dehydroxylation of serpentine group minerals. *American Mineralogist*. 97(4) 666-680.
10. Capitani, G. and M. Mellini. 2004. The modulated crystal structure of antigorite: The m=17 polysome. *American Mineralogist*. 89 (1) 147-158.

High-gravity Carbonation Process using Rotating Packed Bed for Carbon Capture and Waste Utilization in Steelmaking Industry

Shu-Yuan Pan[1], Pen-Chi Chiang[1,2*], Chieh-Li Chen[1], Tai-Chun Chung [2], Chang-Ching Ho[3], and Chin-Jen Hou[3]

[1] Graduate Institute of Environmental Engineering, National Taiwan University, 10673 Taiwan (ROC)

[2] Carbon Cycle Research Center, National Taiwan University, 10673 Taiwan (ROC)

[3] Tung Ho Steel Enterprise Corporation, 368 Taiwan (ROC)

*Corresponding author: pcchiang@ntu.edu.tw

Abstract

In this study, an integrated approach to waste-to-resource technology was developed for CO_2 capture, wastewater neutralization and product utilization in a high-gravity carbonation process, i.e., rotating packed bed (RPB). Steelmaking slags including electric arc furnace slag (EAFS) and basic oxygen furnace slag (BOFS) were gathered for performance evaluation operated under various levels of particle size of slag, reaction temperature, rotation speed, and liquid-to-solid ratio. According to the observations of TGA, XRD and SEM-EDX, the alkaline wastes were found to be successfully carbonated with CO_2 in an RPB, where calcite ($CaCO_3$) was identified as the main product. Furthermore, blended cements containing 10% and 20% replacements of ordinary Portland cement (OPC) with carbonated slag were tested for strength development and autoclave soundness. For instance, compressive strengths of mortars cast into 50 mm x 50 mm x 50 mm molds were tested at 3, 7, and 28 days according to ATSM C109. The results indicated that the cementitious activity and strength contribution of carbonated slag were suitable to serve as supplementary cementitious Materials for cement-based materials, even up to 20% replacement.

Introduction

Application of rotating packed bed (RPB) for accelerated carbonation of alkaline wastes ("high-gravity carbonation", or in short "HiGCarb") should be a promising approach to overcoming BOFS treatment issues; at the same time, mineralizing gaseous CO_2 from industries. A superior CO_2 fixation efficiency can be achieved using BOFS via the HiGCarb process within a few minutes under ambient pressure and temperature.[1-3] Since fine particle size of BOFS (i.e., < 125 μm) is required in HiGCarb for achieving a better CO_2 mineralization efficiency, the disadvantage of BOFS grinding issue may be possibly trade-off to develop a green supply chain with cement industry, if the total energy consumption and operating cost were accepted. According to the bench-scale operation in our previous study,[1] the total energy consumption including BOFS grinding was found to be acceptable (e.g., 707 kWh with a capture capacity of 1.6 kg CO_2/d). However, to accelerate the commercial step of the developed HiGCarb process, an integrated performance evaluation on both CO_2 mineralization and product utilization as supplementary cementitious materials (SCM) performed in industrial plant is needed.

Many activation approaches such as mechanical activation, thermal activation and chemical activation have been used to improve the cementitious activity of BOFS. However, these techniques usually require a great amount of energy and complicated processes. The high strength, particularly high early strength, frequently is achieved through an increase in the cementitious materials content.[4] However, the report from Portland Cement Association (PCA) indicates that the use of BOFS as SCM may increase the later-age strength of the concrete but also may reduce the early-age strength, as compared with a Portland cement-only concrete. Therefore, the utilization of fresh BOFS as a concrete product or a road base material is still restricted by several barriers including (1) hard to grind, i.e., energy intensive and not cost effective, (2) bad soundness because the contains of free-CaO and -MgO may lead to fatal expansion of hardened cement-BOFS paste, (3) potential environmental impacts of heavy metal leaching and high alkalinity, and (4) low strength of blended cements with fresh BOFS, especially for early ages (< 3 days).[5-8]

According to the targets proposed by the U.S. Department of Energy (DOE), a cost-effective CO_2 capture facility should achieve a removal efficiency of 90%, while maintaining <35% impact on the cost of electricity (COE).[9] It is noted that energy consumption is the primary contributor to COE because it reduces power output from a power plant. In this study, an integrated approach to waste-to-resource technology was developed for CO_2 capture, wastewater neutralization and product utilization in the developed HiGCarb Process. The objectives of this study were (1) to determine the CO_2 capture performance by the developed HiGCarb Process, (2) to evaluate the performance of the carbonated steelmaking slags as supplementary cementitious materials (SCM) in cement mortar, and (3) to calculate the energy consumption and cost-benefit analysis by the HiGCarb Process.

Materials and Methods

Steelmaking slags including electric arc furnace slag (EAFS) and basic oxygen furnace slag (BOFS) were gathered for performance evaluation operated under various levels of particle size of slag, reaction temperature, rotation speed, and liquid-to-solid ratio. In addition, the fresh CRW exhibited highly alkaline (pH ~11.8) and complex in compositions, where the major ions were found to be Na^+, K^+, Cl^-, and SO_4^{2-}. Trace amounts of Ca^{2+}, Fe^{3+}, Cu^{2+}, Al^{3+}, Mg^+, and TIC could be also observed in the fresh CRW. Furthermore, the CO_2 source with an average concentration of 30 ± 2% was supplied directly from hot-stove furnace; no capture or concentrated processes were required in advance.

In this study, the packed bed is horizontal rotation with a mean diamter and a height of 465 mm and 199 mm, respectively. The reaction volume of packed bed is about 160 times greater than that used for a lab-scale in previous studies.[10] In addition, the maximal rotation speed of packed bed is designed as 900 rpm to provide a centrifugal acceleration of up to 2,070 m/s^2 (about 210g).

The compressive strengths of cement mortars cast into 50 mm × 50 mm × 50 mm molds were tested at 3, 7, and 28 days according to ATSM C109. For the control group, a mortar cube with 100% Portland cement type I (i.e., no replacement using BOFS) was also prepared. The cubes were demoulded after 24 hr and then put into saturated lime solution until 56 days. A standard flow of 110 ± 5%, as specified by ASTM C 230,[11] was maintained by adjusting the quantity of mixing water to control the mortars with the same workability and to reduce the effect of compaction during casting the mortars into the molds.

Results and Discussion

CO_2 removal performance by HiGCarb Process

Figure 1 shows the effect of the rotation speed and circulation times on CO_2 removal efficiency in hot-stove gas using the HiGCarb process. The rotation speed varied from 150 to 550 rpm, offering a centrifugal acceleration variation from 60 to 770 m/s^2. The results indicate that the efficiency of CO_2 removal moderately increases as the rotation speed increases up to 300–500 rpm. This was attributed to the reduction of mass transfer resistance (i.e., liquid film thickness) by increasing the rotation speed within this range, which was favourable to carbonation reaction. A CO_2 removal of 93% can be achieved by the HiGCarb process at a rotation speed of 350 rpm under a gas-to-liquid (G/L) ratio of 160. However, when the rotation speed further increased, a reduction in CO_2 removal efficiency was observed, indicating that the extent of reduction in mass-transfer resistances at higher rotation speed was compensated for by a reduction of the retention time.

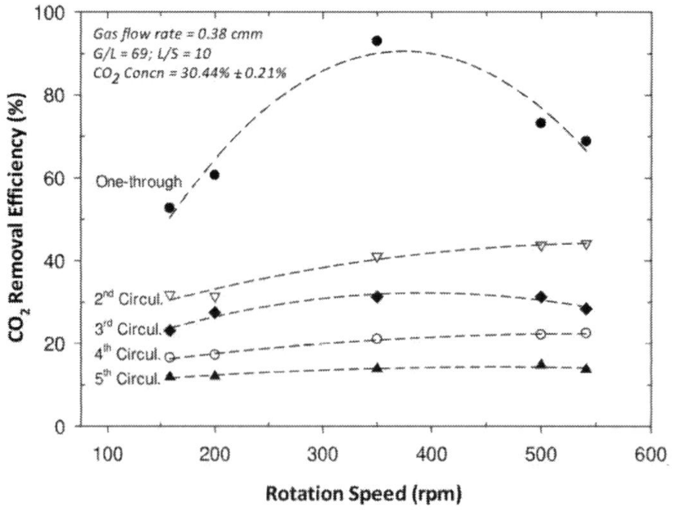

Figure 1. Influence of operation modulus on CO_2 removal efficiency in hot-stove gas using the HiGCarb process

Physicochemical properties of mortar blended with carbonated product

Figure 2 shows the scanning electron microscope (SEM) images of fresh EAFS (F-EAFS) and carbonated EAFS (C-EAFS). It was clearly observed that, before carbonation, the entire F-EAFS was smooth without crystallized precipitates on the surface of slag. After carbonation, the C-EAFS exhibits rhombohedral crystals, with a size of 1 to 3 μm, formed uniformly on the surface of the C-EAFS, which should be calcium carbonate precipitates. In addition, the crystallography of EAFS was also examined by XRD analysis, indicating that the formed product was calcite (i.e., $CaCO_3$). Similar results were obtained in the case of BOFS, which indicated that the steelmaking slags can be successfully carbonated via the HiGCarb process.

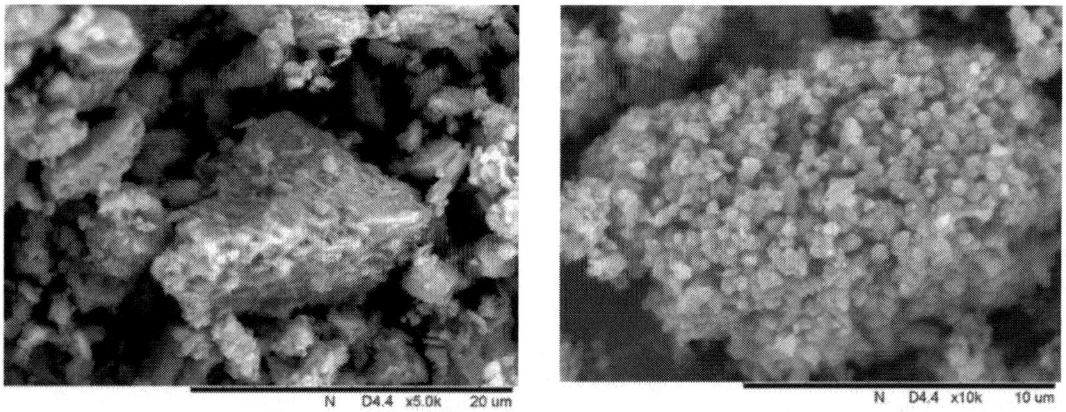

Figure 2. Scanning electron microscope (SEM) images of fresh (left) and carbonated (right) EAFS in the HiGCarb process

In addition, the expansion capacity of mortar was found to increases as the substitution ratio of FBOFS increases due to higher free-CaO contents in mortar. However, after carbonation, the expansion increment of mortar was successfully stabilized, even in the case of up to 20% substitution of CBOFS. This should be attributed to the elimination of free-CaO content in CBOFS.

Furthermore, the compressive strength of blended cement mortar decreased as the substitution ratio of CBOFS in cement increased. The mortar using BOFS with high carbonation conversion exhibited superior 3-day compressive strength to that using pure-Portland cement or fresh BOFS. The hydration of C_3A can be enhanced by $CaCO_3$ to form calcium carboaluminate ($C_3A{\cdot}CaCO_3{\cdot}11H$), as described in Eq. 1, which helps to develop a higher mechanical strength on the early stage:[12]

$$2\,C_3A + 1.5\,CaCO_3 + 0.5\,Ca(OH)_2 + 22.5\,H \rightarrow C_3A{\cdot}CaCO_3{\cdot}11H + C_3A{\cdot}0.5CaCO_3{\cdot}0.5Ca(OH)_2{\cdot}11.5H \quad (1)$$

It suggests that the CBOFS should be suitable for utilizing as high-early strength (HES) cement, specified by ASTM C39,[13] where rapid strength development is desired. The encountered barriers of BOFS utilization as green construction materials could be overcame by the developed HiGCarb to achieve a high early-stage strength development and stabilized expansion properties.

Energy consumption and cost-benefit analysis

In this study, the energy consumptions of the HiGCarb process including air compressors, stirring machines, blowers, pumps, and RPB reactor were measured directly from existing equipment. The grinding power of BOFS was also estimated according to Bond's equation.[1, 14] The results indicate that the overall energy consumption of the HiGCarb process increases with the decrease of CO_2 capture efficiency. The current scale of HiGCarb process was operated at a capture capacity of ~170 kg CO_2 per day, producing ~690 kg of C-BOFS per day which can be utilized as SCM.

According to the cost-benefit analysis as presented in Table 1, the current operating cost for capturing one ton of CO_2 was estimated to be US$18.6. The B/C ratio was about 3.8 if the *i* value is assumed to be 12% for 10 years operation. It was concluded that the developed HiGCarb process exhibited an efficient CO_2 capture performance, which was experimentally determined and mathematically validated, for building up a green waste-to-resource (i.e., slag-to-cement) supply chain.

Table 1. Benefit and cost ratio analysis of HiGCarb process in this study (scale: Q_G=0.38 cmm)

Items	Price (USD/t)	Capacity (t/y)	Annual expense (USD/y)
Direct Cost			
Electricity for CCUM process	18.6	60.9 tons CO_2	1132.8
Total (C_t)			1132.8
Direct Revenue			
Carbon credit (EUR in 2014)	8.1	60.9 tons CO_2	493.0
BOFS treatment cost avoid	10–13.3	192.3 tons F-BOFS	1923.4–2558.2
Carbonated BOFS sale	6	253.2 tons C-BOFS	1519.3
Total (R_t)			3935.7–4570.4
NPV			
IRR (%)			
B/C ratio (i=12%, n=10 yrs)			3.8

Conclusions

In this study, a high CO_2 removal of 93% can be achieved by the HiGCarb process at a rotation speed of 350 rpm under a gas-to-liquid (G/L) ratio of 160. The alkaline steelmaking slag were found to be successfully carbonated with CO_2 in an RPB, where calcite ($CaCO_3$) was identified as the main product. The CBOFS should be suitable for utilizing as high-early strength (HES) cement, where rapid strength development is desired. The encountered barriers of BOFS utilization as green construction materials could be overcame by the developed HiGCarb to achieve a high early-stage strength development and stabilized expansion properties. Furthermore, the current scale of HiGCarb process was operated at a capture capacity of ~170 kg CO_2 per day, associating with an operating cost of USD 18.6/t-CO_2.

Acknowledgements

High appreciation goes to Ministry of Science and Technology (MOST) of Taiwan (R.O.C.) under Grant Number MOST 104-3113-E-007-001 for the financial support. In addition, the authors acknowledge the financial support of Tung Ho Steel Enterprise Corporation under Grant Number 104-S-A17.

References

1. Pan, S. Y.; Chiang, P. C.; Chen, Y. H.; Chen, C. D.; Lin, H. Y.; Chang, E. E., 2013. Systematic Approach to Determination of Maximum Achievable Capture Capacity via Leaching and Carbonation Processes for Alkaline Steelmaking Wastes in a Rotating Packed Bed. *Environ Sci Technol*, 47, (23), 13677-85.
2. Pan, S. Y.; Chiang, P. C.; Chen, Y. H.; Tan, C. S.; Chang, E. E., 2013. Ex Situ CO_2 capture by carbonation of steelmaking slag coupled with metalworking wastewater in a rotating packed bed. *Environ Sci Technol*, 47, (7), 3308-15.
3. Pan, S.-Y.; Chiang, A.; Chang, E.-E.; Lin, Y.-P.; Kim, H.; Chiang, P.-C., 2015. An innovative approach to integrated carbon mineralization and waste utilization: A review. *Aerosol and Air Quality Research*, In press.
4. Caldarone, M. A.; Taylor, P. C.; Detwiler, R. J.; Bhidé, S. B., 2005. Guide Specification for High Performance Concrete for Bridges; *Portland Cement Association: Canada*; p 62.
5. Zhang, T.; Yu, Q.; Wei, J.; Li, J.; Zhang, P., 2011. Preparation of high performance blended cements and reclamation of iron concentrate from basic oxygen furnace steel slag. *Resources, Conservation and Recycling*, 56, (1), 48-55.
6. Monkman, S.; Shao, Y.; Shi, C., 2009. Carbonated Ladle Slag Fines for Carbon Uptake and Sand Substitute. *J. Mater. Civ. Eng.*, 21, 657-665.
7. Wu, H. Z.; Chang, J.; Pan, Z. Z.; Cheng, X., 2009. Carbonate Steelmaking Slag to Manufacture Building Materials. *Advanced Materials Research*, 79-82, 1943-1946.
8. Pan, S.-Y.; Chang, E. E.; Chiang, P.-C., 2012. CO_2 Capture by Accelerated Carbonation of Alkaline Wastes: A Review on Its Principles and Applications. *Aerosol and Air Quality Research*, 12, 770-791.
9. Matuszewski, M.; Ciferno, J.; Marano, J. J.; Chen, S. 2011. Research and Development Goals for CO2 Capture Technology; U.S. Department of Energy: Washington, DC.
10. Chang, E. E.; Pan, S. Y.; Chen, Y. H.; Tan, C. S.; Chiang, P. C., 2012. Accelerated carbonation of steelmaking slags in a high-gravity rotating packed bed. *J Hazard Mater*, 227-228, 97-106.
11. ASTM C 230-98, 2001. Specification for flow table for use in tests of hydraulic cement. In Annual book of ASTM standards.
12. Hawkins, P.; Tennis, P.; Detwiler, R. 2003. The Use of Limestone in Portland: A State-of-the-Art Review; Portland Cement Association: USA.
13. ASTM C39, 2001. Standard Test Method for Compressive Strength of Cylindrical Concrete Specimens. In ASTM American Society for Testing and Materials: West Conshohochen, PA, USA.
14. Bond, F. C., 1961. Crushing and grinding calculations. *Part 1. Br. Chem. Eng.*, 6, 378-385.

Continuous Process for the Aqueous Carbonation of Serpentinite Leachate Derived from Carbonic Acid

T.K. Oliver[1*], B.Z. Dlugogorski[2*] and E.M. Kennedy[1*]

[1] Process Safety and Environment Protection Research Group, School of Engineering
The University of Newcastle, Callaghan, NSW 2308, Australia
[2] School of Engineering and Information Technology Murdoch University
90 South Street, Murdoch, Western Australia 6150, Australia
*Corresponding authors: timothy.oliver@uon.edu.au,
B.Dlugogorski@murdoch.edu.au, eric.kennedy@newcastle.edu.au

Abstract

We present a modelling approach for the continuous aqueous carbonation of serpentinite leachate derived from carbonic acid, suitable for practical application to large-scale CO_2 sequestration. Experiments involved bubbling of gaseous carbon dioxide (CO_2) at a partial pressure of 1 bar into an aqueous suspension of thermally activated serpentinite. Isothermal heating of the ground mineral (-53 μm) at 720 °C for a total period of 30 min resulted in a reactive mineral showing some development of forsterite from the predominantly antigorite basis, with residual hydroxyl content of 53.9 %. An Avrami-Erofe'ev solid state model was fitted to early stage dissolution data (< 10 % Mg extraction) and this was used in conjunction with a kinetic formulation of the carbonate system to model continuous mineral dissolution. Dissolution over a contact time of 5 min was sufficient to produce an alkalised solution that was then degassed in a batch operation at 30°C yielding nesquehonite. X-ray powder diffraction (XRD) and Fourier transform infrared spectroscopy (FTIR) analyses were used to assess changes to the heat activated serpentinite prior to and following dissolution. ICP-OES analysis, and alkalinity measurements were used to estimate magnesium (Mg) and carbon elemental balances.

Introduction

Carbonic acid, formed very simply through the dissolution of gaseous CO_2 in water, is ubiquitous in nature and is uniquely placed in that its reaction with alkaline minerals produces carbonate based alkalinity which is an essential substrate in the formation of mineral carbonates. Although a weak acid, carbonic acid's role in weathering of silicates is of major significance.[1,2,3] In nature, CO_2 degassing and resultant precipitation of carbonates occurs with exposure to ambient conditions of solutions derived from the interaction of carbonic acid with alkaline silicates or carbonate minerals in a subterranean environment.[4] Practical processes for large-scale sequestration of anthropogenic CO_2 must not require additives, catalysts, man-made acids and bases, as economic and thermodynamic considerations prevent recycling of such chemicals. Feedstocks may only comprise naturally-occurring compounds such as silicate minerals, CO_2 itself, water and air. Following recent seminal work at ETH Zurich[5,6,7] this contribution describes development of a new process, based on extraction of Mg by carbonic acid from thermally treated serpentines, followed by precipitation of Mg carbonates as induced by CO_2 degassing, but without a temperature swing.

Methods

Serpentinite samples

Serpentinite (antigorite) sourced from Bingara in the Great Serpentinite Belt in NSW, Australia, was dry crushed to approximately 2 mm and further reduced in size by wet grinding in a rod mill (13 rods, 60% solids, 21 min grind time) specifically to target ground material with 80% passing a 75 µm sieve.[8] The air dried, ground and sieved native serpentinite was further separated yielding a -53 µm sieve fraction using an automatic vibratory sieve shaker (Fritsch Analysette 3 Pro). A Malvern Mastersizer Hydro 2000 SM facilitated laser particle sizing, performed in aqueous media, of the sieved mineral. Isothermal heating of a single sample portion (sufficient for all experimentation) of the -53 µm sieve fraction of the native serpentinite in a muffle furnace (Brother XD-1200N), at a temperature of 720°C for a total period of 30 min, enabled the thermal conditioning of the ground native serpentinite. A Mettler Toledo TGA/DSC 1 machine coupled with a Pfeiffer Thermostar quadruple mass spectrometer (GSD 301 T3) facilitated determination of the extent of dehydroxylation of the thermally activated serpentinite. Alumina crucibles housed the samples which were heated to 900°C at a heating rate of 10°C min^{-1}, under an atmosphere of N_2 flowing at 20 cm^3 min^{-1}.

Mineral dissolution experiments used for subsequent degassing experiments (Expts. D1 to D6)

Dissolution experiments outlined in Table 1 were conducted in order to prepare solutions suitable for degassing experiments (not reported as part of this contribution). Each of the dissolution experiments used a glass 100 cm^3 beaker with the addition of 4.0 g of the thermally activated serpentinite to 50 cm^3 or 80 cm^3 of ultrapure deionised water with electrical resistivity of 18.2 MΩ cm^{-1}. Experiments utilised a heated magnetic stirrer plate open to ambient conditions with the suspension stirred for the duration of each experiment. Food grade (99.5%) CO_2 at a partial pressure of 1 bar was continuously bubbled through the suspension for each experiment via a 2 µm aperture stainless steel diffuser at a controlled flow of 1.39 cm^3 s^{-1} set by mass flow controller (MFC) (Bronkhorst High-Tech; F-201CV) until a stable pH was achieved. The suspension was then heated via the heating plate and the activated serpentinite was added when solution temperature had reached between 50 to 60°C. There was no prevention of water loss during the progress of the dissolution experiments. Solutions were weighed prior to and at the conclusion of each experiment to determine water loss through evaporation. At the end of each experiment the remaining suspension was processed using Whatman Grade 1 filter paper and filtrate subsequently passed through Millex GP 0.22 µm syringe driven filters. Filter cake was dried under ambient conditions and recovered for subsequent analysis.

Table 1. Experimental conditions for batch dissolution experiments (for degassing)

Expt. ID	Activated serpentinite mass (g)/ Mg molarity *(M)*[1.]	Temp. range during heating (°C)	CO_2 flow ($cm^3\ s^{-1}$)	Duration of heating (min)	Total duration of expt. (min)	Initial vol. of soln. (cm^3)
D1	4.0 *(0.80)*	58.4 to 69.3	1.39	5.1	5.1	50
D2	4.0 *(0.80)*	62.7 to 73.0	1.39	4.8	4.8	50
D3	4.0 *(0.50)*	50.9 to 75.9	1.39	62	156	80
D4	4.0 *(0.50)*	60.1 to 76.0	1.39	15.7	77	80
D5	4.0 *(0.50)*	60.0 to 75.8	1.39	90	149	80
D6	4.0 *(0.50)*	60.1 to 77.9	1.39	56	151	80

1. Based on measured data presented in Table 2

Mineral dissolution experiments used for kinetic modelling (Expts. KD1 to KD3)

A second series of dissolution experiments utilising smaller volumes of solution and suitable for estimating kinetic data was also completed. Each experiment, extending over a period of typically 30 min, involved the addition of the thermally activated serpentinite (0.5 g, 1.0 g and 2.0 g lots) to 25 cm^3 of ultrapure deionised water at 55°C. Duplicate experiments were also completed. A flow-through temperature regulated water bath placed over a magnetic stirrer plate allowed for continuous stirring (via PTFE magnetic stirring bar) of the slurry housed in glass 50 cm^3 beakers. The stirring rate was set the same for each experiment with the MFC controlled rate of CO_2 flow to the solution via the diffuser set to 0.69 $cm^3\ s^{-1}$ prior to the addition of the serpentinite. There was no prevention of water loss during the progress of the dissolution experiments, and solutions and filter cake were handled in the same manner as previously described above for the first series of experiments.

Analytical techniques

A Varian 715-ES inductively coupled plasma - optical emission spectrometer (ICP-OES), calibrated using multi-element standard solution, enabled analysis of elemental concentrations of the 0.22 μm filtered solution from each dissolution experiment. Dissolved samples of dried filter cake from selected experiments were analysed using ICP-OES. Dissolution involved adding volumes of 4.5 cm^3 of 65% HNO_3, 4.5 cm^3 of 37% HCl and 3 cm^3 of 50% HBF_4 to 0.1 g of the dried solid residue. Spiking of the samples with 1000 ppm of thulium enabled confirmation of full recovery of digested samples. Samples were heated to 220°C in 20 min in a Milestone start D microwave unit and maintained at this temperature for another 30 min. The clear liquid resulting from digestion was syringe-filtered through a 0.22 μm membrane, diluted and then analysed by ICP-OES. Samples of the dried filter cake were examined using X-ray powder diffraction (XRD) and Fourier transform infrared spectroscopy (FTIR) analytical techniques to assess change in mineral condition. XRD measurements of mineral samples and precipitates were carried out on a Philips X'Pert Pro diffractometer with

Cu Kα radiation operated at a voltage of 40 kV and a current of 40 mA. A goniometer permitted continuous scanning of samples from 5 ° to 90 ° 2θ with a step size of 0.002° or 0.008°. Diffraction patterns were automatched against the International Center for Diffraction Data using X'pert Highscore. FTIR assessment of minerals and precipitates utilised a Perkin Elmer Spectrum Two ATR Perkin-Elmer 100 spectrometer. The range of collected spectra data was 400 to 4000 cm^{-1} wavenumber.

Potentiometric titration of 5 cm^3 aliquots of 0.22 μm filtered samples against standardised 0.01 M sulfuric acid (H_2SO_4) enabled the determination of alkalinity. In-situ pH and temperature measurements for each experiment were automatically logged by a Hanna HI 1131 pH glass electrode (calibrated at working temperature with commercial buffer solutions), HI 3221 meter and HI 92000 software.

Thermodynamic and kinetic modelling

In this study, PHREEQC (Version 2.18) thermodynamic software[9] was used to model the mineral dissolution process. PHREEQC was also used to assess the saturation state of various minerals.

A mathematical model describing the kinetic response of the bulk solution to mineral dissolution was formulated by considering the system of chemical reactions describing homogeneous liquid phase kinetics accounting for species activities and gas-liquid phase kinetics incorporating an open system CO_2 molar flowrate for the transfer of CO_2 from the gas phase to the bulk solution through the gas-liquid interface. The solid state model for dissolution of the activated serpentinite was coupled to the carbonate system to permit modelling of continuous mineral dissolution for aqueous carbonation.

Results and discussion

Mineral characterisation

Results of quantitative elemental analysis of liquid samples of the acid digested -53 μm sieve fraction of the native serpentinite and thermally activated serpentinite used in experiments are shown in Table 2. Calculations of chemical composition were based on the assumption that elements in serpentinites exist as oxides.

Table 2. Chemical composition (by weight) of Bingara serpentinite samples

Items	MgO	Na_2O	SiO_2	Fe_2O_3	Al_2O_3	CaO	NiO	K_2O	Total
Native	37.51	0.44	38.21	8.15	1.76	0.41	0.03	0.52	87.03[1.]
Therm. act.	40.23	0.47	40.21	8.66	1.87	0.42	0.03	0.36	92.25[1.]

1. Structural water accounts for balance

Laser diffraction particle sizing results for the ground native serpentinite of -53 μm sieve fraction and the X-ray diffractogram for the activated serpentinite are shown in Figure 1.

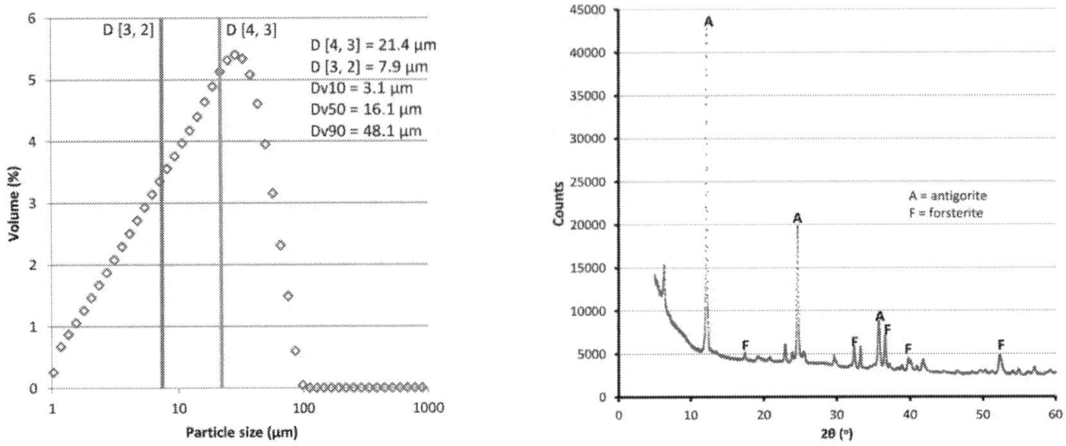

Figure 1. Particle size distribution of native serpentinite samples (-53 µm sieve fraction) used for experiments (left) and X-ray diffractogram of activated serpentinite (right)

The X-ray diffractogram of the native serpentinite showed the mineral as primarily containing antigorite and some magnetite phases. The diffractogram for the activated serpentinite (Figure 1 right) shows that the material has been transformed through heat activation. Antigorite (A) reflection peaks are still evident following heat activation, and forsterite (F) peaks have formed. The degree of dehydroxylation was calculated by comparing the percentage hydroxyl content of the native and heat activated material, assuming the total mass loss correlated to the evolution of water only. We attributed the observed mass loss of 11.5% for the native material to 100% hydroxyl content, and consequently the observed mass loss of 5.3% for the activated sample equates to a residual hydroxyl content for this mineral of 53.9%.

Mineral dissolution

The change in pH and temperature data for selected experiments from the first series, and pH data for the second series of dissolution experiments are shown in Figure 2.

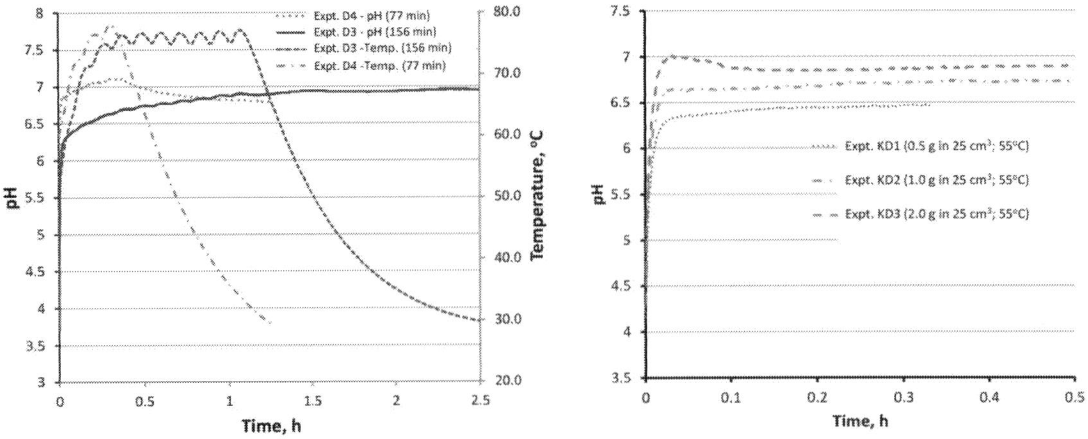

Figure 2. Measured pH and temperature trends for selected batch dissolution experiments (left) measured pH trends for isothermal batch dissolution experiments (right)

pH trends for both series are consistent with the batch configuration of the experiments. Decline in the pH for Experiment D4 after roughly 20 minutes (also evident for Experiments D5 and D6 not shown in Figure 2) is symptomatic of carbonate precipitation[10] and the precipitating phase would be expected to be Mg carbonate[11] although these was no direct evidence through TGA or FTIR analyses of the recovered filter cake for this, possibly as a result of resolubilisation (refer Table 4).

Tabulated experimental conditions and results from solution analyses for the first series of dissolution experiments (Expts. D1 to D6) in Table 3 show stoichiometric proportioning of DIC and Mg suitable for nesquehonite precipitation.[12]

Table 3. Measured concentrations of Mg and Si and calculated DIC for dissolution experiments

Expt. ID	Total duration of dissoln. (min)	Mg content of residual soln. (M)	Si content of residual soln. (M)	DIC content of resid. soln. (M)[3.]	Ratio DIC/ Mg	Ratio Mg/Si	Resid. soln. vol. (cm^3)	Initial soln. vol. (cm^3)	% Mg extrn. (M)[4.]
D1	5.1	4.61×10^{-2}	4.70×10^{-3}	-[1.]	-[1.]	9.82	48.2[2.]	50	5.6
D2	4.8	4.57×10^{-2}	3.77×10^{-3}	0.088	1.93	12.13	48.2	50	5.5
D3	156	6.41×10^{-2}	3.40×10^{-3}	0.141	2.20	18.84	53.7	80	8.6
D4	77	5.06×10^{-2}	3.10×10^{-3}	0.100	1.97	16.35	64.4	80	8.2
D5	149	7.17×10^{-2}	3.21×10^{-3}	0.138	1.93	22.34	44.4	80	8.0
D6	151	6.28×10^{-2}	4.34×10^{-3}	0.109	1.73	14.48	50.1	80	7.9

1. Residual solution diluted with acid prior to alkalinity determination
2. Based on Experiment D2
3. Calculation based on ICP-OES, potentiometric titration analysis and pH using PHREEQC
4. Figures based on ICP-OES of residual solutions, accounting for water loss, relative to total Mg in digested unleached activated mineral

Elevation in pH with batch rather than flow-through configuration impacts Mg extraction for the experiments. The presence of the less soluble phase forsterite as well as particle condition (development of Si passivating layer) in the activated mineral restricts dissolution. Extent of dissolution is also limited by the amount of the amorphous brucite-like phase generated through heat activation in the original sample.

As the mechanism for the formation of alteration layers on silicates during dissolution is commonly attributed to silica leaching accompanied by surface condensation and reconstruction reactions[13], mineral solubility calculations were also completed. Calculated saturation indices (SI, given by log(S) where S=IAP/K_{SP}, where IAP equates to the ion activity product and K_{SP}, the mineral solubility product constant) for amorphous silica and hydromagnesite (HM), likely precipitated phases under experimental conditions using wateq4f database in PHREEQC and shown in Table 4, indicate that for the period of the experiments hydromagnesite transitioned from strongly supersaturated to strongly undersaturated and conversely amorphous silica transitioned from undersaturated to supersaturated. Saturation indices for amorphous silica ($aSiO_2$) and hydromagnesite were calculated for solution states at the end of experiment and at the end of the heating period taking levels of extraction and DIC content shown in Table 3 at the conditions noted in Table 4.

Estimates for amorphous silica saturation show that, over the period of dissolution considered, precipitation of the amorphous silica phase would be expected to impede mineral dissolution for a period following heating. It is also quite probable that hydromagnesite precipitated during heating and was then resolubilised on cooling. These deductions cannot be stated categorically because of uncertainty in thermodynamic data.

Table 4. Calculated amorphous silica ($aSiO_2$) and hydromagnesite (HM) saturation indices (SI) for dissolution Experiments D2 to D6

Expt. ID	SI HM at end of expt. [1.]	SI $aSiO_2$ at end of expt. [1.]	pH at end of expt.	Temp. at end expt. (°C)	SI HM at end of heating [1.]	SI $aSiO_2$ at end of heating [1.]	pH at end of heating	Temp. at end heating (°C)
D2	-1.01 (-0.41)	-0.04 (-0.04)	6.73	73.0	NA	NA	NA	NA
D3	-3.57 (-4.77)	+0.23 (+0.23)	6.95	29.2	+1.20 (+2.28)	-0.10 (-0.10)	6.88	75.9
D4	-5.39 (-6.63)	+0.19 (+0.19)	6.79	29.0	+1.54 (+0.97)	-0.14 (-0.14)	7.05	76.0
D5	-8.81 (-6.30)	+0.18 (+0.18)	6.07	32.9	+0.74 (+2.18)	-0.12 (-0.12)	6.78	75.8
D6	-10.59 (-8.05)	+0.29 (+0.29)	5.87	35.1	+1.20 (+1.58)	-0.00 (-0.00)	6.90	77.9

1. SI for the solution equilibrated with a gas phase of pCO_2 equal to 1 bar are shown in brackets

Differences between the leached and unleached samples were substantiated by XRD analysis (showing enhanced antigorite crystallographic reflections for leached mineral)[14] and differences in FTIR spectra. The -OH stretching vibration region (3700 to 3400 cm^{-1}), centred at 3675 cm^{-1} attributed to OH groups bonded to Mg in octahedral coordination, is retained albeit at reduced strength. More particularly the dipping trend in transmittance within the broader band between approximately 3585 and 3440 cm^{-1} has been demonstrated to be due to the occurrence of interlayer hydrogen bonding, promoted by the substitution for trivalent cations in both tetrahedral and octahedral sites.[15] This trend is evident in native lizardite which has a greater proportion of Al^{3+} and Fe^{3+} and so is always affected by important substitutions for Al^{3+} and Fe^{3+}, whereas antigorites have very low trivalent cations.[15] The same feature for native lizardite is observed for the leached mineral possibly indicating Al^{3+} and Fe^{3+} substitution, (replacing Mg^{2+}) in the mineral during dissolution. The absence of Al and Fe in the leachate from all dissolution experiments may indicate that a proportion of the Al^{3+} and Fe^{3+} ions in octahedral configuration move to tetrahedral positions.

pH data shown in Figure 2 (right) for small isothermal dissolution experiments, similarly observed for replicates, point to a greater rate and extent of dissolution for increased mineral concentrations. There is also apparent proportionality between the amount of activated mineral added and final pH level, and what appears as a slight overshoot and subsequent dipping in pH moving to a stable value for Experiment KD3. The extent of dissolution was measured and also estimated using PHREEQC thermodynamic software.[9] For PHREEQC[9] calculations the dissolved phase was assumed to be brucite ($Mg(OH)_2$) as a result of dissolution results with respect to alkalinity generation and incongruency of dissolution (greatly favouring Mg over Si) (refer Table 3). The measured pH and temperature was used as input to PHREEQC[9] and the amount of brucite necessary to be dissolved to generate the measured pH was calculated for a solution placed in equilibrium with gaseous CO_2 at a pCO_2 of 1 bar reflective of experimental conditions. Listed in Table 5 is the extent of dissolution for each of the isothermal experiments used for kinetic model formulation (Experiments KD1 to KD3) as percentage of total Mg as determined through ICP-OES analysis of residual solutions. Estimated extent of dissolution using PHREEQC[9] based on measured pH at known temperature is also shown. Prior to processing of residual solutions (from isothermal experiments) for ICP analysis, small amounts of precipitate formed in the solutions whilst in sealed sample containers. Acidification of samples resulted in gas evolution and rapid dissolution of the presumed carbonate fraction although trace amounts of a complementary phase, thought to be Mg silicate, persisted.

Table 5. Measured and thermodynamically modelled extent of mineral dissolution for isothermal batch experiments

Expt. ID	Added serpentinite mass, g *(M)*[1.]	Measured extent of dissoln., M *(%)*[1., 2. & 3.]	Extent of dissoln. from PHREEQC model, M *(%)*[1.]
KD1	0.5 *(0.20)*	0.0219 *(11.0)*	0.0183 *(9.2)*
KD2	1.0 *(0.40)*	0.0370 *(9.3)*	0.0367 *(9.2)*
KD3	2.0 *(0.80)*	0.0625 *(7.8)*	0.0587 *(7.4)*

1. Based on measured data presented in Table 2
2. Figures based on ICP-OES of residual solutions, accounting for water loss, relative to total Mg in digested unleached activated mineral
3. Excludes trace amounts of complementary minor phase

Solid state model formulation

Fractional conversion calculations were completed using PHREEQC thermodynamic software[9] for Experiments KD1 to KD3 utilising the observed change in solution pH at a given temperature as described above under the assumption that aqueous equilibrium and gas-liquid phase equilibrium was maintained at all stages during dissolution. This involved inputting to PHREEQC[9] measured pH and temperature values at periodic intervals through the rapid period of dissolution, with the solution equilibrated with a CO_2 gas phase of pCO_2 equal to 1 bar, to arrive at an estimate of extent of brucite dissolution (refer Figure 2 – right). This enabled a time course of estimated extent of mineral dissolution to be generated. In determining fractional conversion (shown for

Experiment KD3 in Figure 3 for example), the full extent of mineral availability for each experimental case was taken as that achieved once the observed rate of pH increase had slowed considerably compared to earlier rates, and this was after an elapsed time of 100 s representing early stage dissolution relevant to rapid ex-situ carbonation applications. Consequently the solid state model does not represent dissolution of the added mineral in its entirety, only the first 100 s of data derived from dissolution experiments. In this study selected solid state models were considered including shrinking core, shrinking disc, 3D and 2D diffusion models and the Avrami-Erofe'ev model described in differential form by Eq. 1[16] and evident in other mineral dissolution studies[17] was found to best fit observed kinetic behaviour for the initial period of dissolution as shown for Experiment KD3 in Figure 3.

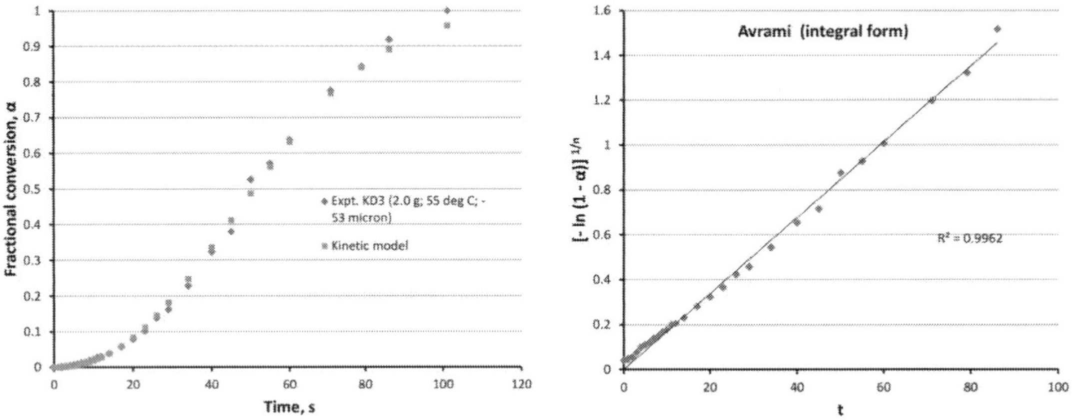

Figure 3. Fractional conversion for Experiment KD3 limited to 100 s (left) solid-state model fit for Experiment KD3 over first 100 s of experiment (right)

$$\frac{d[\alpha]}{dt} = k\ n(1-\alpha)[-ln(1-\alpha)]^{(n-1)/n} \qquad (1)$$

where n and k are coefficients, α the fraction of solid reactant transformed, and t time. The apparent rate constant k is particular to the dissolving particle (size, volume, shape factor, molar density of reactant, and concentration of fluid reactant) and is also dependent on experimental conditions including temperature. This model requires only two fitted parameters, namely the rate constant k and exponent n. Table 6 shows results of fitted Avrami-Erofe'ev rate constants and exponents.

Table 6. Avrami-Erofe'ev apparent rate constants and exponents for isothermal batch experiments

Expt. ID	Avrami-Erofe'ev rate constant, k (s^{-1})	Avrami-Erofe'ev exponent, *n*
KD1	1.94×10^{-2}	1.76
KD2	1.81×10^{-2}	1.91
KD3	1.67×10^{-2}	2.21

Various empirical and surface complexation models (SCM) have been used to describe dissolution kinetics of silicates. The commonly observed 0.5 order dependence of silicate dissolution rate on protonic activity can be equally described by diffusion limitation across a thickening leached layer[18] or proton penetration into the leached layer and adsorption on silica dimers.[19] SCMs can account for surface reactions devoid[20,21] or with complexing agents.[22,23] Analysis using population balance approaches coupled with a mass balance equation to account for the evolution of the particle size distribution (PSD) being a powerful technique accounting for representative mineral feeds and change to dissolution behaviours using a surface reaction control simplification[21] or more complicated approaches accounting for particle structure. Hariharan et al.[7] considered numerous influences in modelling the kinetic behaviour of dissolution of activated lizardite under different modelling scenarios using a SCM and accounting for particle state but with various simplifying assumptions including dissolution of CO_2 and the other associated speciation reactions to be in thermodynamic equilibrium, and showed that observed trends could be matched but with many fitted parameters.

Although particular SCM mechanisms can be formulated for simple molecular surface structures[24,25] capturing active species in dissolution, reactions in actuality may be predicated on other ionic movements and chemical bonding rearrangements within the structure, changing bond strengths accordingly. For example rate controlling exchange reactions between cations of the crystal network which participate in precursor complex formation.[26] These aligned and supporting mechanisms to surface reactions may well impact continuing rates of reaction or aspects of the dissolution mechanism.

Flow-through dissolution studies can provide useful information on the specific influence of factors affecting dissolution, however this approach is limited in that the state of the initial mineral batch varies during the course of the experiment. This is not a significant concern when the mineral sample is mono-sized or with the geometric nature of the particle or particle distribution well defined; of uniform and unvarying composition; of known reactive surface area (both prior to experimentation and after experimentation); and without production of any solid product phase or leached silicate enriched phase which can significantly impede rate of dissolution. However such complicating factors are of considerable importance in the interpretation of Mg silicate dissolution.

In this contribution we outline our approach in using a solid state model fitted to early stage dissolution data for the activated mineral (relevant to rapid ex-situ carbonation) with this model coupled to the carbonate system in order to quantitatively describe carbonate alkalinity generation. This coupled model is then applied to a continuous leaching reactor configuration. This method does not require knowledge of the intricacies of the dissolution mechanism. We emphasise that the fitted solid state model is only representative of the dissolution case to which it is applied, however it embodies all mineral specific and solution specific parameters in its formulation particular to the system including particle condition (particle size and distribution, shape, reactive surface area, molar density of product phase and solid reactant, thickness of product phase, point of zero charge, degree of dehydroxylation and its effect on reactivity/reacting surface area, internal structural change to the mineral and porosity, etc.). It also includes the effects of variable intrinsic properties of the mineral such as number and nature of defects and prevalence of impurities, or any shifts in mechanism throughout the course of dissolution.

Other factors including solution composition (pH, chemical species activities, effect of reducing and complexing agents), temperature (and pressure) dependency of both reaction and any diffusion components of the dissolution mechanism, fluid velocity and turbulence affecting any external mass transfer (diffusion) limiting processes, kinetic aspects of the dissolution of CO_2 and pCO_2 of the system as this directly impacts protonic activity and also the buffering capacity of the system, are also implicit in the model.

Degassing

Solutions from dissolution experiments derived from the first series of experiments (Expts. D1 to D6 noted herein) were degassed at 30°C by sparging with nitrogen (N_2). Whilst details of our experimentation and modelling are not presented in this contribution, nesquehonite was the single phase identified by XRD analysis and confirmed through SEM-EDS analysis. The pH trends exhibited during these degassing batch experiments were similar to that as previously reported[12], and kinetic modelling (incorporating the liquid-solid heterogeneous reaction in the bulk solution) with application to continuous operation has been completed and will be presented in a future publication. Our approach differs from another two-stage low-pressure carbonation aqueous carbonation route currently being researched whereby carbonate precipitation (using leachate derived from the carbonic acid dissolution of heat activated lizardite) is induced through increase in solution temperature which results in decline in solution pCO_2 and decrease in carbonate solubility, with pCO_2 swing without change in temperature (at 90°C) shown as less effective.[6] While this alternative mineralisation strategy was demonstrated as robust, the requirement for large energy input to bring about the necessary temperature shift of process solution through the preferred temperature swing route may limit its effective application to continuous operation in practice.

Kinetic modelling of mineral dissolution

The kinetic modelling of mineral dissolution with alkalinity generation in an aqueous system is facilitated by coupling the derived solid state rate equation with the carbonate system. This is by way of incorporation of the dissolution reaction, given by Eq. 2, in the reaction set for homogeneous reactions comprising the aqueous system. The ODE set was solved to establish trends in speciation and pH as the dissolution reaction proceeded particular to the dissolution case under consideration.

$$Mg(OH)_{2\ (s)} \rightleftharpoons Mg^{2+}_{(aq)} + 2OH^{-}_{(aq)} \qquad (2)$$

The ODE set also incorporated an empirical relationship defining rate of CO_2 mass transfer from the gas phase to the solution. This relationship defines rate of transfer as dependent on gas-liquid interfacial area and difference in CO_2 activity between the gas and aqueous phases. This rate of mass transfer varies during the course of rapid dissolution of the mineral because of varying differential in CO_2 activity such that the actual activity of aqueous CO_2 is not at equilibrium with the gas phase. Also, the rate at which the brucite-like amorphous phase dissolves when first exposed to solution can lead to imbalance in the homogeneous carbonate reactions in the solution. Rapid mineral dissolution in excess of the relaxation time for the carbonate system can lead to measured pH in excess of its equilibrium level as evidenced by overshoot in pH response in Figure 2. By comparing the modelled kinetic response of the system with the observed solid state model, adjustments to the Avrami-Erofe'ev exponent to yield actual dissolution behaviour can be performed, thus accounting for carbonate system kinetics and yielding a true representation of dissolution for the activated mineral. In establishing a suitable kinetic model we consider the system as comprised of Mg and C reflective of incongruency of dissolution during the Avrami-Erofe'ev phase. Other elemental influences are considered minor and consequently the effect of silicic acid and silicic carbonate reactions are excluded.

Continuous dissolution has a variety of important industrial applications.[27] The acid leaching of mineral rich ore configured as a continuous process to extract desired components is a common approach in mineral processing.[28,29] The modelling of dissolution as a continual process involves the continuous addition of the reactant to the system, which in reality equates to a mass flowrate of the bulk mineral, and the extraction of the homogeneous slurry from within the reactor. The dissolution rate proceeds in accordance with the Avrami-Erofe'ev rate equation as defined by the function *dissolve* in Eq. 3 where *Mg(OH)2* is the concentration of brucite at any given time, $Mg(OH)2_{ini}$ is the initial concentration of brucite, Q_{in} and Q_{out} (equal in this case) represent inlet and outlet volumetric flowrates respectively and likewise $Mg(OH)2_{in}$ and $Mg(OH)2_{out}$ represent inlet and outlet concentrations of brucite. Variables in Eq. 4 are as previously defined in Eq. 1.

$$\frac{d[Mg(OH)2]}{dt} = -dissolve * Mg(OH)2_{ini} + Q_{in}[Mg(OH)2_{in}] - Q_{out}[Mg(OH)2_{out}] \qquad (3)$$

$$dissolve = k\ n(1-\alpha)[-ln(1-\alpha)]^{(n-1)/n} \qquad (4)$$

When Eq. 3 is coupled with the ODE set for the aqueous Mg-carbonate system, the extraction of the solution from the reactor necessarily involves the removal of residual solid reactant, along with other species in order to prevent solids build-up within the reactor. The extractive flowrate for the reactant (as given by $Q_{out}[Mg(OH)2_{out}]$) has the net effect of lowering the rate of feed of reagent to the reactor. Furthermore, when the coupled model is considered, the products of the solid-state dissolution reaction (defined by Eq. 2) enter into the system, allowing the determination of the effect of mineral dissolution on the complete system.

As a continuous stirred tank reactor (CSTR), the solution within the model reactor is homogeneous and at any point in time is of uniform composition such that the concentration of the extracted solution mirrors that of the reactor. Accordingly, extractive molar flowrates for all species are included in the model, and the species specific extraction flowrates are defined by the molar concentrations of the species. For the continuous steady flow case, reactor volume is fixed and the effective residence time in the reactor can be varied by the volumetric rate of aqueous extraction. More particularly for a continuous dissolution process, the amount of alkalinity removed from the system with respect to time in conjunction with Mg extraction rates are central to the effectiveness of a subsequent continuous degassing and precipitation process.

The key parameters relating to total Mg extraction flowrate (*mfMgT*) under steady-state, comprising the molar extraction rate for the Mg ion and various Mg carbonate ion pairs are given in Eq. 5. Also shown in similar fashion to that for total Mg extraction flowrate are equations for total steady-state extraction rates for DIC (*mfDICT*) and alkalinity (*mfAlk*) in Eqs. 6 and 7 respectively. Constitutive terms representing the molar extraction rate of respective species are prefixed by *mf*, for example *mfMgOH* for the $MgOH^+$ species. As the reactor liquor would be filtered or clarified of solids in practice, the alkalinity within the mineral is not carried over or considered in calculating alkalinity of the aqueous outflow stream.

$$mfMgT = mfMg + mfMgCO3 + mfMgOH + mfMgCO3aq + mfMgHCO3 \quad (5)$$

$$mfDICT = mfCO2 + mfHCO3 + mfCO3 + mfMgHCO3 + mfMgCO3aq \quad (6)$$

$$mfAlk = mfHCO3 + 2*mfCO3 + 2*mfMgCO3aq + mfMgHCO3 + mfMgOH - mfH + mfOH \quad (7)$$

These key parameters can be used as a means of comparison of cases where the rate of mineral feed to and rate of aqueous extraction from a continuous dissolution reactor based on data derived from batch dissolution experiments with adjusted Avrami-Erofe'ev exponent for a range of scenarios. This analysis will be covered in detail in a subsequent publication.

Conclusions

The use of CO_2 degassing as a method for the production of mineral carbonate for solutions derived from the carbonic acid dissolution of thermally activated serpentinite previously proposed[12] has been demonstrated. Kinetic modelling of both dissolution and degassing processes show that each can be operated on a continuous basis and together as an integrated scheme for the capture and subsequent mineralisation of CO_2. With a continuous arrangement, mineral saturation in both dissolution and degassing process units is maintained at a fixed and optimal level and the precipitated phase of defined classification is extracted with the solution in the degassing reactor.

An overall scheme for a practical application to large-scale CO_2 storage is envisaged which combines dissolution and degassing process units thereby allowing for their individual optimisation, and optimisation as part of an overall integrated mineral carbonation process. Both process units in operation enable rapid capture of CO_2 emissions and subsequent mineralisation of the captured CO_2 with low energy expenditure. Application of the proposed approach is particularly relevant in view of the large deposits of suitable mineral feedstock available world-wide.

Acknowledgements

T.K. Oliver wishes to thank R. Balucan for preparation of native antigorite used in this study, M. Ghoorah for assistance with XRD and SEM analyses and F. Farhang for evaluation of suitable XRD analysis techniques.

References

1. Navarre-Sitchler, A., Thyne, G. 2007. Effects of Carbon Dioxide on Mineral Weathering Rates at Earth Surface Conditions. *Chemical Geology*, 243(1–2), 53-63.
2. Lerman, A., Wu, L., Mackenzie, F.T. 2007. CO_2 and H_2SO_4 Consumption in Weathering and Material Transport to the Ocean, and their Role in the Global Carbon Balance. *Marine Chemistry*, 106(1–2), 326-350.
3. Gaillardet, J., Dupré, B., Louvat, P., Allègre, C.J. 1999. Global Silicate Weathering and CO_2 Consumption Rates Deduced from the Chemistry of Large Rivers. *Chemical Geology*, 159(1–4), 3-30.
4. Pentecost, A., 2005. *Travertine*. Springer-Verlag Pub., Heidelberg.
5. Werner, M., Hariharan, S., Zingaretti, D., Baciocchi, R., Mazzotti, M. 2014. Dissolution of Dehydroxylated Lizardite at Flue Gas Conditions: I. Experimental Study. *Chemical Engineering Journal*, 241, 301-313.
6. Werner, M., Hariharan, S., Mazzotti, M. 2014. Flue Gas CO_2 Mineralization Using Thermally Activated Serpentine: From Single to Double-Step Carbonation. *Energy Procedia*, 63, 5912-5917.
7. Hariharan, S., Werner, M., Hänchen, M., Mazzotti, M. 2014. Dissolution of Dehydroxylated Lizardite at Flue Gas Conditions: II. Kinetic Modeling. *Chemical Engineering Journal*, 241, 314-326.
8. Balucan, R.D. 2012. Thermal Studies of Magnesium Silicates from the Great Serpentinite Belt in New South Wales for CO_2 Sequestration by Mineral Carbonation in Australia, Chemical Engineering, School of Engineering, Faculty of Engineering and Built Environment, The University of Newcastle, Newcastle, Australia.
9. Parkhurst, D.L., Appelo, C.A.J. 1999. User's Guide to PHREEQC (Version 2) - A Computer Program for Speciation, and Inverse Geochemical Calculations. *U.S. Geological Survey Water-Resources Investigations Report 99-4259.*
10. Lin, F.-C., Clemency, C.V. 1981. The Dissolution Kinetics of Brucite, Antigorite, Talc, and Phlogopite at Room Temperature and Pressure. *American Mineralogist*, 66(7-8), 801-806.

11. Hövelmann, J., Putnis, C.V., Ruiz-Agudo, E., Austrheim, H. 2012. Direct Nanoscale Observations of CO_2 Sequestration During Brucite [$Mg(OH)_2$] Dissolution. *Environmental Science & Technology*, 46(9), 5253-5260.

12. Oliver, T.K., Dlugogorski, B.Z., Kennedy, E.M. 2014. Biologically Enhanced Degassing and Precipitation of Magnesium Carbonates Derived from Bicarbonate Solutions. *Minerals Engineering*, 61, 113–120.

13. Schott, J., Pokrovsky, O.S., Spalla, O., Devreux, F., Gloter, A., Mielczarski, J.A. 2012. Formation, Growth and Transformation of Leached Layers during Silicate Minerals Dissolution: The Example of Wollastonite. *Geochimica et Cosmochimica Acta*, 98, 259-281.

14. Tang, J., Valix, M. 2004. Leaching Of Low- Grade Nickel Ores by Fungi Metabolic Acids, in *Separations Technology VI: New Perspectives on Very Large Scale Operations*, C. Fell and Keller II, G.E., Editors, Berkeley Electronic Press.

15. Mellini, M., Fuchs, Y., Viti, C., Lemaire, C., Linarès, J. 2002. Insights into the Antigorite Structure from Mossbauer and FTIR Spectroscopies. *European Journal of Mineralogy*, 14(1), 97-104.

16. Brown, M.E., Maciejewski, M., Vyazovkin, S., Nomen, R., Sempere, J., Burnham, A., Opfermann, J., Strey, R., Anderson, H.L., Kemmler, A., Keuleers, R., Janssens, J., Desseyn, H.O., Li, C.-R., Tang, T.B., Roduit, B., Malek, J., Mitsuhashi, T. 2000. Computational Aspects of Kinetic Analysis: Part A: The ICTAC Kinetics Project-data, Methods and Results. *Thermochimica Acta*, 355(1–2), 125-143.

17. Cornell, R.M., Giovanoli, R. 1993. Acid Dissolution of Hematites of Different Morphologies. *Clay Minerals*, 28(2), 223-232.

18. Brantley, S.L. 2008. Kinetics of Mineral Dissolution, in *Kinetics of Water-rock Interaction*. Springer Science+Business Media.

19. Pokrovsky, O.S., Schott, J. 2000. Kinetics and Mechanism of Forsterite Dissolution at 25°C and pH from 1 to 12. *Geochimica et Cosmochimica Acta*, 64(19), 3313-3325.

20. Prigiobbe, V., Costa, G., Baciocchi, R., Hänchen, M., Mazzotti, M. 2009. The Effect of CO_2 and Salinity on Olivine Dissolution Kinetics at 120°C. *Chemical Engineering Science*, 64(15), 3510-3515.

21. Hänchen, M., Krevor, S., Mazzotti, M., Lackner, K.S. 2007. Validation of a Population Balance Model for Olivine Dissolution. *Chemical Engineering Science*, 62(22), 6412-6422.

22. Prigiobbe, V., Mazzotti, M. 2011. Dissolution of Olivine in the Presence of Oxalate, Citrate, and CO_2 at 90°C and 120°C. *Chemical Engineering Science*, 66(24), 6544-6554.

23. Dove, P.M. 1994. The Dissolution Kinetics of Quartz in Sodium Chloride Solutions at 25 Degrees To 300 Degrees C. *American Journal of Science*, 294(6), 665-712.

24. Pokrovsky, O.S., Schott, J., Castillo, A., 2005. Kinetics of Brucite Dissolution at 25°C in the Presence of Organic and Inorganic Ligands and Divalent Metals. *Geochimica et Cosmochimica Acta*, 69(4), 905-918.

25. Schott, J., Pokrovsky, O.S., Oelkers, E.H. 2009. The Link between Mineral Dissolution/Precipitation Kinetics and Solution Chemistry. *Reviews in Mineralogy and Geochemistry*, 70(1), 207-258.

26. Daval, D., Martinez, I., Corvisier, J., Findling, N., Goffé, B., Guyot, F. 2009. Carbonation of Ca-Bearing Silicates, the Case of Wollastonite: Experimental Investigations and Kinetic Modeling. *Chemical Geology*, 265(1–2), 63-78.
27. Steemson, M.L., White, E.T. 1987. An Analysis of Continuous Dissolution Processes by Population Balance Techniques. *Journal of Crystal Growth*, 82(3), 311-317.
28. Levenspiel, O. 1999. *Chemical Reaction Engineering*, John Wiley & Sons Pub.
29. Crundwell, F.K. 1995. Progress in the Mathematical Modelling of Leaching Reactors. *Hydrometallurgy*, 39, 321-335.

Enzymes and Catalysts for Accelerated Carbonation and Weathering

Roger Aines
Senior Scientist, Atmospheric, Earth, and Energy Division
and Carbon Fuel Cycle Program Leader

Lawrence Livermore National Laboratory

Key rate limiting steps:

1) CO_2 dissolution

2) Silicate dissolution

The transfer of CO_2 into water or other liquids is almost always dominated by chemical reactions at the liquid interface.

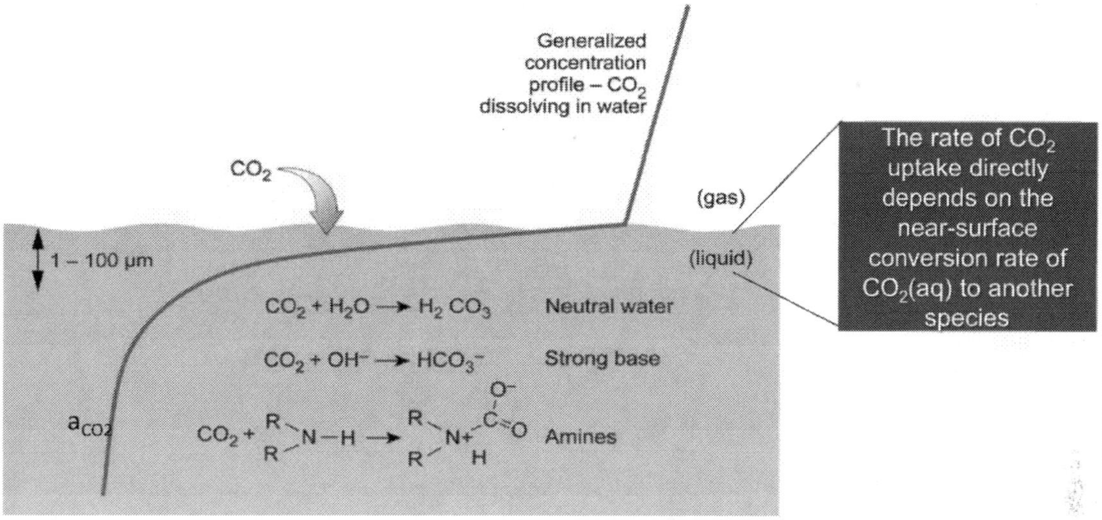

Lawrence Livermore National Laboratory

The reaction rate of CO_2, and energy of that reaction, are strong functions of pH (directly or indirectly)

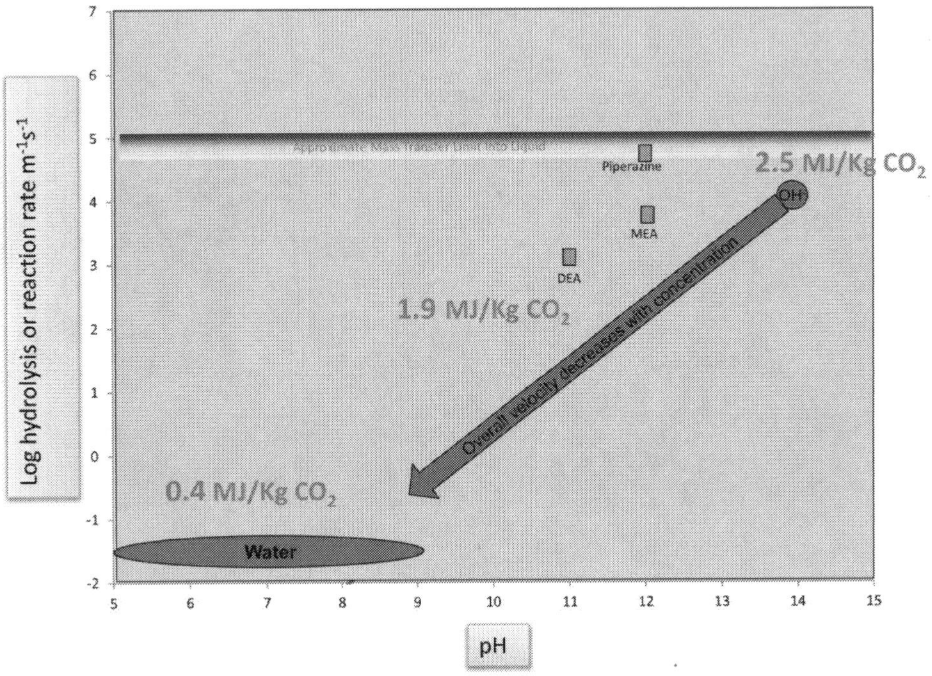

High pH is the single most important factor in transfer rate

Carbonic anhydrase is one of the most rapid enzymes known – it was first discovered in human lungs, where it facilitates CO_2 exhalation

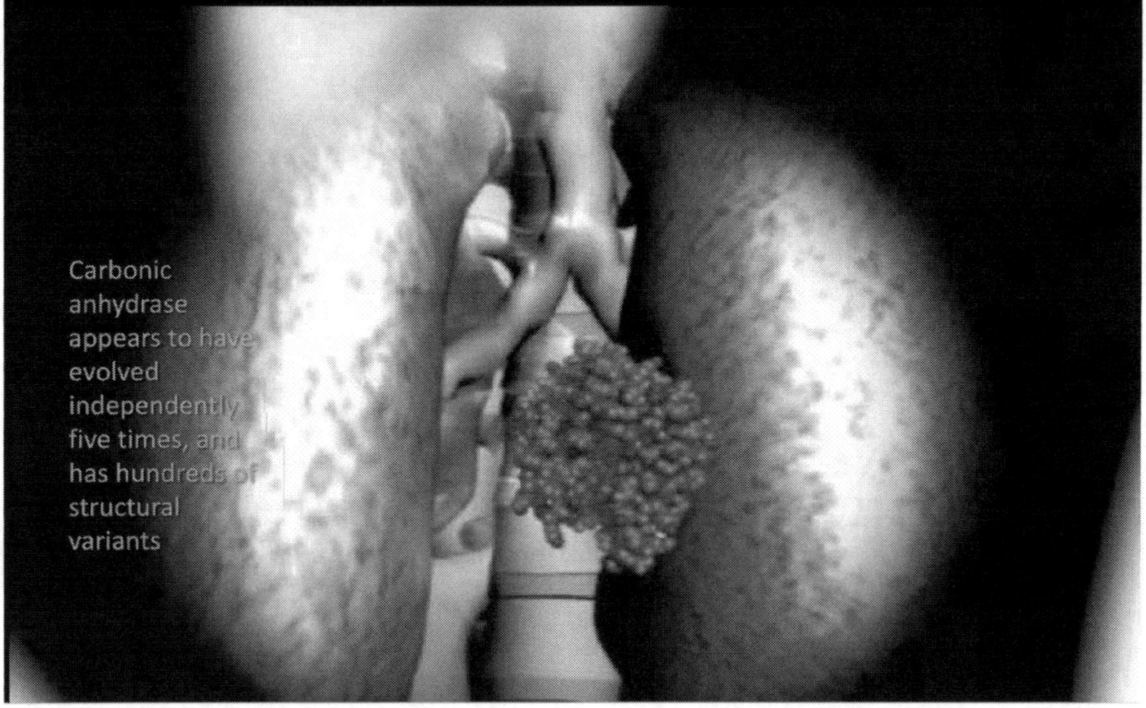

Carbonic anhydrase catalysis of CO_2 hydration is chemically simple

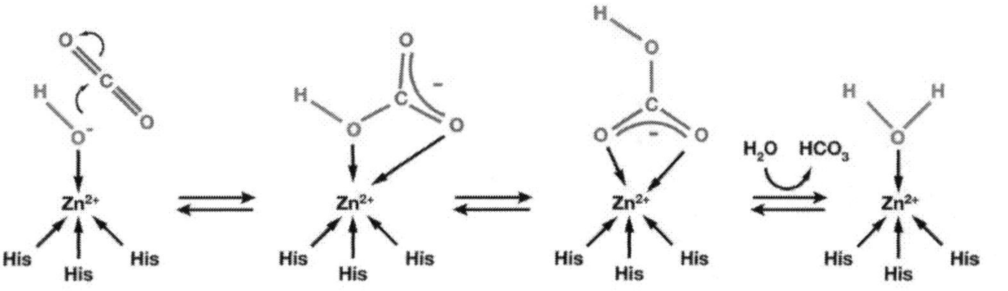

$$H_20 + CO_2 \leftrightarrow H^+ + HCO_3^-$$

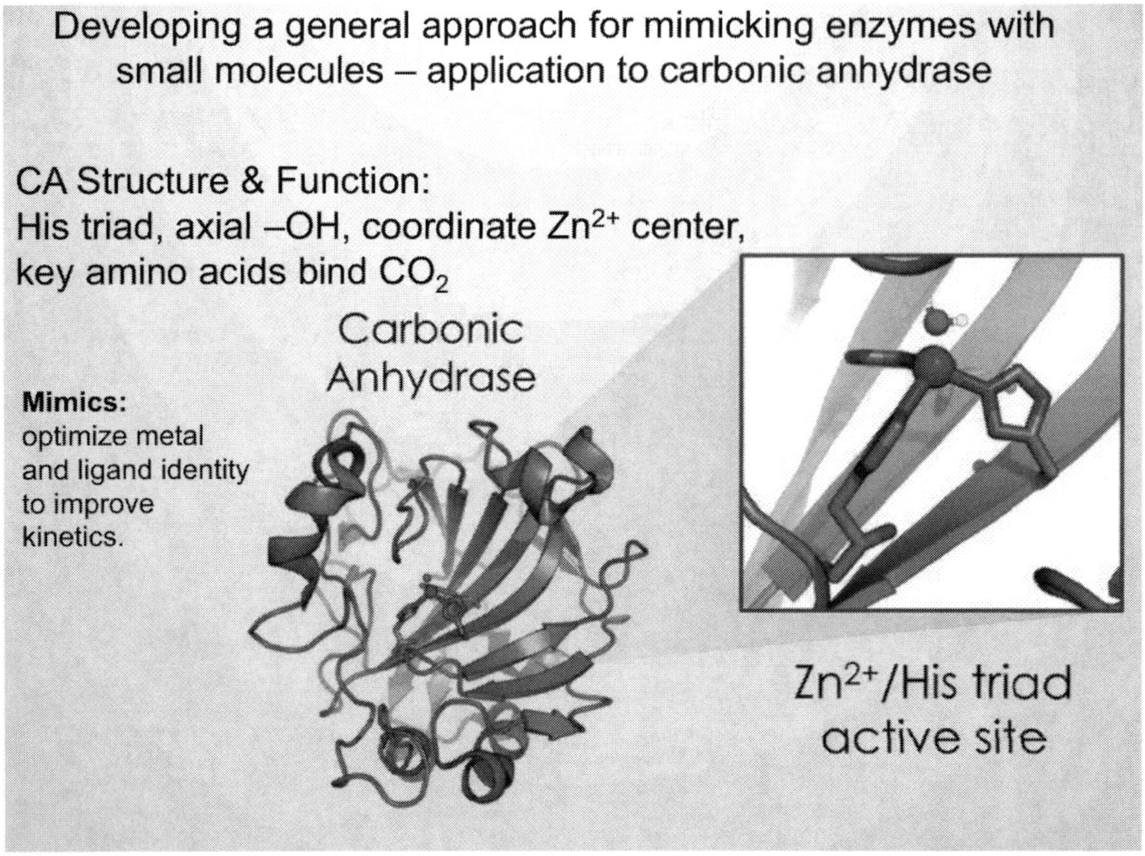
Developing a general approach for mimicking enzymes with small molecules – application to carbonic anhydrase
CA Structure & Function:
His triad, axial –OH, coordinate Zn^{2+} center,
key amino acids bind CO_2
Carbonic Anhydrase
Mimics:
optimize metal and ligand identity to improve kinetics.
Zn^{2+}/His triad active site

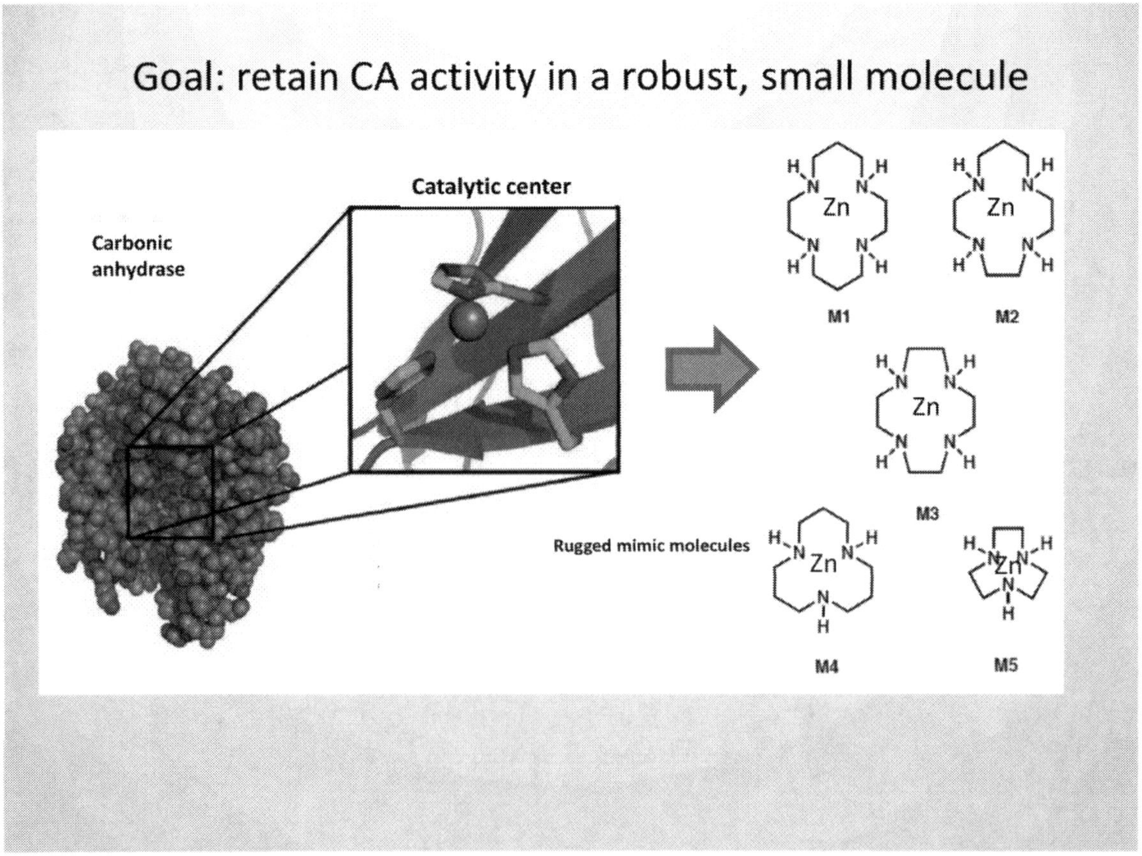
Goal: retain CA activity in a robust, small molecule
Catalytic center
Carbonic anhydrase
Zn
M1
M2
M3
M4
M5
Rugged mimic molecules

We have to understand the full reaction to optimize the design approach.

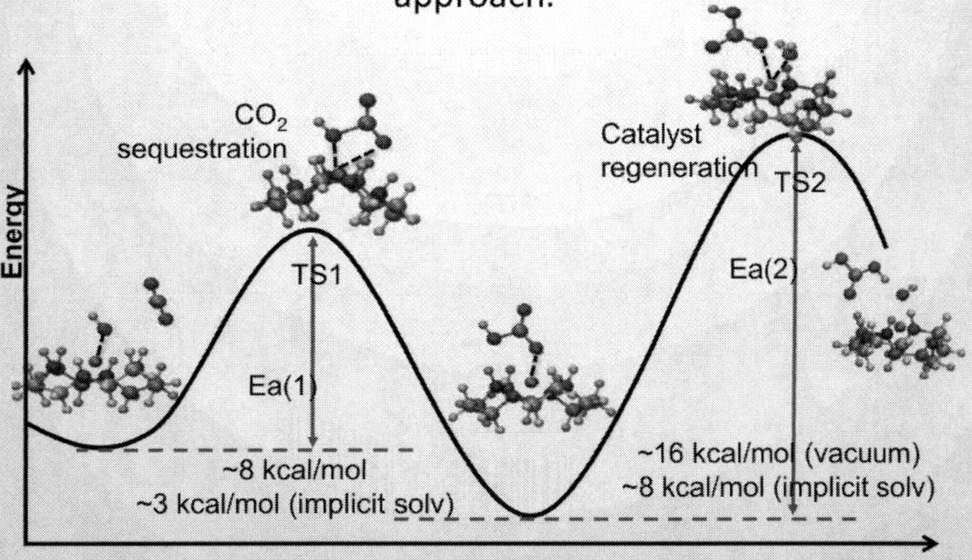

The second transition state is the rate-limiting step

And the winning catalysts are:

Best for:
High T cycles – stable to 120ºC
Low carbonate concentrations

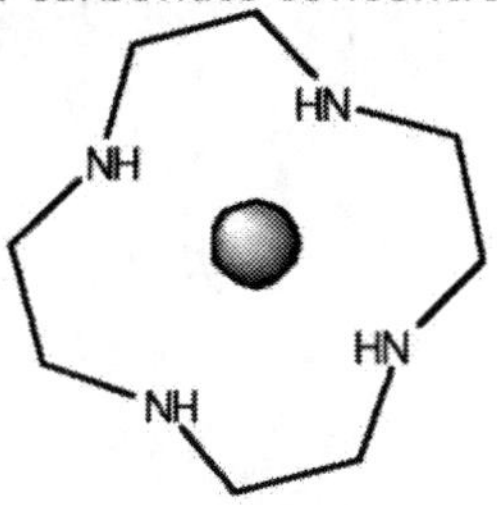

Zn Cyclen
K_{cat} 700 $M^{-1}s^{-1}$

Best for:
High carbonate concentrations
Lower T recycle – to 80ºC

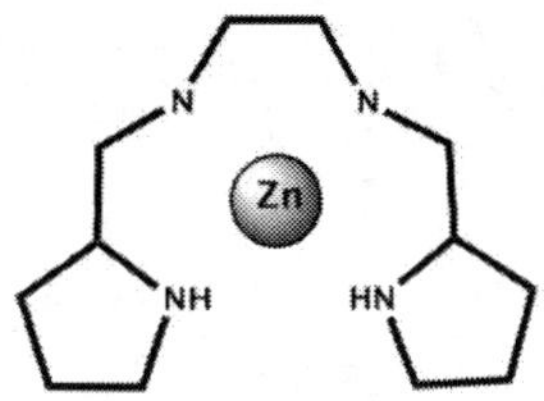

Zn Acyclic Proline
(N,N'-dimethylethylenediamine)
K_{cat} 900 $M^{-1}s^{-1}$

Pressure-drop apparatus: enables catalyst testing in difficult solvents e.g. non-aqueous, high carbonate concentration

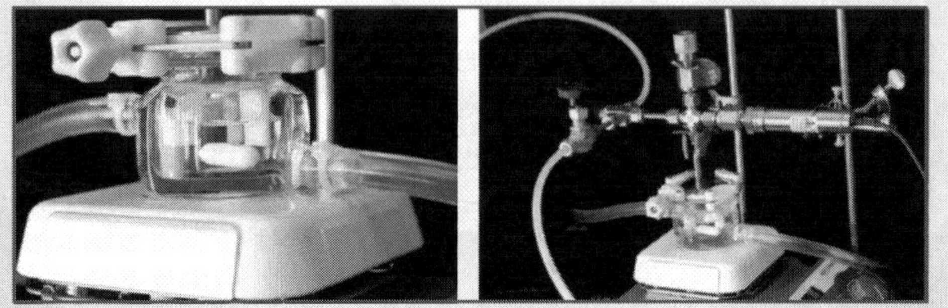

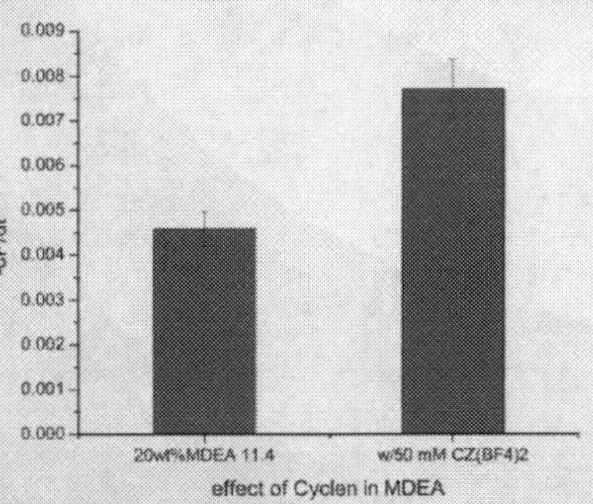

50 mM Zn Cyclen doubles absorption rate in MDEA pH 11.4

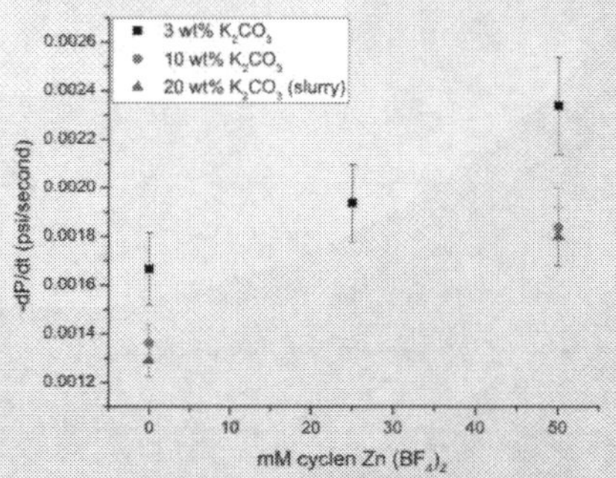

Rate as a function of [Zn Cyclen] in K_2CO_3

Lawrence Livermore National Laboratory

NNSA

Catalysis can improve CO_2 dissolution at moderate pH

Silica spicules form the skeleton of sponges – and are the oldest form of hard tissue

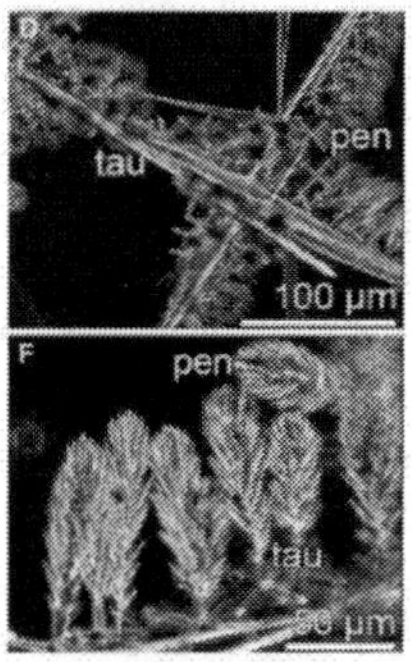

Jochum et al. *Chemical Geology* 2012

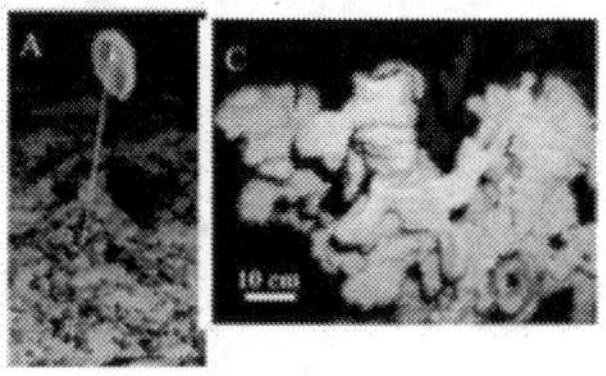

Precipitation is controlled by a framework enzyme, *silicatein,* that remains in the spicule

Dissolution is controlled by *silicase*, which helps control rate and concentration of silica

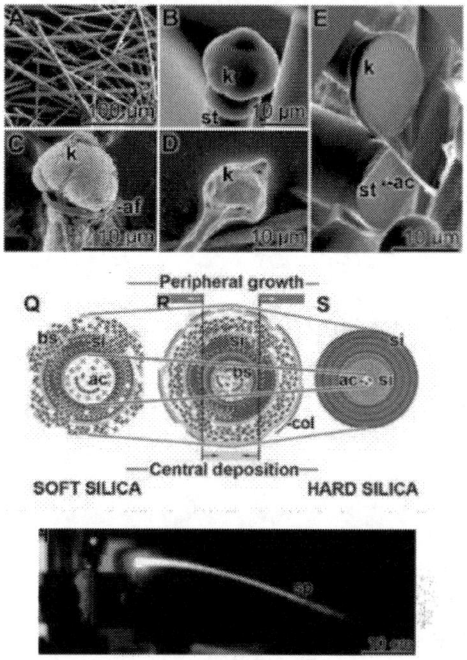

Leys et al, *Advances in Marine Biology*, 2007.

Wang et al. *Soft Matter 2012, DOI: 10.1039/c2sm25889g*

Enzymatic Degradation of Silica

The enzyme s*ilicase*, found in marine sponges, can catalyze the hydrolysis of amorphous or crystalline silicon dioxide to free form silicic acid.

Silicase is a zinc-dependent enzyme that is homologous to alpha-carbonic anhydrase. Both alpha- and gamma-carbonic anhydrase have been shown to exhibit silicase activity.

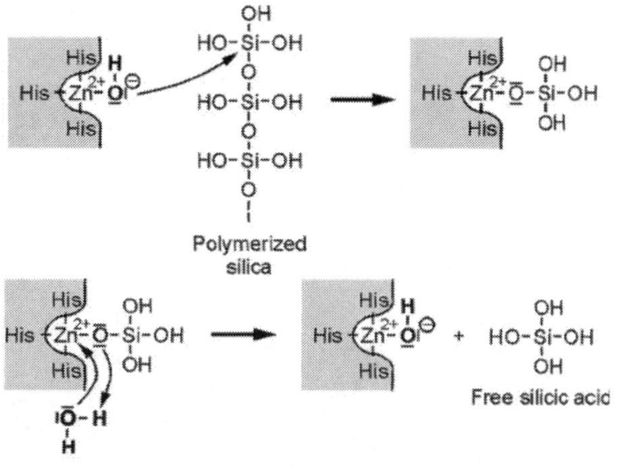

Proposed mechanism for silica hydrolysis by silicase.
Taken from *Chem. Rev.* (2010), **110**, 4656-4689.

Gamma carbonic anhydrase can also double the rate of silica dissolution

Carbonic anhydrase dissolves silica....

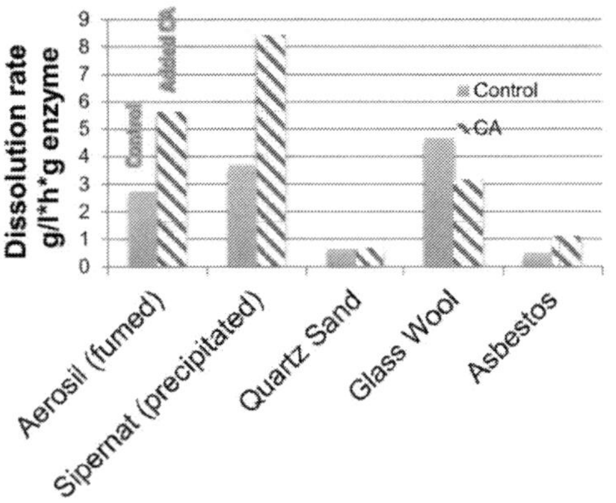

Toender, Janne Ejrnaes; Borchert, Martin; US Patent Application 2012/0021480 A1 Jan 26, 2012

And silicase hydrolyzes esters....

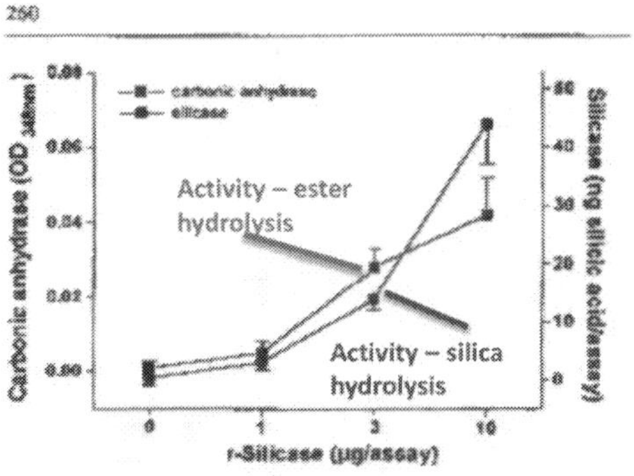

Heinz C. Schroder et al, Progress In Molecular And Subcellular Biology 33 249-268 2003

The reaction mechanism of silicase is assumed to be analogous with carbonic anhydrase and other zinc-dependent enzymes involved in ester-hydrolysis.

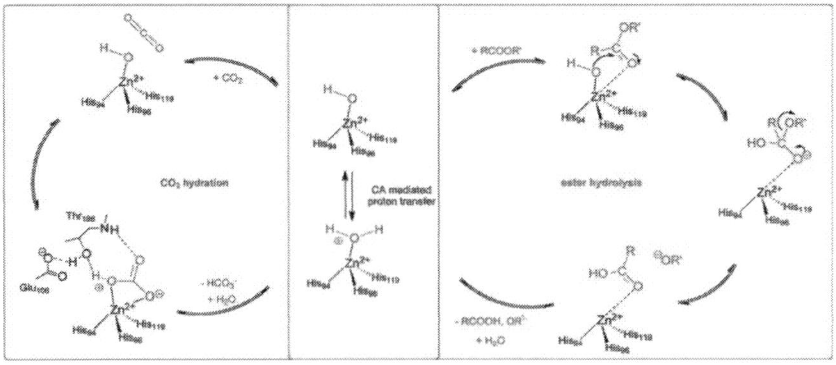

Taken from *J. Am. Chem. Soc.*, (2011), **133**, 18452-18462.

Several of our carbonic anhydrase mimetics have previously been shown to catalyze ester-hydrolysis and are being tested for silicase activity – so far no luck.

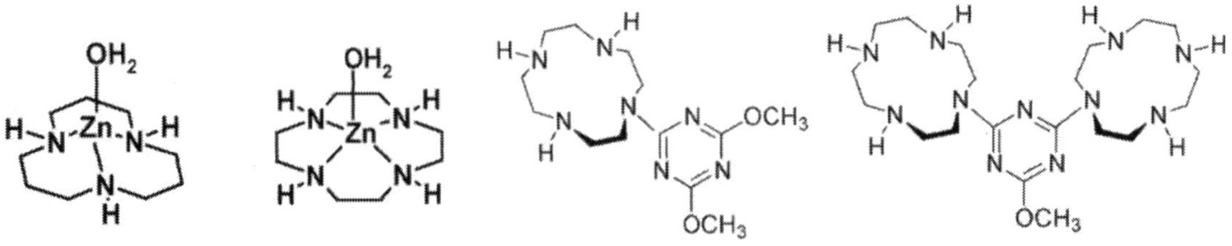

Silicase deserves more research attention – it suffers from having no medical application.

Enzymes and their mimics have clear applications to enhanced mineral carbonation and weathering

→ $CO_2(aq) + H_2O \rightarrow HCO_3^- + H^+$
(fast absorption of CO_2 into solution)

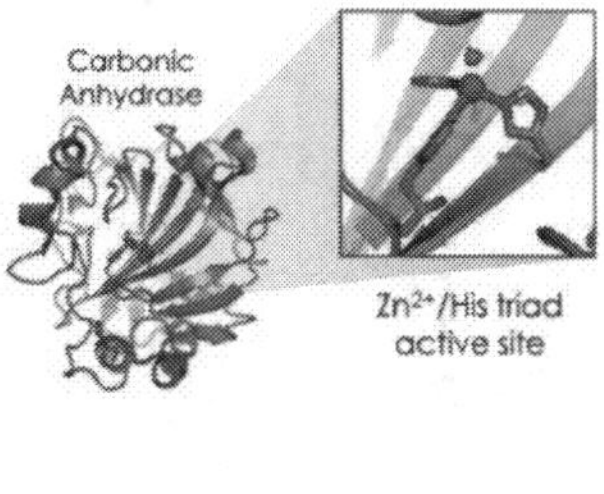

→ $SiO_2(s) + 2H_2O \rightarrow H_4SiO_4(aq)$
(enhanced mineral weathering)

Carbonation-activated steel slag binder for cement replacement and carbon utilization

Y. Shao[1*], M. Mahoutian[1], Z. Ghouleh[1] and A. El-Baghdadi[1]

[1] Department of Civil Engineering, McGill University, Montreal, Canada
*Corresponding author: e-mail address: yixin.shao@mcgill.ca

Abstract
Steel slag is the by-product of steel making process. Because of its chemical composition and high temperature process, steel slag is carbon dioxide reactive. This property can be utilized to develop steel slag binder to replace Portland cement in making precast building products. This paper summarizes a recent study on the carbonation activation of steel slag for binder application. Different types of steel slags are examined for their capacity to develop strength. They include EAF slag, BOF slag and KOBM slag. The slags are compact formed into precast products which are then activated by carbonation. The strength gain and carbon uptake are evaluated. It was found that BOF and KOBM slags are more carbon reactive than EAF slag. The former can develop a strength of 40-50 MPa in two hours reaction while the latter can produce only 10-20 MPa. The strength gain is attributed to the formation of calcium-silicate-hydrates and calcium carbonates through the carbon activation of calcium silicates. Two building products are successfully manufactured using activated slag as primary binder: slag-bond wallboards and slag-bond blocks. The slag-bond products consumes zero natural resources, recycles industry wastes, sequester carbon dioxide and can be made carbon-negative.

Introduction

Steel slag is a major by-product of the steelmaking industry. The steel slag produced during the primary stage of steel production is referred to furnace slag, such as electric arc furnace (EAF) slag, basic oxygen furnace (BOF) slag and Klockner Oxygen Blown Maxhutte (KOBM) slag. KOBM process is considered as a subset of the basic oxygen furnace (BOF) process. Currently the steel slags are often used as aggregates in road construction[1].

As-received steel slags are slowly cooled in air and non-hydraulic, although their chemical compositions are similar to ordinary Portland cement (OPC). In contrast to blast furnace slag, slowly-cooled steel slag do not show cementitious properties because of the highly crystalline nature in slow cooling process. The hydraulic properties of slags however can be enhanced through a fast cooling to form a quasi-glassy structure similar to granulated blast furnace slag[2]. Compressive strength of water quenched BOF slag could reach 29.5 MPa after 28-day hydration, although its 3-day strength was only about 4.5 MPa. In comparison, a three-day strength of OPC could reach 32 MPa. Apparently the treated BOF slag exhibited latent hydraulic behaviour. It can be used as pozzolanic material to be blended with Portland cement, but not as cementing material by itself.

Dicalcium silicate (C_2S) of different polymorphs exists in slowly cooled steel slag, making steel slag reactive with carbon dioxide (CO_2). It is well known that C_2S can be activated by CO_2 to form a strength-contributing carbonate-bonded matrix. The carbonation of dicalcium silicate leads to a rapid strength gain following a short exposure to CO_2 [3-4]:

$$C_2S+(2-x)CO_2+yH \rightarrow C_xSH_y+(2-x)CaCO_3 \quad (1)$$

The stoichiometry of CSH produced by CO_2 activation of calcium silicates was analogous to that resulting from hydration. The continued carbonation will convert CSH to $CaCO_3$ and silica gel, forming a carbonate-bonded matrix. The unique characteristic of C_2S suggests that non-hydraulic steel slag from slow cooling process can be carbon-activated to gain rapid strength. Studies of carbon dioxide reactivity of steel slag under various environmental conditions are scarce when compared to the wealth of information on Portland cement. One cubic meter steel slag blocks were treated by continuous carbonation for 12 days. The carbonated slag exhibited a strength of 10 MPa and a density of 2.2 g/cc[5]. Stainless steel slag was investigated for its solidification under accelerated carbonation. A compressive strength of 9 MPa was obtained after 72 hours of carbonation[6]. Eq. (1) also suggests that CO_2 activation of calcium silicates is a CO_2 consuming process. Carbon dioxide can be sequestered into compounds of calcium and converted to stable calcium carbonates.

The focus of current work is to investigate the feasibility of CO_2 activation of all three types of steel slag obtained from slow cooling process for early strength development and carbon dioxide uptake. Microstructure of CO_2 activated slag is analyzed by SEM and XRD. To promote commercial applications, two building products, slag-bonded wallboards and slag-bonded construction blocks, are fabricated using activated steel slag as primary binder.

Experimental

The chemical compositions of slowly-cooled four types of steel slag are determined by XRF and shown in Table 1. KOBM slag has the lowest iron and highest calcium content than EAF and BOF slag. For the same degree of CO_2 activation, KOBM slag is likely the most reactive to produce strength. EAF and BOF hybrid (EBH) is the mix of EAF and BOF slag. They are collected from a steel mill that simultaneously produces EAF and BOF slag. The as-received slag was crushed to remove metal pieces and pulverized to pass 45 µm. The water to solid ratio (w/s) of 0.15 was selected in terms of promoting the carbonation reaction for C_2S. The slag was mixed with water in a mechanical mixer for approximately 90 seconds. Cylindrical compact specimens of 13 mm in diameter and 25 mm in height were press formed under a pressure of 12 MPa. At least three samples were tested for average of one batch.

Fig. 1 shows the carbonation setup. The slag compacts were placed into the chamber and then exposed to CO_2 for 2 hours and 24 hours, respectively. The effect of carbonation duration on strength gain was studied. The gas was warmed up by a heater and the purity of the gas was about 99.5% to simulate the recovered CO_2 from flue gas at point sources such as thermal power plants or cement plants. A constant supply of gas maintained the chamber pressure at 150 kPa, ensuring that any carbon dioxide consumed by reaction was replaced.

Table 1. Chemical compositions of steel slag

Slag	Cao	SiO_2	Al_2O_3	Fe_2O_3	MgO	Na_2O
EAF	35.2	9.4	10.7	24.2	9.7	0.03
BOF	42.1	13.8	3.1	20.0	10.4	0.05
EAF+BOF hybrid	39.1	12.5	6.8	19.4	10.5	0.01
KOBM	43.7	11.5	2.8	9.9	27.2	0.08

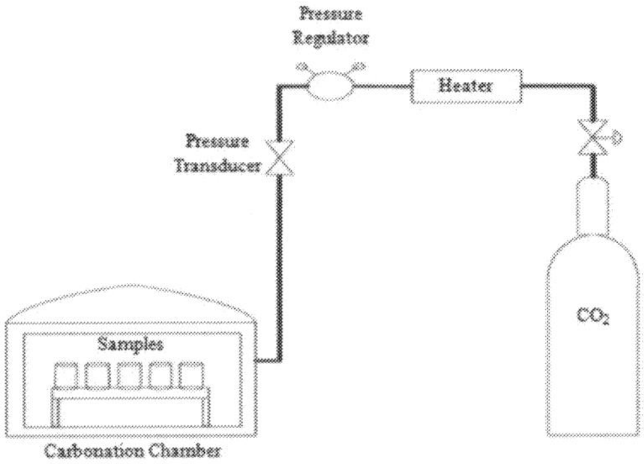

Fig 1: Setup for carbonation activation of steel slag

CO_2 uptake by steel slag through 2-hour or 24-hour carbonation was estimated by mass gain by Eq (2). It was determined by considering the initial slag mass, the final carbonated slag mass (including the water collected due to exothermic evaporation) and the original mass of dry slag binder. CO_2 uptakes by mass gain method is an average of entire batch and is not influenced by initial carbon content. Since water loss from carbonation reaction cannot be fully collected due to the evaporation, mass gain method represents a low bound estimate. The microstructure of activated steel slag was characterized by XRD and SEM.

$$Mass\ gain(\%) = \frac{(Mass_{afterCO2} + Water_{lost}) - Mass_{beforeCO2}}{Mass_{dry\ binder}} = \frac{\Delta Mass}{Mass_{dry\ binder}} \quad (2)$$

Results and Discussion

Carbon Uptake and Strength Gain

Carbon uptake and strength gain of carbonated slag compacts are summarized in Table 2. BOF slag seemed to have produced highest carbon uptake, followed by KOBM slag.

With EAF, either alone or combined with BOF, the slag was less carbon reactive. Prolonged carbonation did not increase significantly the carbon uptake. Most of the reaction occurred in first two-hour reaction. Early strength was measured immediately after carbonation process. The two-hour carbonation strength was highest in BOF slag followed by KOBM slag although the carbon uptake was quite different for the two types of slag. It was indicative that strength gain was not solely dependent on the carbon reactivity. The same phenomenon was observed in EAF and EBH slag in which EBH slag has shown slightly higher uptake but much more increase in strength. It is the quantity and the quality of carbonation reaction products that determine the strength. Both BOF and KOBM slag produced excellent strength in 2 hour reaction due to the high calcium content. Prolonged reaction didn't make significant contribution to strength. The 28-day hydration strength was measured after early carbonation and subsequent hydration up to 28 days in sealed condition. It was interesting to notice that all types of slag had shown latent hydraulic behaviour in subsequent hydration. The strength gain was apparent. The less the early carbonation strength, the more the late hydration strength. The 28-day strength gained by BOF and KOBM slag was comparable to the strength by ordinary Portland cement. Their 2-hour strengths right after carbonation was much higher than that of OPC.

Table 2. CO_2 uptake and strength gain of carbonated steel slag

Slag	CO_2 Uptake (%)		Early carbonation strength (MPa)		28-day hydration strength (MPa)	
	2h carbonation	24h carbonation	2h carbonation	24h carbonation	2h carbonation	24h carbonation
EAF	2.9±0.1	4.6±0.1	7.9±1.9	14.8±3.7	11.7±4.0	20.5±4.1
BOF	12.1±0.2	12.4±0.3	50.3±4.2	53.4±4.2	58.0±0.5	62.3±2.3
EBH	3.3±0.07	4.8±0.25	13.4±2.8	24.3±4.1	23.3±6.6	35.9±11.9
KOBM	7.1±0.2	8.1±0.1	45.0±0.2	49.6±0.7	62.4±3.4	64.1±2.6

Microstructure of Carbon-Activated Steel Slag

Fig. 2 shows typical XRD patterns of KOBM slag before and after carbonation. According to XRD results, the major phase in the original KOBM slag is larnite, β-C_2S. The existence of FeO phase was also noted. The high strength of carbonated KOBM slag is attributed to the carbonation of β-C_2S phase. The C_2S peaks in XRD pattern decreased in intensity in the carbonated slag. At the same time calcite peaks are considerably enhanced. They are the evidence of the reaction. The iron phase in as-received KOBM slag does exist, which is not reactive with CO_2. However the iron phase does not inhibit the calcite growth. Instead it stays as impurity in final reaction product. The comparison of XRD patterns of KOBM slag with that of EAF slag suggests that the relative intensity ratio of calcite to ferrite is much higher in KOBM than that in EAF slag. The low ferrite or high calcium content in KOBM is partially responsible for higher carbon reactivity and more strength gain. The calcium hydroxide phase in as-received slag is completely eliminated after carbonation.

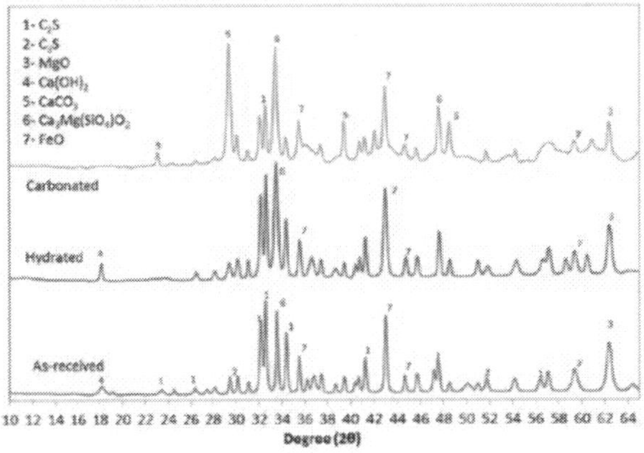

Fig. 2: XRD patterns of as received and carbonated KOBM slag

Typical SEM photographs are shown in Fig. 3. The microstructure of carbonated KOBM slag seems to be more amorphous, and calcium carbonates are intermingled with CSH gel. The traditional carbonate crystals are not distinct even at a high magnification. It is this intermingled poorly crystalline carbonate in amorphous CSH gel that makes KOBM paste a strong cementing material. The SEM photographs also support the assumption that the main reaction product is likely an amorphous calcium silicate hydro-carbonate with impurities. The KOBM compact is much denser and less porous.

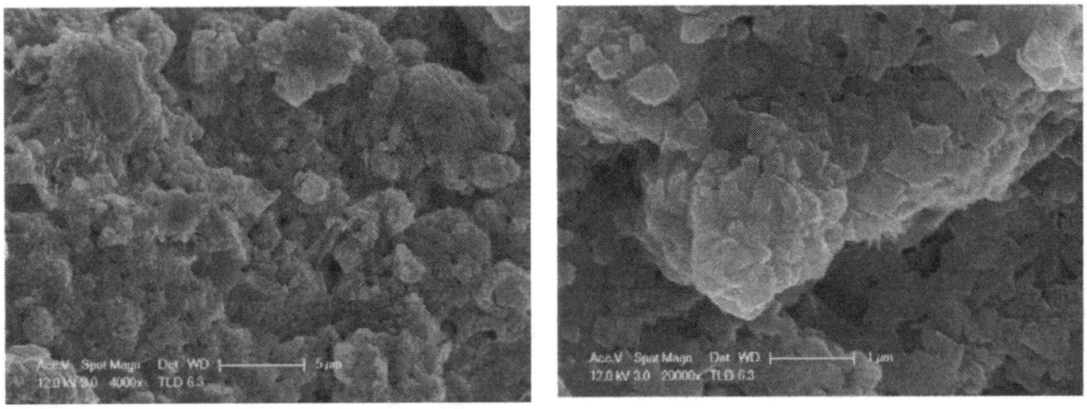

Figure 3: SEM Photomicrographs of carbonated KOBM slag compact (L) x4000; (R) x20000

Slag-Bonded Building Products

Two building products are fabricated using carbon activated steel slag as primary binder: slag-bonded wallboards and slag-bonded construction blocks.

Slag-bonded wallboard. Based on slag-activated slag tests, KOBM slag-bonded sawdust wallboards (127 x 76 x 12 mm) were fabricated using a sawdust as a filler to reduce the weight at a sawdust to slag ratio of 10% by mass and water to slag ratio of 0.15. After mixing, the wallboards were formed under pressure of 12 MPa and carbonated for 24 hours. The final 12-mm thick wallboard is shown in Figure 4.

Figure 4: KOBM slag-bonded wallboard

The results of compressive strength, flexural strength, modulus of elasticity and bulk density are presented in Table 3, along with those of three commercial cement based wallboard products. It was found that the physical properties of the KOBM slag board were comparable to those of the commercial products. The flexural strength of the KOBM slag-sawdust board (6.3 MPa) was higher than that of the cement-mesh board and close to those of the fiber-cement board and cement-EPS board. The compressive strength of KOBM slag-sawdust board was 13.8 MPa, well within the range of strength values of commercial products. KOBM slag-sawdust board had shown the highest density (1.4 g/cm^3) among the four products tested. Slag product is slightly heavier.

Table 3: Comparison of wallboard properties

Wallboard	Compressive strength (MPa)	Flexural strength (MPa)	Modulus (GPa)	Buck density (g/cc)
Slag-sawdust	13.8±0.6	6.3±0.3	1.1±0.1	1.4±0.04
Cement-mesh	13.0±0.5	4.9±0.7	1.9±0.2	1.3±0.01
Fiber-cement	14.8±0.6	6.9±0.2	2.2±0.3	1.2±0.06
Cement-EPS	15.5±0.7	7.9±0.4	2.2±0.1	1.1±0.04

Slag–bonded construction blocks. Slag bonded blocks were made with EBH steel slag (EAF and BOF hybrid) as primary binder. The expanded blast furnace (BF) slag with maximum aggregate size of 5 mm was used as lightweight aggregate to reduce the weight. The density of expanded BF slag was 900 kg/m^3. The water to steel slag ratio was increased to 0.2 to improve the surface quality of the blocks. The mixture of EBH steel slag, expanded slag aggregate and water was compact formed under 12 MPa pressure to make a block of dimension of 127 mm x 76 mm x 30 mm to represent the wall unit of a full-size 20-cm block. After compaction forming and immediate demoulding, the slag blocks were preconditioned for 2 hours in order to reduce the water to slag ratio from 0.2 to 0.1. Afterward, the blocks with the smooth surface and optimal W/S ratio were subjected to carbonation activation for 24 hours at 0.15 MPa gas pressure in the carbonation chamber. The carbon dioxide uptake was determined using the mass gain method. After carbonation, all blocks were kept in the plastic bags for 28 days for subsequent hydration before the mechanical and durability tests. At the same time, tests were also carried out on the commercial cement block for comparison. The wall unit in commercial cement blocks were cut into pieces with same dimensions as that used by slag bonded block (127 mm x 76 mm x 30 mm). Figure 5 shows the carbon-activated construction block with lightweight aggregates.

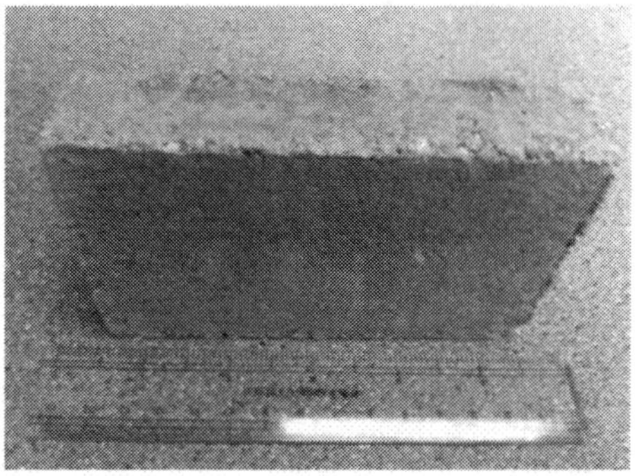

Figure 5: Carbon-activated construction block

The compressive strengths of the commercial cement blocks and activated slag blocks were determined. The commercial cement blocks displayed 23.6 MPa compressive strength and activated slag blocks showed 22.8 MPa. They were of close value. The compressive strength of slag bonded block satisfied the minimum strength requirement for the load-bearing masonry units suggested by ASTM C90 (2014). The freeze-thaw tests were performed to compare the frost resistance of the two blocks according to CSA A23.1 (2009). The freeze and thaw test was performed by submerging the blocks by half of their thickness (30 mm) into the 1% NaCl solution. The experiment was stopped after 20 cycles. Each cycle took 24 hours, 18 hours of freezing and 6 hours of thawing. At the end of each cycle, the weight of the specimens was recorded leading to the calculation of the mass loss as the average of 3 tests. In comparison, the apparent mass loss was observed in commercial cement blocks after 2 cycles and in slag blocks, it was after 7 cycles. Beyond the 7th cycle, both blocks experienced the same rate of mass loss. After 20 cycles of freeze and thaw, commercial cement blocks lost 32% of its mass compared to 17% of mass loss observed in slag bonded blocks. Clearly, block made with activated slag exhibited higher resistance to the freeze and thaw damage. Considering the compressive strength and freeze and thaw resistance, it was clear that the mechanical and durability properties of slag bonded blocks were superior to the commercial cement blocks.

Conclusions

The as-received slowly-cooled steel slags such as EAF slag, BOF slag and KOBM slag are not hydraulic but can be activated by carbon dioxide to gain the strength. The preliminary study has shown that two-hour CO_2 activation can produce a strength of 8

– 50 MPa in two hours and a carbon dioxide uptake varied in a range of 4-12% by slag mass. BOF and KOBM are excellent slag that develops high strength which was directly comparable to Portland cement. EAF slag was less carbon-reactive. High density created by high compact forming force does not reduce the degree of carbonation. Relatively, KOBM slag contains higher calcium and magnesium content and are more reactive with carbon dioxide. KOBM slag also had shown low iron content, indicating that if more ferrite phase could be removed, the slag would show more latent hydraulic behaviour and world be more reactive with CO_2 to promote strength gain. To best use the steel slag for both early and late performance, fast cooling of steel slag at molten stage seems valuable to create latent hydraulic property. With ferrite phase removed and quenching process used, steel slag can be activated by carbon dioxide to obtain high early strength and by subsequent hydration to gain more late strength. Two slag-bonded building products are successfully fabricated. They are slag bonded wallboard to replace cement wallboard and slag bonded blocks to replace cement blocks. The production of slag bonded products can integrate carbon sequestration into commercial process which will show environmental, economical and technical benefits.

Acknowledgements

The financial supports from the Fonds québécois de la recherche sur la nature et les technologies (FQRNT) of Quebec is acknowledged.

References

1. Juckes, L. M. (2003) "The volume stability of modern steel making slags", Mineral Processing and Extractive Metallurgy, Dec, 112, 177-197.
2. Reddy, A. S., Pradhan, R. K. and Chandra, S. (2006) "Utilization of basic oxygen furnace slag in the production of a hydraulic cement binders", Inter. J. of Mineral Processing, 79, 98-105.
3. Young, J. F., Berger, R. L. and Breese, J. (1974), "Accelerated curing of compacted calcium silicate mortars on exposure to CO_2", J. of the American Ceramic Society, 394-397.
4. Berger, R. L., Young, J. K. and Leung, K. (1972), "Acceleration of Hydration of Calcium silicates by carbon dioxide treatment", Nature Physical Science, 240(6), 16-18.
5. Santos, R.M., Van Gerven, T. 2011. Process Intensification Routes for Mineral Carbonation. *Greenhouse Gases: Science and Technology*, 1(4), 287-293.
6. Isoo, T., Takahashi, T. and Fukuhara, M. (2000), "Development of large steelmaking slag blocks using a new carbonation process", Advances in Cement Research, 12(3), 97-101.
7. Johnson, D. C., MacLeod, C. L., and Hills, C. D. (2003), "Solidification of stainless steel slag by accelerated carbonation", Environmental Technology, 24, 671-678.

CO2 capture and storage by mineral carbonation
Hydrotalcites As a Secondary Carbon Sink in Serpentinites
Turvey, C. C., Monash University
Originally presented on: 6/23/2015 10:00:00 - 10:15:00

Hydrotalcite group minerals are commonly found within ultramafic rocks such as serpentinites, which may be used as feedstocks for mineral carbonation programs. Particular members of the hydrotalcite mineral group have the capacity to trap and store CO_2 via anion exchange. Hydrotalcite group minerals, which are also known as layered double hydroxides, undergo anion exchange by replacement of interlayer anions such as Cl^- and OH^- for aqueous CO_3^{2-}. This makes hydrotalcite group minerals a novel secondary sink for carbon within serpentinites.

To better understand the role hydrotalcites may play in CO_2 sequestration, a wide reaching investigation must be undertaken to determine (1) the range of abundance of hydrotalcite minerals in serpentinites, (2) how readily these minerals take up carbonate into their crystal structures and (3) how to fingerprint uptake of atmospheric or industrial CO_2 into these minerals.

We are investigating hydrotalcite mineral abundance and carbonation at a number of serpentinite-hosted mineral deposits. Multiple sites were selected in order to investigate a diverse selection of serpentinites and hydrotalcite minerals: the abandoned Woodsreef Chrysotile Mine (New South Wales, Australia), the Lord Brassey Nickel Mine and several small workings within the Dundas Mineral Field (Tasmania, Australia). Our X-ray diffraction results indicate that samples from these sites are dominated by the presence of serpentine minerals, and many also include magnetite, chromite and brucite. Several of the samples also contain carbonate minerals, including magnesite and dolomite. Pyroaurite [$Mg_6Fe^{3+}{}_2(CO_3)(OH)_{16}{\cdot}4H_2O$], the carbonate-bearing magnesium-iron end-member of the hydrotalcite group is found as a minor phase at Woodsreef with between 0 and 7% abundance, which is consistent with the findings of Oskierski et al. (2013). Pyroaurite appears to have developed on weathering surfaces and may have formed at low temperatures at Woodsreef (Oskierski et al. 2013) and Lord Brassey. Stichtite [$Mg_6Cr^{3+}{}_2(CO_3)(OH)_{16}{\cdot}4H_2O$] the magnesium-chromium analogue of pyroaurite, is common from multiple localities within the Dundas mineral field. Stichtite appears to have formed from replacement of chromite by high temperature fluids. Both of these hydrotalcite minerals are carbonate bearing and represent an existing store for CO_2. However, little is known about the isotopic signatures of anion exchange in hydrotalcite minerals; thus it is difficult to recognise the source of CO_2 in these minerals and whether or not they represent a stable sink for this greenhouse gas.

Anion exchange experiments were conducted to study the rate and isotopic signature of carbonate uptake into a chlorine-bearing hydrotalcite mineral, iowaite [$Mg_6Fe^{3+}{}_2(Cl_2)(OH)_{16}{\cdot}4H_2O$]. Aliquots of iowaite were placed in stirred tank reactors containing deionised water, which was allowed to equilibrate with atmospheric CO_2 over 46 hours. This resulted in complete exchange of the interlayer Cl^- for atmosphere-derived dissolved inorganic carbon (DIC) to form pyroaurite. Stable carbon isotopic analysis of the synthetic pyroaurite shows a marked depletion in ^{13}C relative to atmospheric CO_2, indicative of diffusion fractionation during anion exchange. This apparent preference for light carbon will need to be further investigated as an indicator of CO_2 uptake in natural settings.

Keywords: carbon mineralisation, X-ray diffraction, anion exchange, hydrotalcites

References:

Oskierski H. C., Dlugogorski B. Z. & Jacobsen G. 2013. Sequestration of atmospheric CO_2 in chrysotile mine tailings of the Woodsreef Asbestos Mine, Australia: Quantitative mineralogy, isotopic fingerprinting and carbonation rates. *Chemical Geology* **358**, 156-169.

Ultra-rapid hardening of cement by accelerated carbonation – past, present and future

A. Maries[1,2*] & C. D. Hills[2]

[1] AMSTaR Consultancy, London, UK
[2] University of Greenwich, UK
* Corresponding author: alan.maries@amstar.org.uk

Abstract
The ancient Greeks and Egyptians may have unwittingly employed accelerated carbonation with the polishing technique used to achieve seamless crack-free linings in lime-bound pipes and stucco. However, it was not until the advent of Portland cement in the mid-19th century that carbonation with gaseous CO_2 at atmospheric pressure was proposed as a means of accelerating the setting and hardening of mortar and concrete. In the 1970s the first attempts were made to achieve a more thorough scientific understanding of the process, most notably in the USA by Berger and colleagues, but also by researchers in the USSR, Israel and Sweden. The authors' own development of a carbonation process to accelerate the hardening of precast concrete dates from over 30 years ago, when lowering atmospheric emissions of CO_2 by sequestration was not yet an issue. But today, with global cement production growing relentlessly at nearly 6% per annum to a current total of 4 Gtonnes and releasing almost as much CO_2, reduction of emissions is now a pressing concern. This paper reports on current research and commercial operations involving accelerated carbonation of cement and concrete and will look forward to challenges and opportunities in a low-carbon future.

The Past

Ancient history

Until the advent of Portland cement 150 years ago, lime-based binders in mortar and concrete achieved hardness and strength over time (weeks to years to decades) due to the very slow carbonation of hydrated calcium oxides by the action of atmospheric CO_2, present in those days at a concentration of 280 ppm by volume.

The Roman aqueduct at Caesarea, built nearly 2,000 years ago, was originally 30 km long. The open water canal on the top (Figure 1a) is lined with a lime mortar, some 30 mm thick, consisting of 6 layers, each of a slightly different composition and finished by vigorous hand-polishing. This lining runs for several hundred metres without any apparent shrinkage cracks or expansion joints – a truly remarkable feat of materials engineering!

The Roman architect Vitruvius, in his 10-volume treatise[1] on all aspects of architecture published in Rome at the close of the 1st century B.C., described in detail the technique (already well-established by the Greeks and Egyptians) of 'polishing' partly set lime mortar in order to achieve a seamless crack-free lining in pipe and stucco work:

"... on top of not less than three coats of sand mortar..... a powdered marble mortar ..... so stiff that it does not stick to the trowel *[i.e. low w/c ratio]* ..... is to be applied in 3 more layers ..... and each layer well rubbed down..... The walls will then not be liable to cracks or to any other defect."

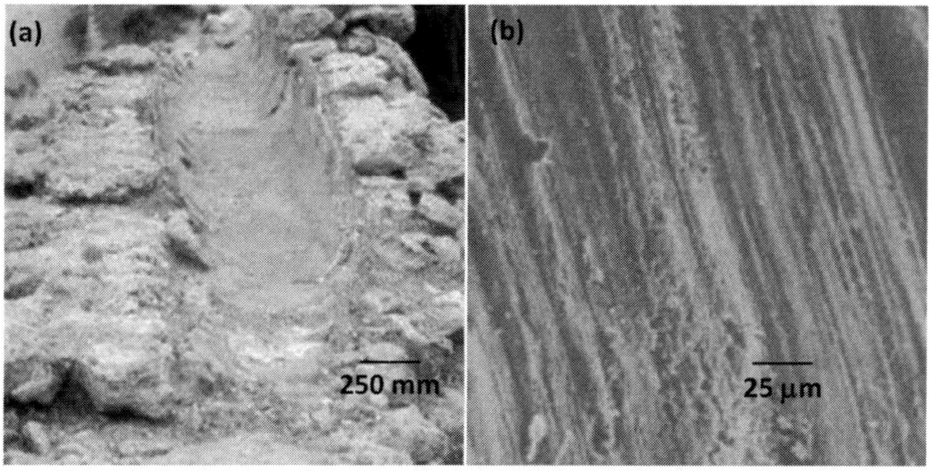

Figure 1. Roman aqueduct at Caesarea (a) lining, (b) micrographic section

Very detailed micro-structural investigations of many such materials were undertaken by Malinowski in the 1970s.[2] He concluded that the composition of the mortar as well as the polishing resulted in the creation of a dense capillary structure parallel to the surface (Figure 1b), producing favourable conditions for intense carbonation and thus accelerating hardening and improving durability.

Many other surfaces in Roman water infrastructure were finished in a similar way, and buildings such as the Pantheon (built a few years later in Rome) were decorated internally with a comparable multi-layer plaster stucco, also very highly polished to look like marble.

As this polishing technique seems to have been inherited from earlier civilizations, accelerated carbonation may have been carried out, unwittingly, for thousands of years!

19th and 20th centuries

This paper is concerned with the application of carbonation to fresh cementitious material in order to induce immediate rapid setting and hardening, (i.e. BEFORE any appreciable hydration of cement has taken place). Carbonation applied AFTER articles have been pre-conditioned by 'normal' curing (often also referred to as 'accelerated carbonation' because it speeds up 'natural' carbonation by atmospheric CO_2) is not covered.

Commercial interest in accelerating the setting and hardening of mortar and concrete by forced carbonation with gaseous CO_2 at atmospheric pressure can be traced back to patents filed in 1855 in Britain and France by Joseph Chaudot,[3] a Parisian architect, who claimed a method for making a stucco coating resembling marble by applying gaseous CO_2 under (unspecified) pressure to a lime paste of low water/solids ratio, with or without added aggregate, compacted into a mould.

A few years later, soon after Portland cement came into common use to replace lime as a binder, James L. Rowland of Milwaukee filed a US Patent[4] describing the application of CO_2 gas to accelerate the setting and hardening of compacted cementitious mixtures. In the late 1800s and early 1900s numerous process patents were issued for making blocks, pipes, columns and slabs by CO_2 curing of various materials including alkaline earth oxides and hydroxides, hydraulic cements and slags, some using increased pressure and others vacuum de-airing. However, high costs and uncertain results in comparison to normally hydrated Portland cement systems prevented much further commercial development and resulted in a decline in interest, with only a few further patent filings of a similar nature appearing in the ensuing decades.[5]

In the 1950s, recirculation of CO_2-rich hot flue gases at atmospheric pressure in curing chambers was found to improve greatly the volume stability of concrete blocks.[6] It was concluded that partial carbonation was induced, which resulted in 'pre-shrinking' of the concrete in much the same way as with the Roman pipe linings described above, but this was not pursued further as a means of accelerating setting and hardening.

First 'scientific' studies

The first systematic and by far the most comprehensive experimental work on accelerated carbonation was undertaken by Berger and co-workers at the University of Illinois during the 1970s.[7,8] Coincidentally, detailed but less extensive studies were published at around the same time by Sorochkin[9] in the USSR and Bentur[10] in Israel, and also by Malinowski[11] in Sweden, all based on mixtures with sand of Portland cement or its principal components $3CaO.SiO_2$ and $2CaO.SiO_2$ (tricalcium and dicalcium silicates), carbonated by gaseous CO_2 at ambient temperature and pressure immediately after mixing with water.

The simple reaction mechanism postulated by Berger[7,15] has been critically reviewed by the authors,[12,13,14] and the overall stoichiometry of the reaction with silicate phases in cement indicates that initial carbonation (Eq. 1a) is accompanied by hydration and is followed, after some delay, by secondary carbonation (Eq. 1b).

For tricalcium silicate:

$$3CaO.SiO_2 + yH_2O + (3\text{-}x)CO_2 \rightarrow xCaO.SiO_2.yH_2O + (3\text{-}x)CaCO_3 \qquad (1a)$$

$$\xrightarrow{+zCO_2} (x\text{-}z)CaO.SiO_2.yH_2O + zCaCO_3 \qquad (1b)$$

Dicalcium silicate reacts similarly, whereas the aluminate phases in the cement appear to be relatively inert to carbonation. Of particular note here is the observation[5] that even non-hydraulic calcium silicates such as gamma-dicalcium silicate may develop considerable strength on being carbonated.

Precast concrete manufacture

The authors' own interest in ultra-rapid hardening of precast concrete dates from the early 1980s, with a challenge presented in a concrete roof-tile plant to develop strength in the fresh mix with sufficient rapidity that tiles could be removed from their aluminium formers within 2 minutes of casting. Accelerated carbonation was found to offer a number of advantages over other chemical methods of rapidly accelerating strength development in Portland cement (for instance by additions of alkali silicates, carbonates or aluminates), which tended to be difficult to control and which generally compromised durability in terms of resistance to frost and (acid) rain.

During development work it was found that the reaction scheme described above (Eq. 1) is in fact an over-simplification of a much more complex process, and fails to account adequately for observed phenomena, such as the rapidity and extent of reaction (up to a quarter of the cement reacting within minutes), its dependence on mortar meso-structure, or the previously unreported effect of ambient temperature. As a result of a detailed experimental investigation of the uptake of CO_2 in sand/cement mortars, a model was developed showing that accelerated carbonation could best be represented as a linked sequence of 9 physical and chemical steps[13] proceeding cyclically through gaseous, liquid and solid phases. The model emphasised the continuous nature of the whole process, where the overall speed is determined by the slowest step, and it was able to explain the influence of ambient temperature and mortar porosity.

The roof tile project developed almost to pilot plant stage, but despite showing commercial promise[16] was not taken further. The requirement for sufficient porosity in the fresh concrete to allow CO_2 to penetrate at atmospheric pressure resulted in the cured tiles being on the borderline of water-tightness. This problem was not resolved at that time, although it is now considered that porosity could be controlled better by adjusting the size distribution of the aggregate and if necessary, incorporating pore-sealing admixtures.

Fibre-reinforced cement

As well as his interests in enhanced carbonation of ancient concrete and in the mechanism of accelerated carbonation, Malinowski[17] patented a process for manufacturing fibre-reinforced panels by applying a vacuum treatment to dewater the fresh material followed by an accelerated carbonation step. This process was described in some detail by Hannawayya[18] and appears to have been marketed for a time by Abetong in Sweden under the product name 'Quickcrete'.[19]

Others who have found that the open structure of fibre-cement makes it particularly susceptible to rapid hardening by accelerated carbonation include Bentur,[10] Schmidt[20] and Pizzol,[21] when the lowering of pH by carbonation gave added protection to glass and other degradable fibres. Moorehead[22,23] investigated the use of cellulose fibre-reinforced lime building elements as a replacement for asbestos-cement sheets, but this process does not seem to have been taken up commercially.

The Present (21st century)

Lime and cement

The use of low-energy binders such as lime is now becoming increasingly attractive to reduce emissions of CO_2 to the atmosphere. Not surprisingly, the renewed interest in these binders has prompted numerous studies of the accelerated carbonation of lime- and magnesia-based pastes and mortars at KU Leuven[24] and elsewhere, many of which have been reported in previous ACEME proceedings. The precise reaction mechanism and product morphology has been found to depend on the type of oxide binder and on its state of hydration.

Waste management, soil stabilisation and aggregate production

Although perhaps not directly relevant to the ultra-rapid hardening of cement, extensive studies have been published by Hills and others[25-29] on the application of accelerated carbonation to immobilise hazardous wastes and stabilise contaminated soils, exploiting the rapid solidification of cement-bound materials.

Moreover, the techniques used to study these phenomena greatly expanded the understanding of accelerated carbonation and paved the way for the later development of a process by Carbon8 (a spin-out company from the University of Greenwich) for producing secondary aggregates entirely by the carbonation of waste materials, with or without added cement.[30] This novel process produces construction blocks from these aggregates that are truly carbon-negative, and has been described in previous ACEME conferences.[31] The latest news of Carbon8's manufacturing facility, now operating on the scale of tens of thousands of tonnes per annum, will be presented in this conference.[32]

Precast concrete

In a study of the properties of CO_2 cured lightweight concrete blocks, Shi & He[33] found that pre-conditioning in a moist environment prior to curing was essential to achieve strengths similar to those of conventionally steam-cured blocks. Other researchers[34] have noted the beneficial effects of such 'pre-carbonation', including reduced setting times, improved compressive strength and enhanced chloride resistance, probably due to deposition of $CaCO_3$ in the pore structure leading to a denser matrix.

The development of accelerated carbonation to achieve rapid hardening in precast concrete masonry blocks has been described in previous ACEME conferences,[35] and a manufacturing process is now being operated commercially by CarbonCure,[36] which can be retrofitted to existing production equipment and systems. Controlled doses of gaseous CO_2 are injected into a mixer feeding into a masonry production machine while the mixing cycle proceeds, with CO_2 uptake measured at around 1.5% w/w cement. The brief gas delivery time allows the CO_2 addition to be simply integrated into a conventional process so that the benefits of carbonation of fresh concrete can be realized in an industrially scalable manner, including greater strength, reduced absorption, improved resistance to chloride permeability, and superior freeze-thaw performance. Additionally, the conversion of gaseous CO_2 into stable, solid, carbonate reaction products effectively serves as permanent storage of CO_2 in the concrete, allowing producers to create sustainable building products with improved performance. This platform technology can be readily integrated with other strategies (for instance, supplementary cementitious materials and secondary aggregates) to reduce the carbon footprint of concrete.

Solidia Technologies have described a process[37] in which a low energy, reduced-lime cement can be manufactured in conventional cement kilns at 250 °C lower than normal, to produce a 'non-hydraulic' powder. When used to make concrete, this cement will carbonate in a CO_2-rich atmosphere at ambient gas pressures and moderate temperatures (20 – 60 °C) to uptake 30% by weight of CO_2 and produce articles with performance similar to or better than traditional products. Furthermore, it is claimed that 70-80 % of water used in concrete production can be recovered, by condensing moisture evaporated during the highly exothermic carbonation reaction.

Passing mention should be made of super-critical carbonation, such as the 'Supramics' process developed in Nevada in the 1990s.[38] However, although this process sequesters CO_2 and confers other environmental benefits, it can realistically only be applied only to treat already hydrated and hardened material as the chemical reaction can be quite violent.

The Future

With current emphasis on the carbonation of waste and sequestration of CO_2 as a means of reducing global emissions, accelerated carbonation of concrete is not seen as making much impact in this respect, particularly as in most applications CO_2 typically combines with less than 15% by weight of cement.

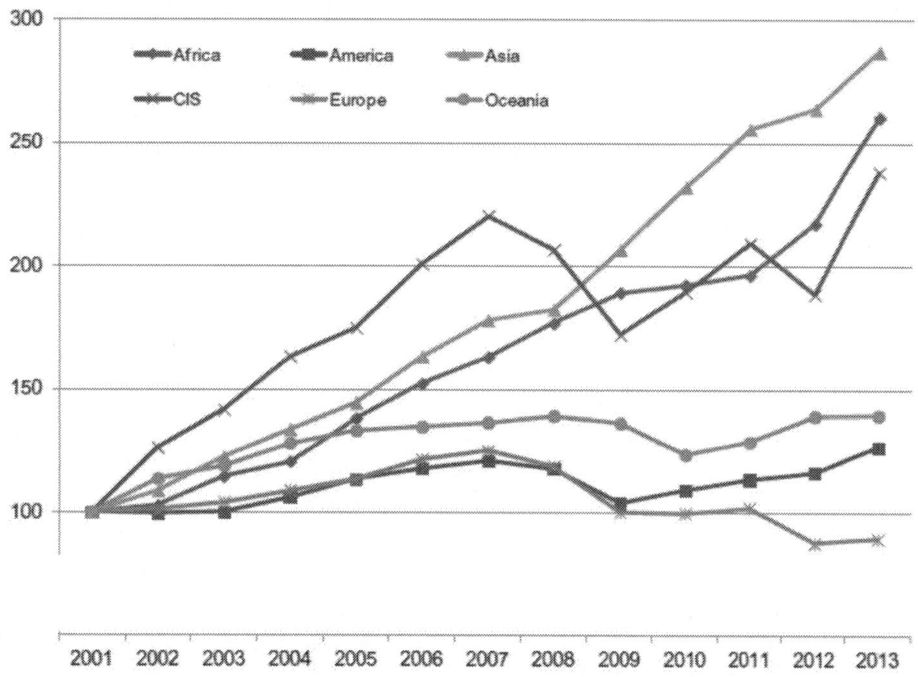

Figure 2. World cement production by region ('100' in 2001)[39]

However, annual cement production now exceeds 4 Gtonnes globally,[40] and the latest projections of growth in cement & concrete production (see, for example Figure 2) indicate no slowing down in the trend established over the past few decades. The sector is continually looking for ways to reduce CO_2 emissions and, as CarbonCure[36] have emphasised, the absorption of CO_2 into fresh concrete offers an attractive new tool for reducing the carbon footprint of the built environment and helping the concrete industry compete with other construction materials in the 'green' building space.

In a world increasingly conscious of the need to pursue low-carbon ideals at every level, accelerated carbonation is capable of providing other advantages. For instance, the control of in-service shrinkage and hydrothermal movement in carbonated concrete blocks has been found to be a commercially beneficial property, and accelerated carbonation seems a particularly attractive process for the manufacture of some product configurations such as fibre-reinforced cement sheets or porous concrete.

It has been noted that accelerated carbonation has been applied extensively to consolidate alkaline waste residues (mainly from thermal processes) such as slags and fly ash, which will carbonate readily if incorporated in concrete as cement extenders. Since the annual global production of such wastes is estimated to be comparable with that of Portland cement, these materials will no doubt be investigated in the future as potential low-energy carbonation-hardened binders.

The ability of carbonation to activate binders such as lime, which are only poorly hydraulic, has inspired a scheme for integrating cement manufacture with concrete production, combining three separated processes.[41] First, a Portland-like cement is made at reduced temperature and cooled so that it self-pulverises without the need for grinding to a cement powder comprised predominantly of the non-hydraulic gamma-polymorph of dicalcium silicate. Second, CO_2-enriched gas is extracted from the cement kiln flue, and this is used in a third part of the scheme to activate the cement in the production of precast concrete products. Considerable energy savings are anticipated, with additional sustainability gains arising from the operation of a closed loop manufacturing process. To date, proof-of-principle has been demonstrated, and scale-up challenges are being addressed. Although the initial focus is on precast concrete production, other potential applications are foreseen for this low-energy cement in areas such as environmental engineering and non-structural concrete.

Accelerated carbonation is a sufficiently rapid technique for most applications,[12] but for some of the less readily carbonatable materials some further quickening may be necessary. Process intensification by catalysis has therefore been studied as a possible way of speeding up the slowest, rate-limiting step in the carbonation cycle, which is considered to be the solvation and hydration of CO_2 in water.[13] Several different chemical 'enhancers' such as sodium hypochlorite have produced promising results in the laboratory, with the initial rate of CO_2 uptake R_{max} increased 5-fold (Figure 3),[14] but there is much more to be done to demonstrate this concept on a commercial scale.

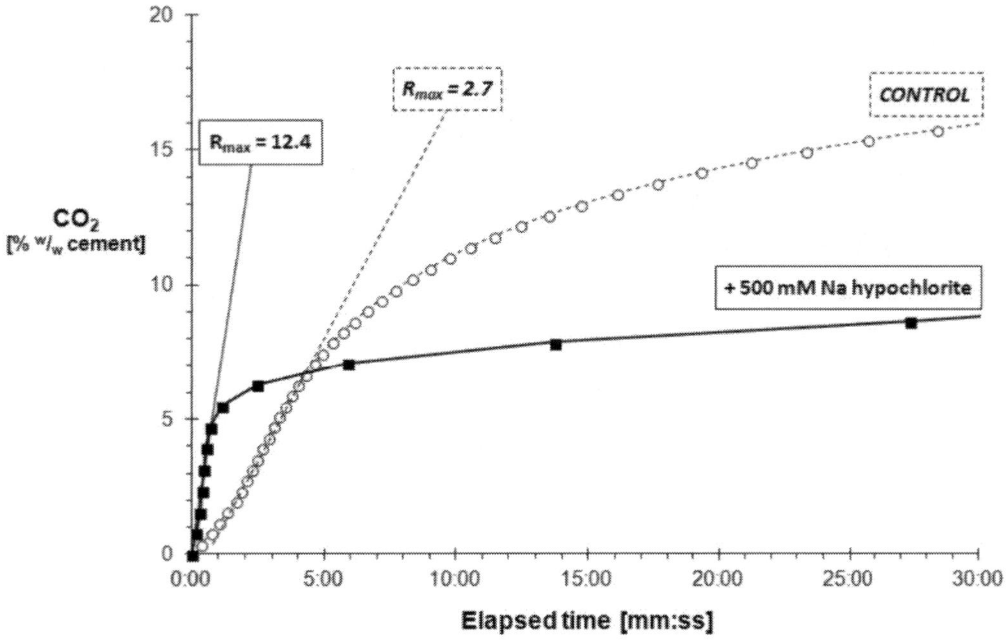

Figure 3. CO_2 uptake in sand:cement mortar enhanced by addition of sodium hypochlorite

Although employing accelerated carbonation to induce rapid strength development is at present restricted to precast concrete, there is no reason why it could not be extended to include site-based processes such as *in-situ* slip-forming or machine-laid masonry mortar. Furthermore, the ability of some cement systems to delay setting until they are activated opens up exciting opportunities for 'set-on-demand' applications in environmental and geotechnical engineering, where rapid strength development could be initiated by injection of CO_2 some considerable time after the concrete has been placed.

Accelerated carbonation could play a potentially significant rôle in improving the additive manufacturing of articles from polymer-cement or concrete, where digitally controlled '3D printing' can be used to make articles at scales from the very small to large load-bearing items for architectural or sculptural use, including intricate shapes otherwise impossible to produce such as lattice designs. Using activation by CO_2 to rapid-harden successive layers between each pass of the machine could speed up what can be a very slow process, if reliant on the normal setting and hardening times of Portland cement. This would enable whole buildings to be constructed on realistic timescales, and could turn into reality Thomas Edison's century-old dream[42] of poured concrete houses filled with concrete furniture, refrigerators and pianos.

References

1. Pollio, M.V., -0015. *De Architectura.* Book VII Chapter III.
2. Malinowski, R., 1979. Concretes and mortars in ancient aqueducts. *Concrete International: Design & Construction* 1(1), 66-76.
3. Chaudot, J., 1855. An improved stucco, *GB Patent* No. 96.

4. Rowland, J.L., 1870. Improvement in the manufacture of artificial stone. *US Patent* No. 109,669.
5. Bukowski, J.M., Berger, R.L., 1979. Reactivity and strength development of CO_2-activated non-hydraulic calcium silicates. *Cem. Concr. Research* 9, 57-68.
6. Toennies, H.T., Shideler, J.J., 1963. Plant drying and carbonation of concrete block. *J. Amer. Concrete Inst.* 60(5), 617-632.
7. Berger, R.L., Klemm, W.A., 1972. Accelerated curing of cementitious systems by CO_2. *Cem. Concr. Research* 2, 647-652.
8. Goodbrake, C.J., Young, J.F., Berger, R.L., 1979. Reaction of hydraulic calcium silicates with CO_2 and water. *J. Amer. Ceram. Soc.* 62(9/10), 488-491.
9. Sorochkin, M.A., Shchurov, A.F., Safonov, I.A., 1975. Study of the possibility of using CO_2 for accelerating the hardening of products made from Portland cement. *J. Appl. Chem. USSR* 48(6), 1271-1274.
10. Bentur, A., Ish-Shalom, M., Grinberg, T., 1981. Cementing properties of oil shale ash: curing in CO_2 atmosphere. *Cem. Concr. Research* 11, 645-650.
11. Malinowski, R., Rodhe, M., 1983. Vacuum carbonation of lightweight aggregate concrete. *Nordic Concrete Research* 2, 166-182.
12. Maries, A., 2008. Measurement of gas consumption during accelerated carbonation of Portland cement mortar. *In: Proceedings of the 2nd International Conference on Accelerated Carbonation for Environmental and Materials Engineering, Rome, October 2008.*
13. Maries, A., 2008. A sequential process model for the accelerated carbonation of Portland cement. *In: Proceedings of the 2nd International Conference on Accelerated Carbonation for Environmental and Materials Engineering, Rome, October 2008.*
14. Maries, A., Hills, C.D., 2013. Homogeneous catalysis of the accelerated carbonation of Portland cement Leuven. *In: Proceedings of the 3rd International Conference on Accelerated Carbonation for Environmental and Materials Engineering, Turku, December 2010.*
15. Young, J.F., Berger, R.L., Breese, J., 1974. Accelerated curing of compacted calcium silicate mortars on exposure to CO_2. *J. Amer. Ceram. Soc.* 57(9), 394-397.
16. Maries, A., Hills, C.D., 1986. Improvement in concrete articles. *UK Patent* No. 2,192,392.
17. Malinowski, R, 1983. A method of accelerating the hardening of concrete. *US Patent* No. 362,679.
18. Hannawayya, F., 1984. Microstructural study of accelerated vacuum curing of cement mortar with carbon dioxide. *World Cement* (Nov) 326-334, (Dec) 378-384.
19. Abetong –Sabema, 1983. 'Quickcrete' brochure, Växjö, Sweden.
20. Schmidt, E., 1986. Rapid setting of cement-bonded fibre composition. *European Patent* No. EP0189127 A2.
21. Pizzol, V.D. et al., 2013. Effect of accelerated carbonation on the microstructure and physical properties of hybrid fiber-cement composites. *Minerals Engineering* 59, 101–106.
22. Moorehead, D.R., 1986. Cementation by the carbonation of hydrated lime. *Cem. Concr. Research* 16, 700-708.
23. Moorehead, D.R., 1980. Method of producing a building element. *GB Patent* No. 2,027,085.
24. Cizer, O. et al., 2012. Real-time investigation of reaction rate and mineral phase modifications of lime carbonation. *Construction and Building Materials* 35, 741–751.

25. Lange, L.C., Hills, C.D., Poole, A.B., 1996. The influence of mix parameters and binder choice on the carbonation of cement solidified wastes. *Waste Management* 16 (8), 749-756, 757-763.

26. Hills, C.D., Sweeney, R.E.H., Buenfeld, N.R., 1999. Microstructural study of carbonated cement-solidified synthetic heavy metal waste. *Waste Management* 19, 325 -331.

27. Hills, C.D., Pollard, S.J.T., 1997. The influence of interference effects on the mechanical, microstructural and fixation characteristics of cement-solidified hazardous waste forms. *J. Hazardous Materials* 52, 171-191.

28. Maries, A., Hills, C.D., Whitehead, K., MacLeod, C.L., 2001. Enhanced cement stabilisation of contaminated clay soils. *Problematic Soils*, Thomas Telford pub., London.

29. Johnson, D.C. et al., 2003. Solidification of stainless steel slag by accelerated carbonation. *Environmental Technology* 24, 671-678.

30. Gunning, P.J., Hills, C.D., Carey, P.J., 2009. Production of lightweight aggregate from industrial waste and carbon dioxide. *Waste Management* 29, 2722–2728.

31. Gunning, P., Hills, C.D., Carey, P.J., 2013. Commercial application of accelerated carbonation: looking back at the first year. *In: Proceedings of the 4th International Conference on Accelerated Carbonation for Environmental and Materials Engineering, Leuven, April 2013*, 185-192.

32. Gunning, P., Hills, C.D., 2015. Negative about carbon: a positive approach. *To be published in: Proceedings of the 5th International Conference on Accelerated Carbonation for Environmental and Materials Engineering, New York City, June 2015.*

33. Shi, C., He, F., 2010. Properties and microstructure of CO_2 cured concrete blocks. *In: Proceedings of the 3rd International Conference on Accelerated Carbonation for Environmental and Materials Engineering Turku, November 2010*, 277-284.

34. Owens, K., Basheer, P.A.M., Sen Gupta, B., 2008. Influence of accelerated carbonation on fresh and hardened properties of cementitious systems. *In: Proceedings of the 2nd International Conference on Accelerated Carbonation for Environmental and Materials Engineering, Rome, October 2008*, 159-168.

35. Monkman, S., Niven, R., 2013. Beneficial carbon dioxide utilization in precast concrete production. *In: Proceedings of the 4th International Conference on Accelerated Carbonation for Environmental and Materials Engineering, Leuven, April 2013*, 309-318.

36. Monkman, S., 2014. Overview of CarbonCure Carbonation Technology. http://carboncure.com/white_paper/carboncure-technology-masonry-white-paper-2014.

37. Solidia Technologies, 2014. White papers. http://solidiatech.com/white-papers.

38. Jones, R.H., 1996. Cement treated with high pressure CO_2. *US Patent* No. *5,518,540.*

39. CEMBUREAU, 2014. World cement production by region: evolution 2001-2013. http://www.cembureau.be/about-cement/key-facts-figures.

40. U.S. Geological Survey, 2015. Mineral Commodity Summaries, January 2015: Cement.

41. Maries, A. et al., 2013. Linked low-carbon manufacture of cement and precast concrete. *Adv. Appl. Ceramics*, 112 (4) 202-6 (2013).

42. American Heritage Society, 1980. Thomas Alva Edison and the concrete piano. *American Heritage 31(5).*

Life Cycle Assessment of the Carbonated MgO Cement- a Green Building Material

Mo, L., Nanjing Tech University
Panesar, D.,
Al-Tabbaa, A., University of Cambridge

Cement industry is one of the main CO_2 emitters, which accounts for up to 7% of the man-made carbon emissions, and thus exists a big demand for cutting its CO_2 emissions. Reactive MgO cement mainly consisting of reactive MgO and industrial wastes was usually considered as a type of low-carbon alternative cement owing to the CO_2 sequestration as a result of the formation of Mg-bearing carbonates during its CO_2 curing process. The accelerated carbonation, microstructure and physical properties of the reactive MgO cement were investigated in some previous studies. However, limited work has been focused on the CO_2 footprint of the carbonated MgO cement particularly considering the varieties of accelerated carbonation techniques, MgO resources, and product types, etc. In this study, the life cycle assessment was employed to evaluate the CO_2 footprint of the carbonated MgO cement product throughout the whole service life. A life cycle assessment model for the carbonated MgO cement was established, and the implications of the carbonation techniques, MgO resources and product type on the CO_2 footprint of MgO cement were analyzed. Results showed that with the accelerated carbonation technology, the reactive MgO cement could be used to prepare green building materials with low-CO_2 emissions, and the CO_2 emissions of the building materials strongly depended on the performance of the products. Therefore, a performance-based design concept for the MgO cement considering the mixes of raw materials, carbonation parameters and performance of product was established. This study provides a reasonable way to estimate the CO_2 emissions of the carbonated MgO cement products and meanwhile gives a basic guidance for the optimum design of green building material with the reactive MgO cements based on the performance requirement.

CO2 capture and storage by mineral carbonation
Increased Stability of CO2-Host Minerals Under Humid Conditions

Morgan, B., CSIRO Mineral Resources
Wilson, S. A., Monash University
Madsen, I. C., CSIRO Mineral Resources
Originally presented on: 6/23/2015 9:45:00 - 10:00:00

Carbon mineralisation provides a safe and effective means of trapping and storing atmospheric CO_2 in highly stable and environmentally benign carbonate minerals, such as magnesite ($MgCO_3$). However, $MgCO_3$ formation is kinetically inhibited under Earth's surface conditions, and while promoting its precipitation to enhance CO_2 trapping efficiency is technologically feasible, it is not currently financially viable owing to low carbon taxes and CO_2 prices.

Hydrated Mg-carbonate minerals, such as nesquehonite ($MgCO_3 \cdot 3H_2O$), precipitate readily at the Earth's surface and are considered promising alternate mineral hosts for CO_2 storage. However, for hydrated Mg-carbonates to be recognised as suitable phases for the capture and storage of anthropogenic CO_2, we need an improved understanding of their stability fields, and subsequent long-term CO_2 security. This is particularly true given the large stockpiles of the minerals that would require long-term storage, or disposal, if the technique were to gain popularity.

Here, we have utilised both *in situ* X-ray diffraction and thermogravimetric analysis to investigate thermal stability of nesquehonite in humid, relatively low-temperature conditions relevant to the Earth's surface. These *in situ* techniques allowed for detailed monitoring of nesquehonite crystallinity and thermal mass loss during short-term decomposition experiments in the presence of water vapour. We found that at both 50°C and 100°C there was a decreased rate of structural decomposition and mass loss when nesquehonite was exposed to humidified gas. We hypothesise that the sorption of water vapour to the mineral surface may impede the outward diffusion of water from the crystal structure, resulting in a higher energy requirement for dehydration to proceed. This has important implications for the disposal and storage of CO_2 in hydrated Mg-carbonate minerals on the millennial timescales required for combating CO_2 pollution. The study is also relevant for tailoring industrial processes to enhance CO_2 storage security via decomposition of nesquehonite to more stable lower hydrates.

Carbon negative cement for atmospheric CO_2 removal

P. Renforth[1*]

[1] School of Earth and Ocean Sciences, Cardiff University, Cardiff, CF10 3AT, United Kingdom

*RenforthP@Cardiff.ac.uk

Abstract

The cement industry emits around 1.6 billion tonnes of carbon dioxide per year (5% of global emissions) to manufacture 4 billion tonnes of cement. The use of cement is ubiquitous in the construction industry, which is essential for economic prosperity. As such, production of cement is likely to rise to at least 4.5 billion tonnes per year by mid-century, emitting even more greenhouse gases. This proposal suggests that it may be possible to not only reduce the emissions of the cement industry to zero but to create an industry that removes carbon dioxide from the atmosphere. To do this, a number of technical and policy innovations are required. Currently, the embodied emissions of cement are created during pre-calcination processes (e.g. mining the raw materials, transport, and comminution), chemical decomposition of limestone, and the fossil fuel energy consumed in the kiln. Usually, life-cycle analysis of cement stops here. However, there is growing evidence that suggests a large proportion of the carbon dioxide emitted by the chemical decomposition of limestone in the kiln is recaptured following the demolition of the structure. Utilising this 'post-use re-carbonation' of cement together with biomass energy and carbon capture and storage, could result in a cement industry that draws around 500 kg of carbon dioxide (NET) out of the atmosphere per tonne of clinker produced or around 3 $GtCO_2\ a^{-1}$ in total. Considering the whole-life cycle of cement in this way is potentially game changing for the industry.

Introduction

The carbon challenge

Once emitted carbon dioxide remains in the atmosphere, and is only removed slowly by natural processes[1]. Methods to accelerate the natural removal of CO_2 from the atmosphere are being investigated should a dangerous concentration be surpassed. The only emission scenario that is likely to result in less than a 2°C global mean surface temperature increase implicitly requires net global atmospheric CO_2 removal towards the end of this century[2]. This scenario requires emissions to peak between 2020 and 2030 with a substantial reduction thereafter. As this is unlikely, a greater amount of CO_2 removal from the atmosphere will be required. The proposed modification to the cement life cycle reduces the emissions of the industry and offers two opportunities for CO_2 removal from the atmosphere (during biomass growth and post-use cement carbonation).

The Cement industry

Global cement production has increased dramatically since the mid-20th century from <500 Mt a^{-1} in 1960 to over 4 Gt in 2013[3] (Figure 1). Predictions for future cement production lie between 3.7 and 5.5 Gt a^{-1} by 2050[4,5]. Cement is manufactured by chemically decomposing limestone in a furnace, creating CO_2 and a number of unstable calcium silicate minerals ('clinker' e.g. the formation of larnite in Eq 1). As this reaction is mostly powered by fossil fuel, the cement industry has a substantial carbon footprint. Between 3.5 and 7.0 GJ are required, and around 0.89 t of CO_2 is produced, for every tonne of clinker[6] (95% of which is from the kiln).

$$2CaCO_3 + SiO_2 \rightarrow Ca_2SiO_4 + 2CO_2 \qquad \Delta G_r \text{ 66.7 kJ mol(carbonate)}^{-1} \text{ Eq. 1}$$

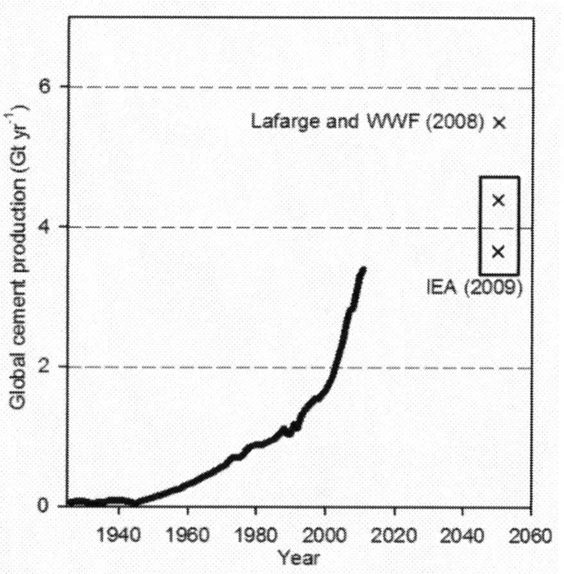

Figure 1: Historic and projected global cement production[3,4,5]

Emissions reduction: Policy and Technology

Realising stringent emission reduction targets in the coming decades is an enormous challenge for the industry. Energy efficiency improvements will contribute, although there is a theoretical limit given the endothermic reaction in Eq1. Current average global energy intensity of cement clinker is 4.2 GJ t^{-1}, which is predicted to decrease to a minimum of 3.2 GJ t^{-1} by 2050 [4] (although best available technology achieves maximum energy efficiencies of 2.7 GJ t^{-1} [6]. Alternative materials may also reduce total emissions, but these are at a very early stage of development and it is uncertain at what scale they can replace conventional production. The most prominent options for emission reduction include lower net carbon intensive fuels (e.g. biomass), and the deployment of carbon capture and storage.

Biomass use in kilns

Currently, biomass use in the cement industry is thought to be around 3%, but has been predicted to rise to 16% by 2050[4]. Based on assumptions relating to human appropriation of biomass, Powell and Lenton[7] estimate approximately 2.4 to 6.0 GtC yr^{-1} of waste biomass may be available for energy generation globally, which corresponds to a lower heating value of 77-186 EJ yr^{-1} (Table 1). The total energy requirement for cement production in 2050 may be between 12 and 18 EJ per year. Therefore, it is notionally possible that the energy demand for cement can be wholly met through biomass combustion without sacrificing food production. Currently, the infrastructure to collect and distribute this type of biomass is underdeveloped, and low energy densities suggest that long distance shipment will not be feasible without prior processing.

Table 2. Potential biomass sources and their heating potential. Adapted from[7]

Feedstocks in 2050 (PgC yr^{-1})	Moisture content (wt%)	Carbon content (dry wt%)	Calorific value (GJ t(dry mater)$^{-1}$)	Low efficiency scenario (Gt yr^{-1})	High efficiency scenario (Gt yr^{-1})	Lower heating* value (EJ yr^{-1})
Crop residues	50	40	14	0.65-0.74	0.48-0.56	14.1-21.7
Manure	70	30	17	0.87-1.07	1.03-1.28	42.7-62.9
Food waste	70	45	15	0.47	0.30	8.5-14.2
Forestry residues	50	40	14	0.39	0.39	11.4-11.7
Bio energy crop	20	45	19		1.66-3.43	0-75.9
Total						**76.8-186.4**

*Assuming a heat of vaporisation of 2.3 GJ $t(H_2O)^{-1}$

A range of fuels can be used in a kiln without sacrificing the quality of the cement, and it may be possible to achieve complete replacement with some biofuels. However, there are limits, which have been reviewed by[8]. These limits include <u>Thermal properties:</u> The volatility of fuels is important for distinguishing where in a kiln it may be used. Highly volatile (low temperature fuels) are more useful for pre-calcination rather than as the main fuel within the kiln. <u>Chemistry</u>: The presence of a number of elements in the fuel can adversely affect the quality of the cement clinker. Alkali metals (Na, K) can cause expansion or reactivity problems in finished concrete due to reaction with silica or sulphate. The presence of Zn (above 0.5%) also alters the reactivity of cement minerals. <u>Environmental impact:</u> Exploiting biomass as a fuel would lead to environmental impacts both as a consequence of creation, harvesting and transport, and also potentially increasing the environmental impact of the production site through the formation of harmful gasses at the kiln. Considerable literature exists on the environmental impact of biofuels. Other than the intrinsic carbon of the fuel, this study externalises all of these other costs (carbon emissions from biomass growth are highly dependent on the source and agricultural practices used). In this study, the carbon draw-down of biomass (and the carbon produced within the kiln) was assumed to be 100 $gCO_2\ MJ^{-1}$ [9], although this will vary depending on the biomass mix.

Carbon capture and storage

Wide scale deployment of flue gas CO_2 capture and storage is predicted to sequester almost 60% of global cement emissions by 2050[5]. Nominally, this involves the separation and compression of CO_2 from the other flue gases for geological storage. Other sequestration methods may be possible (e. g. mineral carbonation[10,11], ocean alkalinity storage[12]) but have yet to be developed beyond laboratory scale. The flue gas in a kiln is typically composed of 50 to 80% by volume CO_2, which could be purified through pre-combustion air separation technologies or post-combustion gas separation. In this study the power consumption for capture and storage was assumed to be 100 kWh tCO_2^{-1}, which is consistent with[13].

Pot-use carbonation

There is growing evidence that cement minerals readily transform back into carbonate minerals when exposed to appropriate conditions (e.g. the hydrated cement mineral jennite in Eq2). This has been demonstrated in controlled reactors[14,15] and discovered in landfill[16,17] at elevated CO_2 partial pressure. However, it is also common, albeit unintentional, in ambient environments receiving demolition waste (urban soils[18]). For instance, following the distribution of concrete demolition waste across a brownfield site in the UK[19,20], almost 60% of the cement minerals had reacted with atmospheric CO_2 within 42 months. Introducing a range of controls on gas and solution mineral interaction through engineered environments (e.g. elevated pCO_2 reactors, landfill cells, or appropriate landscaping) will likely increase this reaction rate and extent.

$$Ca_{1.67}SiO_{1.57}(OH)_{4.2} + 1.67CO_2 \rightarrow 1.67CaCO_3 + SiO_2 + 2.1H_2O$$

ΔG_r -59.5 kJ mol(carbonate)$^{-1}$ Eq2.

Carbon sequestration in cement based waste as a method of creating secondary construction materials has been commercially developed by Carbon8 (http://www.c8s.co.uk/). They have shown that cement rich waste reacts rapidly with CO_2 at elevated partial pressures. Given the likely premium for atmospheric CO_2 sequestration, flue gas capture would be an inefficient use of reactive alkaline waste materials. However, as cement rich wastes readily react with ambient CO_2, flue gas CO_2 may not be necessary. Additional work is required to understand the optimal method of post-use recarbonation.

The maximum global carbon storage potential of this is equivalent to the emissions released from chemical decomposition in the kiln (~0.5 tCO_2 per tonne of clinker). Concrete (in construction and demolition waste) is one of largest waste streams globally, and is likely to grow in the coming decades as the urban environments constructed in the second half of the 20th century renew and develop. The average life span of a building is ~50 years[21], which represents a lag time between production and re-carbonation.

Maximum potential for carbon sequestration

For every tonne of clinker produced approximately 1.2 tonnes of limestone is required (creating 0.543 tCO_2 during chemical decomposition in the kiln). To produce 3.2 GJ of energy per tonne of clinker, approximately 0.21 tonnes of biomass are required which creates (0.352 tCO_2). It is assumed that the 0.895 tCO_2 produced in the kiln represents 95% of the emissions associated with cement production (0.045 tCO_2 are produced from raw material preparation). 80% of the CO_2 produced in the kiln is captured, the rest is assumed fugitive. The purification of the flue gas is achieved through oxy-fuel firing in the kiln, together with CO_2 compression for sequestration adds an additional 580 MJ of electrical energy onto the existing 360 MJ per tonne of clinker (creating 0.235 tCO_2). Finally, post-use re-carbonation of cement is assumed to recapture 95% of the CO_2 lost during chemical decomposition (0.52 tCO_2). This represents a net removal of 0.409 tCO_2 per tonne of clinker created. The aim of this study was to test the carbon draw down as a consequence of changes in biomass fuel substitution, CCS utilisation, and post-use re-carbonation under realistic emission intensity and cement production scenarios.

Methods

Model design

The global cement industry was simplified into the process flow sheet presented in Figure 2, together with a potential route for geological CCS. The main energy consuming activities are included, an outline of the mass and energy flow parameters are included in Table 2.

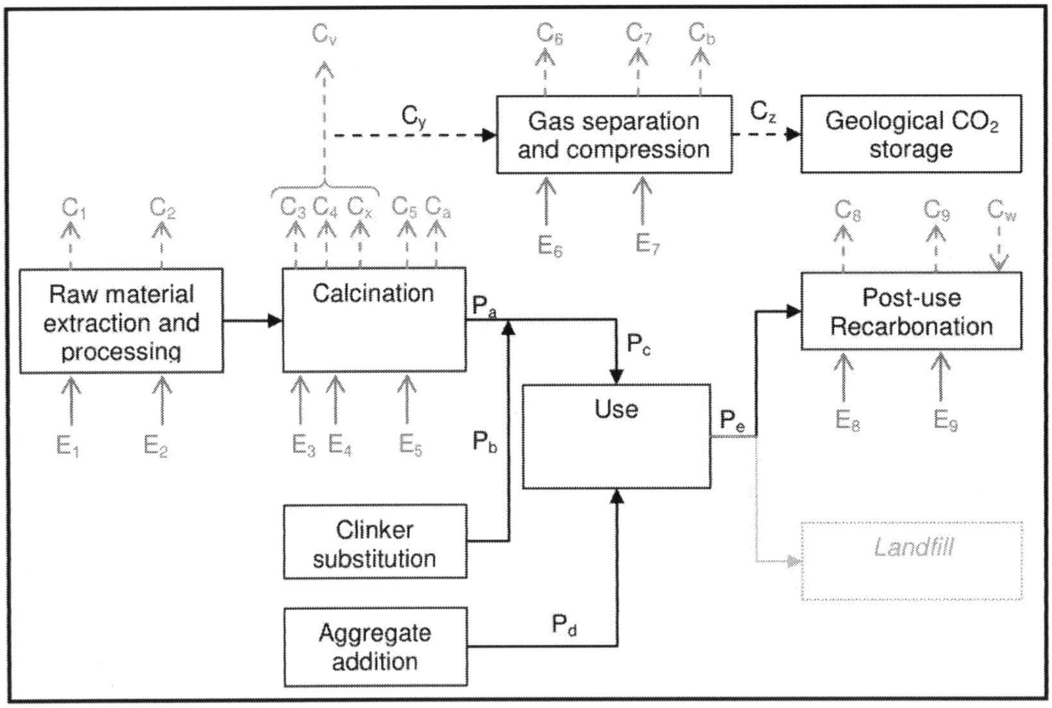

Figure 2. Schematic of cement industry model. Energy flows are denoted by E_x and red arrows, mass flows of clinker, cement, aggregate and waste are denoted by P_x. Carbon dioxide flows are denoted by C_x and blue arrows represent emissions to the atmosphere.

Table 2. Parameters used in the model

Variable	Description	Units	Relationship
E_1, (C_1)	Primary energy requirements for extraction and processing (and associated carbon emissions).	MJ y^{-1} ($MtCO_2\ y^{-1}$)	$C_1 = E_1 \cdot K_F$
E_2, (C_2)	Electrical energy requirements for extraction and processing	kWh y^{-1} ($MtCO_2\ y^{-1}$)	$C_2 = E_2 \cdot K_E$
E_3, (C_3)	Fossil fuel (thermal) energy used in calcination	MJ y^{-1} ($MtCO_2\ y^{-1}$)	$C_3 = E_3 \cdot K_F$
E_4, (C_4)	Biomass energy (thermal) used in calcination	MJ y^{-1} ($MtCO_2\ y^{-1}$)	$C_4 = E_4 \cdot K_B$
E_5, (C_5)	Electrical energy used in calcination	kWh y^{-1} ($MtCO_2\ y^{-1}$)	$C_5 = E_5 \cdot K_E$
E_6, (C_6)	Fossil fuel (thermal) energy used in gas separation	MJ y^{-1} ($MtCO_2\ y^{-1}$)	$C_6 = E_6 \cdot K_F$
E_7, (C_7)	Electrical energy used in gas separation	kWh y^{-1} ($MtCO_2\ y^{-1}$)	$C_7 = E_7 \cdot K_E \cdot C_y$
E_8, (C_8)	Fossil fuel (thermal) energy used in post-use carbonation	MJ y^{-1} ($MtCO_2\ y^{-1}$)	$C_8 = E_8 \cdot K_F \cdot P_e$
E_9, (C_9)	Electrical energy used in post-use carbonation	MJ y^{-1} ($MtCO_2\ t_{Pe}^{-1}$)	$C_9 = E_9 \cdot K_E \cdot P_e$
C_a	Fugitive[1] emissions associated with calcination	$MtCO_2\ y^{-1}$	$C_a = (C_3 + C_4 + C_x) \cdot A$
C_b	Fugitive[1] emissions associated with gas separation	$gCO_2\ y^{-1}$	$C_b = C_y \cdot A$
C_w	The CO_2 draw down from post-use carbonation[2]	$gCO_2\ y^{-1}$	$C_w = L(C_x \cdot X)$
C_x	The CO_2 that is released during the chemical decomposition of the raw materials	$gCO_2\ y^{-1}$	(assumed fixed)
C_y	CO_2 created in the kiln that is available for gas separation	$gCO_2\ y^{-1}$	$C_y = C_a\ (1/A - 1)$
C_z	CO_2 sent for geological storage	$gCO_2\ y^{-1}$	$C_z = C_y - C_b$
P_a	Production of cement clinker	$Mt_{Pa}\ y^{-1}$	$P_a = P_c - P_b$
P_b	Clinker substitution amount3	$Mt_{Pb}\ y^{-1}$	$P_b = P_c \cdot Y$
P_c	Cement production	$Mt_{Pc}\ y^{-1}$	Model boundary condition
P_d	Aggregate addition[4]	$Mt_{Pd}\ y^{-1}$	$P_d = P_c \cdot Z$
P_e	Material fed into post-use activity	$Mt_{Pe}\ y^{-1}$	$P_e = L(P_c + P_d)$
K_B	Carbon intensity of biomass energy	$gCO_2\ MJ^{-1}$	Model boundary conditions
K_D	Carbon intensity of heavy fuel oil	$gCO_2\ MJ^{-1}$	
K_E	Carbon intensity of electricity	$gCO_2\ kWh^{-1}$	
K_F	Carbon intensity of fossil fuel thermal energy	$gCO_2\ MJ^{-1}$	

[1]A is assumed fixed at 1%.
[2]X is the proportion of CO_2 that is released from chemical decomposition that is recaptured during post-use re-carbonation. L is the lag function of use.
[3]Clinker substitution amount is a model boundary condition.
[4]Z is assumed fixed at 4 (to simulate more realistic process requirements during post-use re-carbonation)

Scenario boundary conditions

Two emission intensity scenarios ('High'/'Low') were used for fossil fuel generation (Figure 3A), which correspond to AF1 ('business as usual') and B1 ('extensive decarbonation') used in the SRES emission scenarios[22]. These are similar to RCP 8.5 and between 2.6 and 4.5[23]. Emissions from electricity generation were calculated through a conversion efficiency projection[22]. Two cement production scenarios were created for this ('High'/'Low'; Figure 3B). The high production scenario follows current production estimates, before reaching a constant value in 2050 of 5 Gt y^{-1} (equivalent to the forecast by[5]). The low production scenario assumes increased production before reducing to a constant value of 4 Gt y^{-1} (equivalent to the forecast by[4]). Clinker production was calculated by assuming a substitution scenario that increases from 15% in 1990 to 35% in 2100.

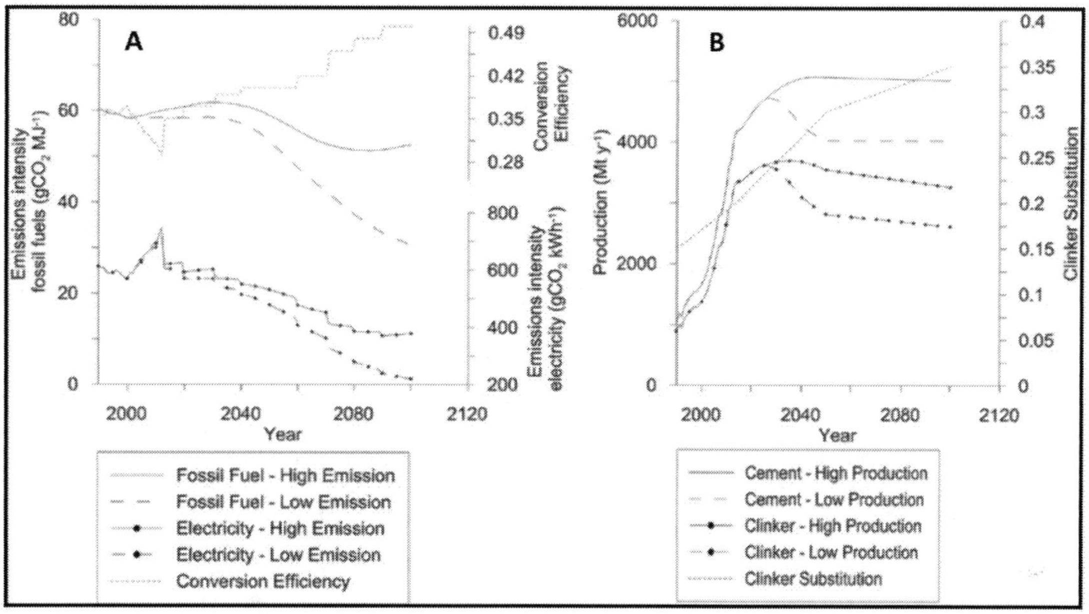

Figure 3. A) Emission intensity and B) cement production used in the simulation

Use of biomass as a fuel source, carbon capture and storage, and post-use recarbonation were tested against each of these emission intensity and production scenarios. It was assumed in each case that current levels remain constant until 2020, after which there is a linear increase until the maximum value is reached by 2050.

Table 3. Current/maximum values for biomass substitution, CCS and post-use recarbonation

Scenario	Current	High	Medium	Low
Biomass fuel substitution	0.03	0.95	0.50	0.10
Carbon capture and storage*	0.00	0.80	0.50	0.10
Post-use recarbonation	0.00	0.80	0.50	0.10

*The values account for the proportion of available CO_2 from the kiln passed to the capture plant. Fugitive emissions are handled separately.

Cost implications
The value added from this activity would reduce the (notional future) penalty associated with atmospheric CO_2 release into the atmosphere (e.g. a carbon tax on emissions). However, it would also create a 'tradable' CO_2 emission reduction from atmospheric CO_2 draw-down. By assuming a relative cost/value of avoided emissions into the atmosphere (~\$60 tCO_2^{-1}) and a value for removal of CO_2 from the atmosphere (~\$100 tCO_2^{-1}). The potential cost savings for exploiting biomass fuel, CCS and post-use recarbonation were investigated.

Results

The model output for the high emission/production and low emission/production scenarios are presented in Figure 4. The 'business as usual' scenarios (including low exploitation of biomass fuel, CCS and post-use recarbonation) still suggest emissions from the global cement industry will peak towards 2050 (as a result of production peaking with population), and decrease in the latter half of the century (mainly due to clinker substitution). In both of the production/emission scenarios, the net emissions of the cement industry are substantially reduced by increasing levels of CCS, biomass or post-use recarbonation (the lowest net emissions corresponding to the highest exploitation mitigation measures given in Table 3). The results indicate that the cement industry could become carbon negative towards the middle of the century with a maximum draw down of around 3 $GtCO_2$ per year.

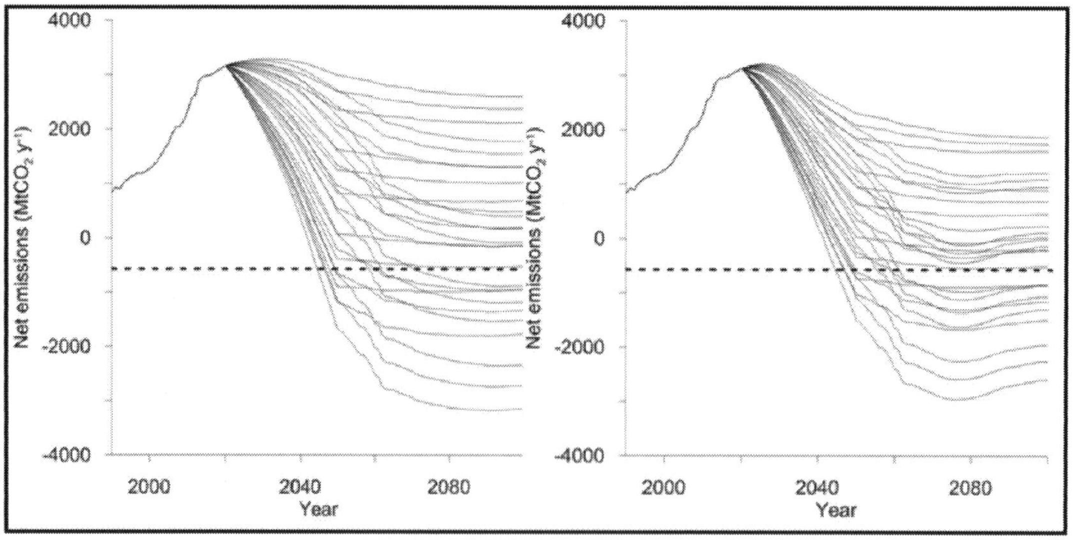

Figure 4. Output from the High Emission-High Production and B) Low Emission-Low production scenario showing the net emissions of the cement industry from 1990 to 2100.

As a result of a \$60 tCO_2^{-1} carbon emission cost, the relative cost of cement may rise by around 65% (if no mitigation policy is implemented). Figure 5 shows the impact of biomass fuel substitution, CCS, and post-use recarbonation on this cost rise (in which atmospheric CO_2 draw down could be traded at \$100 tCO_2^{-1}). A base case of 70%

exploitation for all three options was used. Figure 5A shows that without using biomass energy (while maintaining 70% CCS and post-use recarbonation) the relative increase in the cost of cement is around 20%. The model suggests that increasing the level of biomass use could lower the overall cost increase, providing the cost of the biomass fuel is not sufficiently greater than the cost of conventional natural gas. Exploiting CCS (Figure 5B) may be economical providing the cost of capture and storage is not substantially greater than the cost of emissions. The impact of post-use carbonation seems to have the largest impact on cost reduction (Figure 5C). If a profit can be made from this activity (e.g. through the creation of secondary aggregate or the avoidance of landfill tax on waste) then the overall cost of cement clinker could be reduced by up to 50% even in a world with a \$60 tCO_2^{-1} carbon tax.

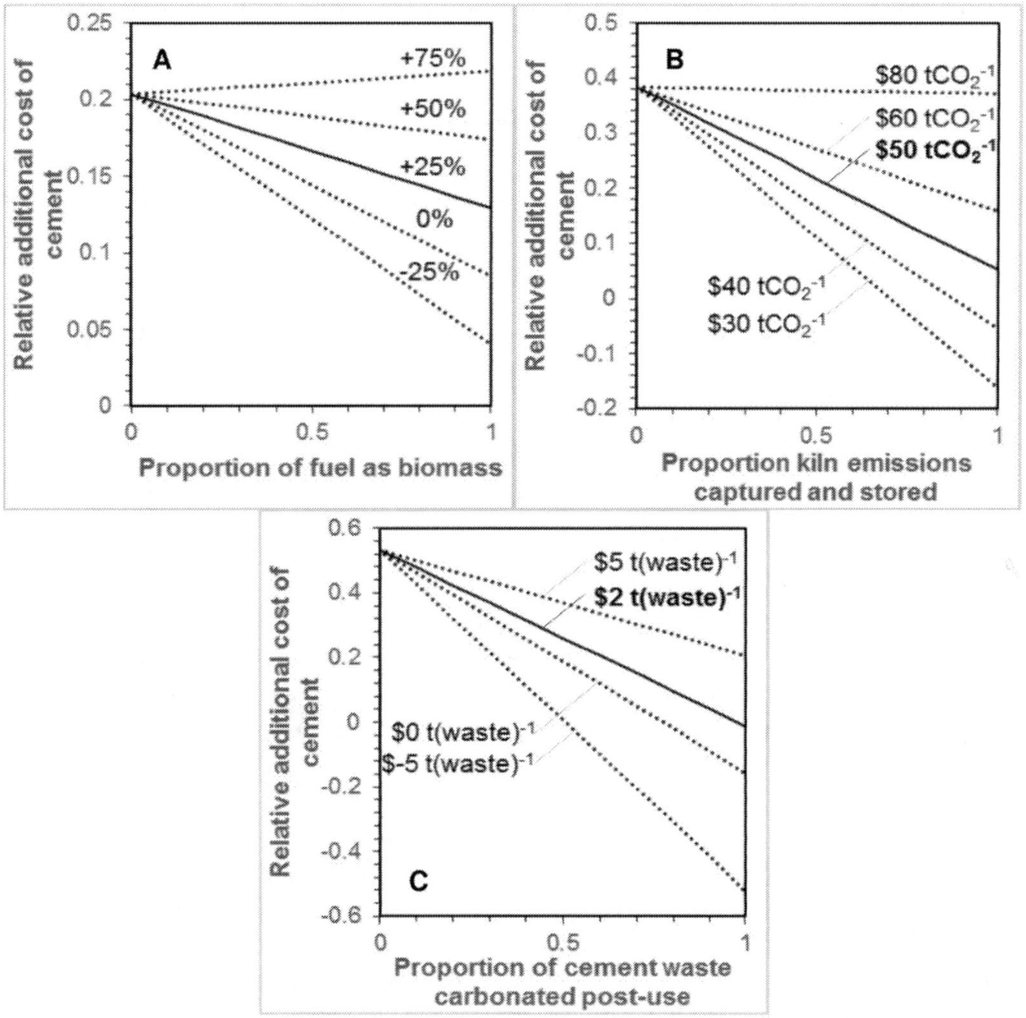

Figure 5. The relative additional cost of cement clinker as a result of A) biomass use as a fuel (lines showing the result in varying the relative cost of the biomass fuel over conventional natural gas fuel), B) use of CCS (the lines showing the cost of CCS), and C) post-use recarbonation (the lines showing the relative cost of cement based waste treatment).

Conclusions

Substituting biomass fuel into the kiln and the use of carbon capture and storage are considered as the primary methods by which the cement industry will decarbonise over the coming decades. However, the results of this study suggest that widespread and extensive deployment of these mitigation measures together with recarbonation of the cement following demolition could result in a carbon negative cement industry that removes around 3 billion tonnes of CO_2 out of atmosphere by the middle of the 21st century. Exploiting these mitigation measures may add value to the cement industry through avoiding costs associated with emissions into the atmosphere and possible tradable 'negative' emissions (carbon dioxide removal from the atmosphere). Further work is required, particularly to:

- Develop a sustainable biomass exploitation strategy for cement manufacturing. This may include the assessment of local biomass waste at each facility. An assessment of the maximum substitution ratio for a particular biomass mixture, and the impact such a fuel switch would have on the quality of the finished clinker.
- Evaluate how CCS can be deployed in the cement industry, including the cost of retrofitting and the location of the production site in relation to potential CO_2 storage repositories.
- Develop optimal strategies for carbonation of cement based waste materials (e.g. natural attenuation, soil addition, or accelerated carbonation under elevated pCO_2).
- Establish appropriate mechanisms by which post-use recarbonation may be valued across the cement supply chain (possibly as 'latent' carbon capture in the sold clinker).

These first two points have gained increasing attention as prominent mitigation measures within the cement industry[4,5]. These need to be holistically integrated with post-use recarbonation[14,18] to maximise the benefit to the cement industry. Furthermore, the implications of this holistic carbon budget approach could be extended for other industries that emit CO_2 through the chemical transformation of their raw materials (e.g. iron and steel, and lime manufacturing). All of these could make substantial gains by recognising the atmospheric carbon capture potential in their sold materials.

References

1. Lowe J.A., Huntingford C., Raper S.C.B., Jones C. D., Liddicoat S.K., Gohar L.K. 2009 How difficult is it to recover from dangerous levels of global warming? *Environmental Research Letters*, 4(1): 014012.
2. Vuuren D., Stehfest E., Elzen M.J., Kram T., Vliet J., Deetman S., Isaac M., Klein Goldewijk K., Hof A., Mendoza Beltran A., Oostenrijk R., Ruijven B. 2011 RCP2.6: exploring the possibility to keep global mean temperature increase below 2°C. *Climatic Change*,. 109(1-2): 95-116.

3. Renforth, P., Washbourne C.L., Taylder J., Manning D.A.C. 2011 Silicate Production and Availability for Mineral Carbonation. *Environmental Science & Technology*, 45(6): 2035-2041.
4. IEA, *Cement Technology Roadmap 2009*. 2009, International Energy Agency.
5. Lafarge-WWF, *A blueprint for a climate friendly cement industry*. 2008.
6. van Oss H.G. and Padovani A.C. 2002 Cement Manufacture and the Environment: Part I: Chemistry and Technology. *Journal of Industrial Ecology*, 6(1): 89-105.
7. Powell T.W.R. and Lenton T.M. 2012 Future carbon dioxide removal via biomass energy constrained by agricultural efficiency and dietary trends. *Energy & Environmental Science,*. 5(8): 8116-8133.
8. Chinyama M.P.M. 2011 Alternative Fuels in Cement Manufacturing, in *Alternative Fuel*, M. Manzanera, (Ed).
9. Hillebrand K., 1993 The greenhouse effects of peat production and use compared with coal, oil, natural gas and wood. *VTT Tiedotteita,* 1494: 50.
10. Lackner K.S., Wendt C.H., Butt D.P., Joyce E.L., and Sharp D.H., 1995 Carbon dioxide disposal in carbonate minerals. *Energy*, 20(11): 1153-1170.
11. O'Connor W.K., Dahlin D.C., Rush G.E., Gedermann S.J., Penner L.R., and Nilsen D.N., 2005 *Aqueous mineral carbonation, Final Report - DOE/ARC-TR-04-002.*.
12. Rau, G.H., 2011 CO_2 Mitigation via Capture and Chemical Conversion in Seawater. *Environmental Science & Technology*, 45(3): 1088-1092.
13. IPCC, *IPCC Special Report on Carbon Dioxide Capture and Storage*, 2005 Prepared by Working Group III of the Intergovernmental Panel on Climate Change [Metz and H.C.d.C. O. Davidson, M. Loos, and L. A. Meyer eds.]: Cambridge, United Kingdom and New York, NY, USA. p. 442.
14. Fernández Bertos M., Simons S.J.R., Hills C.D., Carey P.J., 2004 A review of accelerated carbonation technology in the treatment of cement-based materials and sequestration of CO_2. *Journal of Hazardous Materials*. 112(3): 193-205.
15. Huntzinger D.N., Gierke J.S., Kawatra S.K., Eisele T.C., and Sutter L.L., 2009 Carbon dioxide sequestration in cement kiln dust through mineral carbonation. *Environmental Science & Technology*. 43(6): 1986-1992.
16. Fleming, I.R., Rowe R.K., Cullimore D.R. 1999 Field observations of clogging in a landfill leachate collection system. *Canadian Geotechnical Journal*. 36(4): 685-707.
17. Manning, D.A.C. 2001 Calcite precipitation in landfills: an essential product of waste stabilization. *Mineralogical Magazine*, 65(5): 603-610.
18. Renforth P., Manning D.A.C., Lopez-Capel E., 2009 Carbonate precipitation in artificial soils as a sink for atmospheric carbon dioxide. *Applied Geochemistry*, 24(9): 1757-1764.
19. Washbourne, C.L., Renforth P., Manning D.A.C., 2012 Investigating carbonate formation in urban soils as a method for capture and storage of atmospheric carbon. *Science of The Total Environment,*. 431(0): 166-175.
20. Washbourne C.L., Lopez-Capel E., Renforth P., Ascough P., Manning D.A.C., (in press) Rapid removal of atmospheric CO_2 by urban soils, *Envronmental Science and Technology*.
21. Adalberth, K., 1997 Energy use during the life cycle of buildings: a method. *Building and Environment*, 32(4): 317-320.
22. IPCC, *Emissions Scenarios*, 2000 Nakicenovic N., Swart R. (Eds).. p. 570.

23. Rogelj J., Meinshausen M., Knutti R., 2012 Global warming under old and new scenarios using IPCC climate sensitivity range estimates. *Nature Climate Change*. 2(4): p. 248-253.

Study of CO_2 Capture Capability of Purified Carbide Lime Waste

E. Ruiz-Agudo and C. Rodriguez-Navarro*

Department of Mineralogy and Petrology, University of Granada, Avd. Fuentenueva s/n, 18071 Granada, Spain Belgium
*Corresponding author: carlosrn@ugr.es

Abstract

The recycling of $Ca(OH)_2$ suspensions (carbide lime), obtained as a by-product in the synthesis of acetylene for its use as a building material and/or sorbent for the capture of greenhouse or toxic gases would be a potential, high added-value way to exploit these residues. Here we study the main microstructural characteristics of carbide lime (subjected to a purification process), in order to determine the quality of such solid residues and optimize their use and applications. This waste lime putty is a nanomaterial composed of planar portlandite crystals, 2-100 nm in size, which form non-oriented (and therefore easily re-dispersible) aggregates a few micrometers in size. The small particle size and low degree of aggregation, as well as an essentially planar morphology of the hexagonal portlandite crystals result in surface area values as high as ~40 m^2g^{-1}. These microstructural characteristics give the carbide lime a higher capability for CO_2 capture than other materials of similar composition such as commercial calcitic limes, as determined in both single carbonation experiments and Ca-looping tests.

Introduction

The study and application in civil engineering and construction of industrial mineral wastes is one of the technological and research fields that has experienced a faster progress in recent years.[1,2] In fact, there is a whole new generation of cementitious materials based on the recycling of highly-alkaline silica- or alumina-rich wastes, such as silica fume, cement kiln dust, steel slag, and fly ash, among others.[3-6] On the other hand, in recent years there is a clear trend towards the development and application of sustainable building materials that involve in their manufacture minimum energy cost and emissions, while presenting as added value to be compatible with the environment, or even better, materials that help ameliorating, -over the course of their life-cycle-, pressing environmental problems such as the emission of greenhouse or toxic gases.[7]

In this sense, portlandite ($Ca(OH)_2$) is a cementing material of great potential as, in addition to being compatible with the vast majority of traditional fabric elements (stone, brick), its hardening involves the capturing of atmospheric CO_2.[8,9] Lime (CaO) has been traditionally obtained by calcination of limestone ($CaCO_3$) at temperatures from 800 to 1000 °C.[10] Subsequently, portlandite suspensions (lime putties) have been prepared by hydration using stoichiometric or excess water (slaking).[11,12] An unconventional route to obtain $Ca(OH)_2$ suspensions follows the hydration of calcium carbide during the synthesis of acetylene gas.[13] In this synthesis, CaO is generated, that quickly hydrates and forms a suspension of calcium hydroxide. Normally this lime paste is stored in ponds as a waste, thus representing a problem for the manufacturer, and a potential (environmental) hazard due to its high pH (~12.4) and presence of trace amounts of heavy metals.[14] The recycling of this lime sludge for its use as a building material or for other applications (e.g., conservation of cultural heritage)[15] would be a way to exploit these residues and provide them with an added value. Pre-treatment and conditioning of this waste are nevertheless required in order to remove/fix impurities that give the lime unwanted physical-chemical and organoleptic properties (presence of heavy metals, foul odour associated with sulfides, discoloration).[15]

Another potential alternative for the use of such a waste could be the use of its sorbent capability for the uptake of toxic or greenhouse gases such as SO_2 or CO_2.[16] Capture of CO_2 is critical in both pre-combustion industrial processes, such as steam reforming of methane, and post-combustion CO_2 (and SO_2) capture.[17,18] Ca-bearing solid sorbents have been extensively used to remove CO_2 from flue gas streams, based on the reversible conversion of CaO into $CaCO_3$.[19] In most processes for CO_2 capture, the sorbent solids are used repeatedly (Ca-looping), so that the evolution of their characteristics and their response to cyclic regeneration are critical to determine the feasibility of their use in practical applications.[20,21] A suitable sorbent must meet several requirements such as a high adsorption capacity, long lifetime, fast sorption/desorption kinetics and good mechanical strength.[16,21] In this regard, portlandite is a material of significant potential.[9] Microstructural properties of carbide lime need nevertheless to be studied in detail to determine the quality of carbide limes and to optimize their use and applications. Here we present the results of a comparative study of the main micro- and nano-structural characteristics of carbide and commercial calcitic limes, obtained from the acetylene industry and by calcination of limestone and subsequent hydration, respectively. Importantly, such a comparative analysis allows determining the parameters and criteria that can be critical for the selection of the optimal type of lime for each type of application. Emphasis has been placed on comparing the CO_2 uptake capability of carbide limes with that of commercial calcitic limes, as these latter ones are the most common in industrial and building applications.

Materials and methods

Carbide lime (CL) was obtained from the company AIR LIQUIDE (Seville, Spain) as an aqueous suspension (lime paste) and was further processed (purification pre-treatment, see below) by Trenzametal S.L. (Zamora, Spain). For comparison, a commercial, dry hydrated lime (Andaluza de cales, AC; Seville, Spain) was simultaneously studied. Pre-treatment of the carbide lime to remove sulfides and organic matter in the paste that give the untreated carbide lime an unpleasant odour (of rotten egg) was required. The pre-treatment also involved the fixation as insoluble phases of metal contaminants. Note that these contaminants have been a strong handicap in the past for the widespread use of carbide lime sludge. We has been able to by-pass such a problem designing and applying a pre-treatment which is described below.[22]

Pre-treatment of carbide lime paste

To remove organic matter and sulfides (from the hydrolysis of sulfur phases present in the coal used for the production of calcium carbide) present in the carbide lime suspensions, a fraction of the carbide lime paste was treated with hydrogen peroxide (H_2O_2). 10 ml of H_2O_2 solutions (33% v/v) were added per 100 ml of lime paste. Immediately upon mixing of the solutions, bubbling was observed. This confirms the presence of organic matter, which becomes oxidized to CO_2 in the presence of H_2O_2. After 48 h, bubbling ceases. Additionally, the increased redox potential of the solution results in the oxidation of sulfur residues (possibly in the form of sulfites) to sulfate. To eliminate soluble sulfates (e.g., gypsum) and co-precipitate heavy metal contaminants, the carbide lime waste was treated with $Ba(OH)_2$ dosed in a concentration of 18 mM per 100 g solids. This treatment induces the precipitation of insoluble barite ($BaSO_4$), which in turn co-precipitates (immobilizes) the existing heavy metals (e.g., Zn, Ni, and Cu, present in ppm concentrations).

Characterization of lime pastes

To fully characterize the carbide lime paste, samples were dried at 100 °C during 24 h. The solid volume fraction, ø, of the lime paste was found to be 0.15, slightly lower than that of traditional lime putties used for building purposes (that usually ranges from 0.20 to 0.25). This is possibly related to the massive presence in this lime of particles of size below 100 nm, which gives them the capability of adsorbing a high amount of water. Importantly, once hydrated, these particles tend to stay dispersed, forming a stable suspension that does not easily settle by gravity. This is why the carbide lime pastes have typically more water than conventional lime putty.

Fractions of the samples were ground in an agate mortar and prepared for analysis by X-ray diffraction (XRD, Philips PW 2500 diffractometer) using CuKα radiation (λ = 1.5406 Å). The results of this analysis allowed identifying mineral phases present in each sample. Thermogravimetric analysis (TG) was performed with the aim of determining the concentration of organic matter, portlandite ($Ca(OH)_2$) and calcite ($CaCO_3$) in the carbide lime. The equipment used was a TG Shimahzu. Samples were placed in alumina crucibles and subjected to heating from room temperature up to 950 °C, with a heating rate of 10 °C min^{-1} in flowing air (10 ml min^{-1}). The morphology and texture of the samples was studied at micro- and nanoscale using transmission electron microscopy (TEM, JEOL 400kV) with EDX microanalysis. Before TEM observation, powder samples were dispersed in alcohol and subjected to sonication for 1 minute (for further dispersion). Once dispersed, were deposited on carbon film Cu grids. The surface area was determined by N_2 sorption analysis (adsorption and desorption isotherms) at 77 K using a Micromeritics Tristar 3000 equipment. From the adsorption isotherms the surface area of the samples was determined according to the BET method. Pore size distribution was also obtained by the BJH method. Previously to the sorption analysis, samples were degassed at 200 °C during 3 h.

Testing the CO_2 capture capability of dried lime putties

Once fully characterized, these materials were subjected to two types of CO_2 capture assays, which are described below.

Batch carbonation tests. Solid limes were heated up to 400 °C under vacuum until constant weight (ca. 6 h) and then CO_2 was introduced in the sample chamber (P_{CO2} = 1 atm). Sample weight and surface area were analyzed at regular time intervals.

Cyclic carbonation tests. Solid limes were heated up to 900 °C at 20 °C min^{-1} under flowing nitrogen (100 mL min^{-1}). After this initial step during which dehydroxylation of portlandite takes place, the temperature was lowered to the desired temperature and the sample was carbonated in a stream of pure CO_2 (100 mL min^{-1}). Two different temperatures, 400 and 650 °C, were tested. Then decarbonation of the samples was performed by increasing the temperature again up to 900 °C under flowing nitrogen (100 mL min^{-1}). Carbonation and decarbonation steps were repeated 10 times each.

Results and discussion

Characterization of carbide and commercial calcitic limes

Figure 1 shows the diffraction pattern of carbide lime after drying the untreated lime paste at 100 °C. Diffraction peaks corresponding to portlandite and other minor phases such as calcite or a hydrated calcium aluminosilicate of low crystallinity (zeolite) are detected. This last phase could have formed as a result of the solubilization at very high pH of the alumina and silica impurities present in the carbide lime paste. The presence of this aluminosilicate indicates that the carbide lime has a slight hydraulic character, although it cannot be considered as a hydraulic lime since the proportion of zeolite is below 5 wt %. Finally, diffraction peaks assigned to minor amounts of hydrated calcium sulfite ($CaSO_3 \cdot 4H_2O$) were also detected. As indicated above, this phase could have formed as a consequence of the oxidation of reduced forms of sulfur compounds to sulfite.

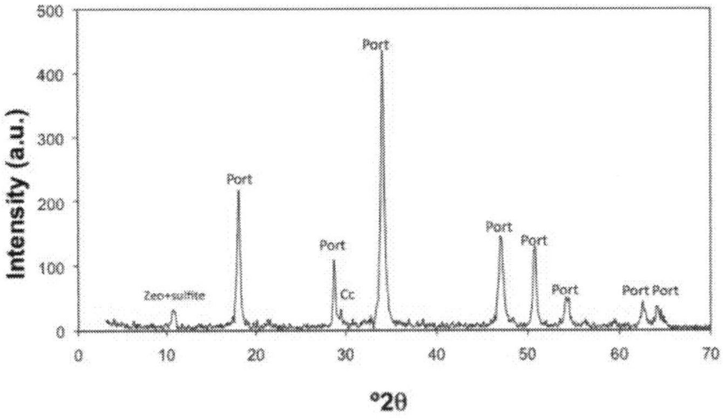

Figure 1. XRD pattern of the carbide lime. Zeo=zeolite;Cc=calcite; Port=portlandite.

After pre-treatment with hydrogen peroxide and $Ba(OH)_2$, no significant changes were observed in terms of abundance and type of crystalline phases detected, except for a reduction in the intensity of the 10.79 °2θ peak corresponding to the mixture zeolite + calcium sulfite. The effect of the treatment with hydrogen peroxide in the variation of the concentration of these phases is possible due to the oxidation of sulfite to sulfate. Nevertheless, the presence of barium sulfate was not detected by XRD, most likely because its concentration was below 5 wt%, i.e., the detection limit of XRD. These results show that the pre-treatment with hydrogen peroxide affects mostly the organic compounds present in the paste.

The phases present in carbide lime are thus very similar to those of a lime obtained by a process of calcination of limestone and subsequent hydration. However, in a calcitic lime usually only diffraction peaks corresponding to portlandite and calcite (5-10 wt%) are seen. However, a significant difference between the diffraction pattern of commercial and carbide limes is that the ratio of (101) to (001) peak intensities is lower in the case of the carbide lime. This is due to the relative overgrowth of basal faces (hexagonal morphology) of portlandite crystals relative to prismatic faces in carbide lime. This plate-like morphology confers a greater plasticity to the carbide limes.[12]

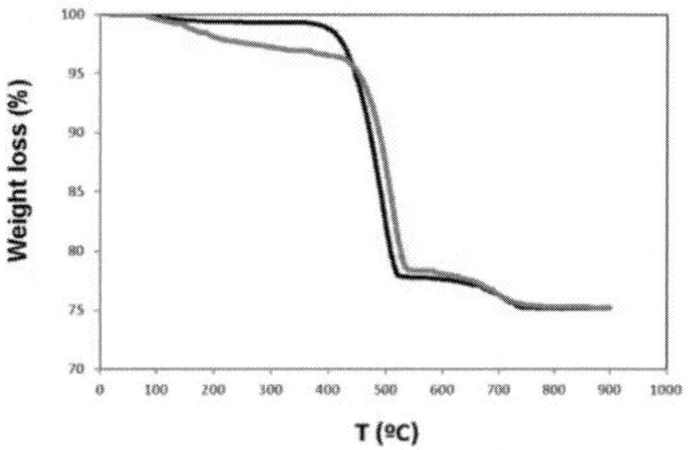

Figure 2. Weight loss curve of CL (red curve) and AC (black curve) limes during thermogravimetric analysis.

Figure 2 shows the results of the TG analysis. CL lime shows an initial weight loss between 100 and 400 ° C, not observed in the case of the traditional lime AC, that corresponds to the oxidation and subsequent dehydration of minor amounts of remaining of calcium sulfite and non-oxidized organic material together with the loss of water present in hydrated calcium aluminosilicates. In the 400-600 °C range, the weight loss is due to dehydroxylation of portlandite. Finally, at 600-800 °C the observed weight loss is associated with the thermal decomposition of calcite initially present as an impurity in the original lime samples. Thermal decomposition of the commercial AC calcitic lime only differs from that of the treated carbide lime CL in the absence of a significant weight loss between 100 and 400 ° C. TG data show that CL lime contains 74.7 wt% portlandite and 7.3 wt% calcite. The rest of the sample (18.0 wt%) accounts for the remaining organic matter, hydrated calcium aluminosilicate and calcium sulfite still present as impurities in the lime. AC lime contains 88.1 wt% portlandite, 6.4 wt% calcite and 5.5% impurities, possibly remnants of silicate residues present in the limestone.

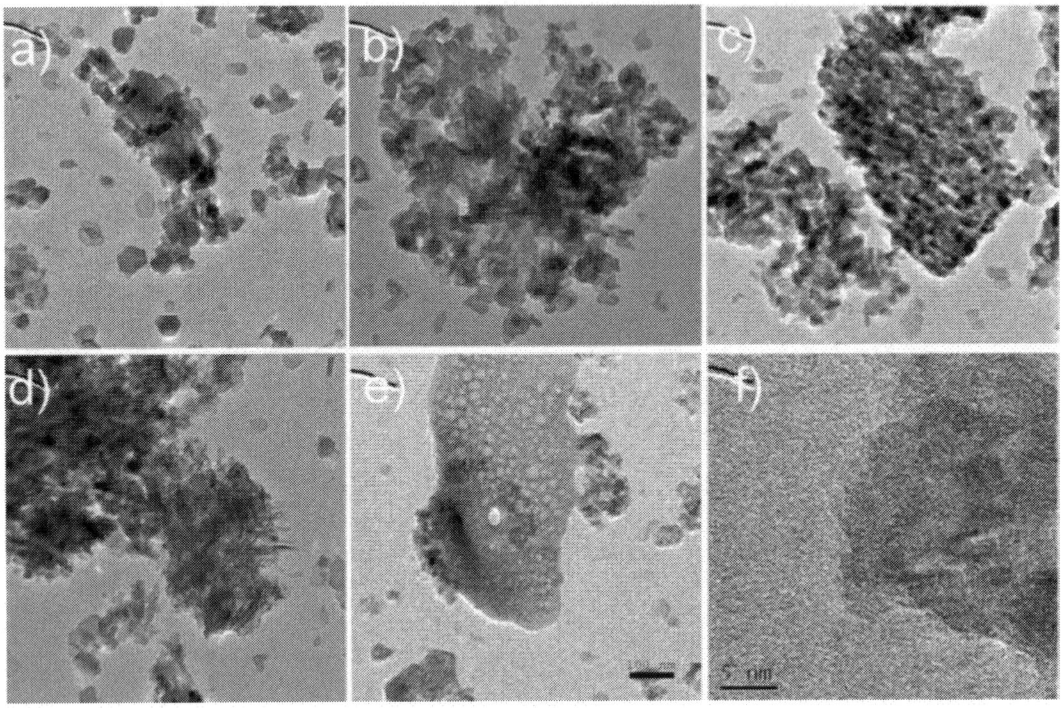

Figure 3. TEM microphotographs of carbide lime samples. a) Planar hexagonal particles of portlandite; b) randomly oriented aggregate of portlandite particles; c) oriented aggregate of portlandite particles; d) aggregate of fibrous particles, possibly zeolite+calcium sulfite; e) amorphous carbonaceous particle with spongy texture after exposure to the electron beam; f) high resolution image of a portlandite crystal formed by an aggregate of nanoparticles with a slight disorientation among them. Figures a) to d) have an image width of 1.5 µm.

TEM observations show that CL is mainly composed of nanometer size portlandite particles with planar, hexagonal morphology (Figure 3). Their sizes range from 5 nm to 100 nm. Overall, the degree of polydispersity of the primary crystals of portlandite is quite small. However, the presence of aggregates of these primary particles contributes to a general increase in polydispersity. The effective size, and thus the reactivity of the lime, is nevertheless controlled by the size of the individual primary particles. Portlandite particles in these micron-sized clusters are typically randomly distributed. These aggregates have a fractal dimension close to 2.4, which implies a process of reaction-limited aggregation and indicates that portlandite particles do not have a high degree of irreversible aggregation. Irreversible aggregation phenomena are due to the oriented aggregation of portlandite primary particles that typically occurs during drying of the lime slurry.[11] A highly oriented aggregation (i) reduces significantly the surface area of the lime, and therefore its reactivity and (ii) adversely affects the rheology of fresh lime putty. However some aggregates in which portlandite crystals stick together with their (001) basal faces parallel are observed in this sample, most likely formed as a result of the drying process at 100 °C (Figure 3). Note that in traditional calcitic limes primary crystals of sizes in the range 200-400 nm are common, whereas aggregates usually have an average size of about 5-20 μm. These sizes are significantly larger than those here observed in the carbide lime, which do not exceed 2 μm. Moreover, the commercial calcitic lime AC has numerous oriented aggregates (irreversible aggregation) caused by drying during stoichiometric hydration of the oxide.[9]

In the TEM images of the carbide lime fibrous aggregates of nanoparticles (Figure 3) are occasionally seen. Such fibrous particles most likely correspond to the aluminosilicate phase (zeolite) identified by XRD. In some areas amorphous carbon particles derived from residues of the carbon used in the manufacture of calcium carbide are detected. Such carbon moieties are responsible for the grayish to bluish color of the carbide lime. These particles volatilize once exposed to the electron beam leading to the development of a spongy, very porous structure. In air limes used in historic mortars it is also common to find carbonaceous particles originated in this case from burning wood used in traditional limekilns. The role that these particles may play in the characteristics of both the fresh lime and in their final state as binder in a lime mortar is currently unknown. However, the effects should be in principle beneficial as for example calcitic lime mortars found in Mayan pyramids, which are preserved in optimal conditions nowadays, include very high proportions of ash residues (carbonaceous particles).[23]

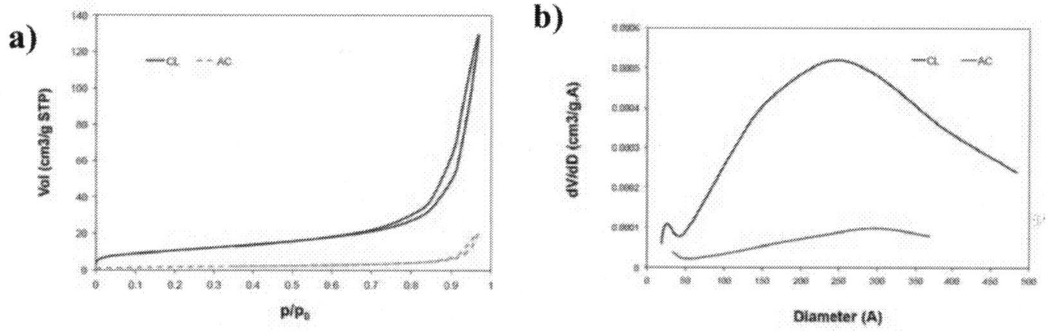

Figure 4. (a) Nitrogen adsorption isotherms and (b) BJH pore size distribution of CL and AC limes.

The N_2 sorption isotherm of the carbide lime paste is of type II (Figure 4a), according to the IUPAC classification (International Union of Pure and Applied Chemistry). Such isotherm is typical of non-microporous solids (pore size > 2 nm). The presence of a H3 type hysteresis loop on the adsorption-desorption curve suggests that such lime has a mesoporous structure (pore size between 2 and 50 nm) with slit-shaped pores. This structure is typical of planar aggregates of particles. These results are in agreement with TEM observations on the morphology and microstructure of the portlandite particles and aggregates formed by these particles in the carbide lime. From the data of the adsorption isotherm the surface area of the carbide lime was determined using the BET method. These limes show surface areas of up to 37 $m^2\ g^{-1}$, a value significantly higher than that typically shown by limes obtained from calcination of limestone, and the subsequent slaking of the resulting calcium oxide (5-20 $m^2\ g^{-1}$).[11,12]

The isotherm of the commercial AC calcitic lime with a surface area of 11 $m^2 g^{-1}$ is also shown in Figure 4a. It can be seen that the amount of N_2 adsorbed by the carbide lime is much higher than that absorbed by the commercial lime. This is possibly due to the larger proportion of particles smaller than 100 nm presents in the CL sample, which also exhibits a lower degree of aggregation than the AC sample.

Figure 4b shows as well the BJH pore size distribution of both CL and AC samples. A higher proportion of mesopores in the carbide lime is observed, with a maximum around 25 nm. This could ensure a remarkable capacity to adsorb gases such as CO_2 and water vapour, which would result in rapid carbonation. Finally, the adsorption isotherm of the carbide lime after pre-treatment with hydrogen peroxide/$Ba(OH)_2$ was also obtained. The treatment slightly reduces the volume of adsorbed gas, and hence the surface area (31 $m^2 g^{-1}$). However these changes do not appear to be relevant, but suggest that a fraction of the surface area of the CL sample could be due to the contribution of the carbonaceous particles, which usually show a significant proportion of mesopores.

Carbonation assays: one-step isothermal dehydration of portlandite and carbonation of lime

Figure 5 show that the surface area of both CL and AC limes decreases significantly with the progress of the carbonation reaction, that is, with increasing degree of conversion of CaO to $CaCO_3$, from 95 to 12 $m^2 g^{-1}$ in the case of CL and from 69 to 4 $m^2 g^{-1}$ in the case of the AC. This decrease in surface area is related to the higher molar volume of $CaCO_3$ compared to CaO that results in the closing of the pores in the starting oxide material. This ultimately results in the blockage of the reaction, as the product layer protects the unreacted lime core from the contact with gaseous CO_2, before reaching a conversion of 100%.

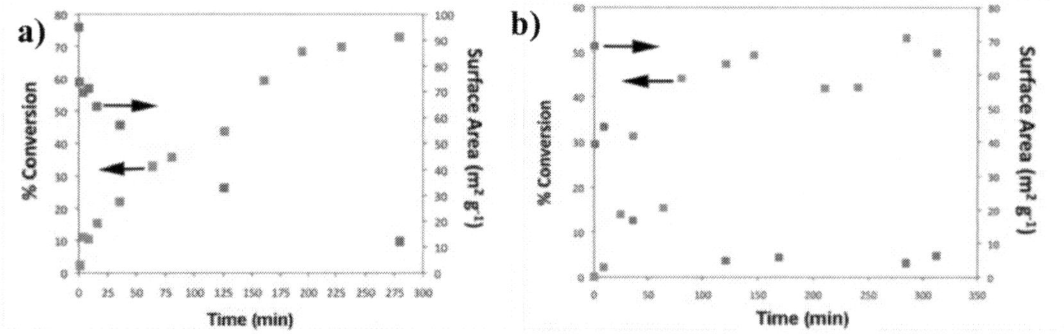

Figure 5. Evolution of the degree of conversion of the carbonation reaction and surface area of (a) carbide and (b) commercial calcitic limes as a function of time.

In this type of tests, with no reuse of the sorbent (i.e., no cycling), a significant increase in the performance of the CL lime in terms of CO_2 capturing with respect to the commercial calcitic lime AC was found. This is related to the better microtextural properties of this lime, primarily its small particle size, low tendency to aggregation and high surface area, which ultimately make this lime more reactive towards carbonation than a commercial lime. This is illustrated in Table 1, where the main parameters that characterize the state of aggregation and porous system of CL and AC limes determined by gas adsorption (nitrogen) before and after carbonation are compared.

Table 1. Characterization of AC and CL porous system by N_2 adsorption.

	CL		*AC*	
	Initial	*Final*	*Initial*	*Final*
Surface area BET ($m^2 g^{-1}$)	95.2	12.3	68.6	4.1
Pore volume ($cm^3 g^{-1}$)	0.307	0.051	0.156	0.0004
Micropore volume ($cm^3 g^{-1}$)	0	0	0.0005	0.003
Mean pore size (Å)	95.7	184.4	65.6	73.1

Carbonation assays: CO_2 looping tests

Figure 6 shows a typical plot of the temporal evolution of the weight percentage gained by the oxide in the cyclic carbonation tests of lime samples. In this plot it can be seen that the weight gain (i.e., the degree of conversion of the carbonation reactions) diminishes significantly with the number of carbonation cycles, an observation typical in this type of tests using calcium-bearing solid adsorbents.[20] This behavior is most likely due to the fact that, simultaneously to the decarbonation of $CaCO_3$, sintering processes take place at 900 °C which cause irreversible aggregation of calcium oxide particles, increasing the particle size and reducing the porosity and surface area of the samples.[21,24] This makes the oxide less reactive as the number of carbonation/decarbonation cycles increases.[20] Regardless of the starting material, it was observed that an increase in the temperature at which the carbonation takes place from 400 to 650 °C results in a significant increase in the performance of the carbonation reaction.

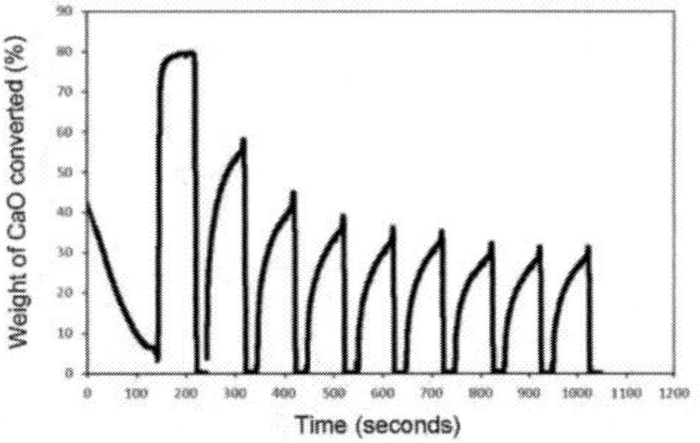

Figure 6. Typical evolution of the weight percentage gained by the oxide in the cyclic carbonation tests of carbide lime samples.

Finally, in Figure 7 the conversion achieved (expressed in wt% of oxide converted) in the case of CL and AC samples is compared with literature data for other two types of calcium-bearing sorbents, including calcium carbonate reagent and calcium carbonate from oyster shells.[25] Although all samples tend to reach a similar value of conversion after 10 carbonation/decarbonation cycles, the CL sample displays a greater CO_2 adsorption capability during the first two cycles of the process. After 10 carbonation/decarbonation loops, this initial better performance of the CL sample represents an increase in the amount of CO_2 captured by ca. 150-180 grams per kg of sorbent with respect to the AC sample. This better performance is mostly related to the optimal characteristics of the starting carbide lime, as mentioned above, in terms of particle size and surface area.

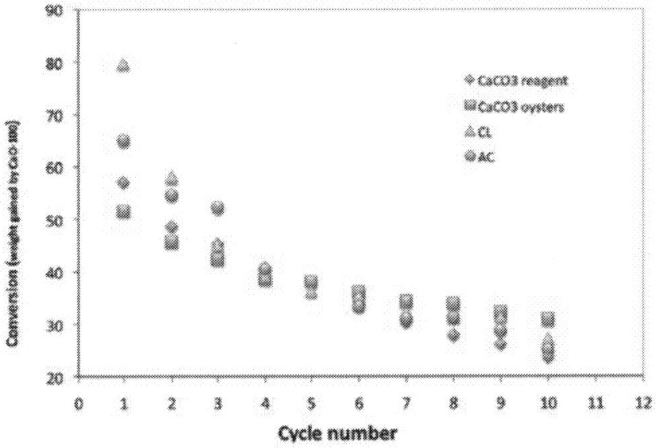

Figure 7. Comparative plot of the degree of conversion of the carbonation reaction (expressed as weight percentage gained by CaO) as a function of the number of carbonation/decarbonation cycles for different Ca-bearing solids.

Moreover, the impurities present in the carbide lime sample, in the form mostly of amorphous carbonaceous particles, do not seem to have a determining effect on the highest performance achieved by the carbide limes, as other materials exhibiting also impurities with a similar composition (as the commercial calcitic lime or the carbonate from oyster shells) have a lower CO_2 uptake capability in the initial cycles of carbonation. Also, the fact that lime samples tested in this study as well as other types of Ca-sorbents used by other researchers eventually yield similar values of CO_2 sorption, regardless their content in carbonaceous impurities, suggest that the reduction of reactive area as a result of the sintering processes that occur during the decarbonation step overrule the potential positive effect of inert impurities in the CO_2 capture performance of Ca-bearing sorbents.

Conclusions

The following conclusions regarding the characteristics, reactivity and CO_2 scavenging capacity of the treated carbide lime at high temperature can be drawn:

• The detailed study, at high resolution, of the microstructure of the carbide lime demonstrate that the carbide lime is a nanomaterial. It is composed of portlandite planar crystals of size ranging from 2-100 nm and non-oriented (and therefore easily redispersible) aggregates of these primary particles with sizes of a few microns. By contrast, commercial calcitic hydrated limes typically include larger aggregates (up to 20 microns) and usually have oriented (and therefore not reversible) aggregation.

• The small particle size and low degree of aggregation, as well as an essentially planar morphology of the hexagonal portlandite crystals in the carbide lime causes surface area values of ca. 40 m^2 g^{-1}, which double or even triplicate the surface area values of commercial calcitic limes.

• These characteristics make carbide limes very reactive and efficient for capturing gases (CO_2 and / or SO_2), for neutralizing acidic waters (e.g. acid mine drainage), and, predictably, and for hydraulic processes in contact with aluminosilicates or industrial wastes (high potential to solubilise silica and alumina). This could have huge advantages both in the preparation of mortars and hydraulic cements, as well as in soil consolidation with carbide lime, both in unstable areas of public works (roads) and in new rural roads or trails in natural parks.

• Carbide lime in fact shows a higher capability for CO_2 capture than commercial calcitic lime in simple single carbonation tests at 400 °C. This material shows as well a better performance in terms of CO_2 capture in Ca-looping tests, particularly in the first two cycles, than other materials of similar composition, including traditional calcitic limes. This is mainly due to the optimal properties of these materials in terms of particle size and aggregation state.

• Furthermore, increasing the temperature at which the carbonation step is performed in looping tests significantly increases the effectiveness of the CO_2 capture process, regardless of the Ca-bearing sorbent used for CO_2 capture.

Acknowledgements

This work was financially supported by the Spanish Government (grants MAT2012-37584-ERDF and CGL2012-35992) and the Junta de Andalucía (Research Group RNM-179 and Project P11-RNM-7550). E.R.-A. acknowledges a Ramón y Cajal grant. We thank the Centro de Instrumentación Científica (CIC; University of Granada) for assistance with TG-DSC and TEM analyses.

References

1. Teramura, S, Isu, N, Inagaki, K. 2000. New building material from waste concrete by carbonation. *J. Mater. Civil. Eng.*, 12, 288–293.
2. Gunning, P.J., Hills, C.D., Carey, P.J. 2009. Production of lightweight aggregate from industrial waste and carbon dioxide. *Waste Manage*. 29, 2722–2728.
3. Huntzinger, D.N., Gierke, J.S., Kawatra, S.K., Eisele, T.C., Sutter, L.L. 2009. Carbon dioxide sequestration in cement kiln dust through mineral carbonation. *Environ. Sci. Technol.* 43, 1986–1992.
4. Bobicki, E.R., Liu, Q., Xu, Z., Zeng, H. 2012. Carbon capture and storage using alkaline industrial wastes. *Prog. Energy Combust*. 38, 302–320.
5. Sanna, A., Dri, M., Hall, M.R., Maroto-Valer, M. 2012. Waste materials for carbon capture and storage by mineralisation (CCSM) – A UK perspective. *Applied Energy* 99, 545–554
6. Sanna, A., Hall, M.R., Maroto-Valer, M. 2012. Post-processing pathways in carbon capture and storage by mineral carbonation (CCSM) towards the introduction of carbon neutral materials. *Energy Environ. Sci.*, 5, 7781–7796.

7. Ortiz, O., Castells, F., Sonnemann, G. *2009*. Sustainability in the construction industry: A review of recent developments based on LCA. *Construction Building Mater.* 23, 28–39.
8. Elert, K., Rodriguez-Navarro, C., Sebastian Pardo, E., Hansen, E.F., Cazalla, O. 2002. Lime mortars for the conservation of historic buildings. *Studies Conserv.* 47, 62–75.
9. Kudlacz, K., Rodriguez-Navarro, C. 2014. The mechanism of vapor phase hydration of calcium oxide: Implications for CO_2 capture. *Environm. Sci. Technol.* 48, 12411–12418
10. Rodriguez-Navarro, C., Ruiz-Agudo, E., Luque, A., Rodriguez-Navarro, A., Ortega-Huertas, M. 2009. Thermal decomposition of calcite: Mechanisms of formation and textural evolution of CaO nanocrystals. *Am. Miner.* 94, 578-593.
11. Rodriguez-Navarro, C., Ruiz-Agudo, E., Ortega-Huertas, M., Hansen E. 2005. Nanostructure and irreversible colloidal behavior of Ca(OH)2: implications in cultural heritage conservation. *Langmuir* 21, 10948-10957.
12. Ruiz-Agudo, E., Rodriguez-Navarro, C. 2010. Microstructure and rheology of lime putty. *Langmuir* 26, 3868-3877.
13. Cardoso, F.A., Fernandes, H.C., Pileggi, R.C., Cincotto, M.A., John, V.M. 2009. Carbide lime and industrial hydrated lime characterization. *Powder Technol.* 195, 143–149.
14. Ramasamy, P., Periathamby, A., Ibbrahim, S. 2002. Carbide sludge management in acetylene plants by using vacuum filtration. *Waste Manag. Res.,* 20, 536-540.
15. Rodriguez-Navarro, C., Suzuki, A., Ruiz-Agudo, E. 2013. Alcohol dispersions of calcium hydroxide nanoparticles for stone conservation. *Langmuir* 29, 11457–11470.
16. Stanmore, B., Gilot, P. 2005. Review - calcination and carbonation of limestone during thermal cycling for CO2 sequestration. *Fuel Process Technol.* 86, 1707–1743.
17. Blamey, J., Anthony, E.J., Wang, J., Fennell, P.S. 2010. The calcium looping cycle for large scale CO_2 capture. *Prog. Energy Combust. Sci.* 36, 260–279.
18. Symonds, R.T., Lu, D.Y., Manovic, V., Anthony, E.J. 2012. Pilot-scale study of CO_2 capture by CaO-based sorbents in the presence of steam and SO_2. *Ind. Eng. Chem. Res.* 51, 7177–7184.
19. Barker, R. 1973. The reversibility of the reaction $CaCO_3 \leftrightarrow CaO + CO_2$. *J. Appl. Chem. Biotechnol.* 23, 733–742.
20. Grasa, G.S., Abanades, J.C. 2006. CO_2 capture capacity of CaO in long series of carbonation/calcination cycles. *Ind. Eng. Chem. Res.* 45, 8846–8851.
21. Liu, W.; An, H.; Qin, C.; Yin, J.; Wang, G.; Feng, B.; Xu, M. Performance enhancement of calcium oxide sorbents for cyclic CO2 capture - A review. *Energy Fuels* 2012, 26, 2751-2767.
22. Bermejo Sotillo, M.A., Rodriguez-Navarro, C., Ruiz-Agudo, E., Elert, K. 2012. *CO_2-capturing binder, production method thereof based on the selection, purification and optimisation of carbide lime, and agglomerates having and environmental activity.* Patent EP2500328; presented May 4th 2010, published September 19th 2012.
23. Hansen, E.F. 2000. *Ancient Maya burnt-lime technology: Cultural implications of technological styles.* Ph.D. Dissertation (Archaeology), University of California, Los Angeles.
24. Sun, P.; Grace, J. R.; Lim, C. J.; Anthony, E. J. The effect of CaO sintering on cyclic CO2 capture in energy systems. *Environ. Energy Eng.* 2007, 53, 2432-2442.
25. Ma, K.W., Teng, H. 2010. CaO powders from oyster shells for efficient CO_2 capture in multiple carbonation cycles. *J. Am. Ceram. Soc.* 93, 221–227.

Accelerated carbonation and performance of steel slag concrete with pressurized CO_2 curing

L. Mo[1*], F. Zhang[1], M. Deng[1,2]

[1] College of Materials Science and Engineering, Nanjing Tech University, Nanjing, China
[2] State Key Laboratory of Materials-Oriented Chemical Engineering, Nanjing, China
*Corresponding author: andymoliwu@njtech.edu.cn

Abstract
The carbonation of various alkali industrial wastes has been widely investigated owing to its potential permanent sequestration of CO_2. To make use of the carbonated products is of great significance, which may not only facilitate to cut the cost of CO_2 sequestration, but also economically utilize the industrial wastes. In this study, steel slag with relatively high contents of MgO and CaO was used as replacements of Portland cement or natural aggregates to cast concrete and the concrete was exposed to CO_2 curing with a CO_2 pressure of 0.10 MPa. The carbonation front, mechanical strength, microstructure, and carbonate products of the concrete were investigated. Results shown that the compressive strength of the slag concrete after CO_2 curing (0.1 MPa CO_2 and 14d) was significantly increased, obtaining an up to five-fold increase in comparison to that of the same concrete cured in normal moist curing condition for 28d. This may attribute to the densification of the concrete microstructure caused by the formation of Ca-bearing carbonates. According to the results obtained in this study, concrete with low-carbon emissions can be produced via the accelerated carbonation of steel slag.

Introduction

Manufacturing of cement, the most widely used construction materials, is an intensive energy consuming and CO_2 emitting process, which accounts for around 5%-8% of the total man-made CO_2 emissions globally [1]. To produce alternative cement with low-carbon emissions is of great significance for the perspective of developing sustainable construction materials. CO_2 curing for construction materials has recently gained increasing attention owing to its advantages with low-carbon emissions and environmental friendly [2, 3]. In some previous studies, binders based on calcium carbonate products were developed via accelerating the carbonation of CaO-rich raw materials [4, 5]. In addition, reactive MgO was also used to prepare magnesia carbonate binders [3]. The presence of Mg enhanced the carbonation of $Ca(OH)_2$ and improved the mechanical strength accordingly [5]. Our previous studies found that, the Mg^{2+} could change the morphology of $CaCO_3$ and make carbonate products aggregated more closely, which was thought to facilitate the strength improvement [6, 7].

Steel slag is a typical industrial wastes which contains high content of CaO as well as MgO. To be used as construction materials, e.g. supplementary cementing materials, aggregates, etc. is one of the utilization ways, which attracts increasing attention [8, 9]. However, as steel slag contains high content of MgO and CaO, particularly in the forms of periclase and free-CaO, it may cause excessive expansion and hence induce volume instability of materials. In recent years, some studies were tried to employ CO_2 curing to carbonate the steel slag to produce construction materials [10]. The formation of carbonate products calcite ($CaCO_3$) caused the densification of the materials. Most of the samples were mortars or pastes precast with pressure at relatively low water-to-solid ratio, and the specimen size was relatively small.

In this study, steel slag was used as replacement of Portland cement to prepare binding materials, and was used as coarse and fine aggregates after being ground into particles with different size. In addition, lime and magnesia were also used as replacements of Portland cement to prepare the binding materials with various MgO and CaO contents. The effects of CO_2 with a pressure of 0.1 MPa on the carbonation front, compressive strengths, and volume stability of concrete prepared with the aforementioned binding materials and aggregates were investigated. This study may provide a potential approach to prepare steel slag concrete, which well integrates the construction materials manufacturing with CO_2 sequestration.

Experimental

Raw materials

Portland cement (type P.II 52.5) manufactured by the Onoda Cement Co., Ltd., Nanjing was used. Lime and magnesia were produced by calcining limestone and magnesite, respectively. Steel slag from Baosteel Co., Ltd., Shanghai was used. Table 1 and 2 show the chemical compositions of Portland cement, lime, magnesia and steel slag. According to the XRD analysis, for the steel slag, the MgO exist mainly in the forms of periclase and solid solutions, while the CaO exists mainly in the forms of free-CaO, C_3S, C_2S, and other solid solutions. Steel slag was ground to into particles with different size ranges. Steel slag powders with particle size smaller than 0.08mm and used as a replacement of Portland cement. Particles smaller than 5mm and used as fine aggregates, and coarse particles with size of 5-16mm were used directly as the coarse aggregates for concrete. For comparison, limestone and sands were used as natural aggregates to cast concrete.

Table 1. Chemical compositions of cement, lime and magnesia /%

	MgO	CaO	SiO_2	Fe_2O_3	Al_2O_3	SO_3	K_2O	Na_2O	IL
Cement	2.35	64.00	19.43	2.96	4.73	2.58	0.68	0.11	2.81
Lime	13.39	82.19	1.63	0.11	0.13	0.04	0.06	0.05	2.15
Magnesia	88.52	3.18	2.76	0.26	0.42	0.61	0.03	0.06	3.62

Table 2. Chemical compositions of steel slag /%

	MgO	CaO	SiO_2	Fe_2O_3	Al_2O_3	SO_3	MnO	P_2O_5	TiO_2	Cr_2O_3	V_2O_5
Steel Slag	7.19	42.42	11.04	27.37	1.61	0.11	2.28	2.09	0.50	0.15	0.13

Preparation of slag concrete

Blended binders were firstly prepared by mixing Portland cement, steel slag, magnesia, and lime. Three types of binders were prepared and their designations are shown in Table 3. In all the binders, 20% Portland cement and 60% steel slag were used, while up to 20% of magnesia and lime in total was incorporated.

Three types of aggregates were used, namely the steel slag aggregates, natural limestone and sand aggregates. Concrete specimens were casted with the same mix proportion of 340 kg/m^3 binders, 1200 kg/m^3 coarse aggregates, 675 kg/m^3 fine aggregates, and 185 kg/m^3 water (water to binder ratio of 0.54). Two types of specimen size of the concrete were used, namely 100×100×100mm and 100×100×400mm. The former was used to test the compressive strength while the latter was used to examine the long term volume stability.

After casting, the concrete specimens were cured in a moist room with 98% relative humidity and temperature of 23℃ for 24±2h, and then demoulded. Two batches of concrete specimens were used. One batch of specimens was placed in the moist room and continuously cured for 3d and 28d. Another batch of concrete specimens after 7d moist curing was exposed to CO_2 curing with a pressure of 0.10 MPa for 1d, 3d, and 14d respectively. At each curing age as mentioned above, the carbonation depth of the concrete was examined by spraying phenolphthalein indicator and compressive strengths of concrete were tested.

The porosity of concrete before and after carbonation was evaluated via examining its water adsorption. The concrete specimens were immersed in water for 24h to fully absorb water, and then dried at 105℃, weighted until obtaining a constant weight, and then the water adsorption was calculated. X-ray diffraction was used to examine the mineral compositions of slag and carbonation products formed in concrete during CO_2 curing.

Table 3. Mix proportions of cementing materials used for slag concrete /%

Mix design	Binding materials				Coarse aggregates	Fine aggregates
	Slag	Cement	Magnesia	Lime		
SCL-S	60	20	0	20	Steel slag	Steel slag
SCLM-S	60	20	10	10	Steel slag	Steel slag
SCM-S	60	20	20	0	Steel slag	Steel slag
SCL-N	60	20	0	20	Limestone	Sands
SCLM-N	60	20	10	10	Limestone	Sands
SCM-N	60	20	20	0	Limestone	Sands

Results and discussion

Carbonation front

Figure 1 shows the optical images of the concrete specimens after being sprayed with phenolphthalein indicator. All the concrete specimens cured in moist condition show obvious pink across the whole section, while exposed to CO_2 the pink area reduced and gradually disappeared with the prolong of exposure time to 14d. This indicates that the CO_2 has gradually penetrated through out the whole sections of specimens and reacted with the relevant MgO and CaO-bearing phases, e.g. Portlandite, brucite, magneisa, lime, etc.. According to XRD analysis, the main carbonate product formed during carbonation is calcite.

Figure 1. Optical images of concrete specimens sprayed with phenolphthalein indicator (1: SCL-S, 2: SCLM-S , 3: SCM-S, 4: SCL-N, 5: SCM-N)

Compressive strength

Table 4 shows the compressive strengths of concrete specimens cured in moist condition and exposed to CO_2 curing. All the concrete specimens cured under moist curing condition exhibit very low compressive strengths at both ages of 3d and 28d, i.e. the compressive strengths of concrete specimens were ranged from 5.1 MPa to 11.5 MPa. This may due to the large replacements of the Portland cement with low hydraulic materials of steel slag, magnesia and lime. With the increase addition of magnesia, the compressive strength increased.

In contrast, after exposing to CO_2 curing for only 1d, the compressive strength of all the concrete specimens have been increased significantly, at 2-3 folds. Moreover, with increasing exposure time, the compressive strengths were increased dramatically. After 14d of CO_2 curing, the compressive strengths of these concrete specimens were 4-6 folds higher than that of concrete specimens cured under moist condition for 7d.

This is attributed to the carbonation taking place in the concrete specimens during CO_2 curing, which produced carbonate products. Regardless of aggregates type, prolong of carbonation time caused more augments in compressive strength of the concrete containing higher content of magnesia. In comparison to the concrete casted with natural limestone and river sands, the concrete prepared with steel slag aggregates exhibited higher strengths. This may due to the carbonation of steel slag aggregates.

Table 4. Compressive strength of concrete specimens / MPa

Mix design	Moist curing	CO_2 curing			
	28d	Before carbonation (moist cured for 7d)	1d	3d	14d
SCL-S	9.9	5.1	31.7	41.2	48.2
SCLM-S	10.5	9.4	33.5	48.4	55.9
SCM-S	11.5	10.7	38.6	52.7	61.3
SCL-N	-	6.5	18.2	23.1	27.2
SCLM-N	9.4	7.9	27.6	36.6	41.2
SCM-N	10.7	9.0	24.6	40.8	46.2

Water adsorption

Table 4 indicates the water adsorptions of concrete specimens. Obviously after 14d exposure into CO_2 with 0.10 MPa pressure, the water adsorptions of all concrete specimens were dramatically decreased, which indicated that the porosity in the concrete was reduced.

Table 4 Water adsorption of slag concrete /%

Mix design	Moist curing for 28d	Before carbonation (Moist cured for 7d)	CO_2 curing for 14d
SCL-S	8.3	7.1	4.2
SCLM-S	7.0	6.9	4.0
SCM-S	6.7	6.6	3.5
SCL-N	-	6.3	5.2
SCLM-N	6.6	6.1	4.3
SCM-N	5.8	5.6	3.3

Volume stability

To examine the long term volume stability of the concrete specimens. Concrete prisms with size of 100 ×100×400mm were prepared. The concrete prisms with or without carbonation curing were then cured in 80℃ water to accelerate the hydration of periclase and free-CaO existing in the prisms, particularly in the steel slag. Figure 2 shows the expansion curves of concrete specimens with or without carbonation treatment followed by continuous curing in 80℃ water. Accordingly, the concrete prisms without carbonation treatment produced obvious expansion when cured in 80 ℃ water. The expansion was mainly caused by not only the hydration of periclase and free-CaO existing in the steel slag but also the lime and magnesia added. The higher the incorporation of periclase, the larger the expansion was produced. In comparison to the concrete prisms casted by using the natural limestone an river sand aggregates, the concrete prisms prepared with steel slag as coarse and fine aggregates produced much higher expansion, and even cracked the specimens (as shown in Figure 3(a) and (c)) due to the excessive expansions, which implies that the hydration of periclase and free-CaO in the steel slag aggregates contributed significantly to the expansion. Regardless of the aggregates type, the CO_2 treatment before the accelerated curing in 80℃ water reduced the expansion dramatically. This may attribute to the contents of periclase and free-CaO in the steel slag, lime and magnesia had been reduced due to the carbonation.

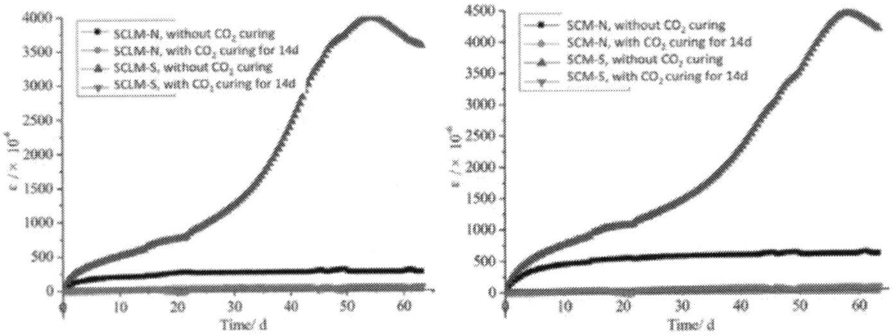

(a) (b)

Figure 2 Expansion of uncarbonated and carbonated concrete specimens followed by curing in 80℃ water: (a) concrete made with SCLM and (b) concrete made with SCM

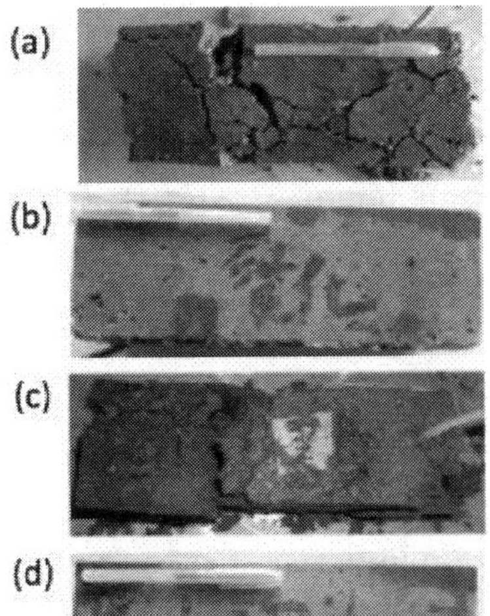

Figure 3. Optical images of the uncarbonated and carbonated concrete specimens after the followed curing in 80℃ water for 60d: (a) SCLM-S, uncarbonated; (b) SCLM-S, carbonated; (c) SCM-S, uncarbonated, and (d) SCM-S, carbonated.

Conclusions

Steel slag was used as binding materials and aggregates to prepare concrete in this study. The effects of CO_2 curing on the carbonation front, compressive strength, water adsorption, and long term volume stability of the concrete made with steel slag were investigated. Some main conclusions can be drawn as follows:

1). Based on the phenolphthalein indicator test, with 0.1 MPa CO_2 curing, the concrete specimens were fully carbonated within 3d. Due to high replacements of PC with 60% of steel slag, and up to 20% of lime and magnesia, the compressive strengths of concrete were very low after conventional moist curing, being 6.5-10.7 MPa at 7d and 5.1-11.5 MPa at 28d of moist curing. However, during the followed CO_2 curing, the compressive strengths of concrete were increased significantly with prolong of exposure time. After 14d of CO_2 curing, the compressive strengths of concrete were increased by two to five folds. This is related to the densification of concrete caused by the formation of carbonate products $CaCO_3$. Concrete prepared with steel slag as the aggregates exhibited higher compressive strengths than that of the concrete casted with natural limestone and river sand as aggregates, which may attribute to the carbonation of steel aggregates.

2) CO_2 treatment on the steel slag concrete improved its long-term volume stability owing to that the carbonation reduced the contents of periclase and free-CaO incorporated in the concrete prisms.

3) In comparison to lime, the magnesia added in the concrete specimens produced greater compressive strength after carbonation curing.

Acknowledgements

The authors are grateful to the financial supports from Jiangsu Natural Science Fund (BK2012427), National Natural Science Foundation of China (51461135003), the Natural Science Foundation of the Jiangsu Higher Education Institutions (14KJB560009), and the Priority Academic Program Development of Jiangsu Higher Education Institutions (PAPD).

References

1. Scrivener, K.L., Kirkpatrick, R.J. 2008. Innovation in use and research on cementitious material. *Cement and Concrete Research*, 28, 128-136.
2. Monkman, S., Shao Y. 2010. Integration of carbon sequestration into curing process of precast concrete. *Canadian Journal of Civil Engeering*, 37, 302-310.
3. Unluer, C., Al-Tabbaa, A. 2014. Enhancing the carbonation of MgO cement porous blocks through improved curing conditions. *Cement and Concrete Research,* 59, 55-65.
4. de Silva, P., Bucea, L., Moorehead, D.R., Sirivivatnanon, V. 2006. Carbonate binders: Reaction kinetics, strength and microstructure. *Cement and Concrete Composites,* 28, 613-620.

5. Chang, J., Li, Y., Cao. M., Fang, Y. 2014. Influence of magnesium hydroxide content and fineness on the carbonation of calcium hydroxide. *Construction and Building Materials,* 55, 82-88.
6. Panesar, D.K., Mo L. 2013. Properties of binary and ternary reactive MgO mortar blends subjected to CO_2 curing. *Cement and Concrete Composites,* 38, 40-49.
7. Mo, L., Panesar, D.K. 2012. Effects of accelerated carbonation on the microstructure of Portland cement pastes containing reactive MgO. *Cement and Concrete Research*, 42, 769-777.
8. Zhu, X., Hou, H., Huang, X. 2012. Enhance hydration properties of steel slag using grinding aids by mechanochamical effect. Construction and Building Materials, 29, 476-481.
9. Mahmoud, A., Ali, B. 2012. Laboratory studies to investigate the properties of CIR mixes containing steel slag as a substitute for virgin aggregates. *Construction and Building Materials*, 26, 475-480.
10. Chang, J., Wu, H. 2010. Study on Carbonation Mechanism of Steel Slag. *Journal of the Chinese Ceramic Society,* 38, 1185-1190.

Understanding Cement Carbonation As A Means Of Carbon Dioxide Utilization

S. Monkman[1*], P. Kenward[2], M. MacDonald[1], G. Dipple[2]

[1] CarbonCure Technologies, Dartmouth, Canada
[2] Department of Earth, Ocean and Atmospheric Sciences, University of British Columbia, Vancouver, V6T 1Z4, Canada
*Corresponding author: smonkman@carboncure.com

Abstract

Industrial use of waste carbon dioxide gas as an upcycled feedstock into concrete has unlocked strength benefits when applied to masonry blocks and ready mixed concrete. The introduction of carbon dioxide to cement paste at the first moments of hydration produce thermodynamically stable carbonate reaction products. The carbonates are effectively formed in the fresh concrete mix and are homogenously distributed throughout the matrix prior to the development of any significant hydration products. The materials science impacts of the carbonation on cement hydration were studied as a means of understanding the physiochemical aspects of this CO_2 utilization approach. A model oil well cement paste was reacted with carbon dioxide gas during mixing. The materials science impacts were assessed through 24 hours of early hydration using isothermal calorimetry, Quantitative X-Ray Diffraction, and scanning electron microscopy. The carbonate reaction products and hydration reactions were characterized. The analysis identified the nature and timing of mineral phase development resulting from carbonation and subsequent hydration of the cement. It was concluded that the formation of calcite does not compromise the formation of the normal hydration products, including amorphous calcium silicate hydrate gel. The initial calcite formation consists of nano-sized particles that may improve the mechanical strength of the fully cured cement paste.

Introduction

Carbon dioxide emissions are recognized as a significant issue relating to cement production and the use of concrete as a building material. The industry has previously recognized a number of approaches to reduce the emissions intensity of the cement produced and used[1]. Further innovative approaches to reducing the industry's carbon emissions are being investigated that can be a part of a portfolio strategy.

One potential method is to upcycle captured carbon dioxide into concrete products. The method involves collecting carbon dioxide, a byproduct from industrial operations, and injecting it into fresh concrete, whereupon it is chemically sequestered. The CO_2 is sourced from industrial emitters. After purification and compression it is transported to concrete production facilities where it is incorporated into concrete during mixing.

The result is that the CO_2 is permanently embedded within the concrete, thereby becoming "sequestered" within the concrete as a mineral. The benefits of incorporating carbon dioxide into concrete is two-fold: (1) the CO_2 is chemically captured thereby reducing greenhouse gas emissions, and (2) the addition of CO_2 can improve the concrete's compressive strength thereby allowing for the production of stronger concrete or the opportunity to reduce the cement content of the mix design. The concept is simple, yet the chemistry is complex.

The main calcium silicate phases in hydrating cement are known to react with carbon dioxide, in the presence of water, to form calcium carbonate and calcium silicate hydrate gel[2]. The calcium silicate hydrate (C-S-H) gel that forms is understood to be intermixed with $CaCO_3$. If the cement hydration has progressed then calcium hydroxide present in the fresh cement paste will react with carbon dioxide. The carbonation reactions are exothermic. The reaction proceeds in the aqueous state when Ca^{2+} ions from the cementitious phases meet CO_3^{2-} ions from the applied gas. The carbonate reaction products are anticipated to form in situ, be of nano-scale and homogenously distributed and influence the subsequent hydration.

Experimental

Industrial scale experiments were performed whereby gaseous carbon dioxide was injected into concrete products during their batching and mixing. The carbonation was integrated with both concrete masonry unit manufacture and ready-mixed concrete production. The goal was to assess the potential of using the carbon dioxide to add, through in-situ chemical reaction during production, nano-scale carbonate reaction products and to determine the attendant material science impacts. The impacts were further investigated through a laboratory study of carbonated cement pastes.

Industrial Work

A tank of liquid CO_2 was connected to a gas control system and manifold. The liquid was expanded to a gas and metered for injection into the mixer (as in the block trial) or mixing drum of the ready-mixed concrete truck. The gas was delivered, at a specified flow rate over a fixed injection interval, whereupon it reacted with the hydrating cement before the mixing was completed.

Masonry Trial Gaseous carbon dioxide was injected, at a rate of 1.5% by weight of the cement, into the concrete mixer once all the dry materials had been loaded. The injection took place over 180 seconds. A medium weight mix design containing 12.7% cement, 70.7% sand and 8.7% gravel was used to make standard 8" (200 mm) concrete blocks.

The fresh concrete was assessed visually and via feedback from the machine (i.e. compaction time) and production personnel. Samples were also processed for quantification of carbon dioxide uptake. The finished blocks were transported to a curing room heated with forced humid air.

The concrete compressive strength was measured at 7, 28 and 56 days at the producer's testing laboratory with 5 blocks for each carbonated condition at each test age. Water absorption and density testing was also conducted on three blocks per condition. Testing was conducted according to ASTM C90 - Standard Specification for Loadbearing Concrete Masonry Units. The control dataset was comprised of two batches contributing 8 strength specimens (5 from the first batch, 3 from the second) and 6 absorption/density specimens.

Ready-mixed Concrete Trial The ready-mixed concrete experiment was conducted at a North York, Ontario ready-mixed concrete producer. The mix design is summarized in Table 1. Concrete was batched into the ready-mix truck and a sample of uncarbonated (control) concrete was removed, assessed, and cast into specimens for testing. The truck was then subjected to three sequential doses of carbon dioxide with assessment and casting of the carbonated concrete between each round.

Table 1. Ready-mixed concrete mix design, per m^3

Component	Quantity	Unit	Fraction
Concrete Sand	Fine aggregate	959 kg	42%
20mm Stone	Coarse aggregate	1080 kg	49%
Slag	Binder	53 kg	2%
Cement	Binder	212 kg	9%
Water Reducer	Admixture	190 ml	
Mix Water	Water	155 kg	

The fresh concrete was assessed via on-site measurement of slump and temperature. The concrete was further cast into 4" x 8" cylinders for compressive strength testing. Concrete performance was assessed through compressive strength testing at 1, 3, 7, 28 and 58 days. All concrete specimens were prepared and tested in accordance with CSA A23.2.

Laboratory work

Laboratory scale experiments were performed on a model system in order to better understand the impacts of the carbon dioxide. The testing used oil well cement due to its low initial calcite content (below detection limits on XRD). Therefore small quantities of carbonate reaction product development could be readily distinguished. Samples were generated by mixing 250 g of water with 500 g of anhydrous oil well cement in a blender for 30 seconds. The blender was flooded with 100% CO_2 gas for a one minute during mixing. Samples were freeze dried to arrest the hydration and carbonation reactions at five distinct times (immediately after the end of mixing, then 5 minutes, 4 hours, 10 hours, and 24 hours after the end of mixing). A parallel set of samples for an uncarbonated (control) system were also prepared. Quantitative X-ray Diffraction (QXRD) was employed to characterize the constituents of the prepared samples. Total inorganic carbon was used to quantify the carbon dioxide.

The microstructural analysis of carbonate reaction products in industrially produced concrete was difficult due to the available sampling approaches and conventional mix designs. An investigation was conducted to characterize the carbonate reaction products through carbonation of a simple cement system. A high degree of carbonation was achieved to allow for direct observation of the crystalline reaction products. The experiment mixed 450 g of GU cement and 50 g of distilled deionized water in an airtight, resealable plastic bag. The materials were hand-agitated through the bag until homogenously blended and the cement was moistened (30 seconds). The bag was inflated with 100% CO_2 gas and sealed. The system was allowed to react until all of the carbon dioxide had reacted (over several minutes) and the bag had deflated. This process was a total of ten times over the course of 1 hour. A parallel control sample set was prepared identically but without no carbon dioxide gas. Carbonate content was quantified by QXRD and the microstructure was imaged using SEM.

Analytical Methods

The carbon dioxide contents of the industrially produced concretes were determined by taking a fresh sample from the production, drying the concrete on a hot plate to remove the water, and subsequently sieving the dried sample through a 160 µm sieve. The carbon contents of the sieved samples were then measured via the combustion infrared detection method using an Eltra CS 800 elemental analyzer.

Laboratory produced paste samples for QXRD analysis were ring milled to a fine powder and then micronized to a particle size of 6.6 □m (volume we
Samples were spiked with 10% fluorite to allow for the quantification of amorphous phases. The micronized samples were mounted and analyzed using a Bruker D8 Focus (0-20, LynxEye detector). Quantitative phase analysis of mixtures was conducted using the Rietveld method on Bruker Topas software.

The comparison between an anhydrous baseline and a system that has increased in mass was achieved by normalizing the QXRD values with respect to the ignited mass as determined through Loss on Ignition (LOI). As additional CO_2 and H_2O mass enters the system it is accounted for by scaling the data with respect to the initial condition appropriately.

Oil well cement samples (anhydrous, hydrated and carbonated) were subjected to quantification of total inorganic carbon (TIC). This accounted for all carbonate in the system, particularly including carbonate minerals that was not detectable through QXRD (due to being xrd-invisible or amorphous). The percent carbon (% C) of the solid samples was determined using a Model CM501C4 CO2 Coulometer from UIC Inc and reported as the %CO2 from carbonate minerals (%carbonateCO2). This method is accurate within ±0.05% by mass with a detection limit of 0.10% C.

Samples of the GU cement pastes were preserved for SEM/EDS analysis. Samples were mounted to aluminum stubs using carbon adhesive tabs. They were then sputter coated with 8 nm of Iridium, using a Leica EM MED020 Coating System, to enhance imaging quality. Imaging was conducted using a Helios NanoLab 650 Focused Ion Beam SEM. Images were collected at 0.5kV and a 25 PA current.

Results and Discussion

Masonry Trial

The first carbonated blocks produced during the masonry test were assessed to have a dry appearance with all other production inputs being equal. The conclusion was that the CO_2 caused an apparent drying effect wherein but less water could be detected in or removed from the mix than had been added during mixing. The blocks were also observed to be underweight.

The operator created a second batch wherein additional mix water was added into the mixer after the CO_2 injection ended and prior to discharge. The water was increased 11% without compromising the appearance or nature of the blocks. The temperature of the blocks was measured when they were sampled for carbon quantification. The control blocks were found to be 26.6 °C. The carbonation treatment increased the temperature 4.2 °C. The temperature rise was consistent with the exothermic nature of the carbonation reaction.

Carbon quantification testing showed that 93% of the supplied CO_2 was absorbed by the water adjusted concrete for an uptake of 1.40% CO_2 by weight of cement. The strength development of the concrete masonry units, in Figure 1, shows the impact of CO_2 on the blocks. The carbonated blocks with the water adjustment were 18-19% stronger than the control at all three test ages.

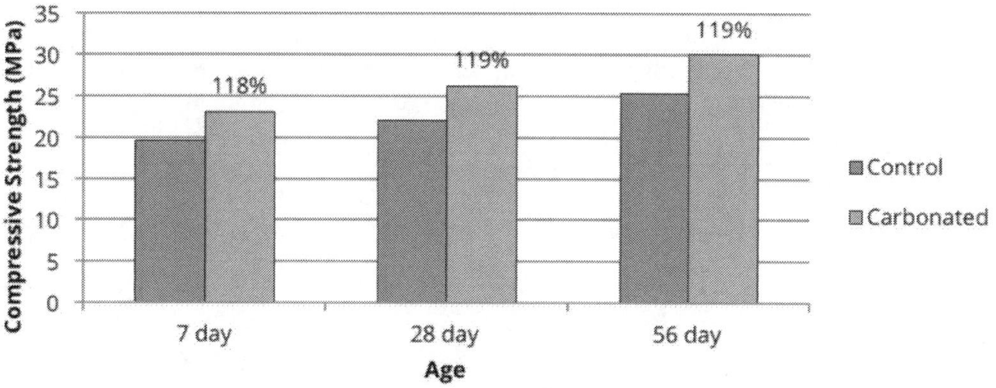

Figure 1 Strength development of carbonated and control concrete masonry units

The dry batch was found to have a density that was 94.9% of the control. The carbonation treatment made the concrete more difficult to compact given that less material was packed into the same mould cavity. Water-adjusted batch improved the density to 99.8% of the control. The average water absorption of the blocks was decreased 18% for the carbonated batch. The fact that the masonry block strength benefit was observed while the block density was unchanged versus the control suggests that the improvement was associated with the carbonation rather than a result of increasing the block density. The reduced water absorption indicates that the carbonate reaction products, or their attendant impact on the hydration, have resulted in an improved pore filling.

Masonry producers lack any internal or external motivation to produce product that has a strength 119% of the conventional product. Instead, an economic gain can be realized by using a mix design that contains less cement and achieves, through help of the carbonation process, 100% of the uncarbonated product strength. A reduced cement mix design would additionally have clear environmental benefits given that Portland cement clinker typically has embodied CO_2 on the order of 866 kg CO_2e/tonne of clinker[3]. If 5% of the cement was removed from the block mix design (about 333 kg/m^3) then the emissions savings would be around 14 kg/m^3 concrete before including any net offset related to the CO_2 absorption.

Ready-mixed concrete trial

The three carbonated batches were numbered 1402, 1403 and 1404 and were treated with 0.1%, 0.3% and 0.6% CO_2 by weight of cement. During the preparation of the carbonated concrete specimens it was observed that the temperature of the mix was increasing and the slump of the mixture was decreasing. A temperature rise can be attributable to the exothermic carbonation reaction. However, given the time involved in conducting the trial it is also possible that the effects reflected hydration progress. A combination of the two influences is likely.

The evolution of heat released from the fresh concrete was measured from 0 to 20 hours after mixing through isothermal calorimetry. The amount of heat released by the concrete in the range of 0 - 24 hours can be used as a proxy for the development of mechanical properties (such as setting or compressive strength) at very early ages[4]. The calorimetry indicated that hydration though the 7 to 14 hours range was accelerated with increasing doses of carbon dioxide. The total energy released through 20 hours was relatively unchanged with results ranging from equivalent (1402) to 6% greater (1403).

The average compressive strength measured for each carbonated condition across the five test ages is summarized for the trial in Figure 3. Each data point represents the average measured compressive strength in MPa with the relative comparison to the corresponding control value displayed as a percentage outside the end of the bar.

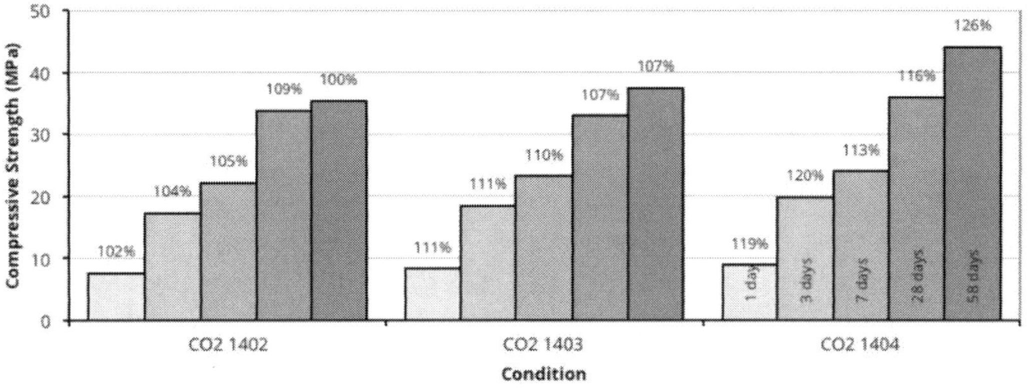

Figure 3 Average compressive strengths of all carbonated ready-mix concrete specimens, further expressed as a percentage of the control value

The results showed an increase in the compressive strength upon carbonation of the concrete mix for the higher two doses. The lowest dose of CO_2 had little impact on the compressive strength ranging between 0 and 9% across the test ages. Batch 1403, with the middle dose of CO_2, was 11% stronger than the control at the two earlier ages, 10% at 7 days and 7% thereafter. The greatest benefit was seen in the final carbonated batch. A 19% increase in strength was observed at 1 day, 16% increase at 28 days and 26% increase at 58 days.

As with the block case ready-mix concrete producers do not typically have an impetus to offer product at 120% of its normal strength when 100% is acceptable. Tangible CO_2 footprint reductions can be realized if the strength benefit from the CO_2 is leveraged towards reducing the cement content in the mix design. The order of a generic benefit is the same as what was proposed for the masonry block case.

Laboratory Study

Oil Well Cement The QXRD results of the oil well cement samples are summarized in Table 3 (hydrated samples) and Table 4 (carbonated series). Results are presented as percentage mass fraction per normalized starting mass. The developments of C3S, calcite, ettringite, calcium hydroxide, and amorphous content were tracked. The Rietveld identification of amorphous content was interpreted, in part, to represent C-S-H gel. While the amorphous content of the anhydrous samples would not adhere to this interpretation, the C-S-H development would generally be associated with the net increase in amorphous content as hydration proceeds.

Table 4. Phase Abundance Summary (wt%) via QXRD for hydrated oil well cement series

Phase	Anhydrous	0 min	5 min	4 hours	10 hours	24 hours
C3S	54.9 ±1.2	54.5 ±1.2	53.4 ±1.2	49.9 ±1.2	46.0 ±1.2	26.4 ±1.0
Calcite	n/d	n/d	n/d	n/d	0.9 ±0.4	1.0 ±0.4
Amorphous	8.6 ±0.7	8.39 ±0.7	7.6 ±0.7	13.5 ±0.8	20.0 ±0.9	34.1 ±1.1
Ettringite	0.6 ±0.3	0.7 ±0.4	1.1 ±0.4	1.4 ±0.4	1.8 ±0.5	4.4 ±0.6
Gyspum	4.0 ±0.6	3.3 ±0.6	4.5 ±0.6	3.6 ±0.6	3.3 ±0.6	0.9 ±0.4
$Ca(OH)_2$	n/d	n/d	n/d	n/d	n/d	6.0 ±0.7

Table 5. Phase Abundance Summary (wt%) via QXRD for carbonated oil well cement series

Phase	Anhydrous	0 min	5 min	4 hours	10 hours	24 hours
C3S	54.9 ±1.2	53.7 ±1.2	52.3 ±1.2	51.6 ±1.2	42.3 ±1.1	27.9 ±1.0
Calcite	n/d	0.8 ±0.4	0.5 ±0.3	1.0 ±0.4	1.5 ±0.4	2.8 ±0.5
Amorphous	8.6 ±0.7	10.4 ±0.8	11.2 ±0.8	13.0 ±0.8	25.4 ±1.0	38.3 ±1.1
Ettringite	0.6 ±0.3	0.6 ±0.3	0.5 ±0.3	0.7 ±0.4	1.1 ±0.4	2.5 ±0.5
Gypsum	4.0 ±0.6	4.1 ±0.6	4.5 ±0.6	4.3 ±0.6	3.0 ±0.6	n/d
$Ca(OH)_2$	n/d	n/d	n/d	n/d	1.0 ±0.4	5.2 ±0.6

The progress of C3S dissolution and reaction is monitored by the change in its relative abundance (decrease) with time. The carbonated case is shown to be parallel the hydrated case wherein the two values are equivalent within the range of error at the initial measurement, 5 minutes and 4 hours. A greater amount C3S has reacted in the carbonated sample (potentially 8% more) at 10 hours but by 24 hours the C3S in the two conditions is again functionally equivalent. The increased reaction of C3S in the carbonated oil well cement at 10 hours as observed by the QXRD agreed with the field ready-mix concrete calorimetry wherein greater energy was released in the 7 to 14 hour interval. This stage of hydration is associated with the end of the acceleration period[5] when the initial silicate hydration starts to slow down. It is possible that the carbonate reaction products are providing a seeding role to boost the hydration or are otherwise affecting the kinetics of the hydration reaction.

Initial concentrations of calcite in the anhydrous cement were below detection limits. In the carbonated sample an increase calcite concentration is observed immediately following the carbon dioxide gas injection. This level of calcite remains relatively constant in the system through the first 4 hours. In the sample at 10 hours the concentration of calcite increases to 1.5% by weight before ultimately reaching its maximum observed concentration of 2.8% at 24 hours. The amount of calcite in the system appears to increase over time but it is recognized that no additional carbon dioxide was added to the system after the initial mixing. The observed increase with time is likely attributable to the calcite reaction products initially being poorly crystalline or too small (xrd-invisible) before developing increased crystallinity or size wherein they could be detected through xrd.

The amorphous content of the carbonated sample is 24% higher than that of the hydrated control immediately after carbonation. At 5 minutes it is 47% greater. It fell 4% behind at 4 hours before accelerating to 27% ahead at 10 hours and 12% ahead at 24 hours. The small lag at 4 hours was mirrored by the C3S content whose consumption was shown to be slightly less for the carbonated paste at 4 hours. The amorphous content, as it would parallel C-S-H content and taken as a proxy for hydration progress, mirrors some of the field observations. The field calorimetry provided evidence of a pivot wherein hydration was slightly behind at 4 hours and notably ahead at 10 hours before showing a strength benefit at 24 hours.

The ettringite content of the carbonated paste was found to be lower than in the hydrated paste. The carbonated paste contained 49% less ettringite at 24 hours. The implication is that the ettringite was slower to form in the carbonated sample. The decrease of gypsum, via consumption during hydration, was greater in the carbonated sample than the hydrated sample. It appeared that all of the gypsum had been consumed in the carbonated sample at 24 hours but less than 80% had reacted in the hydrated sample.

The first detection of calcium hydroxide was in the 10 hour sample, but only in the carbonated paste. At 24 hours the carbonated sample had 88% of the portlandite that was detected in the control paste.

The small dose of carbon dioxide creates nano-calcite but does not prevent the conventional hydration reaction pathways from proceeding. The calcium silicates continue to hydrate, while portlandite and ettringite continue to form.

Total Inorganic Carbon (TIC) measurements were conducted for the three states (anhydrous, hydrated and carbonated). The analysis would account for carbon in amorphous nano-particles of calcite that would be insufficiently crystalline and/or too small to be observed through QXRD. The TIC for the anhydrous cement was 0.098%. Upon the carbonation treatment the carbon had increased to 0.264%. At the equivalent age the carbon content of the hydrated sample was 0.097%. and unchanged from the anhydrous sample. The TIC data proves that carbon dioxide had entered the system even if the QXRD was only detecting some of the ultimate value. A net increase of 0.166% was observed. This represents 0.377% CO_2 by weight of cement.

SEM of GU cement The production of a heavily carbonated paste sample succeeded in increasing the net CO_2 content to 13.6% by weight of the anhydrous cement. This level of carbonation ensures that the reaction products are found in considerably greater abundance than what is achieved in the industrial case but it serves as an effective system for analysis given that the reaction products are readily observed.

Figure 4 Nano-scale (10-150 nm) rhombohedral calcium carbonate reaction products produced through carbonating freshly hydrating cement

The electron microscopy of the carbonated sample revealed that numerous rhombohedral nanocrystals were present in the system. The primary dimension of the particles generally exceeded 10 nm and was predominantly less than 150 nm. The sizes of the particles were too small to allow for an effective direct chemical assessment through EDS. However, given that the particle geometry is consistent with calcite and the QXRD identified the presence of large amounts of calcite, the conclusion can be made that the carbonation process has achieved insitu formation of nano-crystalline calcite.

Conclusions

Concrete blocks were made with carbon dioxide gas injected into the concrete mixer during batching. The absorbed CO_2 improved the strength of the product by 19% once a compensatory water increase was incorporated as part of the production cycle. The carbon dioxide may have had a physiochemical role in improving the strength.

A small dose of carbon dioxide to a ready-mix concrete production accelerated the hydration in the 7 to 14 hour range. The compressive strength was improved up to 19% at one day and 26% at 58 days.

A laboratory study of carbonated cement pastes revealed that the hydration pathways were broadly the same with and without carbonation. The conventional hydration phases formed after the carbonation reaction occurred. The impact of the carbonation may have been to increase the formation of C-S-H in the 10 to 24 hour timeframe.

Strength improvements associated with a carbon dioxide upcycling approach can prompt tangible carbon footprint reductions for the concrete so produced. Producers maybe be motivated to optimize the process and reduce the cement loading in the mix design.

Acknowledgements

Research funding was received from Sustainable Development Technology Canada (SDTC), the National Research Council's Industrial Research Assistance Program (IRAP) and the NRC Engage program. Assistance for the x-ray diffraction analysis was provided by Dr. Mati Raudsepp of the Electron Microbeam/X-ray Diffraction Facility in the Department of Earth, Ocean and Atmospheric Sciences at the University of British Columbia.

References

1. IEA, 2009. *Cement Technology Roadmap: Carbon Emissions Reductions up to 2050.* OECD Pub., Paris.
2. Berger, R. L., Young, J.F., Leung, K., 1972. Acceleration of Hydration of Calcium Silicates by Carbon-Dioxide Treatment. *Nature: Physical Science*, 240, 16–18.
3. Barcelo, L., Kline, J., Walenta, G., Gartner, E. 2014. Cement and carbon emissions. *Materials and Structures*, 47, 1055–1065.
4. Ge, Z.K., Wang, P.J., Sandberg, J., Ruiz, M. 2009. Characterization and Performance Prediction of Cement-Based Materials Using a Simple Isothermal Calorimeter. *Journal of Advanced Concrete Technology*, 7, 355–366.
5. Gartner, E.M., Young, J.F., Damidot, D.A., Jawed, I. 2002. Hydration of Portland Cement. In *Structure and Performance of Cements*, J Bensted and P Barnes eds, CRC Press, New York, 584 pp.

Accelerated carbonation of alkaline materials including industrial wastes, lime, cement and concrete Trace Metal Mobility during Carbon Mineralisation

Hamilton, J. L., Monash University
Wilson, S. A., Monash University
Morgan, B., CSIRO Mineral Resources
Turvey, C. C., Monash University
Jowitt, S. M.,
Southam, G., The University of Queensland
Originally presented on: 6/24/2015 12:00:00 - 12:15:00

Carbon-neutral mines could be made possible by accelerating the natural carbon mineralisation that occurs in mine tailings at the Earth's surface. Ultramafic mine tailings contain an abundance of Mg-rich minerals such as olivine [$(Mg,Fe)_2SiO_4$] and serpentine [$Mg_3Si_2O_5(OH)_4$], which provide an ideal cation source for carbon mineralisation. These minerals naturally react with meteoric or surface water, releasing Mg^{2+} ions, which in turn react with dissolved carbonate (CO_3^{2-}) ions to precipitate Mg-carbonate minerals such as nesquehonite ($MgCO_3{\cdot}3H_2O$) and hydromagnesite [$Mg_5(CO_3)_4(OH)_2{\cdot}4H_2O$]. Carbonation reactions are accelerated with increased mineral surface area, making mine tailings storage facilities an ideal place to observe and promote mineral carbonation.

Natural carbon mineralisation has been documented at several ultramafic mine tailings sites in Australia, Canada and Norway (Wilson *et al.* 2009a; Wilson *et al.* 2009b; Beinlich & Austrheim 2012; Pronost *et al.* 2012; Oskierski *et al.* 2013; Wilson *et al.* 2014). Whilst natural carbonation rates are slow, they may be accelerated *in situ* with relatively inexpensive chemical treatments and changes to tailings management practises. Leaching of industrial and mine wastes with acidic solutions has been widely employed to accelerate mineral dissolution and release of Mg^{2+} for reaction. However, the resulting low pH may enhance metal mobility. Given that mine tailings typically contain significant concentrations of toxic transition metals, the use of acid to accelerate the dissolution of Mg-silicate minerals poses an environmental risk. This concern is also relevant to the *in situ* injection of CO_2 into basaltic or ultramafic rocks, which creates an acidic environment, facilitating metal release (Olsson et al. 2014). As such, we have investigated the mobility of Ni, Cr, Cu, Co and Mn in (1) field samples collected from carbonating mine tailings and (2) laboratory synthesis experiments designed to simulate a worst-case scenario for metal release.

Field samples were acquired from Woodsreef Chrysotile Mine, NSW, Australia, in order to study baseline metal mobility in material known to be carbonating. X-ray fluorescence analyses reveal that both serpentinite ore and carbonate crusts from the Woodsreef Mine contain Ni, Cr, Mn and Co, while mine pit water samples contain no detectable trace metals. These metals are concentrated in chromite and inclusions in magnetite grains within serpentinite and harzburgite. It is expected that these transition metals may substitute for Mg in serpentine minerals and that they may become incorporated into the crystal structures of hydrated Mg-carbonate minerals, such as nesquehonite and hydromagnesite, during carbonation.

Laboratory experiments were conducted to determine whether Mg-carbonate minerals could sequester trace metals from solution during precipitation. Nesquehonite was synthesised in solutions containing up to 100 mg/L of Ni, Cr, Cu, Co and Mn. Our Inductively Coupled Plasma – Mass Spectrometry results confirm that generally >99 wt.% of trace metals were sequestered by nesquehonite. The potential for trace metals to be released from nesquehonite may be influenced by the mechanism controlling their initial sorption to the solid phase. Metals incorporated into crystal structures may be less likely to be released post carbonation than metals sorbed onto crystal surfaces. A combination of X-ray Diffraction and Energy-dispersive X-ray mapping indicates that metals are distributed homogeneously throughout nesquehonite crystals.

Our results suggest that transition metals are substituting for Mg within the crystal structures of the carbonate weathering products that trap and store CO_2 in mine tailings. Furthermore, nesquehonite was able to take up high concentrations of metals, commensurate with conditions found at acid and metalliferous drainage sites, within a timescale of minutes to hours. Consequently, if carbonation of metal-rich industrial wastes or mine tailings were to be accelerated, metalliferous drainage is unlikely to pose an environmental risk. These minerals may therefore represent safe and long-term storage option for both $CO_2$and toxic transition metals.

Keywords: carbon mineralisation, metals, nesquehonite, Woodsreef, mine tailings, CO_2sequestration.

References:

Beinlich A. & Austrheim H. 2012. In situ sequestration of atmospheric CO2 at low temperature and surface cracking of serpentinized peridotite in mine shafts. *Chemical Geology* **332–333**, 32-44.

Olsson J., Stipp S. L. S., Makovicky E. & Gislason S. R. 2014. Metal scavenging by calcium carbonate at the Eyjafjallajökull volcano: A carbon capture and storage analogue. *Chemical Geology* **384**, 135-148.

Oskierski H. C., Dlugogorski B. Z. & Jacobsen G. 2013. Sequestration of atmospheric CO2 in chrysotile mine tailings of the Woodsreef Asbestos Mine, Australia: Quantitative mineralogy, isotopic fingerprinting and carbonation rates. *Chemical Geology* **358**, 156-169.

Pronost J., Beaudoin G., Lemieux J.-M., Hébert R., Constantin M., Marcouiller S., Klein M., Duchesne J., Molson J. W., Larachi F. & Maldague X. 2012. CO2-depleted warm air venting from chrysotile milling waste (Thetford Mines, Canada): Evidence for in-situ carbon capture from the atmosphere. *Geology* **40**, 275-278.

Wilson S. A., Dipple G. M., Power I. M., Thom J. M., Anderson R. G., Raudsepp M.,

Gabites J. E. & Southam G. 2009a. Carbon Dioxide Fixation within Mine Wastes of Ultramafic-Hosted Ore Deposits: Examples from the Clinton Creek and Cassiar Chrysotile Deposits, Canada. *Economic Geology* **104**, 95-112.

Wilson S. A., Harrison A. L., Dipple G. M., Power I. M., Barker S. L. L., Ulrich Mayer K., Fallon S. J., Raudsepp M. & Southam G. 2014. Offsetting of CO2 emissions by air capture in mine tailings at the Mount Keith Nickel Mine, Western Australia: Rates, controls and prospects for carbon neutral mining. *International Journal of Greenhouse Gas Control* **25**, 121-140.

Wilson S. A., Raudsepp M. & Dipple G. M. 2009b. Quantifying carbon fixation in trace minerals from processed kimberlite: A comparative study of quantitative methods using X-ray powder diffraction data with applications to the Diavik Diamond Mine, Northwest Territories, Canada. *Applied Geochemistry* **24**, 2312-2331.

Mineral carbonation of phosphogypsum waste for production of useful carbonate and sulfate salts

H.-P. Mattila[1*], P. Huhtala[1], M. Vahedi[1] and R. Zevenhoven[1]

[1]Thermal and Flow Engineering Laboratory, Åbo Akademi University, Turku, Finland
*Corresponding author: hmattila@abo.fi

Abstract

Phosphogypsum ($CaSO_4{\cdot}2H_2O$) waste is produced in large amounts during phosphoric acid (H_3PO_4) production. Minor quantities are utilized in construction or agriculture, while most of the material is stockpiled, creating an environmental challenge to prevent pollution of natural waters. The stockpiled gypsum waste could be used to capture several hundred Mt of carbon dioxide (CO_2). When gypsum is converted to ammonium sulfate ($(NH_4)_2SO_4$) with ammonia (NH_3) and CO_2, also solid calcium carbonate ($CaCO_3$) is generated. The ammonium sulfate can be utilized in other mineral carbonation processes that use magnesium silicate-based rock as feedstock or as a fertilizer, while calcium carbonate has various uses as e.g. filler material. In this paper the reaction extent of the described process is studied by thermodynamic modeling and experimentally as a function of reactant concentrations and temperature, while the product purity and quality are also followed. Conversion efficiencies of >95% calcium from phosphogypsum to calcium carbonate are obtained. Scalenohedral, rhombohedral and prismatic calcite particles can be produced, though the precipitates contain certain contaminants such as rare earth metals and sulfur. A reverse osmosis membrane cartridge is also preliminary tested as an alternative means of concentrating the ammonium sulfate instead of traditional evaporation of the process solution.

Introduction

Phosphogypsum ($CaSO_4{\cdot}2H_2O$) (PG) is a waste material which is generated during production of phosphoric acid (H_3PO_4) from phosphate rock, mainly apatite ($Ca_5(PO_4)_3OH$), via reaction with sulfuric acid (H_2SO_4). Worldwide, 100-280 Mt/year PG is produced, 85% of which is stockpiled without further treatment.[1] Depending on the origin of the apatite rock, PG may be slightly radioactive, or it may for instance contain some rare earth metals[2]. Besides the huge amount produced, these contaminants often hinder efficient utilization of PG in construction and other industrial applications.[3] Stockpiling causes a risk of pollution to natural waters and is thus not a preferred, although a widely used solution as a PG treatment.

On the other hand, an extensively used production process[4,5] for ammonium sulfate ($(NH_4)_2SO_4$, AS) utilizes calcium sulfate as a raw material (Eq. 1):

$$CaSO_4(s) + 2NH_3(a) + CO_2(a) + H_2O(l) = (NH_4)_2SO_4(a) + CaCO_3(s) \quad (1)$$

AS is currently used as a fertilizer and as a chemical reagent. With an annual production of roughly 20 Mt, it represents a significant share of the nitrogen fertilizer market.[6] It is mainly produced from various process side streams.[7] As can be seen from Eq. 1, the reaction also binds carbon dioxide as solid calcium carbonate ($CaCO_3$, CC). Because carbon dioxide is one of the major greenhouse gases, this reaction could be utilized to mitigate the global climate change.[8-10]

Based on reaction stoichiometry, the annual production of 100-280 Mt PG could bind 26-72 Mt CO_2, and the already stockpiled PG noticeably larger amounts. Moreover, a recently developed mineral carbonation method for magnesium silicate-based rock material (such as serpentinite) utilizes AS as a reagent and requires a make-up stream, although the AS can largely be recycled in the process.[11]

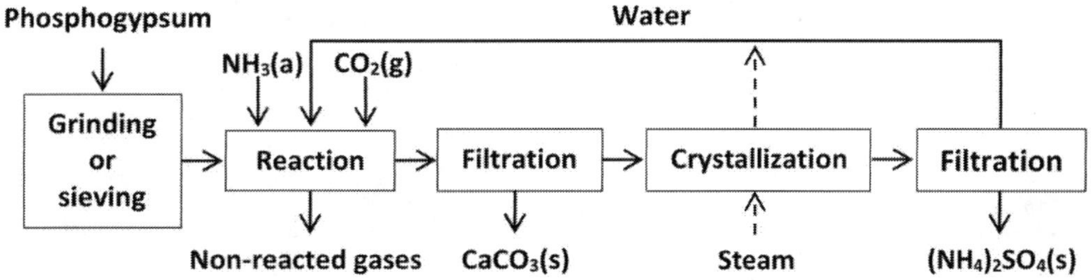

Figure 1. The process concept.

The PG carbonation process is schematically presented in Figure 1, and it has, as mentioned, been broadly used for production of AS. If the aim is specifically in climate change mitigation, it becomes essential to operate the process as environmentally friendly as possible, implying that for example CO_2 emissions caused by the process are to be minimized. In addition, if both the AS and CC products can be obtained in commercial qualities, the possibility of acquiring economic benefit can facilitate a break-through of this process concept as a carbon sequestration technology.

In this paper the emphasis is on an efficient separation of AS and CC products. For this purpose, the presented reaction (Eq. 1) is first modeled with a Gibbs energy minimization software Outotec HSC Chemistry 8.0.3.[12] The effect of different concentrations and reaction temperatures is studied, after which the obtained results are verified in laboratory-scale experiments with synthetic gypsum and an industrial phosphogypsum sample.

The aim of the experimental work is to produce such a high quality CC regarding particle morphology, crystal size and chemical purity, that it can be utilized in commercial applications such as a filler material in papers and plastics.[13] Besides this, a reverse osmosis (RO) membrane cartridge is tested as a means of pre-concentrating the produced AS solution before the AS crystallization, the aim being to reduce the steam consumption during the crystallization and thus to decrease the environmental impact of the entire process.

Thermodynamic modeling

The conversion efficiency of PG to AS and CC at different reactant concentrations was modeled with HSC Chemistry. As a base case, 100 g/L of PG was introduced in a total volume of 200 mL water at ambient (20 °C) temperature and pressure. Figure 2 shows the effect of ammonia (NH_3) and carbon dioxide (CO_2) levels on formation of a) solid CC and b) dissolved Ca^{2+} ions. For comparison, Figures 2 c) and d) present the conversion efficiency of PG to solid CC at 45 and 80 °C, respectively.

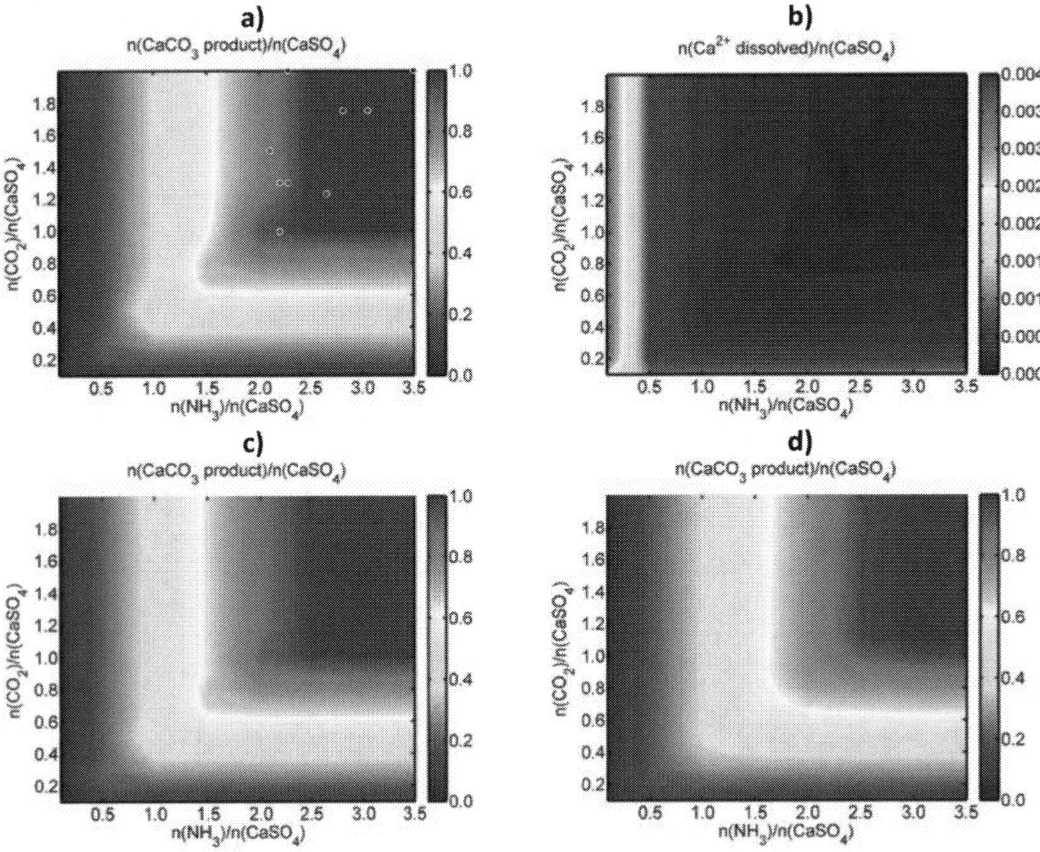

Figure 2. a) Conversion of solid $CaSO_4 \cdot 2H_2O$ to solid $CaCO_3$ and b) to dissolved Ca^{2+} ions at 20 °C. c) Conversion of solid $CaSO_4 \cdot 2H_2O$ to solid $CaCO_3$ at 45 °C and d) at 80 °C. Figures are drawn as functions of initial $NH_3/CaSO_4$ and $CO_2/CaSO_4$ molar ratios with modeling results obtained from HSC Chemistry software. Dots in Figure a) represent experimental results.

As can be seen, at 20 °C a slightly over-stoichiometric $NH_3/CaSO_4$ molar ratio of >2.5 and a $CO_2/CaSO_4$ molar ratio of >1.0 guarantee a high conversion (>99%) of PG calcium to solid CC according to the HSC model. An increase in reaction temperature reduces the absorption of CO_2 gas, whereas it simultaneously increases the evaporation of aqueous NH_3.[7] Somewhat more NH_3 is required to obtain a full conversion at higher temperatures according to the model, while the requirements for CO_2 seem to remain unaffected. Thus, in practice and from the materials efficiency point-of-view, low reaction temperature and a $NH_3/CaSO_4$ molar ratio of 2.5-3.0 are to be preferably used together with a near-stoichiometric CO_2 amount. In addition, the thermodynamic modeling results show that the solid-to-liquid ratio of PG and water has only a minor decreasing effect on the calcium conversion efficiency when larger (>100 g/L) PG-to-water amounts are used. Besides CC, no other solid products were present in the HSC model results, indicating thus an in theory excellent selectivity towards pure CC and AS.

Precipitation experiments

Because it is known from earlier research[13] that the quality of obtained CC depends highly on the used reaction temperature and pH as well as the initial concentrations, laboratory experiments were performed to verify the modeling results and to study the properties of the produced CC in more detail. Two types of gypsum were used; besides synthetic material (Merck KGaA 1.02161.0500), PG was obtained from Yara Suomi Oy, Siilinjärvi, Finland. Thus it was possible to compare the precipitate quality of CC produced with the industrial PG, containing several impurities, to the CC made of synthetic gypsum. Experiments with PG were performed using the material "as delivered", but also after drying it overnight at 105°C. The chemical compositions of both materials are listed in Table 1. Besides the shown elements, the Siilinjärvi PG also contained trace elements including several lanthanides in small (<0.05%) amounts.

Table 1. Chemical composition (% dry solids) of the gypsum samples, based on data from manufacturer, and on XRF and TGA (water content including crystal water, %) analysis.

Sample	$CaSO_4{\cdot}2H_2O$	MgO	Al_2O_3	SrO	Nd_2O_3	P_2O_5	SiO_2	CeO_2	Fe_2O_3	Rest	H_2O_{total}
Synthetic (Merck KGaA)	>99.0	<0.02	-	<0.06	-	-	-	-	0.00	<0.92	21
PG (Siilinjärvi)	97.79	0.58	0.52	0.44	0.17	0.14	0.11	0.07	0.05	0.13	27

Two types of reactions were performed; first a constant amount of reagents (PG, NH_3) was mixed in an aqueous solution (200 mL) with a magnetic stirrer, while CO_2 gas (99.9996%, Oy Aga Ab) was bubbled through this solution at a constant rate. Another tested reaction type was to conduct the carbonation at a constant pH, mixing initially only gypsum and water (200 mL) while the aqueous ammonia solution (100 mL, 25%, Merck KGaA 1.05432.1000) was added during the reaction to maintain the chosen pH value. Also in this case, the CO_2 gas flow was bubbled through the reaction solution at a constant rate (0.07-0.16 L/min). The mixing speed was chosen so that an additional speed increase did not cause any sudden changes in the solution pH. Reaction temperature and pH were recorded during the experiments. After the experiments the solutions were filtered and the collected solids were dried overnight at 105°C. SEM (Scanning Electron Microscopy) pictures and XRF (X-Ray Fluorescence) analysis were taken of the dried samples. Table 2 lists the performed experiments with the utilized parameters.

Table 2. The performed experiments, a-f.[14]

	Sample	Temperature (°C)	pH*	Time (min)	Reaction type**	S/L-ratio*** (g/L)	$NH_3/CaSO_4$ (mol/mol)	$CO_2/CaSO_4$ (mol/mol)
a)	PG	20	-	53	NH_3	100	3.8	1.7
b)	PG	45	8.1	33	pH	160	3.6	1.7
c)	PG	20	7.4	75	pH	160	3.6	4.2
d)	Synthetic	45	7.2	139	pH	150	2.7	5.3
e)	PG, dried, <125 µm	20	7.4	80	pH	130	2.7	3.6
f)	PG, dried, <125 µm	20	8.0	44	pH	85	4.0	2.9

*Average pH during the experiment

**NH_3: a constant amount of reagents, pH: a constant pH-value during the carbonation reaction

***The initial solid-to-liquid ratio; PG-to-water

Results from the precipitation experiments

Within the limits of experimental inaccuracies, the performed experiments confirmed the HSC modeling results for the conversion efficiencies of $CaSO_4$ to solid CC. Overstoichiometric initial conditions regarding both $NH_3/CaSO_4$ and $CO_2/CaSO_4$ molar ratios resulted consistently in calcium conversions of >94%, based on obtained mass conversions and chemical composition analysis. Figure 2 a) represents some of these experimental results as circles, colored according to the conversion efficiency.[14]

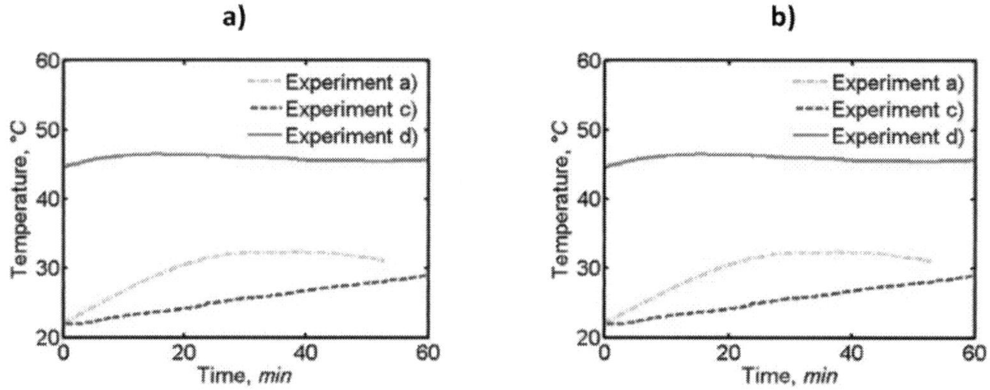

Figure 3. a) Temperature and b) pH during the first 60 minutes of experiments a), c), d).

Figure 3 shows the pH and temperature recordings during experiments a) (constant amount of NH_3), c) and d) (constant pH). As is seen from the graphs, the pH was slightly varying also during "constant pH" experiments c and d. The exothermic nature of the reaction (Eq. 1) caused a temperature increase in the reaction solutions, which may have slightly affected the formation of the CC precipitates. For experiment d) this effect was milder, because the reaction vessel was immersed in a water bath to maintain a temperature of 45 °C. In the other experiments, no water bath was used.

The chemical compositions of the produced precipitates were obtained via XRF analysis. However, because of the small sample sizes and an inaccurate XRF calibration for materials with a high calcium content, the results are also somewhat inaccurate, especially for the trace elements (<0.1 %-wt.). In addition, AS did according to the HSC model not precipitate from the process solutions directly, but small amounts (<3 %-wt.) seemed still to be present in all the samples, because the materials had been dried after the experiments without further washing.

The target of the composition analysis was to estimate the process behavior of the minor impurities which were detected in the original PG (Table 1). For comparison, also one precipitate manufactured from synthetic gypsum (d) was analyzed. As expected, the results showed that this material consisted of CC and unreacted gypsum with even the major impurities (Al, Si) on a 40% lower level than in the PG-based products.

Table 3 presents the $CaCO_3$ content of the products as measured with the gasometric analysis method developed at Åbo Akademi University[15], the calculated conversion efficiency of PG calcium to solid CC, and the estimates of transfer efficiency of different elements from PG to the product precipitate, based on XRF analysis of the solids. Pre-drying of the PG (e-f) lowers the CC formation efficiency by transforming the $CaSO_4{\cdot}2H_2O$ into an apparently more chemically stable $CaSO_4{\cdot}0.5H_2O$.[14]

It might be possible to separate the contaminants from the carbonate material with e.g. density-based separation technologies. Namely, during the experimental work it was observed that small, dark particles were formed and they remained mixed with the otherwise white carbonates. However, due to the small amounts used in these experiments, no separation tests were performed as a part of this study.

Table 3. CC content of the solid products (%), Ca conversion from PG to solid CC (%), and the transfer efficiency of various elements from PG to the solid product material (%).

	$CaCO_3$	Ca conversion	Ca	Mg	Al	Sr	Nd	P*	Si*	Ce*	Fe*	S
a)	97	99	97	<1	6	23	25	135	21	132	184	2
c)	92	97	95	2	5	21	24	123	26	127	173	3
d)	96	98	103	0	-	0	-	-	-	-	-	2
e)	84	94	98	<1	9	24	27	142	19	134	169	6
f)	82	95	98	0	8	24	26	141	18	133	175	5

*Poor accuracy due to small amounts

Figure 4 presents SEM pictures of the precipitates from experiments a)-f), according to Table 2. When the CC precipitation reaction is run at high pH (>8) and with a PG-to-solution ratio above 100 g/L, but with a low $CO_2/CaSO_4$ molar ratio (1.7 mol/mol), the precipitate consists of clustered vaterite particles, so-called framboids (a,b). At 45°C, the vaterite particles form a disc-like surface (b). A longer reaction time (>1h) at 20°C and excess CO_2 (4.2 mol/mol Ca) (c) favor the transformation of vaterite to clustered rhombohedral calcite. A long reaction time (>2h) combined with an elevated temperature (45°C) and a low NH_3 amount (2.7 mol/mol) results into formation of scalenohedral calcite particles (d). In experiment e), CC particles with a crystal form resembling a mixture of rod-like calcium sulfate and scalenohedral calcite were obtained. A moderate reaction time, pH, and CO_2 amount with a somewhat lower solid-to-liquid ratio yielded a prismatic form of calcite precipitate (f). As mentioned above, the precipitates e) and f) contain more unreacted PG than the other products. It seems feasible to utilize PG without any pretreatments such as drying.

These results show that the process can produce CC precipitates in various crystal forms, some of which (scaleno- and rhombohedral calcite) are valuable for commercial purposes.[13] Still, more research is needed to reliably establish the relationships between the different experimental parameters and the crystal forms.

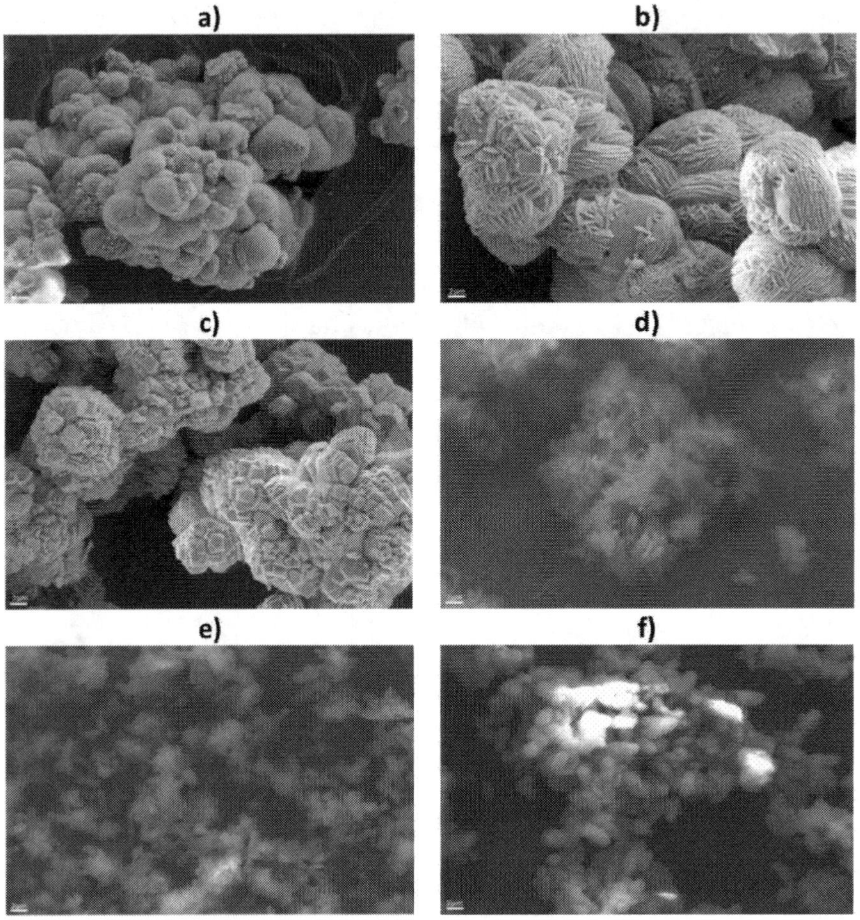

Figure 4. SEM pictures of the produced carbonates. Scale bars 2μm in a)-f).

Treatment of the AS solution using reverse osmosis membranes

Based on the thermodynamic modeling work and the experimental results as well as existing literature,[5,7] AS is obtained from the reaction shown in Eq. 1 as dissolved ions in an aqueous solution. To crystallize the AS from this solution, water is usually evaporated using process steam, which is quite energy-demanding (>2570 kJ/kg when boiled from 25°C). Because AS has a high solubility in water (5.8 mol/L at 25°C),[16] also the amount of water to be evaporated rapidly becomes significant. In the experiments described above, the AS concentration was approximately 0.6 mol/L after the experiment.[14] Thus, 90% of the water could be removed from the process solution before achieving AS supersaturation and initiating the actual crystallization reaction. By using e.g. a RO membrane for water removal, the entire process concept (Figure 1) could be operated more energy-efficiently and also with smaller CO_2 emissions.

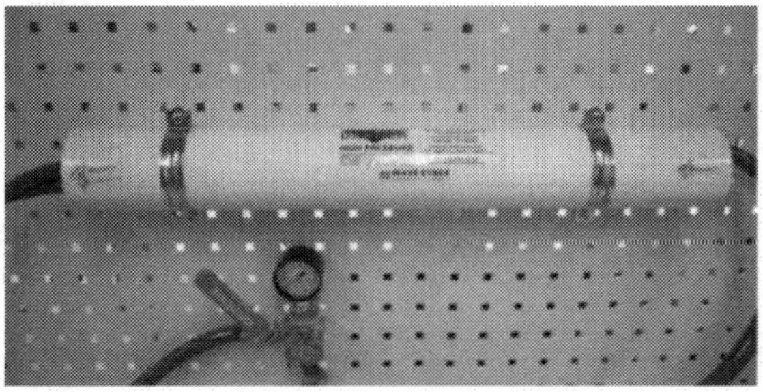

Figure 5. The RO membrane device used in the experiments.

Preliminary experiments to concentrate AS-CC solutions were performed using a Filmtec XLE-2521 RO polyamide thin-film composite membrane cartridge (Figure 5). To standardize the experimental conditions, pre-mixed solutions (10L) of 1) tap water, 2) 0.1 mol/L AS and 3) 0.1 mol/L AS and 0.0002 mol/L CC were prepared using laboratory chemicals (AS: Merck KGaA 1.01217.1000, CC: Sigma-Aldrich 398101-500G) and pumped through the membrane device using a diaphragm pump (KNF-Flodos NF 1.300 KT.18S), as specified in Table 4. Solutions 2) and 3) correspond to the solutions obtained from the above-described precipitation experiments. The conductivity of the permeate flow was measured from samples taken with an interval of five minutes, while the conductivity of the concentrate flow was continuously recorded during the tests to obtain information on the dynamic behavior of the membrane. After experiment 3) the NH_4^+(aq) and Ca^{2+}(aq) concentrations in the taken samples were measured with ion-selective electrodes (ISE, ELIT, Nico2000 Ltd).

Table 4. Experimental results obtained with the reverse osmosis device.

	Solution	Inlet pressure (bar)	Inlet flow rate (L/h)	Permeate flow (%)	Permeate NH_4^+(aq) (% of input)	Permeate Ca^{2+}(aq) (% of input)
1)	Tap water	4	68	53	-	-
2)	0.1 M AS	10	17	34	Not measured	-
3)	0.1 M AS + 0.0002 M CC	10	21	43	3	1

During the experiments, the membrane device reached a steady-state after approximately 10 minutes, after which the tests were continued for another 10 minutes, giving thus a total time of 20 minutes per experiment. According to the ISE measurements (experiment 3), the NH_4^+(aq) concentration was reduced by 97% in the permeate stream. Thus, assuming that no accumulation occurred inside the membrane cartridge, the concentrate stream had a concentration of 0.17 mol/L NH_4^+(aq) after the experiment. Regarding Ca^{2+}(aq), the corresponding values are a reduction of 99% in the permeate stream and a concentration of 0.0004 mol/L in the concentrate stream. The overall conductivity was reduced from 18.6 mS/cm in the inflow to 0.4 mS/cm in the permeate (tap water ~0.8 mS/cm). The concentrate conductivity was 27.6 mS/cm.

Even though the results from this type of "off-line" experiments with an RO membrane device are promising regarding the possible reduction of the energy demand of the PG carbonation process, open questions remain about the long-term operability of RO membranes. If the membrane efficiently prevents both the NH_4^+(aq) and Ca^{2+}(aq) ions from penetrating the barrier, as was now observed to happen, AS and CC will be concentrated in the retentate stream. Eventually, a reverse reaction of Eq. 1 will occur, AS and CC reacting back to solid gypsum, which then blocks the membrane element and requires cleaning or other maintenance. Also, as seen in Table 4, the required inlet pressure increases rapidly with increasing salt amounts, which might cause operational difficulties at higher (>0.1 mol/L) AS concentrations, even though the tested membrane device allows for pressures up to 21 bar.

Discussion and conclusions

It was shown that solid CC (>90% purity) and aqueous AS can be consistently produced from PG with NH_3(a) and CO_2(g) as reagents in an aqueous solution. Both thermodynamic modeling and experimental work confirmed that slightly over-stoichiometric conditions are adequate for an efficient conversion of PG to solid CC. The particle morphology of the CC precipitate can be steered by carefully choosing the precipitation conditions such as concentrations of the starting materials, temperature and reaction time. Future research will focus more on solutions with a higher solid-to-liquid ratio and its possible effects on the CC particle morphology. Still, a better control on the effect of the various parameter combinations is required before a possible larger-scale application can be planned.

Higher quality CC morphologies (scalenohedral, rhombohedral or prismatic calcite instead of vaterite) were obtained in the experiments where a constant pH was maintained in the reactor by feeding both NH_3(a) and CO_2(g) continuously to the system. In practice, if flue gas with <20% CO_2 is blown through the reactor, NH_3(a) evaporation may become an issue. Also regarding this aspect, it is beneficial to feed the NH_3(a) as a continuous stream, reducing thus the amount of free ammonia in the process solution. Otherwise, evaporated NH_3 may require e.g. use of a scrubber of sulfuric acid for the gas stream exiting the reactor.

Several impurities which are present in the PG, including lanthanide trace elements, seem to accumulate in the CC precipitate. A possible topic for further research is to attempt to separate these contaminants, especially the rare earth metals, from the carbonate material. This would, besides purifying the carbonate product, also create a valuable side-product for the process.

It seems that a RO membrane could, in theory, be used to concentrate the AS solution instead of evaporating the water by using process steam. However, other membrane types (or technologies) should be tested to find a material, maybe other than the now tested polyamide, which would also separate ammonium and calcium ions from each other and thus eventually prevent the gypsum formation inside the membrane element. Also, a membrane material with a lower flow resistance with the salt solutions would be beneficial, though this should naturally not compromise the separation efficiency.

In general it can be concluded that the discussed process concept has a significant potential to be utilized for CO_2 capture and climate change mitigation while producing commercially useful carbonate and sulfate salts.

Acknowledgements

This work was carried out with funding from the Environmental Scholarship Fund of Ekokem Oyj (Ekokem Oyj:n Ympäristöstipendirahasto), Finland. Yara Suomi Oy, Siilinjärvi Plant, Finland is acknowledged for providing the material for experiments.

References

1. Choura, M., Maalouf, F., Keskes, M., Cherif, F. 2012. Sulphur matrix from phosphogypsum: a sustainable route to waste valorization. In: Zhang, P., Miller, J. D. and El-Shall, H. (Eds.) *Beneficiation of Phosphates: New Thought, New Technology, New Development.* Society for Mining, Metallurgy and Exploration, Inc.: Englewood, CO, US. 297-302.
2. VTT Technical Research Centre of Finland. 2014. Research collaboration between Finland and Russia to begin recovery of rare metals from a gypsum mountain and industrial side streams. *http://www2.vtt.fi/news/2014/recovery_of_rare_metals_from_gypsum_mountain.jsp?lang=en* (accessed 27.02.2015).
3. Florida Industrial and Phosphate Research Institute. 2004. FIPR Report 2003-2004. Featured Research Topic: Phosphogypsum Use. *http://fipr.us/FIPRreport2003/research-highlights-pg-use.html* (accessed 27.02.2015).
4. Vis, G. N. 1915. Process of manufacturing ammonium sulfate. US patent 1152244.
5. Witte, J. F., Wit, J. J. 1972. Manufacture of ammonium sulfate from gypsum. US patent 3687620.
6. Leip, A., Prud'homme, M. 2013. Integrating nitrogen fluxes at the European scale. In: Sutton, M. A., Howard, C. M., Erisman, J. W., Billen, G., Bleeker, A., Grennfelt, P., van Grinsven, H., Grizzetti, B., (Eds.) *The European Nitrogen Assessment: Sources, Effects and Policy Perspectives.* Cambridge University Press: Cambridge, UK. 345-378.
7. Jang, Y. N., Ryu, K. W., Lee, M. G. 2014. Production method for high-purity calcite and ammonium sulphate by using gypsum. US Patent application 2014/0161692 A1.
8. Lee, M. G., Jang, Y. N., Ryu, K. W., Kim, W., Bang, J. 2012. Mineral carbonation of flue gas desulfurization gypsum for CO_2 sequestration. *Energy*, 47, 370-377.
9. Zhao, H., Li, H., Bao, W., Wang, C., Li, S. 2014. Experimental study of enhanced phosphogypsum carbonation with ammonia under increased CO_2 pressure. *Journal of CO_2 Utilization.* DOI:10.1016/j.jcou.2014.11.004.
10. Pérez-Moreno, S. M., Gázquez, M. J., Bolívar, J. P. 2015. CO_2 sequestration by indirect carbonation of artificial gypsum generated in the manufacture of titanium dioxide pigments. *Chem. Eng. J.*, 262, 737-746.
11. Zevenhoven, R., Fagerlund, J., Nduagu, E., Romão, I., Highfield, J. 2013. Stepwise serpentinite carbonation using the Åbo Akademi route - status and latest developments. In: Nasser, R., Santos, R., Cizer, Ö., van Gerven, T., (Eds.) Proceedings of the 4th International Conference on Accelerated Carbonation for Environmental and Materials Engineering (ACEME); KU Leuven: Leuven, Belgium. 391-400.
12. Outokumpu Research Center, Roine, A., Kobylin, P. 2014. HSC Chemistry 8.0.
13. Mattila, H.-P. 2014. Utilization of steelmaking waste materials for production of calcium carbonate ($CaCO_3$). Doctoral Dissertation. Åbo Akademi University: Turku, Finland.
14. Huhtala, P. 2015. Carbonation of phosphogypsum. M.Sc. thesis, under preparation. Åbo Akademi University: Turku, Finland.
15. Fagerlund, J. 2012. Carbonation of $Mg(OH)_2$ in a pressurized fluidized bed for CO_2 sequestration. Doctoral Dissertation. Åbo Akademi University: Turku, Finland.
16. Lide, D. R. 2006. CRC Handbook of Chemistry and Physics - 87th edition 2006-2007. Taylor & Francis Group, LLC: Boca Raton, FL, US.

Carbon dioxide sequestration by accelerated carbonation of Waelz slag

G. Cappai[1,2], G. De Giudici[3], D. Medas[3] , A. Muntoni[1,2], A. Nieddu[1], G. Orrù[1], M. Piredda[1]

[1] Department of Civil and Environmental Engineering and Architecture, University of Cagliari, 09123, Cagliari, Italy
[2] Institute of Environmental Geology and Geoengineering, National Research Council, 09123 Cagliari, Italy
[3] Department of Chemical and Geological Sciences, University of Cagliari, 09127, Cagliari, Italy
*Corresponding author: gcappai@unica.it

Abstract

This work reports the results of a study on accelerated carbonation of Waelz slag, with a focus on the effects of process conditions on CO_2 uptake and conversion yields. Carbonation experiments were performed at different operating conditions by using different water contents (L/S = 0, 0.2, 0.4 and 1 l/kg), CO_2 pressure (1, 3, 5, 10 and 20 bar) and process duration (up to 240 h). The operating parameters significantly affected the extent of CO_2 sequestration and the evolution over time of the process. CO_2 uptakes benefited from increasing singularly either L/S or CO_2 pressure, with a more relevant effect observed for L/S as compared to CO_2 pressure. The maximum CO_2 uptake of 6.48±1.00% was achieved at a L/S=1 and CO_2 pressure P=20 bar, while the minimum uptake of 1.47±0.56% was achieved for L/S=0 and P=1 bar. The corresponding degrees of carbonation were 18.32 and 4.16%, respectively, indicating that the potential for CO_2 sequestration from Waelz slag was only partially exploited even when the optimal conditions were adopted.

Introduction

Carbon dioxide produced from anthropogenic activity is one of the most important greenhouse gases responsible for global warming. In recent years, mineral carbon sequestration has been proposed as an available and cost effective technology for the containment of atmospheric emissions[1]. The process, consisting in the acceleration of a natural rock weathering process, is based on the storage of CO_2 in a solid phase by neutralization with alkaline earth-metal oxides[1,2].

The use of alkaline industrial residues as sources of carbonate-forming ions has been recently addressed by several studies[3-7]. Waste materials such as metallurgical slag, ashes from combustion processes and cement-based residues exhibit suitable properties toward mineral carbonation which can be used to reduce CO_2 emissions from point sources, since they are generally available at CO_2 emitting plant sites, and possess a high potential reactivity which limits the need for pre-treatment and energy intensive operating conditions[4-7]. This practice could make possible a significant reduction of the CO_2 emissions generated from the process itself, therefore it represents a key factor in view of a full scale application of mineral carbonation. Moreover, mineral carbonation can be seen also as a valorization technique since it may improve the environmental behavior of such residues in view of utilization or disposal [8-10].

One of the alkaline residues that can be considered suitable for the application of the mineral carbonation process is Waelz slag, a by-product of a metallurgical process used for recovering zinc from metallurgical waste, in particular electric arc furnace (EAF) dusts. Indeed, huge amounts of such residue are produced wordlwide; it is estimated that about 500.000 tons/year of Waelz slag would be produced in EU if all the EAF dust were recycled with the Waelz process[11]. Moreover, it possesses a chemical and mineralogical composition suitable for carbonation, since it contains alkaline elements in the form of reactive species [12]. Finally, since the residue is characherized by a relevant metal content which could potentially be leached, an additional benefit in terms of metal mobility reduction is expected[12].

It has been shown that carbonation kinetics and yields, as well as the final properties of the carbonated residues, can be enhanced by appropriately selecting the reaction routes and the operating parameters applied, such as the water content, the CO_2 pressure, the temperature and the eventual residue pre-treatment to be applied. Therefore, it is of crucial interest to identify the combination of all the parameters of concern in order to attain the maximum carbonation yields[7,13,14]. The results of a previous study, where the influence of solid particle size on CO_2 sequestration capacity and on reaction kinetics was investigated, demonstrated that the carbonation yield of Waelz slag could be enhanced considerably by using a grinding/milling pretreatment [12].

In this paper, the influence of operating conditions on CO_2 sequestration capacity and reaction kinetics is presented and discussed. In particular, the effects of parameters such as liquid-to-solid (L/S) ratio (0, 0.2, 0.4, and 1 l/kg), CO_2 pressure (1, 3, 5, 10 and 20 bar) and process duration (up to 240 hr), were evaluated, with the aim of elucidating which combination of operating parameters could allow for the maximum CO_2 uptake over the shortest period of time. The modifications on slag environmental and mineralogical properties were also investigated.

Materials and Methods

Waelz slag were collected from the Waelz Plant located in Portoscuso (CI), Italy, where two furnaces treat EAF dust and other residues, producing about 60.000 t/y of Waelz oxide. Around 200 kg of representative sample were collected from piles containing residues produced over more than 48 hours and under normal operating conditions. The sample was dried at 40°C, homogenized, quartered and stored in closed buckets to prevent natural ageing reactions. The material was crushed to a final particle size below 4 mm. A representative fraction of Waelz slag was used for the physical-chemical and mineralogical characterization.

The content of macro-elements was analized by alkaline fusion with lithium metaborate (1000±25°C) and subsequent dissolution of the molten residue in 2,5%

HNO_3, while minor elements content was assessed according to the EPA 3052 method. Major and minor elements concentration was evaluated by Inductively Coupled Plasma spectrometry (ICP-OES, Perkin Elmer Optima DV 7000). Calcite content was determined by calcimetry analysis, carried out by placing about 0.5 g of dried sample in a Dietrich-Fruehling apparatus, and adding 10 ml of HCl 1:1 v/v. Total carbon content (TC) was evaluated using a CHN analyser (CHN-1000 LECO). The mineralogical composition was investigated by means of XRD analysis, performed with a θ-2θ conventional diffractometer (Phillips X'PERT MPD) with Cu Kα radiation (1.5418 Å). Samples of 200 mg were lightly ground in agate mortar and packed into the sample holder for X-ray diffraction analysis. Peaks for the mineral were attributed according to the Powder Diffraction Cards by using X'Pert Highscore plus software. FTIR spectra were collected by using a Bruker Equinox 55 spectrophotometer.

Several accelerated carbonation tests were conducted at different operating conditions in a pressurized batch reactor equipped with a 1.7 L internal Teflon vessel and an electric jacket for temperature control at 25°C. In particular, different values of parameters such as liquid-solid (L/S) ratio (0, 0.2, 0.4 and 1 l/kg), CO_2 pressure (1, 3, 5, 10 and 20 bar) and process duration (up to 240 h) were applied and the effects on CO_2 uptake as well as on the slag environmental and mineralogical properties were observed. The different operating conditions adopted are summarized in Table 1. In each test, a dry sample of 2.5 grams was mixed with distilled water (0.5, 1 and 2.5 ml, according to the different L/S) in 25x25 mm plastic capsules and placed inside the reactor. As for the test performed at L/S=0, samples at their original water content (1.11%) were used and no water was added. The internal atmosphere was filled with pure CO_2; this was accomplished by flowing pure CO_2 and leaving the gas outlet open for about 1 min to wash out the air atmosphere and, subsequently, closing it to reach a specific CO_2 pressure. Operating pressure was kept constant within ± 0.5 bar of the set-point by replenishment of the consumed CO_2. The contact with a 100% carbon dioxide atmosphere was carried out according to different residence times: 0.5, 1, 2, 5, 10, 24, 48, 72, 144 and 240 hours. A 75% relative humidity was maintained in the reactor by filling the lower part with a saturated solution of NaCl[15].

Table 1. Outline of the accelerated carbonation experiments

Run	L/S (l/kg)	P (bar)	CODE	Run	L/S (l/kg)	P (bar)	CODE
1	0	1	0_1	9	0.4	1	0.4_1
2	0	5	0_5	10	0.4	3	0.4_3
3	0	10	0_10	11	0.4	5	0.4_5
4	0	20	0_20	12	0.4	10	0.4_10
5	0.2	1	0.2_1	13	0.4	20	0.4_20
6	0.2	5	0.2_5	14	1	1	1_1
7	0.2	10	0.2_10	15	1	5	1_5
8	0.2	20	0.2_20	16	1	10	1_10
				17	1	20	1_20

For each operating condition tested, the reaction kinetics were assessed by performing a single batch test, whilst the maximum CO_2 uptake, generally achieved at the longest duration of 240 h, was evaluated in triplicate for all the runs. The CO_2 sequestration capacity was determined by measuring both the gain in dry weight (40°C, to constant weight) and the TC before and after carbonation. Since differences between methods were lower than 5% of the average CO_2 concentration, only the results based on the gain in dry weight will be used for the discussion. CO_2 conversion yields were evaluated by comparing the actual uptake with the maximum achievable, the latter calculated assuming that each mol of Ca, Fe, Mg and Mn initially present in the solid matrix is able to form one mol of carbonate minerals[15,16] (see eq. 1). In the case of Ca, the initial calcite and gypsum content was taken into account. ANOVA test (by using a F-test) was performed in order to assess whether there are any significant differences between the CO_2 uptake assessed at the different operating conditions adopted.

$$\eta = \frac{CO_2uptake[\%]/44}{\frac{Ca_{in}[\%]}{40}+\frac{Fe_{in}[\%]}{55.8}+\frac{Mg_{in}[\%]}{24.3}+\frac{Mn_{in}[\%]}{54.9}-\frac{CO_{3in}[\%]}{60}-\frac{SO_{3in}[\%]}{80.1}} \quad (1)$$

The effect of the operating conditions and their joint interactions on the carbonation yield was assessed by multiple regression analysis[17], expressed by the following second order polynomial function:

$$y = \beta_0 + \beta_1 x_1 + \beta_2 x_2 + \beta_{12} x_1 x_2 + \beta_{11} x_1^2 + \beta_{22} x_2^2 \quad (2)$$

where y is the response variable (CO_2 uptake), βij (i;j= 1, 2) the unknown coefficients of the equation and xi (i= 1, 2) the predictor variables (CO_2 pressure, L/S ratio).

The leaching behaviour before and after carbonation was evaluated through the compliance leaching test (EN 12457-2). At the end of the tests, pH was evaluated and the eluates were acidified with HNO_3 (2%), filtered by a 0.45 µm vacuum system, and the metal content was analyzed by ICP-OES.

Results and discussion

The major elements composition of Waelz slag, given as oxides, show that Fe is the main constituent and accounts for 16,71% (w/w), followed by Ca (16,65%) and Si (7,55%), whilst Mg and Mn were detected at lower concentrations (2,81 and 2,94%, respectively). Among minor elements, potentially critical components such as Zn (9200 mg/kg), Cu (7400 mg/kg), Cr (2600 mg/kg), Pb (2400 mg/kg) were detected in high concentrations. The assessed elemental composition of the Waelz slag (more detailed data are available in Cappai[12]) is in agreement with the available literature data[11,18]. XRD analysis showed that plustite (FeO) is the main iron mineral, Ca can be found in two different phases, larnite (Ca_2SiO_4) and mayenite ($Ca_{12}Al_{14}O_{33}$). Mn was found as independent phase (hausmannite Mn_3O_4), while Mg is likely present as ionic substituent of Fe in plustite.

The results of the carbonation tests performed at the different operating conditions are resumed in Figure 1, where the CO_2 uptake and the conversion yields (mean values and standard deviation) measured at the end of the experiments are shown. It can be seen that by varying the water content and the CO_2 pressure, large variations in CO_2 uptake and conversion yields were obtained. The maximum CO_2 uptake of 6.48±1.00% was achieved at a L/S=1 l/kg and CO_2 pressure P=20 bar, while the minimum uptake of 1.47±0.56% was achieved for L/S=0 l/kg (no additional water other than moisture content) and P=1 bar. The corresponding degrees of carbonation were 18.32 and 4.16%, respectively. ANOVA test showed a statistically significant difference (p-value<0.05) between means CO_2 uptake assessed at the different operating conditions adopted. To determine which means are significantly different from each others, a Multiple Range Test was performed. To discriminate among the means, the Fisher's least significant difference (LSD) procedure was used. Six different homogenous groups were found, identified by letters A, B, C, D, E, and F. Figure 1 evidences that carbonation, as expected, was always less effective during experiments with L/S=0, irrespectively of the CO_2 pressure adopted (see homogenous group A). As for the effect of water content, It is well know that water is necessary for the dissolution of alkaline elements and hydration of oxide/silicates from the solid matrix, whilst non aqueous processes are characterized by very slow reaction kinetics and need very high temperatures to be conveniently speed up[19,20]. However, some studies demonstrated that high water contents limits the achievement of saturation conditions and thus carbonate precipitation, as well as the CO_2 diffusion into the pores of the solid matrix. As a consequence and depending on the kind of alkaline residue, an optimal L/S is suggested in the literature, ranging from 0.2-0.4 for municipal solid waste incineration residues to up to 20 for steel slag[7,14,19]. In the L/S range applied in our study, statistically significant increasing uptakes were found by increasing water content up to L/S =1 l/kg at the same CO_2 pressure. Generally, CO_2 uptake increased by increasing singularly either L/S or CO_2 pressure (See homogenous group E and F).

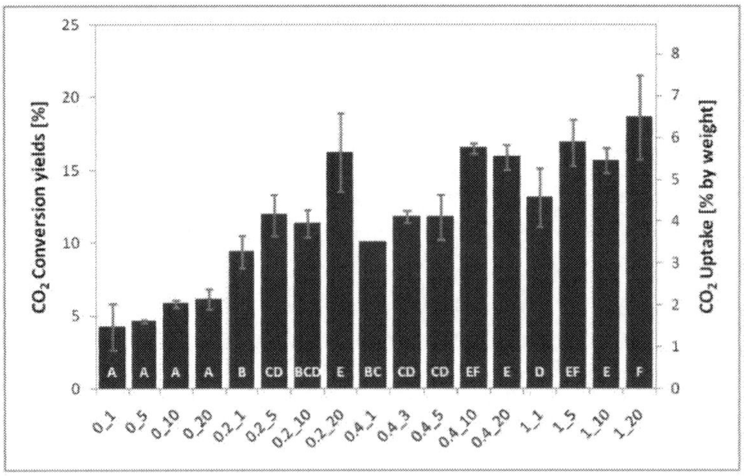

Figure 1. Maximum CO_2 uptakes and conversion yields resulting from the different accelerated carbonation tests. Significant differences between tests (p<0.05) are represented by different upper-case letter labels.

The combined effect of the two considered independent variables was interpreted and described by a multiple regression model. The R-squared obtained accounted for 90%, indicating a statistically significant relationship between the predictor variables. The best-fit regression model was:

$$CO_2Uptake = 1{,}0793 + 0{,}1531 \cdot P_{CO_2} + 10{,}0461 \cdot L/S - 0{,}0217 \cdot P_{CO_2} \cdot (L/S) - 0{,}0035 \cdot (P_{CO_2})^2 - 6{,}6846 \cdot (L/S)^2 \quad (3)$$

As underlined by the coefficients in Eq. 3, it can be highlighted that L/S ratio played a more relevant effect on CO_2 uptake than CO_2 pressure.

The adopted carbonation conditions were also found to influence the kinetics of the process. Different CO_2 uptakes as a function of reaction time were indeed found at the different water content and CO_2 pressures tested in this study. Figure 2 shows CO_2 uptakes as a function of time and L/S ratio, at a fixed CO_2 pressure (up to 20 bar). Kinetics appears to be strongly influenced by water content. As expected, reaction kinetics are slower for the experiment carried out at L/S=0, in particular at low reaction times (8 h), where no carbonation reactions occurred. However, not negligible CO_2 uptakes (from 1.47 at 1 bar to 2.13% at P=20 bar) were obtained at the end of the test. In general, the kinetics show an increase as the L/S ratio increased from 0 to 1 l/kg. For example at P=20 bar, after 48 h, samples where no water was added showed only 0.29% CO_2 uptake (corresponding to 0.82% of conversion of reactive species to carbonates), compared to 3.45% (9.76% conversion), 5.24% (14.83% conversion) and 4.46% (12.62% conversion) observed for L/S ratios of 0.2, 0.4 and 1 l/kg respectively.

As to the influence of CO_2 pressure, Figure 3 shows CO_2 uptakes as a function of time and CO_2 pressure, at a fixed water content (L/S=0.4). It can be stressed that for short reaction times, the intensity of reactive phases conversion into carbonates was mostly similar for the different CO_2 pressure adopted, so that the variation in CO_2 pressure did not appear to affect the carbonation kinetic. Conversely, after a reaction time of about 24 h, the carbonation kinetics slowed down, to an extent that was more pronounced at the lowest pressures, i.e. higher the pressure, faster the CO_2 uptake. The role of CO_2 pressure in accelerated carbonation process has been studied extensively in literature. High CO_2 pressures could lead to higher uptakes due to the increased CO_2 solubility in aqueous phase. Moreover, the dissolution of CO_2 into water leads to a decrease in pH, due to carbonic acid formation which in turn leads to a higher extraction of alkaline elements from the solid residues[7]. Some authors[4,21] found that pressure play a major role in improving the kinetics of the reactions but does not affect the final amount of formed carbonate. In our previous study[12], we've found that during accelerated carbonation of Waelz slag, new phases form as a grain rim, hindering the further reaction of alkaline elements, according to an unreactive core-type model mechanism. In this study, the positive effect on kinetics observed at the higher pressures after the very initial phase of carbonation could be explained taking into account that higher CO_2 pressures could have enhanced the diffusivity of carbon dioxide throughout the particle coatings.

The evolution of the mineralogy during the experiment characterized by the maximum CO_2 uptake (L/S = 1 l/kg and P = 20 bar) are reported in Figure 4, where the XRD patterns of the materials treated according different durations are shown. The I_{100}

peak of larnite decreased in intensity from the sample collected after 0.5 hours to the sample collected after 240 h, suggesting that Ca_2SiO_4 represents the main source of Ca ions for the carbonation process. The grey area in Figure 4 highlights the main peaks attributed to the new phases resulting from carbonation process. XRD analysis, combined to FTIR analysis, allowed us to recognize that the carbonation of the sample led mainly to the precipitation of amorphous carbonate starting from the 24th hour, and of kutnohorite, a double salt of Ca and Mn, after 72 hours. Calcite peaks are recognizable, related to accessory carbonation products.

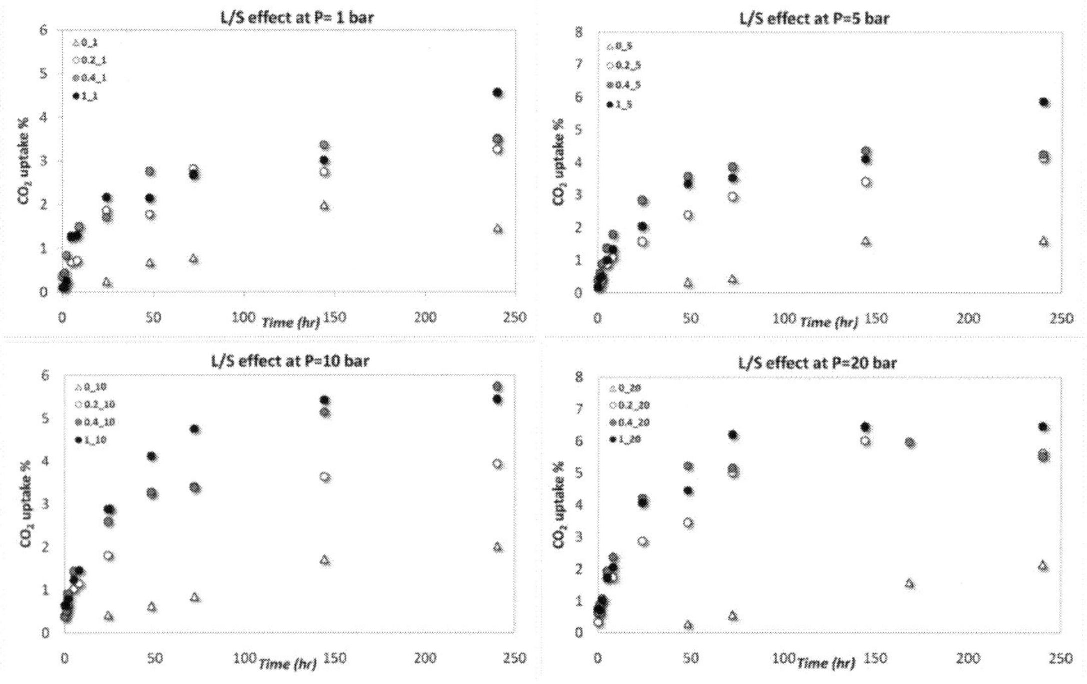

Figure 2. CO_2 uptakes resulting from the different accelerated carbonation tests as a function of time and L/S ratio.

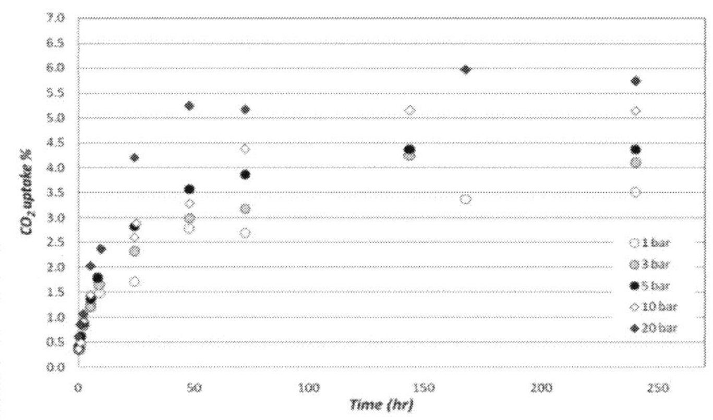

Figure 3. CO_2 uptakes resulting from the different accelerated carbonation tests as a function of pressure and time. L/S=0.4 L/kg, T= 25 °C

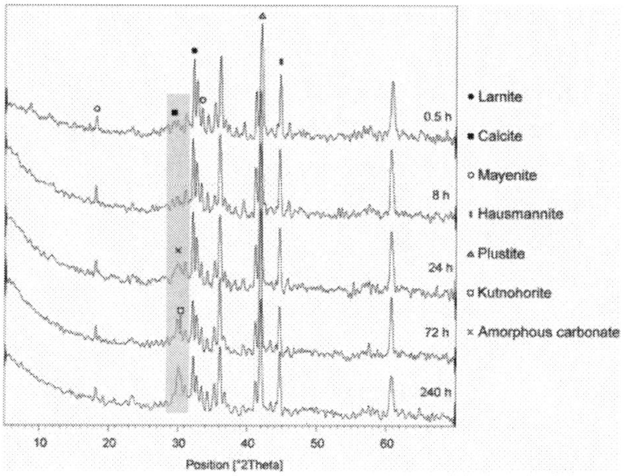

Figure 4. XRD patterns for samples related to the accelerated carbonation experiment performed at L/S = 1 l/kg and P = 20 bar, after 0.5, 8, 24, 72 and 240 h. Symbols marked the I_{100} peak of each phase. The grey area highlights the main peaks attributed to the new phases resulting from carbonation process.

Table 2 shows the results of the EN 12457-2 leaching test performed on the untreated and carbonated (L/S=0 and 1 l/kg, P=20 bar, duration=240 h) Waelz slag. The pH of the residues is shown as well, along with a comparison with the EU regulatory limits for waste disposal in landfills. For the untreated residue, leaching was found to exceed the limit values for disposal in inert waste landfills, in particular for Mo and Sb. Accelerated carbonation at a L/S=1 l/kg resulted in a pH decrease, whilst no variation was assessed for the test performed at L/S=0 l/kg, coherently with the different carbonation extent attained. The carbonation process resulted in a decrease of Ba mobility in both carbonated samples, while the leaching of Sb increased, in agreement with what generally reported in the literature for oxyanion-forming elements. The mobility of the other species considered by the above mentioned EU regulation did not varied substantially after carbonation.

Table 2. Results of the leaching tests EN 12457-2 for untreated and carbonated (P=20 bar, 240 h) Waelz slag and comparison with EU Landfill acceptance criteria. DL=Detection Limit

Element/ Species	Untreated	Carbonated L/S=0 (l/kg)	Carbonated L/S=1 (l/kg)	Inert waste landfills
As (mg/l)	<0.02	<0.02	<0.02	0.05
Ba (mg/l)	0.102±0.048	0.050±0.007	0.060±0.021	2
Cd (mg/l)	<0.002	<0.002	<0.002	0.004
Cr (mg/l)	<0.005	<0.005	<0.005	0.05
Cu (mg/l)	<0.005	0.007±0.002	<0.005	0.2
Mo (mg/l)	**0.379±0.080**	0.247±0.037	0.385±0.009	0.05
Ni (mg/l)	0.009±0.002	<0.005	0.011±0.007	0.04
Pb (mg/l)	<0.02	<0.02	<0.02	0.05
Sb (mg/l)	**0.030±0.002**	0.063±0.003	0.165±0.019	0.006
Se (mg/l)	<0.02	<0.02	<0.02	0.01
Zn (mg/l)	0.029±0.014	0.035±0.009	0.019±0.011	0.4
pH (-)	11.63±0.91	11.65±0.21	9.47±0.05	

Conclusions

Waelz slag could be considered a good candidate for the application of mineral carbonation process, due to its relatively high content of CO_2 reactive minerals phases and the possibility of applying directly at the Waelz plant to reduce the related CO_2 emissions. However, the present study indicates that the process conditions should be carefully selected in order to attain a significant CO_2 uptake. The operating parameters, such as L/S ratio and total pressure, tested in this study appreciably affected the extent of CO_2 sequestration and the evolution over time of the process. CO_2 uptakes and conversion yields were lower for the lower water content and CO_2 pressures applied, and in general increased by singularly increasing either the L/S or the CO_2 pressure. Multiple regression model analysis clearly reveals a more relevant effect on CO_2 uptake played by the L/S ratio than by the CO_2 pressure. The maximum CO_2 uptake of 6.48±1.00% was achieved at L/S=1 l/kg and CO_2 pressure P=20 bar, while the minimum uptake of 1.47±0.56% was achieved for L/S=0 l/kg and P=1 bar. The corresponding degrees of carbonation were 18.32 and 4.16%, respectively, indicating that the potential for CO_2 sequestration was only partially exploited even when the optimal conditions were adopted. This issue could be explained taking into account the coarse particle size and, in turn, the low surface area of the Waelz slag, which does not favour the dissolution kinetics of reactive species. In addition, evidence of particle coatings during carbonate precipitation was observed, which hindered further reactions involving the alkaline elements contained in the inner particle section. In a view of process optimization, a pre-treatment stage based on particle size reduction could contribute to develop a cost-effective sequestration process.

Aknowledgements

This research has been funded by the Regional Sardinian Government, in the framework of "L.R. 7/2007, Promotion of scientific research and technological innovation in Sardinia" (CRP 25 659 project).

References

1. IPCC (Intergovernmental Panel on Climate Change). 2005. *Special report on carbon dioxide capture and storage.* Cambridge University Press, USA.
2. Lackner, K., 2002. Carbonate Chemistry for Sequestering Fossil Carbon. *Ann. Rev. Energy Environment*, 27, 193-232.
3. Johnson, D.C., 2000. Accelerated carbonation of waste calcium silicate materials. *SCI Lecture Papers Series*, Society of Chemical Industry.
4. Sipila, J., Teir, S., Zevenhoven, R., 2008. Carbon dioxide sequestration by mineral carbonation. *Literature review update 2005-2007.* Abo Akademi University, Report 2008-1.
5. Gunning P.J., Hills C.D., Carey P.J. 2010. Accelerated carbonation treatment of industrial wastes. *Waste Manage* 30, 1081-1090.
6. Bobicki, E.R., Liu, Q., Xu, Z., Zeng, H., 2012. Carbon capture and storage using alkaline industrial wastes. *Prog. Energy Combust. Sci.* 38,302-320.

7. Pan, S.Y., Chang, E.E., Chiang, P.C., 2012. CO_2 capture by accelerated carbonation of alkaline wastes: a review on its principles and applications. *Aerosol and Air Quality Research*, 12, 770–791.

8. Lange, L.C., Hills, C.D., Poole, A.B., 1996. The effect of accelerated carbonation on the properties of cement-solidified waste forms. *Waste Manage* 16, 757-763.

9. Shtepenko, O.L., Hills, C.D., Coleman, N.J. Brough, A., 2005. Characterization and Preliminary Assessment of a Sorbent Produced by Accelerated Mineral Carbonation. *Environ. Sci. Technol.* 39, 345-354.

10. Van Gerven, T., Van Keer, E., Arickx, S., Jaspers, M., Wauters, G., Vandecasteele, C., 2005. Carbonation of MSWI-bottom ash to decrease heavy metal leaching, in view of recycling. *Waste Manage.* 25, 291-300.

11. Barna, R., Bae, H.-R., Méhu, J., van der Sloot, H., Moszkowicz, P., Desnoyers, C., 2000. Assessment of chemical sensitivity of Waelz slag. *Waste Manage.* 20, 115-124.

12. Cappai, G., De Giudici, G., Medas, D., Milia, S., Muntoni, A., Orrù, G., Nieddu, A., Piredda, M., 2014. Accelerated Carbonation of Waelz Slag: Effect of Particle Size on CO_2 Sequestration Capacity. *In: Proceedings of Crete 2014, fourth International Conference On Hazardous and Industrial Waste Management*, September 3-5 2014, Chania, Crete, Greece.

13. Costa, G., Baciocchi, R., Polettini, A., Pomi, R., Hills, C., Carey, P., 2007. Current status and perspectives of accelerated carbonation process on municipal waste combustion residues. *Environment Monit. Assess.*, 135, 55-75.

14. Li, X., Bertos, M.F., Hills, C.D., Carey, P.J., Simon, S., 2007. Accelerated Carbonation of Municipal Solid Waste Incineration Fly Ashes. *Waste Manage* 27: 1200–1206.

15. Baciocchi, R., Costa, G., Di Gianfilippo, M., Polettini, A., Pomi, R., Stramazzo, A., 2012. Evaluation of the CO_2 storage capacity of different types of steel slag subjected to accelerated carbonation. *In: Proceedings of Crete2012, 3rd International Conference on Industrial and Hazardous Waste Management*, Chania, Crete (Greece), 12-14 September 2012.

16. Baciocchi, R., Costa, G., Di Gianfilippo, M., Polettini, A., Pomi, R., Stramazzo, A., Zingaretti, D. (2013). Influence of CO_2 partial pressure on direct accelerated carbonation of steel slag. *In: Proceedings ACEME 2013, 4th International Conference on Accelerated Carbonation for Environmental and Materials Engineering*, Leuven (Belgium), 9-12 April 2013.

17. STATGRAPHICS Centurion XV.II, 2005. Ssoftware StatPoint, Inc. Statgraphics Centurion XV, Herndon, VA, USA.

18. Quijorna, N., de Pedro, M., Romero, M., Andres, A., 2014. Characterisation of the sintering behaviour of Waelz slag from electric arc furnace (EAF) dust recycling for use in the clay ceramics industry. *J. Environ. Manag.* 132 (2014), 278-286.

19. Fernández Bertos, M., Li,X., Simons, S.J R., Hills, C.D., Carey, P.J. 2004. Investigation of accelerated carbonation for the stabilisation of MSW incinerator ashes and the sequestration of CO2. *Green Chem.*, 6, 428-436.

20. Baciocchi, R., Polettini, A., Pomi, R., Prigiobbe, V., VonZedwitz, V.N., Steinfeld, A., 2006. CO_2 sequestration by direct gas-solid carbonation of air pollution control (APC) residues, *Energy Fuels* 20, pp. 1933-1940.

21. Rendek, E., Ducom, G., Germain, P., 2006. Carbon dioxide sequestration in municipal solid waste incinerator (MSWI) bottom ash. *J Hazard Mater.* 128, 73-9.

Carbonation-granulation of alkali activated BOF slag for the production of recycled aggregates and CO_2 storage

M. Morone[1], E. Georgakopoulos[2], G. Costa[1*], Ö. Cizer[3], S. Gu[2], R. Baciocchi[1]

[1] Laboratory of Environmental Engineering, Dept. of Civil Engineering and Computer Science Engineering, University of Rome "Tor Vergata", Rome, Italy

[2] Energy and Power Division, Cranfield University, Bedfordshire, UK

[3] Department of Civil Engineering, KU Leuven, 3001 Heverlee, Belgium

*Corresponding author: costa@ing.uniroma2.it

Abstract

This work was aimed at investigating the performance of a combined carbonation and granulation treatment applied to Basic Oxygen Furnace (BOF) steel slag with the aim of producing secondary aggregates and of storing CO_2 in a solid and stable form. In order to improve the mechanical properties of the aggregates, an alkaline solution of sodium silicate and sodium hydroxide was employed as binder instead of water both during the granulation and combined granulation/carbonation tests. The results showed that the granules produced using the alkali activator with or without CO_2 addition, presented a mean size ranging from 1 to 5 mm and adequate mechanical properties for use in civil engineering applications. The maximum CO_2 uptake attained was of 40 gCO_2/kg steel slag for the alkali activated and carbonated granules after 60 min of treatment. As for the leaching behaviour of the produced granules, an increase in the release of Cr and V was found for the product of the granulation/carbonation treatment with alkali activation; nonetheless, the crushed granules attained after a reaction time of 30 min showed to still comply (apart from pH) with the Italian requirements for reuse for land reclamation.

Introduction

Manufacturing processes such as steel production, are typically characterized by the generation of large amounts of solid residues, the recycling or reuse of which is not always feasible, due to the variability of the material characteristics, the lack of a suitable national legislation on industrial by-products management, or the scarce available of low-cost technologies for producing valuable secondary products. As a result, despite the growing awareness on the environmental impacts related to final disposal, for a relevant fraction of these residues no well-established management strategies apart from landfilling have been developed yet. Among the residues generated during steel production, BOF slag, a by-product of the conversion of iron into steel in Basic Oxygen Furnace (BOF) units, is one of the most abundant, amounting for 46% of the total European steel slag production in 2012.[1] Commonly, this material is ground for metal recovery and limitedly reused for low-end applications because some of its properties (i.e. high free lime content, fine particle size and excessive mobility of specific elements) may negatively affect its technical performance and/or environmental behaviour. Different studies have indicated that accelerated carbonation may be an effective treatment for improving the environmental properties of the treated material,[2] as well as its mechanical characteristics by increasing its density.[3] Nevertheless, in order to be used as aggregates in civil works, the carbonated residues should exhibit a specific grain size distribution and adequate mechanical strength. Granulation represents a particle size enlargement process, which is achieved by contacting a mixture of a dry powder and a liquid binder in a dynamic device, such as a disc granulator or a rotating drum.[4] To this aim, previous findings of our research group indicated that the combination of carbonation with a granulation process, using water as binder and applied to a BOF steel slag sample presenting a fine particle size (d_{50}=0.4 mm), allowed to achieve, already after 30 min of reaction, coarse aggregates (d_{50}=4 mm) and significant CO_2 uptakes (120-144 gCO_2/kg).[5] In addition, the pH, and Ca and Ba concentrations measured in the leachates of the uncrushed granules showed a noteworthy decrease compared to those resulting for the untreated slag, while conversely a slight increase of V leaching was achieved.[5] Nevertheless, the mechanical strength of the granules, investigated applying the Aggregate Crushing Value (ACV) test, appeared to be quite poor and inadequate for meeting the requirements for use in civil engineering applications.

In order to enhance the mechanical properties of the granules, in this work we investigated the feasibility of adding alkali additives to the binding solution, thus combining the granulation-carbonation treatment with alkali activation. The latter has in fact shown to be effective for improving the hardening reactions of amorphous residues rich in calcium and silicon, such as blast furnace slag, a by-product of iron manufacturing.[6] Alkali activation is a chemical process in which the raw material, usually amorphous, is mixed with a solution containing sodium or potassium hydroxides, carbonates and/or silicates, in order to favour the dissolution of the reacting species and to enhance hydration reactions leading to the formation of binding phases such as a calcium silicate gel, that yields a compact and strong product.[7]

The performance of the granulation-carbonation treatment combined with alkali activation, as a valorization technique for BOF slag, was compared with that attained by granulation, combined carbonation-granulation performed using only water as a binder and granulation carried out with alkali solutions. In this way, it was possible to assess the influence of the single and combined treatments on the properties of the produced granules, including the particle size distribution, CO_2 uptake, mechanical properties and leaching behaviour.

Materials and methods

Materials

The slag sample employed in this study was collected just downstream of two different Basic Oxygen Furnaces (BOF) units of an integrated steelmaking plant. The material was crushed and subjected to magnetic separation for iron and steel recovery at the plant. The particle size of the untreated material resulted lower than 1 mm, with a mean diameter equal to 0.2 mm.

The elemental composition was determined by alkaline fusion of slag samples with lithium metaborate ($LiBO_2$) in platinum melting pots at 1050 °C for 1 h, followed by dissolution of the molten material with 4% HNO_3 and subsequent ICP-OES analysis of the obtained solution, employing an Agilent 710-ES spectrometer. This analysis was carried out in triplicate. The results, reported in Table 1, showed that Ca and Fe were the main constituents, with concentrations equal to 268 g/kg and 223 g/kg, respectively, followed by Mg (42.5 g/kg), Mn (31.3 g/kg) and Si (15.8 g/kg).
The alkaline activating solution used in this study was a 50:50 by weight mixture of a 2M sodium hydroxide (NaOH) solution, prepared by dissolving NaOH pellets (Fisher Scientific, 99.4% purity) in deionized water, and a commercial solution of sodium silicate (Sigma-Aldrich) whose concentration of silicates in water, evaluated by ICP-OES analysis, resulted equal to 36%.

Table 1. Elemental composition of the untreated BOF slag expressed as g/kg dry weight.

Element	(g/kg)	Element	(g/kg)
Al	8	Fe	222.71
Ba	0.7	Mg	42.55
Ca	268.12	Mn	31.31
Cd	0.17	Si	15.77
Cr	0.73	V	0.47
Cu	0.14	Zn	0.15

Methods

Both the granulation and granulation-carbonation experiments were carried out in a disc granulator reactor, the main features of which are reported in [5]. For both types of tests, a sample of air-dried slag (approximately 500 g) was premixed in a plastic bag with either deionized water or the alkaline activating solution, at set liquid to solid ratios, following the procedure adopted in many previous studies on granulation;[8,9] then the mixture was inserted into the granulator and left to react for a set time. For the carbonation/granulation treatment a 100% CO_2 flow was maintained in the reactor for the entire duration of the experiment.[5] Table 2 reports a general overview of the performed experiments and their respective tested conditions. Each test is identified with a code that is used to present and discuss the results in the next section. As it may be observed, the duration of the granulation tests was of 30 min, while for the combined carbonation/granulation tests also a reaction time of 60 min was adopted. The water to slag ratio (W/S) employed in most experiments was 0.12 l/kg, which was experimentally proven to be the optimal value for the granulation of the slag; for the carbonation/granulation tests with alkali activation a higher W/S (0.14 l/kg) was also tested. The ratios of alkali activation solution to slag (AA/S) employed were 0.16 l/kg (baseline condition) or 0.18 l/kg. The granules produced in each test were cured in a controlled environment (room temperature and 100% relative humidity) for 28 d in order to allow the completion of hydration reactions.

Table 2. Summary of the main parameters adopted for the experiments (W/S: water to slag ratio; AA/S: alkali activation solution to slag ratio).

Experiment code	Reaction time (min.)	Binder type	W/S ratio (l/kg)	AA/S ratio (l/kg)
Granulation tests				
G_W	30	Deionized water	0.12	-
G_AA	30	Alkaline activator	0.12	0.16
Granulation-Carbonation tests				
GC_W	30	Deionized water	0.12	-
GC60'_W	60	Deionized water	0.12	-
GC_AA	30	Alkaline Activator	0.12	0.16
GC60'_AA	60	Alkaline Activator	0.12	0.16
GC_AA_W/S=0.14	30	Alkaline Activator	0.14	0.18
GC60'_AA_W/S=0.14	60	Alkaline Activator	0.14	0.18

The particle size distribution, CO_2 uptake, mechanical strength and environmental behaviour of the cured samples were then assessed and compared (where applicable) to the properties of the untreated slag. Specifically, the particle size distribution was determined by applying the ASTM D422 standard procedure. The CO_2 uptake of the treated slag was assessed on the basis of the results of Inorganic Carbon (IC) analysis that was performed with a Shimadzu TOC VCPH analyser equipped with the SSM-5000A solid sampler. The CO_2 uptake was then calculated by applying Equation 1, in which $CO_{2\ initial}$ and $CO_{2\ final}$ represent respectively the CO_2 content of the untreated and treated sample, as derived by the results the IC analysis. In particular, the initial IC content of the slag resulted equal to 0.03%, which is equivalent to an initial CO_2 concentration of 0.1%. The CO_2 uptake of the granules was assessed immediately after the test and also after 7 and 28 d of curing to evaluate the degree of carbonation occurring during the curing treatment (performed under atmospheric air at ambient temperature).

$$CO_{2uptake}(\%) = \frac{CO_{2final}(\%) - CO_{2initial}(\%)}{100 - CO_{2final}(\%)} \cdot 100 \qquad (1)$$

Batch accelerated carbonation tests with 100% CO_2 were also performed on the as-received BOF slag samples following the experimental procedure adopted in previous works.[10,11] Operating conditions (T=20 °C, p=1 bar, L/S=0.3 l/kg) similar to those employed in the granulation tests were applied in order to evaluate if the different reaction mode adopted (carbonation/granulation instead of static batch carbonation) could affect the resulting CO_2 uptake. Moreover, batch carbonation tests were also carried out on the slag as received and after milling (d<0.125 mm) applying enhanced operating conditions (T=50 °C, p=10 bar, L/S=0.3 l/kg). These specific conditions were selected since in previous works carried out on steelmaking slag they allowed to achieve significant CO_2 uptakes with the thin-film route.[11,12] The CO_2 uptake attained for the batch carbonated samples was determined following the same procedure adopted for the granules.

The mechanical strength of the granules was evaluated by performing the Aggregate Crushing Value (ACV) test, applying the British standard BS 812-110. The leaching behaviour of the granules was evaluated by following the EN 12457-2 standard compliance test that involves grinding of the material presenting a grain size above 4 mm. In addition, the leaching behaviour of the granules obtained after each treatment was also assessed by applying the EN 12457-2 procedure without grinding. The eluates obtained from each test were analyzed by ICP-OES analysis, employing an Agilent 710-ES spectrometer. Leaching tests were carried out in duplicate.

Results and discussion

Particle size distribution

The particle size distribution (PSD) of the granules produced applying the different types of treatments is reported in Figure 1, together with that of the untreated BOF slag. As can be observed in Figure 1a, the granules produced by granulation with water (G_W) or the sodium silicate solution (G_AA) presented a similar PSD with a mean diameter of 10 and 13 mm respectively, over 50 times greater than that of the untreated slag. The PSD of the granules obtained applying the combined carbonation/granulation treatment were significantly smaller, especially when the alkali solution was used as binder (d_{50}=1 mm for test GC_AA). This result may be ascribed to the higher viscosity of the alkali solution compared to water. In fact, an increase in the viscosity of the binder has been indicated to yield stronger granules that tend to deform less during granulation leading to a reduction of the consolidation rate and subsequently of particle size growth.[8]

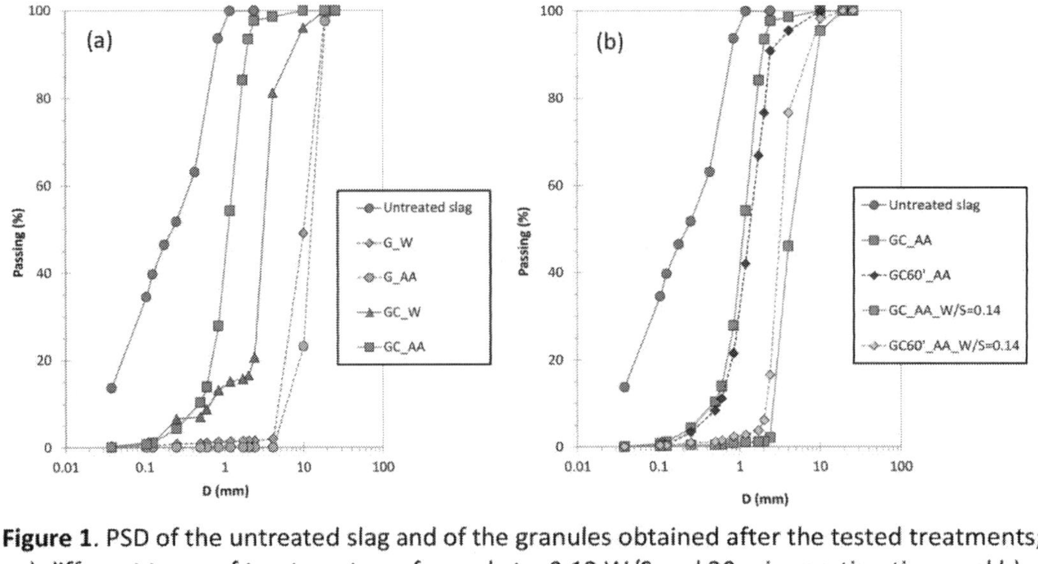

Figure 1. PSD of the untreated slag and of the granules obtained after the tested treatments; a) different types of treatments performed at a 0.12 W/S and 30 min reaction time and b) granulation/carbonation with alkali activation at different W/S ratios and reaction times.

The effect of granulation-carbonation with alkali activation on the size distribution of the granules as a function of reaction time and the W/S is reported in Figure 1b. As can be noted, the W/S appeared to exert a stronger influence than reaction time. In particular, increasing the W/S value from 0.12 to 0.14 l/kg led to a maximum final mean diameter of 4.5 mm, around four times higher than the one obtained at a lower liquid content. Moreover, in both cases, the granule size distribution showed to be quite uniform and not to be significantly affected by a variation in the reaction time. Actually, for a W/S of 0.14 l/kg, a slight decrease of the mean diameter of the granules, passing from 3 to 2 mm, was observed, thus suggesting that breakage may occur during the treatment for prolonged reaction times.

CO_2 uptake

Batch carbonation experiments. The main results of the batch carbonation tests, reported in Figure 2, revealed low differences in the carbonation yield attained at mild and enhanced operating conditions. Indeed, at mild conditions (ambient temperature and 1 bar CO_2), the maximum CO_2 uptake, achieved after 8 hours of treatment, resulted equal to almost 6%, while the corresponding value attained at 50 °C and 10 bar CO_2 was just below 9%. Grain size instead appeared to significantly affect the slag's reactivity with CO_2, in agreement with findings reported in a previous work on stainless steel slag carbonation.[10] Indeed, the carbonation test carried out at enhanced conditions on the milled slag allowed to achieve a maximum uptake of almost 15% already after 2 hours of reaction, comparable to that reported for a similar type of BOF slag, milled below 0.15 mm.[12]

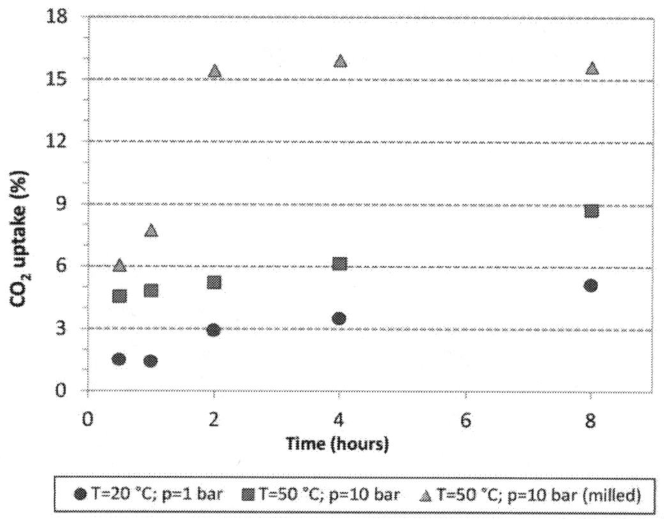

Figure 2. CO_2 uptake kinetics of BOF slag resulting from the batch tests.

Granulation-Carbonation experiments. Making reference to Figure 3a, it can be observed that, regardless of alkali activation, the CO_2 uptake of the granules achieved after granulation under atmospheric air was very low (<0.5%). Instead, after the granulation-carbonation treatment, the average CO_2 uptake of the granules reached values of almost 2% and 3% with water and the alkali solution as binder, respectively.

It is interesting to note that the CO_2 uptake of the carbonated granules with water as binder, which was achieved contacting a considerable amount of material (≈500 g) with a 1 bar CO_2 flow in a dynamic device under ambient T, resulted comparable with that obtained in the batch tests carried out at the same conditions (see Fig. 2), but placing 1 g-BOF slag samples in a static reactor. It should be noted that the BOF slag employed in this study presented a significantly lower CO_2 uptake at the operating conditions of the granulation reactor compared to that exhibited by the slag sampled from the same plant and tested in our previous work on granulation/carbonation.[5] The differences in the degree of reactivity with CO_2 were attributed to the different mineralogy of the two types of slag; the ones analyzed in this work were characterized by a prevalence of Ca silicate-based phases, while the others presented a relevant content of portlandite, that is well known to present a high reactivity with CO_2 also at mild operating conditions.[5] Anyhow, as shown in Figure 3b, the CO_2 uptake showed to increase for a 60 min reaction time, reaching a maximum value of 3.3 % for the test carried out employing the alkali solution with a W/S of 0.12 l/kg. As for the influence of the W/S, for the tests carried out with the higher value (0.14 l/kg) a reduction in the CO_2 uptake (especially for the 30 min test) was achieved. This result may be possibly related to the coarser grain size of the granules formed employing more binder, which may have lead to decrease in carbonation kinetics, as suggested by the results of the batch tests (see Fig. 2).

To evaluate if the carbonation reaction may continue also during the curing treatment carried out under atmospheric air, the CO_2 uptake was estimated also after 7 and 28 d of curing. As shown in Fig. 4, after the first week of curing, a significant increase of the CO_2 uptake of the products of most of the tested treatments (including granulation under atmospheric air) was registered; however, during the next three weeks of curing, no further CO_2 uptake was measured. The additional reactivity of the slag with atmospheric CO_2 may be possibly linked to hydration reactions; hence these results suggest that hydration may achieve completion already during the first week of curing.

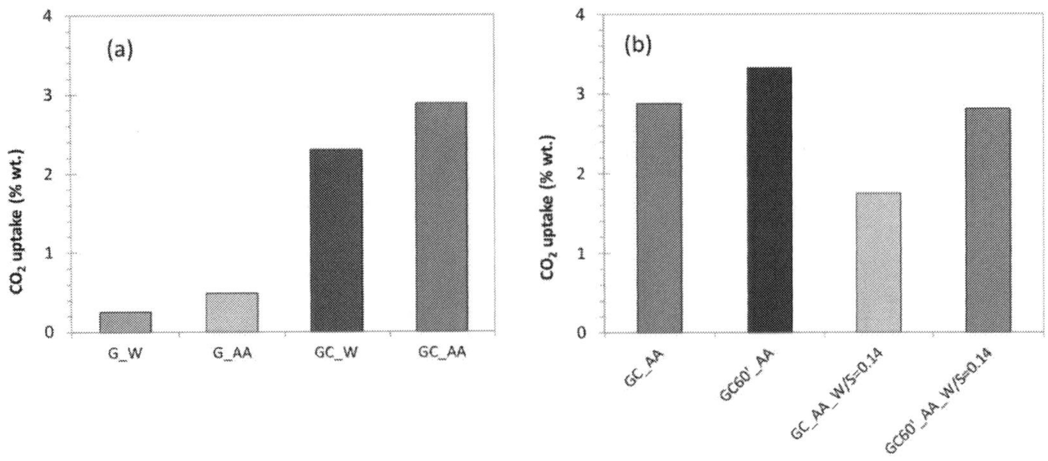

Figure 3. CO_2 uptake of the slag a) for each tested treatment at a 0.12 W/S and 30 min reaction time and b) for granulation/carbonation with alkali activation at different W/S ratios and reaction times.

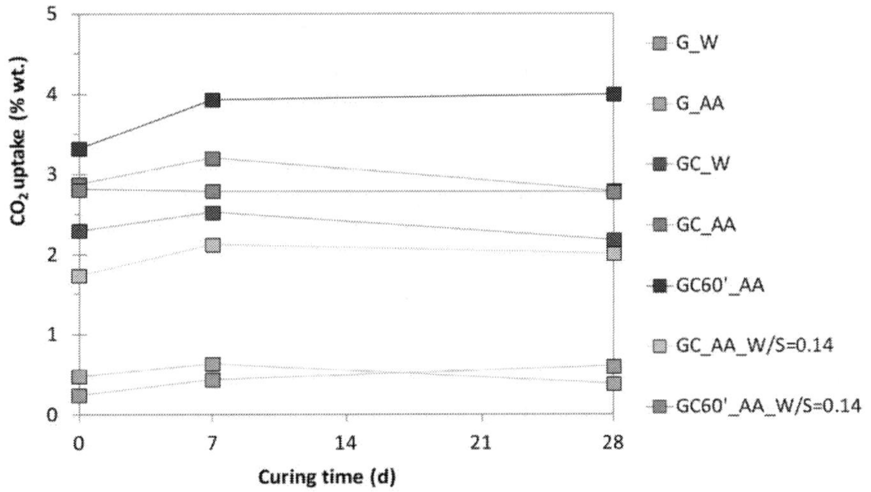

Figure 4. CO_2 uptake with curing time of the granules obtained applying the different types of treatments tested.

Mechanical strength

The results in terms of the ACV of the produced granules are shown in Figure 5 and compared to the average values reported for mixed gravel and blast furnace slag.[13] The mechanical resistance of the granules showed to be affected both by the type of binder used and by carbonation. Namely, the granulation treatment with alkali activation allowed to enhance the strength of the obtained granules, leading to an ACV value of 15%, notably lower than the one obtained using only water as binder (around 45%) and also below the one for mixed gravel. Instead, for the granules resulting from the combined granulation/carbonation treatment, the ACV remained above 35%, indicating a rather poor mechanical performance, even employing the alkali solution (with a W/S of 0.12 l/kg). However, an increase in the amount of alkali binder showed to significantly improve the resistance of the granules; in fact, the granules obtained from test GC_AA_W/S=0.14 were characterized by an ACV equal to 18%, i.e. about half the value achieved using a W/S of 0.12 l/kg. It is also interesting to note that an increase of the reaction time showed to reduce the granules strength for both of the W/S tested. These results may be possibly associated to changes in the physical properties of the granules (e.g. particle size), but also to chemical modifications (regarding e.g. the binder's viscosity or the extent of hydration) related to the CO_2 treatment.

Hence, on the basis of the results achieved, also in comparison to those set by British Standard for mixed gravel or blast furnace slag presenting a comparable grain size,[13] it may be concluded that the granules produced with granulation and alkali activation, or granulation/carbonation and alkali activation at a W/S of 0.14 l/kg, exhibited adequate mechanical properties for use in civil engineering applications; it should however be noted that the maximum ACV is not a fixed value, but instead depends on the specific civil engineering application foreseen for the aggregates.

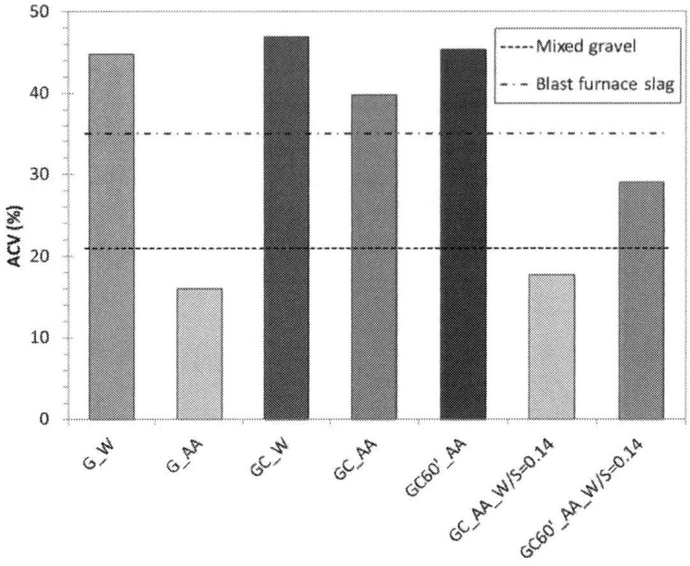

Figure 4. ACV of the granules obtained from the different treatments tested compared to the values reported for mixed gravel (21%) and blast furnace slag (35%).[13]

Leaching behaviour

Table 3 reports the leaching test results of the granules (both crushed and uncrushed) attained from the different treatments tested in comparison to those of the untreated slag. Specifically, the table exhibits the pH values, eluate concentrations of regulated elements that proved higher than instrumental quantification limits (Ba, Cr, V and Zn) and of major constituents (Al, Ca, Na and Si), along with the Italian limits for waste reuse for land reclamation.[14] For the granulation/carbonation with alkali activation treatment, based on the ACV results, it was decided to analyze the leaching behaviour of the granules obtained from the test carried out at a W/S of 0.14 l/kg, which appeared to be the most suitable for construction applications.

As shown in Table 3, the untreated material was characterized by an alkaline pH (13.1), slightly higher than that reported in the literature for this kind of slag,[15] and limited leaching of regulated elements. The product of the granulation treatment carried out with water (G_W) presented a similar behaviour to the untreated slag, although the uncrushed granules exhibited a lower release of some elements (Ba, Cr and Zn) owing to their larger grain size. For the tests performed with the alkali solution (G_AA), the leaching behaviour of the product showed to change significantly especially for the uncrushed granules. In particular, the pH decreased (although remaining above the Italian limit for reuse: pH=12);[14] Cr, Zn and in particular Ba release decreased, as well as Ca, while Si, Al and Na leaching increased. The higher mobility of Na was obviously due to the addition of the Na hydroxide/silicate solution, while the effects on both major and trace constituents' release may be related to the hydration reactions occurring in the material, in particular release control by less soluble phases for Ca and the opposite for Al and Si; for the trace contaminants the reduction may be ascribed both to the lower eluate pH, as well as possible sorption onto hydration reaction

products. It is interesting to observe that the leaching behaviour of the product of the granulation/carbonation treatment performed with water (GC_W) without crushing proved to be quite similar to that of the uncrushed granules of test G_AA (apart from the higher Ca and Ba release), while for the crushed granules to that of the uncrushed product of test G_W. This result can be directly related to the very similar pH values of the eluates of the above reported tests, around 12.5 in the first case and 12.75 in the second one. Hence, in this case the carbonation treatment appeared not to exert a relevant effect on the leaching behaviour of the slag, except for its effect on pH, differently to the findings of previous studies,[5,16] but in accordance to the much lower reactivity with CO_2 shown by the tested slag in this work.
However, the coupling of granulation/carbonation with alkali activation appeared to notably affect the leaching behaviour of the BOF slag. In particular, Ba release further decreased, resulting lower than the instrumental quantification limits for a reaction time of 60 min; correspondingly, however, Cr and V leaching increased, exceeding the reuse limit values after 60 min. With regard to major constituents' release, the effects already observed for the GC_W and G_AA tests (reduction of Ca mobility and increase of Si and Al leaching) were even more evident, especially for the 60 min test. Anyhow, it should be noted that, apart from pH, the leaching behaviour of the crushed granules obtained from test GC_AA for a W/S of 0.14 l/kg that exhibited a high mechanical resistance, showed to still comply with the Italian requirements for reuse for land reclamation.[14]

Table 3. Leaching test results for the untreated slag and the crushed and uncrushed granules obtained from the different tests compared to the Italian limits for reuse.[14]

		pH	Ba mg/l	Cr mg/l	V mg/l	Zn mg/l	Al mg/l	Ca mg/l	Na mg/l	Si mg/l
					Untreated slag					
		13.13	0.36	0.01	<0.003	0.03	0.11	915.2	1.67	0.13
					Granulation					
G_W	Crushed	13.09	0.4	0.01	<0.003	0.03	0.07	1001.3	2.97	0.24
	Uncrushed	12.75	0.2	0.003	<0.003	0.01	0.57	444.35	2.39	0.57
G_AA	Crushed	12.85	0.18	0.003	<0.003	0.02	0.11	480.62	510.2	0.35
	Uncrushed	12.55	0.01	0.005	0.01	0.02	0.58	42.39	429.2	3.86
					Granulation-Carbonation					
GC_W	Crushed	12.75	0.17	0.004	<0.003	0.02	0.48	401.39	2.02	1.11
	Uncrushed	12.52	0.11	0.01	0.01	0.01	0.57	318.91	1.25	2.62
GC_AA_ W/S=0.14	Crushed	12.89	0.03	0.05	<0.003	0.03	0.48	69.17	667	2.49
	Uncrushed	12.71	0.01	0.13	0.09	0.03	1.58	17.83	527.5	13.51
GC60'_AA W/S=0.14	Crushed	12.29	0.003	0.11	1.26	0.07	0.52	14.59	503	101.8
	Uncrushed	12.45	<0.002	0.12	1.16	0.01	0.77	9.67	472.1	69.36
Limit for reuse	It. D.M. 186/2006	5.5-12	1	0.05	0.25	3	-	-	-	-

Another interesting finding of this study was that the differences in the results attained for the uncrushed and crushed granules for each type of test were generally less relevant with respect to those noted in a previous study on granulation/carbonation of BOF slag.[5] In the previous work the formation of an external layer of carbonate product leading to a higher Si release and lower Ca mobility from the uncrushed coarse granules was postulated.[5] In this case, the formation of two distinct layers probably did not occur, both because of the smaller particle size of the granules formed and also of the lower degree of carbonation achieved.

Conclusions

The main outcome of this work was that alkali activation proved to be an effective treatment for improving the mechanical properties of the granules obtained applying granulation or carbonation-granulation processes. In fact, differently from the tests performed using only water as binder, in the new set of experiments performed with a Na hydroxide and silicate solution the ACV of the granules obtained at specific conditions were close to those typically exhibited by natural aggregates. Notably, the best performances were achieved for the granulation treatment (15%) and the combined carbonation-granulation process performed with a W/S of 0.14 l/kg (18%). The latter condition allowed also to achieve a significant uptake of carbon dioxide, with a maximum value of 40 gCO_2/kg steel slag, achieved after 60 minutes of the combined treatment with the alkali activated slag at a w/s ratio of 0.12 l/kg. Moreover, with regard to the effects of the tested treatments on granule particle size, it was observed that also the combined process with alkali activation showed to achieve a remarkable increase in particle size with respect to the starting material. Finally, regarding the environmental behaviour of the produced granules, an increase in the release of Cr and V was retrieved for the product of the granulation/carbonation treatment with alkali activation; nonetheless, the crushed granules attained after a reaction time of 30 min showed to still comply (apart from pH) with the Italian requirements for reuse for land reclamation.
Hence, the results of these preliminary combined tests indicate the potential feasibility of applying the proposed treatments on BOF slag to obtain a product with enhanced mechanical properties. However, the amount and type of binder used as well as the treatment time appear to be critical parameters for the mechanical properties and also for the environmental behaviour of the product.

References

1. Euroslag Statistics, 2012. http://www.euroslag.org/products/statistics/2012/ (accessed on March 2015).
2. Huijgen, W. J., Comans, R. N. 2006. Carbonation of steel slag for CO_2 sequestration: leaching of products and reaction mechanisms. *Environmental Science Technology*, 40, 2790-2796.

3. Lange, L. C., Hills, C. D., Poole, A. B. 1996. The influence of mix parameters and binder choice on the carbonation of cement solidified wastes. *Waste Management*, 16(8), 749-756.
4. Iveson, S.M., Litster, J.D., Hapgood, K., Ennis, B.J. 2001. Nucleation, growth and breakage phenomena in agitated wet granulation processes: a review. *Powder Technology*, 117 (1-2), 3-39.
5. Morone, M., Costa, G., Polettini, A., Pomi, R., Baciocchi, R. 2014. Valorization of steel slag by a combined carbonation and granulation treatment. *Minerals Engineering*, 59, 82-90.
6. Song, S., Sohn, D., Jennings, H. M., Mason, T. O. 2000. Hydration of alkali-activated ground granulated blast furnace slag. *Journal of Materials Science*, 35(1), 249-257.
7. Pacheco-Torgal, F., Castro-Gomes, J., Jalali, S. 2008. Alkali-activated binders: A review: Part 1. Historical background, terminology, reaction mechanisms and hydration products. *Construction and Building Materials*, 22(7), 1305-1314.
8. Iveson, S. M., Litster, J. D., Ennis, B. J. 1996. Fundamental studies of granule consolidation part 1: effects of bonder content and binder viscosity. *Powder Technology*, 88 (1), 15-20.
9. Wauters, P.A.L., van de Water, R., Litster, J.D., Meesters, G.M.H., Scarlett, B. 2002. Growth and compaction behaviour of copper concentrate granules in a rotating drum. *Powder Technology*, 124 (3), 230-237.
10. Baciocchi, R., Costa, G., Polettini, A., Pomi, R. 2009. Influence of particle size on the carbonation of stainless steel slag for CO_2 storage. *Energy Procedia*, 1(1), 4859-4866.
11. Baciocchi, R., Costa, G., Di Bartolomeo, E., Polettini, A., Pomi, R. 2010. Carbonation of stainless steel slag as a process for CO_2 storage and slag valorization. *Waste and Biomass Valorization*, 1 (4), 467-477.
12. Baciocchi, R., Costa, G., Di Gianfilippo, M., Polettini, A., Pomi, R., Stramazzo, A. 2015. Thin-film versus slurry-phase carbonation of steel slag: CO_2 uptake and effects on mineralogy. *Journal of Hazardous Materials*, 283, 302-313.
13. British Standard, 1990. BS 812 Testing aggregates – part 110: Methods for determination of aggregate crushing value (ACV). ISBN 0 580 18825 6.
14. Italian Ministerial Decree 186/2006. Regolamento recante modifiche al decreto ministeriale 5 febbraio 1998 Individuazione dei rifiuti non pericolosi sottoposti alle procedure semplificate di recupero, ai sensi degli articoli 31 e 33 del decreto legislativo 5 febbraio 1997, n. 22. Gazzetta Ufficiale 19 maggio 2006, n. 115) (in italian).
15. Geiseler, J. 1996. Use of steelworks slag in Europe. *Waste management*, 16(1), 59-63.
16. Baciocchi, R., Costa, G., Polettini, A., Pomi, R. 2015. Effects of thin-film accelerated carbonation on steel slag leaching. *Journal of Hazardous Materials*, 286, 369-378.

FACTORS THAT INFLUENCE THE STRENGTH DEVELOPMENT DURING CARBONATION OF MINERAL WASTE COMPACTS

M. Quaghebeur[1]*, P. Nielsen[1], L. Horckmans[1] and Dirk Van Mechelen[2]

[1] VITO, 2400 Mol, Belgium

[2] RECMIX bvba, 3600 Genk, Belgium

*Corresponding author: mieke.quaghebeur@vito.be

Abstract

The carbstone process is developed for the valorisation of non-hydraulic slags and ashes into high quality construction materials. The process makes use of accelerated carbonation by treating various kinds of slags with CO_2 at elevated pressure (5-20 bar) and temperature (up to 140°C) without addition of binders. The carbonates that are produced in-situ during the carbonation reaction act as a binder cementing the slag particles together. The carbonated materials can compete with concrete products (C35/C45) and have excellent environmental and technical properties. The materials have the additional advantage of sequestering 180-200 g CO_2/kg slag. The technology was developed on lab scale by optimisation of the process parameters with regard to the CO_2 uptake rate and the properties of the building materials (chemical, mechanical and leaching). In this paper results will be presented on the factors that influence the strength development during carbonation.

Introduction

Approximately 500 Mton of iron and steel slags were generated worldwide in 2013[1]. The recycling of these slags provides a number of environmental benefits, including the preservation of natural resources (limestone, ...), the recuperation of valuable metals and a reduction in the volume of solid waste. In Europe about 76% of the steel slags were recycled in 2010. The primary application of the iron slag (granulated blast furnace slag) is the use as cement substitute. The recycling of steel slags (Stainless steel slag, Basic Oxygen steel slag) is not as well developed. The main application for these slags is as aggregate in construction applications, e.g. as granulate in asphalt and waterway embankment material. The use of these materials is however, often, hampered by the volumetric instability of the slag aggregate. In Belgium about 0.5 Mton of stainless steel slags are processed each year to recuperate the metal[2]. These steel slags contain on average around 5% of stainless steel. To recuperate the metal the slags are crushed or grinded. After the metal recovery operation about 70% of the slag can be recycled as coarse or fine aggregate, however 25% of the slags are very fine grained (< 0.5 mm) and difficult to reuse due to elevated leaching of chromium, molybdenum and vanadium. Solutions to reduce the volumetric instability and improve the environmental quality of the slags are therefore necessary.

Accelerated carbonation can be used to improve the valorization or reuse of certain waste streams[3,4]. The challenge is here to convert the very fine steel slag aggregates to products with a positive economic value. To accomplish this both the technical and environmental properties of the carbonated products need to comply with stringent product standards and environmental legislation. The carbstone process is developed for the valorization of non-hydraulic slags and ashes into high quality construction materials[5]. The process can hence provide a solution for the valorization of steel slags that are difficult to reuse today. One of the innovations of the technologies is that it combined slag valorization with beneficial use of CO_2. The process makes use of accelerated carbonation by treating various kinds of slags with CO_2 at elevated pressure and temperature. The carbonates that are produced in-situ during the carbonation reaction act as a binder cementing the slag particles together. The carbstone blocks can therefore be produced without the use of traditional binders like for example cement but have the advantage of sequestering up to 200 g CO_2/kg slag. The technology was developed on lab scale by optimization of the process parameters such as pressure, temperature, characteristics and preparation of raw material. Both the technical and environmental properties of the building materials were tested and optimized to be able to comply with the stringent environmental and technical requirements for building products.

Testing of the process using (semi)-industrial process conditions and equipment is however needed to validate the technology. The aim of this study was therefore to investigate:

•The influence of pressure and purity of the CO_2-gas stream on the efficiency of the carbonation process;

• The factors that influence the strength development of the building blocks;

• The technical and environmental quality (leaching behavior) of large building blocks produced via carbonation;

Materials and Methods

Raw Materials

The stainless steel slags used in this study was a fine slag fraction (< 500 μm) produced by the company Recmix (located in Genk, Belgium), which results from the treatment of stainless steel slag to recover the metal. These fine slag particles were used without further treatment. The basic oxygen steel slags (fraction 0-3 mm) were provided by Arcelor Mittal Gent (Belgium). The basic oxygen slags were milled (< 125 μm) before being used in the carbonation process. The chemical composition of the slags was examined using X-ray fluorescence (XRF) or ICP-AES after digestion of the samples with HF/HNO3/HCl. The mineralogy was determined using X-ray diffraction (Philips X'Pert, CuKα, 40kV).

Carbonation process

The carbonation process developed to produce the carbonated building materials consists of 3 main steps: (1) pre-treatment of the slags, (2) shaping of the building blocks by compaction, and (3) curing of the building block by CO_2. A schematic overview of the process is given in Figure 1. The first step (pre-treatment) involves the homogenization of the slag particles and mixing with the appropriate amount of water. For the results presented here, the moisture content of the non-carbonated material was 12 weight%. During the second step, the slag mixture was compacted using a hydraulic press at a compaction pressure between 150-300 kg/cm^2. The compacted samples were subsequently placed in an autoclave and carbonated for up to 16 hours in a CO_2 enriched atmosphere at a pressure between 5 to 75 bar, and a temperature of 140°C. The CO_2 concentration varied between 20% and 100%, the remaining gas was N_2.

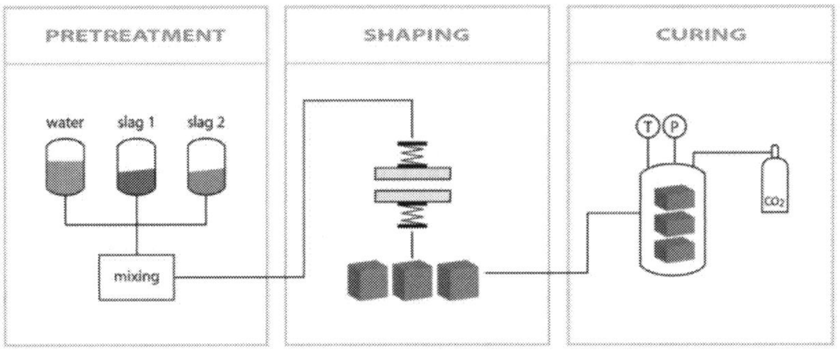

Figure 1. Schematic drawing of the carbonation process on lab scale.

Lab scale carbonation experiments were performed in automated carbonation units consisting of a 3,25 10-3 m³ or 104 10-3 m³ stainless steel Premex autoclave. The autoclave can be heated or cooled using a heating/cooling jacket (3000 W) controlled by a set of Eurotherm control units. The amount of CO2 delivered to the autoclave during a carbonation experiment was measured. The various process parameters were controlled and monitored using an Eurotherm Suite SCADA system.

Characterization of feed and product materials

The carbonate content of the fresh and carbonated samples was determined by measuring the evolved gas by infrared spectrometry (TC) during thermal decomposition of the sample and by measuring the total carbon content with a Ströhlein TOC-analyser C-mat 5500. In addition the amount of carbonate formed over time during a carbonation experiment (kinetics of CO_2 uptake) was automatically monitored in-situ during each carbonation experiment. A detailed description of the method to measure the in-situ carbonation kinetics is described by Quaghebeur et al.[6]. The leaching of fresh and carbonated material was determined using a batch leaching test (EN12457-4) (granular material < 10 mm, L/S 10, 24 hours contact time) and a static diffusion test (CMA2/II/A.9.2, NEN7375) (monolithic material, duration 64 days with renewal of solution). The technical properties of the large (250*100*100 mm) optimised carbonated blocks (16 hours, 20 bar, 140°C) were tested using standard procedures developed for concrete. Compressive strength (NBN EN 12390-3), young's modulus (NBN B15-203), tensile splitting strength (NBN EN 12390-6), shrinkage and swelling (NBN B15-216), frost and thaw resistance (NBN EN 1340) were all tested by a certified testing laboratory.

Results and discussion

Raw Materials

Both slags are rich in Ca (43 and 44% CaO in respectively BOS and SS slags) and have a high alkalinity (pH>12). High concentrations of Si and Mg are observed in the stainless steel slag, while the BOS steel slag is mainly composed of Fe and Si. Both slags are suitable for mineral carbonation. The calculated maximum theoretical CO_2 sequestration capacity is 420 g CO_2/kg for steel slag (when both Ca and Mg are considered) and 337 g CO_2/kg for basic oxygen steel slag (amount of CO_2 expressed on basis of non-carbonated steel slag).

The mineral phases of a representative fresh and non-carbonated steel slag sample was determined using XRD. The main minerals identified in stainless steel slag were Ca-Mg silicates, portlandite ($Ca(OH)_2$), periclase (MgO) and Mg-chromite. The basic oxygen steel slag consisted mainly of portlandite (Ca(OH)2), lime (CaO), various iron oxides (FeO, Fe_3O_4) and a small amount of Ca-Fe silicates and amorphous Ca-silicates.

Carbonate as binder – factors that influence the strength development during carbonation

The objective of the carbonation process is to produce construction materials using steel slags and CO_2 as raw materials without the addition of hydraulic or pozzolanic binders. Strength is developed by the precipitation of carbonate minerals that bind the slag particles together and fill the initial porosity. The lower the final porosity the higher the compressive strength of the compacts will be. The initial porosity (this is the porosity prior to the carbonation step) is a function of the grain size distribution of the slag material, and the compaction force applied during shaping of the compacts blocks. The compaction force should be high enough to develop the necessary green strength to handle the compacts, and to be able to fill most of the pores by carbonate minerals during carbonation. The porosity should however by high enough to allow for CO_2 and water diffusivity within the blocks during the carbonation step. The latter is dependent on the permeability and water content of the pores in the block. The diffusion of CO_2 through the compact is typically controlled by the open porosity of the compact and the water content in the pores. Sample preparation and process conditions should therefore be adapted to avoid water saturation of the pores. The effect of compaction force and grain size distribution can be demonstrated by the reaction kinetics of the carbonation reaction (Figure 2), but also by comparing the developed strength of the compacts in function of grain size distribution and porosity/permeability of the initial compacts (Figure 2). The results obtained show that the reaction rate is significantly affected by the compaction force used during shaping of the compact. A comparable carbonate content (total CO_2 uptake) is achieved after 1000h of reaction for a compaction force of 75-389 kg/cm^2. At high compaction force (609 kg/cm^2) the total carbonate content was significantly lower after 1000h of reaction which resulted in a non-carbonated core. The reaction rate is comparable for the low compaction forces (75-150 kg/cm^2) while at elevated compaction forces (389-609 kg/cm^2) the uptake becomes slower. High compaction results in low porosity which has a significant effect on the CO_2 diffusion through the compact. When compacts are, however, fully carbonated a clear correlation could be observed between the density (or porosity) of the compact before carbonation and the final compressive strength (Fig. 2 right). Both grain size distribution and compaction force can be used to influence the density of the compact before carbonation[7].

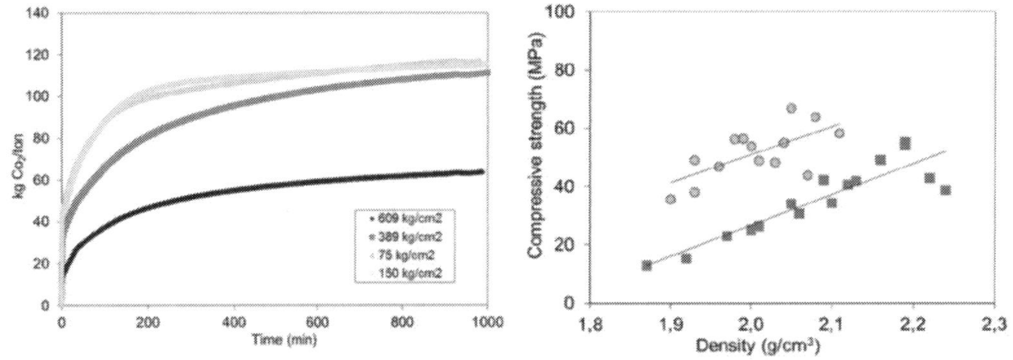

Figure 2. CO_2-uptake versus time for a mixture composed of stainless steel slag compacted at different compaction forces (75 – 609 kg/cm^2) carbonated at 20 bar with 100% CO_2 (left). Compressive strength versus the density of the compacts before carbonation. Different type of steel slags, compaction forces and carbonation conditions were used to produce the building blocks (right).

Carbonation process and influence of pressure and purity of CO_2 gas stream

The objective of the carbonation process is to produce construction materials using steel slags and CO_2 as raw materials without the addition of hydraulic or pozzolanic binders. The carbonates that are produced in-situ during the carbonation reaction act as a binder cementing the slag particles together. Higher CO_2-pressure applied during the carbonation process resulted in a higher amount of carbonates and a higher compressive strength in the carbonated material. Elevated carbonation pressure is therefore advantageous for the strength development of the carbonated materials. Elevated pressure however also results in higher production and investments costs making the process less attractive from an economic point of view. A moderate pressure of 20 bar was therefore chosen as a standard pressure for further development of the carbonation process (Figure 3). Initial process development was done with a gas consisting of 100% CO_2. Many companies are interested in finding a solution for diluted streams of CO_2, since flue gasses generally contain low concentrations of CO_2 and caption of CO_2 is expensive and not yet integrated in existing processes. The impact of the purity of the carbonation gas (with regard to CO_2) in the carbonation gas on the uptake rate and strength development was therefore investigated. The results showed that the uptake rate is lower when less concentrated CO_2 (20%, 50%, 77%) is used (Figure 3). Longer process times are therefore needed when diluted CO_2 is used compared to 100% CO_2 to obtain Carbstone blocks of comparable strength.

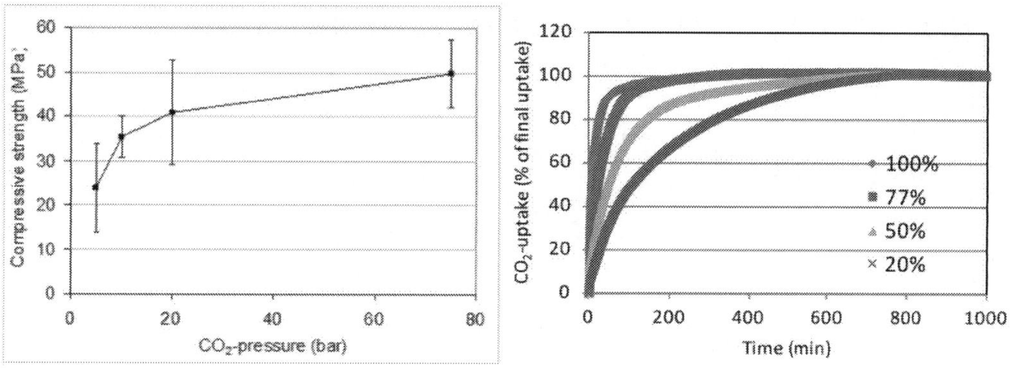

Figure 3. Compressive strength versus CO_2-pressure during the carbonation of stainless steel slags at different pressures (16 hours, 140°C, 100% CO_2)(left) and the CO_2-uptake versus time for a small (minimum dimension of 3.5 cm) compact composed of basic oxygen steel slags carbonated at 20 bar with 20%, 50%, 77% and 100% CO_2. The final CO_2-uptake was approximately 150 kg CO_2/ton.

Environmental and technical quality of the carbonated materials

To evaluate whether the carbonated material is suitable as a construction material a technical characterization of the carbonated blocks was done. An overview of the technical properties for carbonated stainless steel slags (16 hours, 20 bar, 140°C) is given in Table 1. Each characteristic is the average of at least 3 measurements made on different samples. The process can be geared to make construction materials with different technical qualifications. A compressive strength of more than 100 MPa can be obtained. In addition the environmental quality of the fresh carbonated material was evaluated using a maximum availability test[8] and a diffusion test for monolithic material[9]. Critical parameters in stainless steel slag are chromium, molybdenum and vanadium[10]. Carbonation of the stainless steel slags (16 hours, 20 bar, 140°C) resulted in a reduction of the leaching of chromium and molybdenum. To assess the environmental quality with regard to recycling of minerals wastes in Belgium[11] and the Netherlands[12], the leaching properties of the carbonated blocks were tested using a diffusion test [9]. The carbonated blocks made from the basic oxygen slag and stainless steel slags (16 hours, 20 bar, 140°C) complied both to the limit values of the materials decree[11] for reuse of waste in the Flemish region of Belgium and the soil quality decree of the Netherlands[12]. The leaching values are given in Table 2 and compared to the Dutch limit values for monolithic building materials.

Table 1. Overview of the technical properties of compacted and carbonated stainless steel slags (16 hours, 20 bar pCO_2, 140°C), compared to requirements for concrete paving blocks.

	results	requirements[1]
Compressive strength - NBN EN 772-1:2011	134±8 MPa (n=3)	
Frost and thaw resistance - NBN EN 1338 (Annex D)	0.01±0.01 kg/m²	≤ 1.0 kg/m²
Water absorption - NBN EN 1338 (Annex E)	6.9±0.1 weight %	-
Tensile splitting strength NBN EN 1338 (Annex F)	7.1±1.5 MPa (n=3)	> 3.6 Mpa
Abrasion resistance - NBN EN 1338 (Annex G)	21.5±0.8 mm (n=3)	≤ 23 mm
Slip resistance - NBN EN 1338 (Annex I)	88±2 (n=3)	≥ 45
Apparent density (kg/m³)	2370±20 kg/m³	
Apparent density prior to carbonation (kg/m³)	2045±15 kg/m³	

Table 2. Leaching values of the carbonated stainless steel slag compacts (maximum availability test, static diffusion test) compared to the Dutch leaching limit values for monolithic construction materials.

	Maximum Availability (mg/kg)	Diffusion test (E_{64d} in mg/m²)		Dutch limit values (E_{64d} in mg/m²)
As	1.0	< 2.5	< 2.7	260
Cd	0.73	< 0.26	< 0.26	3.8
Cr	110	1.6 - 2.4	2.1 - 2.6	120
Cu	1.2	< 2.5	< 2.5	98
Hg	-	< 0.1	<0.1	1.4
Ni	47	< 2.7	< 3.0	81
Zn	11	< 5.4	< 5.3	800
Sb	< 0.1	<1.2	< 0.86	8.7
Ba	83	<5.2	< 5.0	1500
Mo	10	13	13 - 14	144
Se	0.25	< 1.5	< 0.85	4.8
V	6.2	2.1 - 3.8	3.3 - 4.3	320
Fluoride	6300	480	470	2500
Chloride	370	2500	2400	110000
Sulphate	3600	540-740	570-840	165000

[1] Important performance measures specified in codes of practices in the UK

Conclusions

A carbonation process has been developed for the production of construction materials using slags and CO_2 as raw materials without the need for any additional binder. The process works at moderate pressure (20 bar) and temperature (140°C) and makes use of a unique combination of existing technologies commonly used in the construction industry (concrete and sand lime brick plants). The carbonated blocks show good environmental and technical properties making them suitable to be used as high quality construction materials. Understanding the factors that influence the strength development of the materials during carbonation is needed to have a flexible process with regard to product design. The composition of raw materials, grain size distribution, shaping of materials using different compaction pressures and pCO_2 during carbonation all have a large influence on both the carbonation rate and strength development of carbonated materials. The purity of the CO_2 gas streams used has little effect on the final strength obtained but the uptake rate will be lower when less concentrated CO_2 is used.

Acknowledgements

The authors like to acknowledge financial support from IWT grant 100621 'CarbMAX: Production of high-strength materials with mineral residues and CO_2'

References

1. U.S. Geological Survey (2014) Mineral commodity summaries 2014 (196 p.).
2. Van Mechelen, D., Nguyen, E., & Rijskens, E. (2011) Valorisation of stainless steel slag: zero waste concept. In: *Proc. of the 2nd Intern. Slag Valorization Symp.* (pp. 145), Leuven, Belgium.
3. Nguyen, E., Lecomte, I., & Van Mechelen, D. (2009) New sustainable valorisation routes for fine grained stainless steel slags. In: *Proc. of the Global Slag Conf.* (Paper 9), Brussels, Belgium.
4. Gunning, P.J., Hills, C.D. & Carey, P.J. (2009) Production of lightweight aggregate from industrial waste and carbon dioxide. *Waste Management*, 29, 2722-2728.
5. Patent WO2009/133120 (2009) Production of a mainly carbonate bonded article by carbonation of alkaline materials.
6. Quaghebeur, M., Nielsen, P., & Laenen, B. (2009) Accelerated carbonation to improve quality of recycling methods: a study of the real time kinetics. In: *Proc. of the 7th Int. Waste conference (WASCON)* (pp. 14-15), Lyon, France.
7. Bertos, M.F., Simons, S.J.R., Hills, C.D., & Carey, P.J. (2004) A review of accelerated carbonation technology in the treatment of cement-based materials and sequestration of CO_2. *Journal of hazardous Materials*, B112, 193-205.
8. NEN 7341 (1999) Leaching characteristics of solid earthy and stony building and waste materials - Leaching tests - Determination of the availability of inorganic components for leaching.

9. NEN 7375 (2004) Leaching characteristics - Determination of the leaching of inorganic components from shaped and monolithic material with the diffusion test - Solid earthy and stony materials.

10. Huijgen, W. J. J., Comans, R. N. J. (2006). Carbonation of steel slag for CO_2 sequestration: Leaching of products and reaction mechanisms. *Environ. Sci. Technol.*, 40 (8), 2790–2796.

11. Regulation on Sustainable Management of Material Cycles and Waste Materials (VLAREMA), Decree (2012) OVAM, Mechelen, Belgium (17.02.2012).

12. Soil Quality Decree of the Netherlands (Regeling bodemkwaliteit) (2007) In: *Staatscourant 20 december 2007*, nr. 247 (p. 67).

Evaluation of Activation Efficiency and Quality of Slag Carbonation

F.Trapasso[1*], P.Plescia[1], E.Tempesta[1], G.N. Di Martino[1] and T.M.C. Grasso[1]

[1] Institute of Environmental Geology and Geoengineering (IGAG), National Research Council (CNR), Montelibretti, Italy

*Corresponding author: francesca.trapasso@igag.cnr.it

Abstract

The capacity of some industrial waste to sequester carbon dioxide is well known and it is also noted that to achieve the carbonation reaction it is necessary to activate the precursors overcoming the initial energy barrier through processes of dissolution in an autoclave at high temperature and pressure. To develop the direct carbonation, without autoclaving, our research group uses the mechanical activation of precursor minerals realized through an intense grinding made with friction mill (ring mill). We can estimate the timing of activation of mineral phases using a calculation model realized by us. Knowing the mechanical parameters of the grinding machines, the thermodynamic parameters of the material and the amount of radicalic sites obtained by grinding, we can evaluate the rate of carbonation for each type of mineral. The results show a good correlation between the activation level of the materials and the efficiency of carbon dioxide fixation: at ambient temperature and pressure a yield of up to 98% can be achieved.

Introduction

Mineral carbonation reactions are thermodynamically favorable[1,2]. They begin spontaneously, but generally are very slow: the speed of the carbonation reaction depends on the intermediate reactions which can be unfavourable at atmospheric pressure and temperature. Usually to reduce the reaction time some parameters are modified: the pressure, the temperature, the pH of the solutions, the granulometry of the precursors, the composition of the additives, etc. The formation of precursors by thermal treatments involves calcination to form oxides in hydrothermal autoclave reactor at high pressure and temperature: it is possible to obtain up to 78% efficiency of conversion from silicates to carbonates in 30 minutes[1,2] with the addition of carbonic acid, bicarbonate ions or carbonate ions. Unfortunately this process is very expensive[2].

In the first decade of the new millennium the Department of Energy of the United States embarked on the creation of reactors capable of capturing 90% of the emissions at a cost of fixing that impacted less than 10% on energy costs[3]. The intermediate target place by the DOE was the creation of commercial systems of indirect fixing of CO_2 with a maximum cost of 10$ per ton of carbon. To obtain this result, different solutions have been proposed: one of the most promising regards the use of the mechanochemical activation of minerals[4]. Nelson[5] used a shaking mill to grind for 20 minutes samples of serpentine lizardite in the presence of CO_2 as a dry process gas. Through X-ray diffraction the Author reveals the presence of carbonates in the milled mixture, but a very limited degree, well below which could be obtained by the hydrothermal methods[5]. In our opinion the grinding with simultaneous CO_2 fixation is not energetically convenient, because the mechanochemical action tends to calcine the carbonates and therefore not to encourage the formation of these[6,7]. Haug[8] shows how the direct fixing of CO_2 on olivine increases three times after a process of mechanical activation: the rate of formation of magnesite depends on the speed of dissolution of the precursor (magnesium oxide) in the solution. This speed can be increased acting on the solubility and saturation of the solution.

The effects of mechanical activation are visible through carbonation yields or, in an indirect way, through the X-ray diffraction analysis with which it is observed the crystallinity of the ground precursor. It would be useful to have a direct quantitative method to determine the activation of the precursor and so establish the yield of fixation. For this purpose we undertook a series of studies to determine the correlation between the level of mechanical activation and the percentage of CO_2 fixed on the mineral.

Assuming that for each active site there is one radical, the radicals activity measurement provides an estimate of the number of active sites present in the sample, which is referred to the specific surface of the sample and/or to the weight. Assuming also that the number of active sites of the mineral is comparable to the number of dislocations, we can derive the density of dislocations directly from the measurement of the radicals. From the measurement of the number of defects is determined the correlation with the solubility of the material and so it is possible to "design" the mechanical activations. To realize this measurement, we used a molecule radical scavengers, the compound DPPH (2,2-diphenyl-1-picrylhydrazyl), which binds 1:1 ratio to radicalic sites of the material[9,10]. The DPPH has a particular colour in alcoholic solution, with a main peak at 520 nm: by the reaction of the alcoholic solution of DPPH on the mineral, there is a reduction of the coloration, in particular of the intensity of absorption peaks at 520 nm, in proportion to the number of achieved and annihilated radicals. The reaction is superficial, thus the number of measured radicals refers to the exposed surface of the material. The result can be expressed in sites per square meter or sites on gram of material. Once the relationship between the number of active sites and the leaching of the mineral is determined, it is possible to predict the right time and the intensity of activation.

Methods

The choice of materials for the laboratory tests was performed on the basis of the times of mechanical activation and the presence of alkali metals and alkaline-earth metals. Such materials are MS (secondary metallurgy) blast furnace clear slag and LD (Linz-Donawitz converter) dark slag: these materials, after the chemical analysis, were subjected to a sifting to separate the fraction between 300 and 800 microns and to a chemical analysis in X-ray fluorescence (Bruker S4 Pioneer) to determine the maximum degree of carbonation compatible with the chemical composition.

The mechanical activation was performed using a ring mill with controlled power (Retsch R100) and equipped with temperature probes on the edge of the grinding jar to allow the measurement of the heat produced by the grinding process. The calculation of grinding time (Eq. 1) was carried out adopting an empirical method obtained by the authors of this work and based on the ratio between the enthalpy of formation of the precursor mineral phases and the effective energy ΔE_{ef} in the unit time given by the difference between the total energy delivered from the mill and the energy transferred in the form of heat (Eq. 2):

$$\text{Time (min)} = \Delta H_0 \text{ (KJ)} / \Delta E_{ef} \text{ (KJ/min)} \quad (1)$$

$$\Delta E_{ef} \text{ (KJ)} = \Delta E_t \text{ (KJ)} - \Delta E_h \text{ (KJ)} \quad (2)$$

This simple relation provides a basis of calculation to obtain a level of crystallinity well defined and, in an approximate but effective way, the amount of time needed to reach the complete amorphization of the material. The indication of the attainment of amorphization is the absence of discrete lines in the diffraction spectrum (Figure 1 and Figure 2).

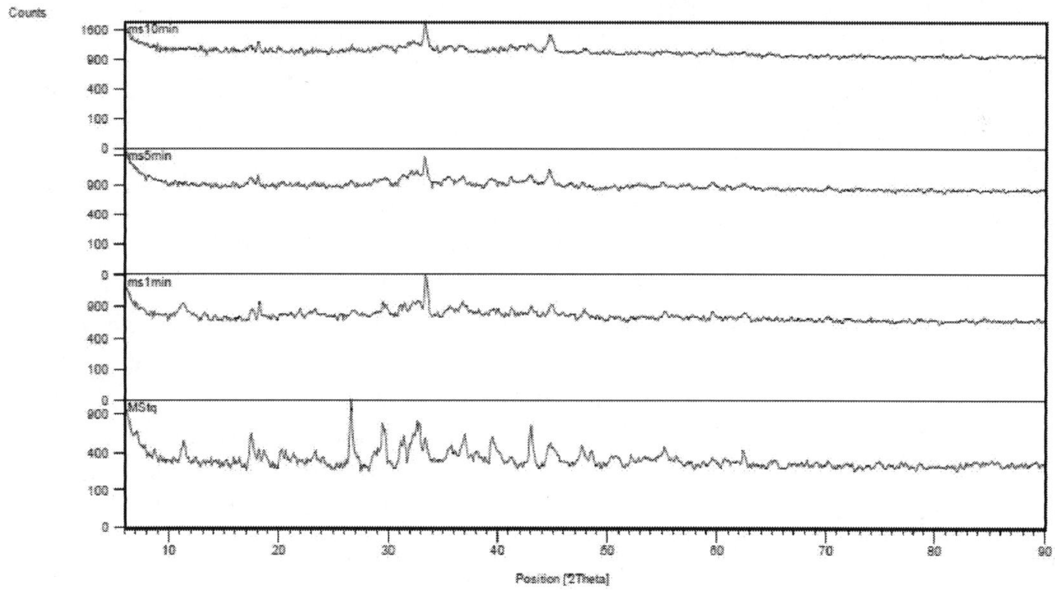

Figure 1. X-ray diffraction patterns of MS slag before and after grinding

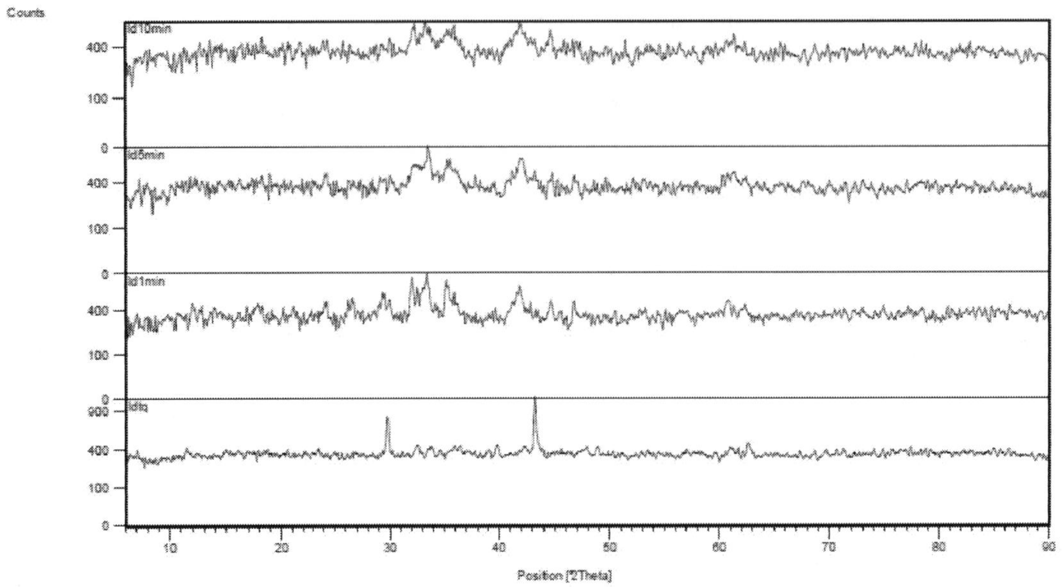

Figure 2. X-ray diffraction patterns of LD slag before and after grinding

The materials were ground for 1, 5 and 10 minutes in a jar of steel coated with tungsten carbide and M/m ratio between the grinding mass (M) and the mill charge mass (m, mass of the material to be ground) equal to 400:1. The needed time for complete amorphization is approximately 24 minutes for the MS slag and 20 minutes for the LD slag for the same amount of energy consumed in the mill. Each material was again subjected to X-ray diffraction analysis in order to evaluate the achieved crystallinity and the final mineralogical composition (by Rietveld analysis).

To complete the description of the characteristics resulting from grinding, the analyses of specific surface area (BET) were performed on the precursor materials, as such and ground, by a nitrogen porosimeter (Micromeritics ASAP 2020).

The ground materials were subjected to quantitative analysis of the radicalic sites, expressed in sites/m^2. This determination was repeated for three consecutive days to determine if the activation decreases over time and in what terms. The analysis of the activation was performed by the method of titration with the compound DPPH (2,2-diphenyl-1-picrylhydrazyl). The chosen concentration of the DPPH for the analysis is 0.01 mM which corresponds to 6.02 x 10^{18} molecules. The amount of solution used is 2.5 ml for 0.1 g of sample. Each ground sample was weighed and imbibed in the solution for one hour while stirring. Then the liquid was separated by centrifugation (8000 RPM, 15 minutes) and analyzed using UV-VIS spectrophotometer (Thermo Evolution 60 Spectrophotometer) in the range 400-700 nm using quartz cuvettes.

The powders were tested for total alkalinity to determine if the materials increase their extractive capacity when they are ground. This was achieved using six Mettler T50 titrators connected by LabX software.

Afterwards the materials were tested for CO_2 fixation for 120 minutes in a beaker in aqueous solution at solid/liquid ratio equal to 1:10 and CO_2 bubbling with a flow of 1 l/min. Then the carbonated samples were subjected to X-ray diffraction and differential thermal analysis (DuPont SDT Q600) to evaluate the content of fixed CO_2.

Results

The Table 1 and Table 2 show the slag mineralogical composition, determined through X-ray diffraction analysis and Rietveld evaluation, and enthalpy of formations[11].

Table 1. MS slag mineral phases analysis

Mineral phase	Enthalpy of formation ΔH_0(kJ/mol)	Molecular weight (g/mol)	Weight (%)			
			Unground	Grinding for 1'	Grinding for 5'	Grinding for 10'
Quartz	- 910	60	0.36	2.86	1.28	1.01
Gehlenite	- 3985	274	15.91	4.51	11.04	19.70
Periclase	- 602	40.3	0.22	0.17	0.04	0.04
Calcite	- 1208	100.1	4.68	1.76	0.57	1.43
Wollastonite-2M	-1635	116.2	45.43	44.17	30.05	14.20
Perovskite	- 1445	100.4	4.61	1.21	0.53	5.50
Magnesioferrite	-1441	199.9	0.86	0.44	0.31	0.17
Glass		450	28	45	56	58
Total			100.07	100.11	99.82	100.04

Table 2. LD slag mineral phases analysis

Mineral phase	Enthalpy of formation ΔH_0(kJ/mol)	Molecular weight (g/mol)	Weight (%)			
			Unground	Grinding for 1'	Grinding for 5'	Grinding for 10'
Troilite	-101	87.9	39.04	10.35	9.70	8.26
Calcite	- 1208	100.1	16.84	6.64	3.05	1.72
Magnetite	-1441	199.9	3.51	2.44	0.97	0.56
Magnesioferrite	-1635	116.2	4.16	1.36	0.24	0.47
Wollastonite-2M	-1441	199.9	1.20	0.84	0.31	0.12
Glass		450	35	78	86	89
Total			99.75	99.63	100.09	99.8

Figure 3 and Figure 4 show the trend of MS and LD slag mineral phases with increasing grinding time. The amount of glassy material produced by the amorphization is higher for the LD slag than for the MS slag considering the same grinding time. The periclase is present in small quantities in the MS slag: it is not shown in Figure 3 to make the graph more legible.

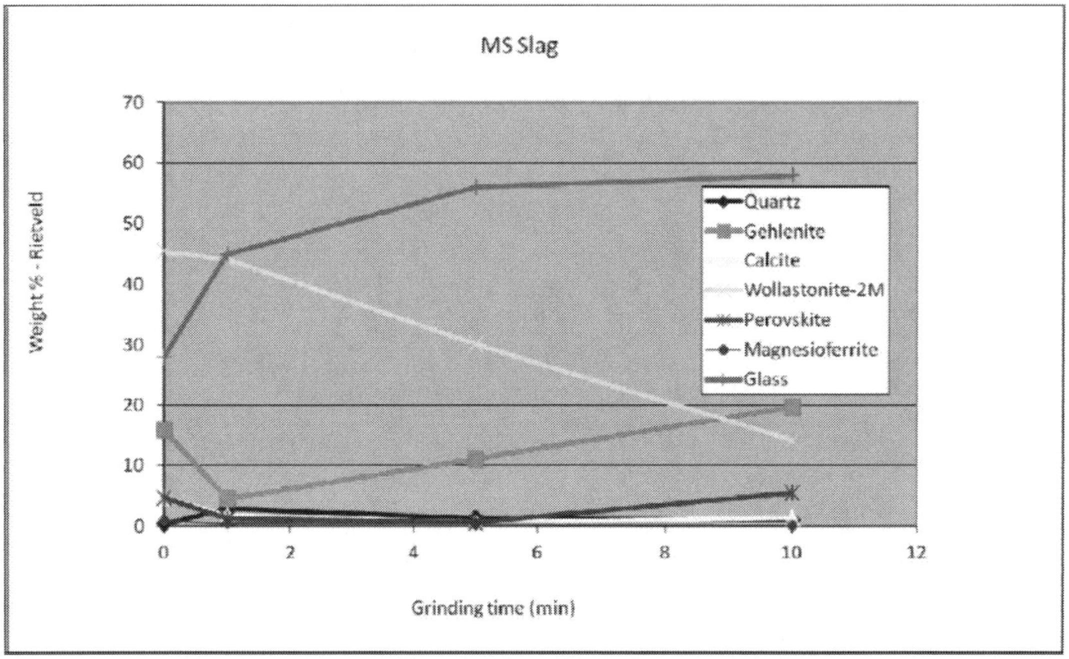

Figure 3. Rietveld evaluation of MS slag mineral phases

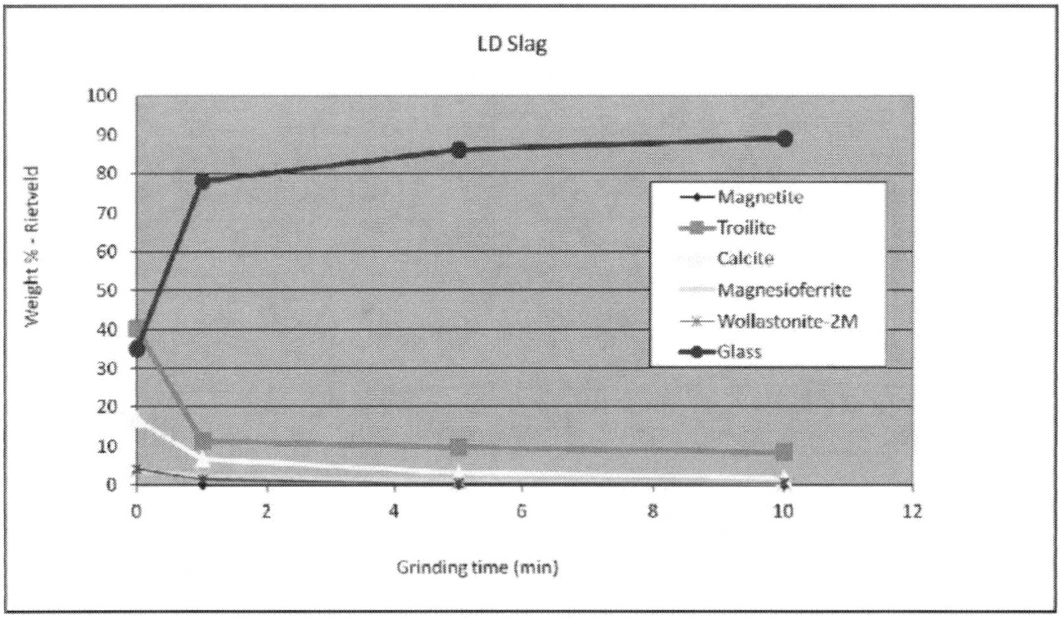

Figure 4. Rietveld evaluation of LD slag mineral phases

The mineral phases that make available the alkalis in the MS slag are 71.2% of the mineral mass (calcite, wollastonite, periclase, perovskite, gehlenite and magnesioferrite), while in the LD slag they are about 24.5% (calcite, wollastonite and magnesium ferrite). However it is possible that alkalis are contained in the glassy fraction.

Table 3 shows the results of X-ray fluorescence quantitative analysis of the slag. The potential amount of alkalis available in the MS slag is approximately 60% (55.59% CaO and MgO 4:26%), while in the LD slag is about 53% (CaO 52.21% and MgO 0.90%).

Table3. Chemical composition of slag (XRF quantitative analysis)

Oxyde	MS	LD
Na_2O	0.33	0.00
MgO	4.26	0.90
Al_2O_3	13.41	1.08
SiO_2	10.96	6.16
CaO	55.59	52.21
K_2O	0.13	0.03
Fe_2O_3	9.47	31.34
MnO	1.31	5.51
TiO_2	0.80	0.36
P_2O_5	0.58	0.83
SO_3	2.54	0.71
Cl	0.44	0.22
Cr_2O_3	0.11	0.46
SrO	0.05	0.03
CuO	0.04	0.00
ZnO	0.00	0.00
NiO	0.00	0.00
V_2O_5	0.00	0.16

Figure 5 shows the trend of the number of radicals generated by grinding with increasing grinding time.

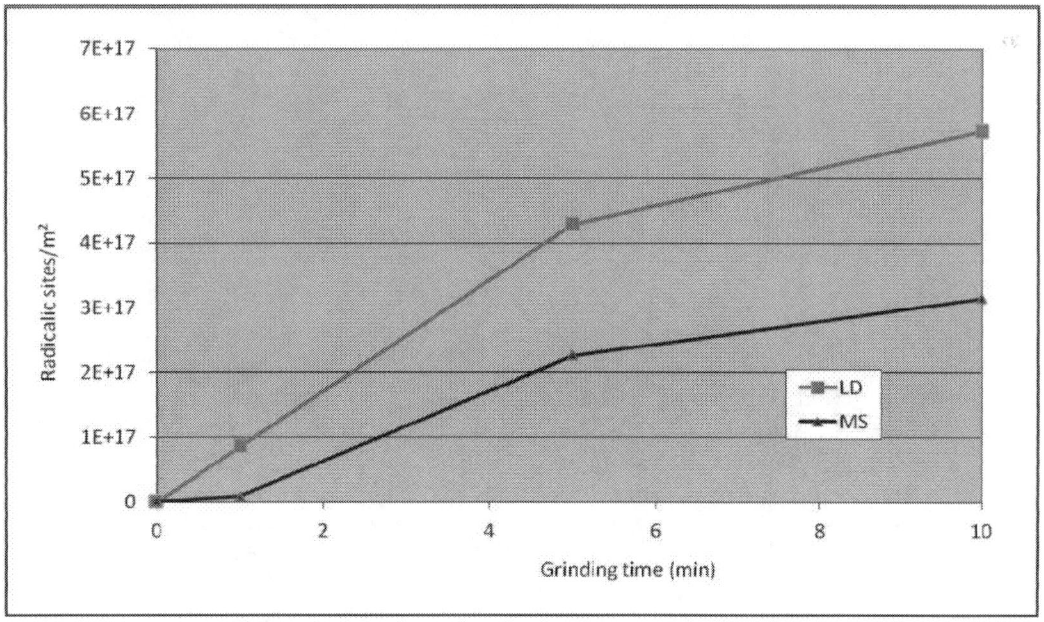

Figure 5. Number of radicalic sites vs. grinding time

Figure 6 shows the graphs of the number of radicals versus the time of ageing, that is time after the grinding. The values changes rapidly: it is observed a decrease of over 30% already 48 hours after the grinding.

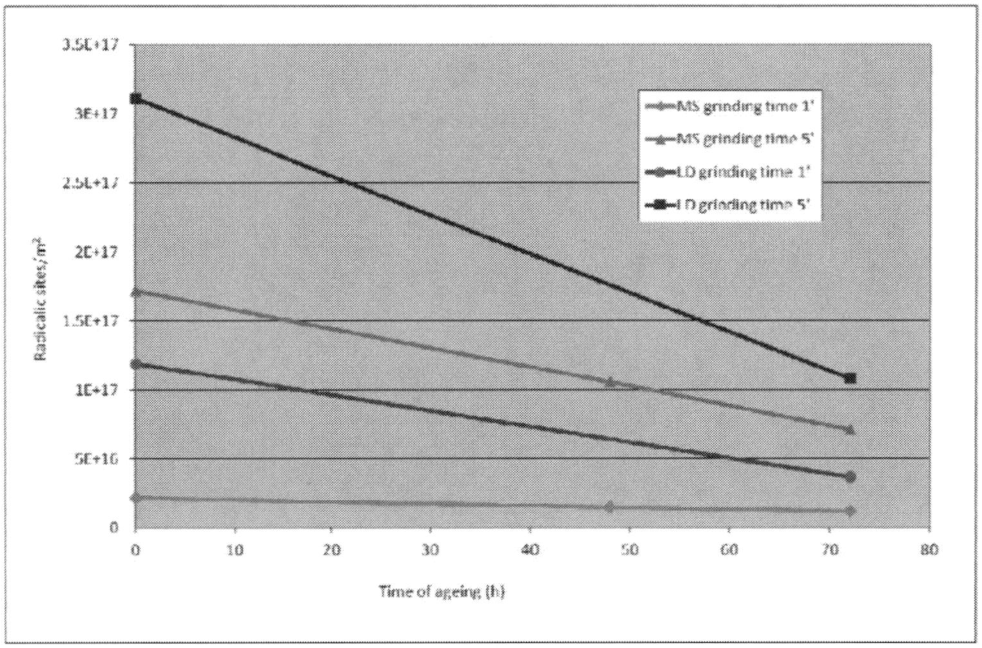

Figure 6. Number of radicalic sites vs. time after the grinding

The correlation between the number of available radicals and the percentage of fixed CO_2 is logarithmic type (Figure 7): the MS slag has a very similar trend to the LD slag. The correlation coefficients are high: it is evident a strong link between the number of radicalic sites and the amount of fixed CO_2.

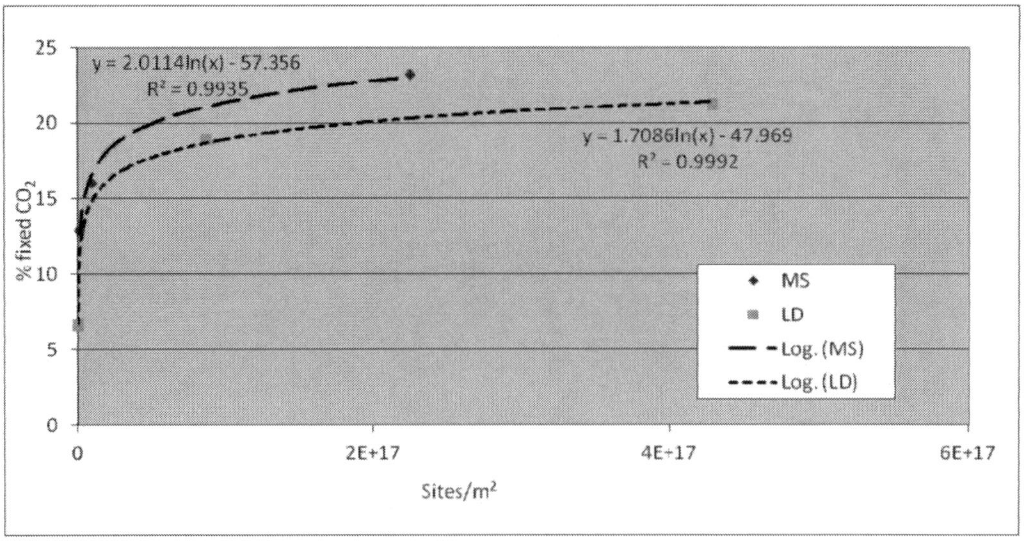

Figure 7. Number of radicalic sites vs. % of fixed CO_2

The analysis of the free alkalis in solution, which were carried out by titration with HCl (0.1 and 1.1 M), show the relationship between the increase of the solubility and the grinding time (Figure 8): the first minute is essential for the activation of the material.

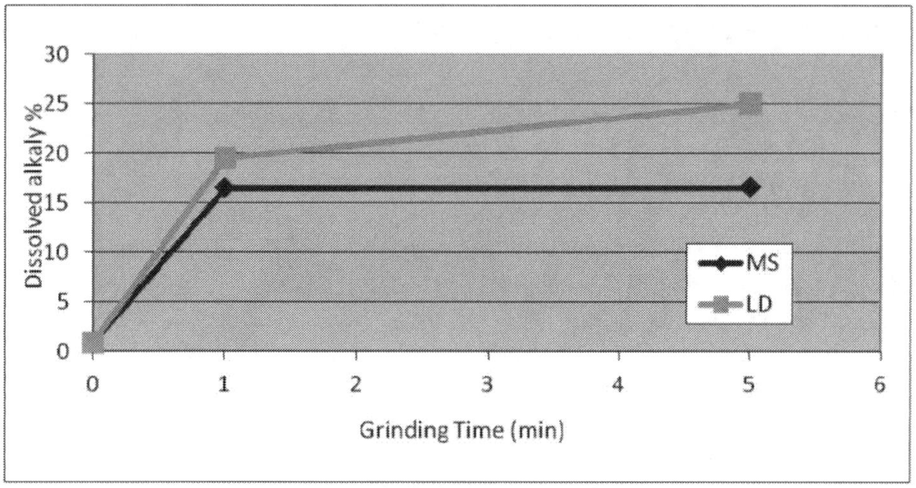

Figure 8. Total alkalinity vs. grinding time

The percentages of CO_2 fixation for the materials, non-ground and ground, subjected to carbonation for 120 minutes were calculated by thermogravimetric analysis (Table 4). The yield of fixation is expressed respect to the theoretical value calculated from the X-ray fluorescence analysis.

. **Table 4.** Slag rate of carbonation

Slag	Grinding time (min)	% fixed CO_2	% carbonate	% of max CO_2 fixation
MS carbonated 120 min	0	12.9	21.5	32.3
	1	21.8	36.4	54.5
	5	23.2	38.7	58.0
LD carbonated 120 min	0	6.6	11.1	18.6
	1	18.9	31.5	53.4
	5	21.2	35.4	76.7

Conclusions

The results demonstrate the possibility of achieving high levels of carbonation without the use of high pressure and temperature treatments, but by simply mechanically activating the materials and controlling the level of activation using a quantitative method. In particular the study shows that it is possible to use the mechanical activation to increase the efficiency of the reactions of CO_2 fixation at atmospheric pressure and temperature up to values which are reached using high pressures and temperatures: the mechanical activation leads to a considerable increase of radicalic active sites, which results in increased efficiency of carbonation.

It is possible to quantify the activation process using a scavenger molecule (DPPH - 2,2-diphenyl-1-picrylhydrazyl): the number of radicalic sites activated by grinding is not stable, but it tends to decrease quickly, probably a result of the neutralization of the active sites.

The protracted grinding leads to a decrease of crystallinity until obtaining the amorphism: these phenomena were modeled[11], together with the increase in the leaching of the minerals, according to the degree of amorphization or the percentage of residual crystallinity. The data we obtained indicate a maximum number of radicalic sites equal to 3.2 E^{17}/m^2, corresponding to the value of defects identified with the equations of Tromans and Mech[11] for prolonged grinding: this suggests that the radicalic sites and the defects induced by grinding are the same thing, at least in the early stages of grinding. The energy due to deformation of the crystal structure during the grinding is stored in the form of the Gibbs free energy and the energy of amorphization[11] that are gradually released increasing the leaching of the ground minerals.

References

1. O'Connor, W.K., Dahlin, D.C., Nilsen, D.N., Rush, G.E., Gerdemann, S.J., Walters, R.P. and Turner, P.C. 2001. Carbon dioxide sequestration: Aqueous mineral carbonation studies using olivine and serpentine. Albany, OR: Albany Research Center.
2. O'Connor, W.K., Dahlin, D.C., Rush, G.E., Dahlin, C.L. and Collins, W.K. 2002. Carbon Dioxide Sequestration by Direct Mineral Carbonation: Process Mineralogy of Feed and Products, *Minerals & Metallurgical Processing*, 19(2), 95-101.
3. Anon. 2003. Carbon Sequestration Technology Roadmap and Program Plan. http://www.globalcarbonproject.org/global/pdf/CsequestrationRoadMapDOE2003.pdf
4. Anon. 2001. HSC Chemistry Chemical Reaction and Equilibrium Software Version 5. 1, Outokumpu Research Oy, Pori, Finland, A. Roine.
5. Nelson, M.G. 2004. Carbon Dioxide Sequestration by Mechanochemical Carbonation of Mineral Silicates, Final Report, University of Utah, Utah, USA.
6. Martinelli, G. and Plescia, P. 2004. Mechanochemical effects on calcite during shear stress, *Physics of The Earth and Planetary Interiors*, 142(5), 205–214.
7. Martinelli, G. and Plescia, P. 2005. Carbon dioxide and methane emissions from calcareous-marly rock under stress: experimental tests results, Annals of Geophysics, 48(1), 167-174.
8. Haug, T.A., Kleiv, R.A. and Munz, I.A. 2010. Investigating Dissolution Of Mechanically Activated Olivine For Carbonation Purposes, *Applied Geochemistry*, 25(10), 1547-1563.
9. Shimada, K., Fujikawa, K., Yahara, K., and Nakamura, T. 1992. Antioxidative properties of xanthan on the autoxidation of soybean oil in cyclodextrin emulsion. *Journal of Agricultural and Food Chemistry*, 40(6), 945-948.
10. Robie, R. A. and Hemingway, B. S. 1995. Thermodynamic Properties of Minerals and Related Substances at 298.15 K and 1 Bar (10^5 Pascals) Pressure and at Higher Temperatures. US Geological Survey Bulletin 2131.
11. Tromans, D. and Meech, J.A. 2001. Enhanced dissolution of minerals: stored energy, amorphism and mechanical activation. *Minerals Engineering*, 14(11), 1359- 1377.

The role of particle size distribution in CO_2 sequestration by BOF slag

A. Polettini*, R. Pomi and A. Stramazzo
Dipartimento di Ingegneria Civile, Edile e Ambientale, Università degli Studi di Roma "La Sapienza", Via Eudossiana 18 – 00184
*Corresponding author: alessandra.polettini@uniroma1.it

Abstract
The paper reports the results of slurry-phase accelerated carbonation experiments using basic oxygen surface steel slag as the source of alkalinity for CO_2 sequestration. In particular, the effect of particle size distribution was specifically investigated here. To this aim, different size classes of the slag were separately subjected to accelerated carbonation, and the CO_2 uptake performance, the mineralogical changes and the final particle size distribution were studied. While the total content of potentially reactive elements in the different size fractions of the slag exerted only a minor influence on the carbonation performance, particle size was found to play a relevant role during the process, and likely in particular during the dissolution stage. Changes of up to two orders of magnitude in CO_2 sequestration (which varied from 4.7 to 465 g of CO_2 sequestered per kg of slag) were observed for the different size classes investigated.

Introduction

Among the proposed solutions to reduce CO_2 emissions, mineral sequestration through accelerated carbonation is a promising alternative to geological storage. A wide variety of natural materials has been tested as the alkaline feedstock for mineral carbonation, as well as a number of residual streams from thermal processes (steel slag[1-13], alkaline ashes from combustion processes[13-15], municipal solid waste ashes[3,14], cement kiln dust[15,16] and cement-based materials[17-19]). Industrial residues have been shown to be more reactive towards CO_2 compared to natural silicates, and accelerated carbonation of alkaline waste materials can achieve high CO_2 sequestration yields under mild operating conditions. Despite accelerated carbonation of industrial residues has been the subject of numerous studies, the relevance of particle size on carbonation kinetics and yields has been mostly underrated so far. While it is commonly acknowledged that CO_2 uptake is inversely correlated with grain size (or directly correlated with the specific surface area) of the material, to the authors' knowledge no specific study has been conducted so far about the influence of particle size distribution of steel slag on the carbonation performance. With this study, we aimed at investigating the effect of particle size distribution on the CO_2 sequestration yield of basic oxygen furnace (BOF) steelmaking slag.

Materials and Methods

Slag samples, collected at the outlet of the basic oxygen furnace of an integrated steelmaking steel plant, were oven-dried at 60 °C to constant weight and separated through dry sieving into the following eight size classes: 850÷1000 μm (C1), 500÷850 μm (C2), 250÷500 μm (C3), 150÷250 μm (C4), 100÷150 μm (C5), 63÷100 μm (C6), 38÷63 μm (C7) and <38 μm (C8). The size distribution of the sieved materials was determined by laser particle size analysis. The characterization of each class also included measurements of the elemental composition, carbonate content and mineralogy, according to the analytical procedures described elsewhere[1-3,17]. Direct slurry-phase accelerated carbonation tests were performed in duplicate in a pressurized stainless steel reactor equipped with an external magnetic stirring device and a heating jacket. The experimental conditions, selected on the basis of previous investigations[1-3,20,21], were: 50 °C temperature, 5 bar total gas pressure, 40% vol. CO_2 content in the gaseous phase (which was a mixture of CO_2 and N_2), 5 l/kg liquid-to-solid (L/S) ratio and 4 h contact time. At the end of the carbonation tests the reactor was immediately degassed and then rapidly cooled down to ambient temperature through immersion in a water bath. Liquid/solid separation was subsequently accomplished through centrifuging at 4000 rpm followed by 0.45 μm filtration. Solid samples were oven-dried at 105 °C and analysed to determine their carbonate content and mineralogy (through XRD analyses), while the liquid solutions were characterized for dissolved IC and metal analysis. To infer about the prevailing chemical equilibria in the aqueous systems, as well as to identify the mineral phases potentially controlling the solubility of elements of interest, geochemical modelling of the liquid solutions obtained at the end of the carbonation experiments was conducted using Visual Minteq[22]. The geochemical speciation code was run using the measured eluate concentrations and pHs as the input data while suppressing precipitation for all solid phases; the potential solubility-controlling minerals were chosen among those displaying saturation indices (SI) between −1.5 and +1.5.

Results and Discussions

The relative proportions of the eight classes in the original BOF slag, along with the main statistical parameters (10th, 50th and 90th percentiles, volume and surface mean diameters) and the elemental composition are reported in Table 1. The statistical data for classes C1 and C2 is not included, since sedimentation of larger particles for these materials made the results of laser particle size analysis poorly reliable. However, for the reasons explained below such two classes were shown to be unsuitable for carbonation, therefore the lack of information on their particle size distribution is not deemed to be detrimental to the interpretation of results. The relative abundance of the different classes did not evidence the prevalence of any specific particle size range. From D_{10} values reported in Table 1 and from the inspection of the size distribution curves (not shown here), it can be inferred that dry sieve separation was not efficient in sharply segregating particles. The presence of particles smaller than the adopted cut size can be explained by physical aggregation or coalescence phenomena of smaller grains into larger aggregates during sieve separation, while the existence of oversize particles is more difficult to justify due to the fact that sonication and dispersion with a chemical agent were adopted during the analysis. A tentative explanation may found considering particle hydration or cementation upon contact with the liquid medium used in the analysis, being the hydraulic properties of steel slag well documented[23-25]. For such reasons, while the measured size distribution do not reflect the actual grain size of the as-received slag fractions, it can be regarded as being representative of the particle size of the material as used during the slurry-phase carbonation experiments. Concerning the elemental composition, a clear difference was observed between the coarser and finer fractions. The major components potentially involved in the carbonation process were more abundant in the finer fractions (C5–C8) compared to the coarser classes (C1–C4) in the case of Ca (28.7–33.0% vs. 22.2–25.1%), Fe (24.1–27.7% vs. 18.1–19.0%), Mg (4.5–8.9% vs. 5.0–5.3%), while the opposite occurred for Mn (2.5–2.8% vs. 3.0–3.2%), Na (0.47–0.55% vs. 0.62–0.70%) and K (173–219 mg/kg vs. 239–279). Metal contaminants (Cu, Ni, Pb and Zn) were enriched in the coarser material by a factor of 1.1–2.6, while Cr displayed 1.0–1.3 times higher concentrations in the finer fractions and V was more concentrated (~1.5 times more than the average) in C8.

Table 1. Relative proportions, particle size distribution parameters and elemental composition of the eight slag classes.

Class	C1	C2	C3	C4	C5	C6	C7	C8
Amount (% dry wt.)	6.7	20.6	17.3	10.8	11.0	14.4	6.5	12.6
D_{50} (μm)	---	---	532.10	200.90	88.40	43.11	31.14	17.48
D_{10} (μm)	---	---	321.20	37.90	11.24	11.81	2.39	6.75
D_{90} (μm)	---	---	746.40	377.00	174.10	87.31	63.26	35.53
D4-3 (μm)	---	---	531.00	209.00	90.12	46.38	31.62	19.48
D3-2 (μm)	---	---	481.00	27.33	15.76	18.68	6.10	9.80
Al (% dry wt.)	0.49	0.42	0.40	0.36	0.35	0.47	0.35	0.64
Ca (% dry wt.)	23.49	25.12	23.24	22.17	32.99	31.05	28.71	28.93
Fe (% dry wt.)	18.10	18.66	19.05	18.14	24.35	27.74	24.26	24.12
Mg (% dry wt.)	5.34	5.17	5.10	5.00	8.91	7.49	4.47	5.21
Mn (% dry wt.)	3.02	3.15	3.05	3.17	2.83	2.82	2.70	2.54
Na (% dry wt.)	0.64	0.62	0.64	0.70	0.55	0.50	0.49	0.47
Si (% dry wt.)	5.42	5.41	5.28	5.13	5.79	5.12	5.05	4.87
Cr (mg/kg dry wt.)	862.87	780.11	797.83	780.22	857.99	990.06	973.97	1018.50
Cu (mg/kg dry wt.)	6.86	6.58	7.98	7.18	5.52	5.65	6.03	6.13
K (mg/kg dry wt.)	238.83	252.39	278.85	270.37	214.50	218.69	173.50	190.97
Ni (mg/kg dry wt.)	25.89	22.41	19.13	24.57	13.11	13.92	14.55	14.32
Pb (mg/kg dry wt.)	20.03	19.27	26.26	24.10	10.01	12.80	11.75	11.02
V (mg/kg dry wt.)	760.40	770.94	758.33	701.43	764.25	830.22	775.24	1152.18
Zn (mg/kg dry wt.)	33.90	29.06	47.25	47.89	32.57	25.45	26.03	25.86

The CO_2 sequestration and the conversion yields attained for the eight size classes are summarized in Figure 1a). The CO_2 uptake, expressed as the mass (in grams) of CO_2 sequestered by 100 grams of as-received dry material, was calculated through equation (1) from the IC contents of carbonated (CO_2final) and untreated (CO_2initial) samples expressed as CO_2 weight percent contents.

$$\text{CO}_2 \text{ uptake } (\%) = \frac{\text{CO}_{2final}[\%] - \text{CO}_{2initial}[\%]}{100 - \text{CO}_{2final}[\%]} \times 100 \tag{1}$$

Considering Ca_{in}, Fe_{in}, Mg_{in}, Mn_{in} and CO_{3in} as the initial weight contents (%) of Ca, Fe, Mg, Mn and carbonates in the as-received untreated dry slag, the conversion yield of the reactive elements into carbonate forms, η, was also calculated according to equation (2):

$$\eta[\%] = \frac{\frac{CO_2uptake[\%]}{44}}{\frac{Ca_{in}[\%]}{40} + \frac{Fe_{in}[\%]}{55.8} + \frac{Mg_{in}[\%]}{24.3} + \frac{Mn_{in}[\%]}{54.9} - \frac{CO_{3in}[\%]}{60}} \times 100 \tag{2}$$

The inclusion of elements other than Ca in the calculation of the conversion yield of the carbonation process is motivated by the experimental evidence gained for the same BOF slag of the formation of Fe, Mg and Mn carbonates, as reported in a previous paper[21] and further confirmed by the present results (see below for details). Equation (2) is based on the implicit simplifying assumption that all the formed carbonate minerals had an equimolar Me/CO_3 ratio (where Me indicates Ca, Fe, Mg or Mn). As shown in Figure 1a, despite the relatively limited differences in the elemental composition of classes C1-C8, the carbonation performance was found to be largely dependent upon the grain size distribution of the material. CO_2 uptake varied indeed by up to two orders of magnitude, from a minimum of 0.47% for C1 to a maximum of 46.5% for C6; the corresponding conversion yield varied between 0.9 (C1) and 72.9% (C7). Thus, in the present case, the surface of particles exposed to the contact with CO_2 was largely more relevant for the carbonation yields than the elemental composition and mineralogy of the solid material. To corroborate this finding, dedicate carbonation experiments were also conducted on the BOF slag that was first ball-milled and then separated into the same size classes defined in Materials and Methods.

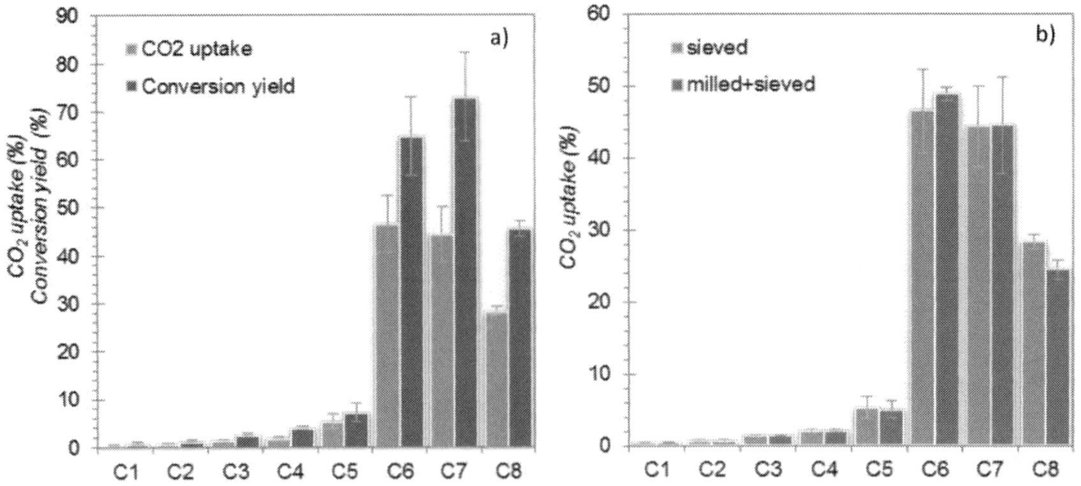

Figure 1. Carbonation performance of the eight slag classes (a) and comparison between the

CO_2 uptakes of the sieve-separated material with and without preliminary milling (b).

Figure 1b) reports a comparison between the CO_2 uptake results for the sieve-separated classes (same data as in Figure 1a)) and the milled + sieve-separated materials. The response upon carbonation was virtually the same for both samples, thus confirming the finding that grain size distribution was the main parameter explaining the different behaviour of the various size classes. From Figure 1, it can be argued that the CO_2 uptake and the associated conversion yield improved as particle size decreased from C1 to C7; further decreasing the grain dimensions (C8) negatively affected the process performance, probably due to the absence of condensation nuclei adequate for carbonate precipitation when the size of the particles was too small. A comparison between the particle size distribution curves for the untreated and carbonated slag classes is provided in Figure 2. It may be observed that the changes in particle size distribution upon carbonation appear to be related to the conversion yield achieved during the treatment. For the C3 and C4 classes, for which only low CO_2 uptakes were attained, the particle size was not modified upon carbonation, showing that both dissolution and precipitation reactions occurred to a very low degree and suggesting that more severe operating conditions should be adopted in order to promote carbonation.

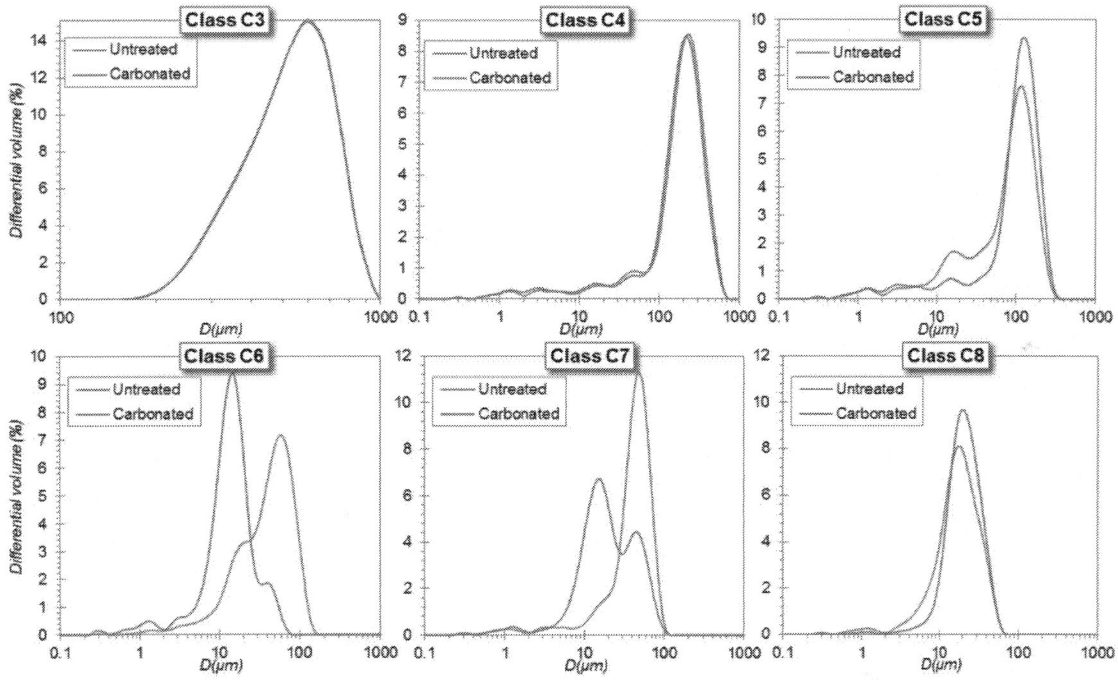

Figure 2. Changes in particle size distribution upon carbonation.

In the case of C5, an increase in the relative amount of >95 μm particles was observed after carbonation, at the expenses of particles in the size range 6-85 μm. Likely, the precipitation of carbonate minerals occurred onto the particle surface, increasing the final grain size of the material and probably hindering further carbonation. Classes C6 and C7, which displayed the highest CO_2 sequestration yield, showed a strong size reduction effect upon carbonation. This suggests that for such samples the dissolution of mineral phases from the solid matrix was the prevalent mechanism affecting particle size. For the finest class (C8) the effect of particle size reduction due to dissolution of the slag minerals was even more evident, with the relative volume of particles below 10 μm being reduced after carbonation, and that of 15–70 μm particles being increased. Surprisingly, the appreciable dissolution effect experienced by class C8 particularly for fine particles was not accompanied by a corresponding gain in CO_2 uptake. It is tempting to hypothesize that the absence of coarser particles, acting as condensation nuclei, significantly hindered the precipitation of new carbonate phases. Another possible explanation may be found in the presence of larger amounts of silica gel resulting from the dissolution if silicate phases, which physically hindered the precipitation of carbonate minerals. The effects of carbonation on the grain size distribution of each class are summarized in Figure 3a) in terms of changes in the D_{50} values before and after the treatment; the observed changes basically reflect the contrasting influence of dissolution and precipitation on particle size. Figure 3b) also clearly shows the negative exponential correlation (R^2 = 0.97) between the CO_2 uptake and the final D_{50} of the slag fractions, thus highlighting the important role of particle size on the process yield.

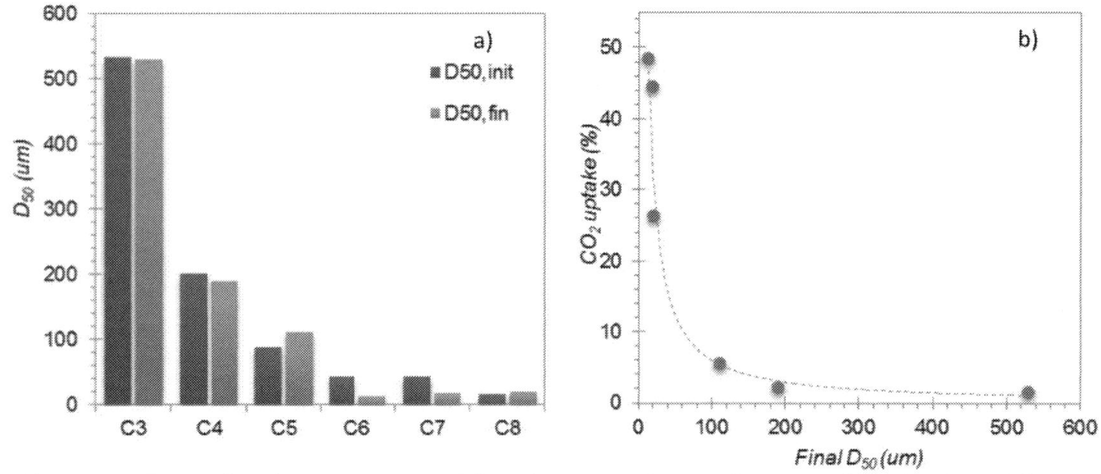

Figure 3. Effects of carbonation on D_{50} (a) and CO_2 uptake as a function of the final D_{50} (b).

The mineralogy of carbonated materials as compared to the corresponding untreated slag fraction was specifically investigated for classes C6 and C8 (data not shown). For both the untreated classes, the main crystalline phases detected included a mixed Ca-Al-Fe oxide ($Ca_2Fe_{1.4}Al_{0.6}O_5$), a mixed Ca-Cr-Fe oxide ($Ca_2Cr_{0.5}Fe_{1.5}O_5$), magnetite ($FeO{\cdot}Fe_2O_3$), Mg silicate ($MgSiO_3$), wüstite, as well as low amounts of portlandite and carbonate phases in the form of calcite, aragonite and Mg-calcite (($Mg_{0.064}Ca_{0.936}CO_3$). For Class C6, the diffraction peaks associated to portlandite disappeared in the carbonated slag, indicating the complete conversion of this phase during the process. The decrease in peak intensity was also relevant for magnetite and $MgSiO_3$, and less pronounced (although still appreciable) for the Ca-Al-Fe and Ca-Cr-Fe oxides, suggesting that the different minerals were variously affected by the carbonation process. For class C8, while again portlandite completely disappeared upon carbonation, the peaks of Ca-Al-Fe oxide, Ca-Cr-Fe oxide and magnetite were still well evident in the carbonated material, indicating that the degree of dissolution reactions were lower compared to C6. Despite the different conversion yields for the two classes, the nature of the newly formed phases appeared to be the same, likely due to the similarities observed in the initial mineralogical composition. The newly formed carbonate phases included calcite, aragonite, Mg-calcite, ankerite ($(CaMg_{0.27}Fe_{0.73}(CO_3)_2)$) and, for C6, kutnohorite ($(Ca_{0.86}Mn_{0.14})(Ca_{0.14}Mn_{0.5}Fe_{0.36})(CO_3)_2$). The diffraction peaks of the newly formed carbonate species were more evident in the C6 sample, thus confirming the larger degree of carbonation attained as evaluated by IC measurements.

The composition of the liquid phase at the end of the carbonation experiments is depicted in Figure 4. The concentrations of largely soluble elements (e.g., Na and K) for the different classes suggest that the dissolution degree of minerals from the original material progressively increased as particle size decreased. Thus, the grain size characteristics of the material are relevant to the carbonation process at the stage of phase dissolution already. For elements potentially subjected to dissolution/ precipitation mechanisms, including Al, Si, Ca, Mg, Fe and Mn among the major components, the interpretation of the measured concentrations in solution should be conducted with care, as multiple contrasting factors were playing a role during the process.

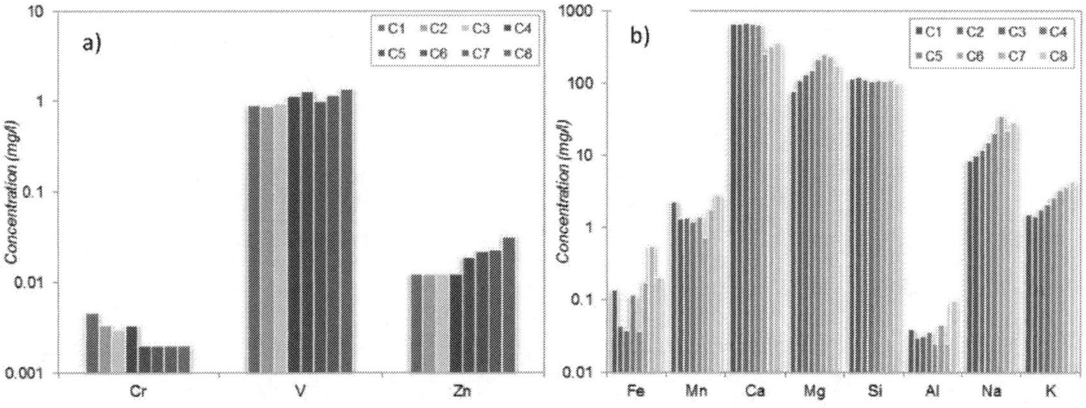

Figure 4. Liquid phase composition of the slurry at the end of the carbonation experiments: a) major components; b) trace metals.

The solution concentration of Si, which may be taken as an indicator of the dissolution of silicate minerals from the solid matrix, was found to be independent from particle size; this may be explained on the basis of the well-known phenomenon of incongruent dissolution, widely documented for natural weathering of silicate minerals[26], which leads to the re-precipitation of new silicate forms and the release of the remaining components into the solution. In the case under concern, geochemical modelling of the liquid phase revealed that the newly precipitated form likely consisted of amorphous silica (see below for details) for all size classes.

In order to separate the effects of metal dissolution from the solid material and metal precipitation in the form of various solid phases, the ratio between the solution concentration of each analysed metal and that of Na (assumed as completely dissolved) was calculated (see Figure 5a) and b). Although the degree of dissolution progressively increased from class C1 to class C8 as discussed above, the extent of solids precipitation for the different classes was found to depend on the specific element of concern. In the case of Si, while the solution concentration was almost the same throughout the various slag classes due to the solubility constraints imposed by the controlling mineral, the precipitated amount increased from classes C1 to C8 as a consequence of progressively enhanced dissolution of the original silicate phases from the slag. The increase in the degree of precipitation with decreasing particle size is also clear for the major components Ca and Mn and the three metal contaminants Cr, V and Zn. On the other hand, Mg precipitation was apparently related to particle size only for classes C6, C7 and C8 (for which Mg concentrations in solution increased to a lesser degree than the corresponding Na concentrations). While the previous considerations provide evidence of the enhanced precipitation of solid phases as particle size decreased, for any conclusive statement about the nature of the precipitated minerals to be derived, further elaboration on the chemical equilibria involved must be performed. To this aim, geochemical modelling of the liquid phases at the end of the carbonation experiments was conducted using the Visual Minteq code. The results are reported in Figure 6 in terms of SI values for Ca, Si, Mn and Mg minerals for the investigated size classes. In the graphs, only the solid phases displaying saturation indices reasonably close to zero are reported.

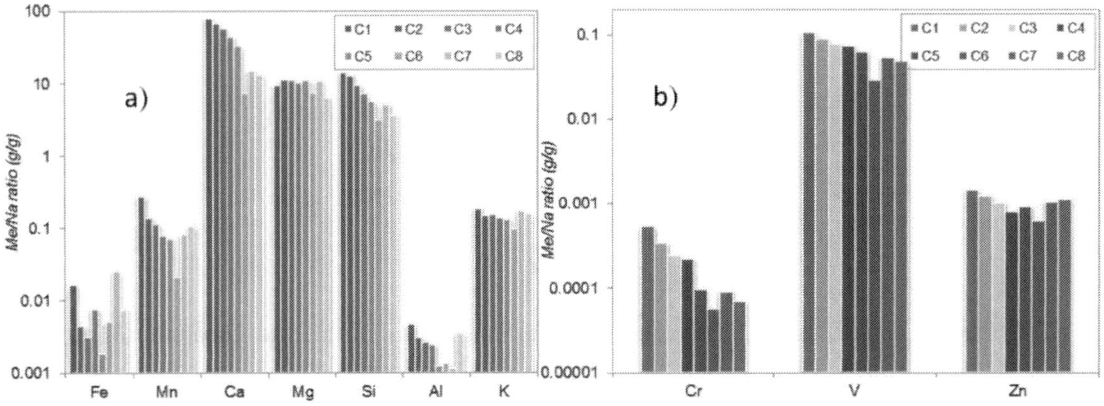

Figure 5. Metal/Na concentration ratio in the slurry at the end of the carbonation experiments; a) major components; b) trace metals.

As briefly mentioned above, the solubility of Si in the carbonated slurries was always clearly found to be controlled by amorphous silica irrespective of the particle size of the tested material. This confirms that SiO_2 is formed as a secondary mineral in all cases as a result of incongruent dissolution of the silicate phases from the original slag fractions. From Figure 6 it is evident that, apart from Si, all the potential candidates for solubility control of the main elements under concern included carbonate minerals, indicating that dissolution/precipitation equilibria of carbonate species largely dominated the carbonated slurry systems.

The most probable phases governing the equilibrium between the solid and the liquid phases included calcium carbonate monohydrate ($CaCO_3 \cdot H_2O$) for Ca, amorphous

$MnCO_3$ for Mn, magnesite ($MgCO_3$) for Mg, siderite ($FeCO_3$) for Fe and possibly $ZnCO_3$

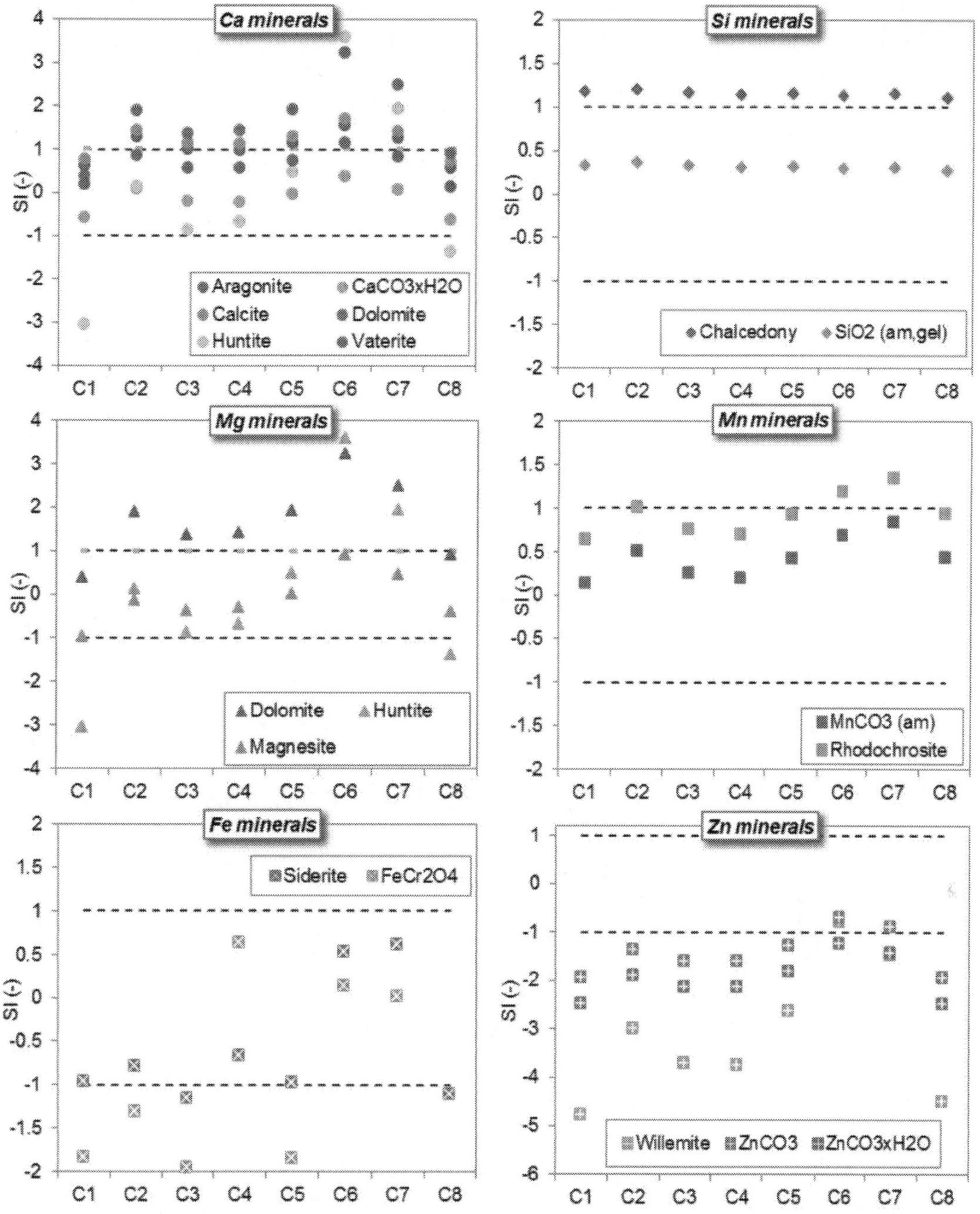

Figure 6. Saturation indices for potential solubility-controlling minerals derived from Visual Minteq results (Legend: Aragonite, Calcite, Vaterite: $CaCO_3$; Dolomite: $MgCa(CO_3)_2$; Huntite: $Mg_3Ca(CO_3)_4$; Chalcedony: SiO_2; Rhodochrosite: $MnCO_3$; Magnesite: $MgCO_3$; Siderite: $FeCO_3$; Willemite: Zn_2SiO_4).

for Zn. These findings further show that not only Ca, but also additional constituents were involved in the carbonate precipitation reactions, thus confirming the evidence of formation of a variety of carbonate minerals derived from mineralogical investigations in this and our previous studies[21]. It was also evident that the nature of the precipitating carbonates was not dependent upon the particle size characteristics of the material, remaining unchanged for the various slag classes investigated. This arguably indicates that the extent of dissolution of the original minerals in the slag, which is directly related to the available surface area, largely determines the carbonation performance.

Conclusions

The CO_2 sequestration capacity of different size fractions of steelmaking BOF-type slag was investigated in the present study through a number of slurry accelerated carbonation experiments. The results showed that particle size plays a very important role in determining the degree of CO_2 uptake by the alkaline material, and is largely more important than the elemental composition. Geochemical modelling of the slurry solutions at the end of the carbonation experiments indicated that a number of carbonate minerals were likely to form as a result of equilibrium with the solid phase, including $CaCO_3xH_2O$, amorphous $MnCO_3$, magnesite, siderite and $ZnCO_3$, as also confirmed by XRD analyses. The nature of the carbonate minerals controlling the equilibrium between the solid and the liquid phase did not vary with the size class, indicating that an increase in the surface area of solid particles enhanced the dissolution of the original minerals from the slag matrix and increased the amount of precipitated solids without changing their nature. A direct implication of this finding is the need of promoting the dissolution of the original matrix through either intensive size reduction pre-treatments or the application of chemical agents while keeping the net CO_2 balance positive.

References

1. Baciocchi, R., Costa, G., Di Bartolomeo, E., Polettini, A., Pomi, R. 2010. Carbonation of Stainless Steel Slag as a Process for CO2 Storage and Slag Valorization. Waste and Biomass Valorization, 1(4), 467–477.
2. Baciocch,i R., Costa, G., Di Bartolomeo, E., Polettini, A. Pomi, R. 2011. Wet versus slurry carbonation of EAF steel slag. *Greenhouse Gases Science and Technoogy*, 1(4), 312–319.
3. Baciocchi, R., Costa, G., Polettini, A., Pomi R. 2009. Influence of particle size on the carbonation of stainless steel slag for CO_2 storage. *Energy Procedia*, 1(1), 4859–4866.
4. Bonenfant, D., Kharoune, L., Sauve´, S., Hausler, R., Niquette, P., Mimeault, M., Kharoune M. 2008. CO_2 Sequestration Potential of Steel Slags at Ambient Pressure and Temperature. *Industrial & Engineering Chemistry Research*, 47(20), 7610–7616.
5. Chang, E.-E., Pan, S.-Y., Chen, Y.-H., Tan, C.-S., Chiang P.-C. 2012. Accelerated carbonation of steelmaking slags in a high-gravity rotating packed bed. *Journal of Hazardous Materials*, 227-228, 97–106.
6. Huijgen, W.J.J., Witkamp, G.-J., Comans, R.N.J. 2005. Mineral CO_2 Sequestration by Steel Slag Carbonation. *Environmental Science and Technology*, 39(24), 9676–9682.
7. Kodama, S., Nishimoto, T., Yamamoto, N., Yogo, K. Yamada, K. 2008. Development of a new pH-swing CO_2 mineralization process with a recyclable reaction solution. *Energy*, 33(5), 776–784.

8. Lekakh, S.N., Rawlins, C.H., Robertson, D.G.C., Richards, V.L., Peaslee K.D. 2008. Kinetics of Aqueous Leaching and Carbonization of Steelmaking Slag. *Metallurgical and Materials Transactions B*, 39(1), 125–134.

9. Santos, R.M., François, D., Mertens, G., Elsen, J., Van Gerven T. 2013. Ultrasound-intensified mineral carbonation. *Applied Thermal Engineering,* 57(1-2), 154–163.

10. Santos, R.M., Van Bouwel, J., Vandevelde, E., Mertens, G., Elsen, J., Van Gerven, T. 2013. Accelerated mineral carbonation of stainless steel slags for CO_2 storage and waste valorization: Effect of process parameters on geochemical properties. *International Journal of Greenhouse Gas Control*, 17, 32–45.

11. Santos, R.M., Ling, D., Sarvaramini, A., Guo, M., Elsen, J., Larachi, F., Beaudoin, G., Blanpain, B., Van Gerven, T. 2012. Stabilization of basic oxygen furnace slag by hot-stage carbonation treatment. *Chemical Engineering Journal*, 203, 239–250.

12. Teir, S., Eloneva, S., Fogelholm, C.-J., Zevenhoven, R. 2007. Dissolution of steelmaking slags in acetic acid for precipitated calcium carbonate production. *Energy*, 32(4), 528–539.

13. Uibu, M., Kuusik, R., Andreas L., Kirsimäe, K. 2011. The CO_2 -binding by Ca-Mg-silicates in direct aqueous carbonation of oil shale ash and steel slag. *Energy Procedia*, 4, 925–932.

14. Gunning, P.J., Hills, C.D., Carey P.J. 2010. Accelerated carbonation treatment of industrial wastes. *Waste Management*, 30(6), 1081–1090.

15. Gunning, P.J., Hills, C.D., Carey P.J. 2009. Production of lightweight aggregate from industrial waste and carbon dioxide. *Waste Manage*ment, 29(10), 2722–2728.

16. Huntzinger, D.N., Gierke, J.S., Kawatra, S.K., Eisele, T.C., Sutter, L.L. 2009. Carbon Dioxide Sequestration in Cement Kiln Dust through Mineral Carbonation. Environmental Science and Technology, 43(6), 1986–1992.

17. Iizuka, A., Fujii, M., Yamasaki, A., Yanagisawa, Y. 2004. Development of a New CO_2 Sequestration Process Utilizing the Carbonation of Waste Cement. *Industrial & Engineering Chemistry Research*, 43(24), 7880–7887.

18. Shtepenko, O., Hills, C., Brough, A., Thomas, M. 2006. The effect of carbon dioxide on β-dicalcium silicate and Portland cement. *Chemical Engineering Journal*, 118(1-2), 107–118.

19. Shtepenko, O.L., Hills, C.D., Coleman, N.J., Brough, A. 2005. Characterization and Preliminary Assessment of a Sorbent Produced by Accelerated Mineral Carbonation. *Environmetal Science and Technology*, 39(1), 345–354.

20. Baciocchi, R., Costa, G., Di Bartolomeo, E., Polettini, A., Pomi, R. 2009. The effects of accelerated carbonation on CO_2 uptake and metal release from incineration APC residues. Waste Management, 29(12), 2994–3003.

21. Baciocchi, R., Costa, G., Di Gianfilippo, M., Polettini, A., Pomi, R., Stramazzo, A. 2015. Thin-film versus slurry-phase carbonation of steel slag: CO_2 uptake and effects on mineralogy. *Journal of Hazardous Material*, 283, 302-313.

22. Gustafsson, J.P., 2012. Visual MINTEQ ver. 3.0. accessed June 2014

23. Shi, C., 2004. Steel Slag—Its Production, Processing, Characteristics, and Cementitious Properties. *Journal of Materials in Civil Engineering*, 16(3), 230–236.

24. Wang, Q., Yan, P., 2010. Hydration properties of basic oxygen furnace steel slag. *Construction and Building Materials*, 24(7), 1134–1140.

25. Belhadj, E., Diliberto, C., Lecomte, A., 2012. Characterization and activation of Basic Oxygen Furnace slag. *Cement and Concrete Composites*, 34(1), 34–40.

26. Ruiz-Agudo, E., Putnis, C. V., Rodriguez-Navarro, C., Putnis A. 2012. Mechanism of leached layer formation during chemical weathering of silicate minerals. *Geology*, 40(10), 947–950.

Mineral carbonation of serpentinite: from the laboratory to pilot scale – the MCi project

G.F. Brent[1*], M.S. Rayson[1], E.M. Kennedy,[2] M, Stockenhuber[2], W.J. Collins[2], J-D Prigge[2], R.G. Hynes[2], T.S. Molloy[2], H. Zulfiqar[2], F.Farhang[2], T.O. Oliver[2], S. Hamblin Wang[3] and M. Dawe[3]

[1] Orica Ltd, NSW, Australia
[2] University of Newcastle, NSW, Australia
[3] Mineral Carbonation International (MCi), ACT, Australia
*Corresponding author: Geoff.brent@orica.com

Abstract

To date, *ex-situ* carbonation of ultramafic ores such as serpentinite has been studied extensively on the laboratory scale. In Australia, the Mineral Carbonation International (MCi) project was launched in 2013, with funding from the New South Wales (NSW) State and the Australian Federal Governments as well as an industry partner, Orica. The project aims to demonstrate the techno-economic viability of mineral carbonation as a sustainable industrial solution for carbon capture, storage and utilisation (CCSU). For the first time, mineral carbonation of silicate ore will be studied at a scale several orders of magnitude greater than laboratory scales hitherto. The MCi project is believed to be the first holistic approach to CCSU via mineral carbonation, examining the full chain of operations from feedstock to final product. It comprises three core themes; the construction and operation of a research pilot plant, a portfolio of research projects and support activities and commercialisation. The research encompasses mineral mapping, characterisation, processing and pretreatment through to process integration with power generation and CO_2 capture, fundamentals of carbonation chemistry, value-adding and final product storage. In line with this holistic approach, Life Cycle Assessments (LCAs) are being conducted.

Introduction

Since the pioneering work by Lackner and co-workers[1] on potential industrial processes for carbonation of ultramafic ores and the intensive study on the aqueous route by the ARC (subsequently NETL) group[2], *ex-situ* mineral carbonation has been advanced by several researchers (see for example the recent review by Sanna *et al.*[3]). While various process routes have been investigated experimentally and also to some extent in desktop studies of industrial-scale processes, actual carbonation of silicate ores has remained restricted to the gram scale in the laboratory. The IPCC Special Report[4] on CCS in 2005 highlighted the absence of pilot scale studies as an impediment to furthering the assessment of the feasibility of mineral carbonation.

Mineral Carbonation International (MCi) is a joint venture between Orica, Newcastle Innovation (the technology transfer arm of the University of Newcastle (UON)) and GreenMag Group. MCi was established in 2013 after funding for a research project was secured in equal parts from the Australian Federal and the NSW State Governments and an industrial partner, Orica. The project aims to demonstrate the technical and economic viability of mineral carbonation of ultramafic ores as a potential carbon capture, sequestration and utilisation (CCSU) solution.

In order to achieve this, we considered it necessary to adopt a holistic approach to mineral carbonation, encompassing the full chain of processes from feedstock to final product. Ultimately, a large CCSU industry will require all elements in the chain to be technically and commercially viable and integrated. A sustainable mineral carbonation industry will require key ancillary operations at both the front-end ore identification and mining phases as well as at the final product disposal phases.

This project is believed to be the first to encompass the full CCSU chain from mineral mapping, characterisation, processing and pretreatment through to integration with power generation and CO_2 capture processes, carbonation, value-adding and final product storage. In line with this holistic approach, quantitative Life Cycle Assessments (LCAs) are being conducted.

A primary objective is to demonstrate carbonation at a much larger scale than hitherto, with a major component of the project being the construction and operation of a world-first carbonation research pilot plant at the Newcastle Institute for Energy and Resources (NIER) at the UON . This research pilot plant (RPP) will go some way to addressing the IPCC's concern in this regard by providing data on scales that are several orders of magnitude larger than before.

A second component of the MCi project is an intensive research portfolio aimed at further understanding and development of the technology required to underpin a viable industrial process. The research includes geological studies, fundamental theoretical and laboratory studies of activation and carbonation chemistry, other pre- and post-carbonation processes as well as a key area that has hardly been addressed in the literature; namely the crucial linkage with fossil fuel-based power generation and carbon capture processes.

The project has primarily been studying the carbonation of serpentine, which is a mineral locally abundant in some major fault zones in NSW, and globally. Since the potential exists for this to become an industrial process on a very large scale, in order to sequester CO_2 emissions from power stations and other stationary sources such as cement plants, we assert that the future feedstock supply needs to be available on the gigatonne scale. The potential for value-adding by-products and linkages with other CO_2-intensive industries also exists as outlined in Brent *et al.*[5]. One of the goals of the MCi project is to determine the potential for utilising the carbonated product in next generation "green" building products. Such a product would represent a world-leading example of CCSU.

A third component of MCi comprises elements required to manage and support the project, along with activities aimed at commercialising the IP developed. The formation of strategic alliances is an important focus.

The major research node is at the Priority Research Centre for Energy at the UON. Other research nodes are at Columbia University in New York and the University of Sydney. The key technical areas of the project are outlined here.

Research Pilot Plant

The benchmark for the MCi project is the well-known single-step aqueous carbonation process developed by the ARC/NETL group. In 2004, the ARC project wound up and the final report concluded that the energy penalty for carbonation of serpentine was prohibitively high. This conclusion was restated in the 2005 IPCC Special Report and we believe it led to a hiatus on work in this area for some years. However, the ARC/NETL analysis had not quantified direct thermal activation of serpentine, instead assuming heat input via electrical energy, leading to an overestimation of the energy penalty by roughly a factor of three. The energy requirement for thermal activation can be further reduced by heat recovery in an integrated process. Direct thermal activation, for example by natural gas, was studied[6] and it was concluded from ASPEN modelling that a suitably integrated process with heat recovery could achieve up to 80% net sequestration of CO_2 with the ARC/NETL process conditions and the highest reported conversions. Other workers[7] have since recognised and pointed out this overestimation and in recent years research in the field of carbonation of serpentine appears to have grown around the world.

The MCi RPP building at the NIER site is shown in Figure 1. The RPP is being constructed in two phases, a batch process followed by a semi-continuous process. The batch plant is designed to carbonate heat-activated serpentine under the ARC/NETL conditions as a benchmark. However, the RPP will have the flexibility to accommodate modified processes and conditions. Several innovations from the MCi research program will be tested on the RPP. In particular, reductions in the overall energy penalty of the ARC/NETL process will be the major objective of the RPP. The production of substantial quantities of carbonated product will also facilitate the studies of by-products, in particular the utilisation of the carbonated product. First production from the RPP is scheduled for quarter three of 2015.

One of the drawbacks of the ARC/NETL process is the use of NaCl in the buffer solution. Chlorides have the potential to induce unpredictable and catastrophic stress corrosion cracking under high temperature and pressure regimes and this potential failure mode was recognised early in the RPP design process.

We urge other workers in the field to be cognisant of this hazard, even under laboratory conditions. Notwithstanding the serious corrosion and cost implications of chlorides for larger scale plants, salt would be a troublesome component in the product, whether it is to be disposed back to the original serpentinite mine or used in concrete. An early objective for us has therefore been to eliminate chlorides from the process.

Figure 1. MCi RPP building on the NIER site at the University of Newcastle

Research programs

Feedstock

The serpentinite feedstock is of vital importance. Mineralogy determines the reactivity of the ore in the mineral carbonation process and constitutes a key process variable. The project aims to utilise ore that is potentially mineable in the foreseeable future on a scale large enough to sequester many millions of tonnes of CO_2. The present focus is on the Great Serpentinite Belt (GSB) along the Peel Fault of NSW, although techniques that are being developed by us are expected to have global applicability.

A key principle held by MCi is the avoidance of any human exposure to hazardous asbestiform fibres. The MCi RPP will not process any material that contains asbestiform fibres. It is well known that chrysotile asbestos is commonly found in serpentinite and occurs within the GSB, as shown in Figure 2. We have an intensive research program within the NSW Institute for Frontier Geoscience at the UON aimed at understanding the factors that govern the occurrence of asbestiform fibres and developing mapping tools to locate regions that minimise the likelihood of encountering fibres. Not only is this important for the MCi project, but we consider that future mining operations would also seek to minimise asbestos exposure. This is despite its occurrence being fairly common in many mines, where it is currently safely managed under appropriate asbestos hazard control plans. Notwithstanding our objective, it should be emphasised that, as pointed out by O'Connor *et al.*[2], the carbonation process readily destroys chrysotile asbestos and hence can fulfil an additional role as an asbestos remediation process.

Figure 2. Serpentine veins containing asbestos in Great Serpentinite Belt, NSW

Apart from being clear of asbestiform fibres, we are targeting feedstock that is highly serpentinised, free of impurities such as clinochlore and preferably with a relatively high content of valuable metals, particularly iron (as magnetite), chromium and nickel. Feedstock characterisation using various analytical tools such as X-ray diffraction (XRD) is an integral part of our work. In this regard we are developing methods for rapid and practical quantitative analysis of serpentinite. Figure 3 shows an XRD peak characteristic of clinochlore impurity in a serpentinite sample from the GSB.

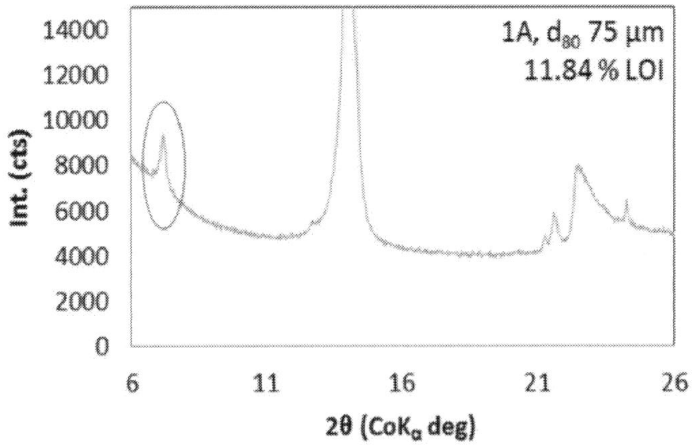

Figure 3. XRD pattern highlighting the clinochlore impurity

Serpentine activation

Activation of serpentine is key to an efficient carbonation process and much of our work has focussed on this aspect. Mann[8] conducted fundamental studies of the kinetics of thermal dehydroxylation to activate serpentine and demonstrated the importance of minimising the formation of crystalline phases such as forsterite. We are collaborating further with the University of Sydney on understanding the fundamental processes and are building on this work to develop processes for enhanced activation that would be practical at larger scale. Fluidised bed roasters are an attractive option for large scale thermal activation of serpentinite owing to excellent heat and mass transfer, and uniform temperature distribution. Figure 4 shows a temperature profile from a fluidised bed roaster using serpentinite from the GSB on the kilogram scale.

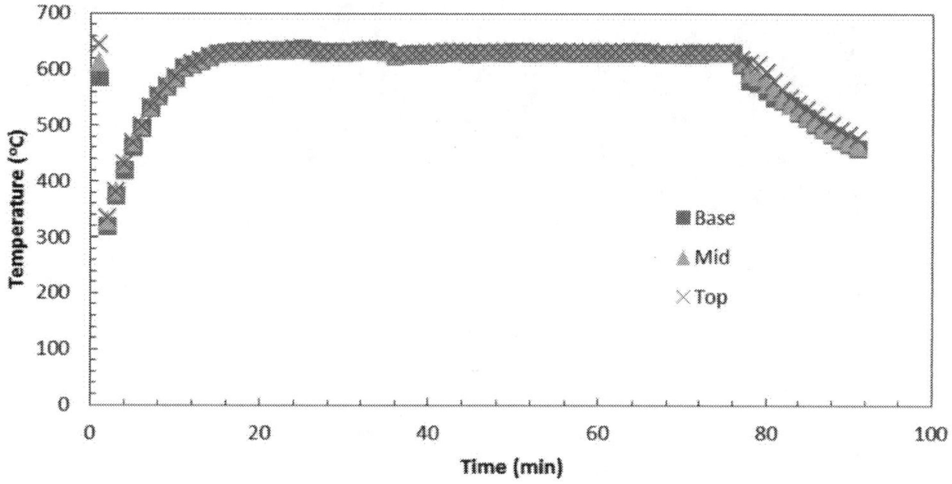

Figure 4. Temperature profile of fluidised bed roaster using GSB serpentine

Carbonation

The ARC/NETL aqueous process is the starting point and benchmark for the MCi process. Although this process was developed over a decade ago and has been studied fairly extensively by other workers since then, it remains poorly understood. Kinetics have yet to be determined and hence there is still scope for further optimisation of process parameters. Our research work is attempting to gain insight into the kinetics as well as understanding unwanted side reactions that may occur within this process. Figure 5 shows two of our research reactors at the UON.

The primary objectives remain to reduce the overall energy and cost penalties of this process. These stem primarily from:

- Elevated pressure (> 100 Bar) and temperature (>150°C), raising energy input and process equipment costs
- Solution additives NaCl and $NaHCO_3$, raising process material costs and reagent costs
- Relatively long residence times (>1 h), raising process equipment costs
- Incomplete conversion of available magnesium to carbonate, raising all process inputs and outputs, energy and equipment costs
- Low solid to liquid ratios (< 30% m/m), raising process equipment and energy input costs

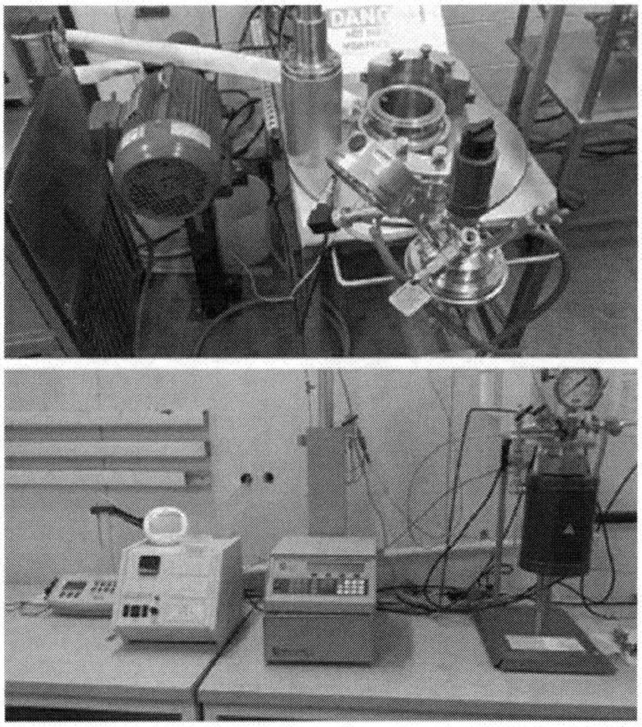

Figure 5. Carbonation research reactors at the UON

Additional collaborative work on carbonation of GSB feedstock is being undertaken in conjunction with MCi by Columbia University. These projects are studying reagents that may have a catalytic effect on carbonation as well as processes that may operate under near-ambient conditions.

Post-carbonation dewatering of the final product is yet another important area of investigation that has not been addressed before. Our program includes work on this.

By-products

A key differentiator for mineral carbonation over conventional underground storage of CO_2 is the potential commercial exploitation of by-products, including metals such as iron, chromium and nickel. We are examining possible routes for the extraction of these metals from the ore. Not only are they a large prospective boost to the overall process economics but, from a sustainability perspective, utilisation of this resource could displace mining activity elsewhere.

In addition, the products of the mineral carbonation process have the potential to provide both economic and sustainability benefits. The carbonated product from the process comprises largely magnesium carbonate and fine amorphous silica. This opens two possibilities for use in concrete construction products. First, the silica can serve as a binder to replace a portion of the cement in concrete. Second, the carbonated product could be simply admixed as filler in concretes. In the former case the product could reduce the net cost of the concrete and avoid the associated upstream CO_2 emissions of the displaced cement. In the latter case, use of this material as a filler displaces the need for mining of sand or aggregate filler elsewhere. In both the former and latter cases, the carbonate represents solid sequestered CO_2, which would form an inherent part of the construction product. It is likely that both silica and magnesium carbonate would be mixed into the concrete product, performing the dual roles of binder and filler. We are examining ways of optimising these roles in concrete products. It should also be noted that magnesium carbonate is a fire retardant and may provide further benefits in certain building products.

Figure 6. Cement test blocks incorporating carbonated product

Synergies and Life Cycle Assessment (LCA)

An area that has hitherto been largely neglected in mineral carbonation studies is that of process synergies with associated power generation and carbon capture processes. In contrast to conventional underground storage of captured carbon where few opportunities for process synergy are readily apparent, synergies could be of crucial importance to mineral carbonation. A prominent example is the utilisation of the exothermic heat of carbonation in the regeneration of CO_2 sorbents. This can provide a substantial reduction in the energy penalty of capture. As an example, the use of the heat of carbonation to regenerate diethanolamine sorbent in a modelling study of an integrated process was reported by Brent *et al.*[5].

We propose that a very significant potential advantage of mineral carbonation lies in the possibility of processing relatively impure streams of CO_2. By contrast, conventional underground injection generally requires CO_2 that is largely free of impurities. Accompanying inert gases such as nitrogen represent an additional compression burden, while oxygen may inhibit injectivity due to oxide formation in pores in mineral bodies underground. Acid gases such as NO_x and SO_x introduce severe corrosion issues to all process equipment and pipelines. Removal of these impurities imposes large cost and energy penalties to carbon capture processes, whether they are post- or pre-combustion. Our process will test the viability of utilising CO_2 streams containing these impurities at various levels. This is expected to lead to a substantial reduction in the costs of capture.

In keeping with our holistic approach to sustainability, we are developing flow sheets of our mineral carbonation processes linked with various fossil fuel power generation and carbon capture technologies to inform cradle-to-grave LCAs. Our key indicators in these studies are energy consumption and net CO_2 per unit of delivered power; however quantification of other important indices and the consumption of resources such as water is also undertaken.

Several potential synergies can be envisaged with other industrial processes, such as iron and steelmaking, construction products, waste disposal and speciality products. Some examples have been outlined and quantified by Brent *et al.*[5] The MCi project is actively pursuing some of these possibilities.

Finally, is should be emphasised that the permanence of the CO_2 sequestration is guaranteed in the thermodynamically stable carbonates, irrespective of acid rain attack as reported by Allen and Brent[9]. This contrasts with the possibility of leakage over the long term from underground injection, and the costs of ongoing monitoring.

Acknowledgements

The authors acknowledge and thank the MCi funding organisations and all those, too numerous to name, who have collaborated with us to date on this project. Our PhD students Leo Anderberg, Ammar Abu Fara and Emad Benhelal are thanked for providing Figures 2, 5 and 6 respectively.

References

1. Lackner, K.S., Wendt, C.H., Butt, D.P., Joyce, E.L. and Sharp, D.H., 1995. Carbon dioxide disposal in carbonate minerals, *Energy* ***20*** *: 1153-1170.*
2. O'Connor, W.K., Dahlin, D.C., Rush, G.E., Gerdemann, S.J., Penner, L.R. and Nilsen, D.N., 2004. Aqueous Mineral Carbonation - Mineral Availability, Pre-treatment, Reaction Parametrics and Process Studies-Final Report, *DOE/ARC-TR-04-002*, Albany Research Center, US DOE.
3. Sanna, A., Uibu, M., Caramanna, G., Kuusik, R. and Maroto-Valer, M. M., 2014. A Review of Mineral Carbonation Technologies to Sequester CO_2, *Chem Soc Rev*, DOI: 10.1039/c4cs00035h.
4. Intergovernmental Panel on Climate Change (IPCC), Prepared by Working Group III of the Intergovernmental Panel on Climate Change: Metz, B., O. Davidson, H.C. de Coninck, M. Loos; L.A. Meyer, eds. *IPCC Special Report on Carbon Dioxide Capture and Storage*. Cambridge University Press: Cambridge, UK and New York, USA, 2005. Available online at: www1.ipcc.ch/ipccreports/special-reports.htm.

5. Brent, G.F., Allen, D.J., Eichler, B.R., Petrie, J.G., Mann, J.P. & Haynes, B.S., 2011. Mineral Carbonation as the Core of an Industrial Symbiosis for Energy-Intensive Minerals Conversion, *J Ind Ecology* ***16*** *(1) 94-104*.

6. Brent, G.F. and Petrie, J.G., 2008. CO_2 Sequestration by mineral carbonation in the Australian context, *Proc Chemeca 2008*, Newcastle, Australia, ICMS.

7. Zevenhoven, R., Fagerlund, J. and Songok, J.K., 2011. CO_2 mineral sequestration: developments towards large-scale application, *Greenhouse Gas Sci Technol* ***1*** *(1) 48-57*.

8. Mann, J.P., 2014. Serpentine Activation for CO_2 Sequestration. *PhD Thesis*, School of Chemical and Biomolecular Engineering, University of Sydney.

9. Allen, D. and Brent, G.F., 2010. Sequestering CO_2 by Mineral Carbonation: Stability against Acid Rain Exposure. *Environ Sci Technol* ***44*** *(7) 2735-9*.

Accelerated carbonation of steel slags using CO_2 diluted sources: CO_2 uptakes and energy requirements

R. Baciocchi[1], G. Costa[1], A. Polettini[2], R. Pomi[2], A. Stramazzo[2] and D. Zingaretti[1*]

[1] Department of Civil Engineering and Computer Science Engineering, University of Rome "Tor Vergata", Rome, Italy,

[2] Department of Civil and Environmental Engineering, University of Rome "La Sapienza", Rome, Italy

*Corresponding author: zingaretti@ing.uniroma2.it

Abstract

This work presents the results of carbonation experiments performed on Basic Oxygen Furnace (BOF) steel slag samples employing gas mixtures containing 40 and 10% CO_2 vol. simulating the gaseous effluents of gasification and combustion processes respectively, as well as 100% CO_2 for comparison purposes. Two routes were tested, the slurry phase (L/S=5 l/kg, T=100 °C and P_{tot}=10 bar) and the wet (L/S =0.3-0.4 l/kg, T=50 °C and P_{tot}=7-10 bar) routes. For each, the CO_2 uptake achieved as a function of the reaction time was analyzed and on this basis the energy requirements associated to each carbonation route and gas mixture composition were estimated considering to store the CO_2 emissions of a medium size power plant (20 MW). For the slurry-phase route, CO_2 uptakes after 24 h ranged from around 8% at 10% CO_2, to 21-29% at 40% CO_2 and 32.5-40% at 100% CO_2. For the wet route, uptakes of 13-19.5% at 40% CO_2 and 17.8-20.2% at 100% CO_2 were attained. The energy requirements of the two analyzed process routes appeared to depend chiefly on the CO_2 uptake of the slag. The minimum overall energy requirements were found for the tests with 40% CO_2 flows (i.e. 1400-1600 MJ/t CO_2 for the slurry and 2220-2550 MJ/t CO_2 for the wet route).

Introduction

In the last decades several studies have shown that accelerated carbonation of alkaline earth metal silicate ores may be an effective method for ex situ carbon dioxide sequestration.[1,2] However, despite minerals such as serpentine, olivine and wollastonite exhibit a high carbon sequestration potential, also in consideration of their widespread availability, the energy intensive operating conditions typically required to accelerate carbonation kinetics still hinders the industrial application of this technology. Alternatively to minerals, the reactivity with CO_2 of alkaline industrial residues characterized by high calcium or magnesium (hydr)oxide or silicate contents has also been extensively investigated.[3,4] These materials have proven to be more reactive than minerals at mild operating conditions.[5] Furthermore, carbonation has shown to affect some of the properties of the tested materials (i.e.: main mineralogy, porosity and mobility of specific elements) improving their long-term technical performance and/or environmental behavior.[3,4] Thus, the application of this treatment to alkaline industrial residues may be indeed considered as a CO_2 storage and utilization option, since accelerated carbonation may be also employed as a valorization technique for widening the reuse options of these materials and also to achieve other environmental benefits. Steel manufacturing plants are typical examples of industrial sites in which relevant flows of both CO_2 and alkaline industrial residues are generated and therefore represent one of the potentially most interesting contexts for the application of accelerated carbonation. Several studies have shown that a number of different types of steel slag present a significant reactivity with CO_2, allowing to achieve, for specific process routes and operating conditions, relevant CO_2 uptakes (see[6] and references therein). Furthermore, several types of residues generated in steel manufacturing plants such as Basic Oxygen Furnace (BOF), Electric Arc Furnace (EAF) and argon oxygen decarburization (AOD) slag, are typically not valorized and generally landfilled, or employed only for low-end applications, owing for their significant content of free calcium and magnesium (hydr)oxides that may result in poor volumetric stability and hence in a low technical performance in construction applications.[7] Therefore, accelerated carbonation of these types of materials may also represent a treatment strategy to improve their properties in view of valorization.

Despite of all of the potential assets mentioned above, industrial-scale applications of accelerated carbonation of residues such as steel slag are also currently still missing owing to uncertainties on the feasibility of this process, especially as far as material and energy requirements are concerned; specifically, with regard to the latter, in order to achieve significant CO_2 uptakes in reasonable timeframes, milling of the residues and operation of the carbonation reactor at enhanced temperature and CO_2 pressure have been reported to be necessary when the process is applied through the slurry phase route.[8] However, it should be considered that promising results in terms of CO_2 uptake (even if lower than those attained through the former route) have been achieved also via the wet or thin film route[6-9] which is operated at lower temperature and CO_2 pressure and also presents the advantage of not generating a liquid by-product that necessitates suitable management and treatment. Furthermore, CO_2 capture from flue gas to a concentrated stream is also an energy intensive process, hence in order to increase the feasibility of carbonation as a carbon capture and storage (CCS) process it would be interesting to directly perform the carbonation step using diluted CO_2 sources, thus lumping capture and storage in a single step.

In this study we performed carbonation tests via both the slurry phase and wet routes on different types of BOF steel slag selecting the operating conditions that allow to achieve significant reaction yields for each route and employing CO_2 diluted gas streams (i.e. containing 40 and 10% vol. CO_2) and also a 100% CO_2 flow as a comparison. Then, based on the CO_2 uptakes achieved in these experiments we estimated the energy requirements associated for storing the CO_2 emissions of a medium size power plant (20 MW). To this aim, the carbonation process layouts corresponding to the slurry and wet routes were considered and the material and energy requirements for each unit operation of the process were assessed and discussed in terms of the type of BOF slag treated, the operating time adopted for the carbonation reaction (from 0.5 to 24 h) and the gas composition considered (10, 40 or 100 % vol.).

Materials and Methods

The slag samples employed in this work were collected from a steel plant employing the integrated steelmaking process both directly at the outlet of the BOF unit (BOF-a and BOF-c) after grinding (to d<1 mm) and at the storage site after grinding (to d< 1 mm) for metals recovery (BOF-b and BOF-d). The slag samples were then oven-dried at 60 °C and ball-milled to a final particle size of less than 150 μm in the case of samples a and b, or only sieved to a particle size below 125 μm, in the case of samples c and d.

The elemental composition of the four BOF slag samples is reported in Figure 1. With regard to the main elements of concern for the carbonation process, the following concentration ranges were detected for the investigated materials: Ca 21.3–28.9 % dry wt., Mg 3.3–6.9%, Fe 13.9–19.6%, Mn 2.4–3.8%. Such elements were present at comparable levels in the four slag samples investigated, although BOF-c and BOF-d exhibited slightly higher Ca and Mg concentrations and somewhat lower Fe contents.

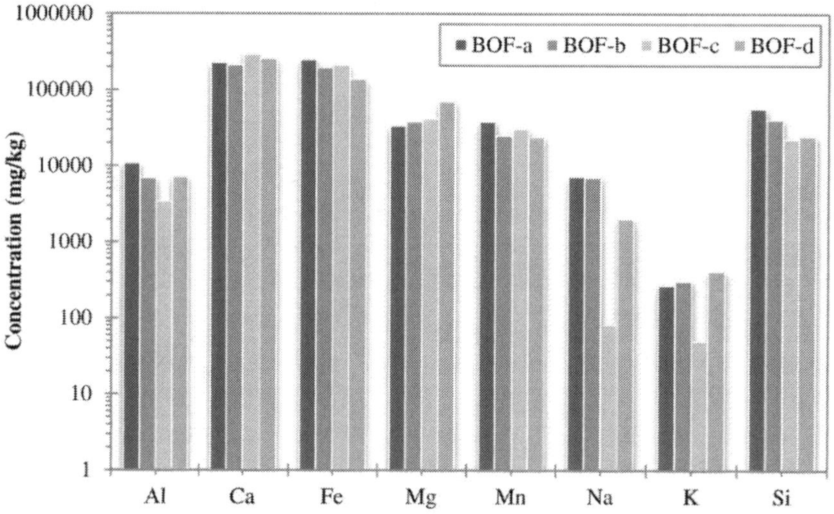

Figure 1. Major Element composition of the BOF slags.

Slurry phase accelerated carbonation tests were performed on BOF slag samples a and b in a pressurized stainless steel reactor equipped with an external magnetic stirring device and a heating jacket. Based on the results of previous investigations[6,9] a liquid to solid ratio (L/S) of 5 l/kg, operating temperature of 100 °C and total pressure of 10 bar were selected. The wet route carbonation tests were carried out on BOF slag samples c and d in a pressurized stainless steel reactor placed in a water bath for temperature control. The operating conditions selected for this route based on previous investigations[6,9] were a L/S of 0.3 l/kg, a temperature of 50 °C and a total pressure of 7-10 bar. For both types of routes different reaction times (ranging from 0.5 to 24 h) and CO_2 concentrations in the gas flow were tested. In particular, the slurry phase tests were performed with 100%, 40% or 10% vol. CO_2 flows to simulate not only a pure CO_2 flow deriving from a preliminary capture step, but also diluted CO_2 flows from gasification or combustion processes. In the wet route only 100% and 40%

vol. CO_2 flows were employed, since previous results obtained for steel slag adopting this configuration indicated that the CO_2 uptakes achievable using a gas composition similar to fluegas were significantly lower than those attained with higher CO_2 concentrations of the treated gas flow.[10]

At the end of the slurry phase carbonation tests the reactor was immediately degassed and then rapidly cooled down to ambient temperature through immersion in a water bath. Then liquid/solid separation was subsequently accomplished through centrifuging at 4000 rpm followed by 0.45 μm filtration. The solid samples obtained from both types of routes were oven-dried at 105 °C and analyzed to determine their carbonate content through inorganic carbon (IC) measurement employing a Shimadzu TOC analyzer equipped with a solids module. The CO_2 uptake, expressed as the mass (in grams) of CO_2 sequestered by 100 grams of as-received dry material, was calculated through Equation (1) from the IC contents of the carbonated (CO_{2final}) and untreated ($CO_{2initial}$) samples expressed as CO_2 weight percent contents.

$$CO_2 \text{ uptake } (\%) = \frac{CO_{2final}[\%] - CO_{2initial}[\%]}{100 - CO_{2final}[\%]} \times 100 \qquad (1)$$

The evaluation of the energy requirements for the carbonation process was carried out using the approach developed in our previous work[10]. Specifically, on the basis of the process layout defined for the slurry-phase and wet configurations (see Figure 2), the energy requirements associated to each route were calculated summing those necessary for all the unit operations foreseen for each configuration (expressed as MJ/tCO_2 stored), on the basis of the experimental data obtained for the type of investigated BOF slag (i.e. chemical composition, mineralogy and CO_2 uptake).

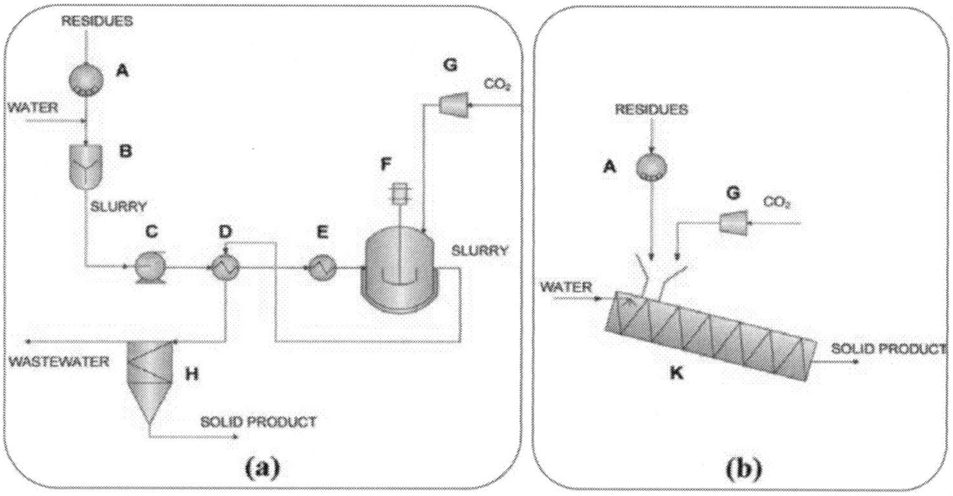

Figure 2. Carbonation process layout for the slurry phase (a) and the wet (b) routes.[10] Legend: A = mill; B = rapid mixing unit; C = pump; D, E = heat exchangers; F, K = carbonation reactors; G= compressor; H = settling tank.

In short, the energy requirements for the slurry-phase route were calculated considering the following unit operations: 1) size reduction of the slag, 2) mixing of the solid material with water, 3) heating of the slurry (assuming to recover the carbonation reaction heat and part of the heat of the slurry leaving the reactor), 4) slurry pumping into the carbonation reactor, 5) CO_2 pressurization, 6) reactor mixing and 7) solid/liquid separation of the carbonated slurry. For the wet route, the energy requirements were assessed making reference to the following unit operations: 1) size reduction of the slag, 2) CO_2 pressurization, 3) heating of humidified material and 4) mechanical operation of the reactor hypothesized as a rotary kiln. For both process routes, when a concentrated CO_2 flow was assumed to be used, the energy demand associated to the preliminary capture step was also accounted for, and overall quantified as 4000 kJ/kg CO_2 captured[11].

Results and Discussion

Carbonation performance

The CO_2 sequestration yield attained from the experiments as a function of the reaction time for the two process routes and the four slag samples investigated is reported in Figure 3.

The CO_2 uptakes measured after 24 h carbonation for the slurry-phase route were around 8.0% at 10% CO_2, 21.1–29.2% at 40% CO_2 and 32.5–40.3% at 100% CO_2, with a generally better performance of the BOF-b slag over the BOF-a sample; the corresponding ranges for the wet route were 12.9–19.5% at 40% CO_2 and 17.8–20.2% at 100% CO_2, with BOF-d slightly outperforming BOF-c. It is interesting to note that, particularly for the wet carbonation route and for short reaction times for the slurry-phase process, the use of a 40% CO_2 flow instead of a pure one did not appear to significantly affect the CO_2 uptake of the slag (especially BOF-b and BOF-d). Instead, for a 10% CO_2 flow the uptake resulted below 10% and appeared not to be appreciably affected by the reaction time.

The higher reactivity exhibited by the slag sampled from the storage site (BOF-b and BOF-d) may be related to its significant content of portlandite which was not detected in the samples collected directly at the outlet of the BOF unit (BOF-a and BOF-c).[6] Although the direct comparison of the two process routes should be made with care owing to the slight differences in composition of the slag samples investigated (see Fig. 1), the observed trends certainly highlight the higher carbonation yields of the slurry-phase process as opposed to the wet route, confirming the findings of our previous study.[6] Compared to the wet process, the slurry-phase conditions are believed to promote the dissolution of CO_2 into the liquid phase and the extraction of reactive metals from the solid phase, as well as to prevent the coating effect exerted by the precipitated carbonate product onto the unreacted core of the slag particles.[12]

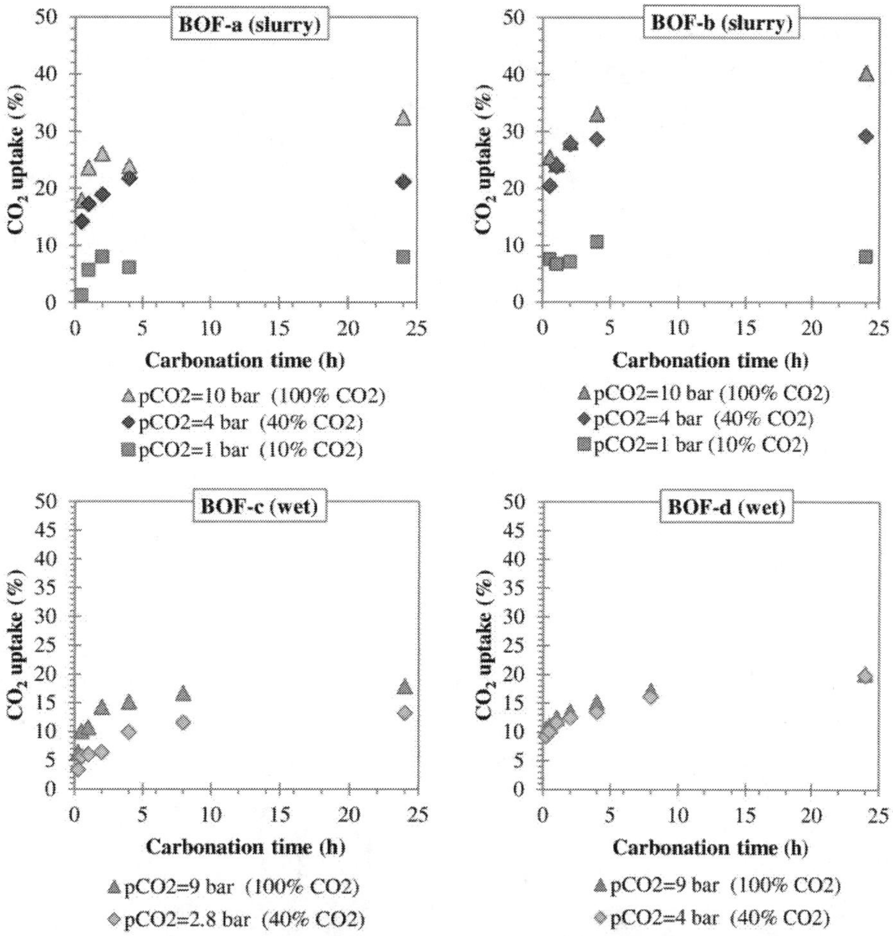

Figure 3. CO_2 uptake attained for the slurry-phase route with BOF-a and BOF-b samples (T=100°C) and the wet route with BOF-c samples and BOF-d samples (T=50°C).

Energy requirements assessment

The specific energy requirements of each unit operation for the slurry-phase route are reported in Figure 4 making reference to the experimental results achieved for the BOF-a and BOF-b slags. It may be noted that the various process units contributed in quite different extents to the overall energy demand. For improved clarity, the relative contributions of the process units to the total energy consumption are reported in Figure 5.

Several interesting features may be derived from the inspection of Figures 4 and 5. Specifically, the carbonation yield attained obviously turned out to play a relevant role on the energy requirements per unit mass of CO_2 sequestered. For a CO_2 concentration in the gaseous phase of 10% (simulating the typical content of combustion flue gases) the notably lower CO_2 sequestration performances achieved implied the use of larger amounts of slag and water for storing the target amount of CO_2 and therefore higher associated energy demands for almost all process units, in particular slurry heating and mixing. Comparing the energy requirements of the slurry-phase process using a combustion effluent or a concentrated CO_2 flow (CO_2 concentration = 100%), it is found that in the latter case the higher energy consumption associated to the preliminary capture stage is counterbalanced by the notably lower consumption of the other process operations. As a consequence, for very diluted gaseous flows a carbonation process including a pre-treatment stage aimed at concentrating CO_2 in the gaseous phase is deemed economically more viable compared to the option without CO_2 capture. On the other hand, for intermediate CO_2 concentrations in the gas phase (CO_2=40% vol., representing the typical CO_2 level in the gaseous effluent from gasification processes), the significant CO_2 uptake potential displayed by the slag, with sequestration yields comparable to those achieved using the concentrated stream, made the carbonation process energetically more competitive since in this case the duties related to the capture step were not considered.

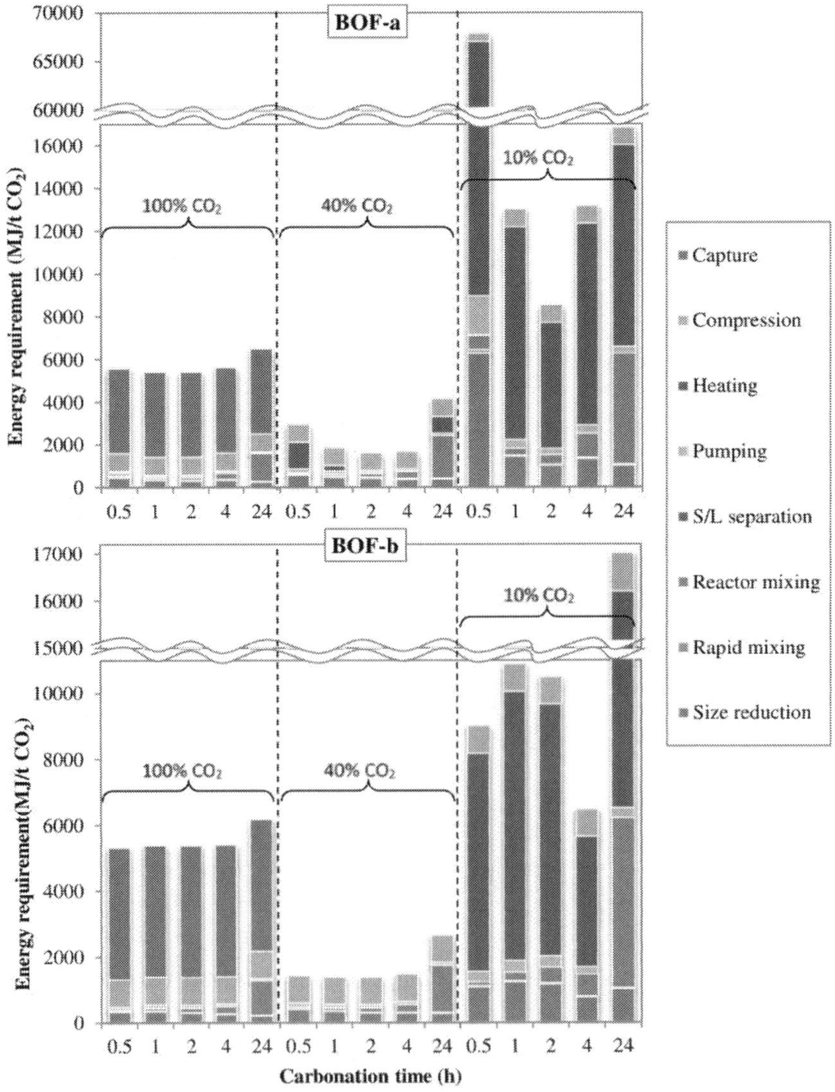

Figure 4. Energy requirement of each unit operation as a function of the carbonation conditions (slurry-phase route).

The relative contribution of the individual unit operations to the total energy consumption was also found to vary with the specific process condition adopted. In particular, at a CO_2 concentration of 10% the most energy-demanding stage was the heating phase, with energy consumptions for BOF-a ranging from 5900 to 58100 MJ/t CO_2 for 2-h and 0.5-h carbonation (corresponding to 69 and 86% of the total energy demand) and for BOF-b ranging from 3900 to 9700 MJ/t CO_2 for 4-h and 24-h carbonation (61 and 57% of the total energy demand). For long carbonation durations (24 h) the relative incidence of the energy consumption associated to slurry mixing inside the reactor became appreciable, with a contribution of ~5200 MJ/t CO_2 for both slag samples (approx. 30% of total consumption). On the other hand, for BOF-a and a short carbonation time (0.5 h), for which the CO_2 uptake was of only 1.3%, the high amount of solid material to be used in the process implied a significant energy consumption (~6300 MJ/t CO_2; 12% of total consumption) for slag milling.

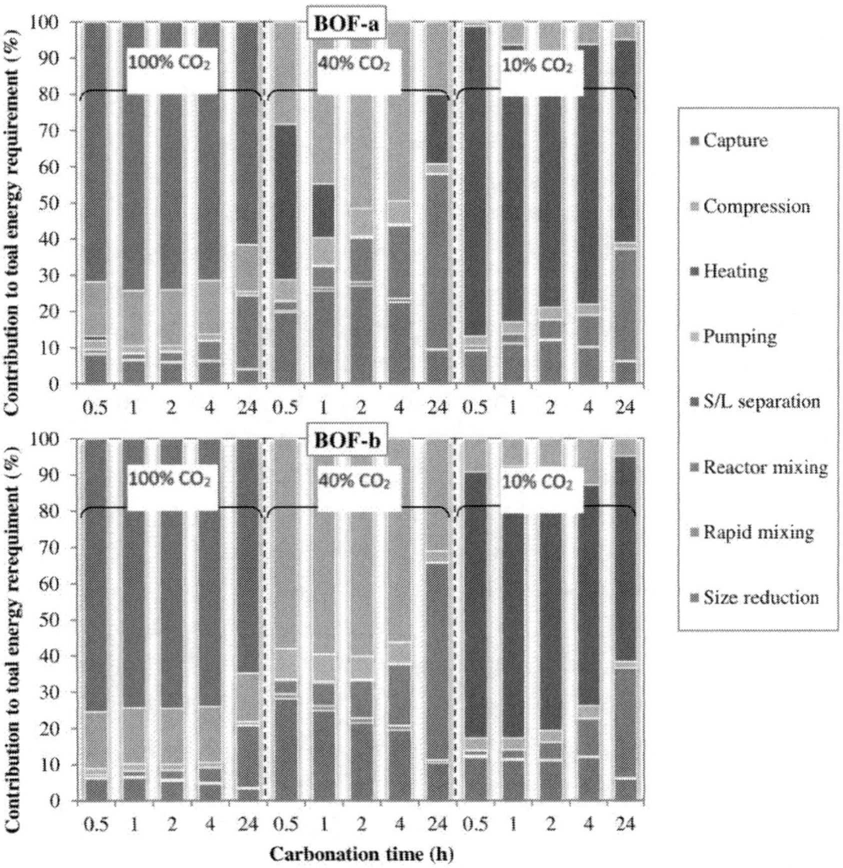

Figure 5. Relative contribution of each unit operation to the total energy consumption as a function of the carbonation conditions (slurry-phase route).

Compared to the results described above, the addition of a preliminary capture stage to the carbonation process layout (i.e. the use of a 100% CO_2 gas flow) resulted in a significant decrease in the energy requirement of all process stages (with the exception of the compression unit), and particularly of the heating stage, which was of several orders of magnitude lower than that observed when working with a gas flow containing 10% CO_2. The overall energy requirement was reduced by 1.2–2.8 times for BOF-b and 1.6–12.2 times for BOF-a compared to the use of a 10%-CO_2 effluent. In absolute terms, the calculated ranges for the total energy consumption were 5400–6500 and 5300–6200 MJ/t CO_2 for BOF-a and BOF-b, respectively, with the highest values being in both cases associated to a carbonation time of 24 h. Using the concentrated gaseous effluent, the most relevant contribution to the overall energy demand was provided by the CO_2 capture stage, which accounted for 62–74% and 65–75% of the total requirements for BOF-a and BOF-b; the gas compression stage had also an appreciable energy duty (~830 MJ/t CO_2), contributing by 12–15% and 13–16% of the overall demand for the two slag samples. The other terms were notably less relevant, except for the contribution of slurry mixing in the carbonation reactor for the longest residence time (24 h), which was of ~1300 and ~1100 MJ/t CO_2 for BOF-a and BOF-b, corresponding to 20 and 17% of the total duty.

When employing a 40% CO_2 concentration in the gas phase, the total energy requirement of the process was found to be considerably lower than that estimated for the two cases discussed above. As mentioned earlier, the improved energetic performance of the process using a 40% CO_2 effluent was due to the fact that, while the reactivity of the slag upon carbonation was comparable to that achieved with pure CO_2, the absence of the capture stage in the process layout allowed to save a significant share of energy. In this case, the overall energy requirements of the CO_2 sequestration process for the BOF-a and BOF-b slags lay within the ranges 1600–4200 and 1400–2700 MJ/t CO_2. Relevant contributions to the energy penalties of the process were provided by the gas compression and size reduction units, while the duty of slurry mixing was only relevant for prolonged contact times; for the BOF-a slag the energy requirement of the heating operation was also appreciable for 0.5-h and 24-h carbonation. In absolute terms, the BOF-a slag had the following associated main energy requirements: ~830 MJ/t CO_2 for gas compression, 380–580 MJ/t CO_2 for size reduction, 70–2000 MJ/t CO_2 for slurry mixing and 0–1300 MJ/t CO_2 for heating. The corresponding values calculated for the BOF-b sample were as follows: ~830 MJ/t CO_2 for gas compression, 280–400 MJ/t CO_2 for size reduction, 50–1400 MJ/t CO_2 for slurry mixing, while heating theoretically required no net energy input thanks to the heat recovered from the exhaust slurry recycled to the first heater (see Figure 2).

The specific energy requirements of each unit operation for the wet route are reported in Figure 6 making reference to the experimental results achieved for the BOF-c and BOF-d slags. The relative contributions of the process units to the total energy consumption are reported in Figure 7. As can be observed, for the case in which carbonation was assumed to be applied on a CO_2 pure stream obtained after capture, the total energy requirement as well as the contributions of the different unit operations for the two BOF slag samples analyzed appeared to be very similar. In particular, for BOF-c slag the overall energy requirements ranged from ~6000 (at 48 h) to 7610 MJ/t CO_2 (at 0.5 h), while for sample BOF-d they varied from 6036 (at 24 h) to

6590 MJ/t CO_2 (at 0.5 h). It should be noted that in this case, differently from the slurry phase route, due to the increase in the CO_2 uptake of the slag for longer reaction times, the energy requirements showed to decrease progressively. This may be related to the fact that in this case the heating requirements resulted significantly lower compared to the slurry-phase route even though heat recovery was not taken into account for the wet route.[10] However, it should be acknowledged with reference to both the slurry-phase and wet route that for an industrial scale application of carbonation as a CCS technique, the adoption of long reaction times (i.e. above 1-2 h) would be unfeasible. As shown in Figure 7, the main contributions to the energy requirements were CO_2 capture (~50-70%), milling and heating of the residues (~10-20%) and CO_2 compression (~13%).

As for the tests carried out with 40% CO_2 gas flows, similarly to what reported for the slurry-phase route, owing to the fact that capture was not assumed to be applied in this case, a notable reduction in the total energy requirements was achieved. In particular, the overall requirement ranged from 2550 to 5920 MJ/t CO_2 for sample BOF-c and from 2220 to 2850 MJ/t $CO_2$for BOF-d, increasing also in this case progressively with the reaction time. The higher requirements resulting for the first type of slag especially for short durations may be related to the lower CO_2 uptakes obtained which in turn may be ascribed to the differences in the mineralogy of the two slags (type c being mainly made up by silicate phases presenting a slower reaction kinetics compared to hydroxides). It should however be noted that for this type of slag the tests with 40% CO_2 were carried out applying a total pressure of 7 bar. As for the contributions of the single unit processes, the most relevant for BOF-c appeared to be milling (40-50%), followed by heating (25-35%) and CO_2 compression (15-30%), whereas for BOF-d the three above mentioned processes appeared to exert a similar weight (30-40%) on the total energy demand.

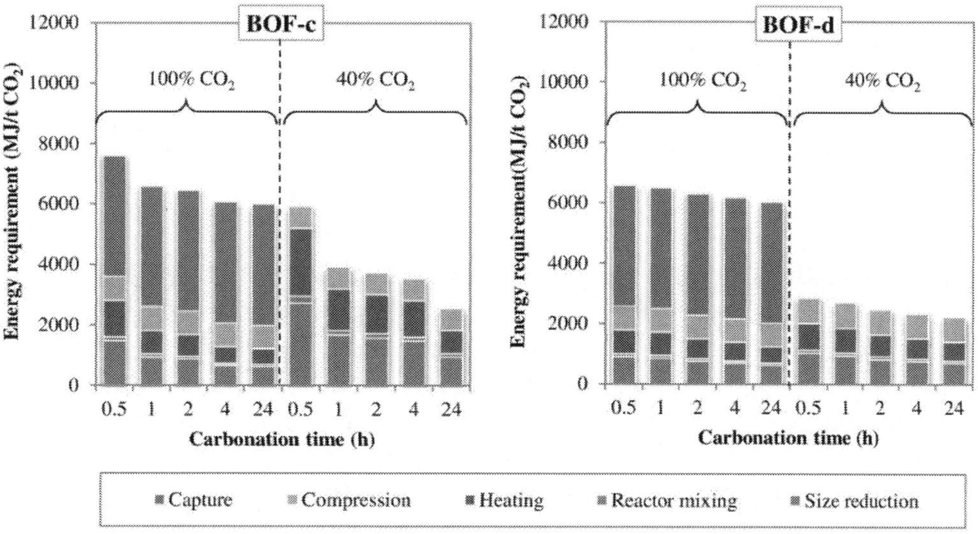

Figure 6. Energy requirement of each unit operation as a function of the carbonation

conditions (wet route).

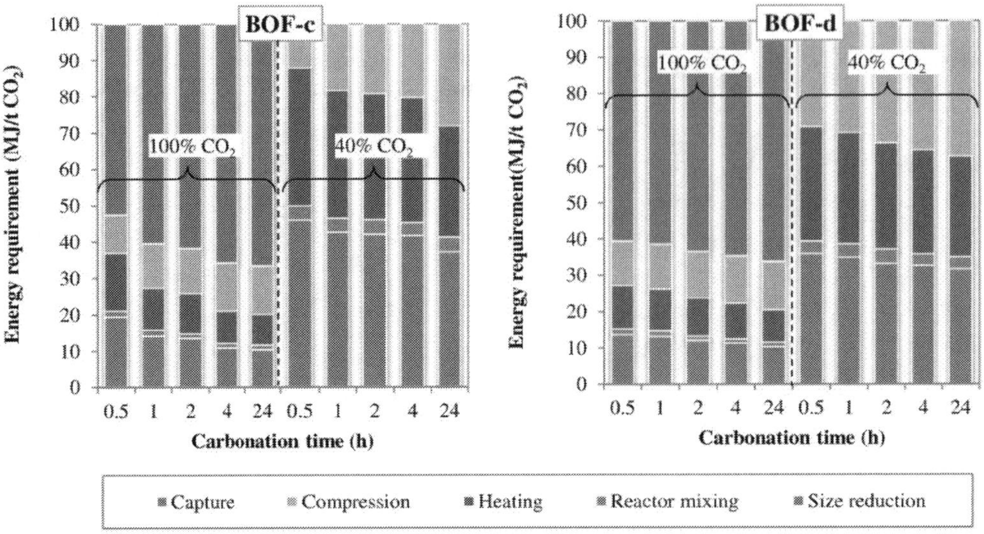

Figure 7. Relative contribution of each unit operation to the total energy consumption as a function of the carbonation conditions (wet route).

In order to derive some considerations of wider applicability concerning the dependence of the total energy requirement of slurry-phase and wet carbonation from the CO_2 uptake of the treated residues, the total energy requirements excluding CO_2 capture obtained for the 4 types of tested slags, the 2 routes and different gas compositions and reaction times tested were all reported together in the same graph as a function of the CO_2 uptake (see Figure 8). The trend of the data points showed to be adequately described by the following theoretical correlation function (see Equation 2) that for a null CO_2 uptake tends to infinity and remains quite constant for CO_2 uptakes close to the maximum achievable ones that depend on the chemical and mineralogical composition of the material:

$$E_{tot} = a + \frac{b}{{CO_2}_{upt}^{\ 1.5}} \qquad (2)$$

With a=1524.2 MJ/tCO_2 and b=101.5 MJ t$CO_2^{0.5}$/tslag$^{1.5}$. This trend highlights the key role of the CO_2 uptake in determining the overall energy requirements related to the carbonation process in particular for low reaction yields (<10 % wt.). It is interesting to note that for the BOF slag samples analyzed in this study, regardless of the type of process route applied and operating conditions selected (apart from the 40% wet and 10% slurry tests) an estimate of the total energy requirement of the process can be directly made on the basis of the CO_2 uptake of the residues.

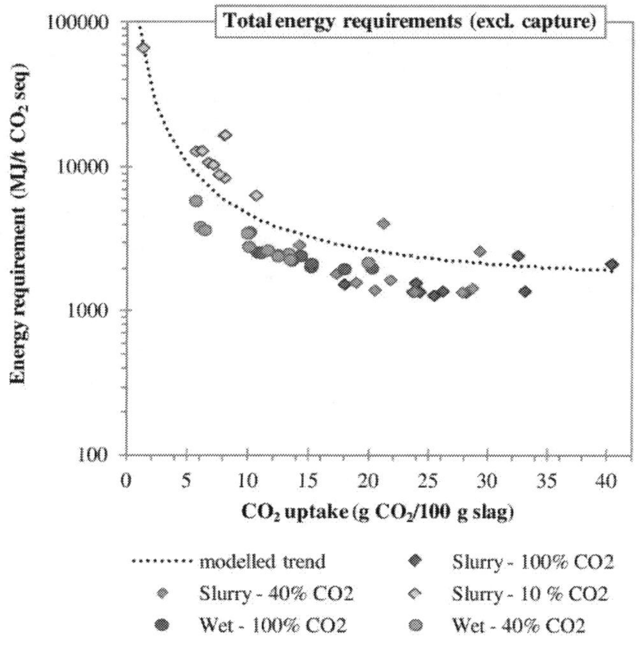

Figure 8. Trend of total energy requirement (excluding CO_2 capture) of the slurry-phase and wet routes as a function of the carbonation yield.

Conclusions

In this study the possibility of directly storing the CO_2 contained in diluted gas flows so to skip the energy intensive capture step was assessed by accelerated carbonation of BOF steelmaking slags both via slurry-phase and wet route batch scale experiments. The maximum CO_2 uptakes obtained with the slurry-phase route ranged from 8% wt. at 10% CO_2, to 21-29% at 40% CO_2 and 32.5-40% at 100% CO_2. For the wet route, maximum uptakes of 13-19.5% wt. at 40% CO_2 and 18-20% at 100% CO_2 were attained. The slurry-phase route hence showed to yield significantly higher uptakes than the wet route. However, it should be noted that this second process operates at lower temperature and makes use of significantly less water than the former one. Hence, the overall performance of these two routes was assessed also in terms of their total energy requirements on the basis of the energy demand of each their unit operations. The results of this evaluation indicated that working with a 40%-CO_2 effluent turned out to be the most energy efficient carbonation strategy for BOF steelmaking slag, at least under the experimental conditions tested, applying either the slurry-phase or the wet route. For the slurry-phase process the most suitable conditions in terms of specific energy requirements were found for both slag samples tested at a carbonation time of 2 h, and implied a total energy duty of 1610 and 1380 MJ/t CO_2 for BOF-a and BOF-b slag samples, respectively. For the wet route, the lowest energy requirements were obtained for the highest reaction times tested (2550 and 2200 MJ/t CO_2 after 24 h for BOF-c and BOF-d slag samples, respectively). In particular, the total energy requirement for both types of routes regardless of the specific properties of the residues and the operating conditions tested showed to be directly related to the CO_2 uptake of the slag. Hence, in conclusion the results of this study suggest the feasibility of applying accelerated carbonation of BOF steel slag as a process to directly store the CO_2 content of diluted gas flows such as syngas or biogas. The most promising reaction route (from an energy requirement perspective) appeared to be the slurry-phase route. However, an overall assessment of the carbonation process should also take into account that the higher CO_2 sequestration performance attained by the slurry-phase route may be counterbalanced by an increased process complexity and the production of a final liquid effluent that needs further treatment prior to discharge. For this reason, prior to the selection of a specific reaction route, additional evaluations that take into consideration potential environmental impacts of the process through a life cycle approach, as well as technical and economic aspects, are required.

References

1. Seifritz, W. 1990. CO_2 disposal by means of silicates. *Nature*, 345-486.
2. Lackner, K.S., Wendt, C.H., Butt, D., Joyce, E.L. Jr., Sharp, D.H. 1995. Carbon dioxide disposal in carbonate minerals. *Energy*, 20, 1153-1170.
3. Bobicki, E.R., Liu, Q., Xu, Z., Zeng, H. 2012. Carbon capture and storage using alkaline industrial wastes. *Progress in Energy and Combustion Science*, 38, 302-320.
4. Pan, S.Y., Chang, E.E., Chiang, P.C. 2012. CO_2 capture by accelerated carbonation of alkaline wastes: a review on its principles and applications. *Aerosol and Air Quality Research*, 12, 770-791.
5. Huijgen, W.W.J., Witkamp, G.J., Comans, R.N.J. 2006. Mechanisms of aqueous wollastonite carbonation as a possible CO_2 sequestration process. *Chemical Engineering Science*, 61, 4242-4251.
6. Baciocchi, R., Costa, G., Di Gianfilippo, M., Polettini, A., Pomi, R., Stramazzo, A. 2015. Thin-film versus slurry-phase carbonation of stee lslag: CO_2 uptake andeffects on mineralogy. *Jounal of Hazardous Materials*, 283, 302-313.
7. Morone, M., Costa, G., Polettini, A., Pomi, R., Baciocchi, R. 2014. Valorization of steel slag by a combined carbonation and granulation treatment. *Minerals Engineering*, 59, 82-90.
8. Huijgen, W.J.J., Witkamp, G.J., Comans, R.N.J. 2005. Mineral CO_2 sequestration by steel slag carbonation. *Environmental Science &Technology*, 39, 9676-9682.
9. Baciocchi, R. , Costa, G. , Di Bartolomeo, E., Polettini, A., Pomi, R. 2011. Wet versus slurry carbonation of EAF steel slag. *Greenhouse Gases Science and Technology*, 1 (4), 312-319.
10. Zingaretti, D., Costa, G., Baciocchi, R. 2014. Assessment of Accelerated Carbonation Processes for CO_2 Storage Using Alkaline Industrial Residues. *Industrial Engineering & Chemistry Research*, 53, 9311-9324.
11. Intergovernmental Panel on Climate Change (IPPC) 2005. *IPCC special report on carbon dioxide capture and storage*. Cambridge University Press, Cambridge, United Kingdom and New York, USA.
12. Santos, R.M., François, D., Mertens, G., Elsen, J., Van Gerven, T. 2013. Ultrasound-intensified mineral carbonation. *Applied Thermal Engineering*, 57 (1-2), 154-163.

Investigation of the effects of alkali activation and CO_2 curing on BOF steel slag

M. Morone[1*], Ö. Cizer[2], G. Costa[1] and R. Baciocchi[1]

[1] Laboratory of Environmental Engineering, Dept. of Civil Engineering and Computer Science Engineering, University of Rome "Tor Vergata", Rome, Italy

[2] Department of Civil Engineering, KU Leuven, 3001 Heverlee, Belgium

*Corresponding author: morone@ing.uniroma2.it

Abstract
The present work reports the results obtained by coupling alkali activation and CO_2 curing as a treatment for BOF slag, with the aim of improving the hydraulic reactivity of the residue while also exploiting its potential as carbon sink. The slag was mixed with two different solutions: NaOH (2M)+Na_2SiO_3 (50:50 by wt.) and NaOH (4M)+Na_2CO_3 (2M) (75:25 % by wt.) and then cured for 3, 7 and 28 days, in either a humidity chamber (RH=90% and T=20 °C) or in a carbonation chamber (RH=75%, CO_2=5%) at both environmental and enhanced conditions (T=50 °C). TGA results, confirmed by XRD analysis, showed an increasing loss in weight corresponding to the decomposition of carbonates by increasing the curing time for both activators and curing conditions, while a significant occurrence of gaylussite at early curing times and low curing temperature was observed, mainly for the sodium carbonate solution. The highest CO_2 uptake (6% wt. slag) was achieved for the BOF slag samples mixed with the sodium hydroxide/sodium carbonate solution after 28 days of curing in the carbonation chamber, regardless of the temperature adopted. The maximum value of the compressive strength (2 MPa) was obtained after 7 days of curing in the carbonation chamber at 50 °C for mortar samples prepared with sodium silicate.

Introduction

The proper management of industrial waste and the reduction of greenhouse gas (GHG) emissions represent two of the main environmental concerns of the industrialized world. This is particularly relevant for the steel making sector, that on the one hand generates large amounts of solid residues, whose reuse is not always feasible, and on the other one accounts for nearly 10% of global anthropogenic CO_2 emissions.[1] One of the most abundant and scarcely reused by-products of the steelmaking process is the BOF (Basic Oxygen Furnace) slag that, referring to EU data of 2012, represents almost 50% of the total steel slag production (23 million tons).[2] The reuse of this residue is still limited or undervalued, owing to both technical (e.g. volumetric expansion, particle size distribution) and environmental aspects (leaching behaviour), that may not allow to meet the compliance criteria for reuse in civil engineering applications or environmental reclamation. In addition, although its mineral composition may suggest possible recycling as cementitious binder in the manufacturing of concrete similarly to blast furnace slag, this type of slag, owing to its high content of iron oxide which is not reactive with water and di-calcium silicate, characterized by a slow hydration kinetics, may present a poor hydraulic activity.[3] In this regard, different experimental studies performed on blast furnace slag, which is typically used in concrete manufacturing as a partial replacement of Portland cement, showed that alkali activation can prove to be an effective treatment to stimulate the material's latent pozzolanic properties and to obtain an improved mechanical performance.[4] Alkali activation is a chemical process in which the raw material, usually amorphous (e.g. blast furnace slag or fly ash), is mixed with specific alkaline solutions (e.g. sodium or potassium alkalis, carbonates or silicates) in order to favour the dissolution of the reacting species and to enhance the hydration reactions leading to the formation of binding phases, such as a calcium silicate gel, that yields a compact and strong product.[5] Recent studies[6,7] found that also crystalline and non-hydraulic steel slag (e.g. continuous casting slag) exhibited cementitious properties if properly activated, developing enough strength to produce a valuable final product. Nevertheless, besides the alkali activation process, as for all cementitious materials, a curing stage is also needed in order to allow the completion of the hydration reactions. Optimizing the curing operating conditions, such as relative humidity, temperature, time of exposition and CO_2 concentration, is required to attain a final product with higher quality.[8] In particular, carrying out the curing treatment in a CO_2 rich atmosphere showed to provide the material a similar compressive strength to that obtained with steam curing, that is the most widely adopted method in the cement field[9,10] but also presents high energy requirements. In addition, different authors have reported that increasing the temperature during the curing phase, regardless of the activator, accelerated the rate of the hydration reactions at early ages, leading to a higher mechanical strength of the activated-slag mortars.[11,12] On the other hand, owing to the fact that the BOF slag has a typically alkaline nature and an abundance of phases containing calcium and magnesium that are potentially reactive towards CO_2, it is possible to take advantage of the required curing step as a valuable technique for storing CO_2 in a thermodynamically stable and solid form. In this regard, aside from the activation of the hydration reactions, alkali activation may contribute in promoting the uptake of CO_2 and carbonation reactions by increasing the alkalinity of the aqueous solution in contact with the slag material.

In this work, we investigate the sequential combination of alkali activation and CO_2 curing applied to BOF steel slag as a valorisation technique for improving the hydraulic reactivity of the residue, while also exploiting its potential as a carbon sink. Specifically, the paper reports the effects of two CO_2 curing temperatures (20 and 50 °C) on the thermal, mineralogical and mechanical properties, along with the extent of CO_2 uptake of BOF slag activated using a sodium hydroxide/sodium silicate and a sodium hydroxide/sodium carbonate solution. The effects of the treatment in terms of the composition and mineralogy of the slag was tested on paste samples, while its influence on the mechanical performance was assessed on mortar specimens.

Materials and methods

Materials

The sample analysed in this study was collected at the outlet of the basic oxygen furnace (BOF) of an Italian steel plant employing the integrated steelmaking process, after crushing and magnetic separation for iron and metal recovery. The chemical characterization of the slag was obtained with the alkaline fusion method, consisting of mixing the milled material with $Li_2B_4O_7$, heating at 1050 °C in a muffle and dissolving the material in a solution of nitric acid, at concentration of 4%; thus, the achieved solution was analysed by ICP-OES. The main chemical constituents, reported in Table 1, were Ca (265 g/kg) and Fe (221 g/kg), followed by Mg (45.7 g/kg) and Mn (41.5 g/kg). Two different alkaline solutions were selected for evaluating the effects of alkali activation on the long-term reactions of the slag. A mixture (50:50 weight ratio) of sodium hydroxide 2M and a commercial solution of sodium silicate (ABCR Gmbh Co. KG, 39-40% silicates in water). The silica modulus of the solution, expressed as the molar ratio of SiO_2 and Na_2O, resulted equal to 2.3. The other alkali solution was prepared by mixing sodium hydroxide 4M and sodium carbonate 2M in 75:25 weight ratio. In the latter case, the molarities of the sodium carbonate solution were set in order to achieve the same sodium oxide content of the sodium hydroxide solutions that resulted equal to 10.5%.

Table 1. Elemental composition of the BOF slag, expressed as g/kg for major elements and as mg/kg for trace elements.

Major elements	Concentration (g/kg)	Element	Concentration (mg/kg)
Al	9.7	Ba	56.5
Ca	265	Cd	193.1
Fe	221.2	Cr	1343.1
Mg	45.7	Cu	273.5
Mn	41.45	K	577.6
Na	1.86	V	530.2
Si	22	Zn	281.1

Methods

The slag pastes (slag + activator) were prepared in a plastic mould of 2.5 x 2.5 x 2 cm^3, using about 15g of the as-received material and an alkali solution to slag ratio equal to 0.2 l/kg. After mixing the slag with the activator solution for three minutes, the pastes were cured for 3, 7 and 28 days in the curing chambers at the following conditions: (i) humidity chamber (HC) at RH=98% and T=22 °C used as reference; (ii) carbonation chamber at 5% CO_2 concentration and 75% relative humidity, but adopting two different temperatures: 20 °C (CC20) and 50 °C (CC50). The relative humidity in the carbonation chamber was kept constant by using a saturated K_2CO_3 solution. After the curing time, the pastes were ground in an agate mortar and dried in a vacuum-drier (Alpha 1-2 LD, Martin Christ) for 2 hours at 0.011 mbar, according to the sublimation curve of water, in order to remove the free water content and stop further hydration reactions. The crushed pastes were analysed in terms of their mineralogy, in order to identify the reaction products obtained for the different alkali solutions and curing conditions tested. The mineralogical compositions of both untreated material and activated slag pastes was evaluated by powder X-ray diffraction (XRD) analysis with Cu Ka radiation using a Philips Analytical PW 1710 diffractometer, operating at 45 kV and 35 mA, with an angular step of 0.02° held for 2s over a range of 5-75 2θ. The thermo-gravimetric analysis (TGA) was carried out with a Netzsch STA 409 PC DSC-TGA equipment in an atmosphere of pure N_2 at 60 ml/min, over a temperature range of 20-1000 °C, at a heating rate of 10 °C/min and the differential thermal gravimetric (DTG) data were used to identify the reaction products. The degree of carbonation of the samples, assessed by measuring their inorganic carbon content (IC) using a Shimadzu TOC VCPH analyser equipped with a SSM-5000A solid sampler, was calculated by applying the following equation:

$$CO_{2uptake}(\%) = \frac{CO_{2final}(\%) - CO_{2initial}(\%)}{100 - CO_{2final}(\%)} \cdot 100 \quad (1)$$

where the IC concentration, expressed in terms of the CO_2 weight percent, of the untreated sample ($CO_{2\ initial}$) and of the treated one ($CO_{2\ final}$) are reported.

The mortar samples for the compressive strength analysis were prepared following the EN 196-1 standard method. Firstly the material was mixed with 0-2 mm sand at a 1:3 slag to sand ratio, and then the activator was added, at a liquid to slag ratio set equal to 0.9 in order to maintain the workability of the mortars. Uniformly, each alkali activated mixture was cast in three steel moulds (4 x 4 x 16 cm^3) and compacted for 3 minutes in a jolting machine so to remove air bubbles. The slag mortars were cured in the carbonation chamber, keeping the same curing conditions and time as for the slag pastes. In fact, based on the crushing resistance of the slag pastes, it was expected that the mortar specimen would not acquire a suitable mechanical resistance for this test if cured in the humidity chamber, that was hence not considered as one of the curing conditions for testing the mortar samples. Once achieved a suitable consistency, the mortar beams were cut in cubes (4 x 4 x 4 cm^3), and then placed back into the CO_2 chamber to complete the curing treatment. The mechanical strength of the mortar cubes was assessed in duplicate by using the Schenck-RM 100 compression testing machine, with a loading rate set equal to 2 mm/min.

Results and discussion

Slag pastes

DTG results. The findings obtained from the thermal analysis showed that different phases may form upon alkali activation and CO_2 curing, whose nature depends on the type of activator solution and curing conditions applied. Figure 1 reports the results obtained for the samples activated with the sodium hydroxide/sodium silicate solution. It can be observed that the samples cured in the humidity chamber and carbonation chamber at 20 °C were characterized by an evident peak at around 100 °C, which corresponds to the dehydration of bound water in C-S-H type reaction products.[13] This phase is an hydration product which is expected to form as a result of the hydration of the crystalline calcium silicate phase present in the slag (see Figure 3), consistently with the results reported in [6] and [7]. Upon carbonation at 50 °C, the formation of a slightly sharp peak at around 100-120 °C was observed, that may be related once again to the formation of C-S-H like phase, or to the decomposition temperature of gaylussite ($Na_2Ca(CO_3)_2{\bullet}5H_2O$), a hydrated sodium carbonate, whose dehydration usually takes place at a temperature below 250 °C.[14] Formation of calcite, identified by the peak at around 650-700 °C, was instead observed at all curing conditions, regardless of the temperature adopted.

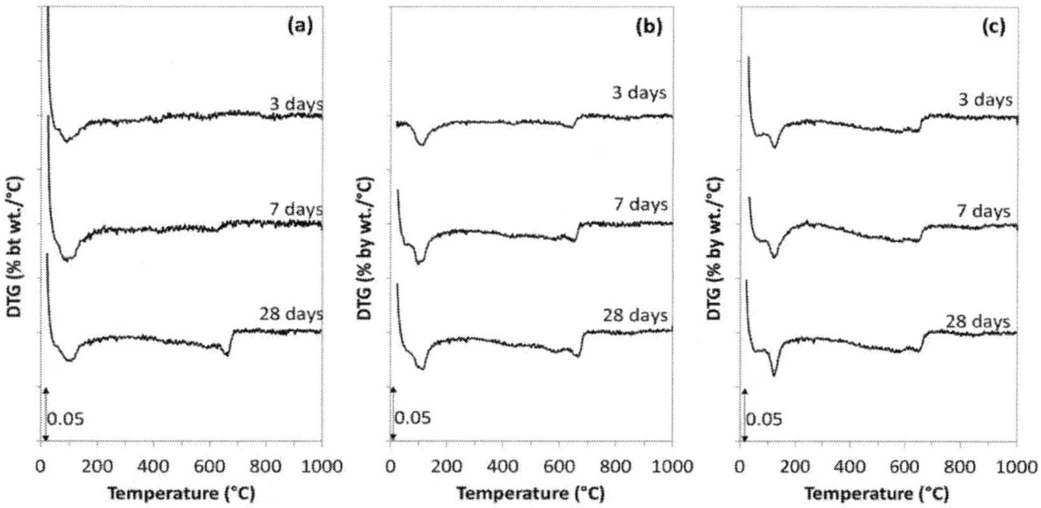

Figure 1. DTG plot of BOF slag pastes activated with NaOH 2M + Na_2SiO_3_50:50 solution and cured, at different times, in (a) the humidity chamber and in the carbonation chamber at a temperature of (b) 20 °C and (c) 50 °C.

The thermal analysis results obtained by mixing the slag with the sodium hydroxide/sodium carbonate solution are reported in Figure 2. It can be seen that most of the mass loss from the samples cured in the humidity and in the carbonation chamber at 20 °C occurred at temperatures below 200 °C. For these curing conditions, an evident narrow peak was observed at around 100-120 °C, whose intensity showed to be highly similar for all the curing times. This peak, whose shape resulted sharper than that observed for the samples activated with the Na-silicate solution, was again associated to gaylussite ($Na_2Ca(CO_3)_2{\bullet}5H_2O$).[13] There is evidence in other literature studies of the occurrence of this phase as a product of the hydration[15] and carbonation[16] reaction of the blast furnace slag activated with a sodium carbonate solution. Its formation may be ascribed to the additional supplying of carbonate ions, resulting from the dissociation of Na_2CO_3, along with those obtained from the hydration of CO_2 that, in combination with the Ca^{2+} obtained from the solid dissolution, forms such a carbonate phase.[15] In addition, it was observed that the intensity of this peak resulted lower at higher curing temperature. In this case, for all the curing times tested, a small shoulder at around 180 °C was also noticed, that can be once again associated to the loss of water from the hydration of the sodium-calcium carbonate phase.[14] Finally, for all the tested curing conditions, a second main peak at around 650 °C was observed, which is related to the decomposition temperature of calcium carbonate. At ambient CO_2 curing temperature (see Figure 2b), the amount of calcium carbonates showed to increase with time for a curing period up to 7 days, remaining constant until 28 days, indicating an increased carbonation of the slag paste with time. Conversely, similar intensities for all the curing times tested were obtained at higher curing temperature, probably related to an enhanced carbonation kinetics with respect to the samples cured in the CO_2 chamber at ambient temperature.

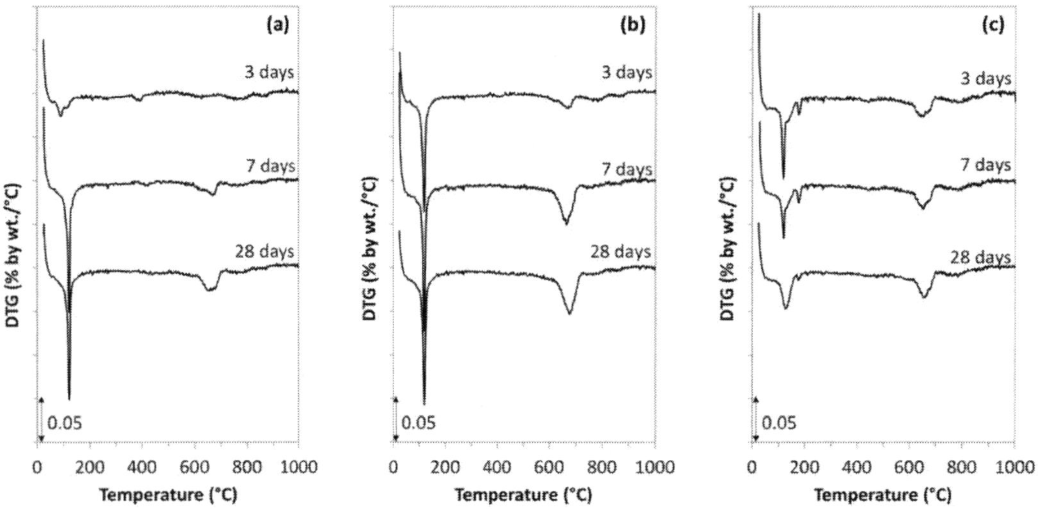

Figure 2. DTG plot of BOF slag pastes activated with NaOH 4M + Na_2CO_3 2M_75:25 solution and cured, at different times, in (a) the humidity chamber and in the carbonation chamber at a temperature of (b) 20 °C and (c) 50 °C.

XRD results. The mineralogy of the samples activated with both solutions, cured for 7 and 28 days at all the tested curing conditions, is reported in Figure 3, and compared with that of the untreated BOF slag. The untreated BOF slag exhibited different peaks of Ca phases, like Ca-containing oxides and Ca-silicate, along with other oxides, such as iron oxide and Ca-Al-Fe oxide. Upon alkali activation with the sodium hydroxide/sodium silicate solution (see Figure 3i), the occurrence of calcite peaks was observed for all curing conditions. The XRD plots were also characterized by a broad hump (44°-52° 2θ), which may be attributed to an amorphous formation, likely a C-S-H type reaction product. Peaks of gaylussite, which based on the DTG plots seemed to form as well in the samples cured in the carbonation chamber at 50 °C, were not retrieved. This result was tentatively ascribed to the higher detection limit of this type of analysis, which may not allow to detect the low intensity peak of gaylussite. Upon activation with the sodium hydroxide/sodium carbonate solution (see Figure 3ii), the occurrence of calcite was noticed for all curing conditions, as reported for the other solution, but with a higher intensity for higher curing temperature. In addition to calcite, different peaks of gaylussite were also observed, in agreement with the results obtained from the DTG analysis. In this case, it is worth noting that these peaks appeared only for two curing conditions, i.e. humidity and carbonation chamber at 20 °C. Moreover, for the latter condition, it resulted that the ratio between the intensity of the peaks of gaylussite and calcite was higher after 28 days of curing than 7 days, indicating that in the former case gaylussite represented the main carbonate phase. Conversely, the main and newly formed peak detected upon activation and curing in the carbonation chamber at 50 °C was only calcite rather than gaylussite, consistently with the peaks' intensity observed in the DTG plots.

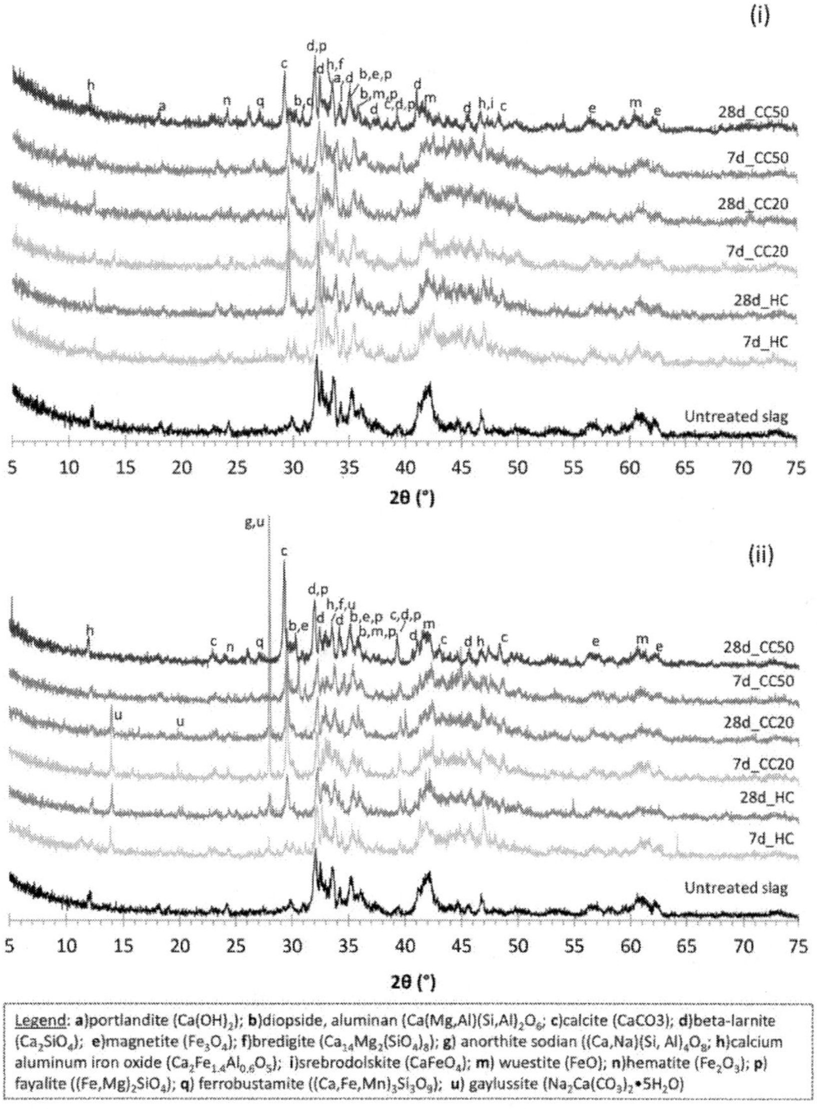

Figure 3. XRD pattern of the untreated BOF slag and the BOF pastes mixed with (i) NaOH 2M + Na_2SiO_3_50:50 solution and (ii) NaOH 4M + Na_2CO_3 2M_75:25 solution, cured in the humidity chamber (HC), carbonation chamber at 20 °C (CC20) and 50 °C (CC50) for 7 or 28 days.

CO_2 uptake. The CO_2 uptake of the activated samples is reported in Figure 4. It can be observed that the degree of carbonation of samples activated with both alkaline solutions and cured in the humidity chamber showed to increase with time, achieving a maximum uptake of 4% after 28 days. Upon CO_2 curing, the uptake values showed to increase for both the curing temperatures tested, with respect to those retrieved for the humidity chamber. In particular, the samples activated with sodium hydroxide/sodium silicate solution (see Figure 4a) showed a maximum CO_2 sequestration of 5.2% upon curing in the carbonation chamber at 20 °C after 28 days, slightly lower than that obtained by mixing the slag with the sodium hydroxide/sodium carbonate solution (see Figure 4b). It is interesting to notice that for both activators, the uptake after 3 days of CO_2 curing at 50 °C resulted higher than that obtained at 20 °C, likely due to the enhancement of the dissolution rate of the reactive minerals (e.g. silicates) at higher temperature. Nevertheless, upon carbonation at 50 °C, a steady CO_2 uptake trend was observed for the samples mixed with both types of alkali solutions, as also reported in the DTG plots where peaks of calcium carbonate of similar intensity were observed for all curing times. It should be underlined that alkali activation provides increased alkaline conditions to promote CO_2 uptake and thereby carbonation reactions. Hence, the low gain of CO_2 uptake or the steady trend observed with curing time at low and high temperature, respectively, may be the result of two possibly combined effects: a limited solubility of Ca-containing minerals present in the slag and/or a limited diffusion of CO_2 due to the rapid precipitation of calcium carbonates that filled the pores of the slag paste.

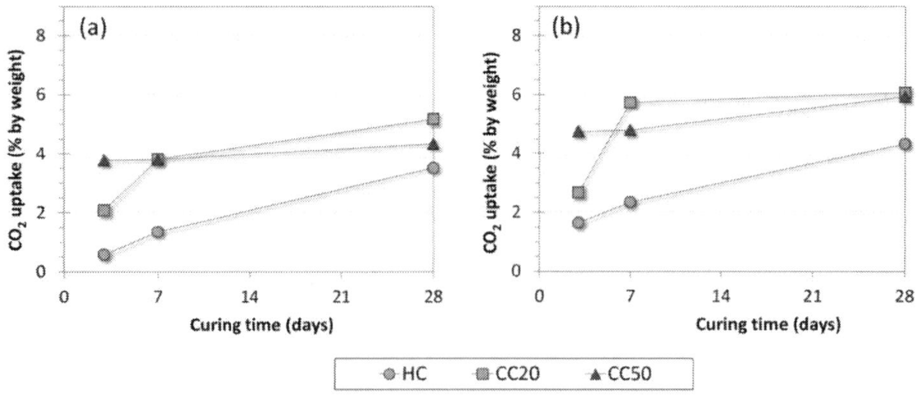

Figure 4. CO_2 uptake with curing time of BOF slag pastes mixed with (a) sodium hydroxide/sodium silicate solution and (b) sodium hydroxide/sodium carbonate solution, cured in the humidity chamber (HC), in the carbonation chamber at 20 °C (CC20) and at 50 °C (CC50).

Slag mortars

Mechanical strength. The compressive strength of the activated BOF slag mortars, reported in Figure 5, showed to be affected by both curing temperature and type of alkaline solution used. Indeed, samples activated with the sodium hydroxide/sodium carbonate solution at 20 °C showed negligible strength for all the curing times. Increasing the curing temperature to 50 °C allowed to achieve a maximum strength of almost 1 MPa after 28 days. On the other hand, activating the slag with the sodium hydroxide/sodium silicate solution allowed to obtain a relevant compressive strength at 20 °C already after 7 days of curing, which increased further to about 1 MPa after 28 days of curing. With this activator, an early strength gain was also noticed at 50 °C, as a strength of 2 MPa was already reached after 7 days. Moreover, it appeared that increasing the curing time did not lead to further strength increase. Working on alkali activation of blast furnace slag, Puertas and Torres-Carrasco[17] observed a similar strength behaviour, i.e. lower values were retrieved by blending the slag with a solution of NaOH and Na_2CO_3, rather than with sodium silicate. The authors associated this result to the nature of the reaction products related to the different nature of the activators. Indeed, they found that the main influence on the final mechanical strength of the mortars was related to the formation of a C-S-H like phase that led to a more condensed structure and hence to a lower porosity of the slag mortar. This result could be also the explanation of the findings attained in this study, as the mechanical performances of the BOF slag mortars prepared with the Na-silicate activator may be associated to the formation of a C-S-H like phase or to a less porous structure. In order to confirm this hypothesis, the micro-porosity of the mortars could be investigated in order to understand the correlation between this parameter and the strength of the samples.

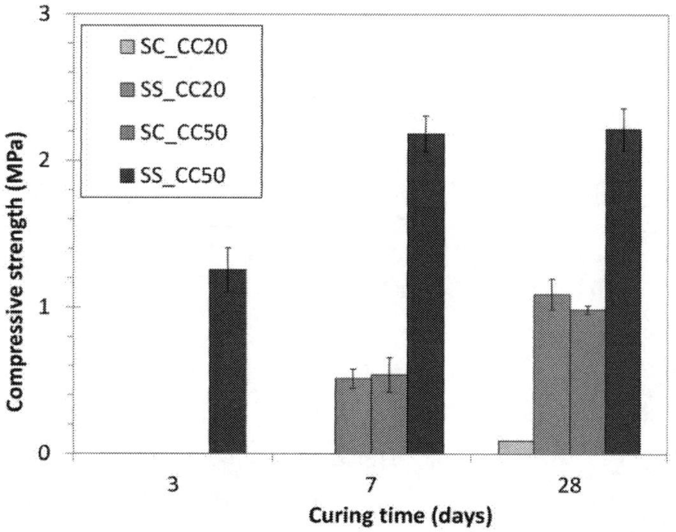

Figure 5. Compressive strength of mortar beams prepared with BOF slag, activated with sodium hydroxide/sodium silicate (SS) and sodium hydroxide/sodium carbonate (SC) solutions and cured in the carbonation chamber at 20 °C (CC20) and 50 °C (CC50).

Conclusions

This paper reports the main results of an investigation carried out to exploit the alkali activation process coupled with CO_2 curing to improve the mechanical performance of BOF slag in mortar formulations, while allowing also to store some CO_2. Two alkaline solutions were tested, i.e. sodium hydroxide (2M) /sodium silicate (50:50 by wt.) and a solution containing NaOH (4M) and Na_2CO_3 (2M), 75:25 by wt. Thermal analysis indicated the formation of calcium carbonate with both types of solutions, while the probable formation of a calcium silicate hydrate like phase was observed for the samples activated with the sodium hydroxide/sodium silicate solution. Formation of a sodium and calcium carbonate, i.e. gaylussite, was observed for the samples activated with the sodium hydroxide/sodium carbonate solution, either cured in the humidity or carbonation chamber at 20 °C. Its formation was ascribed to the additional supplying of carbonate ions, resulting from the dissociation of the activator (Na_2CO_3), along with those obtained from the hydration of CO_2 that, in combination with the Ca^{2+} from the dissolution of Ca containing phases (e.g. larnite) should enhance the formation of this phase. The results in terms of CO_2 uptake showed that similar values (almost 6%) were reached upon 28 days in the CO_2 chamber at 20 °C with both the alkaline solutions, while an increase in the operating temperature from 20 to 50 °C did not seem to enhance the CO_2 uptake. The mechanical strength of BOF slag mortars showed to be affected by the nature of the alkali solution and by the curing temperature. In particular, the highest mechanical performance was achieved with the sodium hydroxide/sodium silicate solution, at a CO_2 curing temperature of 50 °C after both 7 and 28 days of curing.

References

1. Kuramochi, T., Ramírez, A., Turkenburg, W., Faaij, A. 2012. Comparative assessment of CO_2 capture technologies for carbon-intensive industrial processes. *Progress in energy and combustion science*, 38 (1), 87-112.
2. Euroslag Statistics, 2012. http://www.euroslag.com/products/statistics/2012/ (accessed on April 2015).
3. Belhadj, E., Diliberto, C., Lecomte, A. 2012. Characterization and activation of Basic Oxygen Furnace slag. *Cement and Concrete Composites*, 34 (1), 34-40.
4. Shi, C., Jiménez, A. F., Palomo, A. (2011). New cements for the 21st century: the pursuit of an alternative to Portland cement. *Cement and Concrete Research*, 41(7), 750-763.
5. Rashad, Alaa M. A. 2013. Comprehensive overview about the influence of different additives on the properties of alkali-activated slag–A guide for Civil Engineer. *Construction and Building Materials*, 47, 29-55.
6. Salman, M., Cizer, Ö., Pontikes, Y., Vandewalle, L., Blanpain, B., Van Balen, K. 2014. Effect of curing temperatures on the alkali activation of crystalline continuous casting stainless steel slag. *Construction and Building Materials*, 71, 308-316.

7. Salman, M., Cizer, Ö., Pontikes, Y., Snellings, R., Vandewalle, L., Blanpain, B., Van Balen, K. 2015. Cementitious binders from activated stainless steel refining slag and the effect of alkali solutions. *Journal of hazardous materials*, 286, 211-219.

8. Criado, M., Fernández-Jiménez, A., Palomo, A. 2010. Alkali activation of fly ash. Part III: Effect of curing conditions on reaction and its graphical description. *Fuel*, 89(11), 3185-3192.

9. Shi, C., He, F., Wu, Y. 2012. Effect of pre-conditioning on CO_2 curing of lightweight concrete blocks mixtures. *Construction and Building Materials*, 26(1), 257-267.

10. Zhan, B., Poon, C., Shi, C. 2013. CO_2 curing for improving the properties of concrete blocks containing recycled aggregates. *Cement and Concrete Composites*, 42, 1-8.

11. Shi, C., Day, R.L. 1996. Some factors affecting early hydration of alkali-slag cements. *Cement and Concrete Research*, 26 (3): 439-447.

12. Burciaga-Díaz, O., Díaz-Guillén, M. R., Fuentes, A. F., Escalante-Garcia, J. I. 2013. Mortars of alkali-activated blast furnace slag with high aggregate: binder ratios. *Construction and Building Materials*, 44, 607-614.

13. Taylor, H. F. (1997). Cement chemistry. Thomas Telford.

14. Johnson, D. R., Robb, W. A. 1973. Gaylussite: thermal properties by simultaneous thermal analysis. *American Mineralogist*, 58, 778-784.

15. Garcia-Lodeiro, I., Fernandez-Jimenez, A., Palomo, A. 2013. Hydration kinetics in hybrid binders: early reaction stages. *Cement and Concrete Composites*, 39, 82-92.

16. Bernal, S. A., Provis, J. L., Walkley, B., San Nicolas, R., Gehman, J. D., Brice, D. G., Kilcullen, A.R., Duxson, P., van Deventer, J. S. 2013. Gel nanostructure in alkali-activated binders based on slag and fly ash, and effects of accelerated carbonation. *Cement and Concrete Research*, 53, 127-144.

17. Puertas, F., Torres-Carrasco, M. 2014. Use of glass waste as an activator in the preparation of alkali-activated slag. Mechanical strength and paste characterization. *Cement and Concrete Research*, 57, 95-104.

Principles and kinetics of accelerated carbonation
Low Temperature Magnesite Formation

Power, I. M., The University of British Columbia
Kenward, P. A., University of British Columbia
Harrison, A. L., The University of British Columbia
Dipple, G. M., The University of British Columbia
Raudsepp, M., The University of British Columbia
Wilson, S. A., Monash University

Originally presented on: 6/22/2015 13:00:00 - 13:15:00

Mg-carbonate formation is of interest for both *ex situ* and *in situ* CO_2 sequestration strategies, yet precipitation of magnesite [$MgCO_3$], the most stable form, is kinetically inhibited at low temperatures. In the Mg-carbonate system, there are numerous metastable hydrated Mg-carbonate minerals, but magnesite is the most stable and has the greatest carbon density [1], making it a preferred product for sequestering CO_2. However, rates of magnesite precipitation are severely limited at temperatures below 60 °C due to the strong hydration of Mg^{2+} ions in solution [2]. Thus, understanding the rates of and controls on magnesite formation at low temperature remains a challenge. Here, we examine a natural playa environment to understand the long-term viability for storing CO_2within Mg-carbonate minerals and estimate the rates of magnesite formation at near-surface conditions. In laboratory experiments at room temperature, we use carboxylated polystyrene microspheres to accelerate magnesite formation by several orders of magnitude in comparison to the natural environment.

Secure carbon storage within Mg-carbonate minerals requires an understanding of fundamental geochemical processes related to fluid-mineral interactions at large spatial scales and over millennial timescales [3]. Hydromagnesite-magnesite playas (hectare-scale), having formed at the Earth's surface since the last deglaciation, demonstrate the stability of Mg-carbonate minerals for long-term carbon storage [4]. Geochemical, physical and microbial processes have facilitated carbonate precipitation from Mg-rich waters produced from the weathering of ultramafic bedrock [5,6]. Over several millennia, sediment deposition has transitioned from siliciclastic to subaqueous Ca-Mg-carbonate precipitation to subaerial Mg-carbonate deposition [4]. As a consequence of these variable modes of deposition, a complex assemblage of carbonate minerals is present within the playas including metastable hydrated Mg-carbonate minerals and magnesite. Mineralogical analysis in concert with scanning electron microscopy (SEM) show that hydrated Mg-carbonate minerals adhere to the Ostwald's law of phases, with more hydrated phases (e.g., lansfordite, $MgCO_3 \cdot 5H_2O$) transforming to less hydrated and more stable phases such as hydromagnesite [$Mg_5(CO_3)_4(OH)_2 \cdot 4H_2O$]. At surface, magnesite abundance is up to 41 wt.%, while at depth it is present at up to 86 wt.%. Stable, radiogenic, and clumped isotope [7] data demonstrate that the magnesite is modern (<10,000 years old) and has formed at low temperatures (~3-10 °C) through direct precipitation from shallow groundwaters. Magnesite crystal morphology varies with depth suggesting that crystal growth mechanisms also vary along the sediment profiles. Particle size analyses show that increases in magnesite abundance are positively correlated with increasing median particle size, indicating that magnesite formation is likely to be nucleation limited. We estimate the rate of magnesite formation (nucleation + crystal growth) is in the range of 10^{-17} to 10^{-16} mol/cm^2/s with magnesite saturation indices ranging from 1.5 to 3.2, respectively.

Similar to magnesite, dolomite [$CaMg(CO_3)_2$] precipitation is also kinetically inhibited at Earth's surface temperatures making it particularly difficult to precipitate [8]. Previous studies on microbially mediated dolomite precipitation have yielded key insights into the mechanisms of low temperature precipitation of anhydrous Mg-carbonate phases [9-11]. Carboxyl functional groups (R-COOH), found on microbial cell walls, have been shown to cause desolvation of Mg^{2+} ions in aqueous solution and facilitate dolomite formation [10-11]. Using microcosm experiments containing polystyrene microspheres with a high density of carboxyl groups, we were able to precipitate magnesite at room temperature over 72 days. Transmission electron microscopy (TEM) and selected area electron diffraction (SAED) of a thin section obtained using focused ion beam milling were used for mineralogical confirmation. In comparison to natural magnesite formation, this is an acceleration of several orders of magnitude without requiring energy input. Numerous industrial and natural waters (e.g., ocean water) are supersaturated with respect to magnesite and have the potential to serve as feedstock for sequestering CO_2. We postulate that reactive membranes or bioreactors that utilize this reaction pathway could be designed for large-scale magnesite precipitation as part of a strategy for sequestering anthropogenic CO_2.

[1] Lackner et al. (1995) *Energy* **20**: 1153-1170. [2] Hänchen et al. (2008) *Chem. Eng. Sci.* **63**: 1012-1028. [3] Bickle et al. (2013) *Rev. Mineral. Geochem.* **77**: 15-71. [4] Power et al. (2014) *Sedimentology*. **61**: 1701-1733. [5] Power et al. (2009) *Chem. Geol.* **206**: 302-316. [6] Power et al. (2007) *Geochem. Trans.* **8**: 13. [7] Streit Falk and Kelemen, unpublished data. [8] Land (1998) *Aquat Geochem*, **4**: 361-368. [9] Vasconcelos et al. (1995) *Nature* **377**: 220-222. [10] Roberts et al. (2013) *PNAS*. **110**: 14540-14545. [11] Kenward et al. (2009) *Geobiology* **7**: 556-565.

Physical and Chemical Controls on Mineral Carbonation in Mine Tailings from the Pore to Field Scale

Harrison, A., The University of British Columbia
Dipple, G., The University of British Columbia
Power, I. M., The University of British Columbia
Mayer, K. U.,

Mineralization of carbon in solid, stable carbonate minerals through reaction of CO_2 with Mg-rich mining wastes is a promising CO_2 sequestration strategy that could render certain mines greenhouse gas neutral [1]. Here, the physical and chemical controls on rates of and capacity for CO_2 sequestration in systems representative of mine tailings are examined from the pore to field scale using a combination of experimental and reactive transport modeling techniques. Experimental and field data are used to develop a comprehensive reactive transport model that captures the processes governing carbon mineralization in the shallow subsurface. Microfluidic pore scale and decimeter to meter scale column carbonation experiments using brucite [$Mg(OH)_2$] revealed that the primary controls on carbonation include the CO_2 supply rate, the distribution of the reactive phase, the mineral grain size/surface area, and the availability and distribution of water. The rate-limiting step during carbonation varied from CO_2 supply to mineral dissolution depending on the experimental variables [2]. Surface passivation and water-limited reaction resulted in a highly non-geometric evolution of reactive surface area. The extent of reaction was also limited at high water content because viscous fingering of the gas streams injected at the base of the meter scale columns resulted in narrow zones of highly carbonated material, but left a large proportion of brucite unreacted. The incorporation of water consumption, water-limited reactivity, and surface passivation functions into the reactive transport code, MIN3P [3], and the capability to simulate the impact of climate on carbonation rates allows more robust predictions of the CO_2 sequestration rate and capacity at the field scale. This research imparts a better understanding of fundamental mechanisms and chemical processes relevant to CO_2 sequestration in mine tailings, with implications for mineral carbonation in other settings that have greater CO_2 sequestration capacity, such as shallow subsurface formations with similar mineralogy.

[1] Wilson et al. (2009) *Econ. Geol.* **104**: 95-104. [2] Harrison et al. (in press) *Geochim. Cosmochim. Acta*. [3] Mayer et al. (2002) *Water Resour. Res.* **38**: 1174.

Carbon sequestration via steel slag accelerated carbonation: the influence of operating conditions on process evolution and yield

A. Polettini[1], R. Pomi[1*], and A. Stramazzo[1]

[1] Department of Civil and Environmental Engineering, University of Rome "La Sapienza", Via Eudossiana 18 – 00184, Rome, Italy
*Corresponding author: raffaella.pomi@uniroma1.it

Abstract
A factorial study was conducted on steelmaking basic oxygen furnace slag to identify the individual and joint effects of total pressure, CO_2 concentration in the gas phase and temperature on the CO_2 sequestration yield through direct aqueous carbonation. The operating parameters were fixed at three levels according to a 3^3 factorial design, resulting in a set of 27 batch carbonation experiments. The carbonation performance and the changes in the mineralogical characteristics of the slag were investigated in detail. Large effects of the operating factors were observed for CO_2 uptake, ranging from 6.7 to 53.6 g CO_2/100 g slag. The analysis of the solid and liquid phases at the end of the carbonation treatment showed significant changes in the physical, chemical and mineralogical composition of the material. In particular, evidence was gained of other elements (Mg, Fe, Mn, Zn) in addition to Ca being intensively involved in the carbonation reactions, with a variety of carbonate phases being produced in addition to calcium carbonate forms.

Introduction

In recent years the scientific research on CO_2 sequestration through accelerated carbonation of industrial wastes has been focused on a number of residues from thermal processes, including steel slags[1–8], cement wastes[9,10], waste ashes[11–15], coal fly ash[16–18] and others. However, a careful selection of the process conditions is undoubtedly needed in order to enhance the environmental and economic competitiveness of the process. This in turn basically requires that the process is optimized by 1) reducing the overall energy demand, 2) accelerating the reaction rate and 3) improving the degree of conversion.

For complex materials such as industrial wastes, the conversion of the original constituents into carbonates during carbonation results from a vast set of reactions, occurring either in parallel or in series, which are affected by multiple parameters. In aqueous carbonation, the main steps involved include[19–21]: 1) CO_2 solvation into the liquid phase; 2) hydrolysis of CO_2 and formation of carbonic acid; 3) dissociation of H_2CO_3 into HCO_3^- and CO_3^{2-}; 4) dissolution of minerals from the solid matrix and release of reactive metals into the liquid phase; 5) diffusion of ions towards the reaction zone; 6) precipitation of secondary minerals (including carbonates). Numerous factors are recognized to affect one or more process stages, the most relevant of which include the physical and chemical variables of the process (temperature, total pressure and partial CO_2 pressure, liquid-to-solid [L/S] ratio, pH), as well as the solid material properties (particle size, porosity and surface area, mineralogy). While a detailed review of the effects of the operating parameters on the carbonation mechanisms has been reported elsewhere[22,23], Table 1 summarizes the main features of each of the identified important factors during the process. Even from the short summary presented, it is clear that understanding all such effects is a challenging task that requires some thorough and wide investigation, even more considering that the relevant parameters are often interrelated. In the present work a full factorial design approach was adopted to investigate the individual effects and the mutual interactions of some of the relevant operating variables of accelerated carbonation, with the aim of providing a systematic framework for the analysis of process sensitivity to the parameters of concern.

Table 1. Recognized effect of process parameters on carbonation

Parameter	Reaction/mechanism/property affected
Temperature	• Mineral dissolution kinetics • Solubility of minerals (including carbonates) • CO_2 solvation into the liquid phase (→ pH also indirectly affected) • CO_2 diffusion • Viscosity of the liquid phase • Nature of carbonate minerals formed
Pressure	• CO_2 solubility in the liquid phase (→ pH also indirectly affected) • Formation and growth of carbonate crystals
pH	• Mineral dissolution • Mineral precipitation • Morphology of precipitated solid phases
L/S ratio	• Hydration and dissolution of minerals • CO_2 diffusion • Degree of mixing
Surface area	• Mineral dissolution/precipitation equilibria
Mineralogical composition of the solid material	• Mineral dissolution/precipitation equilibria • Nature of carbonate minerals formed

Materials and Methods

Slag samples were collected from a steel plant employing the integrated steelmaking process at the outlet of the basic oxygen furnace (BOF). The slag was separated through dry sieving into various size fractions, and only the 63–100 μm size class was investigated in the present work, since it exhibited the highest carbonation yield[24]. The slag was analysed for elemental composition, carbonate content and mineralogy according to the methods detailed in our previous work[22]. Direct batch accelerated carbonation tests were performed in slurry phase in a 300-ml pressurized stainless steel reactor equipped with magnetic stirring and heating devices. On the basis of previous experiments [2,25–28], the L/S ratio and the carbonation time were set at 5 l/kg and 4 h. A full factorial design in three factors (total pressure [*P*], temperature [*T*] and volumetric CO_2 concentration in the gas phase [*C*]) was adopted to arrange the carbonation tests. The factors were varied over three levels, namely: P = 1, 5, 10 bar; T = 20, 50, 100 °C; C = 10, 40, 100%. The 27 resulting factorial experiments (Table 2) were performed in duplicate in a randomized sequence. After carbonation the slurries were centrifuged at 4000 rpm and filtered on a 0.45-μm membrane. The solid was oven-dried at 105 °C and analysed for carbonate content and mineralogy, while the liquid solution was acidified with 1:1 HNO_3 and analysed for metal content.

Table 2. Experimental conditions for the carbonation runs

Test no.	*P* (bar)	*C* (% vol.)	*T* (°C)	Test no.	*P* (bar)	*C* (% vol.)	*T* (°C)
1	1	10	20	15	10	40	50
2	5	10	20	16	1	40	100
3	10	10	20	17	5	40	100
4	1	10	50	18	10	40	100
5	5	10	50	19	1	100	20
6	10	10	50	20	5	100	20
7	1	10	100	21	10	100	20
8	5	10	100	22	1	100	50
9	10	10	100	23	5	100	50
10	1	40	20	24	10	100	50
11	5	40	20	25	1	100	100
12	10	40	20	26	5	100	100
13	1	40	50	27	10	100	100
14	5	40	50				

The carbonation test results were processed using statistical methods to detect the significant effects and interactions between the operating parameters of concern. To this aim, the t-test was adopted assuming a confidence level of 95%. The experimental results were also used to derive a second-order polynomial relationship expressed by Eq. (1), representing the response surface as a function of the factor levels:

$$y = \beta_0 + \sum_{i=1}^{3} \beta_i x_i + \sum_{i<j,\ j=2} \sum^{3} \beta_{ij} x_i x_j + \sum_{i=1}^{3} \beta_{ii} x_i^2 + \varepsilon \qquad (1)$$

where the response variable y (i.e., CO_2 uptake) is expressed as a linear combination of zero-, first- and second-order terms, plus a random error component (ε); second-order terms are further subdivided into 2-way interactions and pure quadratic terms. In Eq. (1), x_i (i = 1, 2, 3) are the levels of the three factors, β_i (i = 1, 2, 3) the coefficients of the linear component, while β_{ii} (i= 1, 2, 3) and β_{ij} (i= 1, 2; j = 2, 3) those of the quadratic component. The coefficients were estimated by least-square linear regression of the data using the rsm package[29] developed for application with the R software[30].

To investigate the chemical equilibria established in the system and the nature of the mineral phases potentially controlling the solubility of the elements under concern, geochemical modelling of the liquid solutions resulting from the carbonation tests was performed using Visual Minteq[31], as detailed in our previous work[22].

Results and Discussion

Materials characterization

The elemental analysis of the slag revealed that the most abundant elements in the slag included Ca and Fe (with contents of 31 and 27% dry wt., respectively), followed by Mg (7.5%) and Si (5.1%). Among the potentially toxic elements, Cr and V displayed the highest concentrations (990 and 830 mg/kg, respectively).

Effects of the factors on the carbonation yield

The CO_2 uptake yield, expressed as the mass (in grams) of CO_2 sequestered by 100 grams of as-received dry material, was calculated through Eq. (2) from the IC contents of the carbonated (CO_{2final}) and untreated ($CO_{2initial}$) material, both expressed as CO_2 weight percent contents:

$$CO_2\ \text{uptake}(\%) = \frac{CO_{2\,final}[\%] - CO_{2\,initial}[\%]}{100 - CO_{2\,final}[\%]} \times 100 \qquad (2)$$

A summary of the CO_2 uptake results obtained is reported in Figure 1, which shows that all the operating parameters produced some notable effects on the carbonation yield. The CO_2 uptake yield largely varied with the operating conditions tested, ranging from 6.7% (1 bar, 10% CO_2, 20 °C) to 53.6% (5 bar, 40% CO_2, 100 °C). In more than one half of the experiments, the CO_2 uptake exceeded 30%, which is at the upper boundary of documented literature results[1,3–5,32–34].

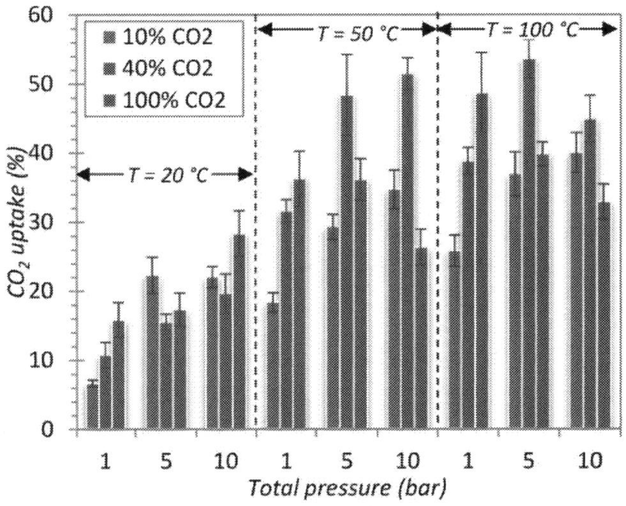

Figure 1. CO_2 uptakes (average values and standard deviations) for the 27 carbonation runs

While the results suggest that more severe carbonation conditions generally promoted CO_2 sequestration, the observed trends with the factor levels were neither linear nor monotonic; this clearly reveals the presence of both higher-order effects of the individual factors and mutual interactions between them. The statistical analysis of the data, which accounted for first-order (**P, C, T**) and quadratic (**P^2, C^2, T^2**) effects as well as two-way (**P×C, P×T, C×T**) interactions, revealed that all terms except the **C×T** interaction were statistically significant. The influence of the three factors on CO_2 uptake is represented by the contour plots in Figure 2, which were derived by cutting the response surfaces derived from Eq. (1) at the three levels for each factor. The curvature of the contour plots confirms that higher than first-order effects of the factors played a role during carbonation.

The maximum CO_2 uptake was found to lie at 5.9 bar, 60.6% CO_2 and 83.7 °C. The curvature of the response surfaces also suggests that sequestration yields in excess of 40 g CO_2/100 g slag can be attained even at low CO_2 levels provided that total pressure and temperature are adjusted accordingly: for instance, at a CO_2 concentration of 10%, for total pressures of 6, 8 and 10 bar, the temperatures required to achieve a yield of 40 g CO_2/100 g slag would be 73, 61 and 58 °C, respectively.

The inspection of Figure 2 a)-c) reveals that, at a constant CO_2 concentration in the gas phase, temperature and pressure affected the carbonation performance to a different extent. In particular, at 10 and 40% CO_2 concentrations, CO_2 uptake increased with both temperature and pressure, although the influence of the latter was more relevant in the lower range of values (up to ~6 bar) and then levelled off. Using a pure CO_2 gas, the contour lines were sub-horizontal in most of the investigated region (Figure 2 c)),

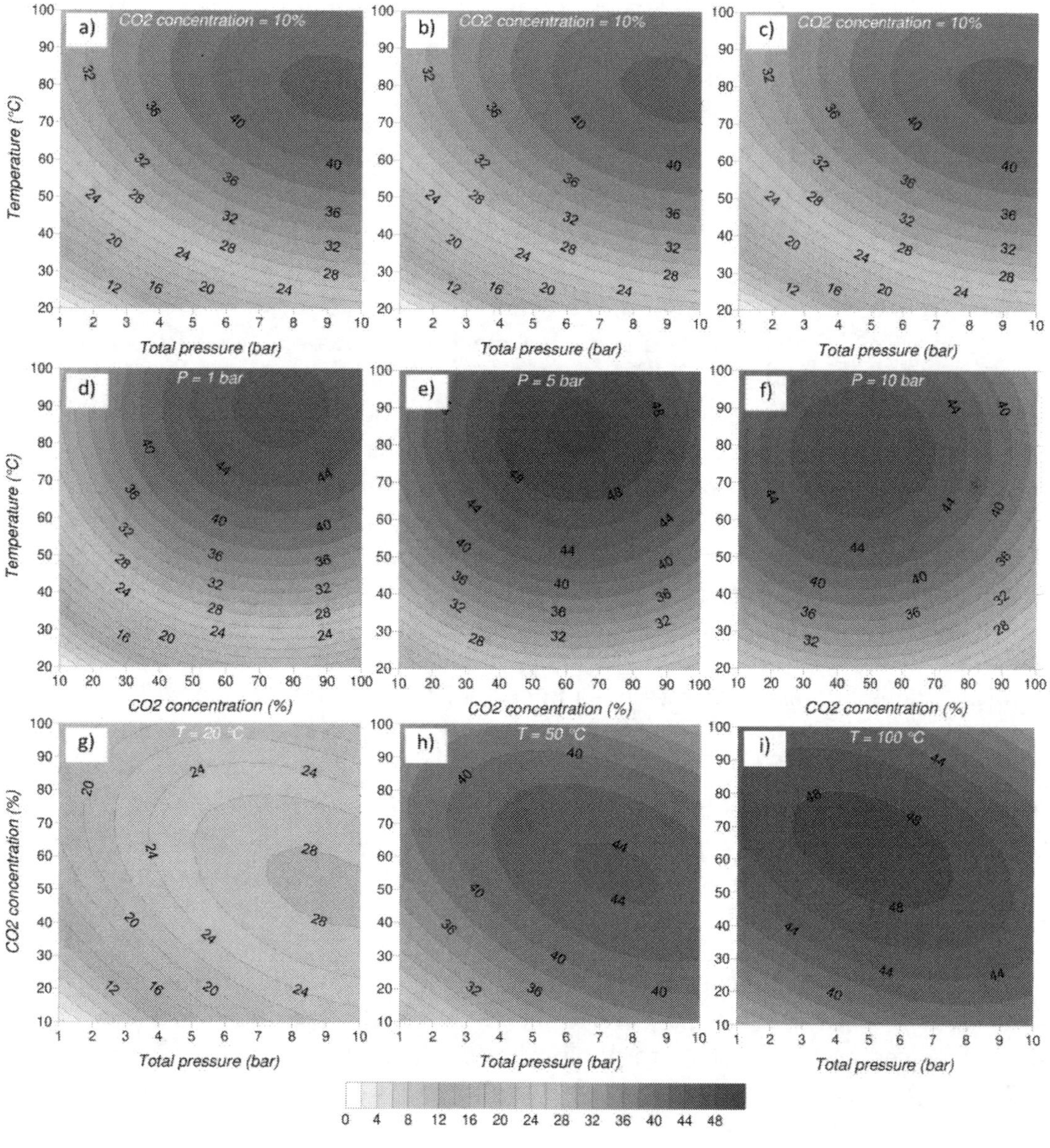

Figure 2. Contour plots for CO_2 uptake (%) as a function of factor levels

indicating a lower influence of pressure on the carbonation yield at high CO_2 contents. The optimum for CO_2 uptake also shifted to lower partial pressures as the CO_2 concentration increased. In principle higher CO_2 pressures – the other operating parameters remaining constant – should promote carbonation by enhancing CO_2 dissolution into the liquid phase. However, this is only true as long as CO_2 transfer is the limiting mechanism of the carbonation process[1,25]. The experimental results here show that increasing the CO_2 pressure beyond the threshold at which it no longer represents the limiting factor for the process, negatively reflects on the carbonation performance, as already noted by other authors[8,35,36]. While some investigators[36] related such a finding to the unfavourable pH conditions for carbonate precipitation at high CO_2 pressures, this was not the case for our study (see below for details). It may on the other hand be hypothesized that the faster precipitation of carbonate minerals induced by higher CO_2 pressures may result in faster clogging of the material's pores and faster formation of the protective carbonate and silica rim around the grains, preventing further contact with CO_2[35]. This may also explain the finding from our experiments that, in the investigated range of experimental conditions, the maximum CO_2 uptake was achieved at lower CO_2 pressures as temperature increased.

As for the influence of temperature, two contrasting effects on CO_2 uptake (i.e., faster dissolution/precipitation of minerals; decreased CO_2 solubility in the aqueous phase) are theoretically expected. Given the general increasing trend of CO_2 uptake with temperature (see Figure 2 a)-c), g)-i)), the results of our experiments show that under the investigated operating conditions the leaching of reactive elements from the slag and their subsequent diffusion to the reaction sites was likely the rate-limiting step of the carbonation process. Previous studies[1,17,19,32] indicated that an optimum exists for temperature beyond which the process kinetics turns from mineral dissolution-controlled to CO_2 transfer-controlled. Obviously, such a threshold depends on the specific type of alkaline material concerned. In our study the existence of an optimum for temperature was not suggested by the experimental results, likely because it fell beyond the investigated range of values, thus indirectly confirming the results of Huijgen and co-workers[1] who found the temperature optimum for BOF slag at ~175 °C.

Chemical equilibria and mineralogical/dimensional changes upon carbonation

The results of geochemical modelling of the slurry solutions at the end of the carbonation experiments are reported in Figure 3 in terms of saturation indices (SI) values for Ca, Mg, Mn, Fe, Si and Al minerals. In the plots, only the solid phases with a SI in the range $-1.5 \leq SI \leq 1.5$ are depicted. It can be noticed that, at CO_2 pressures >0.5 bar (at which CO_2 uptake became appreciable), the potential candidates for solubility control of elements including Ca, Mg, Zn and, under some conditions, Mn and Fe included carbonate minerals. This indicates that dissolution/precipitation equilibria of carbonate species largely dominated the carbonated slurry systems. The most probable minerals governing the equilibrium between the solid and the liquid phases included calcium carbonate monohydrate ($CaCO_3 \cdot H_2O$) for Ca, nesquehonite ($MgCO_3 \cdot 3H_2O$) for Mg and smithsonite ($ZnCO_3$) for Zn; amorphous $MnCO_3$ and either siderite ($FeCO_3$) or the Fe end-member of ankerite ($CaFe(CO_3)_2$) were also identified as possible solubility-controlling carbonates for Mn and Fe, although in some cases the

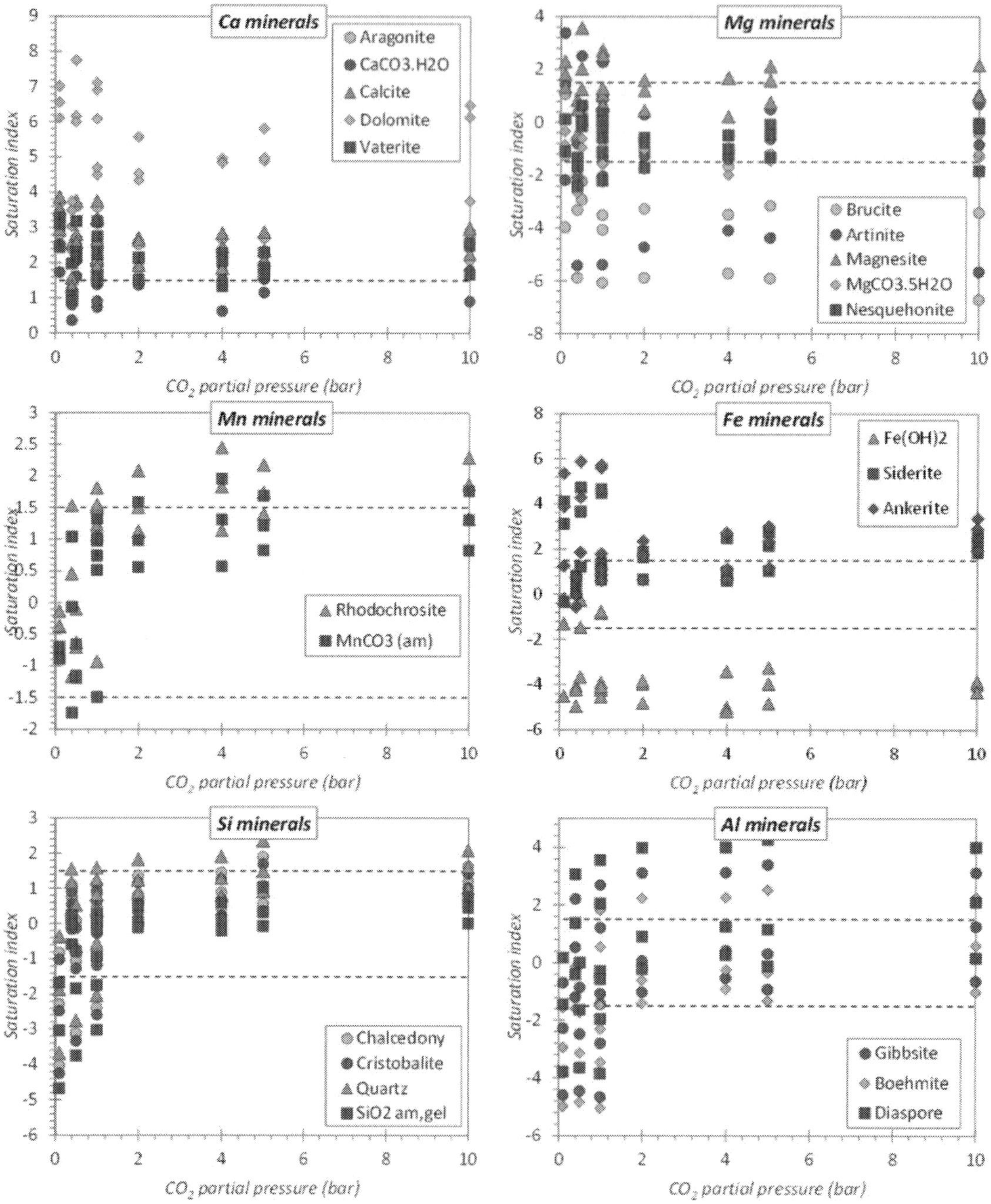

Figure 3. Saturation indices for potential solubility-controlling minerals

solutions were found to be oversaturated in such phases. It was also observed that at CO_2 pressures >0.5 bar, the solubility of Si in the carbonated slurries was always controlled by amorphous silica. This provides evidence of dissolution of silicate minerals from the slag occurring incongruently, leaving a residual Si-enriched leached layer [37,38] that governs the equilibrium between the liquid and solid phases. A similar behaviour was observed for Al, whose solubility was found to be controlled by either hydroxide (gibbiste) or oxy-hydroxide (boehmite, diaspore) forms; the scattering of SI values for the different carbonation conditions tested did not allow to univocally resolve the question as to which of these phases was the most likely to explain Al solubility. Nevertheless, the geochemical modelling results still clearly demonstrate that the Al dissolved upon carbonation from complex minerals partially re-precipitated as a secondary phase, which may have also probably played some role in the formation of the passivating layer around the shrinking reactive core of the slag particles.

The existing equilibria in the liquid phase were of course related to the mineralogical changes induced by carbonation. As reported in Figure 4, a comparison of the XRD patterns between the untreated slag and selected carbonated samples (specifically, samples 1, 14 and 24 in Table 2, which displayed low, high and intermediate CO_2 uptake yields, respectively) shows that the carbonation process had a large impact on mineralogy. The XRD analyses showed that the untreated slag was composed largely by dicalcium silicate, in addition to a number of mixed oxide phases, namely $Ca_2Fe_{1.4}Al_{0.6}O_5$ and $Ca_{12}Al_{113.8}Fe_{0.14}O_{32}$ (and possibly also $Ca_2Cr_{0.5}Fe_{1.5}O_5$, which had similar diffraction peaks compared to the other two oxides), with lower amounts of wuestite (FeO). On the other hand, the XRD patterns of the carbonated materials were dominated by a variety of carbonate minerals, the nature and abundance of which was found to be directly related to the carbonation degree attained. The original minerals in the untreated slag were all found to be highly reactive upon carbonation, so that only minor traces of these (mainly $Ca_2Fe_{1.4}Al_{0.6}O_5$) were retrieved in the carbonated samples, even for low CO_2 uptake yields. The newly formed carbonate phases included calcite in all cases, which was by far the most abundant species identified. Some

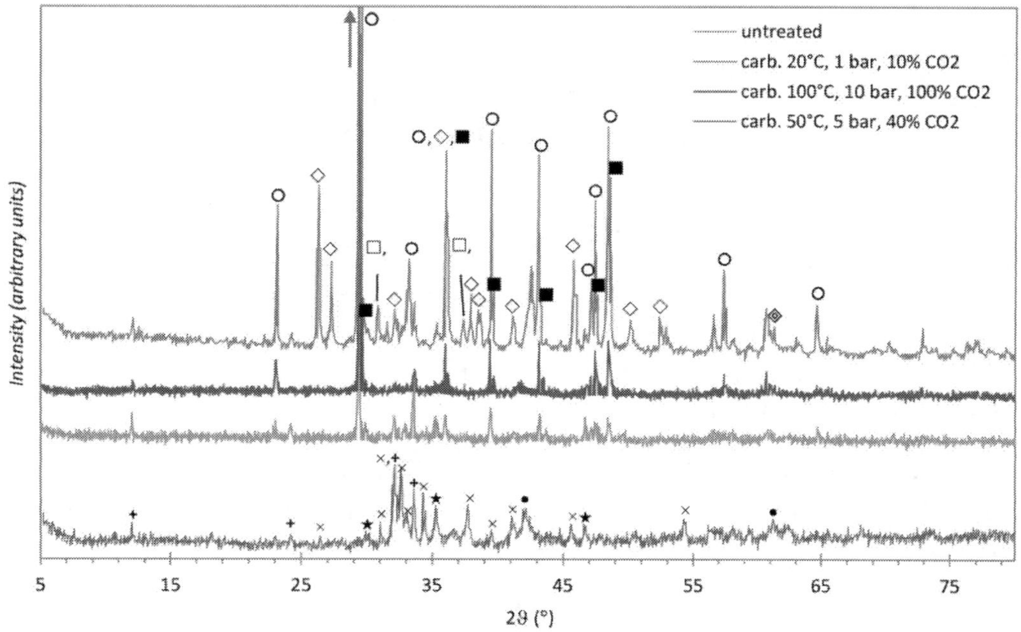

Figure 4. XRD patterns of untreated and carbonated slag samples. Legend: × = Ca_2SiO_4; + = $Ca_2Fe_{1.4}Al_{0.6}O_5$; ★ = $Ca_{12}Al_{113.8}Fe_{0.14}O_{32}$; • = wuestite (FeO); O = calcite; ◇ = aragonite; □ = dolomite; ■ = magnesian calcite ($Mg_{0.064}Ca_{0.936}CO_3$); ◈ = ankerite ($CaMg_{0.27}Fe_{0.73}(CO_3)_2$). Note: for easier identification of peaks the y-axis scale was adjusted cutting the main calcite peak (2ϑ = 29.4°) for the upper curve by 51%

additional carbonates also appeared at increasing concentrations as the carbonation degree increased; for sample 24 (carbonation conditions: 100 °C, 10 bar; 100% CO_2; CO_2 uptake: 26.4%) some magnesian calcite ($Mg_{0.064}Ca_{0.936}CO_3$) was detected, while for sample 14 (carbonation conditions: 50 °C, 5 bar; 40% CO_2; CO_2 uptake: 48.4%) aragonite, dolomite and ankerite were also identified in addition to magnesian calcite.

The mineralogical analyses confirm the findings of our previous studies[28], according to which not only Ca, but other reactive elements are involved in the carbonation process, and the degree of involvement is strongly related to the carbonation yield attained. Compared to the results from other investigators, the present study also clearly demonstrates that the nature and amount of the carbonate minerals formed upon carbonation are strongly dependent on the carbonation conditions applied.

Conclusions

The carbonation study revealed that CO_2 uptake was largely affected by the operating conditions adopted, spanning in the range 6.7–53.6 g CO_2/100 g slag. The best carbonation performance attained was particularly significant compared to previous literature studies on BOF slag, even more when taking into account the relatively mild operating conditions adopted (5 bar, 40% CO_2, 50 °C, 4 h). The statistical analysis indicated that the investigated factors played an important role in the process, both individually and jointly with the other process parameters. The response surface model for CO_2 uptake identified a theoretical maximum within the investigated range of values for the operating parameters, corresponding to 5.9 bar, 60.6% CO_2 and 83.7 °C. The shape of the response surfaces also indicated that, provided that temperature and total pressure are adjusted properly, high carbonation yields (>40 g CO_2/100 g slag) can be attained even at low CO_2 concentrations in the gas phase, thus implying the possibility of removing the CO_2 capture stage upstream of the carbonation process.

The analysis of the carbonated slag showed that the process induced large changes in the physical, chemical and mineralogical characteristics of the material. The observed changes were also quite evidently related to the carbonation degree attained. Mineralogical analyses coupled with geochemical modelling of the slurry systems showed that, although Ca was the main element involved in the process and it largely dictated the composition of the carbonated material, other elements including Mg, Fe, Mn and Zn took also part to the process. While XRD analyses detected the presence of calcite, aragonite, dolomite, magnesian calcite and ankerite as the most abundant carbonate minerals formed, geochemical modelling suggested that additional phases likely formed upon carbonation. Possible additional carbonate phases produced included nesquehonite, amorphous $MnCO_3$, siderite, ankerite and smithsonite.

References

1. Huijgen, W.J.J., Witkamp, G.-J., Comans, R.N.J. 2005. Mineral CO_2 Sequestration by Steel Slag Carbonation. *Environ. Sci. Technol.*, 39(24), 9676–9682.
2. Baciocchi, R., Costa, G., Polettini, A., Pomi, R. 2009. Influence of particle size on the carbonation of stainless steel slag for CO_2 storage. *Energy Procedia*, 1(1), 4859–4866.

3. Santos, R.M., Ling, D., Sarvaramini, A., Guo, M., Elsen, J., Larachi, F., Beaudoin, G., Blanpain, B., Van Gerven, T. 2012. Stabilization of basic oxygen furnace slag by hot-stage carbonation treatment. *Chem. Eng. J.*, 203, 239–250.
4. Chang, E.-E., Pan, S.-Y., Chen, Y.-H., Tan, C.-S., Chiang, P.-C. 2012. Accelerated carbonation of steelmaking slags in a high-gravity rotating packed bed. *J. Hazard. Mater.*, 227-228, 97–106.
5. Chang, E.-E., Chiu, A.-C., Pan, S.-Y., Chen, Y.-H., Tan, C.-S., Chiang, P.-C. 2013. Carbonation of basic oxygen furnace slag with metalworking wastewater in a slurry reactor. *Int. J. Greenh. Gas Control*, 12, 382–389.
6. Lekakh, S.N., Rawlins, C.H., Robertson, D.G.C., Richards, V.L., Peaslee, K.D. 2008. Kinetics of Aqueous leaching and carbonization of steelmaking slag. *Metall. Mater. Trans. B*, 39(1), 125–134.
7. Salman, M., Cizer, Ö., Pontikes, Y., Santos, R.M., Snellings, R., Vandewalle, L., Blanpain, B., Van Balen, K. 2014. Effect of accelerated carbonation on AOD stainless steel slag for its valorisation as a CO_2-sequestering construction material. *Chem. Eng. J.*, 246, 39–52.
8. Santos, R.M., Van Bouwel, J., Vandevelde, E., Mertens, G., Elsen, J., Van Gerven, T. 2013. Accelerated mineral carbonation of stainless steel slags for CO_2 storage and waste valorization: Effect of process parameters on geochemical properties. *Int. J. Greenh. Gas Control*, 17, 32–45.
9. Huntzinger, D.N., Gierke, J.S., Sutter, L.L., Kawatra, S.K., Eisele, T.C. 2009. Mineral carbonation for carbon sequestration in cement kiln dust from waste piles. *J. Hazard. Mater.*, 168(1), 31–37.
10. Gunning, P.J., Hills, C.D., Carey, P.J. 2010. Accelerated carbonation treatment of industrial wastes. *Waste Manage.*, 30(6), 1081–1090.
11. Uibu, M., Velts, O., Kuusik, R. 2010. Developments in CO_2 mineral carbonation of oil shale ash. *J. Hazard. Mater.*, 174(1-3), 209–214.
12. Arickx, S., Van Gerven, T., Vandecasteele, C. 2006. Accelerated carbonation for treatment of MSWI bottom ash. *J. Hazard. Mater.*, 137(1), 235–243.
13. Baciocchi, R., Polettini, A., Pomi, R., Prigiobbe, V., Von Zedwitz, V.N., Steinfeld, A. 2006. CO_2 sequestration by direct gas–solid carbonation of air pollution control (APC) residues. *Energ. Fuel.*, 20(5), 1933–1940.
14. Prigiobbe, V., Polettini, A., Baciocchi, R. 2009. Gas-solid carbonation kinetics of air pollution control residues for CO_2 storage. *Chem. Eng. J.*, 148(2-3), 270–278.
15. Baciocchi, R.; Costa, G.; Polettini, A.; Pomi, R.; Prigiobbe, V. 2009. Comparison of different reaction routes for carbonation of APC residues. *Energy Procedia*, 1(1), 4851–4858.
16. Ukwattage, N.L., Ranjith, P.G., Wang, S.H. 2013. Investigation of the potential of coal combustion fly ash for mineral sequestration of CO_2 by accelerated carbonation. *Energy*, 52, 230–236.
17. Ukwattage, N.L., Ranjith, P.G., Yellishetty, M., Bui, H.H., Xu, T. 2014. A laboratory-scale study of the aqueous mineral carbonation of coal fly ash for CO_2 sequestration. *J. Clean. Prod.*, in press.
18. Montes-Hernandez, G., Pérez-López, R., Renard, F., Nieto, J.M., Charlet, L. 2009. Mineral sequestration of CO_2 by aqueous carbonation of coal combustion fly-ash. *J. Hazard. Mater.*, 161(2-3), 1347–1354.
19. Fernández Bertos, M., Simons, S.J.R., Hills, C.D., Carey, P.J. 2004. A review of accelerated carbonation technology in the treatment of cement-based materials and sequestration of

CO_2. *J. Hazard. Mater.*, 112(3), 193–205.

20. Huijgen, W.J.J., Comans, R.N.J. 2003. *Carbon dioxide sequestration by mineral carbonation: Literature review*. Report no. ECN-C--05-022, pp 1–37.
21. Santos, R.M., Van Gerven, T. 2011. Process intensification routes for mineral carbonation. *Greenh. Gases Sci. Technol.*, 1(4), 287–293.
22. Polettini, A., Pomi, R., Stramazzo, A. 2015. A factorial study of parameters effects during aqueous accelerated carbonation of BOF slag. *J. Clean. Prod.*, submitted.
23. Sanna, A., Uibu, M., Caramanna, G., Kuusik, R., Maroto-Valer, M.M. 2014. A review of mineral carbonation technologies to sequester CO_2. *Chem. Soc. Rev.*, 43(23), 8049–8080.
24. Polettini, A., Pomi, R., Stramazzo, A. 2014. Carbon sequestration capacity of BOF slag as affected by particle size distribution. *In Proc. Venice 2014, 5th International Symposium on Energy from Biomass and Waste*; Venice (IT), 17-20 November 2014.
25. Baciocchi, R., Costa, G., Di Bartolomeo, E., Polettini, A., Pomi, R. 2009. The effects of accelerated carbonation on CO_2 uptake and metal release from incineration APC residues. *Waste Manage.*, 29(12), 2994–3003.
26. Baciocchi, R., Costa, G., Di Bartolomeo, E., Polettini, A., Pomi, R. 2010. Carbonation of stainless steel slag as a process for CO_2 storage and slag valorization. *Waste and Biomass Valorization*, 1, 467–477.
27. Baciocchi, R., Costa, G., Di Bartolomeo, E., Polettini, A., Pomi, R. 2011. Wet versus slurry carbonation of EAF steel slag. *Greenh. Gases Sci. Technol.*, 1(4), 312–319.
28. Baciocchi, R., Costa, G.,Di Gianfilippo, M.,Polettini, A., Pomi, R., Stramazzo, A. 2015. Thin-film versus slurry-phase carbonation of steel slag: CO_2 uptake and effects on mineralogy. *J. Hazard. Mater.*, 283, 302–313.
29. Lenth, R.V. 2009. Response-surface methods in R, using rsm. *J. Stat. Softw.*, 32(7), 1–17.
30. R Development Core Team. 2009. *R: A language and environment for statistical computing*, R Foundation for Statistical Computing, Vienna (AT).
31. Gustafsson, J.P. *Visual MINTEQ ver. 3.0*, http://vminteq.lwr.kth.se/.
32. Chang, E.-E., Chen, C.-H., Chen, Y.-H.,Pan, S.-Y., Chiang, P.-C. 2011. Performance evaluation for carbonation of steel-making slags in a slurry reactor. *J. Hazard. Mater.*, 186(1), 558–564.
33. Pan, S.-Y., Chiang, P.-C., Chen, Y.-H., Tan, C.-S., Chang, E.-E. 2013. Ex situ CO_2 capture by carbonation of steelmaking slag coupled with metalworking wastewater in a rotating packed bed. *Environ. Sci. Technol.*, 47(7), 3308–3315.
34. Chang, E.-E., Chen, T.-L., Pan, S.-Y., Chen, Y.-H., Chiang, P.-C. 2013. Kinetic modeling on CO_2 capture using basic oxygen furnace slag coupled with cold-rolling wastewater in a rotating packed bed. *J. Hazard. Mater.*, 260, 937–946.
35. Bauer, M., Gassen, N., Stanjek, H., Peiffer, S. 2011. Carbonation of lignite fly ash at ambient T and P in a semi-dry reaction system for CO_2 sequestration. *Appl. Geochem.*, 26(8), 1502–1512.
36. Back, M., Kuehn, M., Stanjek, H., Peiffer, S. 2008. Reactivity of alkaline lignite fly ashes towards CO_2 in water. *Environ. Sci. Technol.*, 42(12), 4520–4526.
37. Huijgen, W.J.J., Witkamp, G.-J., Comans, R.N.J. 2006. Mechanisms of aqueous wollastonite carbonation as a possible CO_2 sequestration process. *Chem. Eng. Sci.*, 61(13), 4242–4251.
38. Weissbart, E.J., Rimstidt, J.D. 2000. Wollastonite: Incongruent dissolution and leached layer formation. *Geochim. Cosmochim. Acta*, 64(23), 4007–4016.

Minimization of energy consumption of CO_2 mineralization using recyclable ammonium salts

A. Sanna[1]*, J. Gaubert[1], M.M. Maroto-Valer[1]

[1] Centre for Innovation in Carbon Capture and Storage (CICCS), School of Engineering and Physical Sciences, Heriot-Watt University, EH144AS, Edinburgh, UK
*Corresponding author: A.Sanna@hw.ac.uk

Abstract
Mineral carbonation by aqueous pH swing using recyclable ammonium salts is a promising process to permanently sequester CO_2 into stable carbonates. The ammonium salts are separated from the water solution and thermally regenerated at 300°C. Ammonium salts – water separation by thermal evaporation represents an energy intensive step, which limits the deployment of this technology. To overcome this challenge, the feasibility of using liquid-liquid extraction as an alternative salts/water separation process was evaluated here. The associated energy consumption was calculated, resulting in an energy saving between 26% and to 43% in comparison to water evaporation step depending on the conditions employed, and further research is required to optimise this process.

Introduction

Mineralization of CO_2 using recyclable salts, namely Carbon dioxide Capture and Storage by Mineralization (CCSM,) is considered amongst the most reliable processes to store CO_2 permanently.[1,2] However, the main key challenge to deploy this technology at large scale is related to its high energy requirement.[1]

CCSM involves the reaction of CO_2 with metal oxides to form carbonates. Since the carbonates obtained are stable thermodynamically at ambient conditions, it is not possible for the CO_2 to be released in the atmosphere. Mineralization has the advantage to be an overall exothermic reaction and also, it can take place close to the CO_2 emitters where inorganic wastes (e.g. steel slags) are readily available as carbonation feedstock. Moreover, mineralization can be advantageous if suitable mineral deposits are closely located.[3]

However, at the current stage of development, this process suffers from drawbacks such as its slow kinetics, low efficiency and large energy consumption. As natural carbonation reaction of silicate rocks is very slow because of its inherent low surface available and diffusion limitations, processes based on ex-situ carbonation reactions have been developed to accelerate the reaction. It has been shown that the addition of a mineral dissolution step using chemicals enhances the reaction kinetics. This route is generally referred to as indirect mineral carbonation.[2,4] CCSM by pH swing using ammonium salts is a promising multistep indirect process, which has been extensively described in previous works.[4-7] The process consists of extracting Mg and Ca from the feedstock to produce Mg/Ca-sulphate using an acid solution of ammonium bisulphate and then reacting the Mg/Ca sulphate with CO_2 captured in form of ammonium carbonate by using ammonia. At the end of the process, the remnant ammonium sulphate (AS) is first separated from the water and then regenerated (at 330°C) to NH_3 and NH_4HSO_4, which are re-used in the process. Since the salt/water separation is the most energy consuming step of the process, Wang and Maroto-Valer [5] attempted to reduce the water usage by increasing the serpentine to water ratio. Despite this, the separation of the AS from the water solution remains a problem towards the commercialization of this technology.

To overcome this energy issue, the feasibility of using liquid-liquid extraction as an alternative salt/water separation process has been evaluated and compared to thermal evaporation. The investigated liquid-liquid extraction technique is commercially used in the various processes, which employs methanol as reagent for the separation of AS from different processing fluids.[8,9]

Methodology

A 250mL three necks flask and an oil bath were used to experiment the liquid-liquid extraction technique at ambient pressure. A hot stirring plate was used to control temperature and stirring speed. A water based heat exchanger was connected to the three necks flask to avoid losing vapours from the flask. Finally, a vacuum filter was used to separate the precipitated ammonium salts.

Synthetic mother liquors mimicking dissolution liquors from the CCSM carbonation stage (based on a 100 and 150 $g_{serpentine}$/l, S/L ratio) were prepared and used for all the experiments. When the solution reached the desired temperature, the required amount of methanol was added to the AS solution and left reacting for the chosen time (5, 10 or 30 minutes). The solution containing water, AS and methanol was then vacuum filtered and the filter was then dried at 60°C during 24 hours. The AS was recovered and weighted to establish the % of AS recovered. Thermo-gravimetric analysis (TA Q500) was also used to characterise the precipitated AS. Finally, the energy consumption (kWh/t CO_2) associated to the evaporation and liquid/liquid extraction processes were calculated.

Results and Discussion

A set of different conditions were investigated, including temperature, reaction time, reagents concentration and stirring rate. A summary of the experiments carried out is summarised in Table 1. An AS/water separation of 94% was achieved at 25°C, 10 minutes, 1bar, 100g/l S/L ratio, 115% methanol and 350rpm. The AS separated was 96% using 90% methanol under the same conditions but using a solution containing 150g serpentine per litre.

Table 1. Liquid-liquid extraction experiments at 25°C, 350 rpm, 10 minutes, 100g/l. * 150g/l

	Test 1	Test 2	Test 3	Test 4*
Amount of methanol (%)	58	90	115	90
AS recovered (%)	10.64	67.69	94.00	96.00

Figure 1 shows the energy consumed in the AS salts recovery and regeneration using thermal evaporation and liquid-liquid extraction. The calculations made for the CO_2 capture, mineral dissolution, CO_2 capture using ammonia, pH adjustment, removal of impurities, carbonation, and regeneration of additives (after the salt/water separation stage) were considered equal for both processes. The evaporation of water and methanol was taken into account for the evaporation and liquid-liquid extraction pathways, respectively.

For the 100 g/l case, every 10.5 t of water and 2.7 t of $(NH_4)_2SO_4$ resulting from the carbonation step, a total of 15.2 t methanol (70% of the solution by volume) need to be added. To evaporate this methanol, the total energy required was the sum of the energy to heat the methanol from 25°C to 64.8°C (421 kWh) and its latent heat (4637 kWh). The total energy required to evaporate the methanol was 5058.2 kWh, which is 26% lower than the energy required for the thermal evaporation process (6823 kWh). The energy saving increased to 43% when the initial solid to liquid ratio (serpentine/water) was increased to 150 g/l.

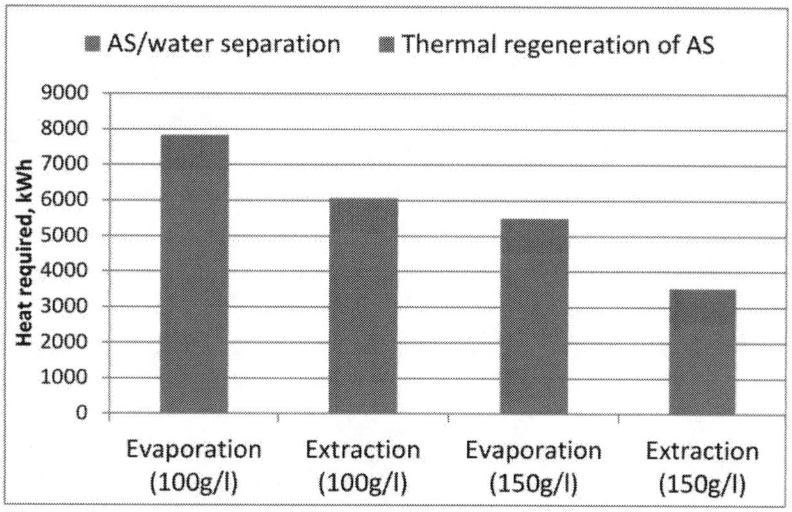

Figure 1. Energy consumption employing thermal evaporation and methanol liquid-liquid extraction under different solid to liquid ratios.

Conclusions

This investigation has indicated that liquid-liquid extraction in presence of methanol is a promising alternative for the separation of AS form the carbonated solution resulting in 26-46% decrease in heat requirement. The heat requirement also decreases with increasing solid/liquid ratio. Further research is required to optimise this technique by evaluating different starting solid-liquid ratios and/or evaluating different solvent to render the process economically viable.

Acknowledgements

The authors thank the Centre for Innovation in Carbon Capture and Storage at Heriot-Watt University (EPSRC Grant No. EP/F012098/2) for support.

References

1. Sanna, A., Uibu, M., Caramanna G., Kuusik R., Maroto-Valer M.M. 2014, Mineral carbonation for geological storage/Eengineering for carbonation, *Chemical Society Reviews*, 43(23), 8049-8080.
2. Park, A., Fan, L.-S. 2004. CO_2 mineral sequestration: physically activated dissolution of serpentine and pH swing process, *Chemical Engineering Science,* 59, 5241-5247.
3. Sanna, A., Hall M.R., Maroto-Valer M.M. 2012. A review of post-processing pathways in carbon capture and storage by mineralisation, *Energy and Environmental Science,* 5, 7781-7796.
4. Wang, X., Maroto-Valer, M.M. 2011. Dissolution of serpentine using recyclable ammonium salts for CO_2 mineral carbonation. *Fuel,* 90, 1229-1237.
5. Wang, X., Maroto-Valer, M.M. 2013. Optimization of carbon dioxide capture and storage with mineralisation using recyclable ammonium salts. *Energy,* 51, 431-438.
6. Hänchen, M., Prigiobbe, V., Baciocchi, R., Mazzotti, M. 2008. Precipitation in the Mg-carbonate system–effects of temperature and CO_2 pressure. *Chemical Engineering Science,* 63(4), 1012–28.
7. Sanna, A., Wang, X., Lacinska, A., Style,s M.T., Paulson, T., Maroto-Valer, M.M. 2013, Enhancing serpentine dissolution by NH4-based pH swing mineral CO_2 sequestration, Minerals Engineering, 49, 135-144.
8. Hoppe, H., 1966, A new method for the production of chlorine-free potash fertilizers. Proc. of the XVIIth International Congress Chemistry Days 1966 on Chemical Fertilizers, Milan, pp. 121 132.
9. Bewsey, J., 2011, Acid mine water, Trailblazer technologies (Pty) Ltd, http://www.nstf.org.za

Computer simulations of the hydration-carbonation processes in reactive magnesia cement systems

M. Wang[1*], A. Al-Tabbaa[2]

[1] Department of Engineering, University of Cambridge, UK
[2] Department of Engineering, University of Cambridge, UK
*Corresponding author: mw567@cam.ac.uk

Abstract
Reactive magnesia cement systems can gain significant strength through their hydration-carbonation processes, providing novel and green construction products. The influencing factors of the hydration-carbonation processes of reactive magnesia cement has been experimentally identified to be the MgO content, water content, porosity and curing conditions, including CO_2 concentration, relative humidity and temperature. It has been found that nesquehonite is formed at a temperature range of 10°C to 50°C under accelerated carbonations of reactive magnesia cement systems. In this paper, a discrete microscopic model is proposed to simulate the microstructure formation of reactive magnesia cement systems under the effect of accelerated carbonation. To be more specific, the initial microstructure of fresh cement before reaction is simulated as a result of digitized particle packing using the technique of Cellular Automata. Then the Lattice Boltzmann Method was applied to simulate the dissolution and precipitation behaviour of MgO under the influence of accelerated carbonation in a diffusion-controlled system to form the final porous microstructure. The percolation threshold of the microstructure was analysed to predict the transportation duration of CO_2 into cement paste. This model provides microstructure with adjustable micro-properties of cement paste, which includes different cement particle size distribution, MgO content, porosity, degree of hydration and degree of carbonation.

Introduction

Computer simulation of the microstructure formation for Portland cement has been intensively studied in the last 20 years. The research can be categorised into discrete modelling [2] and vector-based modelling.[5,11]The development of discrete simulation approach for cement microstructure development appears to have been slower than the development of vector-based models. However, analysis techniques based on the discrete method, such as Cellular Automata (CA) and Lattice Boltzmann Method (LBM), has been widely applied and referred to by vector-based researchers to analyse the microstructural properties such as pore connectivity[15] and transport properties[16], making the boundary between discrete modelling and continuous modelling less and less clear. Within the discrete techniques, Lattice Boltzmann Method (LBM) shows strong potential to simulate the microstructure development of porous material. Recently, Kang, et al.[8] developed a model to simulate the dissolution and precipitation processes for porous materials with a multi-component LBM. Kang's model can be directly applied to the simulation of cement microstructure formation, if the composition and dissolution properties of all the cement phases are known, by simplifying the structure development mechanism into dissolution-precipitation processes.

For both previous discrete and vector-based models, the initial cement microstructure before reaction is generated with Random Sequential Addition (RSA) method. Recent experimental measurement of fresh cement shows that the initial packing fraction is closer to Random Close Packing value of 0.64 [14], which is larger than the RSA value of 0.38. Such observation indicates that it will make the simulation more realistic if particle packing is involved when generating the initial cement microstructure. In terms of packing simulation, Jia & Williams[7] developed a packing algorithm for 2D digitized particles, which was further developed and optimised into 3D by Byholm[6] et al. separately. However their focus was on the packing of particles with arbitrary shapes, and the packing result was slightly lower than 0.64. In this paper, a modified Jia & Williams[7] approach is applied to simulate the initial microstructure formation caused by particle packing before the chemical reaction.

Reactive magnesia cement has been proofed to be more sustainable than traditional Portland cement in term of CO_2 emission, based on previous experimental work[1, 9, 10]. Under accelerated carbonation and ambient temperature, it has been found that the degree of carbonation of MgO for each hydration stage could reach 100% and the major reaction product is $MgCO_3 \cdot 3H_2O$ (nesquehonite).[9,12] Computer simulation for the microstructure formation of reactive magnesia cement is not same as the previous modelling work for Portland cement because the hydration and carbonation processes play an important role in the strength development of reactive magnesia cement rather than the hydration process alone. During the carbonation process of the reactive magnesia cement systems, CO_2 from the surrounding environment is transported to the inside of the cement thorough the connected pores. When depercolation threshold of the pores is reached, the carbonation process will immediately stop or slow down. On the other hand, the depercolation threshold is less important for Portland cement since hydration can still be triggered as long as there is water still left in the isolated pores.

Methodology

Particle Packing

The algorithm of Cellular Automata is applied to simulate the packing of digitized particles as the initial cement microstructure. The initial positions of cement particles are assigned according to the RSA approach as Bentz[2] and Jia & Williams[7]. To be more specific, each particle is randomly placed in a cubic domain one by one from the biggest to the smallest. If overlapping happens, a new random position is chosen to replace the overlapping position. In this work, periodic boundary condition is applied on the four horizontal directions, and solid wall boundary condition is applied on the two vertical directions.

After all the particles are placed in space, random movement is applied to the particles to simulate the influence of vibration. Initially, 26 directions of one single particle are checked to see whether the particle is contacted with each other in these directions. One direction is a possible movement direction for the particle if the contact number of the particles in this direction is zero. Then certain CA rules are applied to choose one of the possible movement directions for the particle to move. In this work, contact force, friction force and gravity is taken into consideration when choosing the movement direction.

After all the particles are placed in space, random movement is applied to the particles to simulate the influence of vibration. Initially, 26 directions of one single particle are checked to see whether the particle is contacted with each other in these directions. One direction is a possible movement direction for the particle if the contact number of the particles in this direction is zero. Then certain CA rules are applied to choose one of the possible movement directions for the particle to move. In this work, contact force, friction force and gravity is taken into consideration when choosing the movement direction.

The relationship between Hertz repulsive force and contact properties are shown with the following equations [13]:

$$\alpha_{ij} = \frac{a^2}{R} = \left(\frac{9P^2}{16RE^{*2}}\right)^{1/3} \qquad (1)$$

where a is the contact area radius; E^* is a constant relevant to the elastic modulus of the contacted material; 1/R is equivalent contact curvature; α_ij is the relative displacement of the centres of two contacted sphere, as shown in Fig.1.

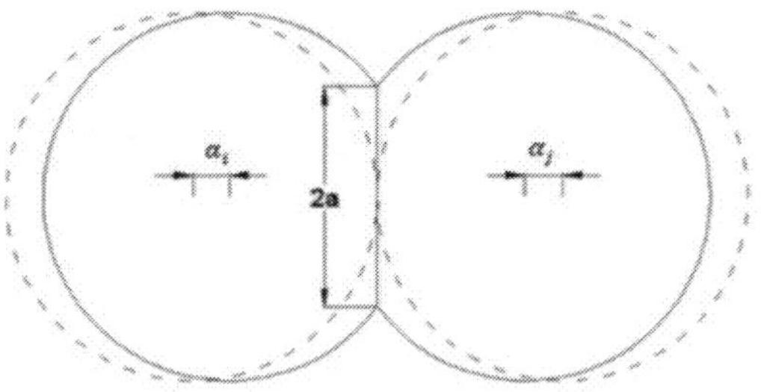

Figure 1. Parameters for particle interaction

This equation indicates that the normal force appears to be a function of equivalent contact curvature, radius of contact area and relative displacement. Assuming that R is equal to the radius of currently being analysed particle, the above function can be converted to the following function:

$$P = P_f(V, A, \alpha_{ij}) \qquad (2)$$

where V is the volume of currently being analysed particle, A is the area value of the contact area. Since α_ij is the relative displacement of the contacted particles during unit computation iteration (time), it is treated as the relative speed in the contact direction. In this work, this value is randomly assigned with the Maxwell-Boltzmann distribution function:

$$f(\alpha_{ij}) = \sqrt{\frac{2}{\pi}}\frac{\alpha_{ij}^2 e^{-\alpha_{ij}^2/2\sigma^2}}{\sigma^3} \qquad (3)$$

where σ is the only control parameter. It was found that the following function can provide a relatively stable final packing fraction for mono-size particle packing:

$$P = V\sqrt{A}\alpha_{ij}(\sigma) \quad (4)$$

where σ is taken as 1.1. With the normal force, the friction threshold in the tangential direction is then calculated:

$$f = \mu P \quad (5)$$

where μ is equivalent to the coefficient of friction, which can be used to adjust the final packing fraction. Gravity proportional to the particle volume is applied on all particles in vertical directions.

The movement in one unblocked direction is possible if the cumulative force in this direction is bigger than the friction threshold. When more than one direction is available for movement, certain priority rules are applied. In this work, one of the available movement directions is randomly chosen with uniform distribution function. For computation efficiency, only the forces in 6 principal directions are calculated. The analysis of the rest 20 directions is considered as a combination of the 6 principal direction calculations. The resolution of this study is set to be 1μm/pixel. More detail about the packing algorithm can be found in the authors' previous pubilication[17].

Fluid flow behaviour in porous material

A modified version of Kang's model[8] is adopted to simulate the dissolution-precipitation process of reactive magnesia cement. It is assumed that the pore space is filled with free water and infinite water is provided from the curing environment. As a result, the ultimate degree of MgO conversion becomes a function of available pore space for the deposition of hydration-carbonation product. The fluid behaviour of the solvent within cement paste is simulated with Lattice Boltzmann equation:

$$f_a(\boldsymbol{x} + e_a\delta t, t + \delta t) - f_a(\boldsymbol{x}, t) = -\frac{f_a(\boldsymbol{x},t) - f_a^{eq}(\boldsymbol{x},t)}{\tau} \quad (6)$$

where f_a is the discrete particles distribution function in a[th] microscopic velocity direction , δt is the time increment, e_a is the microscopic velocity at location x at time t, τ is the relaxation time related to the kinematic viscosity $\gamma = c_s^2(\tau - \frac{1}{2})$, f_a^{eq} is the corresponding equilibrium particle distribution function expressed in the form of a discretization of the continuous Maxwell-Boltzmann equilibrium distribution (Zhang, 2013):

$$f_a^{eq}(\boldsymbol{x}, t) = w_a\rho[1 + \frac{e_a \cdot u}{c_s^2} + \frac{(e_a \cdot u)^2}{2c_s^4} - \frac{e_a \cdot u}{2c_s^2}] \quad (7)$$

where c_s is the lattice speed of sound, w_a are the associated weight coefficient, ρ and $\boldsymbol{u}$ are macroscopic density and velocity which can be calculated with the following equations:

$$\rho = \sum_a f_a \quad (8)$$

$$\boldsymbol{u} = \frac{\sum_a f_a e_a}{\rho} \quad (9)$$

For two-dimensional analysis, nine-speed (D3Q9) Lattice Boltzmann model:

$$w_a = \begin{cases} 1/3 & a = 0 \\ 1/18 & a = 1,2,\dots,6 \\ 1/36 & a = 7,8,\dots,18 \end{cases} \quad (10)$$

$$e_i = \begin{bmatrix} 0 & 1 & -1 & 0 & 0 & 0 & 0 & 1 & 1 & 1 & 1 & 0 & 0 & -1 & -1 & -1 & -1 & 0 & 0 \\ 0 & 0 & 0 & 1 & -1 & 0 & 0 & 1 & -1 & 0 & 0 & 1 & 1 & 1 & -1 & 0 & 0 & -1 & -1 \\ 0 & 0 & 0 & 0 & 0 & 1 & -1 & 0 & 0 & 1 & -1 & 1 & -1 & 0 & 0 & 1 & -1 & 1 & -1 \end{bmatrix} \quad (11)$$

In order to simulate the solute transportation, it is assumed that the solute concentration is sufficient low so that the fluid behaviour of solvent is not influenced by the solute.[8] As a result the distribution function of the solute can be calculated as:

$$g_a^s(x + e_a\delta t, t + \delta t) - g_a^s(x, t) = -\frac{g_a^s(x,t) - g_a^{s,eq}(x,t)}{\tau_s} + w_a q^s \quad (12)$$

$$g_a^{s,eq}(x,t) = w_a C^s[1 + \frac{e_a \cdot u}{c_s^2} + \frac{(e_a \cdot u)^2}{2c_s^4} - \frac{e_a \cdot u}{2c_s^2}] \quad (13)$$

where g_a^s and $g_a^{s,eq}$ are the distribution function and corresponding equilibrium distribution function of solute s, C^s is the concentration of the solute defined as :

$$C = \sum_a g_a \quad (14)$$

Dissolution and Precipitation process

During the simulation of dissolution and precipitation processes, periodic boundary conditions are applied to the cubic computation domain. It is assumed that the rate of deformation for the solid surface is so slow that half-way bounce back condition is applied at the solid-liquid interface. For the part of solid phase where dissolution happens, the solid reaction front is removed and the density of solid phase on the lattice is continuously reduced. The reduction of solid phase is calculated with the following equation:

$$\frac{dC}{dt} = k_s(1 - \frac{\gamma w}{w_o})(C_s - C) \quad (15)$$

where C is the concentration of solute, C_s is saturation concentration of solute, k_s is a constant related to the rate of dissolution, γ is a coefficient of how strong the influence of decreasing amount of water can be on the dissolution rate of the surface pixels, w is the volume of consumed water, w_o is the initial volume of water in the pores of fresh cement. Through previous experimental work[9,12] on reactive magnesia cement, it was observed that 100% carbonation can be achieved in every hydration stage under the curing condition of 20% CO2, yielding the hydration-carbonation product of Nesquehonite:

$$MgO + 3H_2O + CO_2 = MgCO_3 \cdot 3H_2O \quad (16)$$

It is assumed that hydration happened at the same time as dissolution, and the entire hydration product is immediately transferred into nesquehonite($MgCO_3 \cdot 3H_2O$), which is then precipitated as newly formed solid phase as a reverse process of the dissolution with different rate coefficient k_p and saturation concentration C_p. Table 1 list the properties of the relevant substance. The difference in density of MgO and nesquehonite cause the solid expansion as shown in Figure 2.

	Solubility S (g/100g water)	Solubility S_m (mol/ μm^3)	Molar mass M_m (g/mol)	Density ρ (g/cm^3)	N_a/V_m (mol/μm^3)
MgO	8.60×10^{-3}	2.13×10^{-18}	40.31	3.58	8.88×10^{-14}
$Mg(OH)_2$	6.40×10^{-4}	1.10×10^{-19}	58.32	2.35	4.03×10^{-14}
$MgCO_3 \cdot 3H_2O$	1.79×10^{-1}	1.30×10^{-17}	138.37	1.84	1.33×10^{-14}

Table 1 Properties of relevant compounds

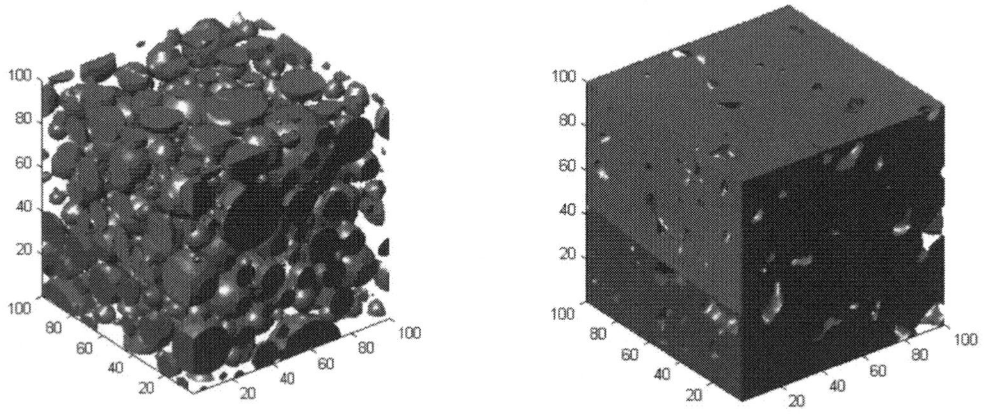

Figure 2. Solid Phase expansion

Results and Discussion

Particle packing

Figure 3 illustrates the packing result of 3000 mono size particles with radii of 6μm in a 100×100×4000 computation domain. The coefficient of friction is set to 0 for this test so that RCP value of 0.64 is aimed. It can be seen that though there is flocculation of cross-sectional packing fraction, the average packing fraction is stabilized to the RCP value of 0.64 when the distance to the bottom boundary is bigger than 25μm. Considering that wall effect is the origin of the Interfacial transition zone (ITZ) , the generated porous structure can be separated into ITZ and bulk cement paste at the point of 25 μm. Bulk cement paste ranging from 200 μm to 300 μm is 'taken out' from the computation domain to form a 100×100×100 sample for further dissolution-precipitation simulation. Figure 4 illustrates the packing fraction result of Random loose packing simulation as a result of increasing coefficient of friction. It can be seen that the fluctuation of cross-sectional packing fraction is more intensive than that of RCP. However the average packing fraction remains stable when the distance to the bottom boundary is more than 25 μm, which is the thickness of the ITZ.

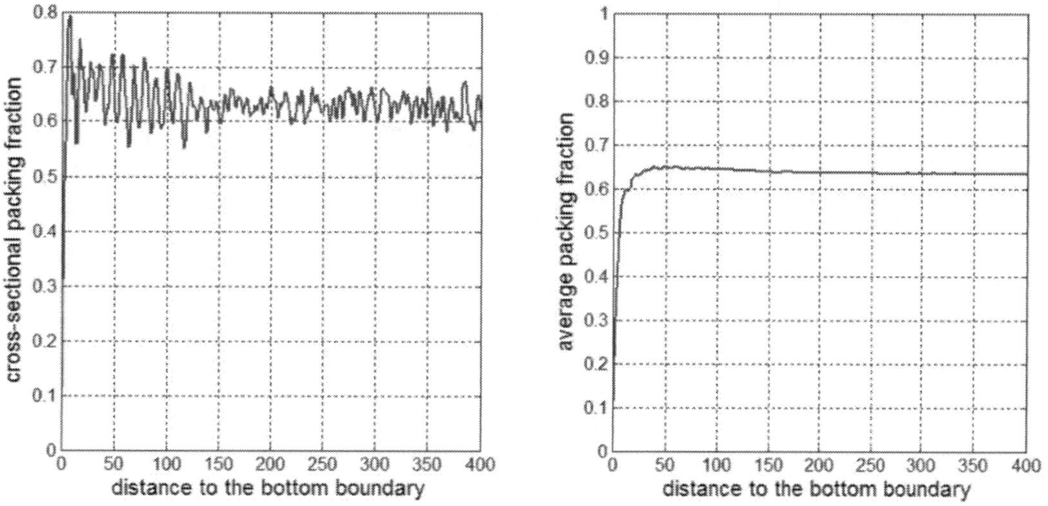

Figure 3. Random close packing fraction

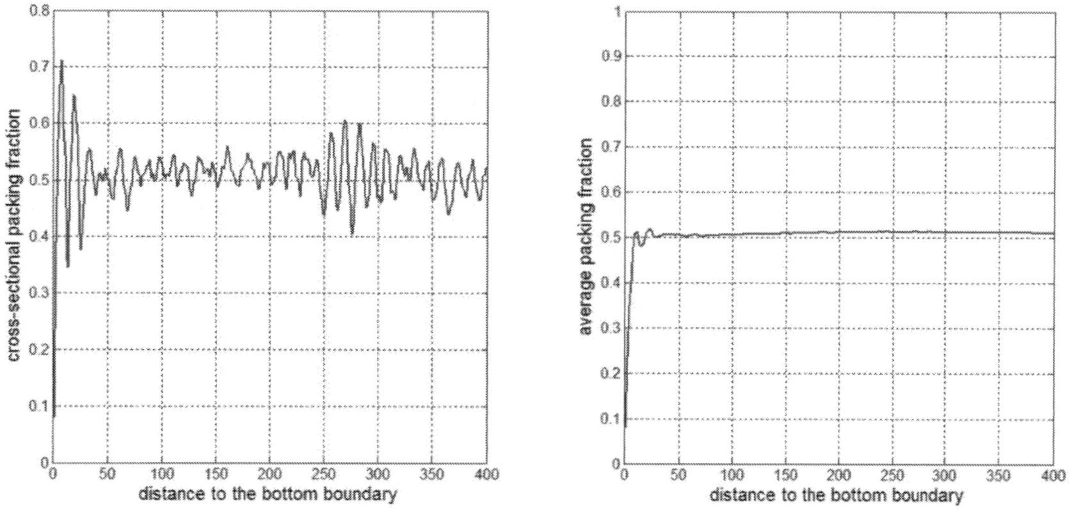

Figure 4. Random loose packing fraction

Pore connectivity

With the bulk cement past generated form the particle packing program, the hydration-carbonation process is simulated with LBM as process of dissolution-precipitation. The pore connectivity is calculated with modified 'burning algorithm'.[3] To be more specific, a 'fire' is initially started on the boundary pore pixels of one side of the simulated microstructure, then their neighbour pixels are checked. If the neighbour pixels are pores, they are classified as 'combustible' and the 'fire' propagate to these pore pixels. In the following interactions, similar neighbour checking is conducted until pore pixels from the other side of the microstructure is fully 'burned' or certain number of computation cycles is reached when depreciation threshold has been achieved. This method can calculate the amount of connected pores and dead end pores which are connected to one end of the cement microstructure.

Figure 5 shows the result of pore connectivity development in terms of porosity in bulk cement paste. It can be seen that the pores are well connected when the porosity is more than 0.05. Similar to the simulation of Ye[15], when the porosity is smaller than the depercolation threshold of around 0.05, the pores become disconnected to each and the connectivity drops drastically. However since the connected pores measurement in this work measures the open pore as well, the pore connectivity does not drop to 0 directly at depercolation threshold. For example at porosity of 0.03, the pore connectivity reaches the value of 0.51. This pore value indicates that there are 51% of open pore are connected to one end of the cement sample. However at porosity value of 0.02, the pore connectivity becomes 0.23. Such reduction in pore connectivity indicates that some of the inner pores, which are part of the measure open pores, are filled by the expansion of the solid phase. Then some of the inner pore become isolated pores since their connection with the open pores are has been filled. As a result, the 51% open pores become 28% isolated pore and 23% open pore which are still connected to one end of the sample. With the burning algorithm used in this study, dead end pore connected to one end of the sample counts as part of the connected pore. As a result, the pores are considered fully isolated when the porosity reaches zero.

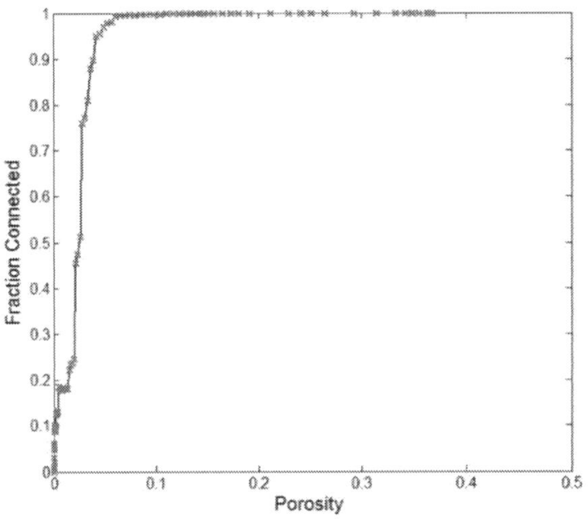

Figure 5. Pore connectivity in terms of porosity

The influence of w/c ratio

Figure 6 illustrates the influence of w/c ratio on the porosity development. It can be noticed that sample with high initial w/c ratio has relative low initial solid packing fraction and high initial porosity. Along with the hydration-carbonation process of the cement, pores are filled by the expansion of solid phase as a result of the formation of nesquehonite. Similar trend was also observed for OPC by Cook & Hover[17] . Since the solid expansion rate for OPC is not as significant as reactive magnesia, the pore-filling speed of OPC is not as fast as reactive magnesia cement.

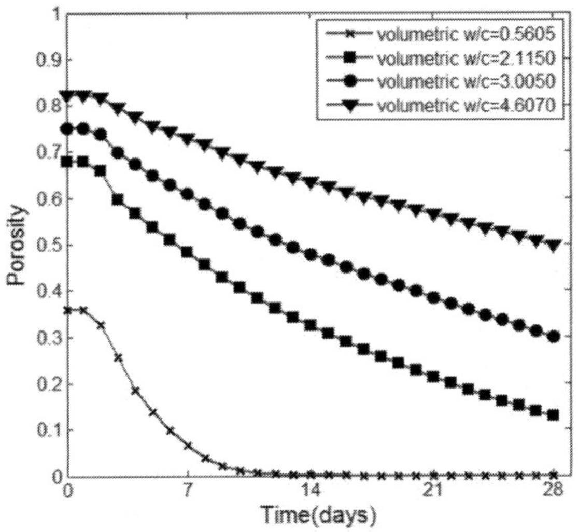

Figure 6. Porosity development for different w/c ratio

Figure 7 shows the degree of conversion result of three 100% MgO sample with different w/c ratio. It can be seen that an incremental volumetric w/c ratio will lead to a faster rate of degree of conversion.

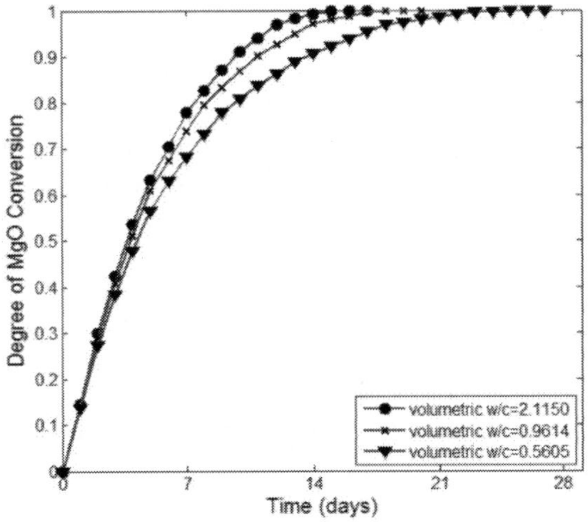

Figure 7. Degree of conversion of 100% MgO samples with different w/c ration

Conclusion

A new 3D model for the microstructure development of reactive magnesia cement systems has been constructed based on the algorithm of Cellular Automata (CA) and Lattice Boltzmann Method (LBM). The initial microstructure generator of this model can generate a structure ranging from RLP to RCP to simulate the micro-structure of fresh cement .The depercolation threshold of 5% obtained from the mono-size particle simulation is in good agreement with the study of Ye[15]. This model simulates the ultimate degree of MgO conversion as a function of available space for the deposition of the product of hydration-carbonation process. It has been simulated that samples with higher w/c ratio have a higher initial and final porosity than samples with lower w/c ratio. It was also observed that MgO in samples with lower initial w/c is converted to nesquehonite more rapidly.

Reference

1. Liska, M and Al-Tabbaa, A. (2009). Ultra-green construction: reactive magnesia masonry products. *Proceedings of the ICE - Waste and Resource Management, 162*(4), 185–196.
2. Bentz, D. (1997). Three Dimensional Computer Simulation of Portland Cement Hydration and Microstructure Development. *Journal of the American Ceramic Society, 80*(1), 3–21.
3. Bentz, D., & Garboczi, E. (1991). Percolation of phases in a three-dimensional cement paste microstructural model. *Cement and Concrete Research, 21.*
4. Van Breugel, K.(1994). Numerical simulation of hydration and microstructural development in hardening cement-based materials:(I) theory. *Cement and Concrete Research, 25*(2), 319–331.
5. Byholm, T., Toivakka, M., & Westerholm, J. (2009). Effective packing of 3-dimensional voxel-based arbitrarily shaped particles. *Powder Technology, 196*(2), 139–146.
6. Jia, X., & Williams, R. . (2001). A packing algorithm for particles of arbitrary shapes. *Powder Technology, 120*(3), 175–186.
7. Kang, Q., Zhang, D., & Chen, S. (2003). Simulation of dissolution and precipitation in porous media. *Journal of Geophysical Research, 108*(B10), 2505.
8. Liska, M. (2009). *Properties and Applications of Reactive Mgnesia Cements in Porous Blocks*. PhD Thesis,University of Cambridge, Cambridge, UK.
9. Liska, M., Vandeperre, L. J., & Al-Tabbaa, A. (2008). Influence of carbonation on the properties of reactive magnesia cement-based pressed masonry units. *Advances in Cement Research, 20*(2), 53–64.
10. Navi, P., & Pignat, C. (1996). Simulation of cement hydration and the connectivity of the capillary pore space. *Advanced Cement Based Materials, 7355*(96).
11. Unluer, C. (2012). *Enhancing the catbonation of reactive magnesia cement-based porous blocks*. PhD thesis, University of Cambridge, Cambridge, UK.
12. Vu-Quoc, L., Zhang, X., & Lesburg, L. (2001). Normal and tangential force–displacement relations for frictional elasto-plastic contact of spheres. *International Journal of Solids and Structures, 38*, 6455-6489
13. Wong, H. H. C., & Kwan, A. K. H. (2007). Packing density of cementitious materials: part 1—measurement using a wet packing method. *Materials and Structures, 41*(4), 689–701.
14. Ye, G. (2005). Percolation of capillary pores in hardening cement pastes. *Cement and Concrete Research, 35*(1), 167–176.
15. Zhang, M. (2013). *Multiscale Lattice Boltzman- Finite Element Modelling of Transport Properties in Cement-based Materials*. PhD thesis, Thechnische University Delft.
16. Cook, R. a., & Hover, K. C. (1999). Mercury porosimetry of hardened cement pastes. *Cement and Concrete Research, 29*(6), 933–943.
17. Wang, M. and Al-Tabbaa, A.(2014) Computer simulation of the packing of digitized reactive magnesia particles. *Proceeding of the RILEM International Symposium on Concrete Modelling.pp.*36-43

Improving technological viability of carbon sequestration technologies with structured agents and industrial wastes

I. Romero-Hermida[1,2], V. Morales-Flórez[1,2], A. Santos[3], L. Esquivias[1,2*]

[1] Departamento de Física de la Materia Condensada, Universidad de Sevilla, Seville, Spain.
[2] Instituto de Ciencia de Materiales de Sevilla (CSIC/US), Seville, Spain.
[3] Departamento de Ciencias de la Tierra, Universidad de Cádiz, Puerto Real, Spain.
*Corresponding author: luisesquivias@us.es

Abstract

The aim of this research is to contribute to resolve some of the major drawbacks of the carbon sequestration technologies under ambient conditions, such as kinetics and costs. We researched the role of the microstructure of calcium-rich wastes in the carbonation kinetics by weathering and we synthesised new efficient carbon sequestration agents from industrial wastes by simple procedures.

Moreover, this work presents the first-to-date combined kinetic and nanostructural research on CaO oriented to carbon sequestration. Complete hydration of the original CaO samples was always found prior to the onset of carbonation, typically, after 300 h of weathering for samples with a specific surface area of 40 m^2/g. Full carbonation (100% efficiency) was obtained in all cases. This experimental research was completed by an empirical description of the weathering reactions in terms of a two-process Random Pore Model. On the other hand, simple procedures were considered to synthesize affordable CO_2 sequesters using phosphogypsum by-product from the phosphoric acid production as calcium source, and an aqueous alkaline Na-rich industrial by-product as basic media for the precipitation of calcium-rich hydroxides from phosphogypsum. These hydroxides were revealed as very efficient sequestration agents under room conditions, as tested by XRD, SEM and TGA analyses.

Introduction

One of the challenges to mitigate the possible harmful consequences of the climate change is to develop and improve technologies for energy production. At the same time, it is also necessary to design and implement technologies to reduce the emissions and the concentration of carbon dioxide and other greenhouse gases in the atmosphere.[1] The mineral sequestration combines in a single step the capture and permanent storage of carbon dioxide in the form of carbonates.[2,3] This process mimics the natural procedure that has regulated atmospheric CO_2 concentration for millennia. It deserves special attention since it is by far the most durable carbon capture and storage (CSS) technology and, indeed, a complementary technology of the already implemented carbon storage facilities. Nevertheless, the low speed of the reaction and the costs, the weakest points of the current mineral sequestration technologies, hinder its general implementation at an industrial scale.

The structure of the carbon sequestration agents and its specific (reactive) surface area in particular, has been identified as a key feature that ultimately determines the kinetics of carbonation[4-6] and, consequently, the industrial viability of the procedure. On the other hand, using industrial wastes with high carbon sequestration potential can contribute to reduce the costs as well as to avoid the ecological impacts of mineral extraction on a massive scale.[7,8] In this framework, maximizing the carbonation efficiency of the designed procedures is mandatory in order to overcome these technological obstacles.

Up to now, the simplest mineral sequestration procedure consists in capturing CO_2 directly from the air[9] (weathering), but it lacks the necessary technological efficiency to be implemented on a large scale. This work explores strategies to overcome the kinetic and efficiency limitations of weathering by using CaO-rich industrial wastes, in order to achieve reaction rates at a human scale under atmospheric conditions (1 bar, 300 K and $[CO_2]$ ~400 ppm). It is worthy to note that the reactions of hydration and carbonation of CaO are key steps for fundamental processes in very different systems, such as in construction materials or carbon capture and sequestration (CCS) technologies,[10-12] but they have been barely studied in the framework of carbon capture under atmospheric conditions. The weathering of CaO implies moisture adsorption and hydration that give rise to calcium hydroxide, and carbonation.[13] Hydration and carbonation reactions may produce a passivating layer[14] which eventually can prevent the exposure of unreacted CaO to gases for further reaction, and therefore, the kinetics can be slowed down. But this retardant phenomenon can be minimized in microstructured samples.[6,15]

The first part of this work is focused on the kinetics of the weathering processes of nanostructured CaO samples obtained from a calcium-rich by-product of the acetylene industry.[16] XRF and XRD analyses of this by-product revealed a composition of calcium hydroxide above 95 wt. % and nitrogen physisorption experiments and scanning electron microscopy showed a high specific surface area (47.5 m^2/g) as well as the presence of submicrometric grains formed by bundles of sticks smaller than 200 nm. The calcium-rich slurry was heat-treated for 1h to obtain nanostructured calcium oxide at 600 °C (samples CaO600) or 800 °C (sample CaO800) to obtain samples with different nanostructures due to incipient sintering.

A simple method to study the kinetics of the weathering reactions consists of accurately registering weight changes in samples exposed to atmospheric gases.[17] For chemically identical samples, the difference between the reaction kinetics and efficiency can be explained in terms of differences in pore geometry or specific surface area. But also, it should be considered that the nanostructure is a dynamic characteristic upon weathering. An exponential form for gas-solid kinetics under controlled atmospheres where the sample mass depends on time as $(1 - e^{-t/\tau})$ has been proposed.[18,19] To describe our results, the Random Pore Model[20-22] (RPM) has been considered in order to discuss the relationship between nanostructure and kinetics.

Moreover, the re-use of industrial wastes for carbon mineral sequestration has a two-fold goal: removing unpleasant wastes and mitigating the emissions of green house gasses in the atmosphere. In this work, industrial wastes with high contents of calcium were considered to synthesize materials with the ability for CO_2 sequestration by simple and cheap procedures. In the second part of this work, phosphogypsum[23] (PG) and Na-rich wastes from the fertilizer industry and aluminum industries, respectively, have been considered as precursors. The industrial wet phosphoric acid production in Huelva (SW Spain) has led to the controversial stockpiling of about 120 Mt of waste phosphogypsum by-products, less than 1 km away from the city. In addition, the aluminum waste consisted in an alkaline liquid waste (pH ~ 14) bearing significant amounts of Na and Al cations. Both industrial wastes were mixed in order to obtain new minerals with the ability of capture CO_2 in simple aqueous sequestration experiments, and their efficiency on carbon sequestration in aqueous media under ambient conditions was checked.

In summary, this work provides results in order to enhance the viability of CO_2 mineral sequestration technologies. It is aimed to determine the role of the nanostructure as one of the most important factors for developing efficient carbon sequestration technologies. In addition, we propose here simple physico-chemical processes to obtain efficient sequestration agents from industrial by-products which are not able to permanently capture CO_2 by themselves.

Results and discussion

Nanostructure and kinetics

Representative curves of the time-dependence of the mass for two samples of calcium oxide obtained at different calcination temperatures (CaO600 and CaO800) and exposed to atmospheric gases are plotted in Figure 1. Real-time mass data were normalized to the conversion reaction parameter $\chi(t)$ spanning from 0 to unity. This assumption of maximum conversion equalling unity is supported by following results which confirmed that the reactions were completed. In all cases, the sample mass increased very fast during the first minutes, revealing extraordinary reactivity. In general, all the curves reproduced analogous behavior with the presence of two clearly separated steps. This feature suggests that weathering of CaO samples occurs in different phases: after initial moisture adsorption, fast hydration occurs (calcium hydroxide precipitation) taking up to several hundreds of hours, and once the hydration of the sample is completed, carbonation takes place (calcium carbonate precipitation) over more than one month.

Moreover, the lower heat treatment of the CaO600 samples led to faster kinetics of weathering with respect to the CaO800 samples, especially in terms of the slope at the initial stages (not shown) and by reaching the plateaus sooner. This was also observed in the onset of carbonation. This feature can be numerically estimated by the change in the sign of the second time-derivative of M* (normalized mass). The estimated carbonation onsets were $t = 300 \pm 100$ h and $t = 550 \pm 140$ h for CaO600 and CaO800 samples, respectively. These differences confirm that reducing the heat treatment by 200 °C led to significant differences in the reactivity, as expected as samples calcined at lower temperature were more reactive due to their expected higher values of specific surface area.

The huge scientific production which refers to calcium hydroxide carbonation by atmospheric CO_2 deals with construction materials involving processes with takes several months or years. This work is one of the first kinetic and nanostructure evolution joint research into calcium oxide oriented to the hydration and subsequent sequestration of atmospheric carbon dioxide. Although standard carbonation mineral sequestration experiments are usually performed at high CO_2 pressures or temperatures, in our case, even working under atmospheric conditions, the samples showed outstanding performance in terms of water adsorption and carbon sequestration.

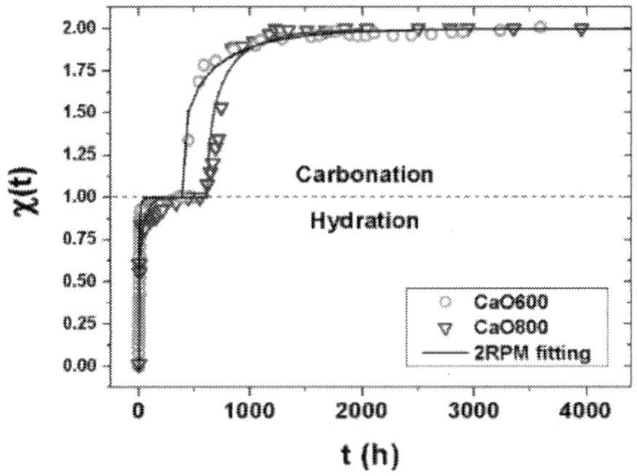

Figure 1. Conversion kinetics during hydration and carbonation for the studied samples and Random Pore Model fitting curves.

Structure and composition

The x-ray diffraction analyses revealed the evolution of different Ca-rich crystalline phases present in the sample during the weathering process. In Figure 2 left, semi-quantitative analyses of the XDR patterns gave an estimation of the time dependences of the relative concentrations of CaO, $Ca(OH)_2$, and $CaCO_3$ for different sets of samples. It was confirmed that the samples were initially composed of CaO, remaining as the major phase up to weathering times below 2–3 h. The patterns obtained at later times revealed the gradual disappearance of CaO leading to $Ca(OH)_2$ precipitation. Then, $Ca(OH)_2$ remained as the most important phase up ~100 h. This result confirms that the first plateau observed in the kinetic data of the normalized mass (Fig. 1) appears after complete hydration of the CaO sample, and prior to carbonation.

Besides, results obtained by thermogravimetric analyses (available at ref. 5) are in good coherence with the expected relative composition regarding the two-step weathering discussed regarding Figure 1 and the quantitative estimations by XRD plotted in Figure 2, left.

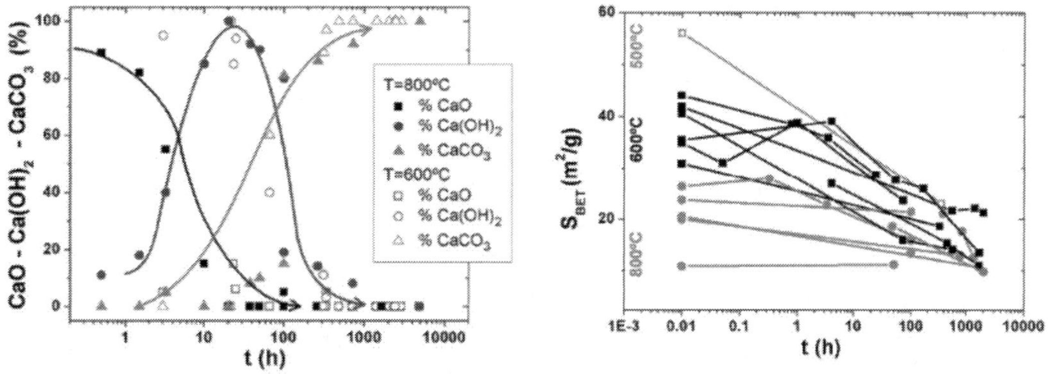

Figure 2. Left: time evolution of the relative presence of crystalline phases measured by semiquantitative analyses of XRD patterns. Open and solid symbols correspond for CaO600 and CaO800 samples, respectively, and curved lines are guides for the eye only. Right: Specific surface area of samples CaO600 (black squares) and CaO800 (red circles).

In view of the specific surface data showed in Figure 2 right, a general decreasing trend with time was obviously concluded. Considering previous analyses, the original CaO grains turned into $Ca(OH)_2$ particles with a lower specific surface area and smaller pore volume. Afterwards, the formation of $CaCO_3$ continued in a similar fashion. Moreover, it is clear that the CaO600 samples had almost doubled the values of the nanostructure parameters of the CaO800 samples. Both kinds of samples experienced a decrease of 40% in the specific surface area due to complete weathering.

These structural data were considered for the two steps of the Random Pore Model, considering hydration and carbonation separately. The very good correspondence between the mathematical model and the experimental values (solid lines in Figure 1) supports the diffusion regime of the Random Pore Model for describing these reactions and the assumptions of this work, and provides empirical relations between the parameters and the structure.

Waste-based sequestration agents

On the other hand, a simple procedure for obtaining carbon sequestration agents[23] has been improved working exclusively with industrial wastes. The experimental process started with the dissolution of the raw phosphogypsum sample in the Na-rich liquid waste under magnetic stirring for 3 h. The phosphogypsum dissolution resulted in the precipitation of a brownish solid phase (sample PGA) and a supernatant liquid, which was discarded. This solid phase was characterized and subsequently considered for

CO_2 sequestration under atmospheric conditions. The carbonation experiments consisted in a flux of CO_2 (1 bar, 20 cm^3/s for 30 min.) into an aqueous media where the sequestration agent was dispersed. The pH during the carbonation experiments dropped down to 6.7. After the carbonation experiments, the sample was centrifuged and the new precipitated solid phase (sample PGAC) was analyzed and the supernatant discarded. X-ray diffraction, ICP-spectrometry, thermogravimetric analyses and scanning electron microscopy were used to analyze the steps of the procedure.

The PG sample was mainly composed by gypsum (96 % wt.), being Na and F the major impurities. Alkaline liquid phase contained mainly Na (110 ± 2 g/L) and Al (52 ± 2 g/L) as revealed by ICP. The raw PG was researched by XRD and the exclusive presence of gypsum was confirmed. The analysis of sample PGA (Fig. 3) confirmed the dissolution of the original gypsum and the precipitation of thenardite (Na_2SO_4), hydrogarnet ($Ca_3Al_2(OH)_{12}$) and portlandite ($Ca(OH)_2$). Finally, the analysis of the sample after the carbonation experiments (sample PGAC) confirmed the successful carbon sequestration process. Hence, different calcium carbonate phases were identified, as calcite, aragonite and vaterite (Fig. 3). Nevertheless, the presence of gypsum also confirmed that the carbonation capture was not optimized and a non-negligible amount of calcium returned into sulfate phases. This result could be originated by the presence of soluble thenardite in the starting PGA which eventually dissolved during carbonation and sulfate anions competed with carbonic anions to form calcium compounds. Future works are targeted to reduce this unwished effect.

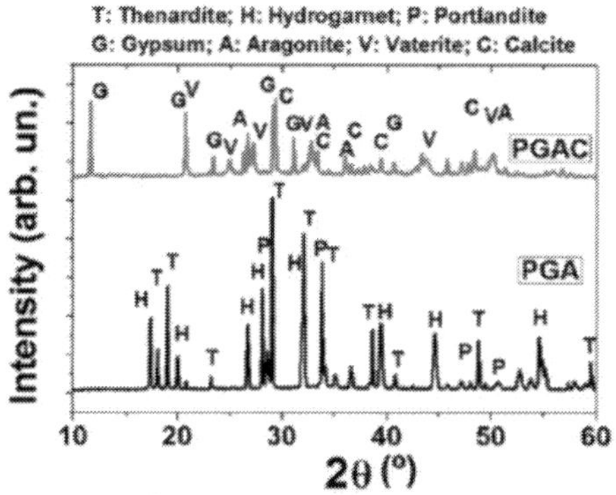

Figure 3. XRD patterns of sample PGA (bottom line, black) and PGAC (top line, red).

The scanning electron microscopy confirmed the previous phase identifications. Thus, hydrogarnet was easily observed, together scarce plate-like morphologies corresponding to portlandite, both of them lying on large gypsum crystals (Fig. 4, left). The carbonated sample was mainly composed by typical calcite rhombohedra, although some spherulitic aggregates composed of framboidal aggregates corresponding to vaterite were also observed.

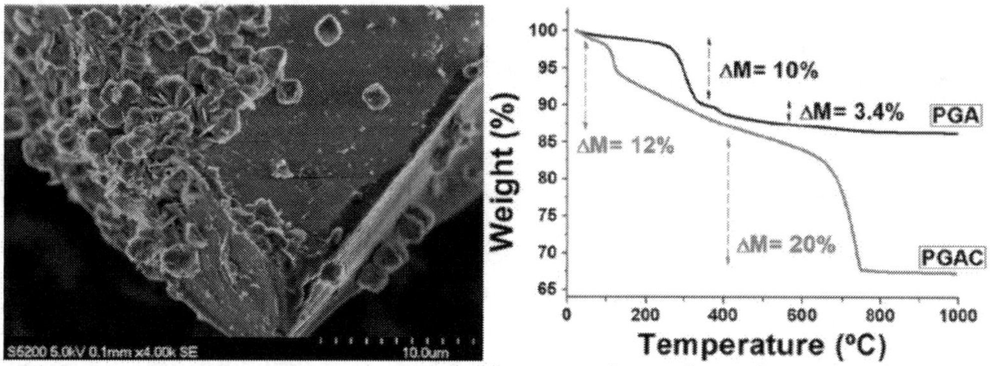

Figure 4. SEM images of sample PGA (left). TGA analyses of samples PGA and PGAC.

The results of the TGA analyses (fig. 4, right) performed to samples PGA and PGAC helped to the quantification of the different phases and to assess the efficiency of the mineral sequestration process. The original sample PGA presented two clear weight losses between 200 °C and 400 °C, which can be interpreted as the release of the structural water of the hydrated phases revealed by XRD. Hence, the first weight loss (10 wt. %) is found at the dehydration temperature of the hydrogarnet, and the second weight loss (3.4 wt. %) can be identified with the dehydration of portlandite. Stoichiometric calculi concluded a chemical composition of the PGA sample of 54 % of thenardite, 45 % of hydrogarnet and 1 % of portlandite.

Regarding the carbonated sample, PGAC, first weigh loss of 12 wt. % at 150 °C was found due to the dehydration of gypsum yielding anhydrite. Finally, from 400 °C to 750 °C the sample mass dropped 20 wt. % due to the decarbonation of the different calcium carbonate polymorphs. The estimated chemical composition corresponds to ca. 55 % of gypsum and 45% of calcium carbonate. This result confirmed that close to 60 % of the calcium present in the sample was successfully involved in carbon sequestration, but the resting 40 % of Ca precipitates as sulfate instead. Thus the carbonation efficiency can be estimated in close to 60 %.

Conclusions

Mineralogical and structural characterizations of the hydration and the carbonation reactions of nanostructured CaO by weathering indicate that both processes take place separately and consecutively. The observed complete carbonation of these samples with a characteristic specific surface area typically above 20 m^2/g suggests the existence of a critical particle size above which the by-product layers will hinder the reactions, which could be estimated above ~90 nm. This feature also highlights the good performance of this sample as carbon sequestration agent by weathering. These results highlight the importance of the structural features of industrial wastes to be considered as carbon sequestration agents.

In addition, this work confirmed that the simple synthesis procedure exclusively based on industrial wastes, namely, the PG waste and the Na-rich alkaline liquid waste from the aluminum industry, successfully provides efficient CO_2 sequestration agents. The efficiency of this new sequestration agent by mineral sequestration under atmospheric conditions in aqueous media was stated at 60 %. Hence, two wastes which are unable to fix CO_2 by themselves can be successfully turned into CO_2 sinks.

Acknowledgements

Thanks to AIR LIQUIDE España, Fertiberia, Verinsur and the technical staff of ICMSE (CSIC/US) and CITIUS (US) for gently supply the raw samples and for their help with the analyses technicques. VMF would like to thank the contract from "Plan Propio de Investigación" of the Universidad de Sevilla (Spain). The funding from the project MAT2013-42934-R from the Spanish Ministry of Economy is also acknowledged.

References

1. The Keeling Curve. A Daily Record of Atmospheric Carbon Dioxide. Available online: http://keelingcurve.ucsd.edu/ (accessed on Feb. 21st 2015); See also: Hansen and 17 more authors, 2013. Assessing "Dangerous Climate Change": Required Reduction of Carbon Emissions to Protect Young People, Future Generations and Nature, Plos ONE 8, e81648.
2. Seifritz, W. 1990. CO_2 disposal by means of silicates. *Nature,* 345, 486.
3. Lackner, K.S., Wendt, C.H., Butt, D.P., Joyce, E.L., Sharp, D.H. 1995. Carbon dioxide disposal in carbonate minerals. *Energy,* 20, 1153–1170.
4. Van Balen, K. 2005. Carbonation reaction of lime, kinetics at ambient temperature. *Cem. Concr. Res.,* 35, 647–657.
5. Morales-Florez, V., Santos, A., Romero-Hermida, I., Esquivias, L. 2015. Hydration and Carbonation Reactions of Calcium Oxide by Weathering: kinetics and changes in the Nanostructure. *Chemical Engineering Journal*, 265, 194-200.
6. Santos, A., Ajbary, M., Morales-Flórez, V., Kherbeche, A., Piñero, M., Esquivias, L. 2009. Larnite powders and larnite/silica aerogel composites as effective agents for CO_2 sequestration by carbonation. *J. Hazard. Mater,.* 168, 1397–1403.
7. Bobicki, E.R., Liu, Q.X., Xu, Z.H., Zeng, H.B. 2012. Carbon capture and storage using alkaline industrial waste. *Prog. Energy Combust. Sci.,* 38, 302–320.
8. Kirchofer, A., Becker, A., Brandt, A., Wilcox, J. 2013. CO_2 Mitigation Potential of Mineral Carbonation with Industrial Alkalinity Sources in the United States. *Environ. Sci. Technol.,* 47, 7548–7554.
9. Stolaroff, J.K., Keith, D.W., Lowry, G.V. 2008. Carbon dioxide capture from atmospheric air using sodium hydroxide spray. *Environ. Sci. Technol.*, 42, 2728–2735.
10. Taylor, H.F.W., 1997. *Cement Chemistry.* 2nd ed., Thomas Telford Pub., London.
11. Lackner, K.S. 2003. A guide to CO_2 sequestration. *Science*, 300, 1677–1678.
12. Power, I.M., Harrison, A.L., Dipple, G.M., Wilson, S.A., Kelemen, P.B., Hitch, M., Southam, G. 2013. Carbon mineralization: from natural analogues to engineered systems. *Rev. Mineral. Geochem.,* 77, 305–360.
13. Van Balen, K., 2005. Carbonation reaction of lime, kinetics at ambient temperature. *Cem. Concr. Res.,* 35, 647–657.

14. Bearat, H., McKelvy, M., Chizmeshya, A., Gormley, D., Nunez, R., Carpenter, R.W., Squires, K., Wolf, G.H. 2006. Carbon Sequestration via Aqueous Olivine Mineral Carbonation: Role of Passivating Layer Formation. *Environ. Sci. Technol.*, 40, 4802–4808.

15. Morales-Flórez, V., Santos, A., Esquivias, L. 2011. Recent insights into xerogel and aerogel mineral composites for CO_2 mineral sequestration. *J. Sol-Gel Sci. Technol.*, 59, 417–423.

16. Morales-Flórez, V., Santos, A., Lemus, A., Esquivias, L. 2011. Artificial weathering pools of calcium-rich industrial waste for CO_2 sequestration. *Chem. Eng. J.*, 166, 132–137.

17. Rendek, E., Ducom, G., Germain, P. 2006. Carbon dioxide sequestration in municipal solid waste incinerator (MSWI) bottom ash. *J. Hazard. Mater.*, B128, 73–79.

18. Beruto, D., Kim, M.G., Searcy, A.W. 1988. Microstructure and reactivity of porous and ultrafine CaO particles with CO_2. *High Temp.– High Press.*, 20, 25–30.

19. Shih, S.M., Ho, C.S., Song, Y.S., Lin, J.P. 1999. Kinetics of the Reaction of $Ca(OH)_2$ with CO_2 at Low Temperature. *Ind. Eng. Chem. Res.*, 38, 1316–1322.

20. Bathia, S.K., Perlmutter, D.D. 1983. Effect of the Product Layer on the kinetics of the CO_2-lime reaction. *AiChe J.*, 29, 79–86.

21. Grasa, G., Murillo, R., Alonso, M., Abanades, J.C. 2009. Application of the Random Pore Model to the carbonation cyclic reaction. *AiChe J.*, 55, 1246–1255.

22. Priggiobe, V., Polettini, A., Baciocchi, R. 2009. Gas-solid carbonation kinetics of air pollution control residues for CO_2 storage. *Chem. Eng. J.*, 148, 270–278.

23. Cárdenas-Escudero, C., Morales-Florez, V., Pérez-López, R., Santos, A., Esquivias L., 2011. Procedure to use phosphogypsum industrial waste for mineral CO2 sequestration. *Journal of Hazardous Materials* 196, 431– 435.

Monitoring CO_2 sequestration by mineral carbonation in experimental milling waste piles (Thetford Mines, Québec, Canada) under natural atmospheric conditions.

K. Lechat[1*], J.-M. Lemieux[1], J. Molson[1], G. Beaudoin[1] and R. Hébert[1]

[1] Département de géologie et de génie géologique, Université Laval, Québec, Canada.
*Corresponding author: karl.lechat.1@ulaval.ca

Abstract
At Thetford Mines, Québec, Canada, the milling wastes consist mainly of poorly-sorted grains and fibers of serpentine with smaller amount of brucite containing a large amount of Mg (up to 25 wt%). This magnesium can react with atmospheric CO_2 by natural and passive mineral carbonation. This reaction is a combination of several complex coupled processes, including air, water and heat transport in the unsaturated porous media and chemical reactions. To improve our knowledge of this natural process under natural conditions, two experimental pilot-scale milling waste cells (200 m^3) were built and instrumented to observe the physico-chemical processes involved during mineral carbonation and to propose a conceptual model for mineral carbonation at this scale. The air depleted in CO_2 from the experimental cells (10 times less than in the atmosphere), the presence of carbonates such as hydromagnesite, and unique chemical conditions (basic pH approximately 10, high alkalinity above 250 mg/L ($CaCO_3$) and the presence of magnesium carbonates) confirm an active mineral carbonation in the mining waste of Thetford Mines. Analysis of seasonal variations in air flow within the pile shows that molecular diffusion is the main process for CO_2 supply within the experimental cells.

Introduction

Mineral carbonation is a process that fixes carbon dioxide in minerals through chemical reactions. In most mineral carbonation reactions, CO_2 reacts with dissolved divalent cations such as magnesium, which is a major constituent of ultramafic rocks, to form magnesium carbonates[1,2]. Natural mineral carbonation is often neglected as a global CO_2 sequestration technology because of its slow reaction rate under ambient conditions. However, mineral carbonation occurs naturally and passively in ultramafic milling and mine waste[3,4,5,6], and has important benefits since it requires no energy nor chemical additives and captures CO_2 in the atmosphere. Moreover, mineral carbonation is considered a sustainable solution to capture and store atmospheric CO_2[7,8,9].

In the region of Thetford Mines, (Québec, Canada) Pronost *et al.*[10] discovered warm air vents capable of melting snow at the upper surface of a large chrysotile milling waste pile. The venting air was depleted in CO_2 (to less than 10 ppm) and it was interpreted to result from exothermic mineral carbonation reactions within the waste pile. In this area, chrysotile mining has produced approximately 0.8 Gt of magnesium-rich milling waste, which mainly consists of poorly sorted grains and fibers of lizardite and chrysotile, with smaller amounts of antigorite, brucite and magnetite. The mining wastes could become significant CO_2 sinks. In the context of a carbon market, it could have a net positive economic impact at the mine operation scale. However, quantification of the amount of CO_2 that can be captured by the wastes must be established.

So far, most mineral carbonation studies in the mining wastes of Thetford Mines have been conducted in the laboratory. Eudiometers have been used to measure the influence of grain size, water saturation and atmospheric CO_2 concentration on the CO_2 capture rate and quantity[11]. The mineral reaction products were also identified as a function of temperature; at a temperature at 21°C, nesquehonite was the main carbonate form, dypingite was at 33°C. Other laboratory experiments, using fixed beds (column tests), have shown the importance of the brucite-chrysotile ratio on reaction kinetics since magnesium is dissolved from brucite faster than from serpentines, yielding the fastest reaction rates for mineral carbonation[12]. The carbonation reaction rates were found to be optimal under water saturation near 30%[13,14]. Partial water saturation is interpreted to provide sufficient water for the dissolution of magnesium and sufficient open pore space for gas circulation and CO_2 supply within the mine wastes.

However, all these experiments have been conducted in controlled laboratory conditions with small sample volumes with a mass inferior to 30 g and over a short duration (up to three weeks), which may not be representative of field conditions. In order to assess the carbonation process under natural conditions, two experimental cells were constructed using milling waste and instrumented to monitor temperature, volumetric water content, gas pressure and composition along with atmospheric conditions (Figure 1). Water geochemistry and waste mineralogy were also monitored over time. The collected transient data are used to observe in situ mineral carbonation and to propose a conceptual model at the scale of the experimental cells.

Figure 1. Experimental cells at the Black Lake Mine (Thetford Mines, Quebec, Canada).

Materials and Methods

Experimental cells 1 and 2 were built in the summers of 2010 and 2011, respectively. They were designed as small scale replicas of the main milling heap (Figure 1 and 2).

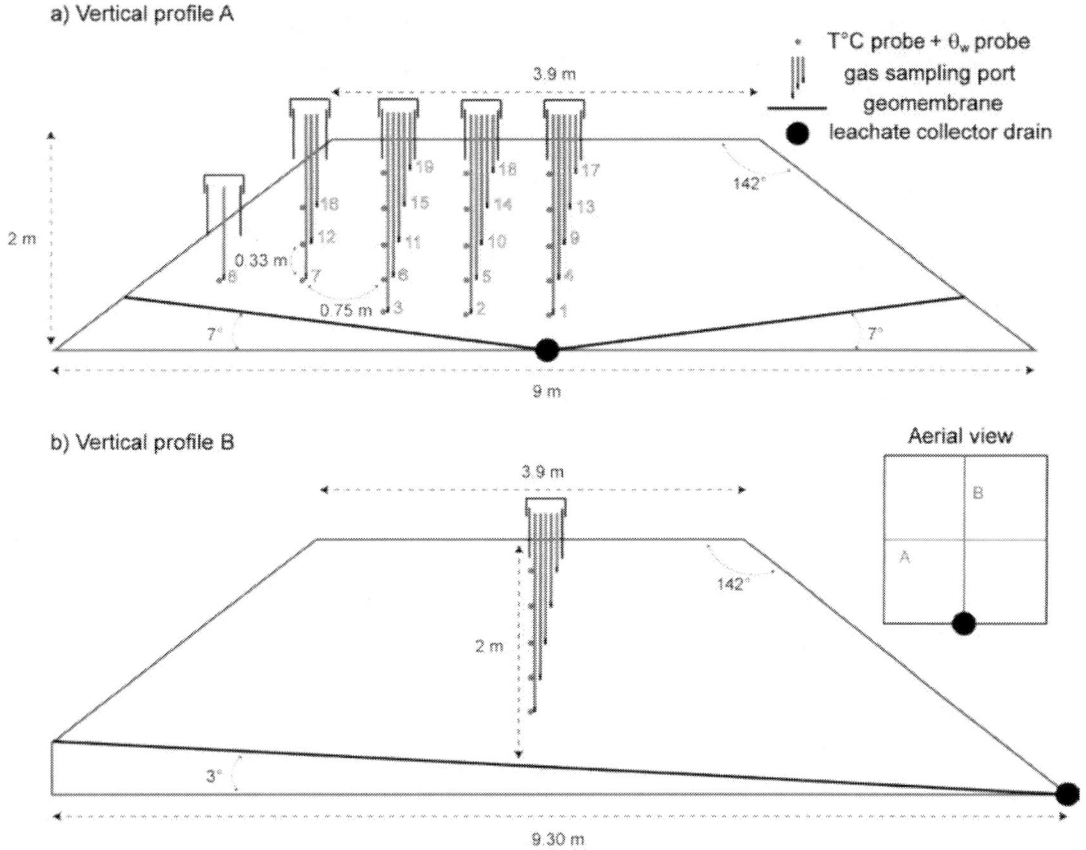

Figure 2. Vertical cross-section profiles of the experimental cells showing dimensions and monitoring instrumentation

Within each cell, five vertical multi-level monitoring stations were emplaced along a symmetrical cross-section, each equipped with multiple sensors for measuring soil temperature (109-BAM temperature probe, Campbell Scientific) and volumetric water content (EC-5 volumetric water content probes, Decagon). The sensors were connected to a Campbell Scientific CR1000 datalogger and two AM16/32B multiplexers powered with a solar panel. Gas sampling ports (made of 0.25'' ID HDPE tubing screened along 5 cm at their bottom) are also located at the same positions as the other instruments in order to measure gas pressure and composition. A geomembrane with a collector drain installed at the base of the experimental cells collects the leachate, which was analysed to determine water chemistry. Ambient conditions were recorded by a meteorological station (HOBO U30 weather station, Onset Computer Corporation), measuring air temperature, precipitation, relative humidity, pressure and snow thickness. Readings are made every five minutes and the average values were recorded every hour.

Gas and water composition

An assembly made of a battery-powered peristaltic pump (Model 410, Solinst) and a flow-through carbon dioxide probe (Model GMP343, Vaisala) mounted on the gas sampling ports is used to measure the CO_2 content air in pores with a precision of 5 ppm for measurements of less than 200 ppm and ± 2 ppm for measurements above

200 ppm. Only ports 1, 2, 3, 7, 8, 9, 10, 11, 16, 17, 18 and 19 were sampled because the CO_2 concentration difference was very low between adjacent ports, and also to prevent gas advection within the porous media due to pumping.

Once or two twice per month, the leachate water was analysed on-site with a multi-parameter probe (Model Professional Plus, YSI) for the primary physico-chemical parameters (pH, conductivity, salinity, redox conditions). Leachate samples were sent to Maxxam Analytics for analysis of dissolved metals (ex: Mg^{2+}, Ca^{3+}, $Fe^{2+/3+}$), alkalinity and sulphur content.

Mineralogy

XRD analysis and SEM analysis were used to identify the minerals in ultramafic mining wastes and the products of mineral carbonation. Carbonate proportions in bulk sample were too low to be identified by XRD. White crusts (suspected carbonates) formed on the surface of rock fragments were scraped with a blade for XRD analysis.

During summer 2014, one of experimental cells was decommissioned and samples were collected along the vertical section where the sensors were located (Figure 1). Twenty samples were collected, and sorted in four grain-size fractions (> 19 mm, 19 - 9.5 mm, 9.5 - 1 mm, < 1 mm). The grain-size fractions were analysed with an Eltra 800 Carbon-Sulfur instrument to measure the total carbon content. The instrument was calibrated with UB-N (%C = 0.106 ± 0.08).

Results and discussion

Data collected from June 2012 to February 2014 in the first experimental cell are presented in Figure 4. No data were collected at the weather station between October 4th and November 6th 2013 due to battery failure. The temperature varied from -25°C to 25°C, which is typical of a moderate and sub-humid regional climate (Figure 3a). The atmospheric CO_2 content is relatively constant, near 400 ppm, but concentrations are higher during winter and lower during summer. Precipitation was regular throughout the year, but rainfall episodes never exceeded 40 mm. During winter, the ground was completely covered by snow that accumulates, reaching more than a meter in thickness (Figure 3b).

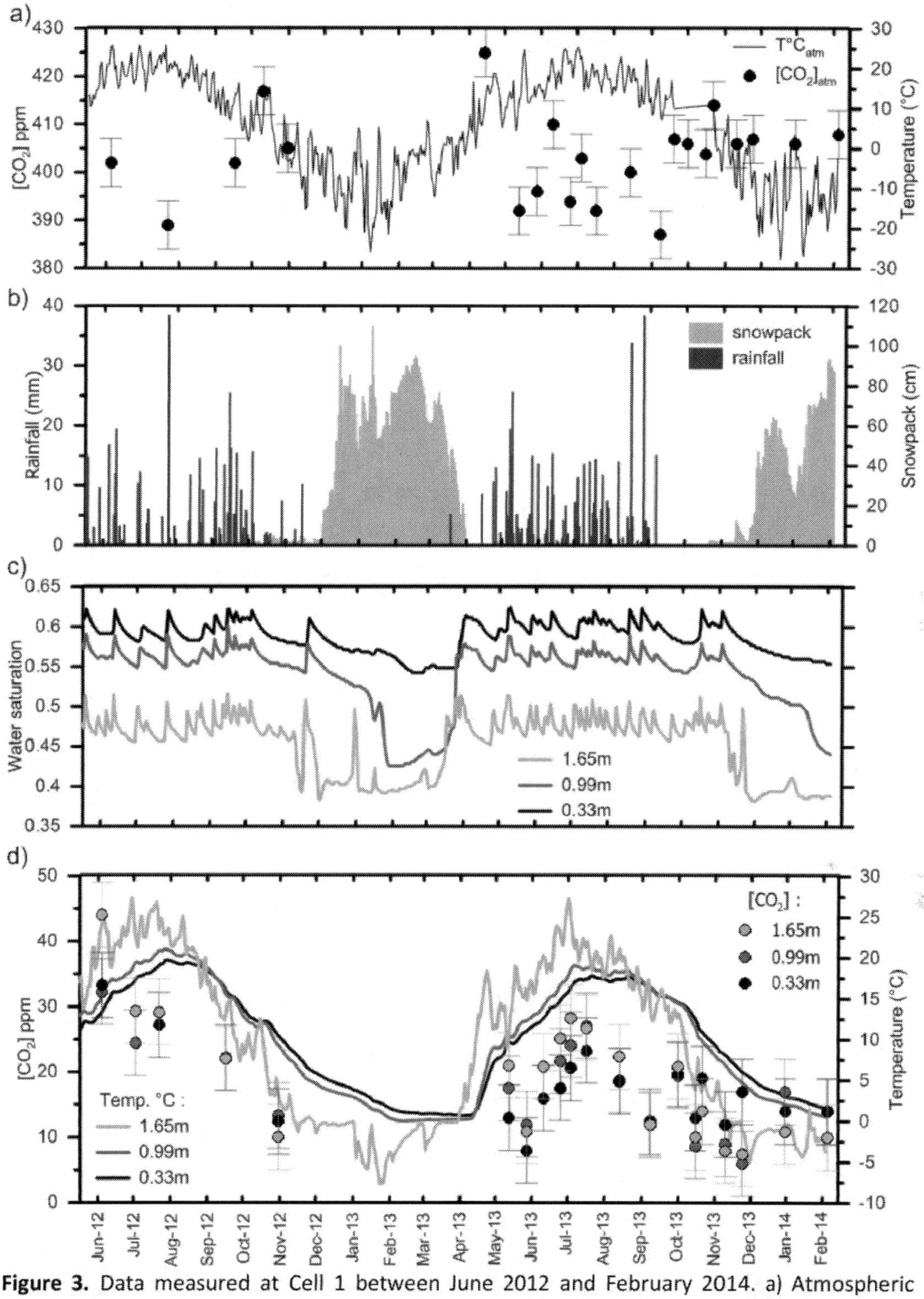

Figure 3. Data measured at Cell 1 between June 2012 and February 2014. a) Atmospheric temperatures and CO_2 concentrations, b) Rain and snowpack, c) Water saturations,d) Temperatures and CO_2 concentrations in the vertical profile in the center of the experimental cell. For c) and d), legend figures represent probe elevations (1, 9 and 17, see Figure 2a for locations).

Figure 3c shows the water saturation in the experimental cell which was obtained from the water content measurements, assuming a porosity of 35 %, as measured in the laboratory. Only values from sampling points 1, 9 and 17 (central vertical profile, see Figure 2), are presented since the measurements from the other profiles are similar. Water saturation increases with depth, the values range from 0.5 at the top to near 0.6 at the bottom in summer, versus 0.4 and 0.55 in winter, respectively.

Figure 3d shows the CO_2 content in the interstitial air within the experimental cell. The CO_2 content changes between 5 and 50 ppm depending of the date and the depth, which is at least eight times less than in the atmosphere (Figure 3a). The concentrations decrease with depth in summer, whereas they increase with depth in winter. The CO_2 depletion of air in ultramafic milling wastes is consistent with the conceptual model proposed by Pronost[10]. The porous medium is never saturated, allowing the gas to circulate and supply the cell with CO_2 from the atmosphere.

Chemical analysis of the leached water shows that the pH is most often above 10, magnesium concentrations range from 85 to 140 mg/L, and total alkalinity from 260 to 300 mg/L (Figure 4). The silica concentrations are less than the detection limit (1×10^{-3} mg/L).

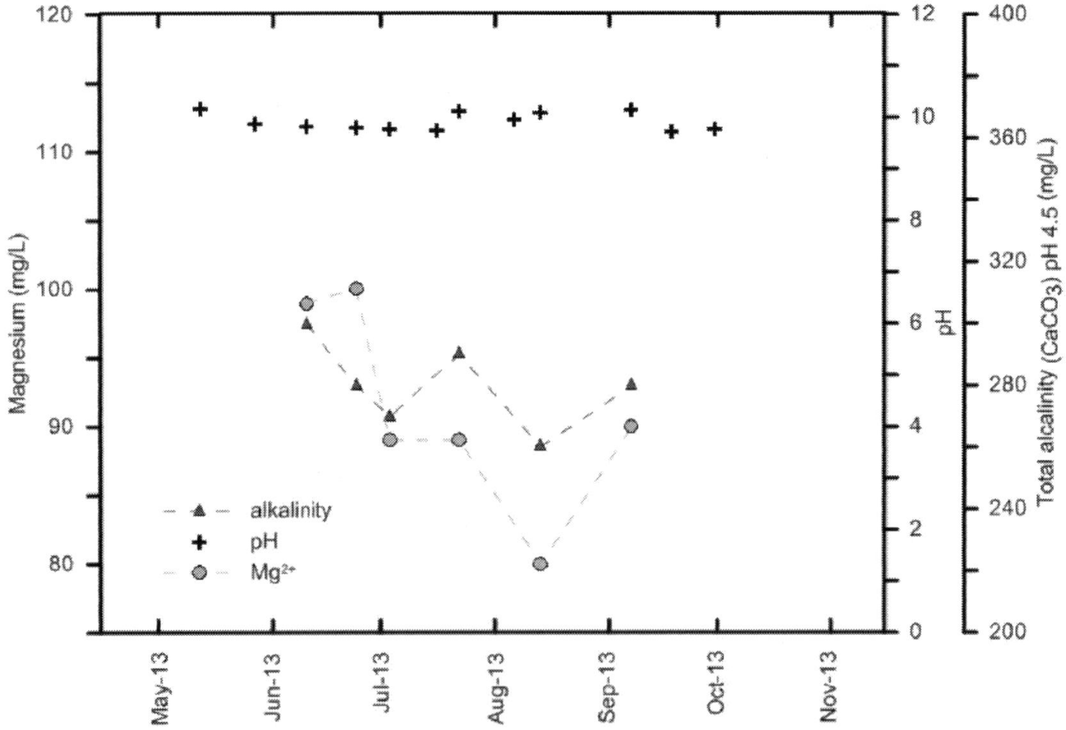

Figure 4. Observations of Mg^{2+}, pH and alkalinity in the leachate

These results are consistent with active CO_2 mineralization reactions within the cell. The slightly acidic rain water (pH 6 to 6.5) hydrolyses the magnesium minerals (particularly brucite) which liberates the magnesium and yields an alkaline leachate. CO_2 dissolution can explain the high alkalinity of the leached water. These processes are supported by the highly depleted CO_2 concentration in the interstitial air within the cell.

The XRD analyses show that the mining waste is composed of chrysotile, lizardite, antigorite magnetite, brucite and chlorite. These results are similar to the results obtained by Assima *et al.*[12], Pronost *et al.*[10, 11], and Huot *et al.*[15], where only the minor mineral phases vary like albite, phlogopite and chromite. Hydromagnesite is the main carbonate mineral phase, while sjögrenite and pyroaurite have also been found. SEM analyses confirmed the presence of carbonates. However, during the SEM analyses, only hydromagnesite was observed, which exhibited a lamellar texture on cemented grain surfaces (Figure 5).

Figure 5. Carbonates with lamellar texture (left) and observations of the contact between carbonates and lizardite (right). Images were obtained from SEM analysis.

The carbon content for each particle size class is presented in Table 1. Values are between 0.29 % and 0.56 %, with a minimum of 0.05 % and a maximum of 1.2 %, without any relation to the sample depth, suggesting that the reaction is occurring everywhere in the cell. The range in values is much larger for coarser grain-size subsamples, showing significant heterogeneity.

Table 1. Carbon content in mining waste

Grain size (mm)	Mass fraction (%)	Average %C	Std. deviation %C	Min. value %C	Max. value %C
> 19	36	0.29	0.35	0.05	1.21
9.5 - 19	13.68	0.38	0.2	0.1	0.97
1 - 9.5	30.31	0.39	0.03	0.31	0.44
< 1	19.99	0.56	0.04	0.48	0.63

These observations have helped to develop a conceptual model for mineral carbonation in the wastes (Figure 6) which will be used to calibrate a reactive transport model.

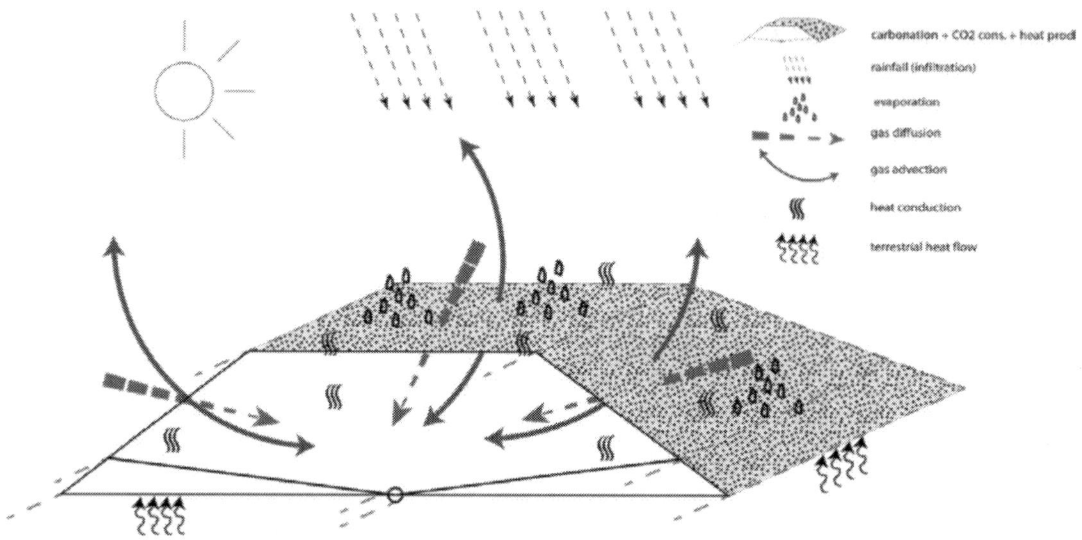

Figure 6. Conceptual model of the physical processes implied in mineral carbonation within the experimental cell of ultramafic mining residues.

For this conceptual model, we suggest that molecular diffusion is the main process for CO_2 supply within the experimental cells. Diffusion in the pile is facilitated by the fact that the milling wastes are only partially saturated, so the gas can circulate which supplies CO_2 within the cell. Lower CO2 concentration within the pile during the winter is attributed to the snowpack which reduces CO2 supply. Although it is suspected that carbonation occurs mostly at shallow depth below the surface of the pile (due to a major drop in CO2 concentration), carbon analysis on samples collected during the decommisionning of of the pile do not support this hyposthesis. The next step of this study will be to test the conceptual model with a reactive transport model.

References

1. Gerdemann, S.J., O'Connor, W.K., Dahlin, D.C., Penner, L.R., Rush, H., 2007. Ex situ aqueous mineral carbonation. Environ. Sci. Technol. 41, 2587–2593. doi:10.1021/es0619253
2. Sipilä, J., Teir, S., Zevenhoven, R., 2008. Carbon dioxide sequestration by mineral carbonation, Litterature review update 2005-2007. Abo Akademi University, Turku.
3. Beaudoin, G., Hébert, R., Constantin, M., Duchesne, J., Cecchi, E., Huot, F., Vigneau, S., Fiola, R., 2008. Spontaneous carbonation of serpentine in milling and mining waste, southern Québec and Italy. Presented at the Proceedings of Accelerated Carbonation for Environmental and Materials Engineering (ACEME2008), Rome, Italy, pp. 73–82.
4. Power, I.M., Wilson, S.A., Small, D.P., Dipple, G.M., Wan, W., Southam, G., 2011. Microbially mediated mineral carbonation: roles of phototrophy and heterotrophy. Environ. Sci. Technol. 45, 9061–9068. doi:10.1021/es201648g
5. Wilson, S.A., Dipple, G.M., Power, I.M., Thom, J.M., Anderson, R.G., Raudsepp, M., Gabites, J.E., Southam, G., 2009. Carbon dioxide fixation within mine wastes of ultramafic-hosted ore deposits: examples from the Clinton Creek and Cassiar Chrysotile deposits, Canada. Econ. Geol. 104, 95–112. doi:10.2113/gsecongeo.104.1.

6. Wilson, S.A., Dipple, G.M., Power, I.M., Barker, S.L.L., Fallon, S.J., Southam, G., 2011. Subarctic weathering of mineral wastes provides a sink for Atmospheric CO_2. Environ. Sci. Technol. 45, 7727–7736. doi:10.1021/es202112y
7. Lackner, K.S., 2003. CLIMATE CHANGE: A guide to CO_2 sequestration. Science 300, 1677–1678. doi:10.1126/science.1079033
8. Lackner, K.S., Wendt, C.H., Butt, D.P., Joyce, E.L., Sharp, D.H., 1995. Carbon dioxide disposal in carbonate minerals. Energy 20, 1153–1170. doi:10.1016/0360-5442(95)00071-N
9. Seifritz, W., 1990. CO2 disposal by means of silicates. Nature (345), 486.
10. Pronost, J., Beaudoin, G., Lemieux, J.-M., Hebert, R., Constantin, M., Marcouiller, S., Klein, M., Duchesne, J., Molson, J.W., Larachi, F., Maldague, X., 2012. CO2-depleted warm air venting from chrysotile milling waste (Thetford Mines, Canada): Evidence for in-situ carbon capture from the atmosphere. Geology. doi:10.1130/G32583.1
11. Pronost, J., Beaudoin, G., Tremblay, J., Larachi, F., Duchesne, J., Hébert, R., Constantin, M., 2011. Carbon sequestration kinetic and storage capacity of ultramafic mining waste. Environ. Sci. Technol. 45, 9413–9420. doi:10.1021/es203063a
12. Assima, G.P., Larachi, F., Beaudoin, G., Molson, J., 2012. CO_2 sequestration in chrysotile mining residues—Implication of watering and passivation under environmental conditions. Ind. Eng. Chem. Res. 51, 8726–8734. doi:10.1021/ie202693q
13. Assima, G.P., Larachi, F., Beaudoin, G., Molson, J., 2013a. Dynamics of carbon dioxide uptake in chrysotile mining residues – Effect of mineralogy and liquid saturation. Int. J. Greenh. Gas Control 12, 124–135. doi:10.1016/j.ijggc.2012.10.001
14. Assima, G.P., Larachi, F., Molson, J., Beaudoin, G., 2014a. Comparative study of five Québec ultramafic mining residues for use in direct ambient carbon dioxide mineral sequestration. Chem. Eng. J. 245, 56–64. doi:10.1016/j.cej.2014.02.010
15. Huot, F., Beaudoin, G., Hébert, R., Dipple, G.M., Raudsepp, M., 2003. Le piégeage du CO_2 anthropique dans les parcs à résidus d'amiante du sud de Québec: concept et valorisation. Laval University, Qc, Canada.

CO2 capture and storage by mineral carbonation
Utilization of Chrysotile Mine Tailings and Microbially Accelerated Carbonate Mineral Precipitation Reactions As a Strategy for in Situ Carbon Sequestration

McCutcheon, J., The University of Queensland
Wilson, S. A., Monash University
Dipple, G. M., The University of British Columbia
Southam, G., The University of Queensland
Originally presented on: 6/23/2015 11:00:00 - 11:15:00

Carbonate mineral precipitation was successfully completed in chrysotile mine tailings collected from the abandoned Woodsreef Asbestos Mine in New South Wales, Australia. This was achieved by implementing microbially accelerated mineral precipitation reactions as a method of carbon sequestration.

A preliminary leaching experiment was first completed to determine the ability of the chrysotile tailings to act as a magnesium source for carbonate precipitation. Tailing samples (5 g) were leached in five, 100 mL sulfuric acid [H_2SO_4] solutions with acid:serpentine ratios of 0.25, 0.5, 1, and 2, with water as a control. Tracking the pH of each leaching system monitored chrysotile dissolution/acid neutralization. The 0.25 system had a final pH of 6.3 and 16 % of the magnesium was leached from the chrysotile, producing 1800 ppm Mg^{2+} in solution. These results were used to design the reaction conditions for the microbial carbonate mineral precipitation trial. Cultures of cyanobacteria enriched from the Woodsreef Mine pit waters and endolithic, carbonate-rich habitats at the mine were capable of producing highly alkaline waters (up to pH 12) when grown in the laboratory. The alkalinity generated by these phototrophic microbes was used to accelerate magnesium carbonate mineral precipitation in acid treated tailings at ambient temperature and pressure. This was achieved by leaching tailings in columns constructed in 60 cc syringes, after which three columns were inoculated with the alkalinity generating cyanobacteria consortium. Two additional columns were not inoculated to act as controls. The microbes were given 4 weeks to grow in the columns, at which point the columns were sampled for characterization of their mineralogy (micro X-ray diffraction), water chemistry (inductively coupled plasma atomic emission spectroscopy), and microbial colonization within the leached tailings (scanning electron microscopy).

Micro X-ray diffraction of the surface material collected from the inoculated columns identified dypingite [$Mg_5(CO_3)_4(OH)_2 \cdot 5H_2O$], hydromagnesite [$Mg_5(CO_3)_4(OH)_2 \cdot 4H_2O$], nesquehonite [$MgCO_3 \cdot 3H_2O$], and chrysotile [$Mg_3Si_2O_5 \cdot (OH)_4$]. When examined using scanning electron microscopy, the carbonate mineral precipitates were observed directly on the filamentous cyanobacteria. This is a result of the microbial cells becoming surrounded by microenvironments containing high concentrations of adsorbed magnesium and photosynthetically generated carbonate ions. When these localized 'pockets' of reactants are combined with the ability of the cell exterior and extra-cellular polymeric substances to act as nucleation sites, magnesium carbonate precipitation under ambient conditions becomes possible.

This study demonstrated the ability of microbially enabled carbonate mineral precipitation to act as a method of sequestering carbon dioxide (CO_2) in chrysotile mine tailings, initiating the formation of regolith carbonate. The ability of this process to use atmospheric CO_2 rather than a point source of carbon makes it suitable for utilization at a range of remote abandoned and active mine sites that generate ultramafic (Mg-rich) mine tailings.

Carbon Dioxide Sequestration Using Waste Serpentine

Hsin Wang*, **Ta-Wui Cheng, Yung-Chin Ding, Yu-Hsu Chang, Wei-Hao Lee, Tsai-Chia Chung**

National Taipei University of Technology

Institute of Mineral Resource Engineering

1, Sec. 3, Zhongxiao E. Rd., Taipei 10608 Taiwan, R.O.C.

peppy@ntut.edu.tw

Abstract

In Taiwan, more than 0.54 million tons of serpentine waste are produced per year from three main sources: well-developed joints in the ore body after mining operations, dimension stone cutting, and the requirements of particle size by iron making company. The purpose of this study is the reutilization of waste serpentine for carbon dioxide mineral sequestration. By using an attrition scrubbing method at 80°C for HCl acid leaching waste serpentine powders, an 88% magnesium ion leaching rate can be reached. After the magnesium-rich leached solution for carbonation process, hydromagnesite was formed and the transferring carbonation degree was 86%, as determined by DTA results. Additionally, the highly reactive amorphous silica residue resulting after acid leaching was then synthesized to 99% β-phase silicon carbide using waste graphite. The process stated above is feasible and provides a potential use for serpentine waste in the future.

1. Introduction

Serpentine is an important resource in Taiwan. The major applications for serpentine are in dimension stones and flux materials. However, there are 0.54 million tons of serpentine waste produced each year [1] due to the well-developed joints in the serpentine orebody after mining operations, as well as the strict requirements of the particle size (6 mm ~ 80 mm) and chemical composition of serpentine by China Steel. The waste also creates considerable environmental problems. There are several ways to recycle serpentine waste for particular purposes, such as: fertilizer production, extraction of amorphous silicate and pure magnesium compounds, synthetic forsterite refractory materials [2], and synthetic silicon carbide

[3]. However, due to serpentine's theoretical composition of Mg-rich (MgO = 44.1%), it can be a great raw material for carbon dioxide sequestration.

In earlier studies, serpentine was pre-treated to 630°C for dehydroxylation as it increased magnesium extraction rate, however, heating it required energy of 326kWh/t_{rock} [4], which is unsuitable for carbon dioxide reduction. Although the heat pre-treated serpentine sample had a higher extraction rate, silicon gel was also produced, which led to subsequent filtration difficulty. Therefore, the main focus here is on using non-pre-treated serpentine extraction under 100 °C, attrition scrubbing leaching, and not pressurizing with CO_2 gas during the carbonation system. Conversely, the main focus is is trying to totally recycle the serpentine leached amorphous silica residue for synthesized silicon carbide.

2. Experiment

Waste serpentine samples were obtained from Hualian, Taiwan. The particle sizes were between 1.3 and 9.6 μm, and D_{50}=6.2 μm. The major mineral phase of the serpentine sample was antigorite (Figure 1), which was determined by using XRD for analysis. The chemical composition of the serpentine sample is shown in Table 1.

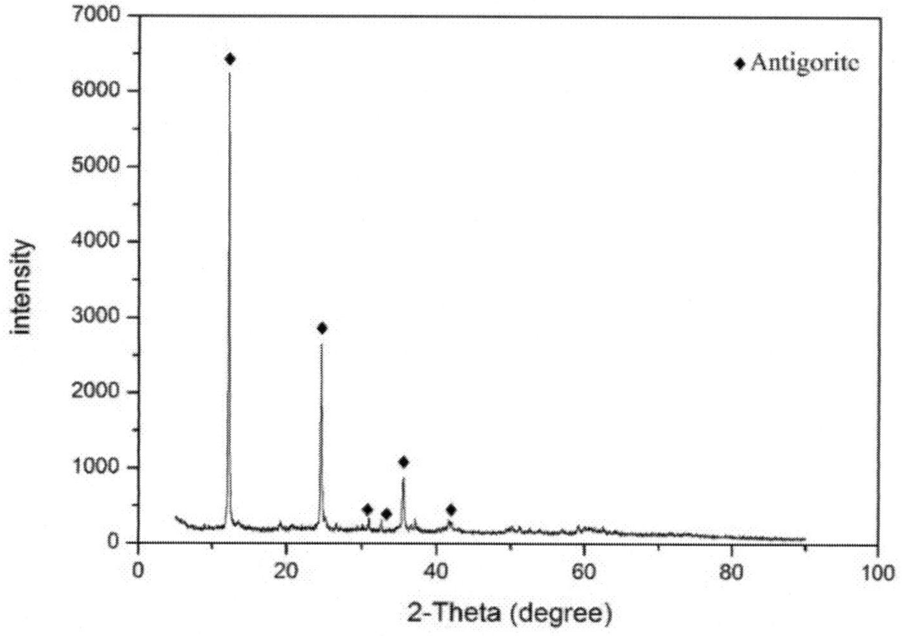

Figure 1 XRD pattern for waste serpentine in this study

Table 1 Chemical composition of serpentine sample

Composition	SiO_2	MgO	Fe_2O_3	CaO	Al_2O_3	Cr_2O_3	Ig. Loss
(wt.%)	34.7	33.4	6.4	2.4	0.8	0.3	14.9

Figure 2 illustrates the treatment process for this study. The serpentine sample was leached by 7M HCl at 80°C with a L/S ratio of 20 for 2 h by using the attrition scrubbing method. Attrition scrubbing is a mineral processing technology generally used for separating minerals when the clay or silts are tightly bound to the silica grains, or when clay particles are similar in size to the silica sand grains. For appropriate attrition scrubbing, it is important that the solid content of the slurry reach a high solid percentage, i.e. 50-80%. At those proportions, there are good particle-to-particle collisions and it is easier for the fine particles to separate from the coarse particle surfaces. In the case here, after HCl leaching, a silica skeleton layer can form on the serpentine surface. From an earlier study, it was found that removing the diffusion limiting SiO_2 layer and exposing the inner Mg layer is important to promoting further dissolution of the mineral [5]. Therefore, attrition scrubbing was conducted to improve the leaching rate. Then, the sample was filtered. After filtration, the residue remaining consisted of highly reactive amorphous silica that can be synthesized into silicon carbide using waste graphite. After the Mg-rich solution was distilled at 95 °C 2.5 h for the recovery of HCl, the purified Mg-rich solution was adjusted the pH to 4.5 by using NaOH to remove $Fe(OH)_2$. The solution pH was adjusted again to a pH of 6 and then CO_2 was introduced to form hydromagnesite. The white precipitates of hydromagnesite were dried in a low-temperature oven, and analyzed using an X-ray diffraction (XRD) and thermal gravimetric analyzer (TGA). The transferring carbonation degree of carbonated samples was determined with a Thermal Gravimetrical Analysis system (TGA). Samples (30 mg) were heated in platinum crucible under a nitrogen atmosphere from 25 to 1000 °C at a temperature increase ratio of 20 °C/min. Weight loss was measured by the TGA. Three weight fractions were distinguished: (1) 25-105 °C, (2) 105-400 °C, and (3) 400-800 °C. These fractions represented the decomposition temperature of (1) moisture, (2) organic, and (3) carbonate, respectively. The third weight fraction of the TGA curve was used as the magnesium carbonate content, expressed in terms of CO_2 (wt. %):

$$CO_2(wt\%) = \frac{\Delta m_{400^\circ C\sim800^\circ C}}{m_{105^\circ C}} \times 100\% \quad (1)$$

The transferring carbonation degree (δMg) can be determined from equation 2. In equation 2, it is assumed that only Mg was carbonated during the carbonation process.

$$\delta_{Mg} = \frac{\frac{CO_2(wt\%)}{100-CO_2(wt\%)} \times \frac{MW_{Mg}(kg/mol)}{MW_{CO_2}(kg/mol)}}{Mg_{total}(kg/kg)} \times 100 \qquad (2)$$

As

Mg_{total} = the total Mg content of the serpentine,

MW_{Mg}= the molar weights of Mg

MW_{CO2}= the molar weights of CO_2

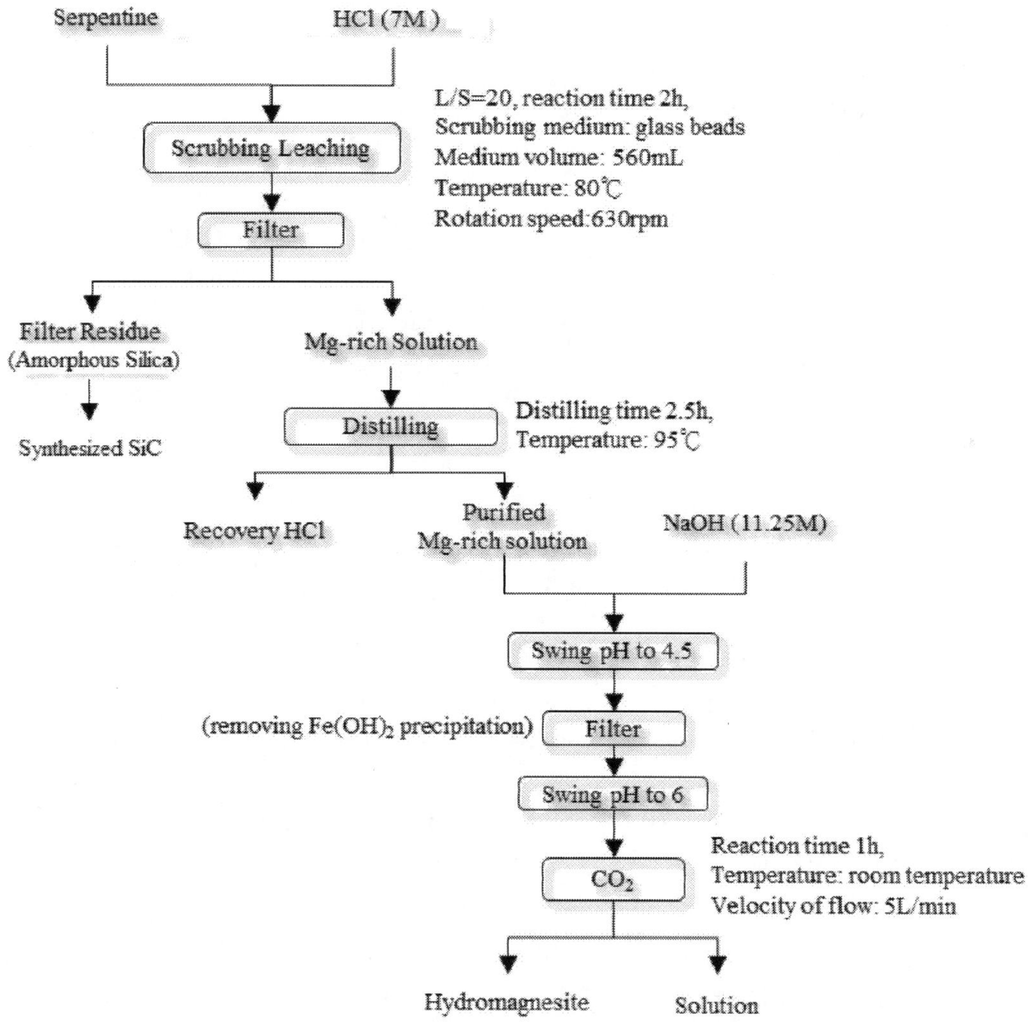

Figure 2 Treatment process of this study

3. Results and Discussion

3.1 Leaching Test using Attrition Scrubbing Technology

Figure 3 shows the effect of the serpentine Mg extraction ratio at 7M HCl on different attrition scrubbing conditions. It was found that the Mg extraction ratio increased as the scrubbing time increased and the higher extraction ratio was reached after 2 h in each condition. The extraction rate with attrition scrubbing at 80 °C was reached around 90%. This was because attrition scrubbing removed the extracted SiO_2 layer exposing the inner Mg layer and promoted further dissolution.

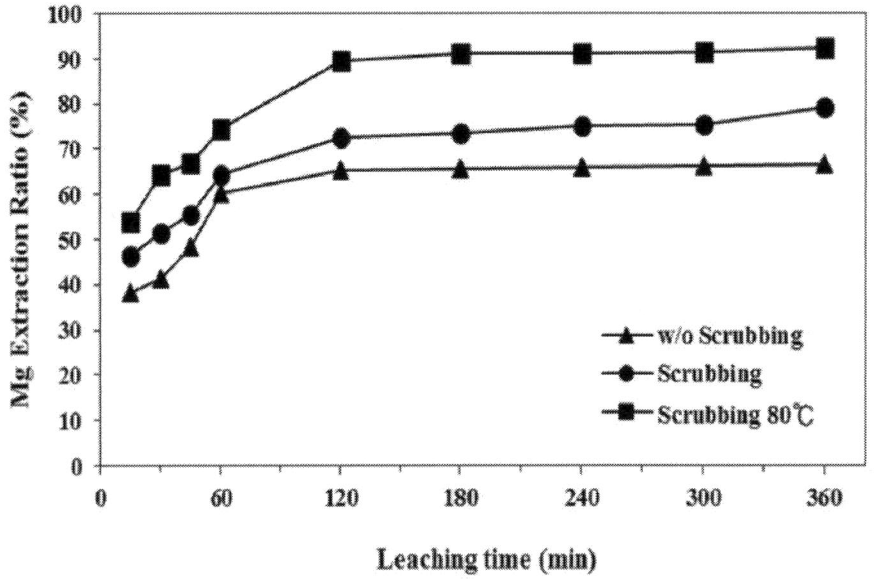

Figure 3 Effect of Mg extraction ratio at 7M HCl on attrition scrubbing time

(L/S : 20ml/g, glass bead Φ=2mm, glass beads volume = 560mL)

3.2 SiC Synthesis Using Leached Serpentine and Waste Graphite

After HCl leached the serpentine sample, high reactive amorphous silica can be obtained from the residual parts. It can be a great raw material for the synthesis of SiC. Figures 4 and 5 were the XRD patterns for the synthesis of SiC using leached serpentine mixed with waste graphite on the ratio of C/silicate=3 at 1400 °C and 1600 °C. As show in below, after the 1400 °C and 1600 °C thermal carbon reactions, SiC, Fe_5Si_3 and residual carbon graphite signals were found (Fig. 4(a) and Fig. 5(a)). However, after the synthesized SiC samples were treated again with thermal treatment at 800 °C, the un-reacted carbon was removed and the Fe_5Si_3 intensity was reduced (Fig. 4(b) and Fig. 5(b)). Subsequently, the synthesized SiC samples were washed by

HF resulting in the Fe_5Si_3 being removed and a 99% β-phase silicon carbide was obtained (Fig. 4(c) and Fig. 5(c)). The SEM micrograph of synthesized SiC is shown in Figure 6.

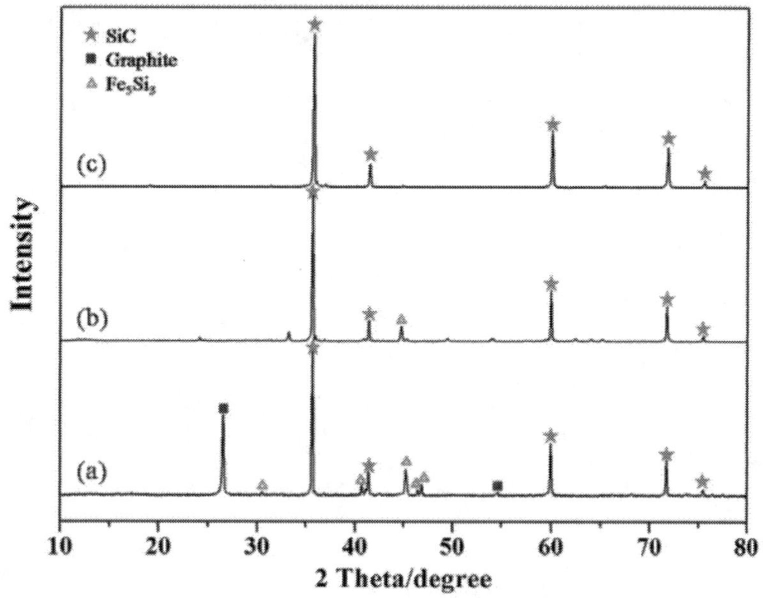

Figure 4 XRD Pattern of synthesized SiC (a) waste graphite and leached serpentine (C/silicate=3) at 1400 °C for 3 h (b) after heat treatment at 800 °C (c) after HF washed

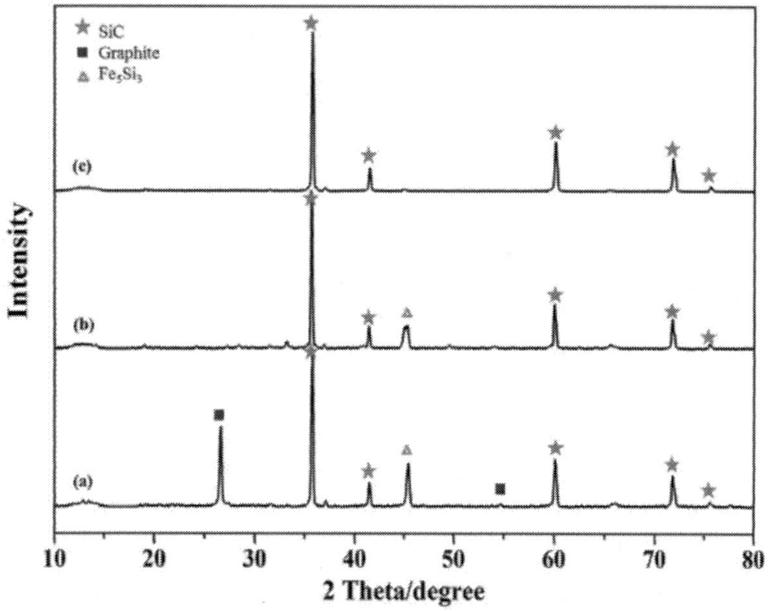

Figure 5 XRD Pattern of synthesized SiC (a) waste graphite and leached serpentine (C/silicate=3) at 1600 °C for 3 h (b) after heat treatment at 800 °C (c) after HF washed

Figure 6 SEM micrograph of synthesized SiC

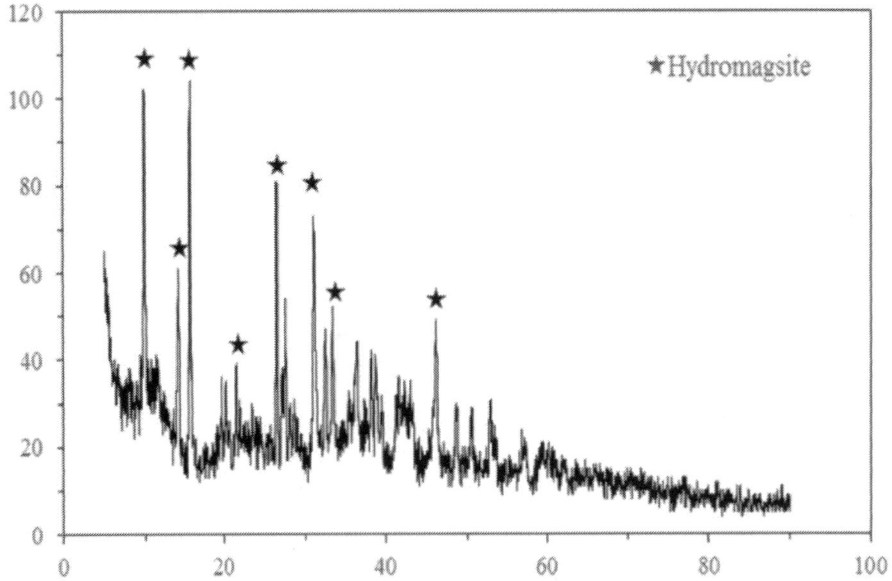

Figure 7 XRD results for white precipitation material

Figure 8 SEM micrograph of hydromagnesite

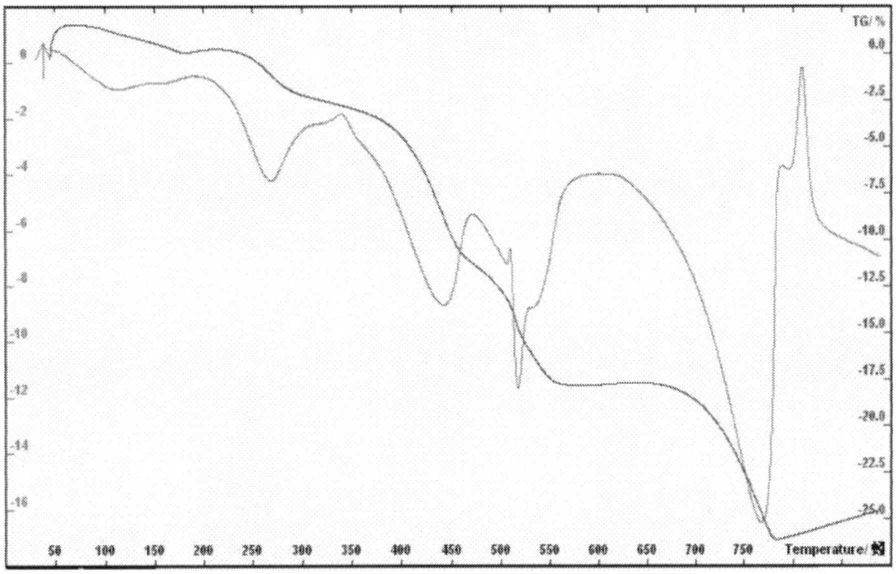

Figure 9 DTA results for hydromagnesite

3.3 Carbonation Tests

From earlier work, it was found that after leaching by HCl, the Mg-rich aqueous phase was fairly acidic. Therefore, it is important to maintain the pH value to 4.5 by adding NaOH to remove $Fe(OH)_2$, and to adjust the Mg-rich solution to pH 6.5 for the carbonation process afterwards. The carbonation process was inducing carbon dioxide to react Mg-rich solution for 15 min. and white precipitation of hydromagnesite ($Mg_5(CO_3)_4(OH)_2 \cdot 4H_2O$) was formed. The XRD results for the white precipitation material are shown in Figure 7 and the SEM micrograph of hydromagnesite is illustrated in Figure 8.

The transferring carbonation degree of carbonated samples was determined with a thermal gravimetrical analysis system (TGA), as shown in Figure 9 (blue line). The calculation of the transferring carbonation degree followed equations (1) and (2). The CO2% reached 27% and the transferring carbonation degree was 86%.

4. Summary

1) Attrition scrubbing is an efficient extraction method. The best extraction ratio can be reached 90% (80℃, 7M HCl, 2h).
2) Synthesis of SiC by using leached serpentine residue mixed with waste graphite on the ratio of C/silicate=3 at 1400 °C and 1600 °C, can result in 99% β-phase silicon carbide being obtained.
3) For carbonation tests, a white precipitation material of hydromagnesite was formed. Carbonated samples were determined by TGA and equations and the transferring carbonation degree was 86%.

5. References:

[1] Ministry of Economic Affairs, (1996), The application of limestone and serpentine wastes. Ministry of Economic Affairs, Republic of China, Taipei.

[2] T.W. Cheng, Y.C. Ding, J.P. Chiu, (2002), A Study of Synthetic Forsterite Refractory Materials Using Waste Serpentine Cutting, Minerals Engineering, Vol. 15, No. 4. , pp. 271-275.

[3] T.W. Cheng, and C.W. Hsu, (2006), A Study of Silicon Carbide Synthesis from Waste Serpentine, Chemosphere, 64(3), pp. 510-514.

[4] George Alexander, M. Mercedes Maroto-Valer, Parvana Gafarova-Aksoy, (2007), Evaluation of reaction variables in the dissolution of serpentine for mineral carbonation, Fuel, 86, p. 273-281.

[5] A.-H. Park, R. Jadhav, L.-S. Fan, (2003), CO2 Mineral sequestration: chemically enhanced aqueous carbonation of serpentine, Canadian Journal of Chemical Engineering, 81, p. 885–890.

In Situ Studies of "Mineral-CO_2" Interactions relevant to Carbon Sequestration

Andrew V.G. Chizmeshya and Jason Diefenbacher#*

*Department of Chemistry and Biochemistry,
Department of Physics,
Arizona State University, Tempe AZ

#Freescale Semiconductor,
Sensor Solutions Division, Chandler AZ

June 22, 2015

ARIZONA STATE UNIVERSITY Argonne NATIONAL LABORATORY NETL

ARIZONA STATE UNIVERSITY

Presentation outline

- Introduction and background
- Development of microreactors for in-situ studies of CO_2 sequestration "reactions"
- *Interaction of supercritical CO_2 with wet/dry clay minerals*

ARIZONA STATE UNIVERSITY

CO_2 sequestration research at ASU

Mineral Sequestration Working Group
NETL, ASU, ARC (Albany), LANL,SAIC
(1998-2005):

- CO_2 **mineral carbonation** reaction studies
- *in situ* process investigations
- Thermodynamic & ab initio simulations
- Initial multi-phase fluid studies

ASU CO_2 Sequestration Group
(2006-present):

- Optimization of **mineral carbonation** reactions
- Fluid/solid reaction mechanisms (stirred)
- Diffusion rates of carbon-bearing species in solution and porous minerals
- Pressure and temperature dependence of key solution species' activities. Formation kinetics of solid carbonate phases and their aqueous complexes
- "Ex-situ" vs "in-situ" comparisons of aqueous CO_2-mineral interactions in **geological sequestration** setting

Support: DOE-NETL, Argonne National Laboratory

ASU Arizona State University

Development of microreactors for in-situ studies

1st generation ASU in-situ microreactor technology:

- Moissanite (single crystal SiC) window technology enables visual and spectroscopic access to chemical processes up to T=400 °C and P=350 atm.
- Large external ballast to maintains **constant chemical activity** as reactants are consumed (e.g., unlike closed autoclave systems)...simulates "infinite" $scCO_2$ source (μ-reactor reaction volume ~ 100 μL)
- "*in situ*" mineral-fluid interactions can be efficiently and accurately mapped at **typical aquifer conditions** (P,T, aqueous chemistry, etc).

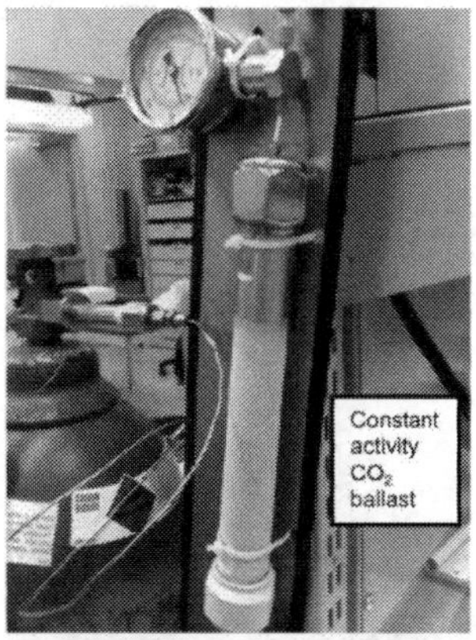

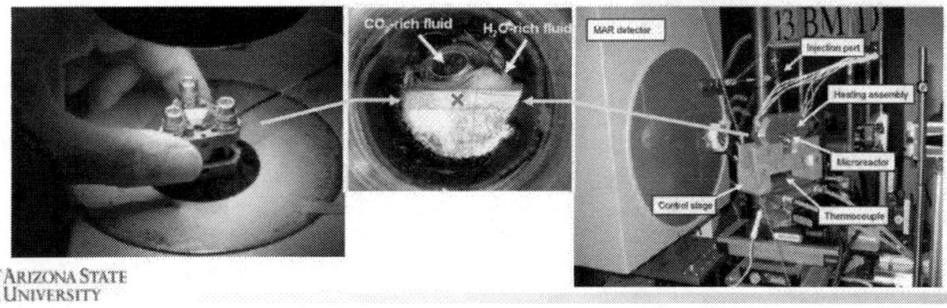

ASU Arizona State University

Development of microreactors for in-situ studies

1st generation ASU in-situ microreactor technology:

Lizardite carbonation in H_2O-CO_2 ($T=180°C$, $P_{CO2}=150$ atm)

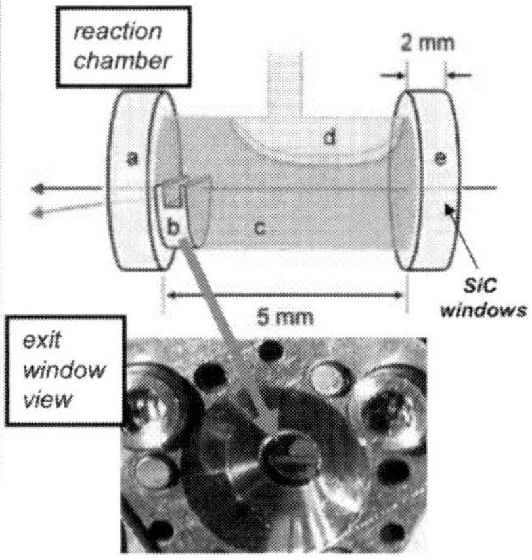

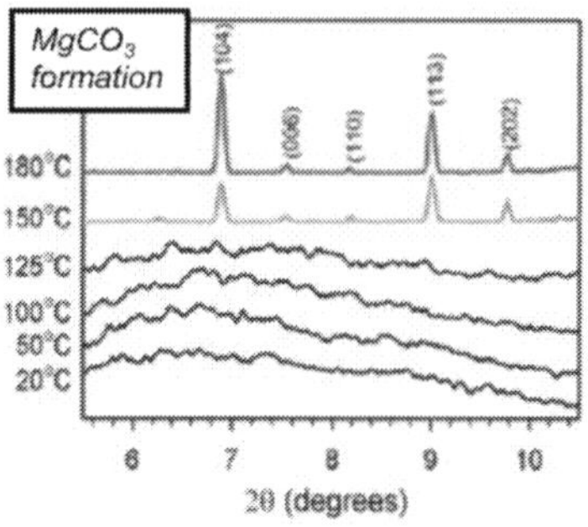

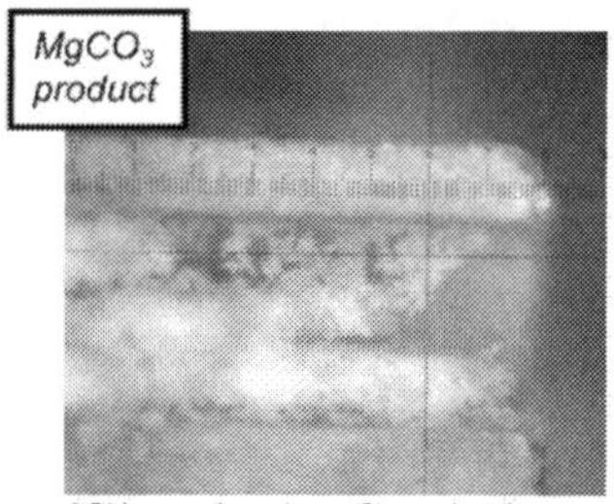

ASU experiments performed at the GeoSoilEnviroCARS synchrotron beamline (Argonne National Laboratory)

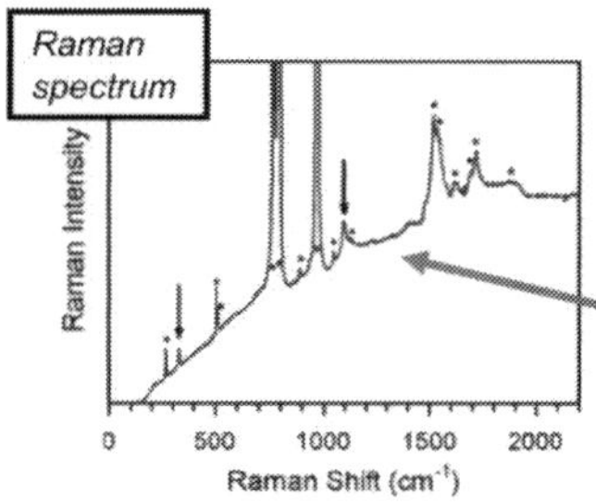

Formation of crystalline $MgCO_3$ confirmed by Raman (331 cm^{-1}, 1094 cm^{-1})

Technique applicable to both crystalline and "non-crystalline" systems

ARIZONA STATE UNIVERSITY

Application to Below-ground CO_2 Sequestration

Interaction of supercritical CO_2 with wet/dry clay minerals

ARIZONA STATE UNIVERSITY

Geological Sequestration…

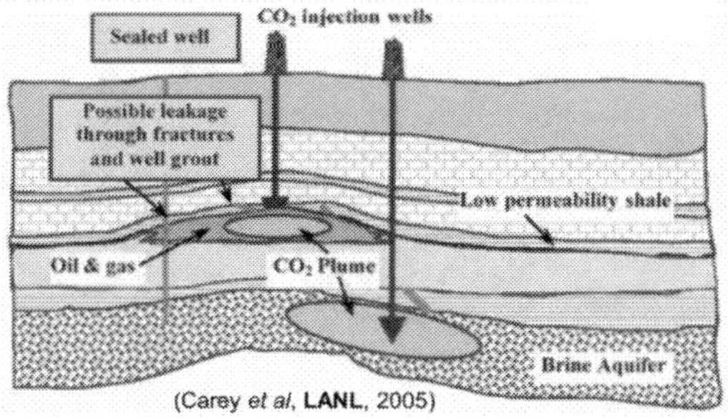

(Carey *et al*, **LANL**, 2005)

Mechanisms:

(1) **structural trapping** due to low-permeability

(2) **solubility trapping** via $scCO_2$ dissolution into oil and/or water

(3) **mineral trapping reactions** producing permanent mineral products

Factors affecting the redistribution/capacity of injected CO_2 ⇨
- miscibility
- viscous fingering
- reaction kinetics
- gravity segregation
- leakage → natural/artificial pathways

Stability: Leak rate less than 0.05 %/yr considered acceptable (IPCC report)

⇨ Catastrophic CO_2 release would produce a major disaster.

⇨ Natural CO_2 release from volcanoes & Lake Nyos (Cameroon) 1986: **2000 dead**.

ASU Arizona State University

Motivation – relevance to DOE programs

- Practical implementation of safe, long-term Geological Sequestration of CO_2 requires a thorough understanding of the interaction of buoyant $scCO_2$ with caprock minerals.
- Clays and clay-rich mineralogy responsible for low permeability of most caprocks and engineered barriers (Lin et al, 2008)
- Recent studies suggest that the **integrity** of "shaly" caprock may **decrease** with $scCO_2$ exposure (Busch et al, 2009). Impact on long-term geological storage?
- Clay-H_2O-CO_2 reaction studies are typically done in autoclaves (closed systems): ⇒ *ex situ* analysis **after quenching**! (Credoz et al 2009)
- Are "wet" clay minerals in post-injection core samples **altered by the quenching** associated with their removal? If so, how?

What is the state of the clay-H_2O-CO_2 system "***in situ***"?

ASU Arizona State University

Project objectives

Fundamental Scientific goals

- Use in situ studies to develop fundamental knowledge-base for the response of clay minerals to "dry" $scCO_2$, "wet" $scCO_2$ and H_2O-rich fluid component.
- Target representative clay minerals such as Ca- and Na-montmorillonites (STx-1, SWy-1) at T and P conditions spanning those encountered in typical aquifers (T=30-100 °C, P_{CO2}=70-200 atm).
- Map out swelling/shrinkage characteristics in "wet" and "dry" $scCO_2$ via desiccation or CO_2 absorption, respectively, and possible carbonation reactivity.

...develop new experimental technology along the way

- Adapt first generation ASU microreactor technology to enable systematic "in situ" synchrotron-based x-ray structural and chemical response of clays.
- Develop second generation microreactors to take advantage of alternative low-energy synchrotron x-ray sources → expand their versatility, throughput and reliabillity (ease of use).
- Develop tandem reactors to allow simultaneous parallel (comparative) reaction studies and/or long-term reaction studies, for the first time.

ASU Arizona State University

Clay-CO_2 Interactions: Baseline characterization...

- Standard clay materials in 75-150 mm ground powder form were obtained form the SESE-ASU mineral repository (Prof. P. Buseck). Several samples were obtained in coarse (gross) form and ground into powders, as needed.

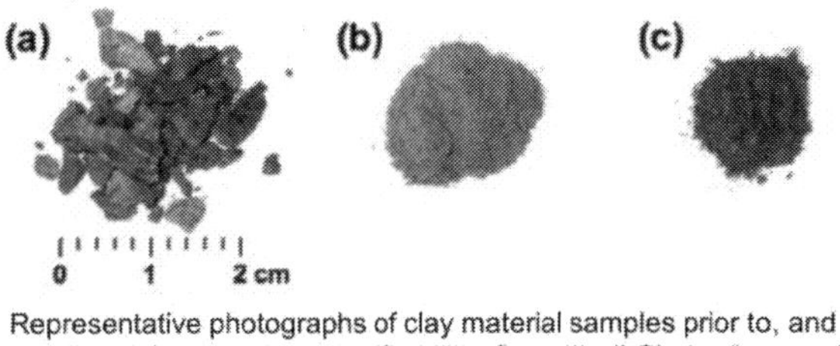

Representative photographs of clay material samples prior to, and after grinding: (a) natural interstratified Illite-Smectite (I-S) clay "as-received", (b) I-S clay ground to 150 mm, (c) Glauconite clay sample ground to 150 mm

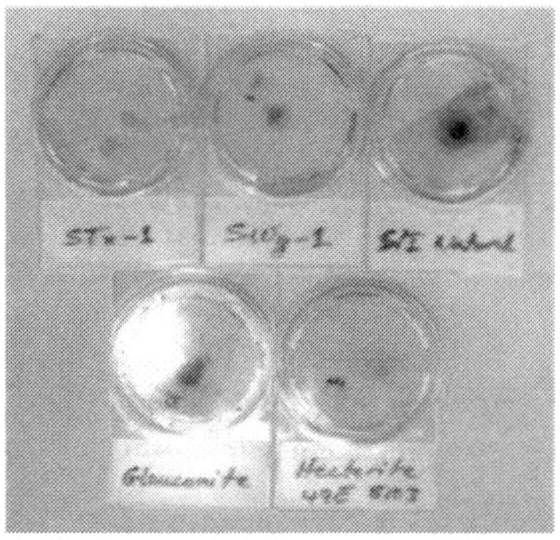

Most clay samples could be easily hydrated by exposure to liquid water or water vapor. Representative photographs of hydrated clay material samples, tape mounted and stored in translucent dishes.

ASU Arizona State University

Clay-CO_2 Interactions: Baseline characterization…

- Compositions/structure characterized using NMR, TGA, FTIR, SEM, EDAX and XRD.

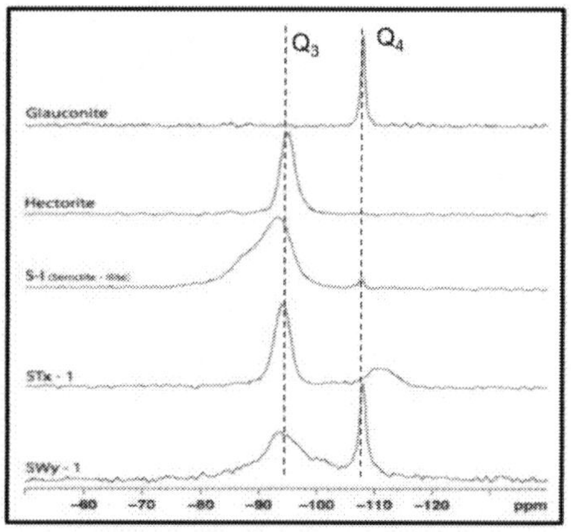

^{29}Si NMR spectra of the clay samples showing predominantly Q_3 and Q_4 Si species. STx-1 and Hectorite exhibit the most characteristic lamellar structure (Q_3). The solid state spinning spectra were obtained by spinning the samples at a rate of 10 kHz.

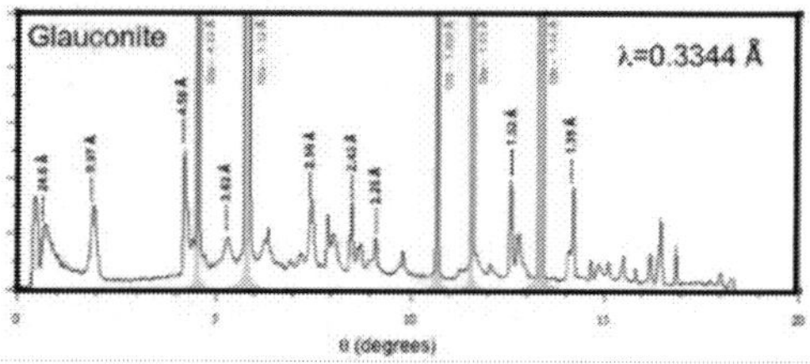

Background subtracted synchrotron XRD spectrum of Glauconite confirm the presence of a substantial α-quartz phase fraction.

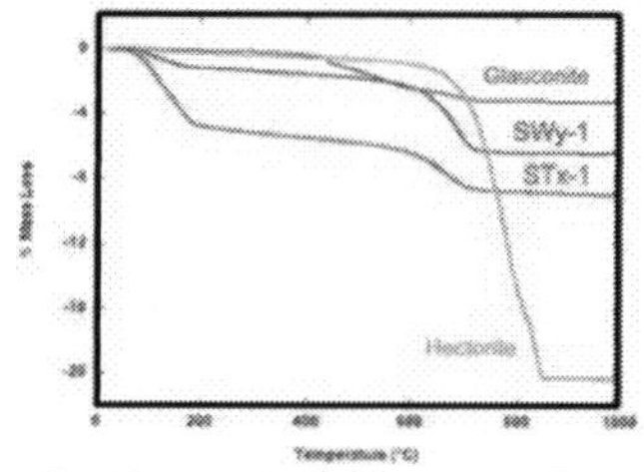

Thermogravimetric analysis (%mass loss vs T) at constant heating rate show similar trends of initial mass loss at low temperature, and complete dehydration around 600-700 °C.

ARIZONA STATE UNIVERSITY

Clay-CO_2 Interactions: Baseline characterization…

Morphological characterization in an Environmental Scanning Electron Microscope

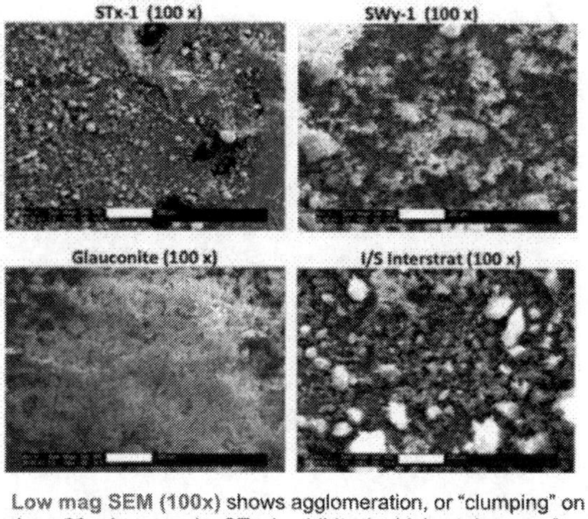

Low mag SEM (100x) shows agglomeration, or "clumping" on the < 20 micron scale. STx-1 exhibits the highest degree of homogeneity (10-50 μm particles). SWy-1 and ilite/smectite mixture (I/S) materials show moderate agglomeration. Glauconite displays large scale clumping on the mm scale.

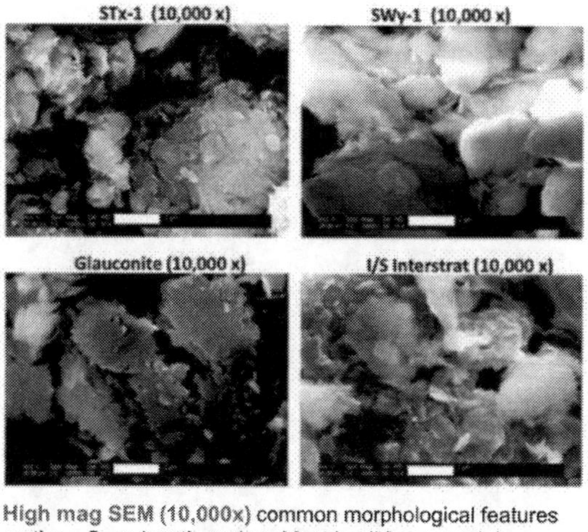

High mag SEM (10,000x) common morphological features on the ~ 2 μm length scale evident in all four sample's micrographs. At even higher magnification levels (50,000-150,000x) the lamellar features characteristic become observable only in the STx-1 and Glauconite .

ARIZONA STATE UNIVERSITY

Clay-CO_2 Interactions: Baseline characterization…

In situ Energy Dispersive Spectroscopy (EDS) in the E-SEM

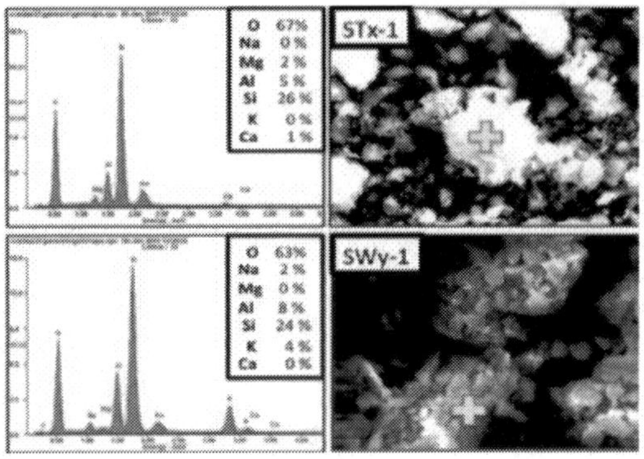

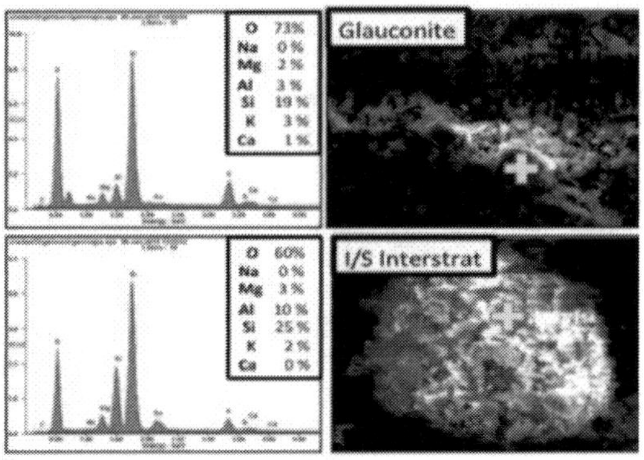

- STx-1 and SWy-1 materials contain the expected Ca/Mg- and Na/K- signatures
- Glauconite spectra indicate a depletion of Al and Si in relation to the standard montmorillonites
- Illite-Smectite (I/S) enhanced in Al and Mg, suggesting illite component predominant.

ARIZONA STATE UNIVERSITY

Clay-CO_2 Interactions: Synchrotron studies

- Initial synchrotron studies corroborated **known hydration behavior** of STx-1 (Ca-montmorillonite), and confirmed viability of using microreactors for further studies.

Characteristic d_{100} spacings for various hydration states in STx-1 clay (Monsalvo et al, 2006) containing 0-3 layers of water (W):	0W (9.5–9.7 Å) 1W (11.2–12.5 Å) 2W (14.6–15.5 Å) 3W (18.0–19.2 Å)

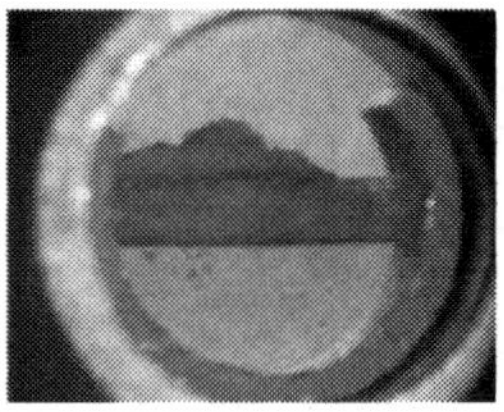

Microscope image of clay powder wafer mounted within the microreactor

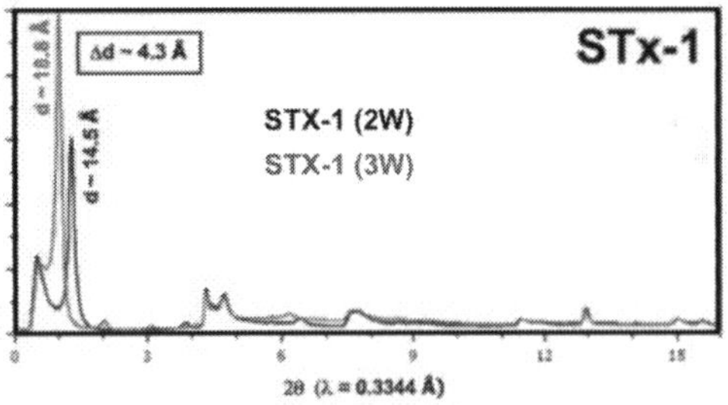

Removal of one layer of water (3W to 2W) obtained by simple heating of a hydrated sample to ~ 150 °C in a "dry" microreactor.

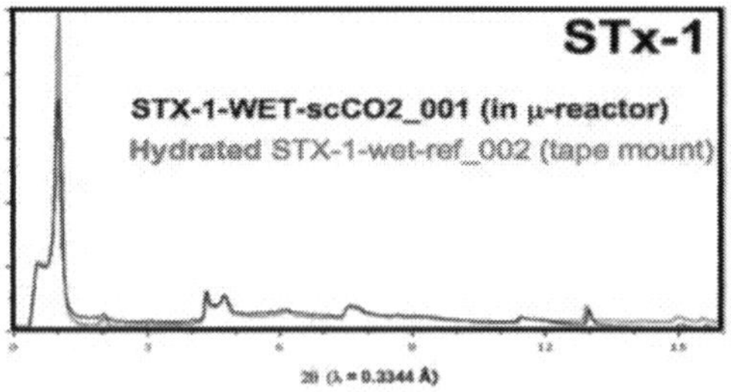

Comparison of STx-1 diffraction patterns obtained for a "tape mounted" sample and one collected through the microreactor, demonstrating excellent S/N (following background subtraction).

ARIZONA STATE UNIVERSITY

Clay-CO_2 Interactions: Synchrotron studies

- Our first observation of $scCO_2$ induced desiccation was made in a pilot study (Feb 2010) on STx-1 clay, at P_{CO2}~140 atm and T~ 40 °C, by monitoring the d_{100} basal reflections.
- Similar desiccation processes were subsequently observed over a broad range of (P,T) conditions.
- Drying behavior is a complex function of the hydration state of the clay, and that of the $scCO_2$,

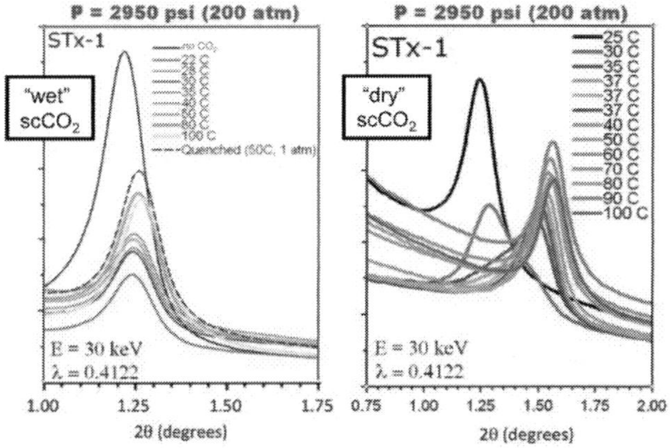

Representative results at P_{CO2}~200 atm: Interlayer d_{100} reflections in STx-1 clay as a function of temperature and P_{CO2}~200 atm. The 2θ~1.24° position corresponds to the 3W hydrated clay d-spacing of ~19.1 Å.

Key observations...

- Only a slight expansion (~0.7 Å) is observed in "wet" $scCO_2$ (not fully quenchable). Possible (minor) intercalation of CO_2 at P_{CO2}~200 atm?
- A very abrupt dehydration transition occurs near 32 °C in "dry" $scCO_2$.
- Final d-spacing of ~15.5 Å obtained at 100 °C corresponds to the 2W hydration volume change of ~19% relative to initial state).
- Net volume change of ~19% relative to initial state.

ARIZONA STATE UNIVERSITY

Clay-CO_2 Interactions: Synchrotron studies

- Water exits the clay only if the $scCO_2$ can accommodate it...

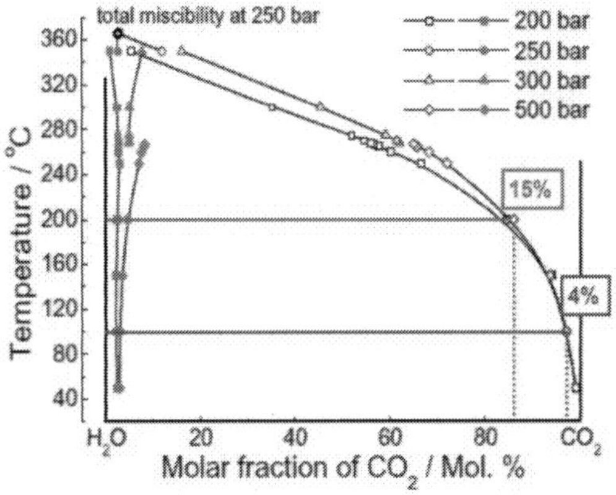

Phase diagram of H_2O-CO_2 at constant pressure (based on a plot by Tassaing et al. (2005)).

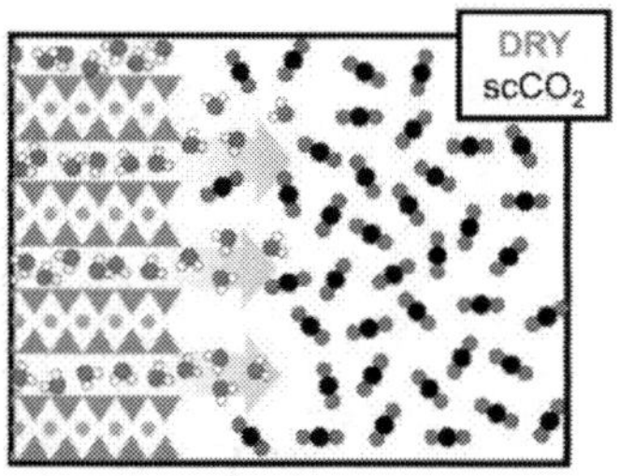

Hydrated clay in dry $scCO_2$
- desiccation promoted

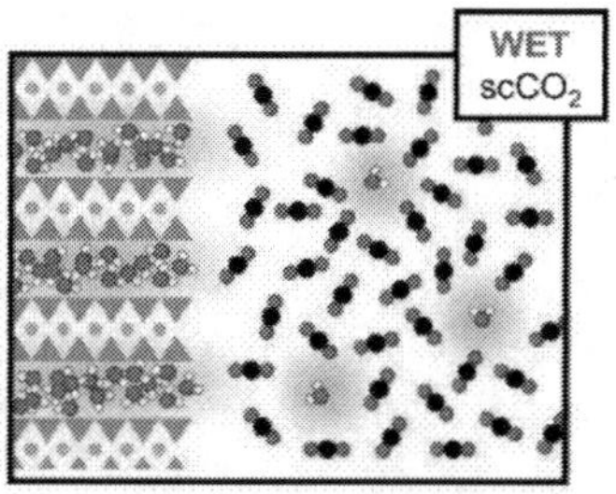

Hydrated clay in H_2O saturated $scCO_2$ (~ 4% H_2O @ 150 atm and 100 °C)
- desiccation inhibited

ARIZONA STATE UNIVERSITY

Clay-CO_2 Interactions: Synchrotron studies

- STx-1 clay exhibits desiccation behavior over a broad range of P_{CO2} pressures…

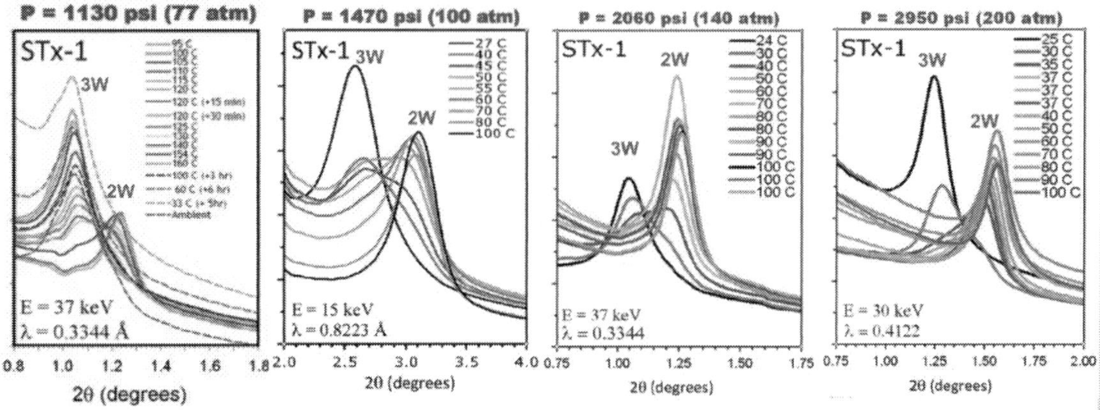

Interlayer d_{100} reflections in STx-1 clay as a function of temperature for a range of pressures spanning P_{CO2}~77-200 atm. In all cases the lowest angle reflection corresponds to the 3W hydrated clay d-spacing of ~19.1-19.2 Å, while the lower intensity reflection at higher angles corresponds to the 2W hydration state with d_{100} ~ 15.1-15.5 Å.

Note: The different horizontal scales reflect the use of various x-ray energies accessible during separate visits to the BMC and BMD beamlines (GSECARS).

Arizona State University

Clay-CO_2 Interactions: Synchrotron Studies

- Analysis of the diffraction data (from peak fitting) yielding 3W → 2W transition temperatures and critical volume changes corresponding to each pressure.

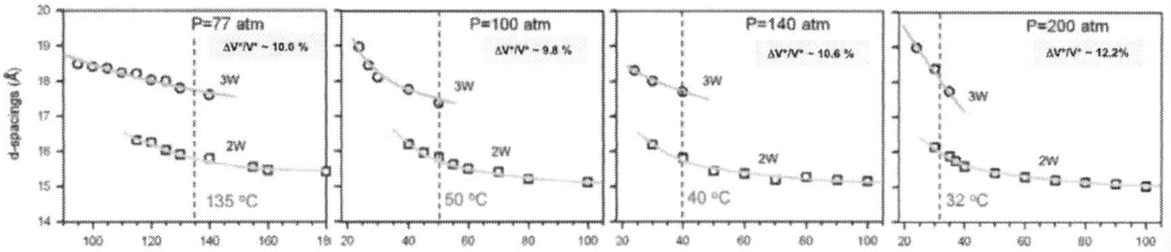

- D-spacings vs temperature for a range of pressures (77-200 atm) indicate an abrupt 3W → 2W dehydration behavior at critical temperatures (vertical dashed lines) of 135, 50, 40 and 32 °C.

- Data shows a progressive decrease in the critical desiccation temperature with increasing $scCO_2$ pressure, with a limiting high-pressure value of the desiccation temperature approaching the critical temperature of CO_2 (~31 °C).

Arizona State University

Clay-CO_2 Interactions: Synchrotron Studies

- Dynamics and quenching behavior…look at low-pressure case (just above P^*_{CO2}~73 atm)

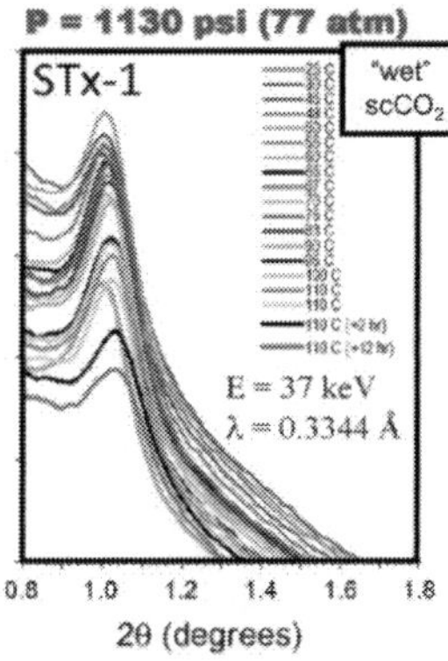

Temperature dependence of the (001) "hydration" diffraction peak in STx-1 clay in "wet" $scCO_2$. The bold green line (25 °C) corresponds to 3W hydration state (d_{100}~ 19.2 Å).

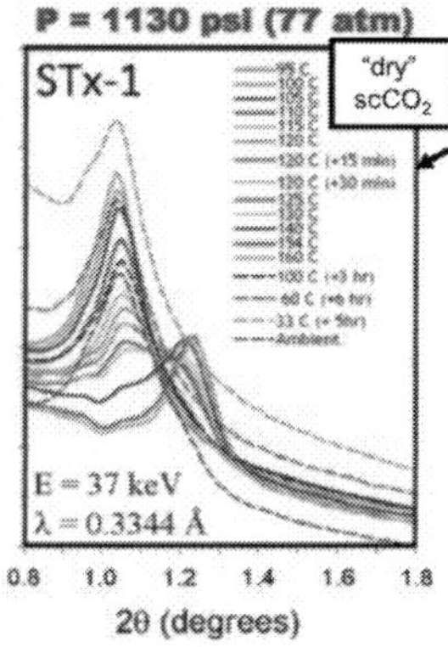

Temperature dependence of the (001) "hydration" diffraction peak in STx-1 clay in "dry" $scCO_2$ (transition T~135 °C).

Spectra of STx-1 hydration state in the vicinity of the transition (T~120 °C) acquired throughout the course of ~ 1 hour.

Equilibration occurs within the span of about 30 minutes, with no further observable change in the diffraction peak profile.

Quenching:

Cool to 100 °C, hold for 3 hours.
Cool to 60 °C, hold for 6 hours.
Cool to 33 °C, hold for 5 hours.

At low pressures STx-1 appears to re-hydrate

However…

STx-1 quenched from high pressures (P_{CO2}> 100 atm) does NOT rehydrate.

Pressure dependent mechanism?
At low pressures, the water liberated upon the 3W→ 2W desiccation transition remains "wetted" (remains in the vicinity of the clay surface). Significantly higher pressures are required to "activate" the promotion of water into the bulk $scCO_2$.

ARIZONA STATE UNIVERSITY

Clay-CO_2 Interactions: Main outcomes…

- The "desiccation" transitions occur on the scale of minutes-hours, and involve the release of one water layer (3W→ 2W).
- The volume change in the clay associated with the transitions is about 9-12%, and up to ~ 19% in relation to the initial volume of the 3W clay.
- Quenching to ambient conditions from low-pressures leads to re-hydration in STx-1 (■ square), suggesting that the diffusion barrier for H_2O into bulk $scCO_2$ may be pressure-dependent.
- Desiccation behavior NOT observed in "wet" $scCO_2$, or in corresponding saline solutions containing 1-3 M NaCl.
- At high pressures (~ 200 atm) we occasionally observed minor expansion of the clay (Δd_{100}~0.2-0.7 Å) suggesting the possible intercalation of CO_2 into the clay (further studies are needed).
- Virtually identical "desiccation" behavior was observed, with smaller volume changes (~5-9%), in SWy-1 (Na-montmorillonite) clay at P_{CO2}~140 atm (T_D~40 °C).
- No evidence of carbonation reactions (carbonate formation)

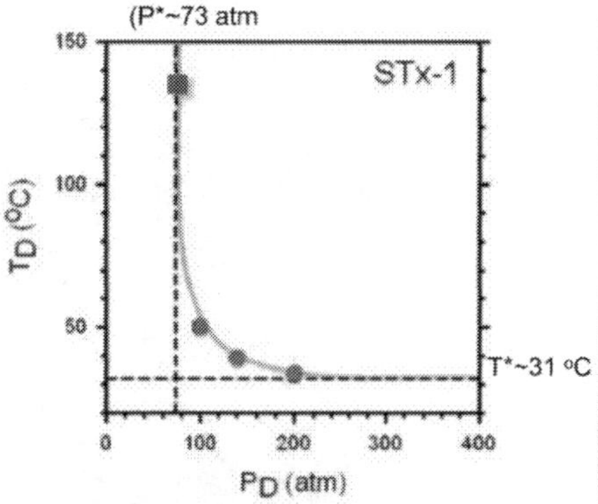

Why this plot is useful:

Post-injection equilibration of aquifer puts almost pure $scCO_2$ in contact with caprock (due to buoyancy). If the temperature of the aquifer is known (T_D), the critical pressure (P_D) at which desiccation is likely to occur can be interpolated form the graph above. Our study provides estimates of the associated volume changes in the clay fraction, which can be integrated into geophysical models to assess leakage potential.

ARIZONA STATE UNIVERSITY

$Clay\text{-}CO_2$ Interactions: Technology Development

Tandem μ-reactor rig with dual-redundant pressure ballast system

Motivation and design highlights:

- tandem reactors with common pressure control and thermal sink allow simplified setup for **differential reactivity studies**
- Compact design with dual-redundant pressure ballast system ideal for **long-term "constant activity" reaction studies**

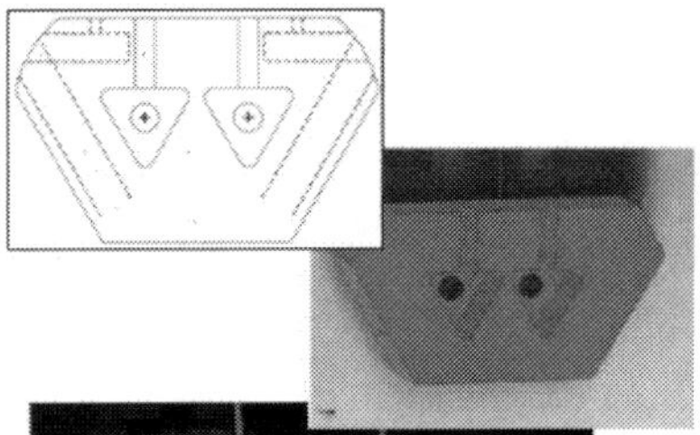

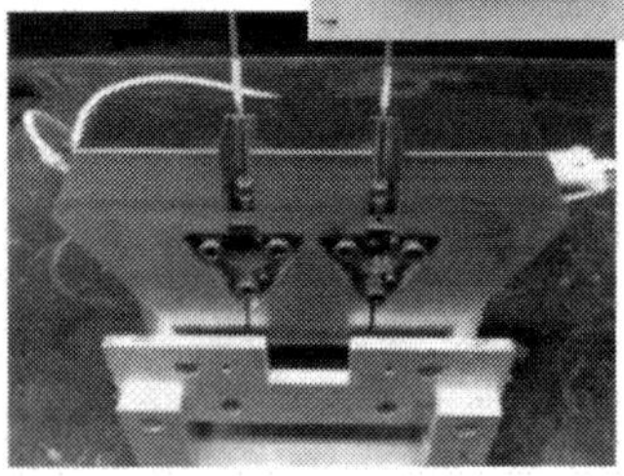

Prototype (May 2012)

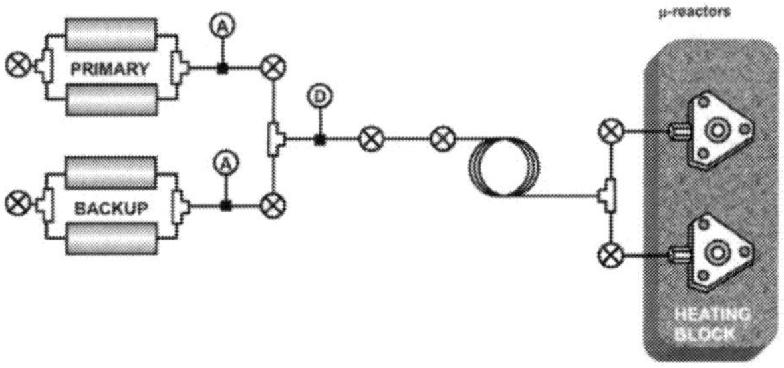

ARIZONA STATE UNIVERSITY

$Clay\text{-}CO_2$ Interactions: Technology Development

Tandem μ-reactor rig with dual-redundant pressure ballast system

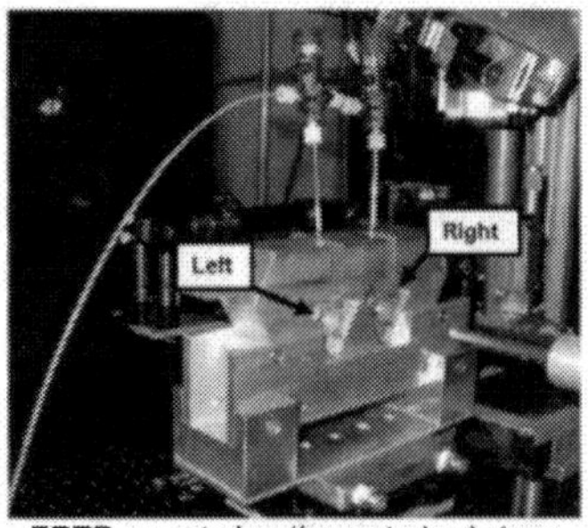

ERTR mounted on the motorized stage in the Sector 13-BMD hutch, connected to high-pressure plumbing

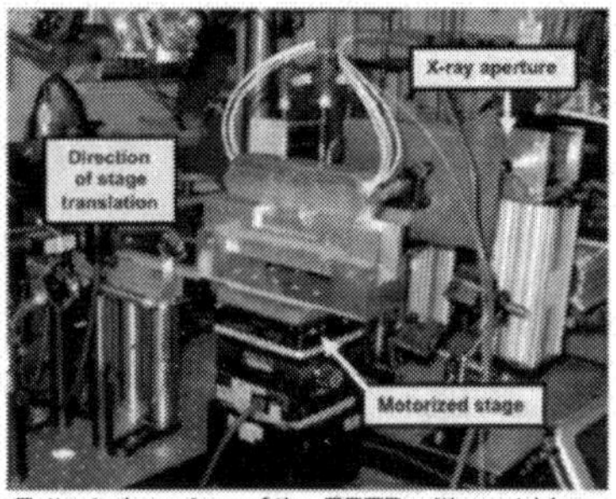

Exit window view of the ERTR with cartridge heaters inserted, indicating its position relative to the x-ray source aperture

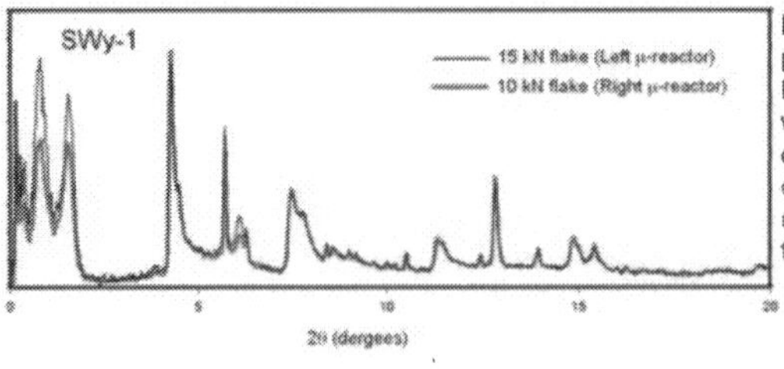

BMD/GSECARS: Diffraction patterns obtained from the ERTR for two SWy-1 powder wafers prepared by slightly different methods, showing an **excellent level of reproducibility between the two independent reactors**

ARIZONA STATE UNIVERSITY

Accelerated Carbonation of Cement Kiln Dust and Lime Kiln Dust for Permanent Storage of CO2

Gadikota, G., Columbia University
diDonato, M., University of Rome "La Sapienza"
Yip, S. W., Columbia University
Naibryf, M., Columbia University
Polettini, A., University of Rome "La Sapienza"
Park, A. H. A., Columbia University

According to the International Energy Agency, cement production contributes to 7% of the global CO_2 emissions. A significant reduction in the carbon footprint of the cement industry is expected to reduce the detrimental impacts on the carbon cycle. One of the ways to reduce CO_2 emissions from the cement industry is to react the produced CO_2 with alkaline wastes produced during the manufacturing of cement such as cement kiln dust (CKD) and lime kiln dust (LKD). CKD and LKD are fine-grained, unreacted wastes obtained from the calcination of finely ground limestone fed into cement kilns. Therefore, these materials have a high content of CaO which can react with CO_2 to form carbonates. In order to accelerate the dissolution of CKD and LKD to allow for faster conversion to carbonates, weak organic chelating agents such as 0.1 M Na-acetate and 0.1 M Na-gluconate are tested. Direct carbonation experiments are then performed with the optimal chelating agents at varying reaction temperatures (Tmax = 90 °C) in the presence of the best chelating agent and 0.64 M $NaHCO_3$ which acts as a pH buffer and carbon carrier, and at P_{CO2} = 150 atm. The extents of carbonation and formation of different phases are determined using Thermogravimetric Analysis (TGA), Total Carbon Analysis (TCA), Total Inorganic Carbon Analysis, and X-Ray Diffraction (XRD). Morphological alterations in CKD and LKD were identified using BET pore volume distributions, particle size analyses, and Scanning Electron Microscopy (SEM).

Carbonated Silicate Minerals and Industrial Wastes As Sustainable Construction Materials

Zhou, X., Columbia University
Gadikota, G., Columbia University
Jang, S. H.,
Yin, H.,
Kawashima, S.,
Park, A. H. A., Columbia University
Originally presented on: 6/22/2015 17:30:00 - 17:45:00

Given that the cement industry is responsible for 7% of global CO_2 emissions, various CO_2 mitigation strategies need to be employed to reduce the overall carbon footprint of cement and building materials. One such method to reduce CO_2 emissions during the production of cement and building materials is to incorporate carbonated materials and their by-products into concrete. Since silicate minerals such as serpentine ($(Mg, Fe)_3Si_2O_5(OH)_4$) and olivine ($(Mg,Fe)_2SiO_4$) contain more than 35% of magnesium and silicon, they can be reacted with CO_2 to form carbonates and other by-products which can then be used as cementitious materials. In addition, industrial wastes such as steel slag also have a significant composition of calcium and silicon, which are key constituents in building materials. In this study, various materials are synthesized from the alkaline rich minerals such as olivine and industrial wastes such as steel slag to produce silica, magnesium carbonates, and carbonated steel slag. The synthesis of these materials is via various routes which include direct carbonation of alkaline material-CO_2-reaction fluid, and indirect carbonation which involves dissolution of the alkaline content of the mineral followed by the formation of carbonates. The composition of these materials is determined using X-Ray Fluorescence (XRF), X-Ray Diffraction (XRD), Thermogravimetric Analysis (TGA), Total Carbon Analysis (TCA), and Total Inorganic Carbon Analysis (TIC). Morphological features were characterized using BET pore volume distributions, particle size analyses, and Scanning Electron Microscopy (SEM). Varying amounts of these materials were then incorporated into concrete and their compressive and mechanical strengths were tested over a period of time to determine the optimal composition of these materials to achieve the highest compressive and mechanical strengths of these novel composites.

Carbon isotope evidence for passive mineral carbonation of mine wastes from the Dumont Nickel Project (Abitibi, Quebec) _ ACEME 2015

A. Gras[1*], G. Beaudoin[1], J. Molson[1], B. Plante[2], B. Bussière[2], J.M. Lemieux[1] and B. Kandji[2]

[1] Department of Geology and Geological Engineering, Université Laval, Québec, Canada
[2] Research Institute of Mining and Environment, Université du Québec en Abitibi-Téminscamingue, Rouyn-Noranda, Canada
*Corresponding author: antoine.gras.1@ulaval.ca

Abstract
Natural weathering of ultramafic rocks in mine tailings allows the sequestration of atmospheric CO_2 through the formation of magnesium carbonates. In this context, the Dumont Nickel Project (DNP) is of particular interest as it will generate 1.7 Gt of ultramafic residues. A field experiment has been conducted at the DNP site in order to estimate its CO_2 sequestration potential and to evaluate the impact of carbonation reaction on the quality of mine drainage. Two experimental cells were built and instrumented. A decrease of the interstitial CO_2 concentration in both cells, from atmospheric values (~390 ppmv) near the surface to ~100ppmv near the bottom, reflects active CO_2 consumption by the residues. The total carbon content of the weathered DNP mine waste ranges from 0.2 wt% to 3.5 wt% C. Mg-Fe carbonates, nesquehonite, dypingite and hydromagnesite precipitated during passive mineral carbonation. In situ measurements using Wavelength-Scanned Cavity Ring Down Spectroscopy reveal an increase of $\delta^{\square}C_{(air)}$ from -8‰ to ~2‰ that is correlated with the decrease in CO_2 concentration. This trend is explained by kinetic carbon isotope fractionation during dissolution of atmospheric CO_2 in interstitial water ($\Delta_{DIC\text{-}Co2}$= -11‰). These observations are used to constrain a conceptual model of the carbonation reaction in the residues.

Introduction

Many deposits within mafic and ultramafic rocks are mined worldwide for economic minerals such as asbestos, nickel or diamond. Natural weathering these in mine tailings and waste rocks allows the sequestration of atmospheric CO_2 through the formation of magnesium carbonates [1,2,3].During the carbonation reaction, minerals such as brucite, olivine or serpentine release magnesium in leachate. Magnesium ions in drainage water can react with carbonate (CO_3^{2-}) and bicarbonate (HCO_3^-) ions to form carbonates minerals. Passive mineral carbonation of large quantities of ultramafic rocks accumulated from mining could reduce the net anthropogenic carbon released from mining operations[4,5,7]. The carbonation reaction has been extensively studied at historical and active mining sites [4,7,8,9,10].Several newly formed magnesium carbonates have been observed including lansfordite, nesquehonite, dypingite or hydromagnesite.

The Royal Nickel Corporation (RNC) plans to mine a serpentinised ultramafic body, located near Amos in northwestern Quebec (Canada), which contains 0.27 % nickel [11]. The exploitation of the Dumont Nickel Project (DNP) deposit will generate approximately 2.2 Gt of waste rock and tailings among which 1,7 Gt will be ultramafic rocks[12]. Previous laboratory experiments have evaluated the carbon sequestration capacity of the DNP residues, at high CO_2 partial pressure [13,14,15,16,17]. Furthermore the influence of several parameters (mineralogical composition, water saturation, watering) has been tested in order to optimize the carbonation process and increase CO_2 sequestration [18,19]. The DNP mining wastes are currently being studied in field experiments to estimate the CO_2 sequestration potential of the future mining operations and to evaluate the impact of mineral carbonation on the quality of mine waste drainage water. In order to improve our evaluation of the CO_2 sequestration capacity and to develop useful planning tools for the mining industry, field experiments are necessary.

Using two experimental cells, this study aims at understanding the key factors of the carbonation reaction and to propose a conceptual model, which can be used as a basis for numerical modeling. The carbon and oxygen isotope compositions (□^{13}C and □^{18}O) of Dissolved Inorganic Carbon (DIC) and newly formed carbonates were used to characterize the carbonation process [6,8,9,10,20]. A conceptual model of passive mineral carbonation in DNP mining wastes is developed using both geochemical and mineralogical analyses.

Materials and Methods

Experimental cells

Two experimental cells were built and instrumented in July 2011, to reproduce the dynamic processes in tailings storage facilities and waste rock piles. The first cell (EC-1) contains approximately 104 tonnes of heterogeneous mine waste rock ranging in size from blocks to silt. The second cell (EC-2) is constructed from approximately 2.1 m^3 of homogeneous processed waste (called tailings) with an average grain size of 175µm . The bottom of each cell is lined by a geomembrane (PVC Solmax 240) and a drain collects the leachate for sampling. A tipping bucket pluviometer is connected to the drain of each cell and records the flow of leachate. The mineralogical composition of mining wastes filling the two cells was measured using XRD and XRF instruments [14,15]. The main minerals in these wastes are chrysotile, lizardite, brucite, chlorite and magnetite, while Calcite and millerite were found in minor amounts. The amount of brucite, the key mineral for carbonation [14,15,18,19,6], is up to 10.18 wt% by GC-TGA[15]. Both cells contain interstitial gas sampling ports, located at different depths. In EC-1, the sampling ports are located at 28, 70, 109, 150 and 190 cm below the surface of the cell, whereas in cell EC-2, they are 5 and 25 cm beneath the surface. Cell EC-2 also contains probes for monitoring the hydrogeological parameters. (1) The changes in volumetric water content and temperature are measured at the surface and at 25 cm below and (2) suction is measured using a dielectric water potential sensor near the surface. A meteorological station was installed between the two cells to log solar radiation, temperature, wind speed and direction. Additionally, a tipping bucket rain gauge provides information on precipitation.

Interstitial gases

The interstitial gases were sampled twice a month using a peristaltic pump. The gases were analysed in situ with a Vaisala Carbo probe (Model GMP343; ±5 ppmv). The carbon isotope composition ($\delta^{13}C$) of interstitial gases was measured, in-situ, using a Wavelength-Scanned Cavity Ring Down Septroscopy (WS-CRDS) instrument (Picarro G-1101-i). Analyses were performed 5 times between March and November 2014. Results are reported using the delta notation ($\delta^{13}C$) relative to Vienna Pee Dee Belemnite (VPDB) in per mil (‰). In EC-1, the WS-CRDS instrument is connected to the sampling port and the interstitial gas is analysed for 20 min for CO_2 concentrations (+/- 1 ppmv) and carbon isotope ratios (+/- 0.3‰). A reference gas ([CO_2]= 420 ppmv +/- 8 ppmv; $\delta^{13}C$= -14 ‰ +/- 0,1 ‰) is measured after three gas analyses to correct for instrumental drift. In cell EC-2, the internal pumping system of the WS-CRDS instrument does not allow sufficient gas flow for C isotope analysis. Interstitial gases are pumped with a peristaltic pump for 3 minutes into Tedlar bags. The bag is then connected to the WS-CRDS instrument and the gas is analysed following the same method.

Newly formed minerals

White crusts were found on rock fragment surfaces in EC-1. On the surface of cell EC-2, scattered white crusts indurated the tailings at the edges of the cell. The white crusts were regularly sampled from both cells. In addition, cores were sampled in the center of EC 2 and split into 5 to 7 sub-samples depending on the length of the core. To provide a better identification of the carbonation front, the samples are thinner at the surface than at the bottom.

Qualitative X-ray diffraction

Samples were crushed using an agate mortar and analysed by X-ray diffraction (XRD) using a Siemens D5000 X-ray diffractometer (Cu-Kα radiation). XRD spectra were collected with a scanning step at 1°/min (0.02° step size) over the 5-40° scattering angle range. The Jade computer program was used to identify the mineral phases.

Scanning Electron Microscopy

A JEOL 840-A scanning electron microscope (SEM) equipped with an energy dispersive X-ray spectrometer (EDS) was used to observe mineral habits and textures and for mineral identification. Selected samples of carbonate coatings and sub-sample cores were impregnated with epoxy under vacuum to prepare thin sections.

Electron Probe Micro-Analysis

Selected thin sections were analysed using a CAMECA SX- 100 Electron Probe Micro-Analyser with an accelerating voltage of 15 kV, a current of 20 nA. Quantitative analyses and element maps were recorded for Si, Fe, Ni, Ca, Mg, Al and O. Calibration was performed using natural and synthetic silicates and oxide standards.

Total Carbon

The carbon content was measured with an Electra CS 800 carbon sulphur analyser. The samples were crushed in an agate mortar. The ceramic crucibles were then placed in a furnace at 950°C to destroy residual carbon. The powders were weighed and mixed with 0.18 cm^3 of tungsten and 0.18 cm^3 of iron. Two detectors allowed identifying high and low-level carbon. High carbon content was calibrated using the AR 4018 (1.30 +/- 0,05 wt% C), whereas we used UBN serpentine ANRT (0.106 +/- 0.08 wt% C) for low carbon contents.

Results and Discussion

CO_2 concentration within the cells

A decrease of CO_2 concentration in the waste rock cell, from atmospheric concentrations (~390 ppmv) at the surface of the cell to ~75 ppmv near the bottom has been recorded since 2011 (Figure 1-A). The high CO_2 concentration measured in sampling port 5, at the bottom of the cell, is caused by air infiltration in the drain used to sample leachate water. Despite the smaller thickness of the tailings materials (30 cm) in cell EC 2, the CO_2 concentrations decrease from atmospheric values (~390 ppmv) to ~50 ppmv (Figure 1-B).

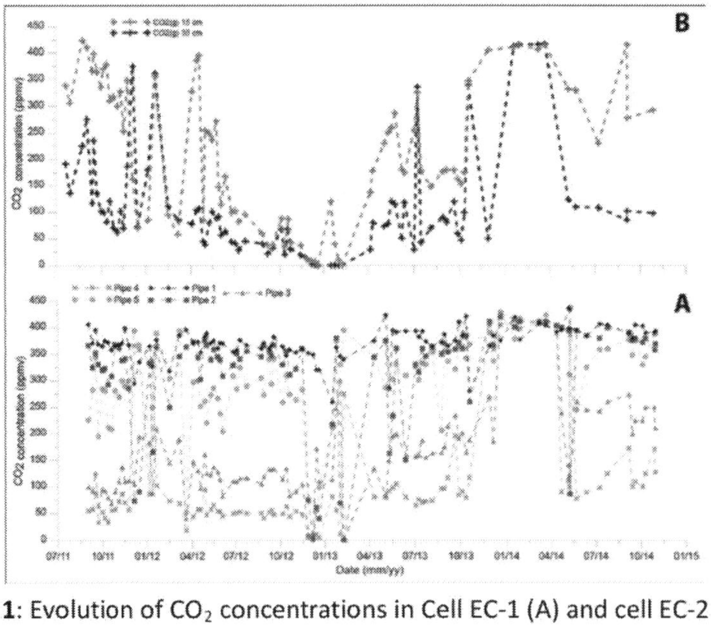

Figure 1: Evolution of CO_2 concentrations in Cell EC-1 (A) and cell EC-2 (B)

Since 2012, the CO_2 concentrations at one position have tended to increase from year to year in both cells. During the winter, sampling is disrupted by cold weather. The decrease of CO_2 concentrations in both cells reflects active CO_2 consumption by the mining wastes. The decrease of $CO_{2(g)}$ concentration is greater in cell EC-2 than in cell EC-1 (Figure 1-A,B).

This difference between EC-1 and EC-2 could reflect (1) differences in CO_2 availability or (2) a higher rate of $CO_{2(g)}$ consumption by mine tailings. (1) In cell EC-2, the mine tailings have a homogeneous grain size, about 175 μm. Laboratory measurements of show that the tailings material in cell EC-2 has a porosity and hydraulic conductivity, similar to that of a silty-clay porous medium (Table 1), compared to the waste rock of cell EC-2, which has properties analogous to a gravel-silt porous medium. Differential pressure measurements show that the difference between the atmosphere and the interstitial gases in the cells is less than 1 atm. These results, and other studies on similar experimental cells[7], indicate that advection is not the main mechanism for gas transport in both experimental cells. Therefore, the supply of $CO_{2(g)}$ is controlled by diffusion in the cells. The effective CO_2 diffusion coefficient ($D_{e\text{-}CO2}$) depends on the hydrogeological parameters (porosity, water content, tortuosity). Due to cell-specific hydrogeological conditions, diffusion of $CO_{2(g)}$ is slower in cell EC-2 than in cell EC-1. (2) The mineral surface area of the mine tailings in cell EC-2 is larger than that of mine wastes in cell EC-1, which could suggest a higher CO_2 consumption rate in cell EC-2. Further analyses are in progress to evaluate the CO_2 consumption capacity in the laboratory. The preliminary results show that mine the tailings have a greater CO_2 consumption capacity than the waste rock[16,21]. On one hand, CO_2 gas diffusion is more effective in cell EC-1 than in cell EC-2 . On the other hand the rate of $CO_{2(g)}$ consumption by mine tailings (EC-2) is higher than in cell EC-1. The differences of CO2 concentration decrease between the two cells can be explained by a combination of these two processes.

The increase in $CO_{2(g)}$ concentrations in both cells from year to year likely reflects a reduction of the carbonation reaction rate and, therefore, a decrease of carbon capture capacity with time.

Table 1. Hydrogeological parameters of the experimental cells

Hydrogeological parameters	EC-1 (Waste rock)	EC-2 (Tailings)
Porosity (%)	35*	45
Water content [1] (%)		30
Hydraulic conductivity (ms^{-1})	$10^{-3} - 10^{-5}$*	$10^{-8} - 10^{-9}$

* Parameters estimated after Aubertin et al. 2008. ,

[1] Average water content recorded between March and November since 2011

Total carbon content and newly formed carbonates

In order to describe the carbonation reaction of the DNP wastes, the products of the reaction were analysed. The total carbon content varies, in the white crusts on waste rock fragments and at the surface of the mine tailings, between ~1 wt.% and ~3.5 wt.% C. In core samples, the total carbon content ranges from ~0.8 wt.% at the surface to

~0.2 wt.% C at the bottom of the cell (Figure 2). This demonstrates a carbonation front in EC-2. Seasonal variations of total carbon content (Figure 2) at the surface of EC-2 yield no net increase in total carbon content over time. The absence of an increase in C concentration from year to year is not consistent with the continuous weathering of the residues. One hypothesis is that the newly formed carbonates are episodically dissolved and transported with the leachate during intense rain events in the Spring and Fall. Further investigations are in progress to explain these observations.

The low abundance of carbonate minerals in the weathered residues makes it difficult to identify the newly formed phases. Studies of crusts and cores by XRD, SEM and EPMA reveal nesquehonite, dypingite, hydromagnesite and several Mg-Fe carbonates, including brugnatellite, and pyroaurite-sjögrenite. XRD data does not allow an accurate identification of the low-abundance Mg-Fe carbonates. Mg-Fe carbonates such as pyroaurite, found in earlier studies in chrysotile mines tailings were interpreted to be bedrock-derived [8,20]. In this study, unweathered waste rock and mine tailings did not contain Mg-Fe carbonates. However, the precipitation of Mg-Fe carbonates (brugnatellite, pyroaurite-sjögrenite) could store the carbon content in the tailings core samples, which increases from 0.09 wt% to 0.2 wt%. The newly formed carbonates identified by SEM had a platy texture, which suggests dypingite and hydromagnesite. Nesquehonite was only found in EC-2 using XRD. Epoxy-impregnated thin sections allowed studying the contact between rock particles and newly formed carbonates. SEM and EPMA data indicate that the type of carbonate mineral does not vary with the composition of the rock particles on which it has precipitated. Progressive replacement of serpentine was not observed. Further investigations are in progress to confirm these preliminary results.

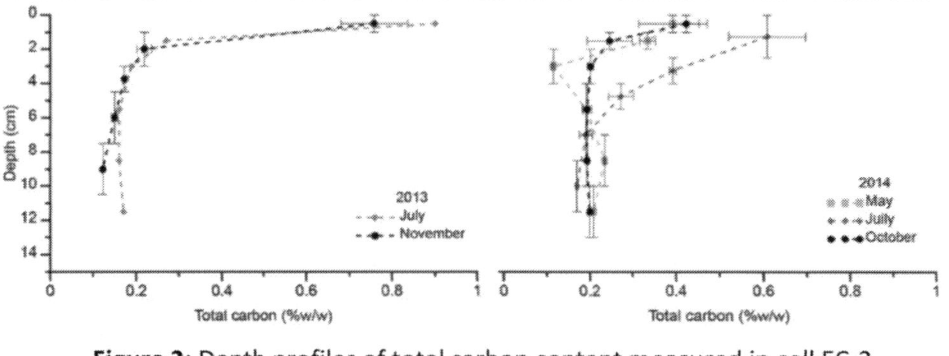

Figure 2: Depth profiles of total carbon content measured in cell EC-2

Carbon isotope composition of interstitial gases

The WS-CRDS analyses yielded the CO_2 concentration and carbon isotopic composition ($\delta^{13}C$) of the interstitial gas. The samples collected in EC-1 show that the decrease in CO_2 concentration from ~390 ppmv to ~100 ppmv is associated with an increase of $\delta^{13}C(_{CO2(g)})$ from circa -10 to ~1 ‰ (Figure 3). The high CO_2 concentration and low $\delta^{13}C$ values at the bottom of the cell reflect the infiltration of air through the leachate drain.

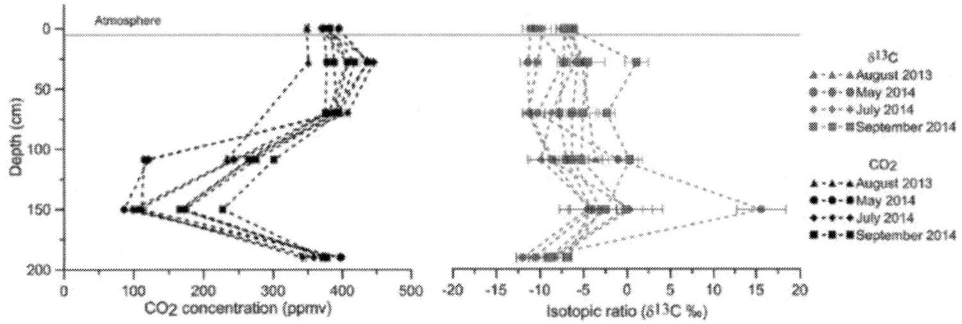

Figure 3: Depth profiles of CO_2 concentrations and $\delta\ ^{13}C$ in EC-1

The WS-CRDS measurements in EC-2 also show a decrease in CO_2 concentrations with an associated increase in $\delta^{13}C_{(air)}$ from -10 ‰ to ~3 ‰ (Figure 4).

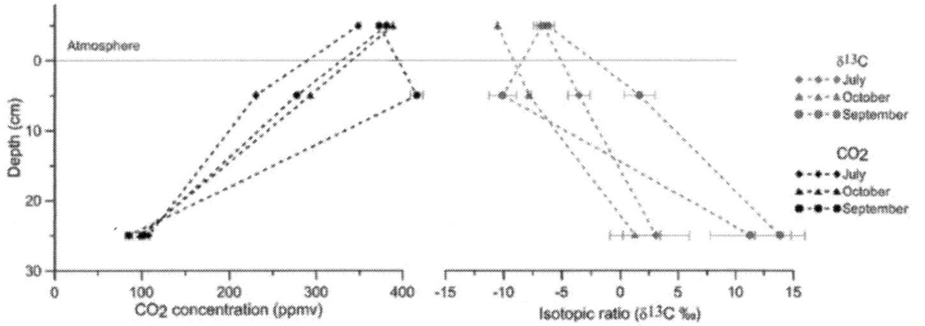

Figure 4: Depth profiles of CO_2 concentrations and $\delta^{13}C$ in EC-2

The fractionation between DIC and $CO_{2(g)}$ is a temperature-dependent process [22]. The average atmospheric temperature between March and November 2014 was ~8°C. Inside cell EC-2, the average temperature during this time interval was ~10°C, 15 cm below the surface. Using a Rayleigh distillation model and Mook et al.[22] equilibrium carbon isotopic fractionation ($\Delta^{13}C_{DIC\text{-}CO2(g)}$ = 9.6 ‰ at 10°C), a $\delta^{13}C$ decrease from -6.5 ‰ to -28 ‰ in the interstitial gases is predicted. Instead, an increase of $\delta^{13}C$ was measured in the cells. This suggests that $CO_{2(g)}$ was not in isotopic equilibrium with interstitial water. Kinetic isotope fractionation during dissolution of $CO_{2(g)}$ in water [6,20,23] yields a carbon isotope fractionation of $\Delta^{13}C_{DIC\text{-}CO2(g)}$ = -11.2 ‰. Using the kinetic isotopic fractionation between $DIC_{(aq)}$ and $CO_{2(g)}$ in a Rayleigh distillation model, and the range in on-site $\delta^{13}C$ values of atmospheric CO_2 (-6.5 ‰ and -11 ‰) allows modelling the measured carbon isotope compositions in the cells. Figure 5 presents the change in $\delta^{13}C$ in gases as a function of the remaining fraction of $CO_{2(g)}$. CO_2 concentrations higher than typical atmospheric values were measured in the upper sampling tube in EC-1, which yields a value greater than 1 for the remaining $CO_{2(g)}$ fraction in models using an atmospheric CO_2 concentration of 390 ppmv. Figure 5 shows that dissolution of CO_2 in the interstitial pore water controls the carbon fractionation under conditions where the CO_2 supply is limited by diffusion from the atmosphere. The slow dissolution of CO_2 into the interstitial water causes the water to become depleted in ^{13}C. Assuming that Mg-carbonate minerals precipitate under isotope equilibrium with the interstitial water[23], the $\delta^{13}C$ values of the newly formed minerals should follow the ^{13}C depletion trend in water.

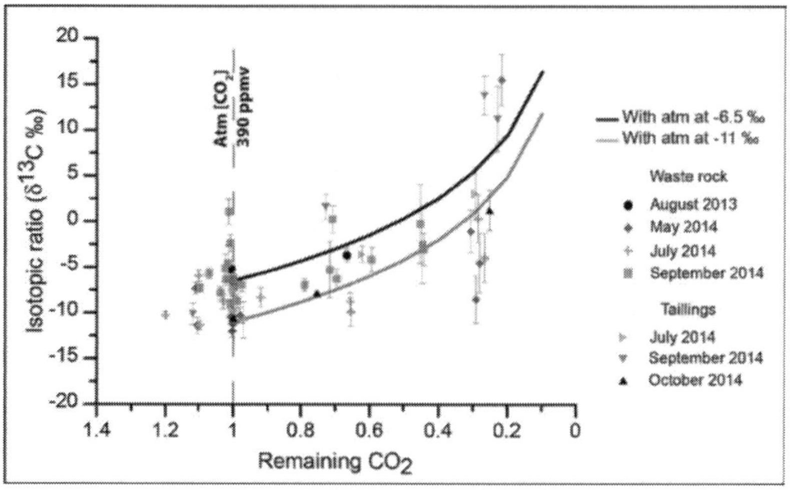

Figure 5: Theoretical fractionation of $CO_{2(g)}$ in the Rayleigh model and measured $\delta^{13}C$ in interstitial gases

In order to evaluate the CO_2 sequestration capacity it is important to constrain the source of the CO_2 sequestered. The $CO_{2(g)}$ in the experimental cells may be derived from three different sources having distinct $\delta^{13}C$ signatures: (1) atmosphere (~ -8‰), (2) dissolution of bedrock carbonates and (3) oxidation of organic matter by microorganism respiration (~-15‰). The isotopic signature of the $CO_{2(g)}$ derived from carbonate dissolution and microorganism respiration can be approximated using carbon isotope equilibrium fractionation. The oxidation of organic matter by microorganisms will produce $CO_{2(g)}$ highly depleted in ^{13}C. Considering the $\delta^{13}C$ measurements described above, $CO_{2(g)}$ released by microorganism respiration (~-15‰) is unlikely to have contributed to the carbon isotope composition of interstitial gases in the experimental cells. The initial carbon isotopic signature of carbonate minerals controls the $\delta^{13}C$ signature of gas evolved from carbonate dissolution. The $\delta^{13}C$ values of bedrock carbonates (measurements in progress) will allow testing for the potential dissolution of bedrock carbonates. Previous investigations of the radiogenic carbon content of newly formed carbonates [8,9,10,20], have yielded values of $F^{14}C$ (expressed as a fraction of modern carbon) greater than or equal to 1, which indicates a dominant source from modern carbon.

We propose a process-based conceptual model for carbon isotope fractionation during the carbonation reaction of the DNP wastes, as shown in Figure 6. As discussed above, three steps control the carbon cycle in the experimental cells: (1) diffusion of atmospheric CO_2 into the porous medium, (2) $CO_{2(g)}$ dissolution in interstitial water, which enriches the interstitial gases in ^{13}C and (3) precipitation of Mg-carbonates from the ^{13}C depleted interstitial water. According to this conceptual model, the limiting step in the carbonation process is CO_2 dissolution in interstitial water, which is then the major limiting step to reach the maximum carbon sequestration capacity. This conclusion was also reached in laboratory studies [6,23].

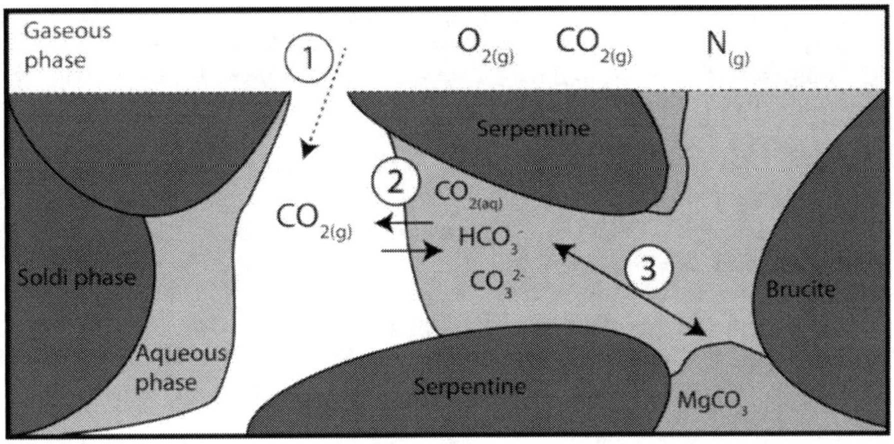

Figure 6: Conceptual model of carbon isotope fractionation during mineral carbonation in DNP residues.

Conclusion

In summary, in situ measurements and analysis of samples collected from field experimental cells (EC-1 and EC-2) have provided evidence for passive mineral carbonation and carbon sequestration in DNP mining wastes at an atmospheric partial pressure of CO_2. At least 4 carbonate minerals precipitate in the cells, which increases the total carbon content in the weathered DNP mining wastes. In spite of the differences between the two experimental cells (hydrogeological parameters, grain size and surface area) the results suggest that the same processes occur in both cells during passive mineral carbonation. WS-CRDS measurements reveal an enrichment of ^{13}C in the interstitial gas associated with a decrease in CO_2 concentration within the cells. As gas advection is slow in the partially saturated porous medium of the experimental cells, CO_2 input is driven by diffusion. CO_2 dissolution in interstitial water under limited CO_2 input causes an enrichment of ^{13}C in the interstitial air. The ^{13}C enrichment is explained by kinetic isotope fractionation during dissolution of atmospheric CO_2 in interstitial water ($\Delta_{DIC\text{-}Co2}$= -11‰), which is the limiting step for the carbonation reaction in the experimental cells.

Acknowledgements

The authors thank Royal Nickel Corporation and the Natural Sciences and Engineering Research Council of Canada Collaborative Research and Development (NSERC CRD) program for their support. We thank U. Laval and Royal Nickel Corporation staff for their help in the field and in the laboratory, and for insightful discussions. The anonymous reviewers are thanked for their helpful and constructive suggestions.

References

1. Lackner, K.S., Wendt, C.H., Butt, D.P., Joyce Jr, E.L., Sharp, D.H., 1995. Carbon dioxide disposal in carbonate minerals. *Energy 20*, 1153-1170

2. Wilson, S.A., Raudsepp, M. & Dipple, G.M. 2006. Verifying and quantifying carbon fixation on in minerals from serpentine-rich mine tailings using the Rietveld method with X-ray powder diffraction on data, *American Mineralogist* 91:1331–1341. doi:10.2138/am.2006.2058.

3. Beaudoin G., H.R., Constantin M., Duchesne J., Cecchi E., Huot F., Vigneau S., Fiola R., 2008. Spontaneous carbonation of serpentine in milling and mining waste, southern Québec and Italy., in: *2nd International Conference on Accelerated Carbonation for Environmental and Materials Engineering* (ACEME 2008) (Ed.), 1-3 Oct. 2008, Rome, p. 10 p.

4. Hitch, M., Ballantyne, S.M. and Hindle, S.R. 2010. Revaluing mine waste rock for carbon capture and storage, *International Journal of Mining, Reclamation and Environment*, 24 (1): 64-79.

5. Mills, S.J., Wilson, S.A., Dipple, G.M. & Raudsepp, M. 2010. The decomposition of konyaite: importance in CO_2 fixation in mine tailings, *Mineralogical Magazine*, 74: 91-917.

6. Harrison, A. L., Power, I. M., & Dipple, G. M. (2012). Accelerated carbonation of brucite in mine tailings for carbon sequestration. *Environmental Science & Technology*, *47*(1), 126-134.

7. Lechat K., Lemieux J.M., Beaudoin .G., Hébert R., Molson J. , 2014 Monitoring CO_2 Sequestration by Mineral Carbonation in Mine Tailings at Thetford Mines, Quebec, Canada. *Poster session AGU 2014* V23A-4774

8. Oskierski, H. C., Dlugogorski, B. Z., & Jacobsen, G. (2013). Sequestration of atmospheric CO 2 in chrysotile mine tailings of the Woodsreef Asbestos Mine, Australia: Quantitative mineralogy, isotopic fingerprinting and carbonation rates. *Chemical Geology*, *358*, 156-169.

9. Wilson, S.A., Dipple, G.M., Power, I.M., Thom, J.M., Anderson, R.G., Raudsepp, M., Gabites, J.E. & Southam, G. 2009a. Carbon dioxide fixation within mine wastes of ultramafic-hosted ore deposits: Examples from the Clinton Creek and Cassiar chrysotile deposits, Canada, Economic. *Geology*, 104:95–112. doi:10.2113/gsecongeo.104.1.95.

10. Wilson, S.A., Raudsepp, M. & Dipple, G.M. 2009b. Quantifying carbon fixation in trace minerals from processed kimberlite: A comparative study of quantitative methods using X-ray powder diffraction data with applications to the Diavik Diamond Mine, Northwest Territories, Canada, *Applied Geochemistry*, 24 (12): 2312-2331.

11. Sciortino, M., Mungall, J. E., & Muinonen, J. (2015). Generation of High-Ni Sulfide and Alloy Phases During Serpentinization of Dunite in the Dumont Sill, Quebec. *Economic Geology*, *110*(3), 733-761.

12. Royal Nickel Corporation. 2013. Technical Report on the Dumont Project, Launay and Trécesson Townships, Québec, Canada. 2280-RPT-000, 24 October 2013, 432p.

13. Pronost J., B.G., Constantin M., Duchesne J., Hébert R., Larachi F., 2010. Evaluation of the mineral carbonation potential of mining residues from Royal Nickel Dumont Nickel Project, Amos (Québec, Canada) in: géologique, D.d.G.e.d.G. (Ed.), Université Laval.

14. Pronost, J., Beaudoin, G., Tremblay, J., Larachi, F., Duchesne, J., Hébert, R., Constantin, M., 2011. Carbon Sequestration Kinetic and Storage Capacity of Ultramafic Mining Waste. *Environmental Science & Technology* 45, 9413-9420.

15. Assima, G. P., Larachi, F., Molson, J., & Beaudoin, G. (2013). Accurate and direct quantification of native brucite in serpentine ores—new methodology and implications for CO 2 sequestration by mining residues. *Thermochimica Acta, 566*, 281-291.

16. Plante, B., Kandji, E.H.B., Bussière, B., Awoh, A.S., Ibrahim-Saib, B.A., Beaudoin, G., Gras, A., Molson, J., Dupont, P.P. (2014). Geochemical behavior of carbon-sequestrating mine wastes: Dumont project, Royal Nickel Corporation. Proc. Tailings and mine waste, Keystone, Colorado, 5-8 October 2014.

17. Kandji, E.H.B., Plante, B., Bussière, B., Awoh, A.S., Ibrahim Saïb, B.A., Dupont, P.P. (2015). Prediction of drainage water quality from mining waste with carbon sequestration potential. *Proc. (IMWA/ICARD)*, Santiago, Chili, 21-24 April 2015.

18. Assima, G. P., Larachi, F., Beaudoin, G., & Molson, J. (2013). Dynamics of carbon dioxide uptake in chrysotile mining residues–effect of mineralogy and liquid saturation. *International Journal of Greenhouse Gas Control, 12*, 124-135.

19. Assima, G. P., Larachi, F., Molson, J., & Beaudoin, G. (2014). Emulation of ambient carbon dioxide diffusion and carbonation within nickel mining residues. *Minerals Engineering, 59*, 39-44.

20. Wilson, S. A., Harrison, A. L., Dipple, G. M., Power, I. M., Barker, S. L., Mayer, K. U., ... & Southam, G. (2014). Offsetting of CO 2 emissions by air capture in mine tailings at the Mount Keith Nickel Mine, Western Australia: rates, controls and prospects for carbon neutral mining. *International Journal of Greenhouse Gas Control, 25*, 121-140.

21. Awoh, A.S., Plante, B., Bussière, B., Mbonimpa, M. (2014). Measurement and prediction of the CO2 effective diffusion coefficient in unsaturated media. *Conférence Canadienne de Géotechnique*, Montréal, Canada.

22. Mook, W. G., Bommerson, J. C., & Staverman, W. H. (1974). Carbon isotope fractionation between dissolved bicarbonate and gaseous carbon dioxide. *Earth and Planetary Science Letters, 22*(2), 169-176.

23. Wilson, S. A., Barker, S. L., Dipple, G. M., & Atudorei, V. (2010). Isotopic disequilibrium during uptake of atmospheric CO2 into mine process waters: Implications for CO_2 sequestration. *Environmental Science & Technology, 44*(24), 9522-9529.

CO2 capture and storage by mineral carbonation
Experimental Investigation of the Pressure of Crystallization of Ca(OH)2: Implication for the Reactive-Cracking Process

Lambart, S.,
Savage, H., Lamont-Doherty Earth Observatory
Kelemen, P., Columbia University
Originally presented on: 6/23/2015 8:30:00 - 9:00:00

Carbonation of peridotite may be an important sink in the global carbon cycle. Some natural systems attain 100% carbonation. In these systems, all Mg, Ca and Fe is extracted from silicate minerals to form carbonate minerals, perhaps as a result of reaction-driven cracking processes that maintain or enhance permeability and reactive surface area. Without reaction-driven cracking, in situ mineral carbonation could fill pore space and armor reactive surfaces, limiting reaction progress and leaving much of the silicate unaltered (Kelemen and Hirth, 2012).

The reactive-cracking process will occur if stress resulting from anisotropic volume changes during fluid-rock interactions, i.e., the "crystallization pressure", is sufficient to fracture rocks. The crystallization pressure is usually described as:

$P' = -\Delta Gr/\Delta Vs$ (1)

where P' is the crystallization pressure, ΔGr is the Gibbs Free Energy of a reaction, and ΔVs is the change in solid volume resulting from this reaction (Scherer, 2004; Steiger, 2005). Kelemen and Hirth (2012) proposed that use of Helmholtz Free Energy of reaction ΔFr may be preferable, yielding:

$P' = -\Delta Fr/\Delta Vs = -\Delta Gr/\Delta Vs + P\Delta Vr/ \Delta Vs$ (2)

where P is the confining pressure and ΔVr is the volume change of a stoichiometric reaction including fluid components.

However, the conditions most favorable for reaction-driven cracking are poorly understood, especially at confining pressures relevant to CO_2 storage. Fundamental uncertainties remain in estimating the crystallization pressure that drives reaction-driven cracking and, so far, experiments on peridotite hydration or carbonation have not produced reactive cracking, possibly due to limited reactive surface area in low porosity samples. To address this, we explore crystallization pressure in simple systems under controlled laboratory conditions.

We have performed experiments on reaction (1): solid $CaO + H_2O =$ solid $Ca(OH)_2$, with a 100% increase in the solid volume and high initial porosities (Φ). Using equation (1), the calculated pressure of crystallization of $Ca(OH)_2$ is 4.1 GPa. However, equations (1) and (2) do not incorporate energy sinks such as exothermic heating and/or thermal diffusion, implicitly assuming that all chemical potential energy is converted into stress.

Thus, both equations yield an upper bound for P'.

We cold-pressed CaO powder to form ~20 mm long cylinders 4 to 12 mm in diameter, with initial Φ ranging from 0.36 to 0.53. These cylinders were confined in steel, and compressed with an axial load of 0.1 to 27 MPa while water was introduced through a micro-porous frit. Without expansion of the total volume, the reaction would stop when $\Phi = 0$, producing $\Phi = 0, 2\cdot\Phi\cdot V_{Ca(OH)2} + (1\text{-}2\cdot\Phi)\cdot V_{CaO}$.

Instead, in all experiments the volume of cylinders increased with time, maintaining $\Phi >$ 20%. Hence, the stress exerted by the crystallization of $Ca(OH)_2$ is higher than the maximum axial load (27 MPa). 27 MPa is also higher than the tensile strengths of most of rocks, suggesting that, at least at low confining pressure, reactive-cracking can happen with this system. The evolution of ΔV is best fit by a power law. The reaction slows over time, either because of a reduction of permeability or consumption of the reactant. The fact that all experiments show the same slope ($\Delta V = bt^m$ with m ~ constant) whether the porosity was decreasing or increasing during the reaction, suggests that the latter explanation is correct. Experiments were stopped at reaction extents from 42 to 100% but all would likely reach 100% at longer durations.

To investigate the conditions leading to reaction-driven cracking in this system, we've begun a series of higher pressure experiments in a triaxial deformation apparatus. We will map stress and strain resulting from CaO hydration at geologically relevant combinations of confining pressure (10 to 100 MPa), temperature (20 to 400°C), and volume change. The configuration that we chose is a CaO cylinder embedded in a relatively inert, porous, sandstone (i.e., mainly SiO_2). CaO solid + H_2Ofluid = solid $Ca(OH)_2$ and CaO solid + CO_2 fluid = solid $CaCO_3$. Both have a big volume change. The configuration with has already been tested without confining pressure, in an experiment in which the first macroscopic fractures appeared after only 3 min. Additional fractures continued to form during the next 15min, after which the sample fell apart.

References:

Kelemen PB, Hirth G, 2012, Reaction-driven cracking during retrograde metamorphism: Olivine hydration and carbonation: Earth Planet. Sci. Lett., v. 345-348, p. 81–89. Scherer GW, 2004, Stress from crystallization of salt: Cement and Concrete Research, v. 34, p. 1613-1624. Steiger M, 2005, Crystal growth in porous materials – I: The crystallization pressure of large crystals: J. Crystal Growth, v. 282, p. 455-469.

Pilot and full-scale applications

Quantification of Carbon Mineralization with Reactive and Non- Reactive Tracers

Stute, M., Columbia University
Matter, J., University of Southampton

Originally presented on: 6/24/2015 15:15:00 - 15:30:00

Environmental tracers are useful tools for tracking the fate of CO_2 injected and stored in the subsurface, including the quantification of mineralization. Tracers in various ways tend to preferentially partition into gas, liquid, or solid phases and thus allow for the quantification of these processes. Of particular importance are ^{13}C and ^{14}C because they are the only ones tracking the CO_2 molecule itself. The potential of these environmental tracers will be illustrated using laboratory experiments and field studies. One of the case studies we will focus on is the Carbfix site in Iceland. Pure CO_2 and CO_2/H_2S mixtures from the Hellisheidi geothermal power plant near Reykjavik, Iceland were dissolved downhole in groundwater and injected into a permeable confined basalt formation at about 500m depth below ground. We are using non-reactive (sodium fluorescein, SF_5CF_3, and SF_6) and reactive (^{14}C and ^{13}C) tracers in the project to characterize subsurface CO_2 transport and in situ CO_2-water-rock reactions. In January 2012, 170 tons of pure CO_2 tagged with ^{14}C and SF_6 were injected followed by 73 tons of a CO_2/H_2S mixture starting in June 2012. Mass balance calculations based on the tracer breakthrough curves suggest that about 90% of the injected CO_2 has been mineralized over a period of only 2 years.

Directed Formation of Hydrated and Anhydrous Magnesium Carbonates and Enhanced Water Gas Shift Reaction

Fricker, K., Columbia University
Park, A. H. A., Columbia University
Originally presented on: 6/22/2015 14:15:00 - 14:30:00

The integration of carbon capture and storage (CCS) technologies with energy conversion processes will increase the sustainability of using carboneous fuels. Combined CCS via carbon mineralization, where CO_2 reacts to form solid, stable mineral carbonates represents an attractive solution that avoids the energy costs of sorbent regeneration and public and scientific inquiries surrounding the safety of long-term geological storage of CO_2 in subsurface reservoirs. Magnesite is the most desirable phase within the magnesium carbonate family for carbon storage for a number of reasons: magnesium efficiency, omission of additional crystal waters and thermodynamic stability. Unfortunately, the formation of metastable hydrated magnesium carbonate phases (e.g. $MgCO_3 \cdot 3H_2O$ and $Mg_5(CO_3)_4(OH)_2 \cdot 4H_2O$) interferes with the production of anhydrous magnesite under a variety of reaction conditions because magnesite crystals are slower to both nucleate and grow compared to the hydrated carbonate phases. Furthermore, the reaction conditions required for the formation of each magnesium carbonate phases have not been well understood with conflicting literature data. In this study, the effects of both magnesite ($MgCO_3$) and inert (Al_2O_3) seed particles on the precipitation of magnesium carbonates from a $Mg(OH)_2$ slurry were explored as well as the implication of the findings in terms of enhanced water gas shift reaction.

Utilization of Flue Gases for Waste Paper ONP/OCC/OMG Recycling - Semi Pilot Scale Demonstration

Ji Whan, A., Korea Institute of Geosciences and Mineral Resources(KIGAM)
Thriveni, T., Korea Institute of Geosciences and Mineral Resources(KIGAM)
Sun Yu, K., Korea Institute of Geosciences and Mineral Resources(KIGAM)

In the present work, we report a novel microstructure of synthesized colloidal calcite PCC (precipitated calcium carbonate) without any additives by a simple and ecofriendly carbonation process carried out in a liquid-gas system on the crystal growth of the colloidal calcite phase. Especially, the process parameters of $Ca(OH)_2$ concentration and CO_2 flow rates were investigated to enhance the sensitivity of the process. The colloidal pcc calcite was obtained at pH 6.0, temperature 25^0C, $Ca(OH)_2$ concentration 0.2M, CO_2 flow rate 80mL/min, and total volume 1L. The utilization of flue gas-derived CO_2 from coal fired power plants could be a form of carbon capture and storage using carbonation reaction with colloidal PCC. Finally, the synthesized colloidal PCC was used as a filer for ONP/OCC/OMG recycling. We demonstrated all these experiemnts at semi pilot plant scale.

Utilization of municipal solid waste incineration (MSWI) landfill real flue gases mixed with KR-slag for CO_2 sequestration-Pilot scale demonstration

Ahn Ji Whan[1*], T. Thriveni[1] and Yujin Jegal[2]

[1] Korea Institute of Geosciences and Mineral Resources, Daejeon, South Korea
Mineral Processing Division, Yuseong gu, 305 350, Daejeon, Korea
[2] Department of Resources Recycling, University of Science and Technology, Daejeon, South Korea
*Corresponding author: ahnjw@kigam.re.kr

Abstract
Municipal solid waste is being produced in large quantities, in keeping with a dramatic population increase and the growth of industrialization. Most of this waste is sent to landfills in South Korea. However, current treatment methods, including incineration and recycling is limited owing to insufficient landfill space and the generation of leachates and soil contamination. This study seeks to determine the degree of influence of several main factors, such as the optimum temperature, solid-to-water ratio and CO_2 concentration, on the process of accelerated carbonation of the landfill flue gases (7% concentration) mixed with KR slag. This paper presents the experimental results obtained from a pilot scale demonstration of landfill real flue gases mixed with KR slag. This demonstration was performed with the capacity of 20 tons/hour.

Introduction

In general, solid waste arises from lot of human activities such as domestic, agricultural, industrial, commercial, waste water treatment, construction, and mining activities etc. If the waste is not properly handled and treated, it will have a negative impact on the hygienic conditions in urban areas and pollute the air with greenhouse gases (GHG) and surface and ground water, as well as the soil and crops. Currently, wide variety of smart technologies available for MSWI management in developed countries. Recycling is the other major alternative process for MSWI landfill issues.

In the case of South Korea as a whole, the waste disposal rate on a per capita basis dropped by 40% from 1995 with the remaining waste treated by incinerators and then landfilled(Fig. 1(a) and (b)).[1,2] The capital city Seoul, with a population of more than 10 million, has built four incinerators for treating MSW with a daily capacity of 2,850 tonnes in total.[3]

In 2004, 94% of greenhouse gas emissions come from energy consumption and industrial processes in Korea. To achieve the 5% reduction that was mandated to Korea in 2004, 30 million tons of CO_2 gas emissions must be reduced, and such reduction will cost domestic industry approximately 4 trillion won every year. With the recent highlights of CO_2 reduction and resource recycling, various policies have been delivered to achieve greenhouse gas reduction and improved competitiveness of the national resource industry. According to the results of the UNFCCC, the attempts of each country to secure certified emission reductions and greenhouse gas reduction technology have attracted a great deal of attention due to the enormous economic profit from CDM business.

Current domestic greenhouse gas emissions constitute 1.8% of global emissions and it is highly probable that Korea will be classified as a greenhouse gas reduction mandatory subject for the 2nd commitment period (2013-2017). Therefore, corresponding preparations are urgently required. Developed countries are nurturing green environment industries based on the traditional green technology of 21st century, and they are promoting open innovations, grafting the recycling industry of waste byproducts of the past with novel business paradigms. Regarding technology in response to climate change, the world has set national agendas that take resource recycling as the basis of not only every industry but the social economic system as a whole. Lead the development of an important future industry with high global potential and provide a resource recycling action plan as a model fused with energy.

The International Energy Agency (IEA) has predicted that approximately 22% of carbon dioxide emitted in 2050 can be reduced with CCS technology. Industries with high energy consumption occupy the majority in Korea. Therefore, it is predicted that approximately 40% or more of carbon dioxide emissions can be reduced with CCS technology. A demonstration plant that will stably conduct a solid-solution treatment on independently condensed CO_2 from landfill gas, with core technology completed from the research mentioned above, and recycle the resource is required.

Based on the waste composition of 2007, municipal waste, business and facility wastes, and construction wastes were 15.0%, 34.0%, and 51%, respectively. Construction waste accounted for the largest proportion. The trend analysis of municipal solid waste generations revealed that municipal solid waste incineration bottom ash, construction waste, byproducts from the steel industry, and other inorganic wastes constituted 70% of total municipal solid waste generation. According to the Ministry of Environment, most inorganic wastes are easily recycled but most wastes are nevertheless buried as bank materials. Most municipal solid waste incineration bottom ash, construction waste, fine cement powders, and others are buried. Although a small portion of the wastes is recycled, most is used in low quality purposes, and high value addition is required.

Therefore, the construction of a CO_2-utilizing resource treatment and CO_2 fixation demonstration plant by establishing a treatment technology that recycles inorganic wastes through the fixation of CO_2 and CO_2 carbonation for CO_2 greenhouse gas reduction and recycling of inorganic wastes is required. The present objective of our study is to characterize the municipal solid waste incineration bottom ash and utilization of MSWI landfill real flue gases mixed with KR-slag for CO_2 sequestration at pilot scale demonstration.

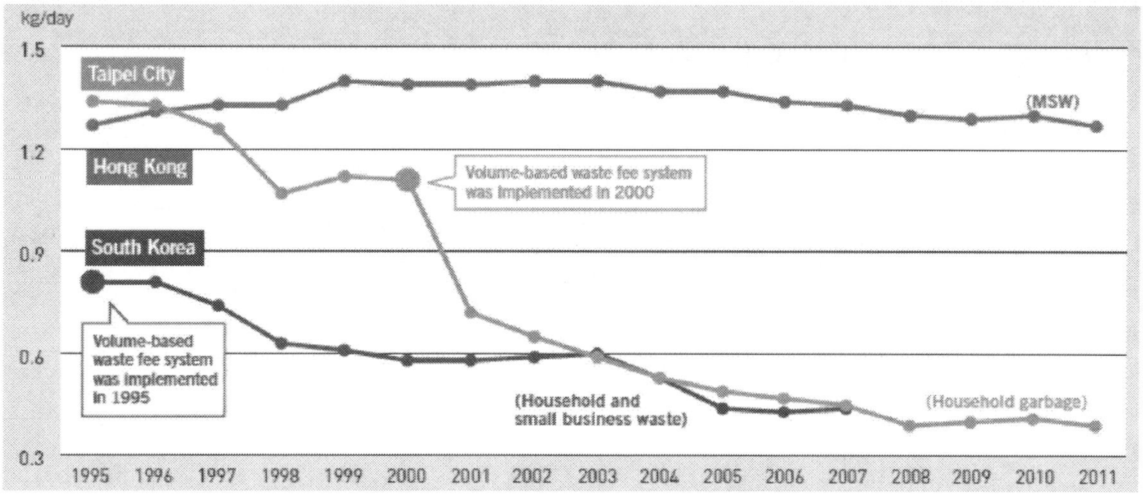

Figure 1. (a). Asia MSWI production

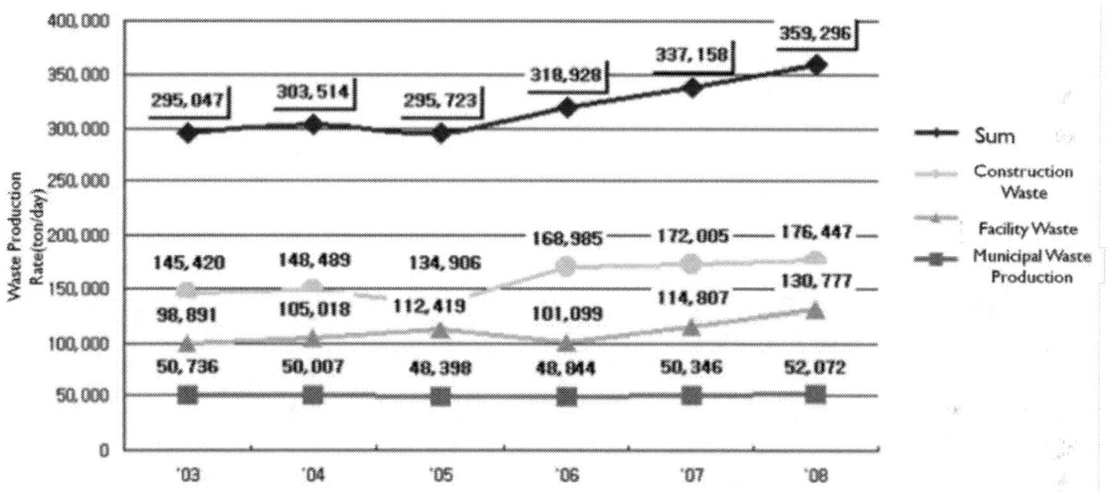

Figure 1. (b). MSWI production in Asia

Waste Treatment Process

The South Korean Government has been promoting the 3R (Reduce, Reuse and Recycle) policy in recent years. Major policy initiatives implemented on the part of waste recycling include[4]:

(a) Waste Charge System (applicable to products not easy to recycle or containing hazardous materials); and

(b) Extended Producer Responsibility System. Disposal of municipal wastes (i.e. wastes other than industrial wastes) in South Korea is by way of landfilling, recycling, composting and incineration. The waste treatment process is illustrated in (Fig.2).

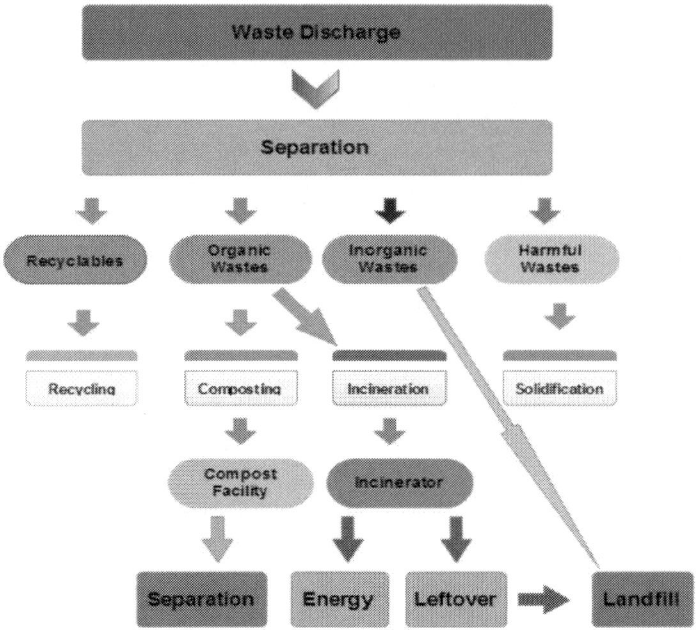

Figure 2. Waste treatment process in South Korea

MSWI bottom ash (Practical Approach, Pilot plant 20 tons/hour)

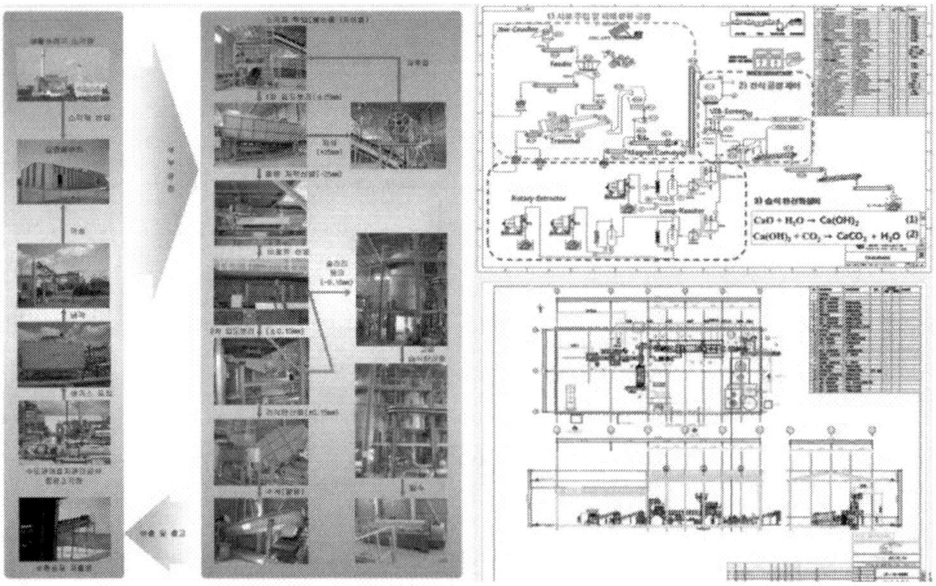

Figure 3. Waste treatment process in South Korea

i) By-product that is produced during the combustion of municipal solid waste in solid waste combustor facilities (Fig.4 (a) bottom ash and KR Slag (Fig.4 (b)). (Table-1).
ii) Category : Bottom ash(Grate ash, siftings), fly ash(boiler ash, precipitator or bag house ash)
iii) Bottom ash
- Approximately 75 to 80 percent of the total combined ash stream

- Similar in appearance to a grayish, silty sand with gravel
- Consists primarily of glass, ceramics, ferrous and nonferrous metals, and minerals.

Figure 4(a). Municipal Solid Waste Incineration Ash

Figure 4(b). KR slag

MSWI bottom ash waste treatment process

State-of-the-art plants produce approximately 200 kg great or bottom ashes per 1000 kg of burnt waste. In semi-dry or dry systems the amount of residues is significantly increased because of not reacted additives. Figure 5 shows the flow diagram of bottom ash treatment process.

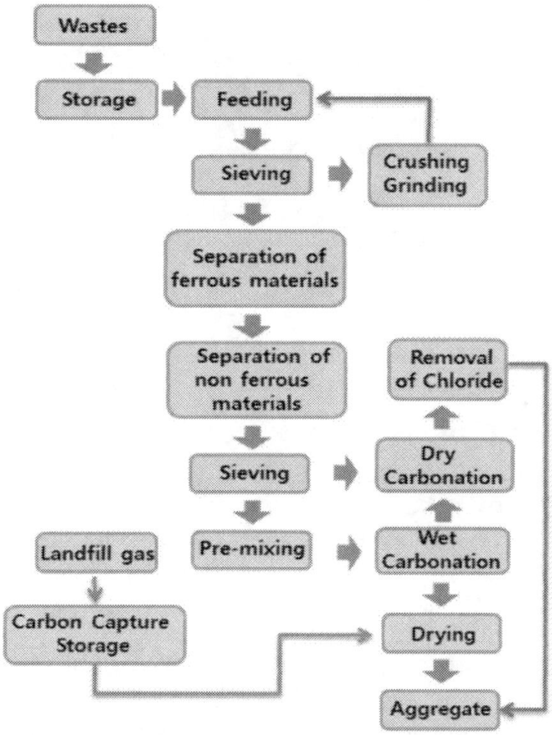

Figure 5. MSWI bottom ash treatment process flow sheet in KIGAM

Pretreatment Process

In general, the mixture of MSWI bottom ash was stabilized as the following pretreatment processes: water washing, phosphate stabilization and carbonation.[5] After the optimum processing condition for each pretreatment was determined, the performance of each pretreatment was examined using pH stat leaching tests (pH 6,9, 12) and availability tests (Figure 6). For the performance tests, fly ash alone was used, instead of a mixture of MSWI residues, to observe leaching differences among the pretreatments more distinctly.

In addition, these pretreatment techniques were usually used for the MSW of high calorific value. However, the relatively high water content of waste (74%), a common characteristic of the refuse produced in Asian countries, will lead to low calorific values.[6]

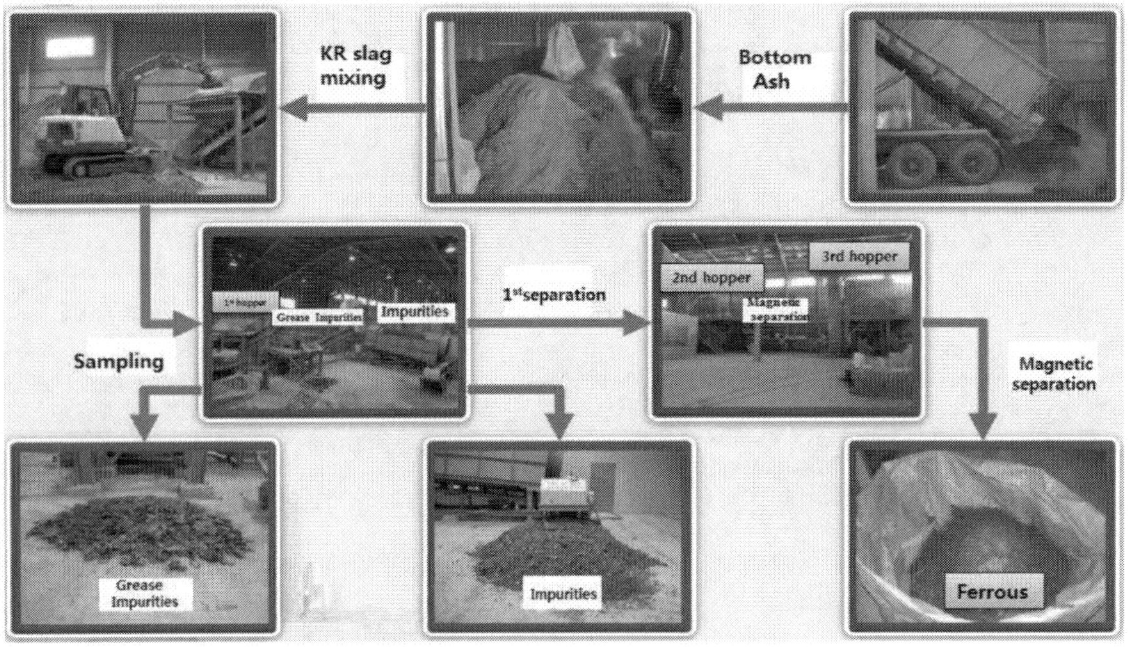

Figure 6. Pretreatment process of MSWI bottom ash

Accelerated carbonation of MSWI bottom ash

Accelerated carbonation has been developed and it has been commercialized· CO_2 is artificially injected into the target material, the waste, to accelerate the carbonation reaction and the treatment is not limited by space. Therefore, time and space are saved compared to natural carbonation[7-11]. In addition, the main culprit of the greenhouse effect, CO_2, is produced in many areas. Using CO_2 to accelerated carbonation is a significantly effective method from the perspective of CCS.

Major factors in accelerated carbonation include temperature, water content, CO_2 density, and pressure, and these factors are closely related to reaction rate. The reaction rate increases with increasing temperature. However, high temperature has a negative effect on high CO_2 content. The actual CO_2 employment obtained from the process demonstration in pilot plant scale using the municipal waste was 26.0kg-CO_2/ton-waste, and 35.0kg-CO_2/ton-waste less than 4.75mm. Considering the domestic municipal solid waste incineration bottom ash production, CO_2 emission reduction of 16,500ton/year is expected.

The highest carbonation rate was observed in 4.75mm or smaller bottom ash at 20°C, solid-liquid ratio of 0.3, and 30% of CO_2, proving the efficiency of the utilization of incinerator exhaust gas. The phase-boundary control reaction was dominant in the initial rate-determining step of dry carbonation. However, it was later revealed that the product layer diffusion control reaction becomes the rate-determining step as the reaction proceeds. The result also showed effective reduction of Pb and Cu, which exceed the elution standard, among the heavy metals of domestic municipal solid waste incineration bottom ash. The heavy metal stabilization of municipal solid waste incineration bottom ash seems to be comprehensively influenced by the generation of stable compounds, such as carbonate, due to the decrease in pH, physical blocking with capsulation, and the generation of absorbents from the decomposition of hydrates.

MSWI bottom ash landfill flue gases with KR slag

After carbonation, the employment of CO 31.38 wt% of KR slag was obtained. Treatment of a ton of KR slag is predicted to employ approximately 457.3kg of carbon dioxide. Construct and operate a recycle treatment demonstration plant capable of treating 50,000ton/year of inorganic wastes, such as municipal solid waste incineration bottom ash, and capture 3,000 ton/year of CO_2(Figure 7). This research aims to develop and demonstrate an environmentally friendly waste treatment process by integrating CCS and resource recycling technology. The technology fixes greenhouse gas (CO_2) in various inorganic wastes, such as incineration ash and others, which are traditionally buried, to manufacture safe cyclic aggregates by preventing the elution of harmful heavy metals. The current demonstration plant mainly treats municipal solid waste incineration bottom ash and KR slag produced during the steel-making process.

The current demonstration plant mainly treats municipal solid waste incineration bottom ash and KR slag produced during the steel-making process. The imported wastes go through impurities removal, primary size sorting, iron separation, and other pretreatment processes. The wastes are finally categorized into cyclic aggregates for manufacturing processes and CCS processes through secondary size sorting. The fine powder wastes under a 100mesh (0.15mm) are made into a slurry of an appropriate solid-liquid ratio in a mix-tank, and combined with exhaust gas (CO_2) in a loop-reactor to collect CO_2 in large amounts, and then discharged after the dehydration process.

Particles larger than a 100mesh (0.15mm) go through dry carbonation, dechlorination, and other stabilization processes that generate carbonate layers on the particle surfaces in an extractor. The particOles are then recycled and stored as cyclic aggregates. Such pretreatment, cyclic aggregates production, CCS, and other processes of the demonstration plant are composed as an automated central system controlled by a small number of people. The system operation was stabilized and the plant processed 15 ~ 20 tons of waste every hour. In CO_2 capturing, bottom ash powders, KR slag, and combination of the bottom ash and KR slag in a 10:1 ratio yielded stable carbonation rates of approximately 5%, 31% and 19% respectively in CO_2 capture.

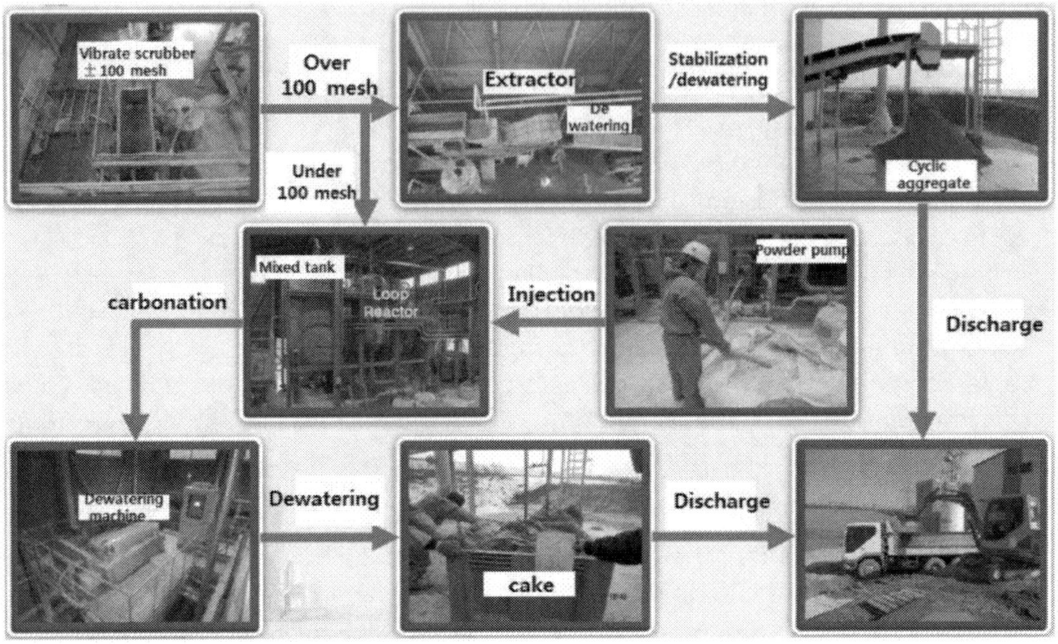

Figure 7. Carbonation process of MSWI bottom ash

Table 1. XRF analysis of MSWI bottom ash samples

	Al_2O_3	CaO	Fe_2O_3	K_2O	MgO	MnO	Na_2O	P_2O_5	SiO_2	SO_3	L.O.I
Ash (Wt%)	15.66	39.36	0.88	0.20	1.72	0.14	0.68	1.86	6.28	1.19	29.8

Effect of CO_2 pressure before and after carbonation

CO_2 pressure change ($\Delta P-\Delta P_{blank}$) in the case of using CO_2 in 10 and 20 bar ranged from 0.296 to 0.300 bar. Interestingly, the pressure drop was not shown in the case of using CO_2 in 30 and 40 bar. In the case of using 10 and 20 bar of initial CO_2 pressure, CO_2 pressure drop ($\Delta P-\Delta P_{blank}$) was 0.296 and 0.300 bars. However, in the case of using 30 and 40 bar of CO_2, the pressure rather increased after the carbonation, indicating no carbonation effect in those conditions.

Sequestrated CO_2 as mol was calculated by the ideal gas equation, and then the amount of sequestrated CO_2 as g/kg was max. 25.85 g/kg. Carbonation efficiency (CE) was also calculated from the ratio of real and theoretical CO_2 sequestration amount, and then the value was max. 7.88% (Figure.8).

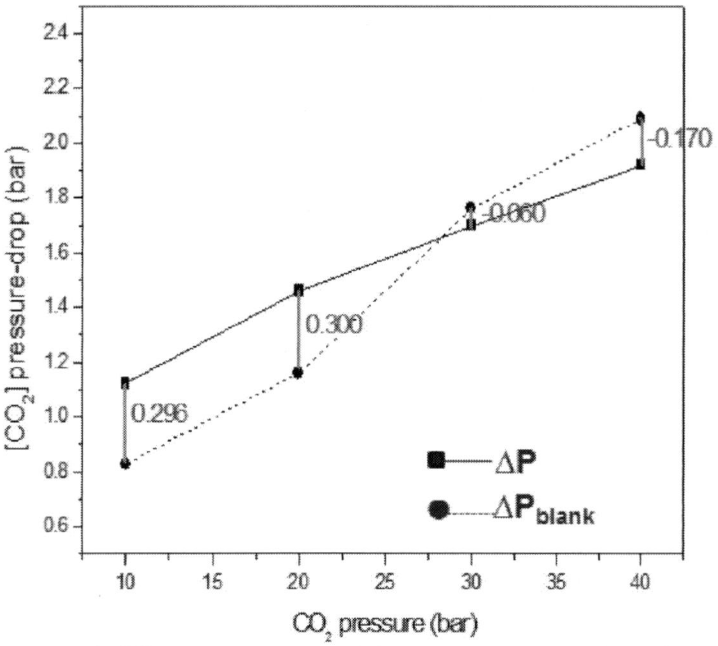

Figure 8. Effect of CO_2 pressure before and after carbonation

Conclusions and Outlook

The accelerated carbonation of solid wastes containing alkaline minerals such as Ca and Mg before their landfill treatment is effective for decreasing the mobility of heavy metals by adjusting pH to below 9.5 at which their solubility is lowest. In the worldwide trend towards encouraging carbonation treatment of MSWI bottom ash as described above, our research results shows the carbonation studies using MSWI bottom ash and CO_2 in order to increase the recycling percentage of the ash and reduce the concentration of atmospheric CO_2 in Korea. We established a pilot plant successfully for multi-processing of MSWI ash in South Korea. The capacity of this pilot plant is 200 kg/hour. This pilot plant was the first one in the waste recycling field in Korea and even throughout the Asian countries, and includes multi-processes of particle separation, heavy metal stabilization, chloride removal, and green aggregate/concrete manufacture. The higher capacity 20 tonns / hour plant (Fig.3), for all inorganic waste treatment by accelerated carbonation and utilized direct flue gases from landfill sites, mixed with KR slag for CCS model in South Korea.

Acknowledgements

This research was conducted in 2014 with support from the Energy Technology Development Project [20141010101840] of the Korea Institute of Energy Technology Evaluation and Planning, financed by the Ministry of Trade, Industry and Energy.

References

1. The Republic of Korea Year Book 2011. Ministry of Environment of South Korea.
2. Kwang-yim Kim., 2008. Performance of Waste Management Policy in Korea-Volume-based Waste Fee System and Packaging Waste EPR,. Korea Environment Institute.

3. Korea Zero Waste Movement Network (KZWMN) http://www.waste21.or.kr/html/eng.asp.
4. Korea Development Institute, 2012.
5. Sang Yul Kim., Toshihiko Matsuto., Nobutoshi Tanaka., 2003. *Evaluation of pretreatment methods for landfilling municipal solid waste incinerator residues*, Thesis, 1-24.
6. Wen-Bing Li., Jun Yao., Zaffar Malik., Gen-Di Zhou., Ming Dong., Dong-Sheng Shen., 2014. *Impact of MSWI bottom ash codisposed with MSW on landfill stabilizawith different operational modes,* Biomed Research International, 2014, 1-10.
7. Nam Il Um, Seong Young Nam., Ji Whan Ahn., 2013. *Effect of accelerated carbonation on the leaching behaviour of Cr in municipal solid waste incinerator bottom ash and the carbonation kinetics.* Materials Transactions, 54(8), 2013, p.1510-1516.
8. Eva Rendek., Gaelle Ducom., Patrick Germain., 2006. *Carbon dioxide sequestration in municipal solid waste incinerator (MSWI) bottom ash.* Journal of Hazardous Materials, B128, 2006, p.73-79.
9. Rafael, M.S., Gilles, M., Muhammad, S., Ozlem, C., Tom Van Gerven., 2013. *Comparative study of ageing, heat treatment and accelerated carbonation for stabilization of municipal solid waste incineration bottom ash in view of reducing regulated heavy metal/metalloid leaching.* Journal of Environmental Management, 128, 807-821.
10. Jong, S. C., Soon, M.K., Hee, Dong, C., Gun, W. H., Chang, H. L., 2011. *Carbon dioxide capture with accelerated carbonation of industrial combustion waste*, International Journal of Chemical Engineering and Applications, 2(1), 60-65.
11. Lovat, E., 2013. Municipal solid waste incineration bottom ash: from production to stabilization, Thesis, 1-130.

NATIONAL ENERGY TECHNOLOGY LABORATORY

NETL

Ab initio Thermodynamic Approach to Identify Solid Sorbents for CO_2 Capture Technology

Yuhua Duan

DOE-National Energy Technology Lab.

Pittsburgh, PA 15236

Yuhua.duan@netl.doe.gov

ACEME 2015

(2:00-2:15 pm, Jun. 22, 2015)

(Columbia University, Davis Auditorium)

Collaborators

- Drs. Dan C. Sorescu, David Luebke, H. W. Pennline, Bryan Morreale, Jonathan Lekse (NETL)
- Drs. Xianfeng Wang, Bingyun Li (WVU)
- Drs. Keling Zhang, Xiaohong S. Li, David King (PNNL)
- Drs. Jinling Chi, Lifeng Zhao, Yunhan Xiao (CAS)
- Drs. Heriberto Pfeiffer, Brenda Alcantar-Vazquez (NAUM)
- Drs. Bo Zhang, J. Karl Johnson (Pitt)
- Prof. K. Parlinski (IFJ, Poland)
- Prof. J. Woods Halley (UMN)

NATIONAL ENERGY TECHNOLOGY LABORATORY

Motivation

- CO_2 emission causes Global Warming.

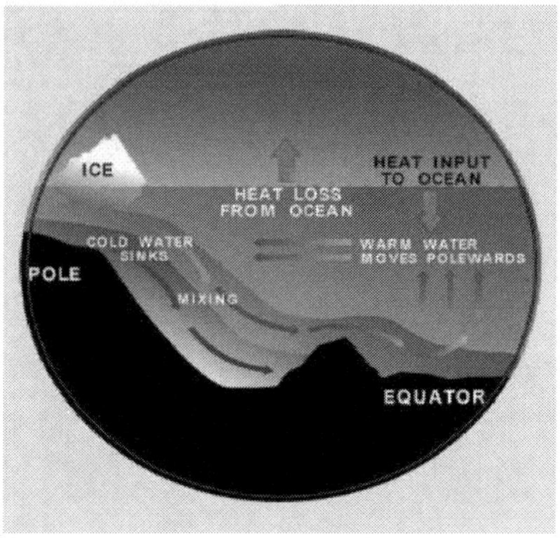

- To solve such environmental issue, instead of emitting into atmosphere, the CO_2 must be captured and sequestrated.

Carbon capture & geological storage

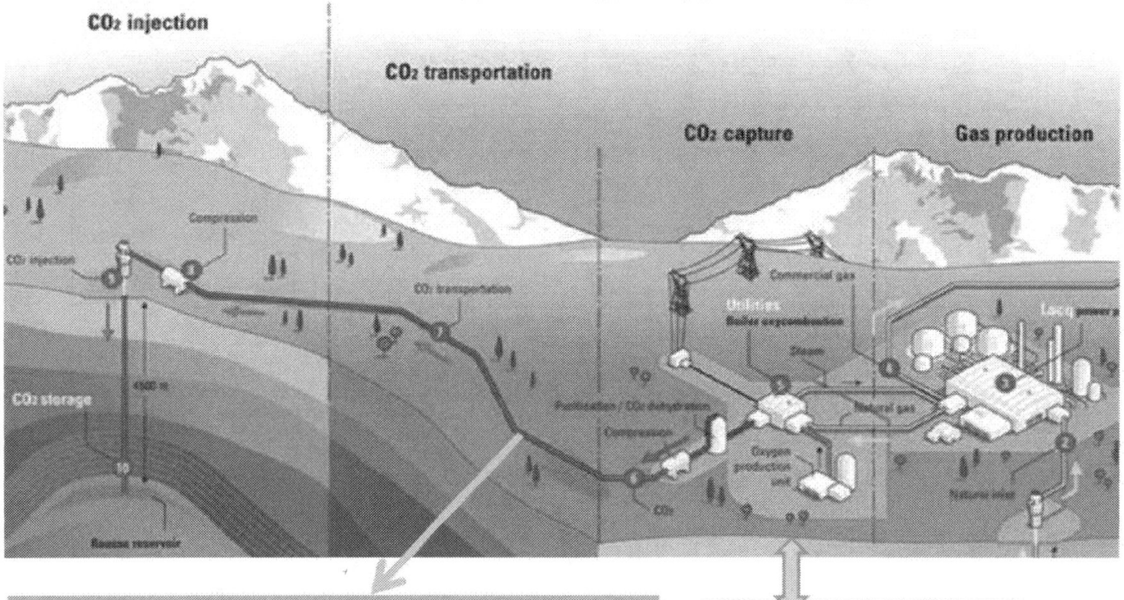

Utilization: (only can use very small portion <5%)
- Enhanced oil recovery;
- Chemical resource, *etc.*

Capture technologies are developed.

Current CO_2 Capture Technology

CO_2 Capture Technology for Power Plants						
Process	**Schematic**	**Main separating Gas**	**Desired working T (°C)**	**P_{CO2} (atm)**	**Possible technology**	**Capture goal**
Post-combustion	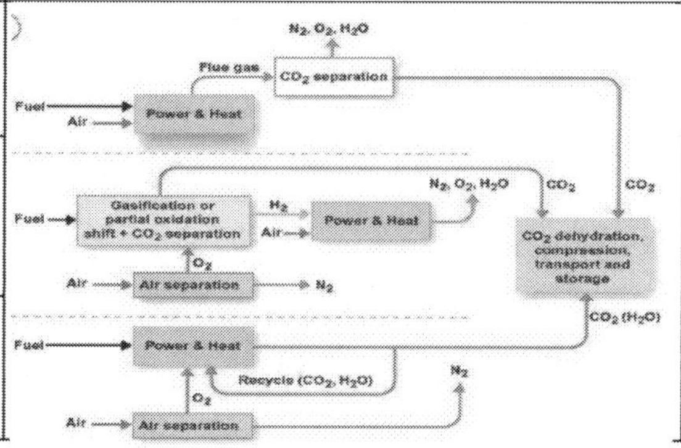	CO_2/N_2	<100 or >600	0.1~0.2	**Solvent:** MEA Solid: CaO **Membrane**	**Capture** >90% **Cost** <35%
Pre-combustion		CO_2/H_2	40~260 (post WGS) 260~500 (concurrent WGS-separation)	10-20	**Solvent:** Selexsol Solid: salts, oxides **Membrane** **Chemical-looping**	**Capture** >90% **Cost:** <10%
Oxyfuel		O_2/N_2		0.2 (P_{O2})	**Cryogenics** **OTM** **IGCC**	

- Current CO_2 sorbents have big limitations due to large energy usage → high operating costs. Need to identify CO_2 sorbents with optimal energy usage. Theoretical simulations are powerful tools for selecting good candidates of CO_2 sorbents.

- J. D. Figueroa *et al*, **Int. J. Greenhous Contr. 2**(2008)9-20
- **Applied Energy**, 102(2013)1439-47

Objectives

- To capture CO_2, we need materials with optimal performance and low costs;
- We are establishing ***a theoretical procedure to*** identify most potential candidates of CO_2 solid sorbents from a large solid material databank;
- Computational synthesis new materials to fit industrial needs;
- To explore the optimal working conditions for the promised CO_2 solid sorbents, especially from room to warm T ranges with optimal energy usage.

Our Modeling Approach (1)

▪For reaction, Solid_A+CO_2+[H_2O]↔Solid_B+[solid_C]+[H_2O], where [...] is optional, the chemical potential is:

$$\Delta\mu(T,P) = \Delta\mu^0(T) - RT\ln\frac{P_{CO_2}}{P_{H_2O}^{\pm 1}}$$

If no H_2O involved, P_{H2O} term vanished

●Search literature and known database. If the thermodynamic properties of all solids involved are known,

$$\Delta\mu^0(T) \approx \Delta G^{solid}_{product}(T) - \Delta G^{solid}_{reactant}(T) - G_{CO_2}(T) \pm G_{H_2O}(T)$$

●Reaction Heat:

$$\Delta H(T) = \Delta H^{solid}_{product}(T) - \Delta H^{solid}_{reactant}(T) - \Delta H_{CO_2}(T) \pm \Delta H_{H_2O}(T)$$

- ***Proc. of 7th Ann. Conf. on Carbon Capture & Sequestration***, 2008
- ***Phys. Rev. B* 77**(2008)045332, **84**(2011)104113; **79**(2009)014301
- ***J. Chem. Phys.* 133**(2010)074508, **136**(2012)064516; ***Int. J. Clean Coal & Energy*, 1**(2012)1-11

NATIONAL ENERGY TECHNOLOGY LABORATORY

Our Theoretical Approach (2)

▪For reaction, Solid_A+CO_2+[H_2O]↔Solid_B+[solid_C]+[H_2O], where [...] is optional, the chemical potential is:

If the thermodynamic data of solids are not available

$$\Delta\mu(T,P) = \Delta\mu^0(T) - RT\ln\frac{P_{CO_2}}{P_{H_2O}^{\pm 1}}$$

where

$$\Delta\mu^0(T) \approx \Delta E^{DFT} - G_{CO_2}(T) \pm G_{H_2O}(T) + \Delta E_{ZP} + \Delta F^{PH}(T)$$

VASP

Ideal gas Statistics physics

For solids Phonon dynamics

Reaction Heat:

$$\Delta H^{cal}(T) = \Delta\mu^0(T) + T*(-S_{CO_2} \pm S_{H_2O} + \Delta S_{harm})$$

- ***Proc. of 7th Ann. Conf. on Carbon Capture & Sequestration***, 2008
- ***PRB* 79**(2009)014301, ***J. Chem. Phys.* 133**(2010)074508; ***Int. J. Clean Coal & Energy*, 1**(2012)1-11

NATIONAL ENERGY TECHNOLOGY LABORATORY

What Info. Can We Obtain?

A. Thermodynamic Properties of Capture Reaction: $\Delta H(T)$, $\Delta G(T)$, $\Delta S(T)$, etc.

B. Chemical Potential ($\Delta\mu(T, P)$) , T, P_{CO2} Relationship to Identify Suitable Sorbent Materials

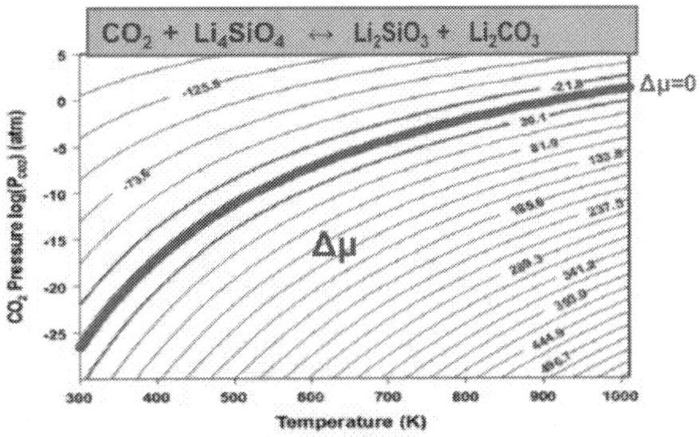

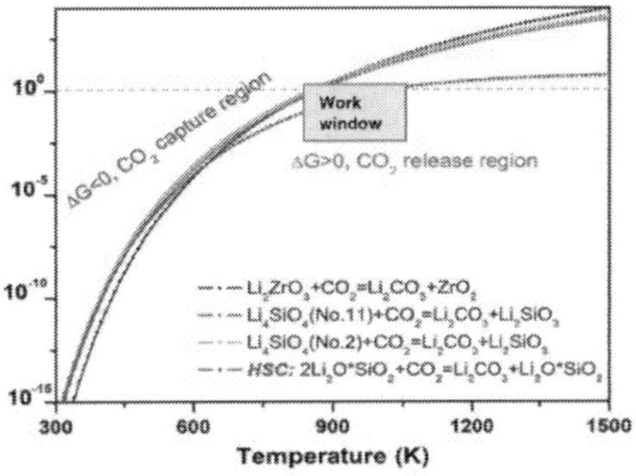

- For each solid, provide its working conditions (T, P_{gas})
- At given work window, select suitable solid materials.

C. Reaction pathways & Energy Cost at Different Capacity

D. Provide basic information for CFD & Process Modeling

New Sorbents by Mixing Solids

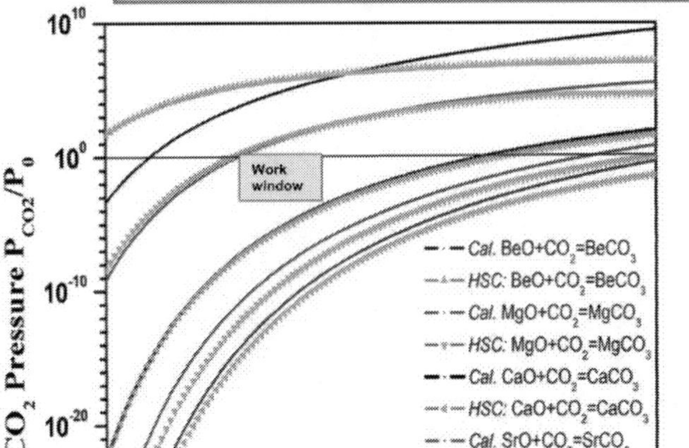

- In many cases, some solids P-T ($\Delta\mu$=0) curve above the desired "WORK WINDOW", while some other solids below the "WORK WINDOW". Is it possible to form a new sorbent by mixed two or more solids?
- In other words, if we only have those solids which behaviors are off the "work window". What can we do?

•Generally speaking, if two solids (A, B) mixed to form a new solid C, the thermodynamic properties of C are located between those of A and B.

- *PRB* **84**(2011)104113, **79**(2009)014301
- *J. Renewable Sustainable Energy,* **3**(2011)013102, **4**(2012)013109
- **Phys. Chem. Chem. Phys. 15**(2013)9752-9760; 13538-13558
- **J. Phys. Chem. C118**(2014)19822-32; **116**(2012)14461-70
- **RSC Advances, 5**(2015)34157-65

Mixing Scheme

- **For a promising sorbents, its CO_2 capture turnover temperature T_t is fixed. In order to use them in different CO_2 capture technologies under specific working conditions, we need to adjust their T_t.**

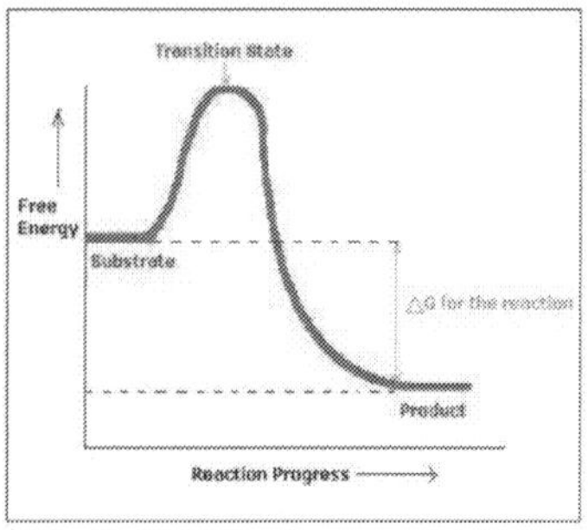

Reactant + CO_2 = Product-CO_2

$$\Delta\mu(T,P) = \Delta\mu^0(T) - RT\ln\frac{P_{CO_2}}{P_{H_2O}^{\pm 1}}$$

$$\Delta\mu^0(T) = \Delta\mu_{product}{}^0(T) - \Delta\mu_{reactant}{}^0(T)$$

Goal on T_t	$\Delta\mu^0(T)$	$\Delta\mu_{product}{}^0(T)$	$\Delta\mu_{reactant}{}^0(T)$
Decrease	**Less negative**	↑, destablize	↓, stablize
Increase	**More negative**	↓, stable, lower energy level	↑, destablize, increase energy level

- We proposed a mixing scheme to shift T_t.

NATIONAL ENERGY TECHNOLOGY LABORATORY

Mixture Sorbents

•Single material may not work at desired conditions.
•By mixing two (A and B) or more solids to form a new sorbent may fit our needs.

❑Assume A is a strong CO_2 sorbent with turnover temperature T_A, B is a week CO_2 sorbent with turnover temperature T_B. Obviously, $T_A>T_B$;
❑For mixing A and B to form sorbent C with turnover temperature T_C, $T_A>T_C>T_B$. We have typically three scenarios:

A. $T_A>T_B$, using A component to capture CO_2. In this way, to reduce T_A to T_C since $T_C<T_A$. For example, Li_2O/SiO_2, Li_2O/ZrO_2

B. $T_A>T_B$, using B component to capture CO_2. In this way, to increase T_B to T_C since $T_C>T_B$. N_2O (N=Na, K, Cs), CaO-promoted MgO

C. Using both A and B components to capture CO_2. In this way, the turnover T_C does not change much as $T_A \approx T_C \approx T_B$, the capacity does not decrease.

- PCCP 15(2013)13538-13558
- Phys. Rev. Applied 3(2015)044013

NATIONAL ENERGY TECHNOLOGY LABORATORY

Mixture A/B Sorbents (I): $T_A > T_B$ using A

A: Li_2O: strong CO_2 adsorbent
B: SiO_2: does not absorb CO_2.

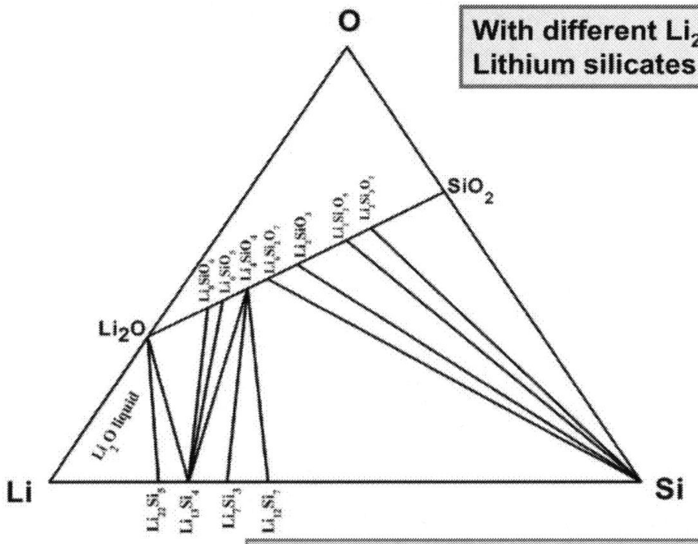

With different Li_2O/SiO_2 ratios, a series of Lithium silicates can be formed.

Silicates	Li_2O/SiO_2 ratio
Li_2O	1:0
Li_8SiO_6	4:1
Li_4SiO_4	2:1
$Li_6Si_2O_7$	3:2
Li_2SiO_3	1:1
$Li_2Si_2O_5$	1:2
$Li_2Si_3O_7$	1:3
α–SiO_2	0:1

***PRB* 84(2011)104113, *79*(2009)014301; *Phys. Chem. Chem. Phys.* *15*(2013)13538-13558**

Calculated thermodynamic properties of the reactions for lithium silicates capture CO_2

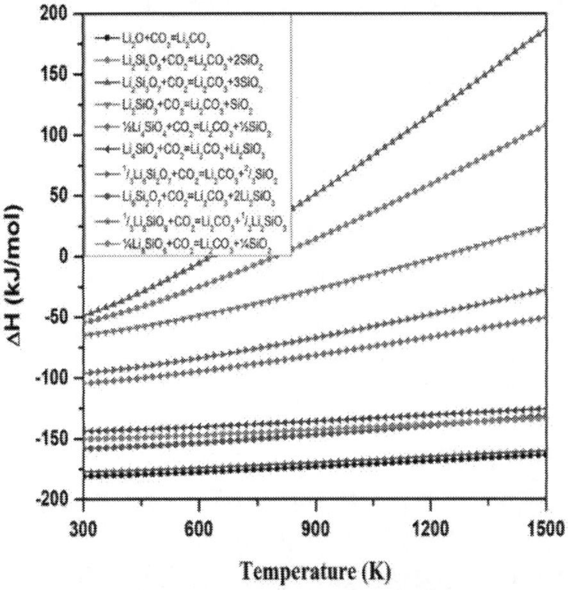

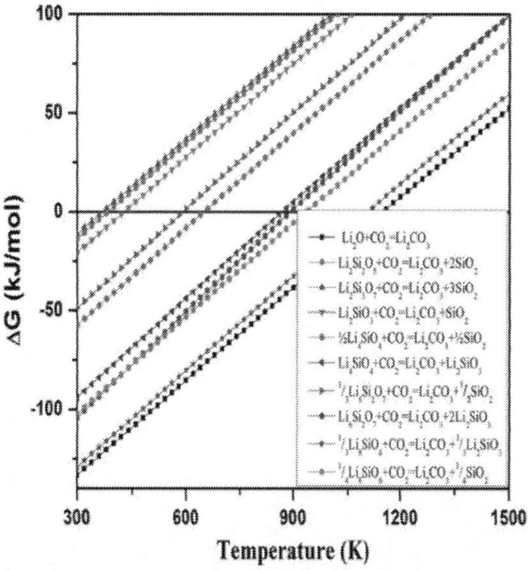

• ***Phys. Rev. B* 84(2011)104113, 79(2009)014301; *PCCP* 15 (2013)9752-9760, 13538-13553**

Relationship of chemical potential, temperature, and CO_2 pressure

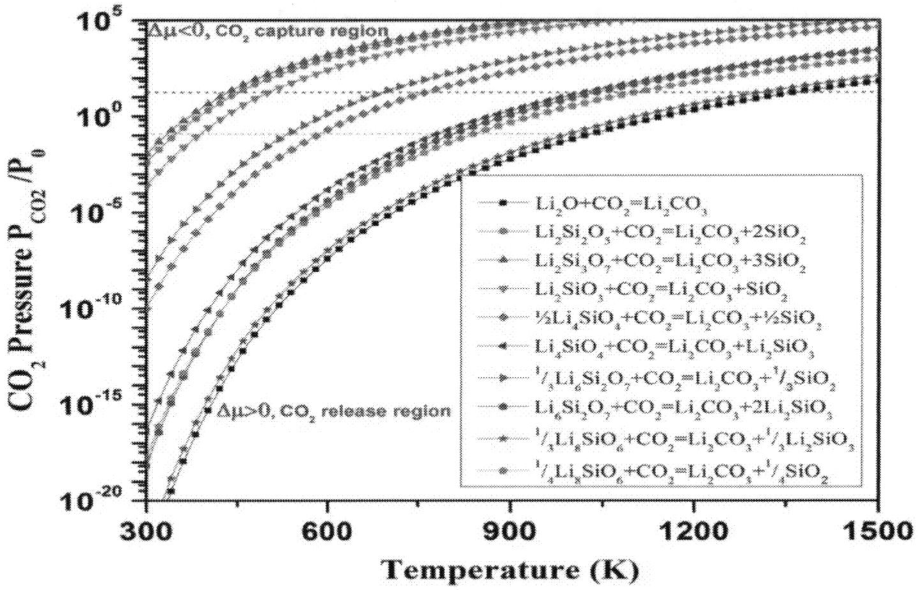

- *Phys. Rev. B* **84**(2011)104113, 79(2009)014301; *PCCP.* **15** (2013)9752-9760, 13538-13553

NATIONAL ENERGY TECHNOLOGY LABORATORY

Li_2O/ZrO_2 Mixed Lithium Zirconates

Zirconates	Li_2O/ZrO_2 ratio
Li_2O	1:0
Li_8ZrO_6	4:1
$Li_6Zr_2O_7$	3:2
Li_2ZrO_3	1:1
ZrO_2	0:1

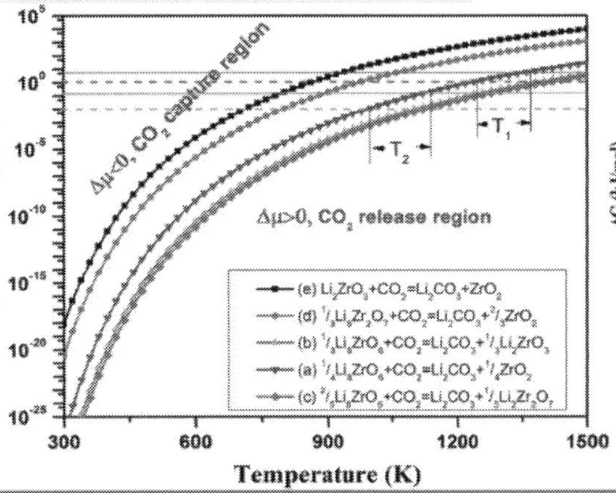

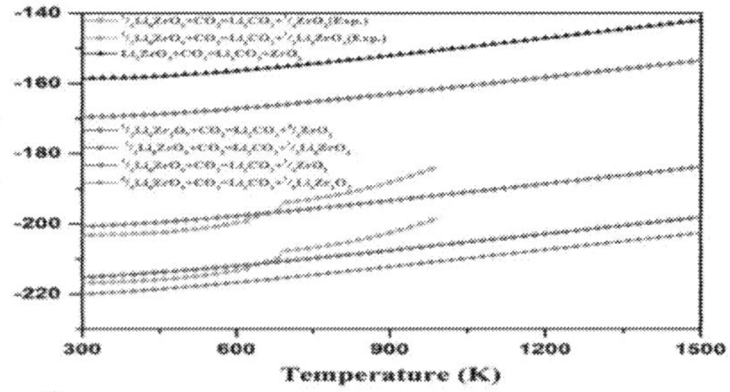

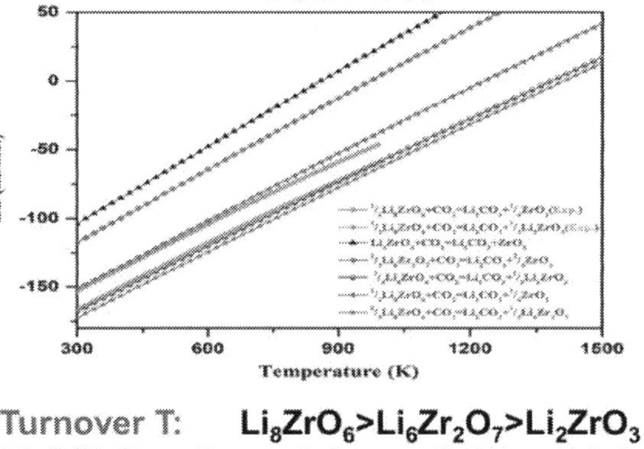

Turnover T: $Li_8ZrO_6>Li_6Zr_2O_7>Li_2ZrO_3$

Li_2O/ZrO_2 ratio: 4:1 3:2 1:1

- *J. Renewable Sustainable Energy,* 3(2011)013102, **4**(2012)013109; *Phys. Chem. Chem. Phys.* **15**(2013)9752-9760

Synthesis of new lithium silicate materials by adjusting the Li_2O/SiO_2 or Li_2O/ZrO_2 ratio

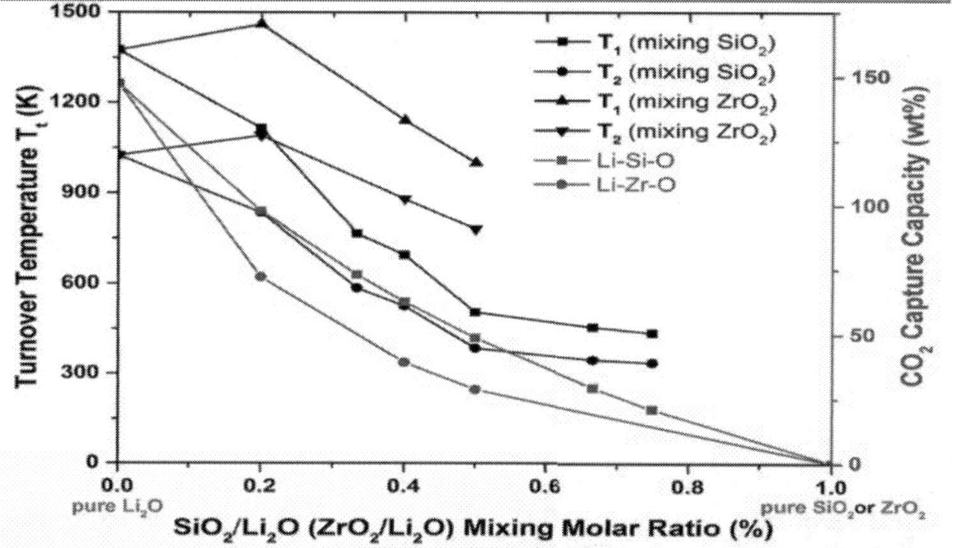

- *Phys. Chem. Chem. Phys.* **15** (2013)9752-9760; 13538-13553
- *J. Renew. Sustain. Energy*, **3**(2011)013102, **4**(2012)013109
- *Phys. Rev. B* **84**(2011)104113; *Phys. Rev. Applied*, **3**(2015)044013

NATIONAL ENERGY TECHNOLOGY LABORATORY

Mixture Sorbents: Case II

$T_A > T_B$ using B as active CO_2 capture component. By mixing to increase turnover T_B

A: Na_2O (or Na_2CO_3)
K_2O (or K_2CO_3)
CaO (or $CaCO_3$)
B: **MgO** active for CO_2 capture

Capture reactions:

$$N_2CO_3 + MgO + CO_2 \xleftrightarrow{T,P} N_2Mg(CO_3)_2 \quad (N = K, Na)$$

$$CaCO_3 + MgO + CO_2 \xleftrightarrow{T,P} CaMg(CO_3)_2$$

NATIONAL ENERGY TECHNOLOGY LABORATORY

Na_2CO_3-promoted MgO sorbent

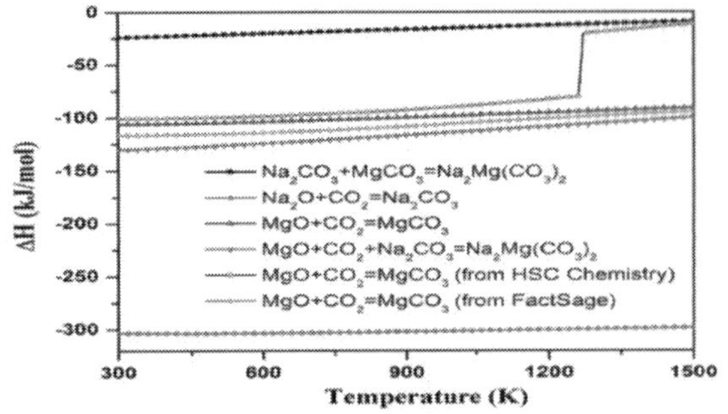

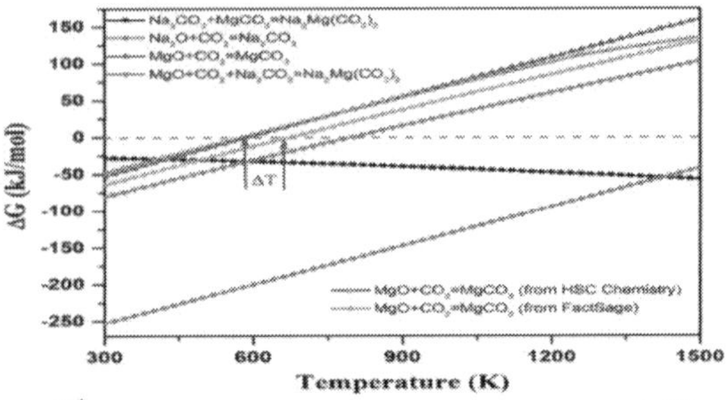

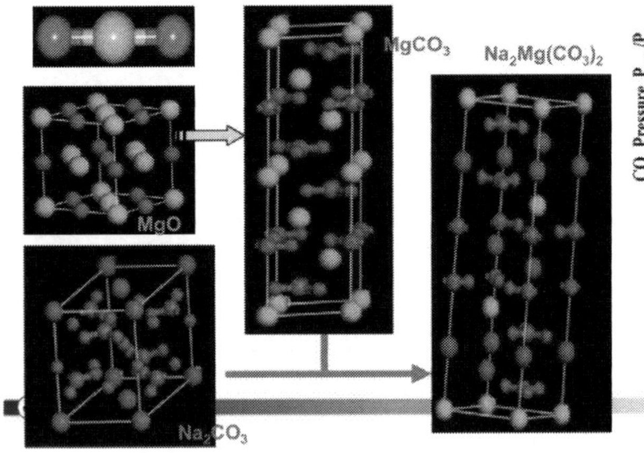

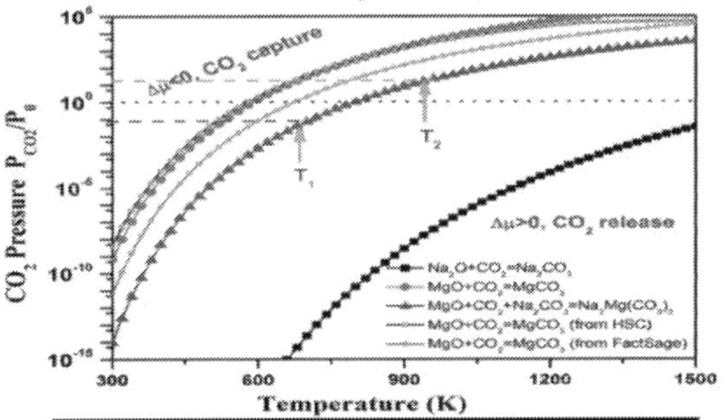

Reaction	ΔH (kJ/mol)
$MgO+CO_2=MgCO_3$	-92.51
$MgCO_3+Na_2CO_3=Na_2Mg(CO_3)_2$	-27.68
$MgO+CO_2+Na_2CO_3=Na_2Mg(CO_3)_2$	-120.19

Na_2CO_3-promoted MgO sorbent

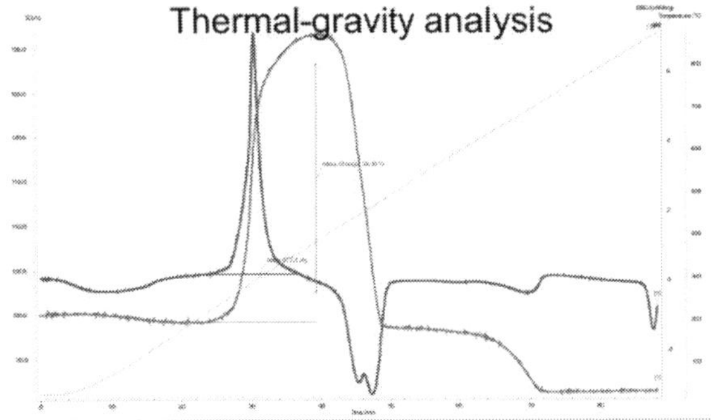

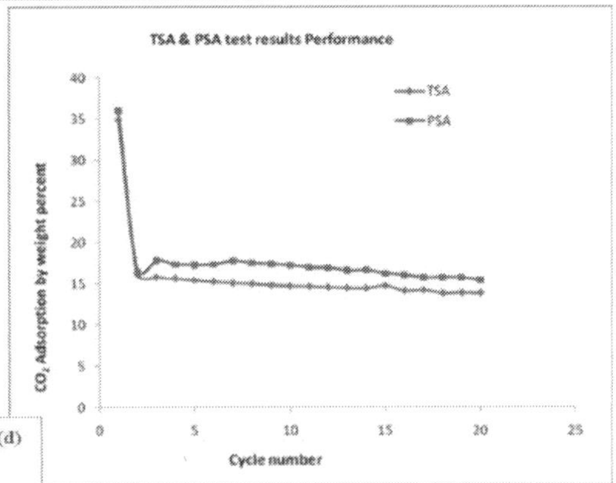

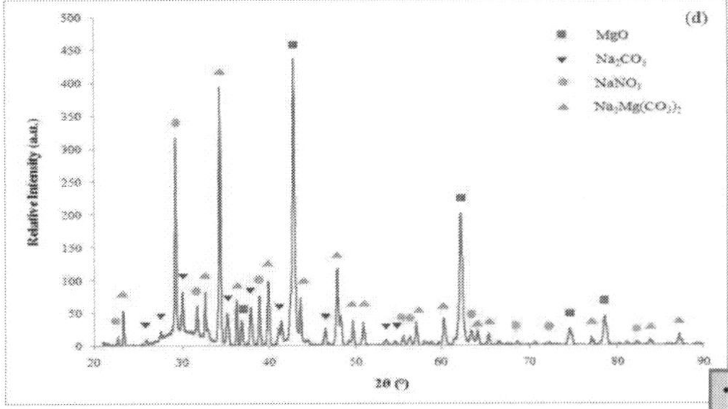

- Adsorption process is through forming $Na_2Mg(CO_3)_2$ double salts;
- CO_2 capture capacity: ~15%(wt)
- Easily regenerated both through TSA & PSA.

- **Int. J. Greenhouse Gas Control 12**(2013)351-358,
- **Adv. Mater. Interfaces 1**(2014)1400030

Turnover T for salt-promoted MgO sorbents

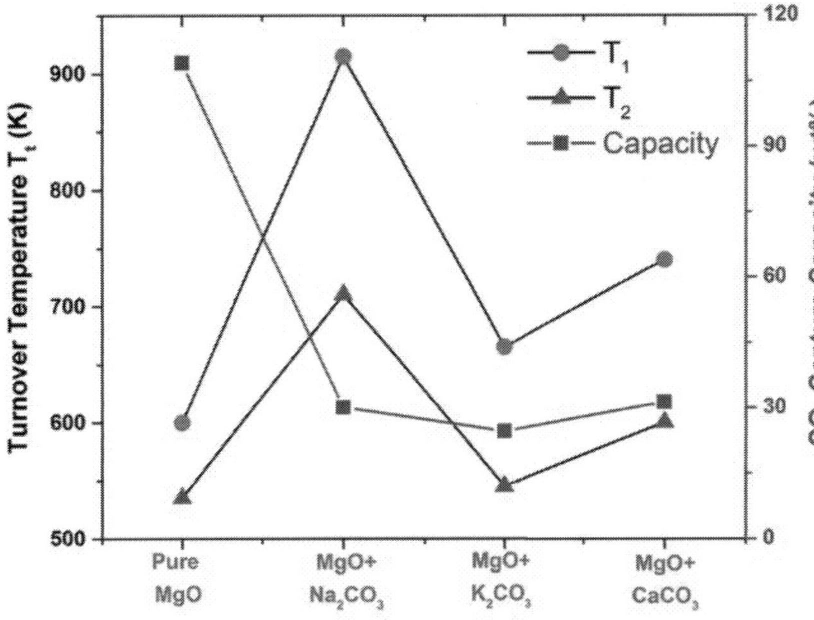

NATIONAL ENERGY TECHNOLOGY LABORATORY

Mixture Sorbents: Case III

Using both A and B as CO_2 capture components

Doped systems: Li_2ZrO_3 doping with M=Na, K to form $Li_{2-x}M_xZrO_3$
Na_2ZrO_3 doping with M=Li, K to form $Na_{2-x}M_xZrO_3$
Li_2MeSiO_4: Me=transitional metal

Such doped system can be treated as mixing of 3 or more oxides:

- $Li_{2-x}M_xZrO_3 \leftrightarrow (2\text{-}x)Li_2O+(x)M_2O+ZrO_2$
- $Na_{2-x}M_xZrO_3 \leftrightarrow (2\text{-}x)Na_2O+(x)M_2O+ZrO_2$
- $Li_{2-x}M_xMeSiO_4 \leftrightarrow (2\text{-}x)Li_2O+(x)M_2O+MeO+SiO_2$

Capture CO_2 reactions:

$$LiMZrO_3 + CO_2 \leftrightarrow \frac{1}{2}Li_2CO_3 + \frac{1}{2}M_2CO_3 + ZrO_2 \quad (M = Na, K)$$

$$Na_{2-\alpha}M_\alpha ZrO_3 + CO_2 \leftrightarrow \frac{2-\alpha}{2}Na_2CO_3 + \frac{\alpha}{2}M_2CO_3 + ZrO_2 \quad (M = Li, K; \alpha = 0, 0.5, 1.0, 1.5, 2.0)$$

$$M_xN_{2-x}MeSiO_4 + CO_2 \leftrightarrow \frac{x}{2}M_2CO_3 + \frac{2-x}{2}N_2CO_3 + MeSiO_3 \quad (M, N = Li, Na, K; Me: \text{transition metal})$$

NATIONAL ENERGY TECHNOLOGY LABORATORY

Mixture Sorbents: Case III

$$LiMZrO_3 + CO_2 \leftrightarrow \frac{1}{2}Li_2CO_3 + \frac{1}{2}M_2CO_3 + ZrO_2 \qquad (M = Na, K)$$

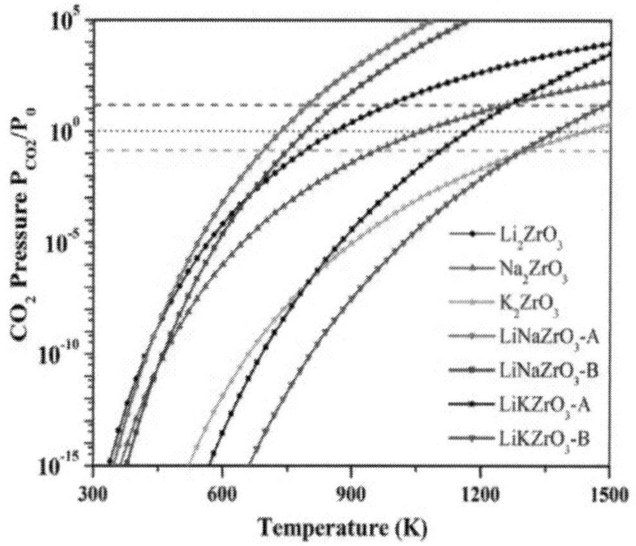

Doped system.

- Such doped system can be treated as mixing 3 oxides:

$LiMZrO_3 \leftarrow\rightarrow Li_2O+M_2O+ZrO_2$

- Using Li_2O & M_2O as CO_2 capture components.
- Here the ratio is $Li_2O:M_2O:ZrO_2=1:1:2$

Using both Li_2O and M_2O as CO_2 capture components

• *ScienceJet* 3(2014)56

NATIONAL ENERGY TECHNOLOGY LABORATORY

$Na_2O+M_2O+ZrO_2$ Mixing sorbents

$$Na_{2-\alpha}M_{\alpha}ZrO_3 + CO_2 \leftrightarrow \frac{2-\alpha}{2}Na_2CO_3 + \frac{\alpha}{2}M_2CO_3 + ZrO_2 \quad (M = Li, K; \alpha = 0, 0.5, 1.0, 1.5, 2.0)$$

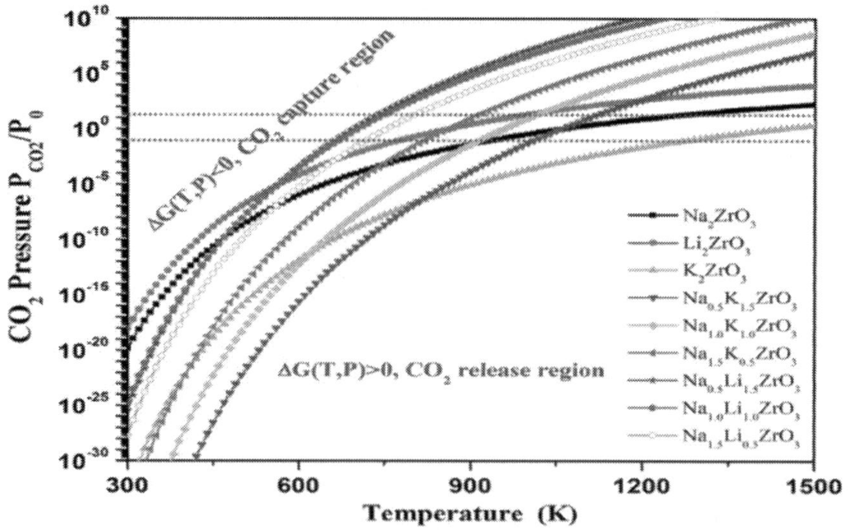

• *Phys. Rev. Applied* 3(2015)044013

NATIONAL ENERGY TECHNOLOGY LABORATORY

Doping level versus Turnover T

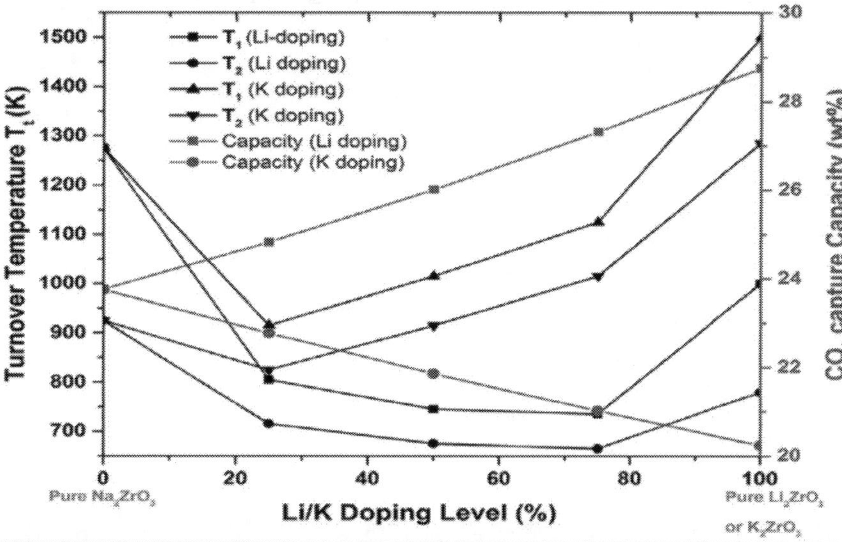

Doping Li into Na_2ZrO_3, T_t decrease with doping level
Doping K into Na_2ZrO_3, T_t first decrease then increase with increasing K-level

• *Phys. Rev. Applied* ***3***(2015)044013

Conclusions

- Our methodology can predict thermodynamic properties of solid materials and their CO_2 capture reactions.
- Single solid may not satisfy the industrial operating conditions as CO_2 sorbent, however, by mixing two or more solids, the new formed solid may satisfy the industrial needs.
- Depending on the needs of specific capture technology, by different mixing schemes, it's possible to move turnover T of newly formed sorbent to the desired operating T range.
- These results provide guidelines to synthesize sorbent materials by mixing different solids with different ratio.

Disclaimer

This report was prepared as an account of work sponsored by an agency of the United States Government. Neither the United States Government nor any agency thereof, nor any of their employees, makes any warranty, express or implied, or assumes any legal liability or responsibility for the accuracy, completeness, or usefulness of any information, apparatus, product, or process disclosed, or represents that its use would not infringe privately owned rights. Reference therein to any specific commercial product, process, or service by trade name, trademark, manufacturer, or otherwise does not necessarily constitute or imply its endorsement, recommendation, or favoring by the United States Government or any agency thereof. The views and opinions of authors expressed therein do not necessarily state or reflect those of the United States Government or any agency thereof.

NATIONAL ENERGY TECHNOLOGY LABORATORY

Thank You!

NATIONAL ENERGY TECHNOLOGY LABORATORY

In-situ X-ray Diffraction Carbonation of Cementitious Silicates

Email: dkopp@scarletmail.rutgers.edu
Presenting Author: Daniel Kopp
Co Authors:
Ryan Anderson, rsa78@scarletmail.rutgers.edu
Kevin Blinn, ksblinn@scarletmail.rutgers.edu
Haohan Wu, haohanwu@gmail.com
Michael Sender, msender98@gmail.com
Herbert T Schaef, todd.schaef@pnnl.gov
Richard Riman, riman@rci.rutgers.edu

Affiliations:
Department of Materials Science and Engineering, Rutgers, The State University of New Jersey, Piscataway, NJ USA
Pacific Northwest National Laboratories, 902 Battelle Boulevard, Richland WA

Abstract
Carbonated mono-calcium silicate or carbonate cement has the potential to bring the cement industry to the next level by simultaneously providing a more stable alternative to traditional Portland cement and by reducing the carbon footprint of cement production. Carbonate cement replaces the key strength-imparting hydrating minerals, di-calcium (Ca_2SiO_4) and tri-calcium (Ca_3SiO_5) silicate, with mono-calcium silicate ($CaSiO_3$), which utilizes carbon dioxide instead of water to chemically bond the $CaSiO_3$ grains. In this work, we analyze the curing kinetics of carbonate cement by monitoring the crystallization of calcium carbonate (the product of $CaSiO_3$ carbonation) via in-situ x-ray diffraction We assess the kinetic differences of carbonation between synthetic and naturally occurring $CaSiO_3$. Semi-quantitative x-ray diffraction techniques are combined with thermal analysis measurements to accurately measure reaction progression.

Introduction

Calcium silicate cement (carbonate cement) has recently gained traction as a viable alternative to portland cement due to its impact on global sustainability[1,2,3,4]. Unlike the constituents of portland cement (Ca_2SiO_4 and Ca_3SiO_5), carbonate cement ($CaSiO_3$) does not react with water during the curing process and instead requires humid CO_2 as a curing agent to form a dense, uniform, and strong structural material[1,2,3,4,5]. Carbon sequestration through cement curing combined with efficient production can heavily offset the CO_2 emissions typically imposed by the hydraulic cement industry. Specifically, CO_2 production can be reduced by 545 kg per tonne by replacing portland cement with carbonate cement technology[3,4].

The binding mechanism relies on the following carbonation reaction:

$$CaSiO_3 + CO_2 \rightarrow CaCO_3 + SiO_2$$

The $CaSiO_3$ grains are chemically bonded together by the expansive growth of $CaCO_3$ and SiO_2 within the porosity of the matrix[1,2,5].

The kinetics of this carbonation reaction are important in understanding how the processing variables could affect the curing rate of the cement. This manuscript focuses on introducing a heated and pressurized in-situ x-ray diffraction (XRD) cell aimed at monitoring the carbonation kinetics of $CaSiO_3$.

Experimental Methods

A heated and pressurized x-ray transparent vessel was constructed in collaboration with PNNL to monitor the kinetics of the $CaSiO_3$ carbonation reaction (Fig 1). The vessel was designed with stainless steel to withstand temperature and pressure specifications of 100°C and 30 psig (CO_2) respectively and to maintain uniform heating on the sample. A film of polyimide was chosen as the x-ray transparent window due to its high temperature and pressure durability, and low replacement cost. An x-ray diffractometer (Bruker D8 Discover, Cu-Kα Radiation: λ=1.514 Å, Horizontal Goniometer, Vantec 1 Detector) with parallel beam x-ray geometry was utilized throughout the study. The scan parameters consisted of consecutive 20-40° 2-theta scans, with a 0.018° step size, and 0.1 second dwell time. An x-ray spot-size reduction slit was used when determining optimum scan parameters.

Sample preparation consisted of mixing the $CaSiO_3$ powder with 20 wt% de-ionized H_2O with a mortar and pestle. The wet powder was then lightly compacted into the plastic sample holder and placed into the cell.

Pressurized CO_2 was bubbled through a heated (80°C) water bath and flowed into the reaction cell to establish a flow-through humid-CO_2 atmosphere. Carbonation rates can be adjusted by changing the temperature, pressure, and material wetting. In this study, the inlet CO_2 pressure remained constant at 15 psig. The outlet valve was slightly opened to establish 10 psig inside the cell. This pressure differential resulted in approximately 450 ml/min of CO_2 flow. The pre-reaction powder wetting content was fixed at 20 wt% H_2O. The reaction chamber temperature was varied between 20, 40, 60, and 80°C to determine its effect on the reaction kinetics.

Qualitative carbonation was observed by the evolution of the $CaCO_3$ peak in the XRD pattern. The degree of carbonation was obtained from post-reaction thermal decomposition of $CaCO_3$ in the TGA-MS (Netzsch STA449F3 (TGA/DSC) coupled with Aëolos 403D Mass Spectrometer) system.

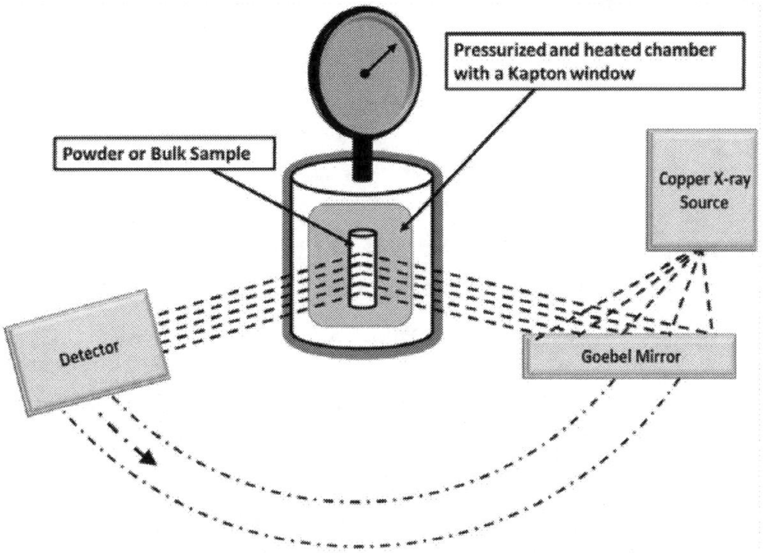

Fig 1. Schematic of the in-situ XRD carbonation vessel

Results

The carbonation of naturally occurring $CaSiO_3$ (Wollastonite 2M, PDF #98-000-0463), was compared to synthetic $CaSiO_3$ (Pseudowollastonite, PDF #97-012-3129). The chosen reference XRD reflections (hkl) for the natural $CaSiO_3$ were located at 28.8° (202) and 29.9° (320), while the synthetic material reference reflection was located at 31.9° (134). Fig 2 shows typical XRD patterns for the carbonation reaction for both synthetic and natural materials. The irregularities in the intensity of the diffraction patterns may be due to H_2O condensation interfering with the incident x-rays. Carbonation of both materials began almost immediately upon exposure to humid CO_2 as noted by the evolution of the $CaCO_3$ peak (Calcite, 29.2°, (104) PDF#97-016-4935). Carbonation occurred rapidly in the synthetic $CaSiO_3$ while the natural $CaSiO_3$ did not exhibit the same obvious kinetics. Longer time trials did not show additional calcite formation in the natural mineral.

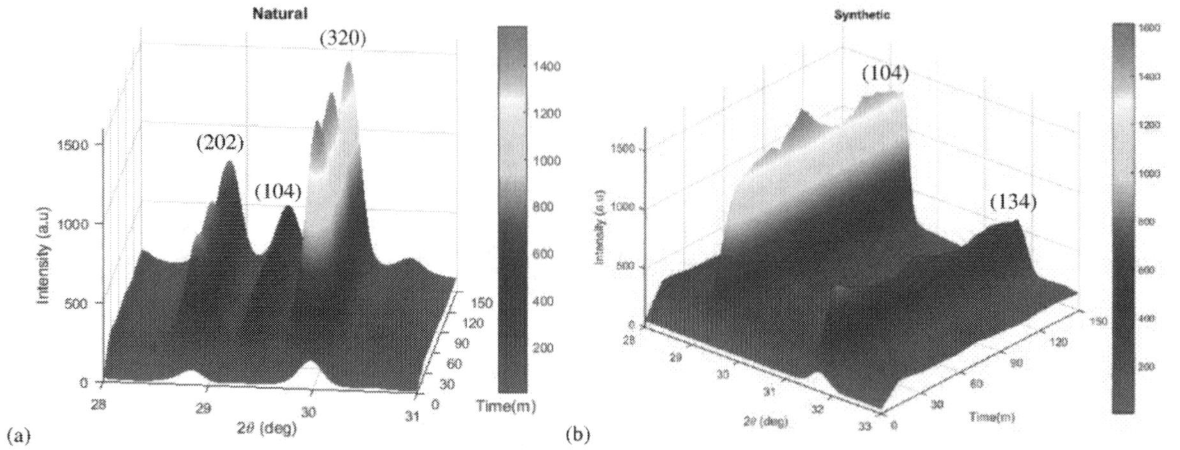

Fig 2. In-situ XRD carbonation at 80°C and 10psi CO_2 for 150 min of the (a) naturally occurring and (b) synthetic $CaSiO_3$

Post-reaction TGA-MS results provided quantitative carbonation by thermally decomposing all of the precipitated and adventitious $CaCO_3$. The TGA weight loss data was converted to carbonation percentage via the following equation:

$$\text{Carbonation } \% = \left(\frac{m_{CO_2}}{m_s} * \frac{1}{0.275}\right) * 100$$

Where, m_{CO2} is the mass loss attributed to CO_2 evolution from $CaCO_3$ decomposition, m_s is the mass of the sample, and 0.275 is the wt% of CO_2 present in the product of a 100% $CaSiO_3$ carbonation reaction ($CaCO_3$ + SiO_2). Table 1 summarizes all of the experiments performed in the study and lists the TG determined carbonation percentage. Raising the temperature directly enhanced the carbonation rate for both systems. It can be seen for the synthetic system that carbonation at 20°C for 570 min produced 39.4%, which is approximately equivalent to 44.6% observed after 75 min at 60°C. The natural mineral did not exhibit extensive carbonation in the in-situ XRD pattern, but TGA showed that substantial carbonation (>20%) did occur at temperatures above 40°C.

Table 1. Quantitative reaction analysis from post-reaction TGA-MS of the newly formed $CaCO_3$. Listed values are the carbonation percentage calculated from the mass-loss associated with decomposition of $CaCO_3$. "-"represents omitted experiments.

Temp (°C)	*% Carbonation*					
	Synthetic (%)			Natural (%)		
	75 min	*150 min*	*570 min*	*75 min*	*150 min*	*660 min*
20	27.3	-	39.4	7.8	9.5	-
40	34.8	42.0	-	14.8	-	-
60	44.6	48.8	-	25.3	-	-
80	45.6	44.9	-	16.9	23.7	26.8

Discussion

Our in-situ XRD monitoring tool can observe carbonation reaction kinetics. The results in Table 1 show a correlation between reaction kinetics and temperature for both systems. The natural mineral however deviated slightly from this correlation and exhibited less carbonation at 80°C than 60°C. This may be due to rapid initial carbonation forming a $CaCO_3$ or SiO_2 passivation layer that inhibits kinetics.

Other polymorphs of calcite (Aragonite and vaterite) may also be crystallizing during carbonation, and therefore higher definition and broader 2θ scans are necessary to fully resolve those phases.

Future in-situ XRD data interpretation will incorporate reaction quantification via Rietveld refinement and crystallite size calculations via the Scherrer Equation.

Conclusion

A heated and pressurized in-situ x-ray diffraction cell was constructed to monitor the curing kinetics of carbonate cement ($CaSiO_3$). Naturally occurring (Wollastonite 2M) and synthetic $CaSiO_3$ (Pseudowollastonite) powders were carbonated in-situ at 20, 40, 60, and 80°C at 10 psi CO_2. The crystalline product of carbonation for both powders was Calcite ($CaCO_3$). A clear trend between reaction temperature and kinetics was evident from post-reaction TGA-MS.

References

1. R.E. Riman, Q. Li, S. Gupta, C. Vakifahmetoglu, V. Atakan, L. Tang, and A. Krugler, Carbonate Ceramics by CO_2 Sequestration (In submission)
2. Riman, R. R. & Atakan, V. Systems and methods for carbon capture and sequestration and compositions derived therefrom. Patent No.: US 8114367 B2 (2012).
3. Solidia Technologies Inc. Solidia Cement™, Part One of a Two-Part Series Exploring the Chemical Properties and Performance Results of Sustainable Solidia Cement™ and Solidia Concrete™. Http://solidiatech.com/white-papers/. Solidia Technologies Inc., 17 Dec. 2013. Web.
4. Solidia Technologies Inc. Solidia Concrete™, Part Two of a Series Exploring the Chemical Properties and Performance Results of Sustainable Solidia Cement™ and Solidia Concrete™. Http://solidiatech.com/white-papers/. Solidia Technologies Inc., 19 Feb. 2014. Web.
5. Riman, R. R. & Atakan, V. Method of hydrothermal liquid phase sintering of ceramic materials and products derived therefrom. Patent No.: US 8,313,802 B2 (2012).

Rationalizing the effect of accelerated carbonation and accelerated ageing on transition zone of cellulose fibers in cement based composites

G.H.D. Tonoli[1], V.D. Pizzol[1], G. Urea[2], S.F. Santos[3], L.M. Mendes[1], V. Santos[2], M. Frias[4], H. Savastano Jr.[2],

[1] *Department of Forest Science, Universidade Federal de Lavras, C.P. 3037, 37200-000 Lavras, MG, Brazil.*

[2] *Department of Bio-systems Engineering, Faculdade de Zootecnia e Engenharia de Alimentos, Universidade de São Paulo (USP), Avenida Duque de Caxias Norte, 225, Pirassununga, SP, 13635-900, Brazil.*

[3] *Universidade Estadual Paulista "Julio de Mesquita Filho"– UNESP, Guaratingueta/SP, Brazil*

[4] *Eduardo Torroja, Institute for Construction Science (CSIC), C/Serrano Galvache, 4, 28033 Madrid, Spain*

**Corresponding author: gustavotonoli@yahoo.com.br; Tel:+55 (35) 3829-5236*

Abstract

The objective of the present work is to show the effect of accelerated carbonation on the transition zone of cellulose fibers in cement based composites. Accelerated carbonation increased the content of calcium carbonate ($CaCO_3$) polymorphs and consumed the calcium hydroxide ($Ca(OH)_2$), and other less stable cement phases. The scanning electron microscopy (SEM) images showed that absence of S- and Al-rich needles around the cellulose fibers in the carbonated composites, and confirmed by the energy dispersive spectrometry (EDS) mapping. The $CaCO_3$ formed from the carbonation reaction is precipitated in the pore structure of the matrix, decreasing diffusion of Si, S an Al during curing and secondary hydration, because act as a binder and refine the pore size distribution. The interface between the cellulose fibers and the cement matrix in the carbonated composites was improved, decreasing the typical transition zone around the cellulose fibers that prejudice the fiber-cement performance at long term. Additionally, the consequence of carbonation at early age is the volume stabilization, as indicated by the lower drying shrinkage. The use of this technology during the fiber-cement production consists of an interesting procedure to prematurely decrease the alkalinity of the cement matrix, which is potentially harmful to those reinforcing fibers that are vulnerable to the alkali attack. It also improves the microstructure of the material and is an initiative to CO_2 sequestration and partial replacement of petroleum-based fibers as is the case of cellulose pulps.

Keywords: A. Fibers; A. Hybrid; B. Fiber/matrix bond; E. cure, F. interface

Introduction

Early age carbonation had been successfully used in partial replacement of steam in concrete-pipe production to improve durability (Rostami et al., 2011) and can result in a number of other benefits in the pre-cast concrete production, including the improvement of the initial strength, dimensional stability, cement savings and less product waste (Rostami et al. 2012; Soroushian et al. 2012; Neves Jr. et al., 2015). Durability is improved because relatively insoluble crystalline calcium carbonate ($CaCO_3$) polymorphs are formed from the soluble $Ca(OH)_2$.

Fiber-cement materials have been widely used in the world due to their versatility for manufacturing construction elements, such as flat boards for walls, corrugated sheets and tubes, present in most of houses in developing regions as well as in agricultural and industrial buildings (Ikai et al., 2010). Cellulose fibers are commonly used as reinforcement of fiber-cement composites, and present low density, renewable character, biodegradability and availability at low cost and in a wide variety of morphologies and aspect ratios. One of the drawbacks in the use of cellulose fibers in cement based composites is their mineralization and degradation of their constituents in the high alkalinity of the Portland cement matrix, with the pH around 13. This low durability of the cellulose fibers in the cementitious matrix causes the reduction of the mechanical performance of the fiber-cement composites due to the fiber enbrittlement and loss of adhesion between fiber and matrix (Tonoli et al. 2010; Tonoli et al., 2013).

According to Soroushian et al. (2012), CO_2 curing of cellulose fiber-cement has shown to enhance productivity and engineering properties (including dimensional stability) of the final product. Furthermore, the consequences of carbonation are the stability of the mechanical behavior (Soroushian et al., 2012), densification of the cementitious matrix and reduction of its permeability (capillary) and porosity (Lesti et al., 2013). In a previous study, Tonoli et al. (2010) and Pizzol et al. (2014a) showed the necessity to anticipate the fast carbonation for initial periods of cure, which would not allow time for the alkali to attack the cellulose fibers during the cement hydration. The morphology and distribution of the hydration products formed during the carbonation curing cannot yet be controlled with the same precision as for metals and other engineered materials, but the better understanding may permit new strategies for predicting the evolution of microstructure and properties, and ultimately engineering microstructure and overall performance. Furthermore, there is still a lack of information about the changes promoted by the carbonation at early age and accelerated ageing cycles on the mineralogical phase composition at the transition zone of cellulose fibers and microstructure of the fiber-cement composites. In this context, the aim of this work was to analyze the changes in the transition zone of cellulose fibers promoted by the use carbonation at early ages and accelerated ageing of fiber-cement composites reinforced.

Material and methods

Sampling and carbonation at early age

The fiber-cement samples with nominal thickness of 4 mm were obtained from the Hatschek process. The Hatschek process is the most employed in the fiber cement industry worldwide, and consists of producing fiber-cement boards by assembling thin layers (up to 1 mm) made of sub-layers (approximately 0.3 each) that are formed in rotary sieves and progressively accumulated in the filtering felt from a suspension of Portland cement, mineral additions (limestone filler for example), cellulose and synthetic fibers and water. Vacuum is applied to remove the excess water from the layer before it is transferred to the formation cylinder (resembling a paper production process) where the layers are stacked. Then, the green fiber-cement sheet is cut, shaped (e.g. corrugating) and cured.

The average formulation is composed of around 70% of Portland cement type II-E (NBR 11578, 1991), 24% of calcium carbonate filler, 4% of eucalyptus cellulose pulp and 2% of polyvinyl alcohol (PVA) fibers. The calcium carbonate filler (limestone) is normally used in the industry for partial substitution of Portland cement in order to reduce costs concerning the production of fiber-cement.

Preconditioned (water-saturated air) samples were submitted to 10 h of accelerated carbonation at around 60°C and 90% relative humidity (RH). The CO_2 content into the climate chamber was monitored to around 15% by volume. Then the carbonated composites were taken out and maintained in water-saturated air at around 25°C until the characterization tests. The conditions were the same for the non-carbonated specimens, however without applying CO_2. The fiber-cement samples were cut into test specimens (160 mm x 30 mm x 4 mm) with a diamond disk cooled with water.

Soak and dry accelerated ageing cycles

Composites were successively immersed into water at 20 ± 5°C for 170 min and, after resting for 10 min, they were heated to a temperature of 70 ± 5°C for 170 min in a ventilated oven. An interval of 10 min (at room temperature) is usual prior to the beginning of the next cycle, as recommended by EN 494 (1994) standard. Each soak and dry set represents one cycle, which was repeated 200 times, i.e., 200 ageing cycles as used in Tonoli et al. (2013). This method simulates natural ageing in severe conditions.

Scanning electron microscopy

A back-scattered electron (BSE) image detector operating at around 15.0 kV and 20.0 kV was applied for viewing cut and polished surfaces. The BSE imaging was used to study the transition zone of the cellulose fibers. The preparation of specimens for BSE was accomplished with vacuum (80 kPa gauge) impregnation using epoxy resin. BSE samples were manually polished with silicon carbide abrasive paper with sequential grit sizes of 320, 600, 1200 and 2000 for 2 min each, using alcohol as lubricant. A final polishing was carried out using in turn 8–4, 4–2 and 1–0 μm diamond polishing compound during 4, 2 and 1 min each size respectively. Polished samples were carbon coated before being analyzed in a FEI Quanta 600 (FEG) microscope. Energy dispersive spectrometry (EDS) mapping analysis was performed in order to localize the carbon (C), calcium (Ca), silicon (Si) and sulfur (S) atoms around the cellulose fibers on the polished surface of cross-section of the specimens. Around twenty SEM micrographs were obtained for each composite treatment and just the typical images of the microstructure observed were used in this manuscript.

Textural study of the surface of fracture of the samples was performed using secondary electrons (SE) detector on a Hitachi S-4800 with tungsten source and detector Bruker X- Flasb 5030, which operates at a voltage of 1-30 kV with a range of between 10 and 80000x magnification, at a work distance from 1 to 70 mm. Around six SEM micrographs were obtained for each composite treatment and just the typical images of the observed microstructure were used in this manuscript.

X-ray diffraction (XRD)

Diffractometer Bruker D8 Advance model consisted of a high voltage generator 3 kW X-ray tube with copper anode (CuKa radiation of 1.54 Å), usually operating at 40 kV and 50 mA, LYNXEYE detector, carrying an anti-scatter slit of 3 mm and a filter of Ni K-beta (0.5%) and no monochromator. The semi quantitative mineralogical composition was determined by the method of reflectant powers. The areas under the reflection for each mineral were determined through Gaussian fitting where the baseline intensity was subtracted by a conventional program. The percentage of error by this method is 10%.

Thermogravimetry analysis (TG)

To check the carbonation extension, thermogravimetric (TG) analysis was recorded with a Netzsch equipment with STA 409 simultaneous analysis system using 50-mg samples and a dynamic nitrogen stream (flow rate = 100 cm^3/min) at a heating rate of 10°C/min.

Drying shrinkage

Drying shrinkage was measured following recommendations of standard procedures (NBR 15498, 2007). It was measured the length variation (mm/m) of the composites in relation to the initial condition (water vapor saturated). The values of length decrease during drying were divided by the initial length of the specimens in order to find the length variation in mm/m. Specimens were 10 days of total curing when they were submitted for this first drying.

Results and discussion

Changes in the hydration products due to early carbonation

The carbonation is the reaction of cement hydration products with carbon dioxide (CO_2). The carbonation is an acid-base neutralization reaction of the alkaline nature of cement. CO_2 first dissolves in the pore solution and then reacts with the OH^-, forming CO_3^{2-} (Eq. 1). Consequently, the pH of the pore solution decreases. As the solution is saturated in calcium ions, due to the C_3S and C_2S hydration, (pH of 12.5, Eq. 2), the precipitation of calcium carbonate ($CaCO_3$) polymorphs starts (Eq. 3), according to the following reactions (Frias et al., 2013):

$$CO_2 + 2OH^- \Leftrightarrow CO_3^{2-} + H_2O \quad \text{(Eq. 1)}$$

$$Ca(OH)_2\downarrow \Leftrightarrow Ca^{2+} + 2OH^- \quad \text{(Eq. 2)}$$

$$CO_2 + Ca(OH)_2\downarrow \Leftrightarrow Ca^{2+} + CO_3^{2-} + H_2O \Leftrightarrow CaCO_3\downarrow + H_2O \quad \text{(Eq. 3)}$$

After 10h of carbonation several changes occur in the matrix structure. The XRD results (Figure 1) show important changes in the cementitious phases formed in the two different curing conditions (control and carbonated). Calcite (Ca) is the main mineralogical phase observed in the control sample (non-carbonated), due to the presence of the calcium carbonate filler in the composition. Calcium hydroxide (portlandite - P) is detected in the control samples, and is formed from the calcium silicates hydration. Also, in lower proportion the remaining cement anhydrous phases, such as alite (C_3S), belite (C_2S), tetracalcium alumino ferrite (C_4AF) and very few traces of tricalcium aluminate (C_3A) are detected both after 10 days of standard cure and after the ageing cycles. Finally it is also observed the formation of ettringite (AFt), monocarboaluminate (Mc) and monosulphate (AFm) due to the presence of gypsum as a setting regulator.

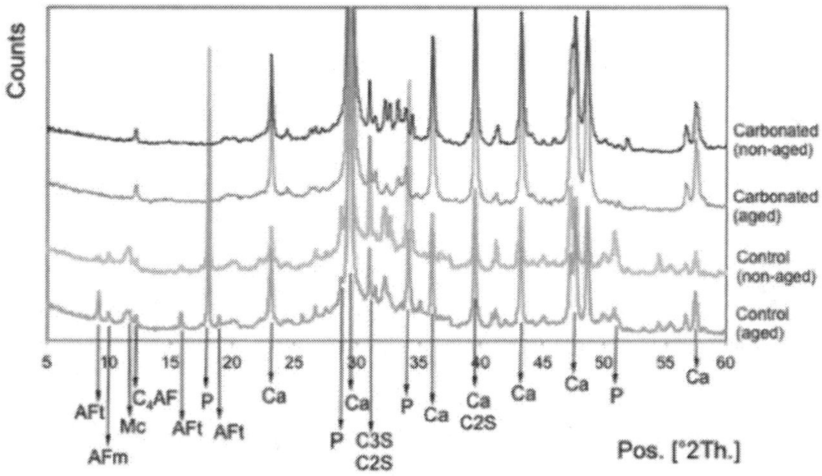

Figure 1. X-ray diffraction patterns of the control (non-carbonated) and carbonated composites before (non-aged) and after 200 soak and dry cycles (aged). Ca = calcite; AFm = monosulfoaluminate; AFt = ettringite; P = portlandite; Mc = monocarboaluminate; C_4AF = tetracalcium alumino ferrite; C_2S = belite and C_3S = alite.

The presence of CO_2 during the curing period (early carbonation) led to disappearing of calcium hydroxide (portlandite - P). In the carbonated composite, it was not detected the presence of this mineralogical phase, which indicates that all calcium hydroxide was converted to calcium carbonate (mostly Ca), whose XRD peaks have increased in intensity. The calcium carbonate polymorphs is still not fully elucidated. Furthermore, the AFt, AFm and Mc phases were not identified as crystalline phases in the carbonated composite, as they were decomposed by CO_2 curing. It is worth to note the higher peak intensity for C_3A, which is an anhydrous phase that was less hydrated in the carbonated composites. The consequence is the lower content of Al-rich hydrated phases in the carbonated composites.

Figures 2 and 3 show the EDS mapping for different chemical elements (identified with different colors) in the cut and polished cross-section of control (non-carbonated) and carbonated composites, respectively. These samples were analyzed after 200 soak and dry ageing cycles.

Figure 2 shows that Si-rich phases are widely distributed in the matrix of control (non-carbonated) composite, except in the fiber transition zone. During curing and secondary hydration during ageing of the control composites, the Si ions firstly migrate by solution diffusion to empty spaces in the bulk matrix, and less to transition zone (Figure 2). Those hydration ions re-precipitate in the pores (bulk or transition zone) when conditions are adequate for nucleation. In the case of the transition zone, the main hydration products reported in literature are etringite - AFt, monosulfoaluminate - AFm and calcium hydroxide. Figure 2 show the S- and Al-rich phases around the cellulose fibers. These hydration products can also penetrate into the fiber cell wall (nucleating there when adequate conditions), and penetrate into fiber cavity (lumen), leading to the re-precipitation of ettringite/monosulphate and calcium hydroxide into the fiber.

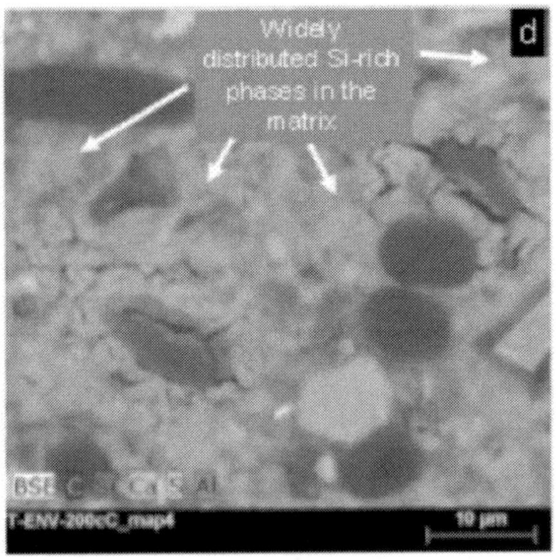

Figure 2. SEM/BSE micrographs of the cut and polished surface control (non-carbonated) fiber-cement composites, usin colored EDS mapping for C, Si, Ca, S and Al. Arrows also show the Si-rich phases widely distributed in the bulk of the fiber-cement matrix.

Figure 3 shows that Si-rich phases are still concentrated in the alite (C_3S) and belite (C_2S) anhydrous grains, with low diffusion in the matrix and fiber transition zone. The carbonation of $Ca(OH)_2$ is accompanied by an increase in the volume of solids [(Johannesson et al., 2001). $CaCO_3$ is precipitated in the pore structure of the composite matrix, also acting as a binder, filling the voids and thereby blocking the free intake of water due to the decrease of pore size (Hyvert et al., 2010). Then, these pores become almost closed instead of the initially predominant capillary pores, and less permeable to silicates, decreasing diffusion of Si and S ions during curing and secondary hydration. Figure 3 show that S is widely distributed in the carbonated matrix, while Al-rich phases are concentrated in the C_3A (tricalcium aluminate) or C_4AF (tetracalcium alumino ferrite) anhydrous grains, and have not been solubilized and re-precipitated around the cellulose fibers. Therefore, less Aft, Afm and Mc phases are formed around the fibers because of lower diffusion of S- and Al-rich phases. These S-rich phases are responsible for peeling off cellulose chains, degrading its structure.

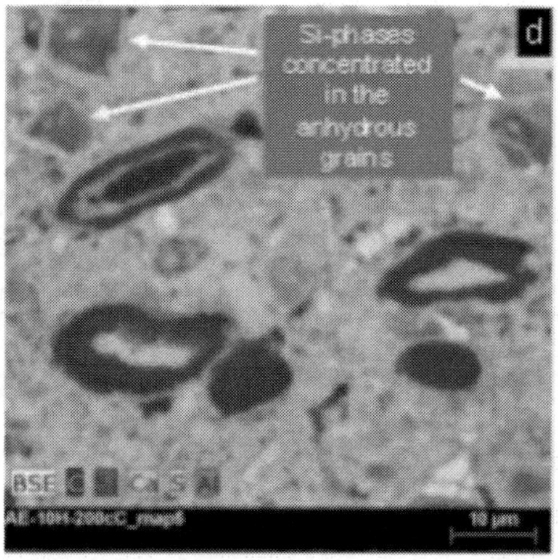

Figure 3. SEM/BSE micrographs of the cut and polished surface of early carbonated fiber-cement composites, using colored EDS mapping for C, Si, Ca, S and Al. Arrows also show the Si-rich phases are concentrated in the alite (C_3S) and belite (C_2S) anhydrous grains.

Figures 4 and 5 depict the differential weight loss (DTG) of the composites at the different carbonation conditions, before and after ageing. Endothermic events can be identified in the three main zones of decomposition: 50-250°C (hydrated phases: C-S-H, AFm, Mc and Aft), 400-450°C ($Ca(OH)_2$ or portlandite – P) and 500-850°C (stable and less stable carbonates). These results indicate that the carbonation rate occurred in a high extension, reducing by 58% the weight loss in the range between 50-250°C of the carbonated composites. The C-S-H is progressively decalcified, being converted to $CaCO_3$ and S-H (silica hydrates). Dissolution of

$Ca(OH)_2$ generally takes place more readily, while decalcification of C-S-H occurs when $Ca(OH)_2$ is inaccessible or locally reduced.

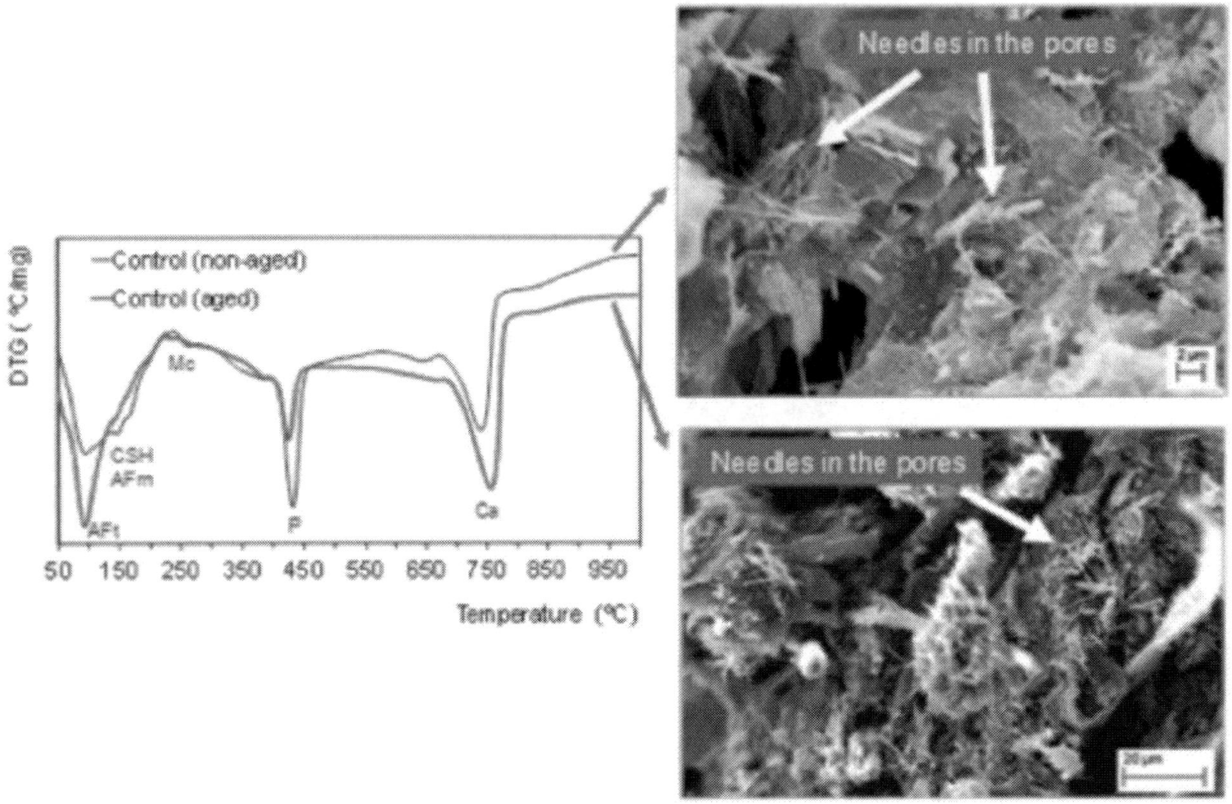

Figure 4. (Left) DTG of control composites before and after soak and dry ageing AFt = ettringite; AFm = monosulfoaluminate; P = portlandite; C-S-H = calcium silicate hydrate; Mc = monocarboaluminate; and Ca = calcium carbonate polymorphs. (Right) SEM/SE of the fractured surface of control composites - arrows show the needles (probably AFt).

Figure 4 also shows the changes in the endothermic peaks between 450 and 500°C related to the decomposition of the $Ca(OH)_2$. The mass loss in this range decreases as the carbonation occurs. No mass loss at this temperature range was observed for the carbonated sample (Figure 5). Finally, the accelerated carbonation increased the mass loss at the range between 500 and 850°C, which are related to the carbonates polymorphs. The decomposition ranging between 500 and 850°C for non-carbonated (control) composites is related in great part to the carbonate content in as-received cement and of calcium carbonate from calcareous filler, the supplementary material applied in the fiber-cement formulation, in order to reduce production costs.

The 200 soak and dry cycles decreased the content of hydrated phases (AFt, AFm, Mc and C-S-H) and mainly of portlandite [$Ca(OH)_2$] in the case of the control sample (non-carbonated). This decrease is resultant of the natural carbonation occurred during the drying stage of the accelerated ageing cycles.

Arrows in Figure 4 show needles of AFt formed in the matrix pores and around the cellulose fibers before and after soak and dry cycles of the control (non-carbonated) composites, respectively. This was particularly evident in the EDS semi-quantification analysis in previous work (Pizzol et al., 2014b), where sulfur (S) is present, with high S:Ca ratio (between 0.06 and 0.22) and Ca:Si ratio (between 12 and 40). The needles were not observed in the accelerated carbonated composites (Figure 5). In this case the microstructure is compact and formed by layered structures (arrows in Figure 5), probably related to the $CaCO_3$ polymorph phases, where very little content of S is detected. These results corroborate with the lower (or null) content of $Ca(OH)_2$, C-S-H, AFt, AFm and Mc observed in the XRD patterns (Figure 1) and weight loss (DTG – Figure 5) of the accelerated carbonated composites.

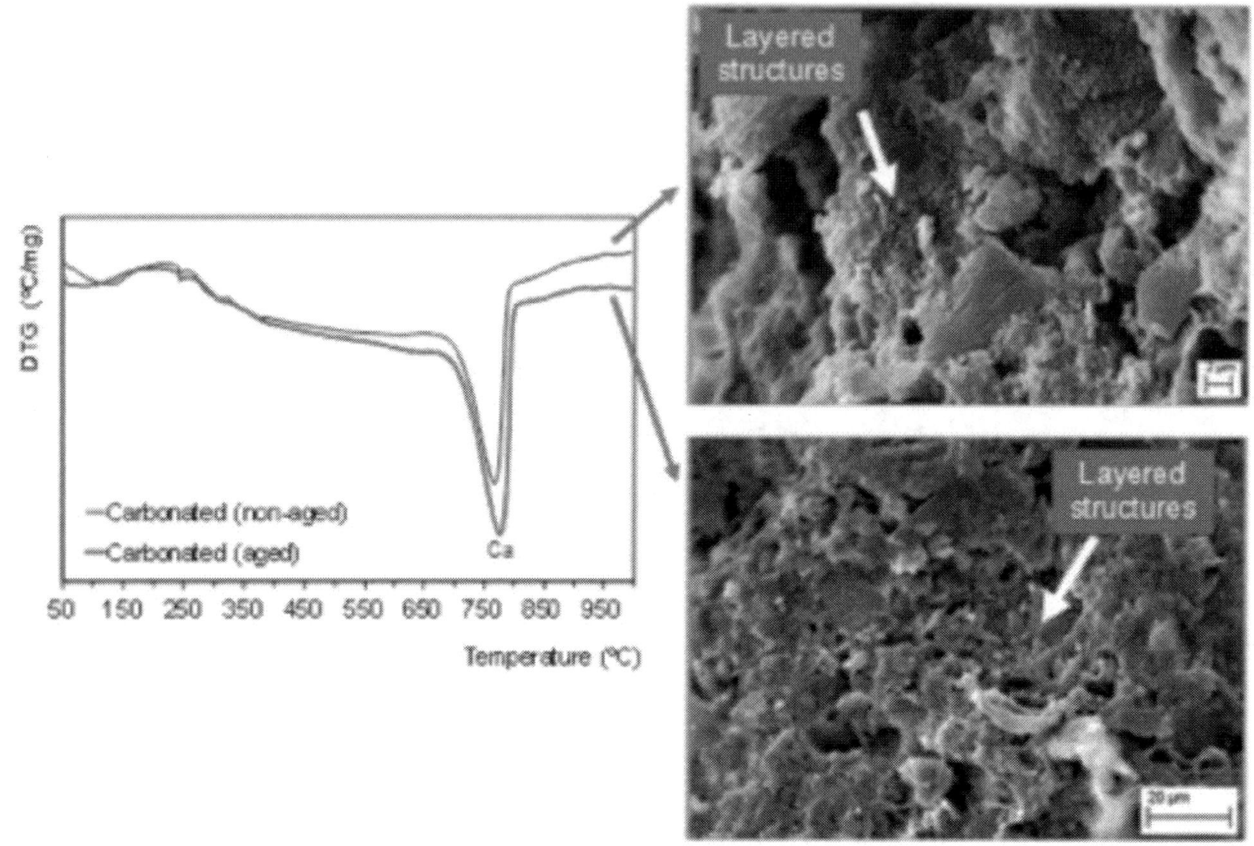

Figure 5. (Left) DTG of early carbonated composites before and after soak and dry ageing cycles; Ca = calcium carbonate polymorphs. (Right) SEM/SE of the fractured surface of control composites - arrows show the layered carbonates.

Consequences of the early carbonation on the drying shrinkage

When the relative humidity decreases porous materials undergo a so-called drying shrinkage. Indeed, the evaporation of the interstitial liquid water and the consequent decrease of the liquid saturation give rise to capillary pressure effects that led to contraction of the solid structure of the composite (Coussy et al., 2004). Figure 6 shows that carbonated composites presented lower drying shrinkage than control composites. The lower pores in carbonated composites increases interfacial energy during drying shrinkage as relative humidity is reduced in relation to the initial condition. The cement matrix of the carbonated composite is denser and compact, which can improve the contact between the fibers and cement matrix to form a rigid structure with lower drying shrinkage. In the opposite (for control) the large porosity provokes large loosing of water, which leads to great volume change in the microstructure of the non-carbonated composite.

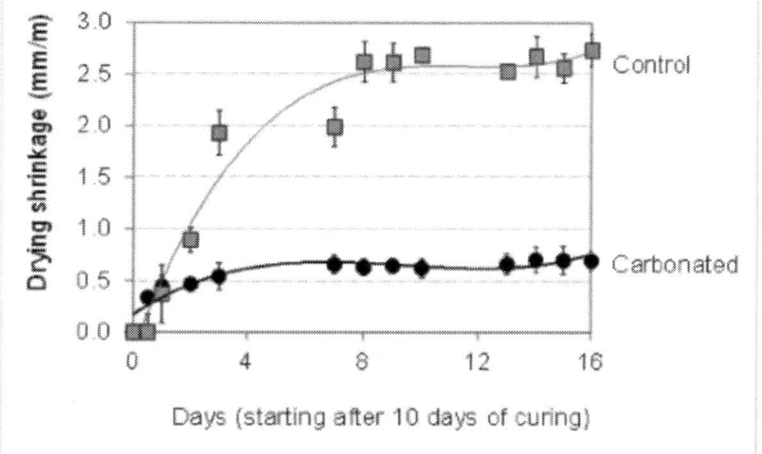

Figure 6. Drying shrinkage, of the control (non-carbonated) and carbonated composites. These values are length decrease of the composites in relation to the initial condition (water vapor saturated).

Conclusion

Accelerated carbonation is presented here as an interesting strategy to prematurely decrease the alkalinity of the cement matrix by consumption of ions Ca^{2+} and OH^- in the pore solution. The absence of C-S-H, AFm, AFt and Mc was also observed during this carbonation. The SEM micrographs show that the $CaCO_3$ polymorphs formed from the carbonation reaction is precipitated in the pore structure of the matrix, filling the voids, and resulting in the densification of the matrix. Then, in carbonated composites the Si-rich phases are still concentrated in the alite (C_3S) and belite (C_2S) anhydrous grains, with low diffusion in the matrix and fiber transition zone. Therefore, matrix becomes more chemically stable with lower diffusion of ions from curing and secondary hydration. It improves the contact between fibers and cement matrix and thus preventing the ingress of water due to the decrease in the pore size network and diffusion of degradation products (e.g. Ca and S ions) into the composite. Otherwise, the presence of S- and Al-rich in the fiber-matrix transition zone was particularly evident in the EDS mapping of the non-carbonated composites. The S- and Al-rich needles (probably AFt and AFm) were not observed in the accelerated carbonated composites, and S was detected in very low content around the cellulose fibers. Additionally, the consequence of carbonation at early age is the volume stabilization, as indicated by the lower drying shrinkage. Accelerated carbonation can be an effective procedure to mitigate the degradation of cellulose fibers in fiber-cement composites for indoor and outdoor exposure applications, encouraging the partial replacement of the expensive plastic (e.g. PP or PVA) fibers by vegetable pulp fibers in the fiber-cement production.

Acknowledgments

Thanks to Conselho Nacional de Desenvolvimento Científico e Tecnológico (CNPq), Fundação de Amparo à Pesquisa do Estado de São Paulo (Fapesp), Fundação de Amparo à Pesquisa do Estado de Minas Gerais (FAPEMIG), Coordenação de Aperfeiçoamento de Pessoal de Nível Superior (Capes), Infibra S. A., Imbralit Ltda., Research Nucleus on Materials for Biosystems - Biosmat (FZEA/USP) and Rede Brasileira de Compósitos e Nanocompósitos Lignocelulósicos (RELIGAR), in Brazil. Thanks also to the collaboration agreement between the University of São Paulo (FZEA / USP, Brazil) and the State Agency for National Research Council (IETcc-CSIC, Spain) that benefited this present partnership.

References

Associação Brasileira de Normas Técnicas. NBR 15498: placa plana cimentícia sem amianto, requisitos e métodos de ensaio. Rio de Janeiro, 2007. 26 p.

Coussy, O.; Dangla, P.; Lassabatere, T.; Baroghel-Bouny, V. The equivalent pore pressure and the swelling and shrinkage of cement-based materials. Materials and Structures / Concrete Science and Engineering, 37, 2004, pp 15-20.

European Standard, EN 494. Fibre–cement profiled sheets and fittings for roofing – products specification and test methods; 1994.

Frías, M. and Goñi, S. Accelerat6ed carbonation effect on behavior of ternary Portland cements. Composites: Part B, 2013, 48, 122-128.

Hyvert, N.; Sellier, A.; Duprat, F.; Rougeau, P. and Francisco, P. Dependency of C–S–H carbonation rate on CO_2 pressure to explain transition from accelerated tests to natural carbonation. Cement and Concrete Research, 2010, 40, 1582-1589.

Ikai, S.; Reichert, J.R.; Rodrigues, A.V. and Zampieri, V.A. Asbestos-free technology with new high toughness polypropylene (PP) fibers in air-cured Hatschek process. Construction and Building Materials, 2010, 24 (2) 171-180.

Johannesson, B. and Utgenannt, P. Microstructural changes caused by carbonation of cement mortar. Cement and Concrete Research, 2001, 31, 925-931.

Lesti, M.; Tiemeyer, C. and Plank, J. CO_2 stability of Portland cement based well cementing systems for use on carbon capture & storage (CCS) wells. Cement and Concrete Research 45 (2013) 45-54.

Neves Jr., A; Toledo Filho, R.D.; Fairbairn, E.M.R.; Dweck, J. The effects of the early carbonation curing on the mechanical and porosity properties of high initial strength Portland cement pastes. Construction and Building Materials 77 (2015) 448–454.

Pizzol, V.D.; Mendes, L.M.; Frezzatti, L.; Savastano Jr., H.; Tonoli, G.H.D. Effect of accelerated carbonation on the microstructure and physical properties of hybrid fiber-cement composites. Minerals Engineering 2014a 05/2014; 59:101–106.

Pizzol, V.D.; Mendes, L.M.; Savastano Jr., H.; Frías, M.; Davila F.J.; Cincotto, M.A.; John, V.M.; Tonoli, G.H.D. Mineralogical and microstructural changes promoted by accelerated carbonation and ageing cycles of hybrid fiber–cement composites. Construction and Building Materials 68 (2014b) 750–756.

Rostami, V.; Shao, Y.; and Boyd, A.J. Durability of concrete pipes subjected to combined steam and carbonation curing. Constr. Build. Mater., 2011, 25(8):3345–3355.

Rostami, V.; Shao, Y; and Boyd, A.J. Carbonation Curing versus steam curing for precast concrete production. J. Mater. Civ. Eng. 2012. 24:1221-1229.

Soroushian, P.; Won, J-P. and Hassan, M. Durability characteristics of CO_2-cured cellulose fiber reinforced cement composites Construction and Building Materials, 2012, 34, 44-53.

Tonoli, G.H.D.; Santos, S.F.; Joaquim, A.P. and Savastano Jr., H. Effect of accelerated carbonation on cementitious roofing tiles reinforced with lignocellulosic fibre. Construction and Building Materials, 2010a, 24, 193-201.

Tonoli, G.H.D.; Santos, S.F.; Teixeira, R.S.; Pereira-da-Silva, M.A.; Lahr, F.A.R., Pescatori Silva, F.H., Savastano Jr., H. Effect of eucalyptus pulp refining on the performance and durability of fibre-cement composites. Journal of Tropical Forest Science. 25(3): 400–409, 2013a.

Tonoli, G.H.D.; Mendes, R.F.; Siqueira, G.; Bras, J.; Belgacem, M.N.; Savastano Jr., H. Isocyanate-treated cellulose pulp and its effect on the alkaly resistance and performance of fiber cement composites. Holzforschung (Berlin. Print).67(8), p.853-861, 2013b.

Performance Evaluation of High-gravity Carbonation Process using Electric Arc Furnace Slag for CO_2 Fixation and Cement Replacement Materials

Chieh-Li Chen[1], Pen-Chi Chiang[1,4*], Chang-Ching Ho[2], Chin-Jen Hou[2], Yi-Hung Chen[3], Shu-Yuan Pan[1]

[1] Graduate Institute of Environmental Engineering, National Taiwan University, Taiwan (ROC)
[2] Tung Ho Steel Enterprise Corporation, Taiwan (ROC)
[3] Department of Chemical Engineering and Biotechnology, National Taipei University of Technology, Taiwan (ROC)
[4] Carbon Cycle Research Center, National Taiwan University, Taiwan (ROC)

*Corresponding author: pcchiang@ntu.edu.tw (PC Chiang)

Abstract

In this study, an integrated process of waste-to-resource technology was developed. Carbon dioxide (CO_2) sequestration using the accelerated carbonation of electric arc furnace slag (EAFS) under various operational conditions in a high-gravity rotating packed bed (RPB) was investigated. Moreover, wastewater neutralization and product utilization were investigated as well. The performance of CO_2 capture by EAFS were evaluated under various levels of reaction time, reaction temperature, rotational speed and liquid-to-solid ratio. The samples of reacted slurry were analyzed quantitatively and qualitatively using thermogravimetric analysis (TGA) and X-ray diffraction (XRD), respectively. The morphology and microstructure of samples were also examined using scanning electron microscopy with energy-dispersive X-ray spectroscopy (SEM-EDX). The utilization of carbonated EAFS as Portland cement replacing materials was examined. Both fresh and carbonated EAFS were blended in the cement mortar with the replacement ratios of 5, 10, 15 and 20% of Ordinary Portland Cement (OPC) were tested for compressive strength. According to ATSM C109 regulation, the cement mortars were tested at 3, 7 and 28 days of curing age with the 50 mm x 50 mm x50 mm molds. The physico-chemical properties and surface structure of the hardened cement mortars were characterized. The results indicated that the carbonated EAFS was suitable as a filler in the OPC-EAFS mortars, which the compressive performance was superior to that using fresh EAFS in cement mortars.

Introduction

Authors Nowadays, people are concerning about global warming, it is imperative to reduce greenhouse gas emissions such as carbon dioxide. The increasing atmospheric concentration of CO_2 attributed to fossil fuel combustion is a serious problem leading to global warming. The atmospheric concentration of CO_2 isϕ_{CO2} = 0.028% in preindustrial times, and it is around ϕ_{CO2} = 0.0396% today.[3] However, the main attribution of CO_2 emission is steelmaking and cement industries. At the same time, industry wastes are generated during the process of manufacturing, for instance, electric arc furnace. Therefore, those kind of slags could be controlled and manage properly as resources, the utilization of wastes could be highly beneficial and developed the byproduct can be developed as "green materials".[1,2,4]

The byproducts produced from the process of combining the CO_2 emission with industry wastes have high economic potential. Not only CO_2 could be removed from the atmosphere and reach the goal of reducing greenhouse gas emission but also industrial wastes could be transfer into utilized materials. At the same time, some of the byproduct produced from the carbonation process could be used as cement replacing materials, which has great benefit to the environment and economy.

CO_2 capture experiments were conducted by accelerated carbonation in a rotating packed bed (RPB), which is shown in figure 1. EAFS was selected as feedstock in this research due to its high calcium content which is useful in CO_2 sequestration. Response surface methodology (RSM) has been used to determine the optimum condition of experiment, which is a statistical method based on the multivariate non-linear model. Therefore, the objectives of this research were focus on (i) Evaluating the effects of operating factors including temperature, rotating speed, liquid and solid ratio on EAFS carbonation in an RPB; (ii) Investigate the properties and hydration of cement with both fresh and carbonated slag as partial replacement by 5%, 10% and 20% binder weight; (iii) Determining the optimum conditions of carbonation of EAFS on RPB by using response surface methodology.

Results and Discussion

Carbonation

In the study, the result shows that at the beginning of experiment, pH value is high due to high alkalinity of EAFS, which is around 11 to 12. When the reaction begins, pH value will drop along with the proceeding experiment. The longer the reaction is, the higher carbonation conversion. As the result indicated that it has the highest carbonation conversion at 60 min during a 60-minutes experiment. According to several factors, including liquid-to-solid ratio, rotating speed of RPB and temperature have influence on carbonation conversion. The liquid-to-solid ratio also has influence on carbonation conversion. The experiment under L/S ratio of 20 ml/g has higher carbonation conversion than the experiment under 10 L/S ratio. It also indicates that the higher amount of EAFS doesn't has positive influence on carbonation conversion. The effect of temperature has a significant influence on the carbonation conversion. The higher the temperature is, the better carbonation conversion.

Cement Replacement

As results show, the compressive strength of 5%, 10%, 15% or 20% replacement in the early age are mostly over the national regulation. The compressive strength of mortar with 5% replacement of EAFS is slightly higher than the original Portland cement. According to CNS 61-R2001 requirements, 3-day, 7-day and 28-day compressive strength must reach 1800 psi (≒12.4 MPa), 2800 psi (≒19.6 MPa) and 4000 psi (≒ 28.0 MPa), respectively. However, national regulation has no compressive strength requirements for 56-day and 90-day of cement. The cement with 20% replacement has lower compressive strength than the national regulation until the mid-age of cement. However, 5% and 10% of replacement still satisfy the national regulation. 28-day compressive strength must reach 4000 psi (≒28.0 Mpa). The cement mortar is shown in figure 2.

Conclusions

The highest carbonation conversion was found to be 57.1% under 60℃, an L/S ratio of 35 ml/g, 1100 rpm and 31.5% of CO2 concentration by RPB process. The higher the operational temperature is, the better conversion will be expected until it reaches the limitation of carbonation. The compressive strength of 5% and 10% cement replacement satisfy national regulation in all curing age. However, Cement with 15% and 20% replacement can only satisfy the national regulation at the early age such as 3-day and 7-day. The compressive strength of 20% replacement at 28-day was below the regulation.

Figures

Figure 1. Overview of rotating packed bed (RPB) reactor

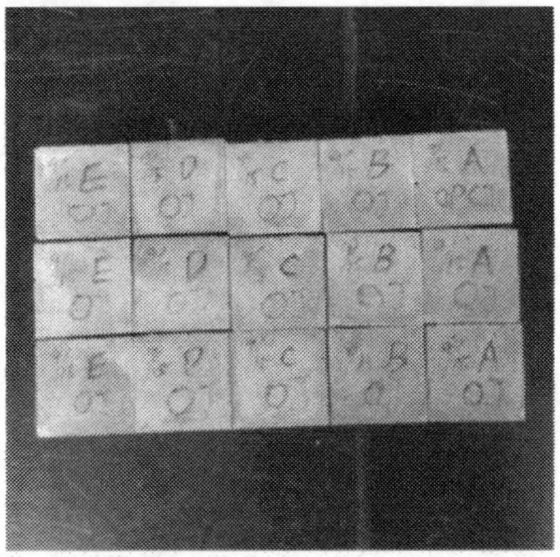

Figure 2. Photo of cement mortars for compressive strength tests

Acknowledgements

This study is supported by Tung Ho Steel Enterprise Corporation, Kaohsiung, Taiwan (ROC). In addition, the Ministry of Science and Technology (MOST) of Taiwan (ROC) is highly appreciated by authors for the financial support under Grant No. MOST-104-E-3113-007-001.

References

1. Kelly, K.E., Silcox, G.D., Sarofim, A.F., Pershing, D.W. 2011. An evaluation of ex situ, industrial-scale, aqueous CO_2 mineralization. *International Journal of Greenhouse Gas Control,* 5, 1587-1595.
2. Khoo, H.H., Bu, J., Wong, R.L., Kuan, S.Y., Sharrant, P.N. 2011. Carbon capture and utilization: preliminary life cycle CO_2, energy, and cost results of potential mineral carbonation. *Energy Procedia.* 4, 2494-2501.
3. Metz, B.D., editor. 2005. IPCC Special Report on carbon dioxide capture and storage. Prepared by working group III of the intergovernmental panel on climate change. Cambridge UK: Cambridge University Press.
4. Pan, S.-Y., Chiang, P.-C., Chen, Y.-H., Chang, E.-E. 2013. Ex situ CO_2 capture by carbonation of steelmaking slag coupled with metalworking wastewater in a rotating packed bed. *Environmental Science & Technology,* 47, 3308-3315.

Carbonation of Calcium-Silicate-Hydrate Gel: Elucidation of Atomic Structure Mechanisms and Reaction Kinetics Using Pair Distribution Function Analysis

Antoine E Morandeau, Claire E. White*

*Assistant Professor
Department of Civil and Environmental Engineering, and Andlinger Center for Energy and the Environment

June 22, 2015

andlinger center
for energy+the environment

PRINCETON UNIVERSITY

INTRODUCTION

Ph.D.

The University of Melbourne
Australia

Department:
Chemical and Biomolecular Engineering

"Atomic structure evolution in amorphous geopolymer precursors and gels"

Post Doc

Lujan Center / T-1 / CNLS
Los Alamos National Laboratory

- Multiscale modeling of materials
- Combining theory and experiment
- Low-CO_2 cements (geopolymers), aluminosilicates and carbonates

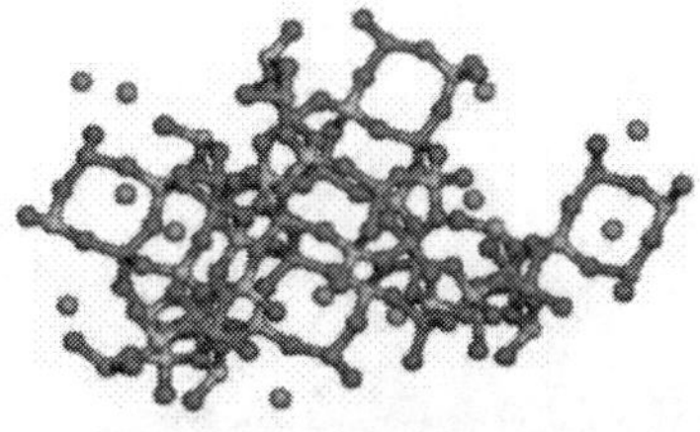

andlinger center
for energy+the environment

PRINCETON UNIVERSITY

SOME FACTS ABOUT CEMENT

THE HIDDEN DETAILS

- Cement manufacture makes up 5-8% of all CO_2 (3rd highest emitter)
- Second most used material next to WATER
- ~ 0.8-0.9 ton of CO_2 per 1 ton of cement
- Market to double by 2050

Global Cement Production 1970 - 2050

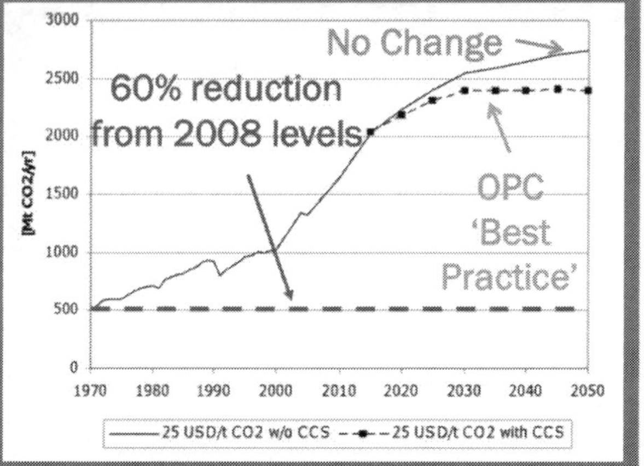

Alternatives to ordinary Portland cement (OPC) are required to meet the VAST MAJORITY of cement in concrete.

ORDINARY PORTLAND CEMENT VERSUS GEOPOLYMER

Hydrated ordinary Portland cement

- Silica chains ("dreierketten")
- Break two Si-O-Si bonds to remove a silicon atom
- Bound water of hydration

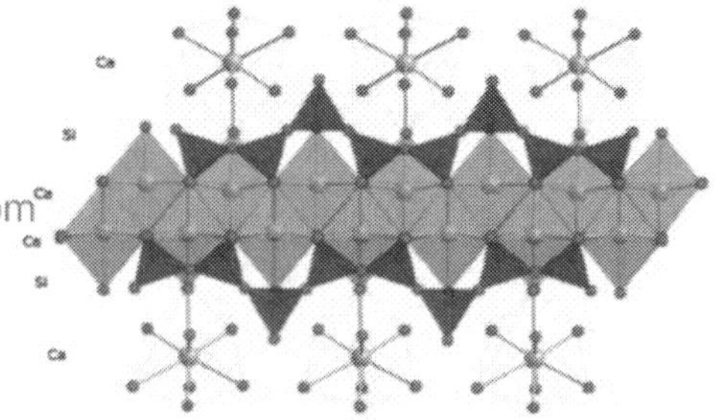

Geopolymer cement

- Highly crosslinked gel structure (pseudo-zeolitic)
- Break four Si-O-(Si,Al) bonds to remove Si or Al
- Little chemically bound water
- Precursors: Fly ash and metakaolin

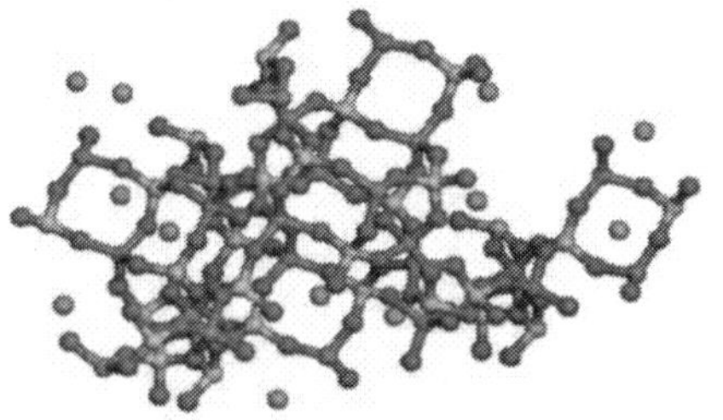

C-S-H image from: I.G. Richardson, *Cement and Concrete Research*, **2004**, 34(9): 1733-1777.

CONTEXT AND CHALLENGE FOR THIS INVESTIGATION

- Promote low clinker (OPC) content binders
- Develop new binders that are not clinker based

  → Durability issues

- Need to understand phase stability and reaction kinetics occurring during exposure of calcium-silicate-hydrate (C-S-H) gel to CO_2

CARBONATION MECHANISM

- Dissolution of gaseous CO_2 in the pore solution of concrete leads to a drop in the pH

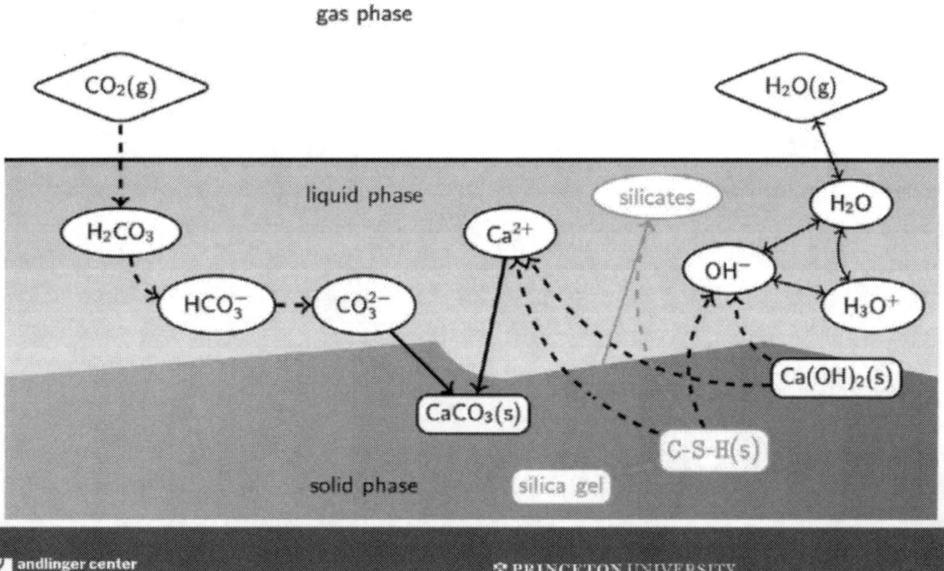

MATERIALS AND METHODS

- Synthetic C-S-H (double decomposition of Na_2SiO_3 and $CaNO_3$)
- 100% CO_2 gas flow
- Data set every 30 sec to 2 min
- 11-ID-B, APS, ANL

andlinger center
for energy+the environment
PRINCETON UNIVERSITY

THE PAIR DISTRIBUTION FUNCTION (PDF)

A sine Fourier transform of the total (Bragg and diffuse) scattering components

The structure factor: $S(Q)$

$$Q = \frac{4\pi \sin(\theta)}{\lambda}$$

The pair distribution function: $G(r)$

$$G(r) = \frac{2}{\pi}\int_0^\infty Q[S(Q)-1]\sin(Qr)dQ$$

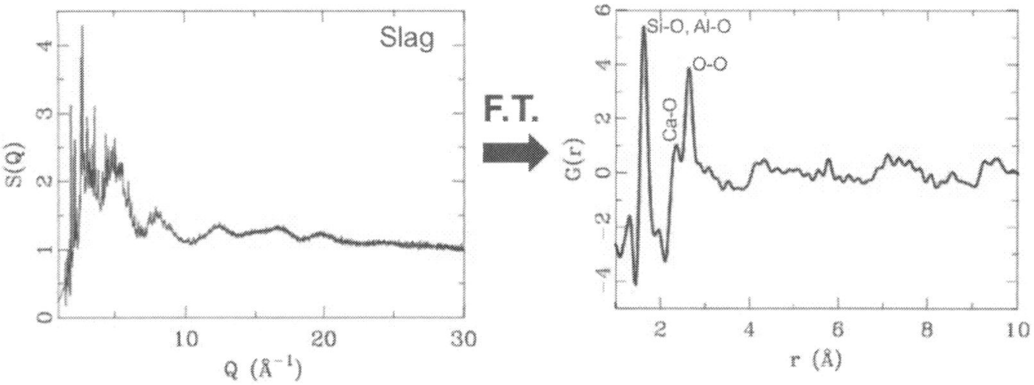

T. Egami and S.J.L. Billinge, *Underneath the Bragg peaks: structural analysis of complex materials*, Pergamon Press Elsevier, Oxford, England, 2003.

andlinger center
for energy+the environment
PRINCETON UNIVERSITY

IN SITU X-RAY DIFFRACTION

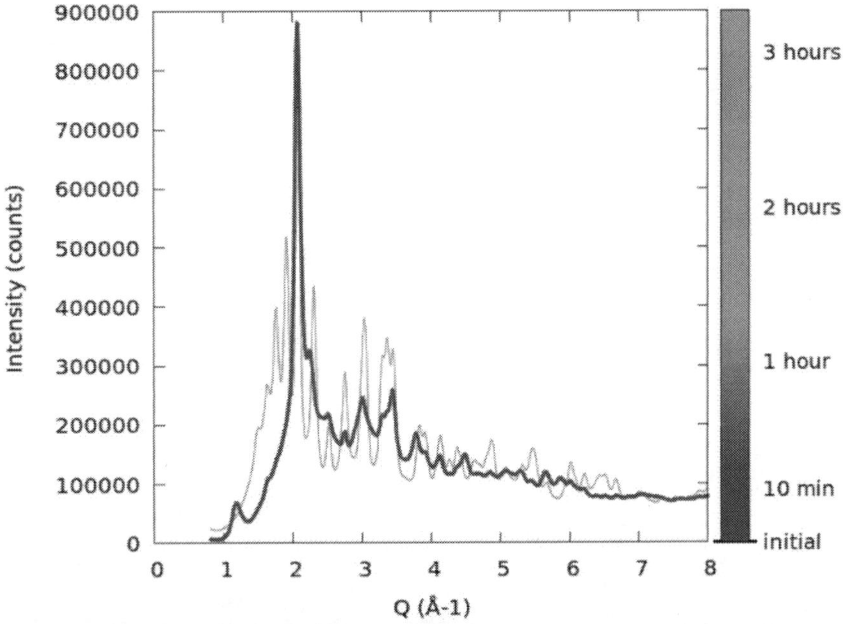

andlinger center for energy+the environment PRINCETON UNIVERSITY

A.E. Morandeau, C.E. White, J. Mater. Chem. A, **2015** 3: 8597-8605.

IN SITU X-RAY DIFFRACTION

- Formation of calcite and vaterite over time

andlinger center for energy+the environment PRINCETON UNIVERSITY

A.E. Morandeau, C.E. White, J. Mater. Chem. A, **2015** 3: 8597-8605.

PHASE EVOLUTION – RIETVELD ANALYSIS

andlinger center for energy + the environment

PRINCETON UNIVERSITY

IN SITU X-RAY PAIR DISTRIBUTION FUNCTIONS

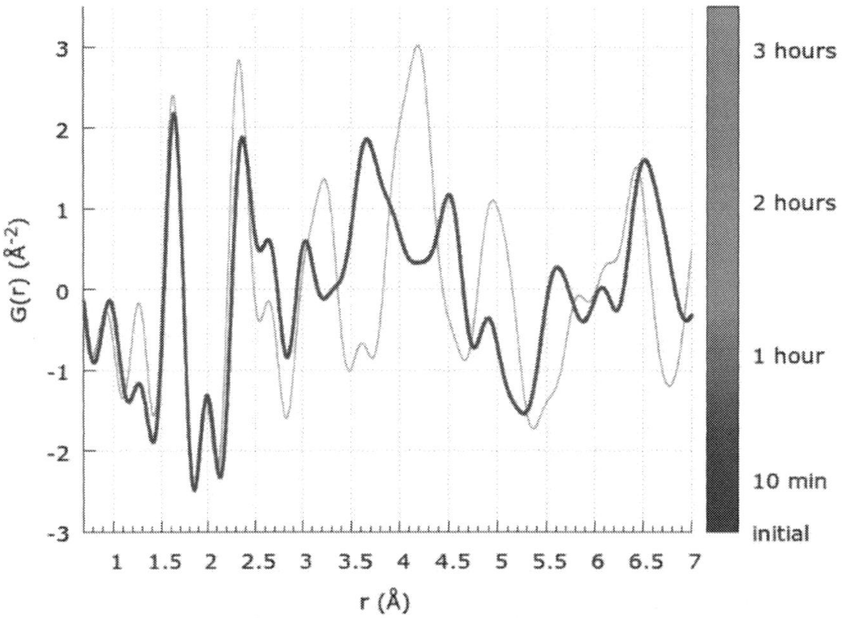

andlinger center for energy + the environment

PRINCETON UNIVERSITY

A.E. Morandeau, C.E. White, J. Mater. Chem. A, **2015** 3: 8597-8605.

IN SITU X-RAY PAIR DISTRIBUTION FUNCTIONS

- Decalcification of the C-S-H gel

andlinger center
for energy+the environment

PRINCETON UNIVERSITY

A.E. Morandeau, C.E. White, J. Mater. Chem. A, **2015** 3: 8597-8605.

ISOLATION OF THE DECALCIFYING GEL

andlinger center
for energy+the environment

PRINCETON UNIVERSITY

A.E. Morandeau, C.E. White, J. Mater. Chem. A, **2015** 3: 8597-8605.

EVOLUTION OF THE DECALCIFYING GEL

- Emergence of an amorphous silica gel and potentially amorphous calcium carbonate (ACC)

andlinger center for energy+the environment

PRINCETON UNIVERSITY

PHASE EVOLUTION: INFLUENCE OF DATA TREATMENT

Real space

- No evidence of vaterite dissolution

Reciprocal space

- Overestimates the amount of vaterite precipitation due to lack of accurate models describing the decalcifying gel

andlinger center for energy+the environment

PRINCETON UNIVERSITY

CONCLUSIONS

- The phase evolution and reaction kinetics occurring during carbonation of C-S-H gel can be accurately followed using *in situ* X-ray pair distribution function analysis
- Removal of the crystalline phases from the PDFs reveals the exact atomic structural changes that occur as nanocrystalline C-S-H gel transforms to amorphous silica gel
- Calcite formation is the rate limiting factor
  - Excess calcium and carbonate ions in the supersaturated solution lead to the formation of vaterite (and potentially ACC)

andlinger center for energy+the environment   PRINCETON UNIVERSITY

ACKNOWLEDGMENTS

Funding

Financial support by the Princeton E-ffiliates Partnership award, Andlinger Center for Energy and the Environment (Princeton University)

People/Facilities

Karena Chapman and Kevin Beyer for their help with the gas cell

The 11-ID-B beam line is located at the Advanced Photon Source, an Office of Science User Facility operated for the U.S. DOE Office of Science by Argonne National Laboratory, under U.S. DOE Contract No. DE-AC02-06CH11357

Argonne NATIONAL LABORATORY   PRINCETON UNIVERSITY

andlinger center for energy+the environment   PRINCETON UNIVERSITY

Techno-economic assessment and environmental impacts of Mineral Carbonation of industrial wastes and other uses of carbon dioxide

Evgenia D. Mehleri[1,2], Amit Bhave[4], Nilay Shah[2], Paul Fennell[3] and Niall Mac Dowell[1,2,*]

[1] Centre for Environmental Policy, Imperial College London, UK
[2] Centre for Process Systems Engineering, Imperial College London, UK
[3] Department of Chemical Engineering, Imperial College London, UK
[4] cmcl innovations, Sheraton House, Castle Park, Cambridge CB30AX, UK
*Corresponding author: niall@imperial.ac.uk

Abstract
In this contribution, we present the results of an in-depth techno-economic analysis of some leading CO_2 capture and utilisation (CCU) and conversion (CCC) options. Specifically, we consider CO_2 conversion to methanol, formic acid and urea (CCC) in addition to mineral carbonation of industrial wastes (CCU). We compare the CCC and CCU options using a range of key performance indicators (KPIs), including 2nd law efficiency, CO_2 avoided and $tonne_{CO2}/tonne_{Product}$.
The results indicate that CCU and CCC technologies are unlikely to provide a significant contribution to mitigating anthropogenic climate change. The primary bottleneck to industrial scale deployment of CCC technologies is likely to be the cost effective availability of low carbon-hydrogen in the case that the conversion option requires hydrogen. Further, we find that mineral carbonation may have niche applications in the context of industrial waste remediation but the large scale deployment of this technology as a substitute for the geological sequestration of CO_2 is unlikely to be either cost effective or scalable. Moreover, although formic acid offers attractive economic profiles, we note that this process is at a lower TRL (TRL 4-5). Thus, we conclude that CCC and CCU technologies are only likely to be viable at scale in the event that substantial subsidies are available to offset the high costs associated with producing renewable hydrogen and the thermodynamic cost associated with processing such a stable molecule.

Introduction

Global emissions of carbon dioxide (CO_2) from fossil fuels have been increasing by 2.7% annually over the past decade and are now 60% above 1990 levels, the reference year for the Kyoto Protocol. By contrast, it is estimated that the CO_2 emissions should be reduced by approximately 50 % to limit the rise of the global average temperature to 2°C by 2050[1]. The International Energy Agency (IEA) suggests that CCS is an essential part of the technology portfolio needed to achieve deep reductions in global CO_2 part of the technology portfolio needed to achieve deep reductions in global CO_2 emissions with CCS expected to contribute something in the range of 14-10% of total reductions in emissions by 2050[2]. More recently, a related alternative - carbon capture and utilisation (CCU) has started to attract attention worldwide under the narrative the CCU and CCC can provide a route to convert waste CO_2 emissions into valuable products such as chemicals and fuels, while at the same time contributing to climate change mitigation. Before starting and describing the different technologies for CCU and CCC, it is important to define the different terms used and be clear about the purpose. CCU which is the abbreviation of CO_2 capture and Utilisation, refers to the capture of CO_2 from a large, fixed point source and subsequent utilisation in industrial purposes, principally CO_2 Enhanced oil Recovery (CO_2-EOR). CCC on the other hand, is the capture of CO_2 from a large fixed point source and subsequent conversion of the CO_2 to a "high value" end product, i.e. chemicals or fuels. It is then argued that CO_2 capture and utilisation (CCU) and conversion (CCC) may have a potential role to play as a complementary step to large scale CO_2 capture and storage (CCS) deployment. The potential of CCU as a CO_2 abatement option has been discussed previously from various perspectives[3,4]. The novelty in this contribution is in the application of a detailed engineering-based techno-economic analysis which both ranks the CCU options and identifies knowledge gaps, highlighting key areas for future work. However before it is deployed at a large scale there are some other issues that need to be addressed. Firstly, we need to consider that, owing to its thermodynamic stability, the conversion of CO_2 to various products will be energy intensive. Secondly, the economics associated with the deployment of the different technologies at large scale, needs also to be considered. Moreover, the current global demand for chemicals does not have the capacity to sequester enough CO_2 emissions to contribute significantly to meeting the carbon reduction targets. For example, the annual production of urea and methanol, two of the most important chemicals, would consume only 0.5% of the current 34.5 Gt/yr of the anthropogenic CO_2 emitted globally[5]. Furthermore, the utilisation of CO_2 in this manner, only delays its emission and does not provide storage for longer timescales, needed for mitigating climate change.

In an attempt to contribute to the debate on this field, this paper provides a techno-economic analysis of some leading CCU and CCC options, namely, CO_2 conversion to methanol, urea and formic acid and the mineral carbonation of industrial wastes. For each of the aforementioned processes, the technical process design was initiated with a literature review to identify the most salient CO_2 conversion routes to the desired chemical products, based also on the technology readiness level (TRLs) definitions from the EU Horizon 2020 framework for each of the processes[6]. The four processes are evaluated in terms of key performance indicators (KPIs): process efficiency, energy requirements, avoided CO_2 emissions, amount of CO_2 utilised, capital and operating costs and average levelised production costs (LPC).

To set the context, the paper starts by a description of the process flow diagrams proposed for each option. This is followed by the results of the energetic and economic analysis for the different processes, and their evaluation based on the different key performance indicators (KPI's). Finally conclusions are presented in the last section.

Description of CCU and CCU alternatives

In this section, we present a detailed overview of each of the process flow diagrams (PFDs) proposed for each technology. For each option, a brief introduction to the process is presented, followed by a PFD and the associated unit operations and streams. All processes were modelled in the Aspen Plus and Aspen Hysys platform. In all cases, we assume CO_2 is available from a CO_2 transport network, i.e., at high pressure (110 bar) and purity (>95% CO_2, < 5% N_2 and essentially dry and deoxygenated), in line with the ZEP recommendations[7]. Similarly, wherever H_2 was required, as in the case of methanol production, it was assumed that the H_2 was obtained via the alkaline electrolysis of H_2O using wind energy thus rendering the hydrogenation process as "renewable".

Mineral carbonation

Mineral carbonation is considered as one of the options for permanently storing CO_2 by converting calcium and magnesium-bearing minerals and alkaline industrial wastes into environmentally benign, insoluble and thermodynamically stable solid magnesium and calcium carbonates. It is worth noting that in terms of availability of alkaline wastes, there is 130 Mt/yr of fly ash, 15 Mt/yr cement kiln dust, 10 Mt/yr steel slag and 5 Mt/yr bauxite residue produced in the US and this could account for 200-300 Mt/yr[8]. This is a large number in relation to current CO_2 conversion and utilisation options (200 Mt/yr Globally), small relative to the US' CO2 emissions of 5.9 GtCO2 in 2013[9] . Therefore, there might be some benefit to the carbonation of, e.g., fly ash, in some contexts, but as much of this material is already used in industrial applications (e.g., concrete production, primarily[10]) the counter factual would also need to be explored in detail before one could come to a view that the mineral carbonation of industrial wastes is a viable option from a whole-systems perspective. The highly heterogeneous nature of these waste streams, both in composition and morphology serves to considerably complicate this analysis. The process of mineral carbonation is thermodynamically favoured, but kinetically limited - CO_2 will spontaneously react with alkali materials, just relatively slowly under normal conditions. In this study, we have focused on the ACT (Accelerated Carbonation Technology) process, as developed by the Carbon8 company[1]. This has the dual benefits of producing building products from otherwise landfilled wastes and avoiding the use of aggressive chemicals such as ferrous sulphate. In this paper, a dry one-step mineralisation process was simulated at a scale of 80,000 t/yr of end-product.

In Figure 1, the process flow diagram of the mineral carbonation process is presented. In this process, Flyash containing CaO, MgO and inerts is fed to the reactor with CO_2 where the carbonation takes place at a temperature of 25°C and pressure of 1 bar. Prior to being introduced to the reactor, the CO_2 is expanded from transport conditions to 1 bar. The next steps after the carbonation reaction are the gas and solid separations and the mixing of the carbonated ash with water to form a slurry and the carbonated final product (pellets). CO_2 is fed in a stoichiometric ratio to the concentration of CaO in the fly ash feed. As the CaO is not 100% converted to $CaCO_3$

[1] Carbon8 company: http://c8a.co.uk/

under these conditions, this equates to a slight excess of CO_2 with unreacted CO_2 being recycled back to the reactor. The water used in the binding process is 1:1 to the end-product. Finally, the carbonated product and the solid residue, in the form of pellets leaves the process to be transported and used.

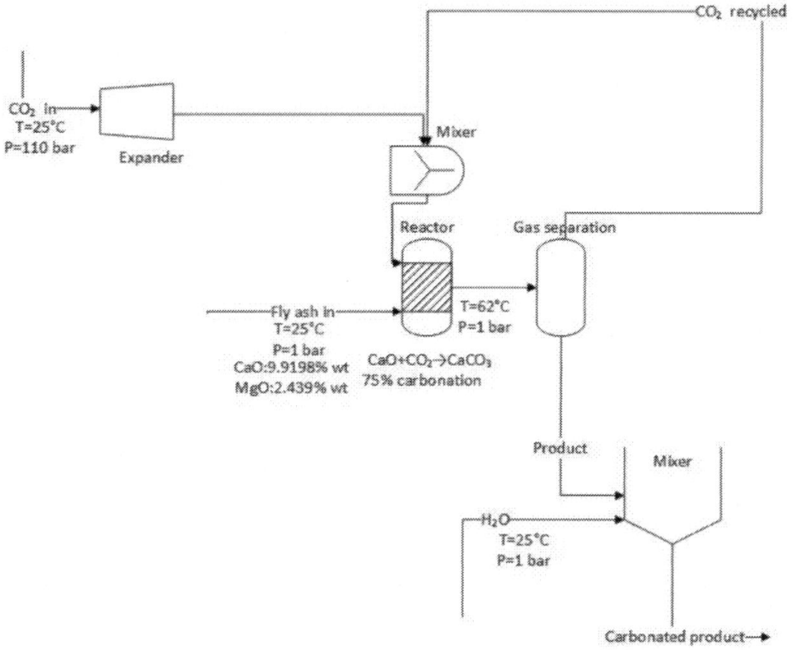

Figure 1. PFD of mineral carbonation process

Methanol

The catalytic hydrogenation of CO_2 is estimated to be at TRL 5-6 as it has been demonstrated at pilot scale and as the catalyst materials continue to be tested and optimised. This process has been commercialised by CRI in Iceland, and therefore this process was chosen as the basis for this study[11].

The methanol production process was simulated at a scale of 400,000 t/yr and the NRTL model was used to calculate the thermodynamic properties of the stream, while the kinetics scheme for the reactor model and the amount of catalyst were calculated using the work of Van-Dal and Bouallou[12]. The process flow diagram as implemented in Aspen Plus is illustrated in Figure 2. The two streams are mixed and subsequently combined with the recycle stream. The stream is then heated to 210°C and injected into the fixed bed adiabatic reactor. The stream leaving the reactor is separated via a flash separator and the vapour portion of this stream is recycled. The pressure of the liquid stream is reduced in two stages to approximately 10 bar and then 1.2 bar in S3 and S4 respectively. The vapour fraction is removed via a further flash separator and the residual liquid portion is heated to 80°C prior to being fed to a distillation column which operates in the temperature range 70 – 100 °C at a pressure of approximately 1 bar. After exiting the distillation column, the stream rich in methanol is compressed to 1.2 bar and then cooled to 40°C. The liquid fraction rich in methanol is removed via a flash separator.

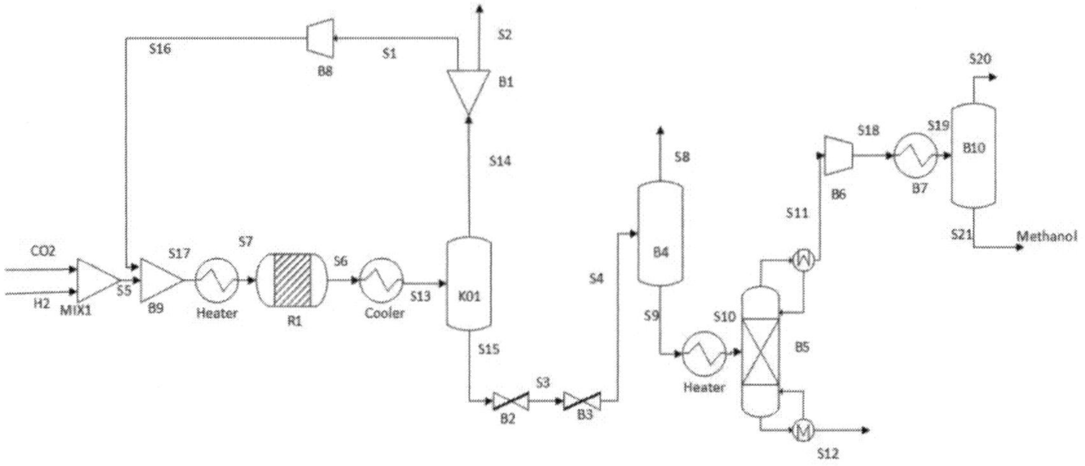

Figure 2. PFD of CO_2 to methanol process

Formic acid

Formic acid can be produced from CO_2 with three different routes: CO_2 hydrogenation in a two-phase liquid–liquid system (diol/[NHex3•HCOOH]—NHex3) using liphophilic ruthenium complexes such as [Ru(H)2(PnBu3)4], CO_2 hydrogenation using a [C1Rh(TPPTS)3] catalyst (Wilkinson's catalyst) and through electrochemical reaction. The electrochemical process is currently being demonstrated at the pilot plant level by DNV and Mantra (TRL 5)[13]. The homogeneous catalytic hydrogenation of CO_2 route has been backed by the worldwide market leader BASF and offers a process comprising two reactor steps and one distillation step. The homogeneous catalytic hydrogenation of CO_2 route offers further benefits in terms of lower energy input requirements when compared with the diol-based hydrogenation process[14]. The ultimate performance will be governed by the underlying thermodynamics for the specific route and hence the one with the most promising overall thermodynamics was selected for further investigation. This is important because CO_2 use must be demonstrated to be energetically sensible for society to perceive it to be sustainable.

The formic acid process was simulated at a scale of 180,000 t/yr and the NRTL fluid package was used for the calculation of the thermodynamic properties. The PFD of the process is presented in Figure 3. In this process, CO_2 and H_2 are assumed available at 60 and 30 bar, respectively. The two streams are mixed and combined with the recycle stream. The stream is then cooled down to 25°C before entering the conversion reactor. The liquid stream leaving the reactor is heated to 60°C before entering the distillation column (T-100). In the distillation column the liquid stream rich in formic acid is separated from the vapour stream which is recycled back to the mixer. After exiting the distillation column, the stream rich in formic acid is cooled down to 25°C.

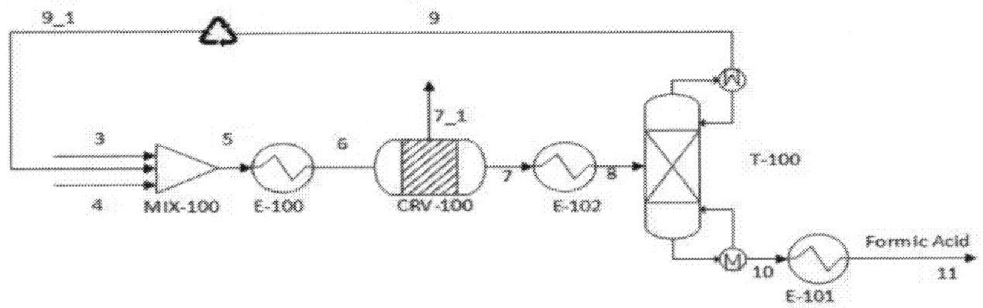

Figure 3. PFD of CO_2 to formic acid process

Urea

CO_2 reuse for urea manufacture is considered to be a mature technology (TRL 9), as the conventional process already utilises CO_2 and is well-established on an industrial scale. Conventional urea synthesis is from natural gas-derived ammonia and carbon dioxide. Commercial urea production is carried out on a very large industrial scale (250,000 to 700,000 t/yr). Although several technologies are available for urea manufacture, the ammonia stripping urea manufacture process, i.e. Snamprogetti process[15] and the CO_2 stripping urea manufacturing process licensed by Stamicarbon[16] together represents more than 75% of the urea production technology market share (TRL 9). Therefore the Stamicarbon process has been considered for further investigation. The urea process was simulated at a scale of 600,000 t/yr and the SR-POLAR model was used to describe the thermodynamic properties of the system, as described in Figure 4.

In this process, CO_2 gas is fed to the CO_2 stripper, to strip the urea solution coming from the reactor. In the stripper, ammonium carbamate decomposes, liberating more NH_3 and CO_2 to be stripped out. The outlet liquid solution from the stripper is rich in urea and goes to the downstream section for urea purification. In the adiabatic urea reactor, an aqueous solution of NH_3 and CO_2 and vapours flow upward through the reactor. Remaining gases condense and the carbamate decomposes in the reactor to provide heat for the slightly endothermic reaction of carbamate to urea. The urea solution flows from the top of the reactor back to the stripper, while the unreacted gases pass out from the top of the reactor. These unreacted gases pass to the scrubber where recycled carbamate solution from the evaporation/recirculation section is passed over the top of a packed bed and fills the tube side section of vertical tubes. The gases rise up through the tubes and pass up through the packed section before leaving the top of the vessel. The contact with the carbamate solution absorbs the unreacted NH_3 and CO_2, while the inert gases of O_2, N_2, and others vent out from the top. Carbamate solution flows out of the vessel. The solution together with the top vapour stream from the stripper is fed to the carbamate condenser through the use of an ejector, where the ammonia feed serves as the pumping fluid. Ammonium carbamate forms in this condenser. The vapour-liquid mixture falls through tubes and the heat of reaction is removed by generation of 3.5 bar steam on the shell side. The mixture goes to the bottom of the reactor for urea production. Finally, the bottom stream from the stripper is sent to the downstream section to recover urea. The recycled stream after recovering the urea is passed back to the high pressure scrubber to complete the loop.

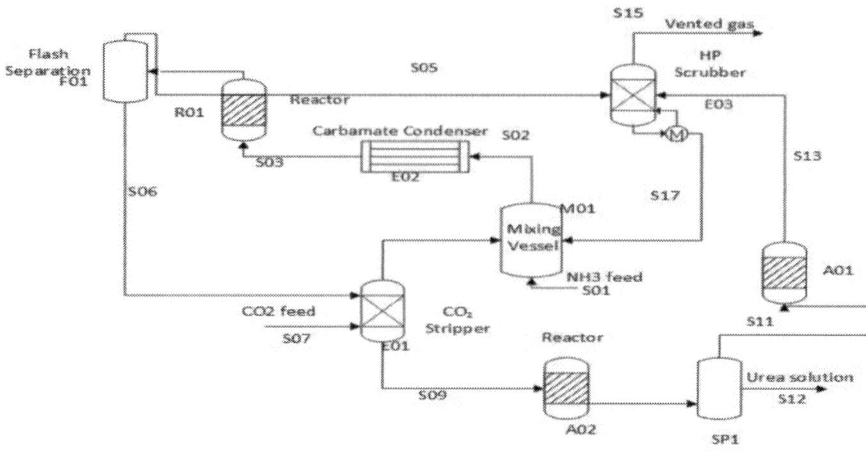

Figure 4. PFD CO_2 to urea process

Results and discussion

The four processes as described above, were evaluated based on key performance indicators as described in Table 2. To determine the capital expenditure and the production costs, a common framework has been applied across the five technology options. In the framework, the first step involved the evaluation of the total Purchase Cost of Equipment (PCE). These costs were calculated for each of the four processes by accounting for the process units within the corresponding PFD. The mass and energy balance streams data was used for sizing the equipment and the main assumptions regarding the materials, the cost factors and the empirical cost-correlations were adopted from the validated and established guidelines reported by Couper[17] and using the Aspen integrated economic modelling tool. Specifically for the reactor cost in the mineral carbonation, we have used the two standard different types of mineral carbonation reactors based on different pressures and for the process scale[18]. In order to evaluate the operating costs (OPEX), first the variable costs were determined using the costs associated with the raw materials, miscellaneous material and utilities. The raw materials and the utilities were obtained from the mass and energy balance streams whereas the miscellaneous material costs were assumed at 10% of the fixed maintenance costs. The fixed costs comprising maintenance, operating labour, supervision, plant overhead, capital charges, insurance, local taxes and royalties were also evaluated as a proportion of the fixed capital costs. The capital, operating and levelised costs were calculated at CO_2 credit price 0 €/t, in other words, the CO_2 was available free of charge, but nor was a credit available for its utilisation. In Table 3 the results of the different KPIs calculated are presented.

Table 2. Key Performance Indicators (KPIs)

KPIs	Description
Process efficiency	The efficiency of CO_2 retention in the final product, per unit CO_2 fed in to the process : $\text{Process efficiency} = \frac{(CO_{2,IN} - CO_{2,OUT})}{CO_{2,IN}}$
Energy requirements	The total (heating, cooling, electrical) energy used in the process.

Avoided CO_2 emissions	The sum of the CO_2 that would have been emitted had the product been produced via a conventional process and the CO_2 that is required to drive the process.
Amount of CO_2 utilised	The total amount of CO_2 utilised in the process. This is calculated as the nominator of the first KPI (Process efficiency).
Capital cost	The total capital expenditure associated with the construction of this process.
Operating cost	The total operational expenditure associated with the operation of this process.
Levelised production cost	This is defined as $LPC = \frac{\sum_{t=1}^{n}\left(\frac{I+O_t}{(1-r)^t}\right)}{\sum_{t=1}^{n}\left(\frac{P_t}{(1-r)^t}\right)}$ where I is the total capital investment, O_t is the operations and maintenance cost in a given year t and finally P_t is the total annual product in a given year t.

Table 3. KPI results for the different processes

KPIs/Processes	Mineral carbonation	Methanol	Formic acid	Urea
TRL	9	5-6	4-5	9
Scale (t/yr)	80,000	400,000	180,000	660,000
Process efficiency (%)	75	99.6	95.16	80
Energy requirements ($GJ/t_{product}$)	0.3	2.5	1.843	2.83
Avoided CO_2 emissions ($t_{CO2}/t_{product}$)	0.795-0.84	4.52-4.98	1.65	0.77
Amount of CO_2 utilised ($t_{CO2}/t_{product}$)	0.786	3.68	1.27	0.77
Capital cost (€/$t_{product}$)	115	1742	457	87
Operating cost (€/$t_{product}$)	111	1892	493	252
Levelised production cost (€/$t_{product}$)	181	3015	788	287

Based on the model assumptions made and the conditions studied, the costing model for CCC to methanol indicated that the CAPEX and production costs of renewable hydrogen plant contributed to more than half of the costs of the overall plant.

For the mineralisation process, the amount of CO_2 utilised is relatively low, and is a strong function of the fly ash composition, i.e the amount of CaO (in this case is 9.9198 wt%) and also in the CO_2 conversion efficiency of the process. The mineralisation process has low energy requirements compared to the other processes and also the capital cost is mainly driven by the cost of the reactor.

The results from the evaluation of the three CCC and one CCU processes, have shown that for the CO_2 price credit across all the five CCU routes set at zero (base case); urea and formic acid were the two production routes that yielded positive cumulative cash flow whereas methanol and the mineral carbonation routes considered here rendered a negative (loss-making) cash flow profile. In order to make the methanol process cost effective, a CO_2 price credit of -€400/t is required. For the formic acid process, at a CO_2 price credit of -10 €/t, the cumulative cash flow has a break-even after about 7 years of initial investment.

For the mineral carbonation, the economic viability is strongly influenced by not only the CO_2 price but also the landfill tax credit. Taking as a reference the selling price of the pellets at €71[2] and considering a 2% yearly increase in the price of the pellets for the period of 30 years, we examined the influence of the CO_2 and landfill tax credit on the cumulative cash flow. This analysis, indicated that the landfill tax credit is the most important parameter. When the current landfill tax credit is considered as constant over the 25-year period operation of the plant, at a price of -20 €/t the break-even point for the cumulative cash flow was less than five years. On the other hand, if the landfill credits are not considered, then a CO_2 price credit of -20 €/t credit is necessary for the overall CCU mineralisation process to be financially viable and sustainable in terms of cumulative cash flow in less than five years. It may be worth noting that contemporary flue gas desulphurisation (FGD) processes operate on a similar principle and, in some cases, are a decidedly positive asset to the economics of the base power plant.

For formic acid and mineralisation existing gaps in the know-how, were identified. Although formic acid offers attractive economic profiles, it is highlighted that this process is at lower TRLs (i.e. TRL 4-5). Also, the uncertainties and gaps in the techno-economic information relating to the process catalysts are high. The urea process yielded a positive cash flow with a relatively longer break-even period of 13 years for the base case, however, given its higher penetration levels (TRL9), it offered the highest potential for CO_2 avoided (24.7 Mt by 2030).

Conclusions

In this contribution, we present the results of an in-depth techno-economic analysis of some leading CO_2 capture and utilisation (CCU) and conversion (CCC) options. Specifically, we consider CO_2 conversion to methanol, formic acid and urea (CCC) in addition to mineral carbonation of industrial wastes (CCU). We compare each of the CCC and CCU options using a range of key performance indicators (KPIs). Owing to a relative difference in scales, we content that CCC or CCU routes are unlikely to provide a material contribution to the mitigation of climate change. However, some CO_2 conversion or utilisation routes may be important to bootstrapping early stages of CCS deployment – particularly if they equate to the effective sequestration of CO_2, as in the case of mineral carbonation of industrial wastes. It is, however, vital that the CCS/CCU routes do not unduly increase the cost of carbon sequestration, thus highlighting the value of studies such as this. We have found that, in the context of methanol and formic acid production, the primary bottleneck to industrial scale deployment for methanol and formic acid process is likely to be the cost effective availability of low carbon/hydrogen. For mineralisation, the influence of the landfill tax credits is important in order to make the process economically viable.

Acknowledgements

The authors gratefully acknowledge the financial support from the EPSRC under grant EP/M001369/1 MESMERISE-CCS.

[2] Based on personal communication with industrial suppliers

References

1. IPCC, Climate Change 2013: The Physical Science Basis, Intergovernmental Panel on Climate Change, 2013 available at: https://www.ipcc.ch/report/ar5/wg1/.
2. Centre for Low Carbon Futures, Carbon Capture and Utilisation in the Green Economy: Using CO_2 to Manufacture Fuels, Chemicals and Materials, 2011.
3. Von der Assen N., Voll P., Peters M. and Bardow A., Life cycle assessment of CO_2 capture and utilization: a tutorial review, Chemical Society Reviews, 2014.
4. International Energy Agency, 2014, Energy Technology Perspectives, Harnessing Electricity's Potential.
5. PBL and JRC, Trends in Global CO_2 emissions: 2013 Report, PBL Netherlands Environmental Assessment Agency and European Commission Joint Research Institute for Environment and Sustainability.
6. EC (European Commission), Horizon 2020, Work Programme 2014-2015, Technology Readiness Levels (TRL), General Annexes, 2014, URL: http://ec.europa.eu/research/participants/data/ref/h2020/wp/2014_2015/annexes/h2020-wp1415-annex-g-trl_en.pdf (Accessed on 06/05/2015).
7. Zero Emissions Platform (ZEP), http://zeroemissionsplatform.eu
8. Greeshma Gadikota, Ah-hyung Alissa Park. 2015. Accelerated Carbonation of Ca- and Mg-Bearing Minerals and Industrial Wastes Using CO_2. Book Chapter (Carbon Dioxide Utilisation: Closing the Carbon Cycle).
9. BP Statistical review of world energy, 2014.
10. American Coal Ash Association. "Coal Combustion Products Production & Use Statistics" and http://www.fhwa.dot.gov/infrastructure/materialsgrp/flyash.htm.
11. Carbon Recycling International (CRI), 2012. Renewable Methanol Pilot Plant, Based on 5 million liters CH_3OH/year. Retrieved from: http://cri.is/index.php?option=com_content&view=article&id=14&Itemid=8&lang=en , accessed 06/05/15.
12. Van-Dal E.S. and Bouallou C. 2013. Design and simulation of a methanol production plant from CO2 hydrogenation, *Journal of Cleaner Production*, 57, 38-45.
13. DNV (Det Norske Veritas), Sridhar et al., Electrochemical Production of Chemicals: Applicability to CO_2 Conversion, 2012. URL: http://www.arpa-e.energy.gov/sites/default/files/documents/files/3_Narasi_Sridhar_DNV.pdf (Accessed on 06/05/2015).
14. Dinjus E. 1998. Organometallic Reactions with CO2 – Catalyst Design and Mechanisms, *Advances in Chemical Conversions for Mitigating Carbon Dioxide, Studies in Surface Science and Catalysis*, 114, 127-140.
15. Snamprogetti. URL: http://www.intsok.no/style/downloads/Snampr-Inc---Snamprogetti-Highlights.pdf and http://www.snamprogetti.ro/Projects (Accessed on 06/05/2015), 2014.
16. Stamicarbon, 2014. URL: http://www.stamicarbon.com/ (Accessed on 06/05/2015).
17. Couper J. R., Penney W. R., Fair J. R., Chemical process Equipment: Selection and Design, Elsevier, 2010.
18. Kirchofer A., Brandt A., Krevor S., Prigiobbe V., Wilcox J. 2012. Impact of alkalinity sources on the life-cycle energy efficiency of mineral carbonation technologies, Energy and Environmental Science, 5, 8631-8641.

Åbo Akademi University

Fixation of CO_2 in Carbonates of Calcium or Magnesium

Hannu-Petteri Mattila
Ron Zevenhoven

Thermal and Flow Engineering Laboratory
Åbo Akademi University, Turku, Finland

Åbo Akademi University

CCUS/CCS in/for Finland

		In Finland	Near Finland and abroad
"CCS"	**Geological storage**	*Not possible*	Baltic sea ?
	Ocean storage	*Not possible*	*Not possible*
	Mineral sequestration	Large potential Projects ongoing	Projects ongoing PT, SG, LT, (CA?, ZA?)
"CCUS*"	**CO_2 utilization**	Several applications (PCC, CO_2 solvent...) Projects ongoing	Projects ongoing US

**Carbon capture, utilization and storage, Carbon capture, use and sequestration, etc.*

CO_2 mineralization in Finland

- The only CCS option in/close to the country is CO_2 mineral sequestration, at least 2-3 Gt capacity
- Interest currently limited to metal /limestone/mineral processing sector
- Producing valuable solids is an important economic driver
- Co-operation with countries in a similar situation: Singapore, Portugal, Lithuania (future: South Africa, Chile, South Korea)

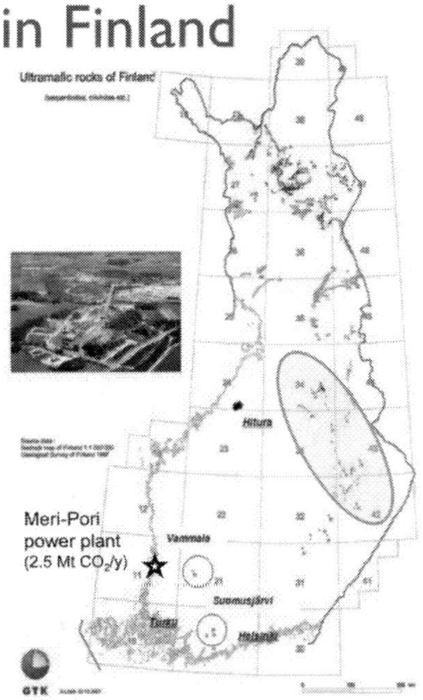

Mineral resources; source-sink links

Picture from: http://www.co2geonet.com/UserFiles/file/ESF/10-5-Teir_Application-of-CCS-in-finland.pdf

Åbo Akademi University

Mineral resources; source-sink links

- Vammala Mg-silicate resources (~200 Mt rock)

↕ ~85 km

- Meri-Pori power plant (2.5 Mt CO_2/y)

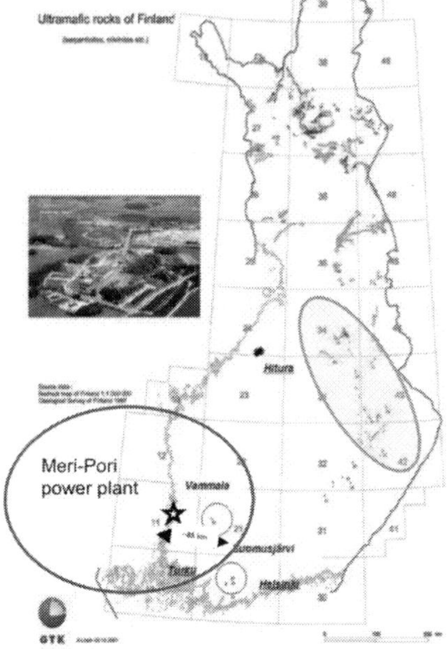

Åbo Akademi University

Process concepts

- Mg: serpentinite rock carbonation
  - *"ÅA route": $Mg(OH)_2$ production from rock + carbonation in a gas/solid reaction*
  - *Alternatively: $MgSO_4$ production from rock + carbonation in aqueous solution*
  - *Large-scale application; direct operation on flue gas*
- Ca: steelmaking slag carbonation
  - *Slag2PCC concept*
  - *See also afternoon presentations by Said and Teir*
- Ca: phosphogypsum carbonation
  - *Yesterday: presentation by Mattila*
- Conclusions

Åbo Akademi University

Serpentinite rock carbonation

Åbo Akademi University

Mg-silicate carbonation: the ÅA route

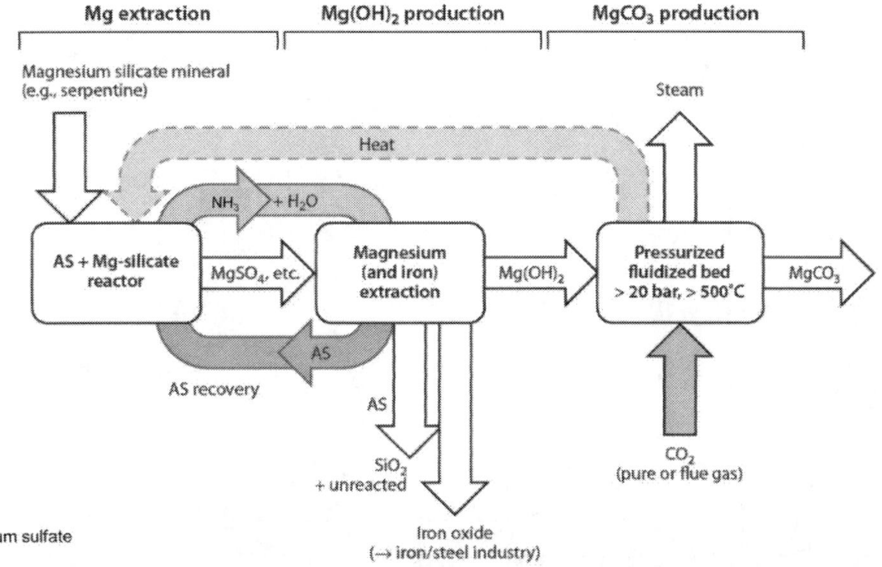

AS = ammonium sulfate

Åbo Akademi University

PFB carbonation of $Mg(OH)_2$

T: up to ~ 600°C
p_{CO2}: up to 80 bar

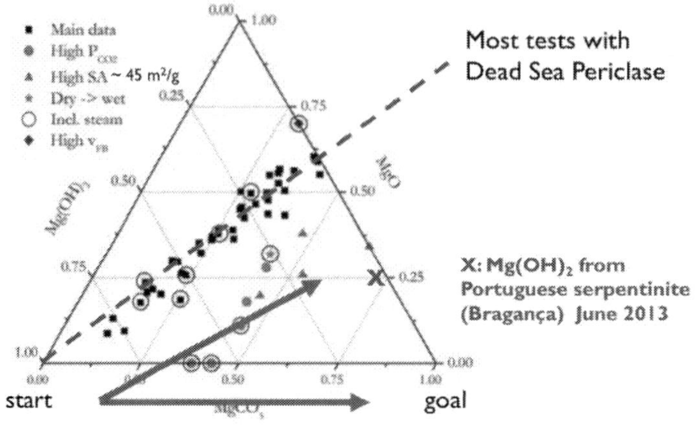

- Competition between de-hydroxylation and carbonation

Original Figure Fagerlund, dr. thesis ÅA 2012

Åbo Akademi University

Progress towards >90% conversion

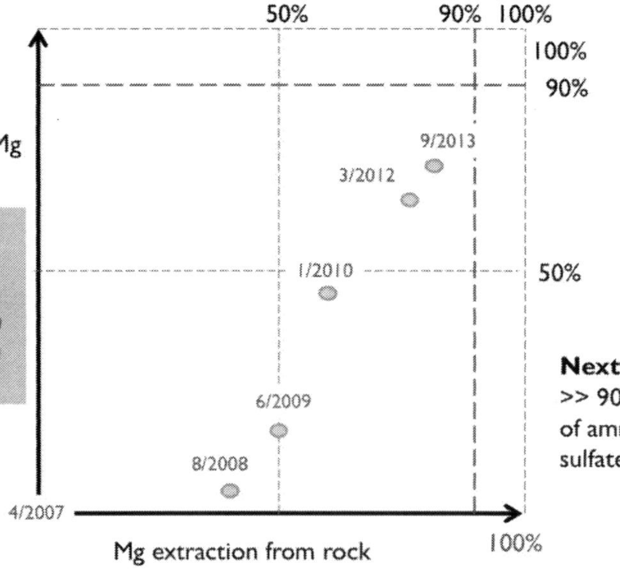

Since summer 2013:
- *Focus on alternative (wet) route for ($MgSO_4$) carbonation*
- *Processing a diopside – type material*

Next challenge:
>> 90% recovery of ammonium sulfate salt

Åbo Akademi University

Serpentinite rock carbonation: alternative route

Åbo Akademi University

Comparison of the routes

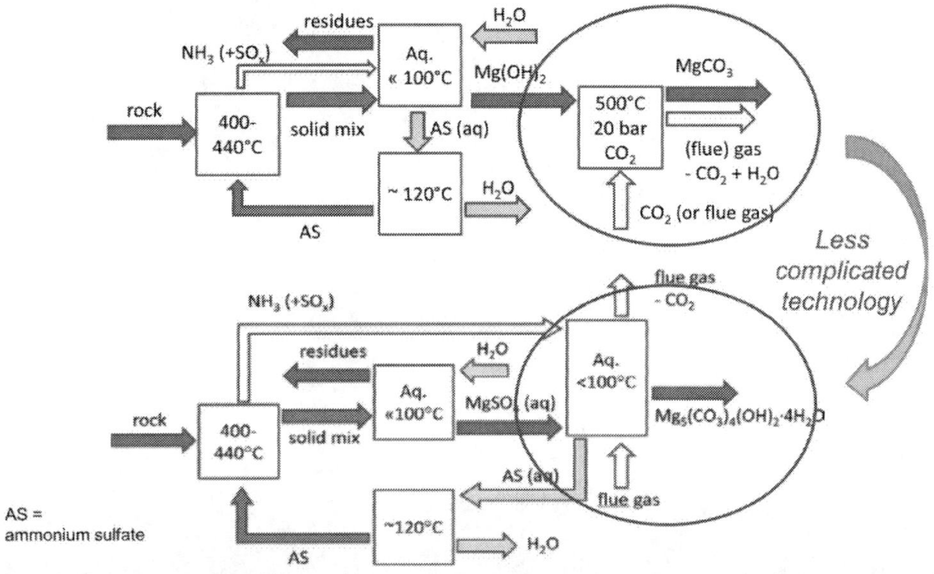

Alternative route: $MgSO_4$ carbonation in aqueous solution

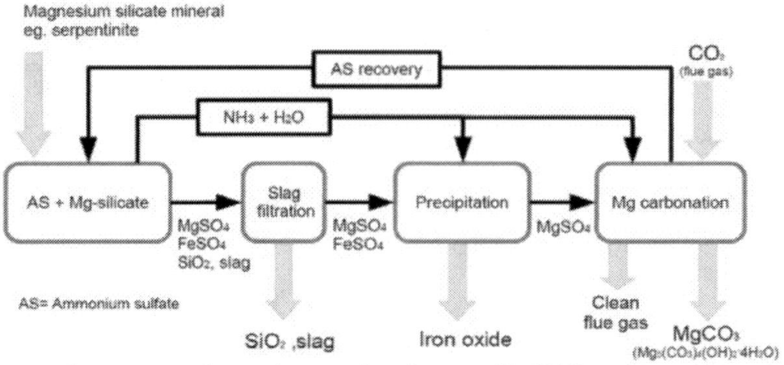

+ : *No need to separately produce and carbonate $Mg(OH)_2$*

+ : *Still, no separate CO_2 capture stage*

- : *Carbonation heat cannot be recovered*

- : *Product is hydromagnesite ~ only 80% of Mg carbonated*

Application at a lime kiln ~ 200 kg CO_2 / h

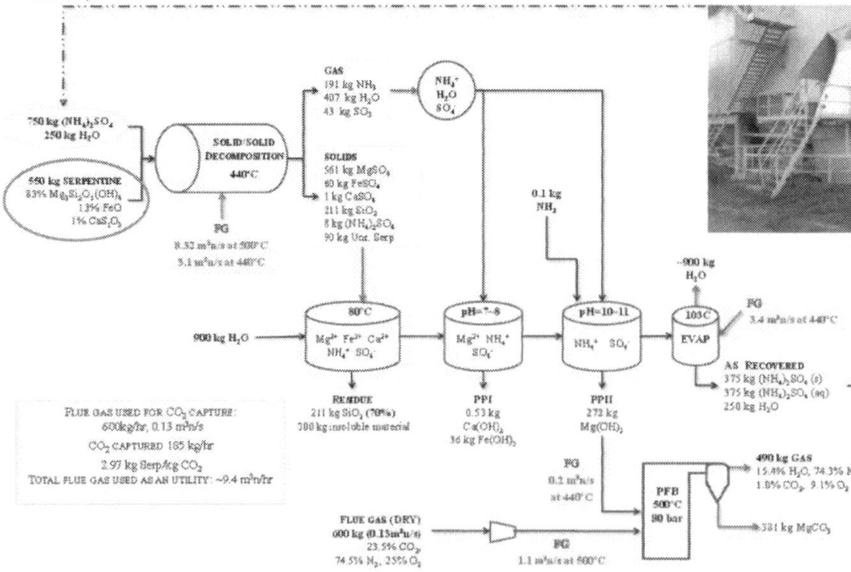

Challenge: Diopside as raw material

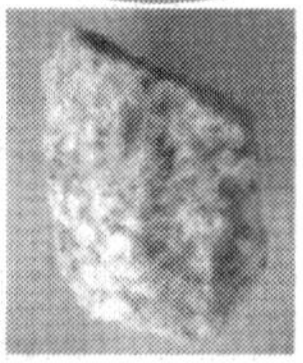

Application at a lime kiln ~ 200 kg CO_2 / h

Integration at an industrial lime kiln,
No separate CO_2 capture,
full flue gas compression
to 80 bar ~ 20 bar CO_2,
0.2~0.25 t CO_2/h fixed,
heat (kiln gas) 2.6 GJ/t, power 0.9 GJ/t CO_2

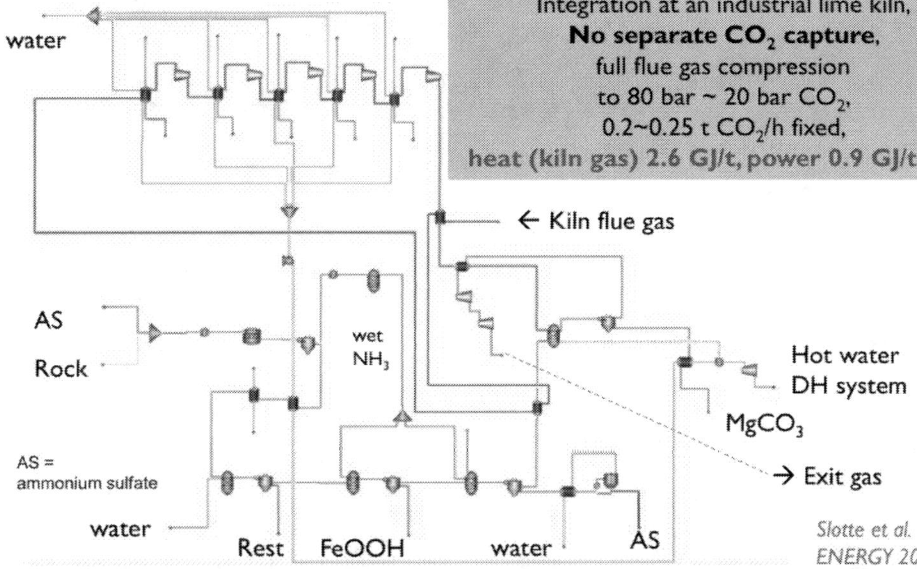

Slotte et al.
ENERGY 2013

Production of precipitated calcium carbonate from steelmaking slags (Slag2PCC)

Åbo Akademi University

Steelmaking

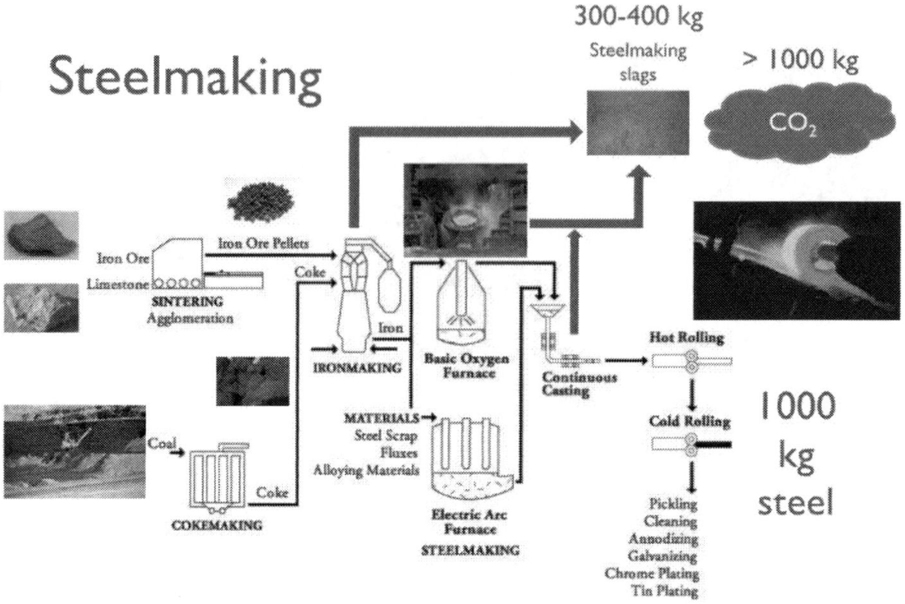

Pictures from Wikipedia &
http://images.businessweek.com/ss/08/12/1211_most_influential/18.htm

Åbo Akademi University

The Slag2PCC concept

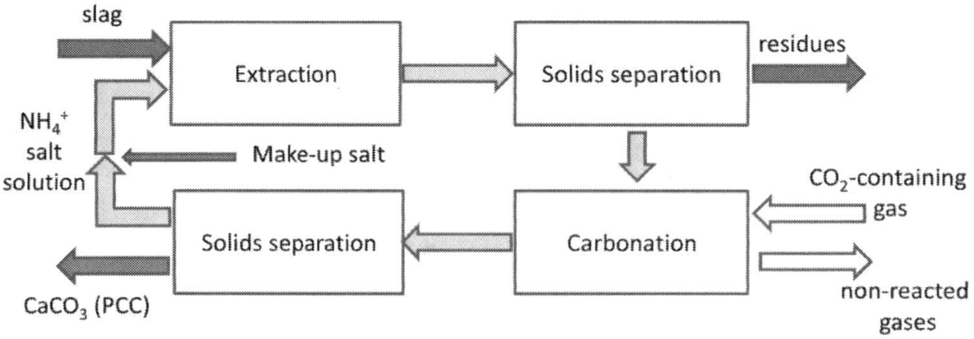

$$CaO(s) + 2NH_4X(aq) + H_2O(l) \rightarrow CaX_2(aq) + 2NH_4OH(aq) \quad (R1)$$

$$2NH_4OH(aq) + CO_2(g) \leftrightarrow (NH_4)_2CO_3(aq) + H_2O(l) \quad (R2)$$

$$(NH_4)_2CO_3(aq) + CaX_2(aq) \leftrightarrow CaCO_3(s) + 2NH_4X(aq) \quad (R3)$$

"X" in reactions (R1)–(R3) represents CH_3COO^-, NO_3^- or Cl^- ions, (acetate, nitrate or chloride)

The Slag2PCC concept

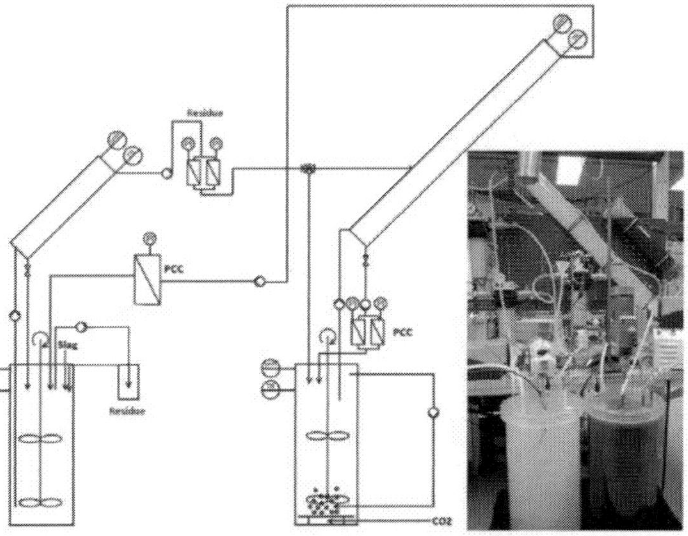

Precipitated Calcium Carbonate (PCC): product quality

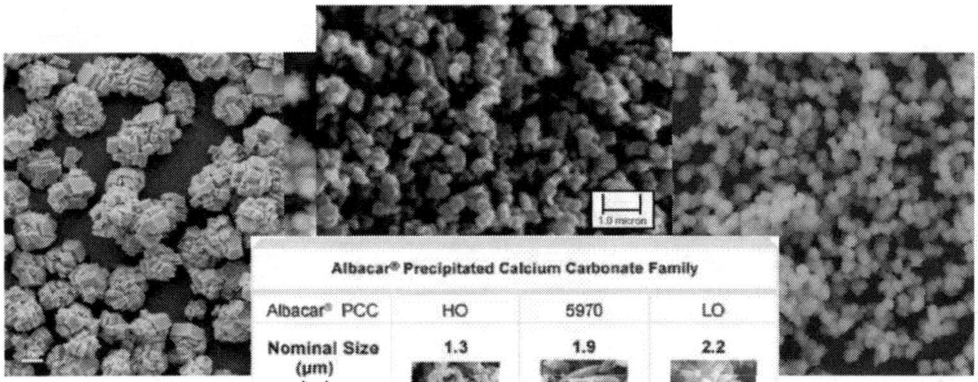

Albacar® Precipitated Calcium Carbonate Family

Albacar® PCC	HO	5970	LO
Nominal Size (µm) 1 µm	**1.3**	**1.9**	**2.2**
Surface Area (m²/g)	**12**	**7**	**6**
Papermakers Use It For...	**Optical Properties**	⟷	**Strength Caliper**

Pictures from HPM, Wikipedia & http://www.specialtyminerals.com/paper/products-and-applications-for-paper/paper-applications/albacar-pcc-family/

Precipitated Calcium Carbonate (PCC) manufacturing: Life cycle assessment (LCA) of the production methods

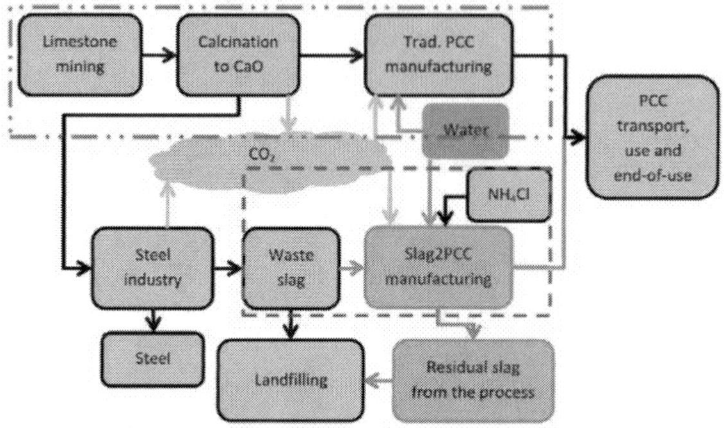

Picture from DOI: 10.1016/j.jclepro.2014.05.064

Precipitated Calcium Carbonate (PCC) manufacturing: Life cycle assessment (LCA) of the production methods

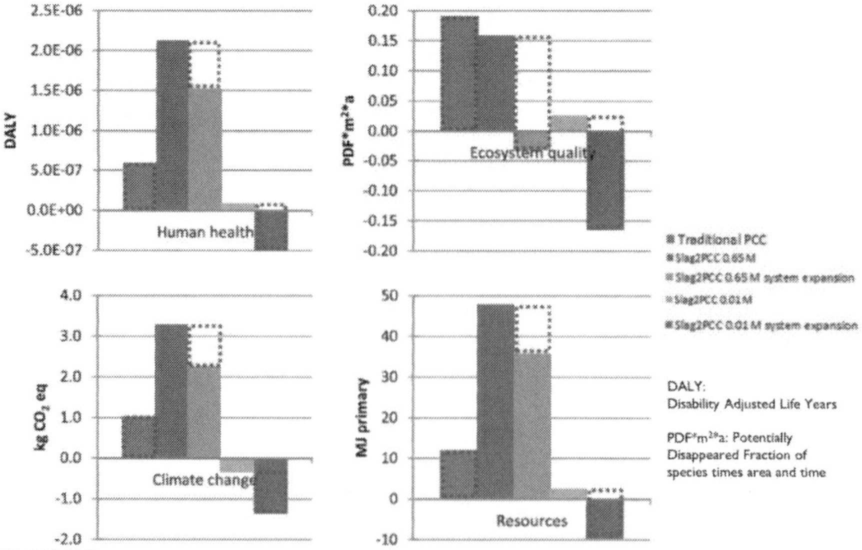

DALY:
Disability Adjusted Life Years

PDF*m^{2}*a: Potentially Disappeared Fraction of species times area and time

Picture from DOI: 10.1016/j.jclepro.2014.05.064

Summary: steel slag carbonation

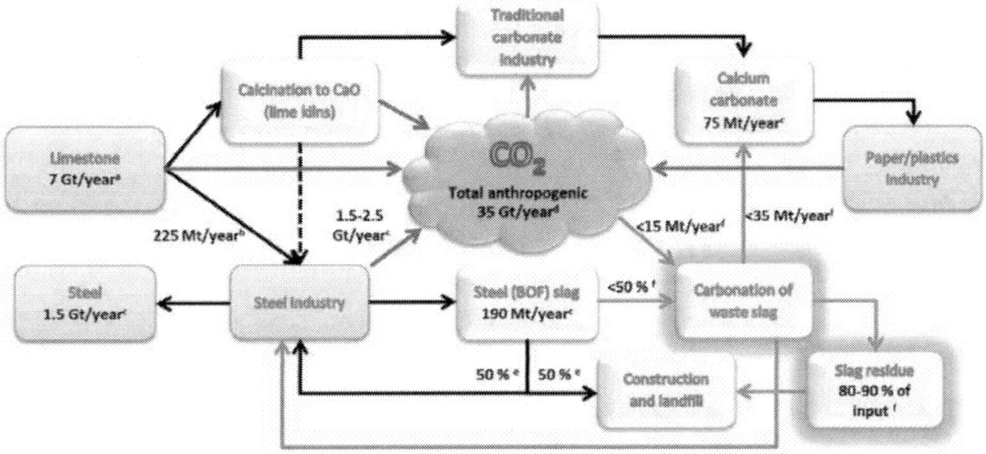

Picture from: http://urn.fi/URN:NBN:fi-fe201409244878

Steel slag carbonation: challenges

- Process economics and investment payback time (more at 3:45 PM)
- Further development regarding process equipment and verification of the product applicability for various purposes is required
- Utilization of slag residue, control of particle morphologies etc.

- The lack of large-scale references of mineral carbonation processes (more about pilot-scale demonstration at 2:15 PM)

General conclusions

- CO_2 mineral sequestration offers a leakage-free alternative for CO_2 underground storage worldwide; only option ***in*** Finland
- CO_2 capture from oxygen containing gases isn't "taking off", and capture is more expensive than economically viable CC(U)S!
  →no capture step, operate mineralization directly on flue gas
- ÅA staged process route: ~ 90 % Mg extraction from serpentinite + fast (~10 min) $Mg(OH)_2$ carbonation ~ 70%
- Energy input stand-alone ÅA route **~3 GJ/t ≈ CO_2 capture**
- Application at lime kiln, no separate CO_2 capture step:
  ~ 0.9 GJ/t CO_2 power in
- An alternative route with $MgSO_4$ aqueous carbonation appears simpler, though no heat recovery from carbonation
- Industrial waste – based processes provide important possibilities of feasibility demonstration for mineral carbonation, even if their total CO_2 sequestration potential is limited

Thermal and Flow Engineering Laboratory | Åbo Akademi University | Turku Finland 10.7.2015 25

Acknowledgements

ÅBO AKADEMI
Aalto University
Tekes
ACADEMY OF FINLAND
Institute of Chemical and Engineering Sciences
A*STAR
ccsp
Carbon Capture and Storage Program
SusEn
Sustainable Energy
GSCE
Nordkalk
Outotec
ruukki
PART OF SSAB

& support from

Educação e cultura
Leonardo da Vinci
Mobilidade
UNIVERSITY OF COIMBRA
UNIVERSITY OF CALGARY

Thermal and Flow Engineering Laboratory | Åbo Akademi University | Turku Finland 10.7.2015 26

Mineralization with capture integration

Mischa Werner
Institute of Process Engineering, ETH Zurich

ACEME15
NYC, USA, June 21-24, 2015

The origin of CO_2 Mineralization…

CO_2 disposal by means of silicates

SIR—If fossil fuels are to be used on a massive scale and the greenhouse effect avoided, carbon dioxide emissions into the atmosphere must be reduced. One way to achieve this is to collect the CO_2 and to dispose of it in empty natural gas fields or the deep oceans. But such a disposal system would need a substantial amount of energy. It would be advantageous if there were an abundant mineral to which the CO_2 could be bound chemically via an exothermic reaction to form a stable, permanent substance.

It is known that about 100 million tons of carbon per year are bound by silicate-weathering throughout the world. This natural process is of a scale that would consume the CO_2 inventory of the atmosphere in 7,000 years. There are other substances to which CO_2 can bind, but the weathering reaction of calcium silicate is slightly exothermic whereas most of the others are endothermic.

$$2CO_2 + H_2O + CaSiO_3 \rightarrow Ca^{2+} + 2HCO_3^- + SiO_2 - 29.6 \text{ kcal}$$

One could imagine a process by which pressurized CO_2 is pumped into a closed container containing a suspension of pulverized silicates in water. The resultant hard water and SiO_2 could be drained into the deep ocean and the process repeated, thus disposing of CO_2. This idea may not be practicable because the energy required to dig up and grind the calcium silicate may be too great and the kinetics of the above reaction may be too slow. Experimental work is necessary to determine whether this sort of scheme would be feasible in practice.

W. SEIFRITZ

Chapfstrasse 4,
CH-5200 Windisch,
Switzerland

NATURE · VOL 345 · 7 JUNE 1990

The origin of CO_2 Mineralization…

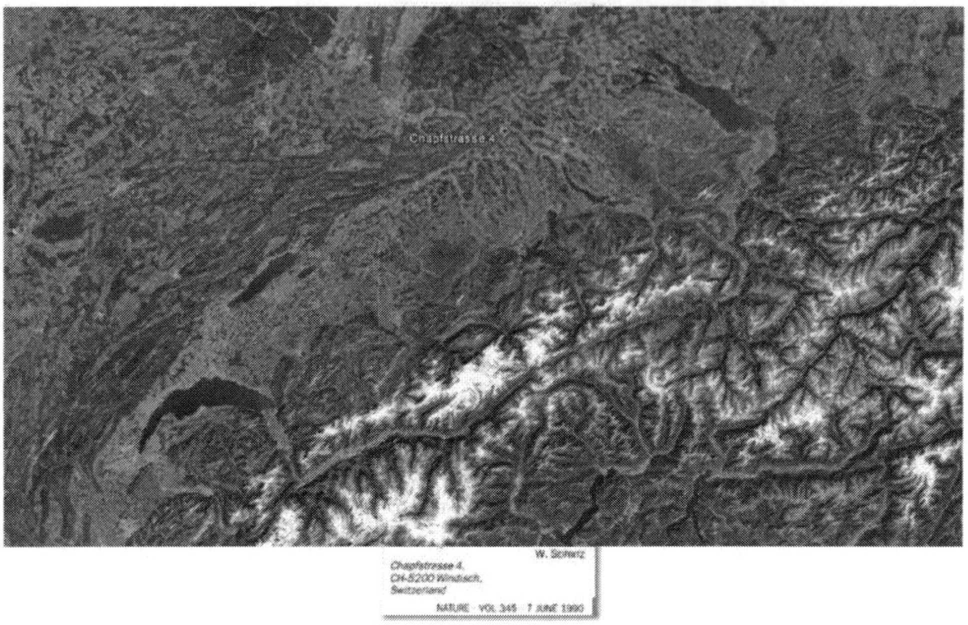

The early days of CO_2 Mineralization research at ETH

- Kinetic study of dissolution of San Carlos olivine, gemstone quality olivine, at high pCO_2

Standard CSTR setup

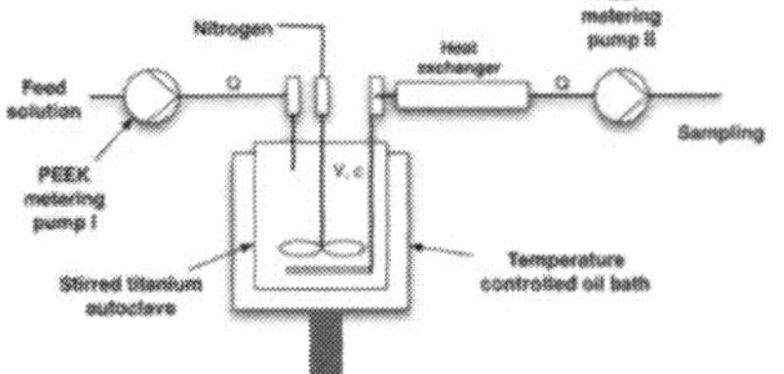

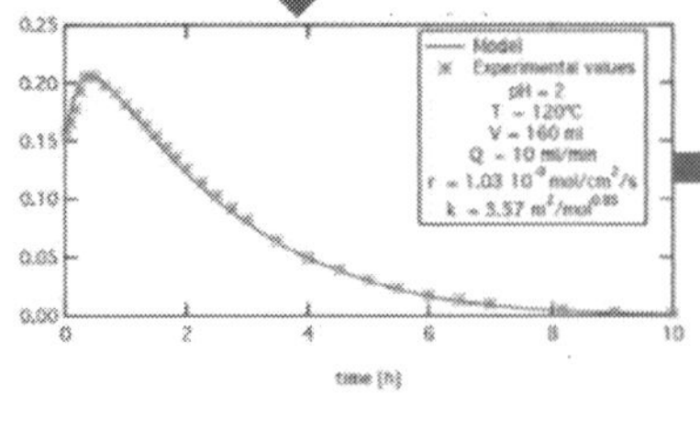

Why olivine?

Hänchen et al., *GHG Control Technologies* (2005)

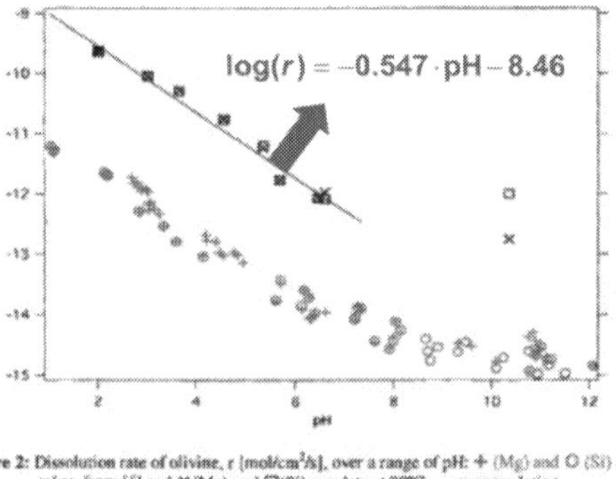

Figure 2: Dissolution rate of olivine, r [mol/cm²/s], over a range of pH: + (Mg) and ○ (Si) at 25°C taken from [5] and ×(Mg) and □(Si) our data at 90°C, — our correlation.

ETH zürich Mischa Werner | Mineralization with capture integration | werner@ipe.mavt.ethz.ch | 5

Pioneering work at ARC

INTERIM REPORT

AQUEOUS MINERAL CARBONATION

Mineral Availability, Pretreatment, Reaction Parametrics, And Process Studies

September 30, 2004

Table IX. Optimum carbonation conditions, by mineral

Mineral	Carbonation conditions		
	T, °C	P_{CO_2}, atm	Carrier solution
Olivine	185	150	0.64 *M* $NaHCO_3$, 1 *M* NaCl
Wollastonite	100	40	Distilled water
HT serpentine	155	115	0.64 *M* $NaHCO_3$, 1 *M* NaCl

Feed Material History				Pretreatment Energy Consumption, kW•h/ton								
						Grinding						
Ore mineral & grade	Pretreatment methodology	R_{CO_2}	R_X, %	Crush.	Bene.	Stage 1	Stage 2	Stage 3	Heat treatment	Total	Ore/conc., kt/day	Pct total plant energy
Olivine, 100%	Ball mill (-200 mesh)	1.8	16	2		11				13	286	15
	Ball mill (-400 mesh)	1.8	61	2		11	70			83	75	26
	SMD mill	1.8	81	2		11	70	150		233	56	55
Serpentine, 100% (lizardite)	Ball mill (-200 mesh)	2.5	9	2		11				13	706	37
	Heat treatment (-200)	2.5	40	2		11			326	339	158	222
Olivine, 70%	Ball mill (-200 mesh)	1.8	16	2	2	15				19	286	22
	Ball mill (-400 mesh)	1.8	61	2	2	15	70			89	75	28
	SMD mill	1.8	81	2	2	15	70	150		239	56	56
Serpentine, 100% (antigorite)	Ball mill (-200 mesh)	2.1	12	2		11				13	445	24
	Heat treatment (-200)	2.1	62	2		11			293	306	86	110
	Heat treatment (-400)	2.1	92	2		11	70		293	376	58	90
Wollastonite, 50%	Ball mill (-400 mesh)	2.8	43	2	4	21	70			97	165	67
	SMD mill	2.8	82	2	4	21	70	70		167	87	61

experimental carbonation data. The basis for this calculation was a 1 GW coal-fired power plant, burning coal with an as-fired heat value of 12,500 Btu/lb and carbon content of 74%, at 35% plant efficiency.

Why olivine?

ETH zürich Mischa Werner | Mineralization with capture integration | werner@ipe.mavt.ethz.ch | 6

Pioneering work at ARC

$$Mg_3Si_2O_5(OH)_4 \xrightarrow{heat} 1.5\ Mg_2SiO_4 + 0.5\ SiO_2 + 2\ H_2O$$

- **Isothermal, 2 hours at 630°C → full dehydroxylation?!?**
- **Sensible heat: 206 kWh/ton$_{sh}$** c_p = 89 cal/mol/K, King et al., 1967
- **Latent heat: 120 kWh/ton$_{sh}$** Govier,Arnold, 2004, ARC internal rep.

326 kWh/ton$_{sh}$ ➡ **$$$/ton$_{sh}$** Electricity as heat source, no heat recovery

inclusion of a heat-treatment step. However, thermal activation of serpentine, while extremely effective, results in more CO_2 generated than is sequestered by the process, making it an impractical methodology. These

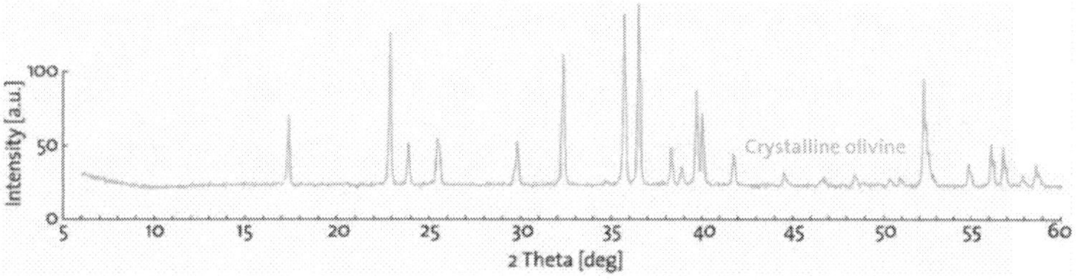

Heat treating (HT) serpentine: room for improvement

$Mg_3Si_2O_5(OH)_4 \xrightarrow{heat} Mg_3Si_2O_{6.5}(OH) + 1.5\ H_2O$

- Only partial dehydroxylation for maximum reactivity
- Sensible heat: 206 kWh/ton$_{sh}$ c_p = 89 cal/mol/K, King et al., 1967
- Latent heat: 120 kWh/ton$_{sh}$ Govier, Arnold, 2004, ARC internal rep.

326 kWh/ton$_{sh}$ ➡ \$\$\$/ton$_{sh}$ electricity as heat source, no heat recovery

- **Balucan et al.'s optimal HT strategy: 143 kWh/ton$_{sh}$** ➡ **1.15 A\$/ton$_{sh}$** nat. gas as heat source, accounting for secondary emissions

Pro-grade heating, 20% residual hydroxyls, 80% heat recovered.
Balucan et al., 2013, Int J GHG control (reported value: 570 MJ/t)

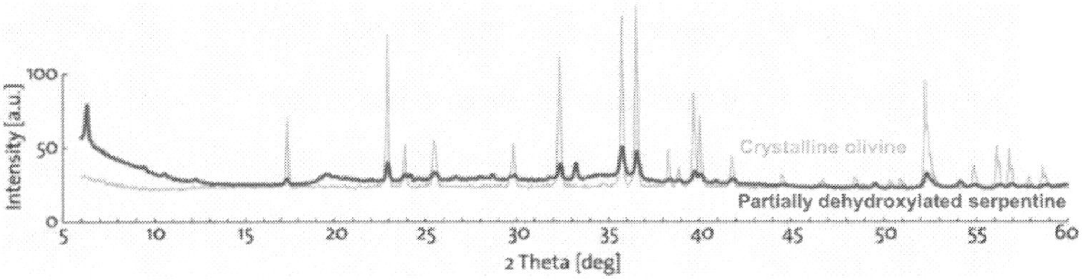

Heat treating (HT) serpentine: room for improvement

$Mg_3Si_2O_5(OH)_4 \xrightarrow{heat} Mg_3Si_2O_{6.5}(OH) + 1.5\ H_2O$

- Integrate the HT step, e.g. like in Shell's 2007 patent:

(19) World Intellectual Property Organization
International Bureau
PCT
(43) International Publication Date
31 May 2007 (31.05.2007)
(10) International Publication Number
WO 2007/060149 A1

- **Use** syngas from coal gasification **(e.g. CTL plant) for HT**
- **Cooled syngas and water is subjected to** water-gas-shift **conversion**
- **The CO_2 obtained from the water-gas-shift reaction is contacted with the HT serpentine**

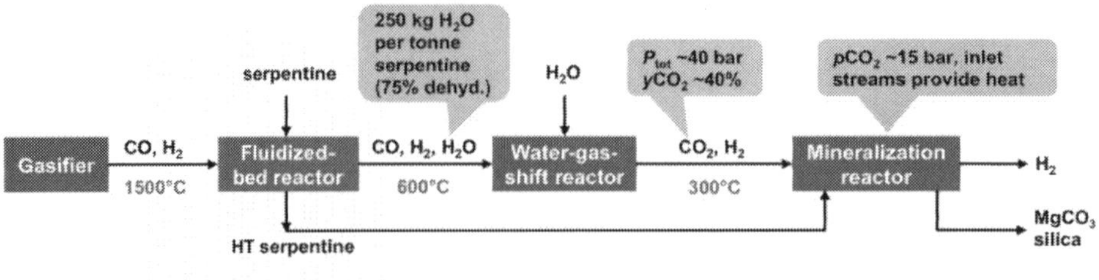

Heat treating (HT) serpentine: room for improvement

$$Mg_3Si_2O_5(OH)_4 \xrightarrow{\text{heat}} Mg_3Si_2O_{6.5}(OH) + 1.5\ H_2O$$

- Integrate the HT step, e.g. by exploiting thermal energy storage (TES) cycles in a concentrated-solar power plant: **(just a thought)**

Day:

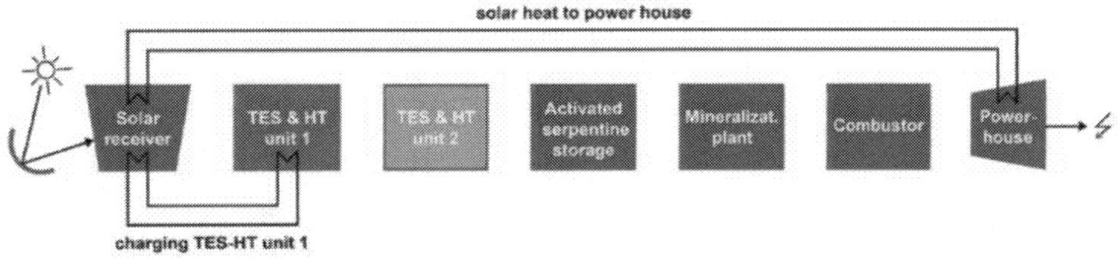

Heat treating (HT) serpentine: room for improvement

$$Mg_3Si_2O_5(OH)_4 \xrightarrow{\text{heat}} Mg_3Si_2O_{6.5}(OH) + 1.5\ H_2O$$

- Integrate the HT step, e.g. by exploiting thermal energy storage (TES) cycles in a concentrated-solar power plant: **(just a thought)**

Evening:

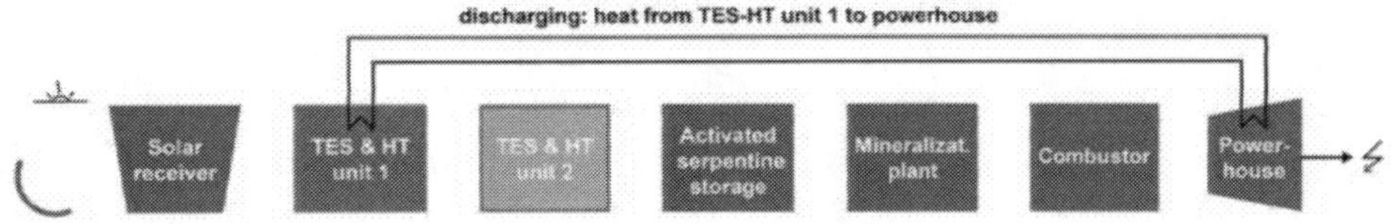

ETH zürich   Mischa Werner | Mineralization with capture integration | werner@ipe.mavt.ethz.ch | 11

Heat treating (HT) serpentine: room for improvement

$Mg_3Si_2O_5(OH)_4 \xrightarrow{heat} Mg_3Si_2O_{6.5}(OH) + 1.5\ H_2O$

- Integrate the HT step, e.g. by exploiting thermal energy storage (TES) cycles in a concentrated-solar power plant:

(just a thought)

Night:

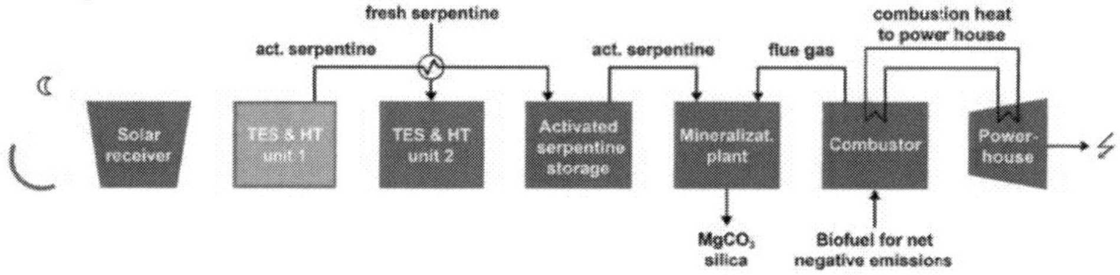

ETH zürich   Mischa Werner | Mineralization with capture integration | werner@ipe.mavt.ethz.ch | 12

Heat treating (HT) serpentine: room for improvement

$Mg_3Si_2O_5(OH)_4 \xrightarrow{heat} Mg_3Si_2O_{6.5}(OH) + 1.5\ H_2O$

- Integrate the HT step, e.g. by exploiting thermal energy storage (TES) cycles in a concentrated-solar power plant:

(just a thought)

Next day:

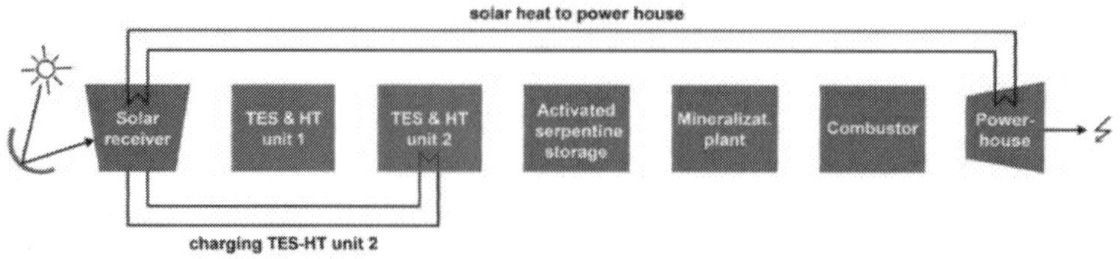

Heat treating (HT) serpentine: room for improvement

$$Mg_3Si_2O_5(OH)_4 \xrightarrow{heat} Mg_3Si_2O_{6.5}(OH) + 1.5\ H_2O$$

- Integrate the HT step, e.g. by exploiting thermal energy storage (TES) cycles in a concentrated-solar power plant:

(just a thought)

Applied Thermal Engineering

High-temperature thermal storage using a packed bed of rocks – Heat analysis and experimental validation

Markus Hänchen, Sarah Brückner, Aldo Stein...

DEVELOPMENT OF A NOVEL CONCEPT OF SOLAR RECEIVER/THERMAL ENERGY STORAGE SYSTEM BASED ON COMPARTMENTED DENSE GAS FLUIDIZED BEDS

R. Chirone[a], P. Salatino[b*], P. Ammendola[a], R. Solimene[a], M. Magaldi[c], R. Sorrenti[c], G. De Michele[d], F. Donatini[d]

[a]Istituto di Ricerche sulla Combustione - Consiglio Nazionale delle Ricerche; ... Napoli Federico II

International Conference on Concentrating Solar Power and Chemical Energy Systems, SolarPACES 2014

Using sand and other small grained materials as heat storage medium in a packed bed HTTESS

D. Schlipf[a*], P. Schicktanz[a], H. Maier[a], G. Schneider[a]

ScienceDire...

Parametric analysis of a high temperature packed bed thermal storage design for a solar gas turbine

P. Klein[a,b,*], T.H. Roos[b], T.J. Sheer[a]

Design of packed bed thermal energy storage systems for high-temperature industrial process heat

G. Zanganeh[a], A. Pedretti[b], A. Haselbacher[a,*], A. Steinfeld[a,c]

Back to Shell

(19) World Intellectual Property Organization
International Bureau

PCT

(43) International Publication Date
26 January 2006 (26.01.2006)

(10) International Publication Number
WO 2006/008242 A1

(51) International Patent Classification[7]: C01F 5/24, 11/18

(21) International Application Number: PCT/EP2005/053258

(81) Designated States *(unless otherwise indicated, for every kind of national protection available)*: AE, AG, AL, AM, AT, AU, AZ, BA, BB, BG, BR, BW, BY, BZ, CA, CH, CN, CO, CR, CU, CZ, DE, DK, DM, DZ, EC, EE, EG, ES, FI,

„An aqueous slurry is contacted with CO_2 containing gas…"

„Preferably, the CO_2 containing gas is an industrial off-gas, for example an industrial flue gas."

(71) Applicant *(for all designated States except US)*: **SHELL INTERNATIONALE RESEARCH MAATSCHAPPIJ B.V.** [NL/NL]; Carel van Bylandtlaan 30, NL-2596 HR The Hague (NL).

(72) Inventors; and

(75) Inventors/Applicants *(for US only)*: **GEERLINGS, Jacobus Johannes Cornelis** [NL/NL]; Badhuisweg 3, NL-1031 CM The Hague (NL). **VAN MOSSEL, Gerardus Antonius Franciscus** [NL/NL]; Badhuisweg 3, NL-1031 CM Amsterdam (NL). **IN 'T VEEN, Bernardus Cornelis Maria** [NL/NL]; Badhuisweg 3, NL-1031 CM Amsterdam (NL).

GM, KE, LS, MW, MZ, NA, SD, SL, SZ, TZ, UG, ZM, ZW), Eurasian (AM, AZ, BY, KG, KZ, MD, RU, TJ, TM), European (AT, BE, BG, CH, CY, CZ, DE, DK, EE, ES, FI, FR, GB, GR, HU, IE, IS, IT, LT, LU, LV, MC, NL, PL, PT, RO, SE, SI, SK, TR), OAPI (BF, BJ, CF, CG, CI, CM, GA, GN, GQ, GW, ML, MR, NE, SN, TD, TG).

Published:
— *with international search report*

For two-letter codes and other abbreviations, refer to the "Guidance Notes on Codes and Abbreviations" appearing at the beginning of each regular issue of the PCT Gazette.

(54) Title: PROCESS FOR PRODUCING CACO3 OR MGCO3

Mineralization, a business-case?

... ONLY IF: a separate capture step is avoided **... AND started from the** most abundant precursor

- **Low pCO_2 carbonation seems to work for gas-solid route, starting from alkaline residues, e.g. Baciocchi et al.**

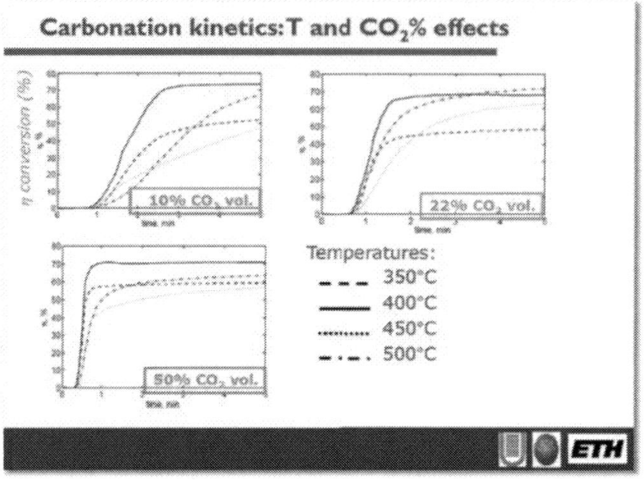

"High-temperature Gas-Solid Carbonation of Air Pollution Control Residues"
Baciocchi et al., 2008, ACEME08, Rome, Oct 08

Mineralization, a business-case?

... ONLY IF: a separate capture step is avoided **... AND started from the** most abundant precursor

- **Low pCO_2 carbonation seems to work for gas-solid route, starting from alkaline residues, e.g. Baciocchi et al.**
- **But alkaline residues ≠ abundant, e.g Hänchen et al. 2013 (Internal report for ALSTOM)**
- **theoretical CO_2 sequestration potential using** ALL residues produced worldwide = ~2% of current global emissions

	Source material properties			
	Global resources estimates [Mt]	Production [Mt/yr]	$R(CO_2)$	CO_2 sequestration potential [Mt]
Natural Minerals				
Serpentine (ultramafic rock)	2,900,000 - 43,000,000[a]	-	2	1,450,000 - 21,500,000[a]
Olivine	-[b]	4-8 [155],c	1.8	-[b]
Wollastonite	90 - 270[199]	0.6[199]	2.64	34-102
Waste Materials				annually [Mt/yr]
Steel Slag	-	383[d]	2.45	157[d]
Fly Ash	-	808[d]	11.9	68.2[d]
CKD	-	32.8[d]	8.08	4.06[d]

[a] Includes olivine (dunite), data taken from Table 2.2.

"Mineral Carbonation Assessment"
Hänchen et al., 2013, internal report on behalf of ALSTOM SCHWEIZ (AG)

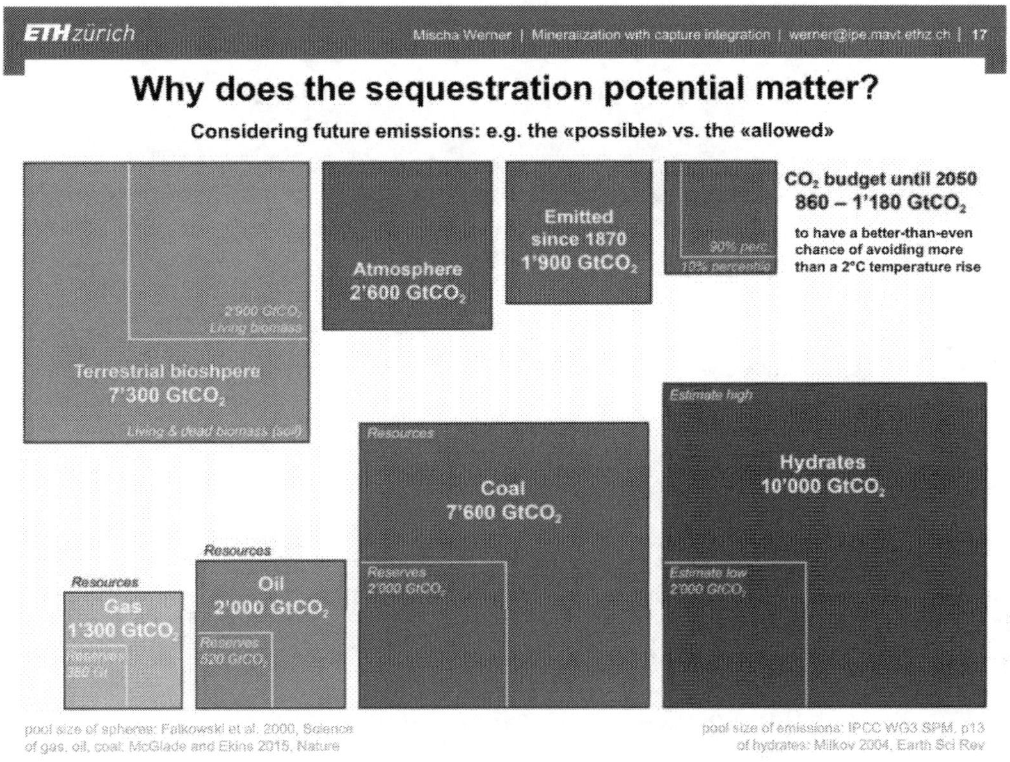
ETH zürich
Mischa Werner | Mineralization with capture integration | werner@ipe.mavt.ethz.ch | 17
Why does the sequestration potential matter?
Considering future emissions: e.g. the «possible» vs. the «allowed»
Terrestrial bioshpere
7'300 GtCO2
2'900 GtCO2
Living biomass
Living & dead biomass (soil)
Atmosphere
2'600 GtCO2
Emitted
since 1870
1'900 GtCO2
90% perc.
10% percentile
CO2 budget until 2050
860 – 1'180 GtCO2
to have a better-than-even chance of avoiding more than a 2°C temperature rise
Resources
Gas
1'300 GtCO2
Resources
Oil
2'000 GtCO2
Reserves
520 GtCO2
Resources
Coal
7'600 GtCO2
Reserves
2'000 GtCO2
Estimate high
Hydrates
10'000 GtCO2
Estimate low
2'000 GtCO2
pool size of spheres: Falkowski et al. 2000, Science
of gas, oil, coal: McGlade and Ekins 2015, Nature
pool size of emissions: IPCC WG3 SPM, p13
of hydrates: Milkov 2004, Earth Sci Rev

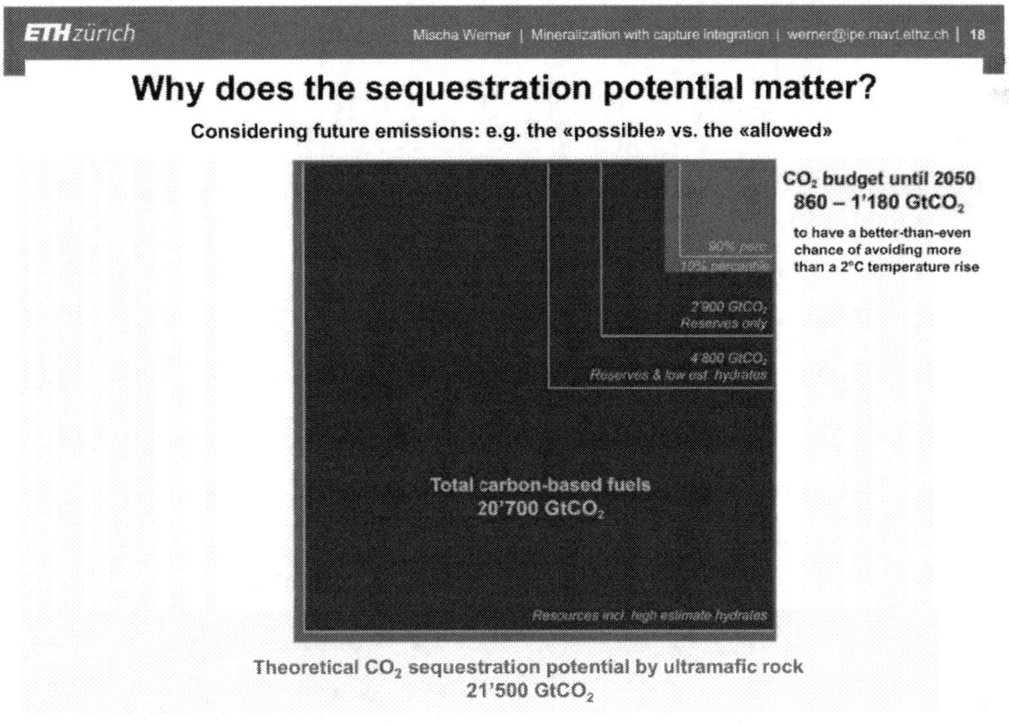
ETH zürich
Mischa Werner | Mineralization with capture integration | werner@ipe.mavt.ethz.ch | 18
Why does the sequestration potential matter?
Considering future emissions: e.g. the «possible» vs. the «allowed»
CO2 budget until 2050
860 – 1'180 GtCO2
to have a better-than-even chance of avoiding more than a 2°C temperature rise
90% perc.
10% percentile
2'900 GtCO2
Reserves only
4'800 GtCO2
Reserves & low est. hydrates
Total carbon-based fuels
20'700 GtCO2
Resources incl. high estimate hydrates
Theoretical CO2 sequestration potential by ultramafic rock
21'500 GtCO2

Why does the sequestration potential matter?

Considering future emissions: e.g. from coal fired power

- **World's yearly capacity build up of coal plants:**
  - **1990-99:** 33 GW per year
  - **2000-09:** 66 GW per year
  - **2010-12:** 89 GW per year
- **With all the coal plants built by the year 2012, we are committed to:**
  - 192 Gt CO_2 **if plants run for** 30 y **on average**
  - 307 Gt CO_2 **if plants run for** 40 y **on average**
  - 578 Gt CO_2 **if plants run for** 60 y **on average**

➡ Compare this to the 860-1180 GtCO_2 budget to stay below 2°C warming (IPCC WG II, AR5)

➡ Carbon lock-in, and it keeps increasing

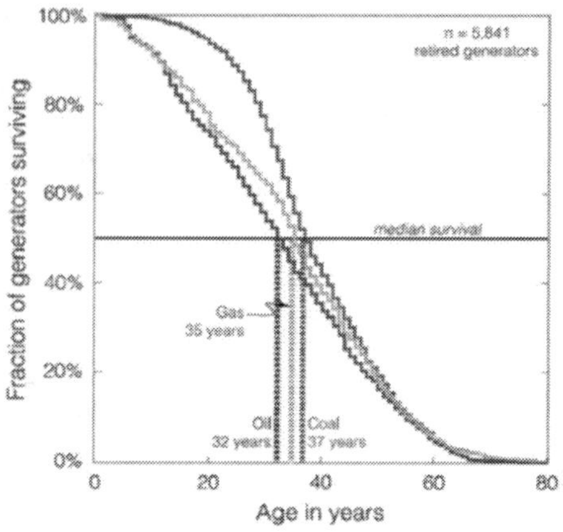

Historical survival of generators by fuel type (power sector only)

Davies & Socolow, 2014, Environ. Res. Letters

Why does the sequestration potential matter?

Considering future emissions: e.g. from coal fired power

- **World's yearly capacity build up of coal plants:**
  - **1990-99:** 33 GW per year
  - **2000-09:** 66 GW per year
  - **2010-12:** 89 GW per year
- **With all the coal plants built by the year 2012, we are committed to:**
  - 192 Gt CO_2 **if plants run for** 30 y **on average**
  - 307 Gt CO_2 **if plants run for** 40 y **on average**
  - 578 Gt CO_2 **if plants run for** 60 y **on average**

➡ Compare this to the 860-1180 GtCO_2 budget to stay below 2°C warming (IPCC WG II, AR5)

➡ Carbon lock-in, and it keeps increasing

➡ Hence, why not thinking «large-scale» from the start?

➡ Then, serpentine becomes the precursor material of choice

- **In 2012: 350 coal plants at «some stage of development in China»** World Resource Institute
- 560 GW **new-built capacity**

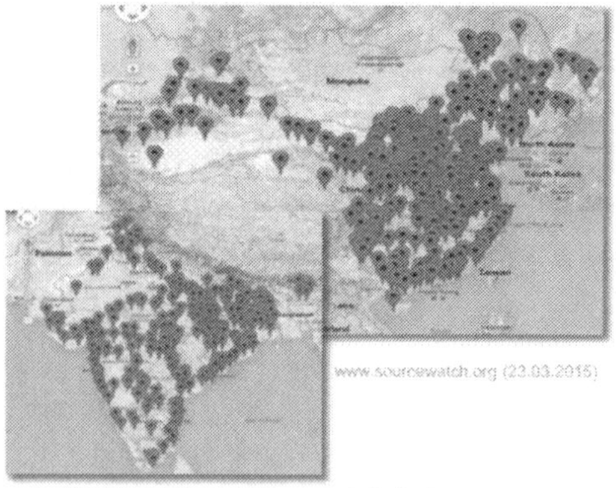

www.sourcewatch.org (23.03.2015)

- **In 2011: 554 proposed projects in India** Prayas Energy Group
- 620 GW **new capacity**
- **Compared to US in 2011:** 320 GW **total coal capacity**

Change of focus at ETH: use serpentine, integrate capture

- Two options: 1) afford the heat treatment OR 2) afford the recovery of enabling additives

- Use Shell's activated serpentine
- "Put a camera on these particles" to understand their dissolution mechanism at flue gas conditions
- Called for a substantial upgrade of our setup
  - Liquid and gas flow-through at variable *P* and *T*
  - Zero-tolerance to changes in liquid level
  - Stand-alone operation
  - Avoid initial «blind-spot» typical for pressurized CSTR experiments

Standard CSTR setup for $P_{tot} > 1$ atm

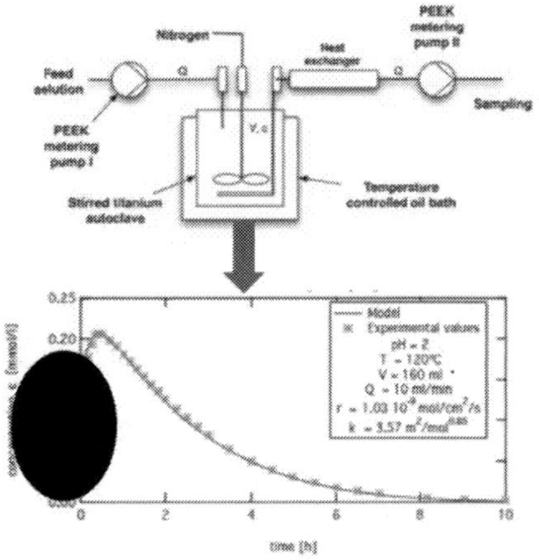

"Putting a camera on these particles"

1. Flow sheet design 2. Layouting 3. Construction 4. Automation

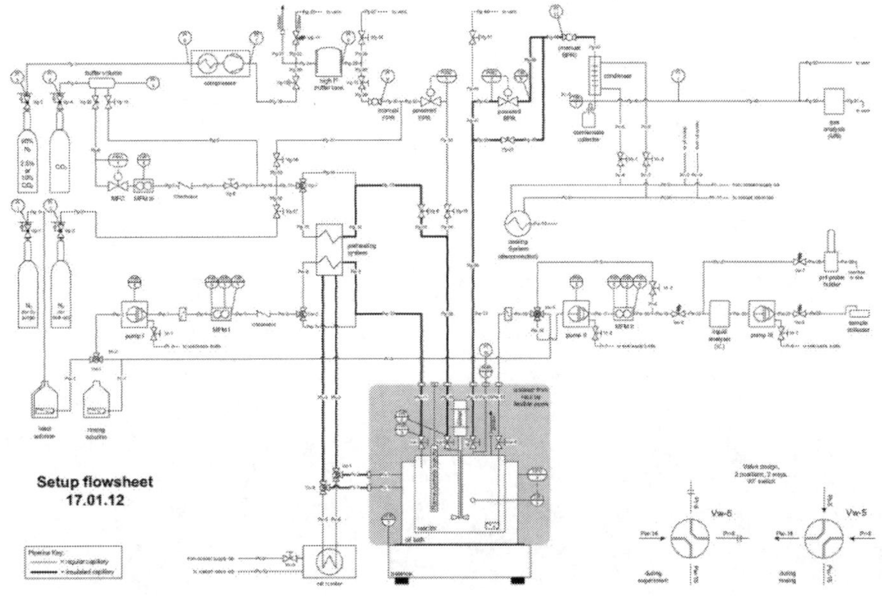

ETH zürich — Mischa Werner | Mineralization with capture integration | werner@ipe.mavt.ethz.ch | 23

"Putting a camera on these particles"

1. Flow sheet design **2. Layouting** 3. Construction 4. Automation

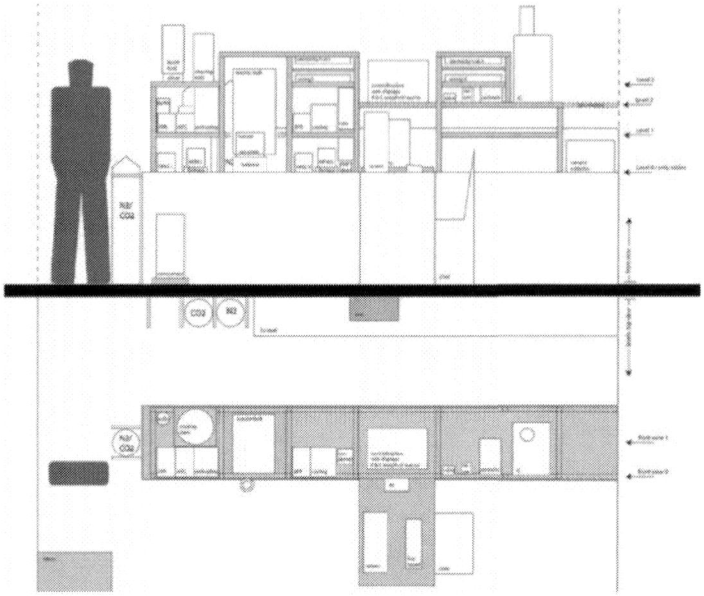

ETH zürich — Mischa Werner | Mineralization with capture integration | werner@ipe.mavt.ethz.ch | 24

"Putting a camera on these particles"

1. Flow sheet design 2. Layouting **3. Construction** 4. Automation

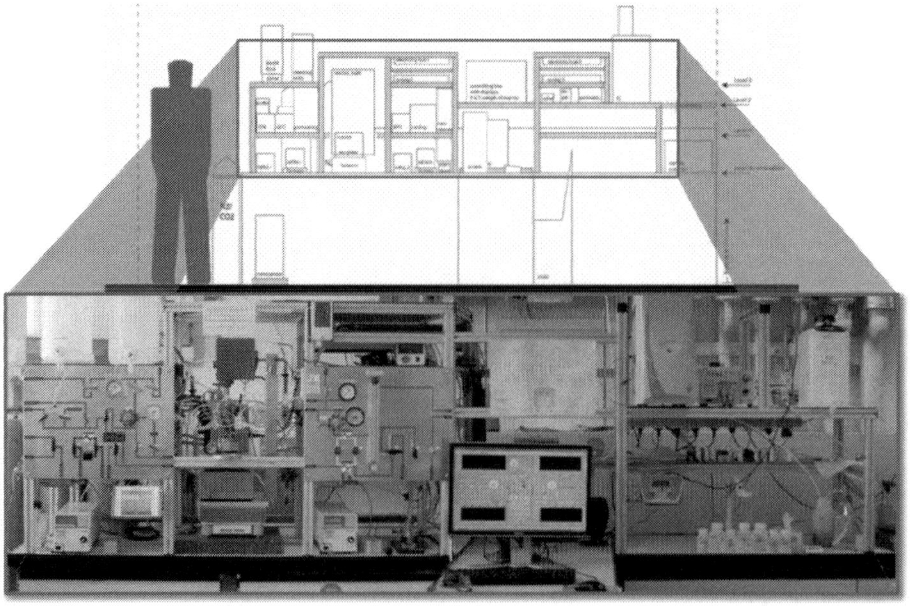

"Putting a camera on these particles"

1. Flow sheet design 2. Layouting 3. Construction **4. Automation**

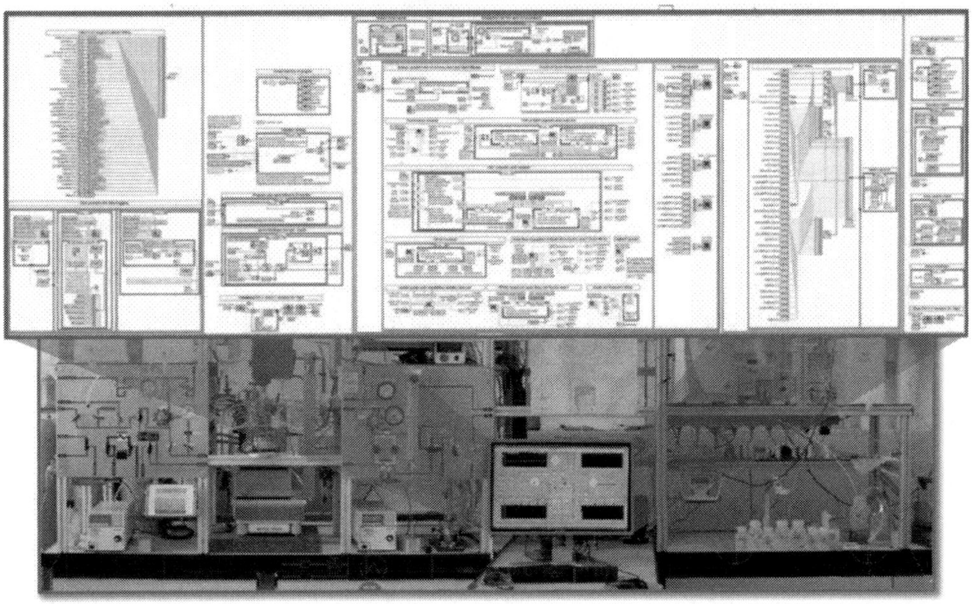

"Putting a camera on these particles"

Avoiding the «blind spot»: Solid sample injector

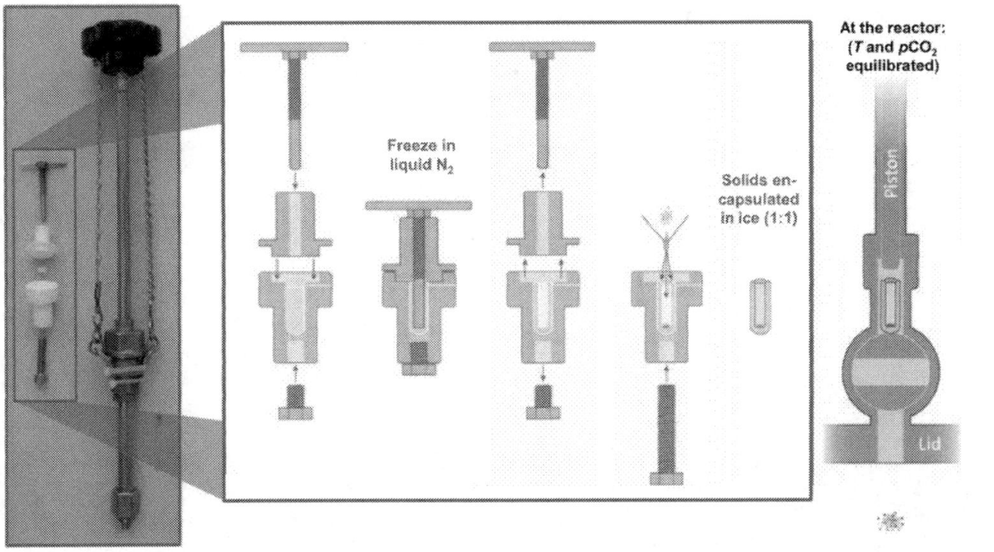

▶▶ Fast forward

- **Dissolution of dehydroxylated lizardite at flue gas conditions: I. Experimental study** Chem. Eng. J. *241*, 301-313
- **Dissolution of dehydroxylated lizardite at flue gas conditions: II. Kinetic modeling** Chem. Eng. J. *241*, 314-326
- **Dissolution of dehydroxylated lizardite at flue gas conditions: III. Near equilibrium dynamics** in preparation

➡ **check near equilibrium highlights on our POSTER**

- **Flue gas CO_2 mineralization using thermally activated serpentine. From single- to double-step carbonation** PhysChemChemPhys *16*, 24978-24993

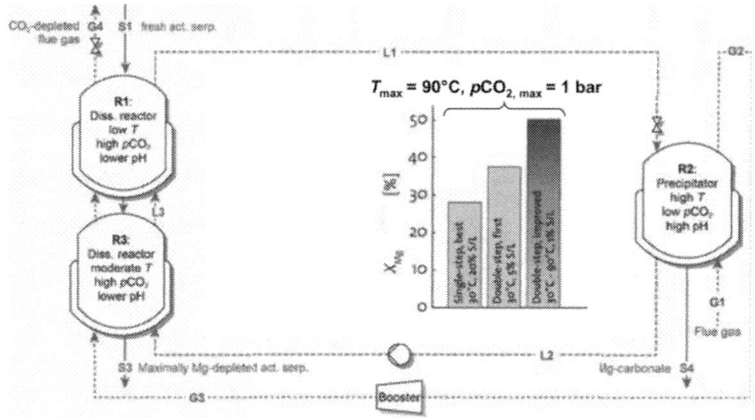

▶▶ Fast forward

- **Dissolution of dehydroxylated lizardite at flue gas conditions: I. Experimental study** Chem. Eng. J. *241*, 301-313
- **Dissolution of dehydroxylated lizardite at flue gas conditions: II. Kinetic modeling** Chem. Eng. J. *241*, 314-326
- **Dissolution of dehydroxylated lizardite at flue gas conditions: III. Near equilibrium dynamics** in preparation

➡ **check near equilibrium highlights on our POSTER**

- **Flue gas CO_2 mineralization using thermally activated serpentine. From single- to double-step carbonation** PhysChemChemPhys *16*, 24978-24993

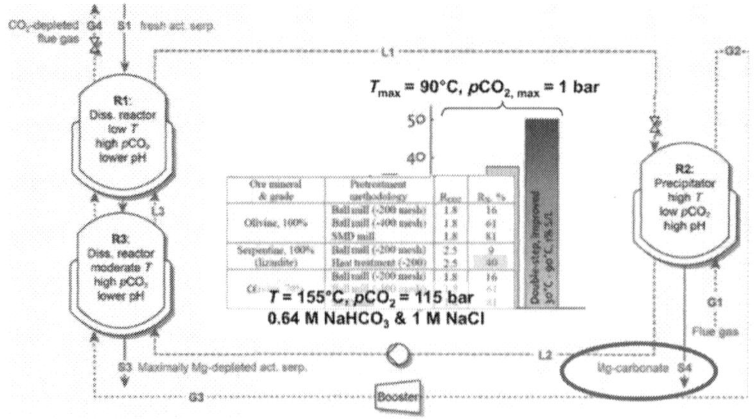

ETH zürich Mischa Werner | Mineralization with capture integration | werner@ipe.mavt.ethz.ch | 29

Fast forward

- **Dissolution of dehydroxylated lizardite at flue gas conditions: I. Experimental study** Chem. Eng. J. *241*, 301-13
- **Dissolution of dehydroxylated lizardite at flue gas conditions: II. Kinetic modeling** Chem. Eng. J. *241*, 314-26
- **Dissolution of dehydroxylated lizardite at flue gas conditions: III. Near equilibrium dynamics** in preparation

➡ check near equilibrium highlights on our POSTER

- **Flue gas CO_2 mineralization using thermally activated serpentine. From single- to double-step carbonation** PhysChemChemPhys 16, 24978-24993
- **Precipitation kinetics of Mg carbonates at 90°C** Victor Amado-Blanco: Semester thesis, 2015

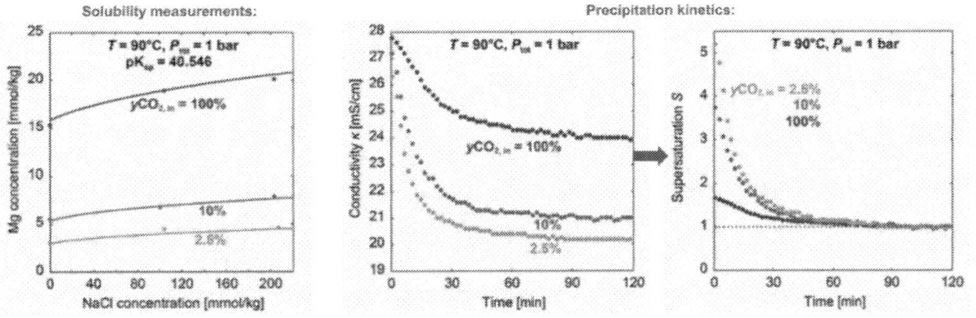

ETH zürich Mischa Werner | Mineralization with capture integration | werner@ipe.mavt.ethz.ch | 30

Fast forward

- **Dissolution of dehydroxylated lizardite at flue gas conditions: I. Experimental study** Chem. Eng. J. *241*, 301-13
- **Dissolution of dehydroxylated lizardite at flue gas conditions: II. Kinetic modeling** Chem. Eng. J. *241*, 314-26
- **Dissolution of dehydroxylated lizardite at flue gas conditions: III. Near equilibrium dynamics** in preparation

➡ check near equilibrium highlights on our POSTER

- **Flue gas CO_2 mineralization using thermally activated serpentine. From single- to double-step carbonation** PhysChemChemPhys 16, 24978-24993
- **Precipitation kinetics of Mg carbonates at 90°C** Victor Amado-Blanco: Semester thesis, 2015
- **Double-step process for flue gas CO_2 mineralization** Clemens Leopold: Master's thesis, ongoing

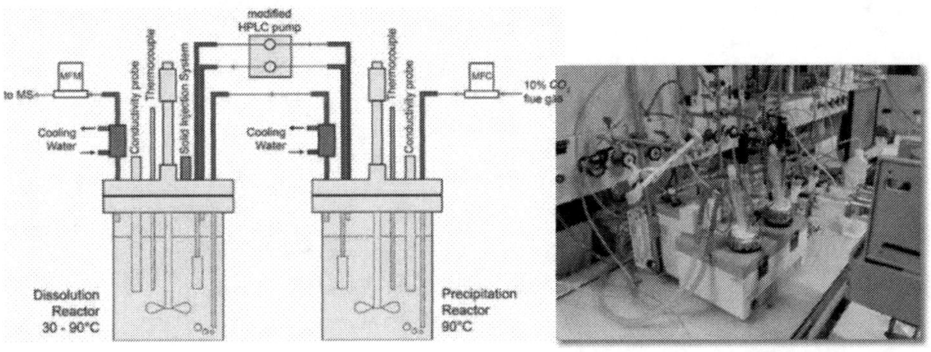

▶ Play: The way ahead

The process – as envisioned – offers many degrees-of-freedom for optimization:

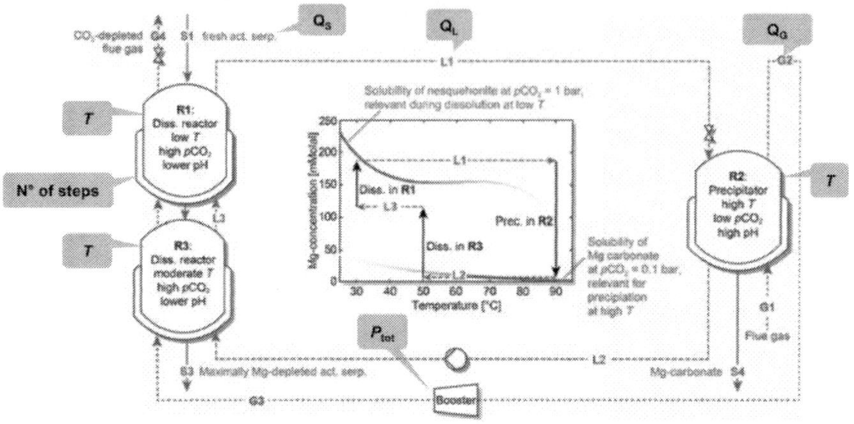

- **Our results so far have convinced us that this design space offers promising opportunities**
- **Our kinetic models will allow us to explore these opportunities by simulations**
- **Our experience in the lab will allow us to turn the best opportunities into reality**

Thanks for your attention!

The process – as envisioned – offers many degrees-of-freedom for optimization:

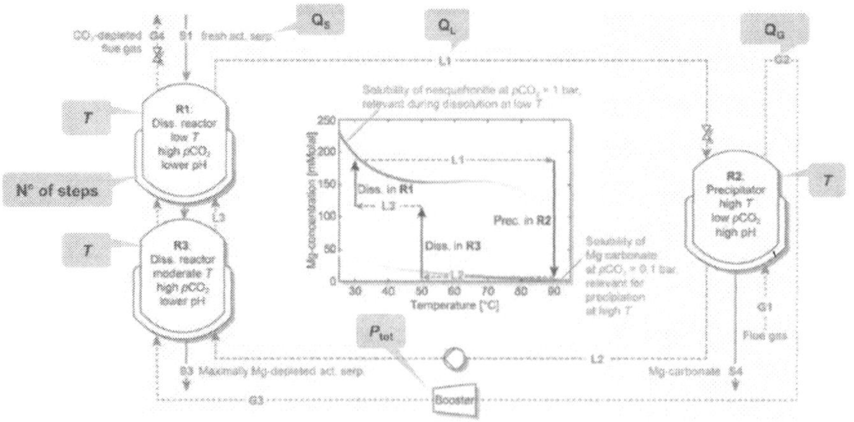

- **Our results so far have convinced us that this design space offers promising opportunities**
- **Our kinetic models will allow us to explore these opportunities by simulations**
- **Our experience in the lab will allow us to turn the best opportunities into reality**

Carbon Dioxide Sequestration in the Staten Island Serpentinite

Nothaft, D., Earth and Environmental Sciences
Kelemen, P., Columbia University

The Intergovernmental Panel on Climate Change has suggested carbon dioxide capture and storage (CCS) as a method for mitigating climate change resulting from anthropogenic greenhouse gas emissions. The Staten Island Serpentinite has potential for permanent carbon storage via *in situ* mineral carbonation. This study proposes to dissolve CO_2 sourced from the Fresh Kills Park landfill gas processing plant into groundwater pumped from a saline aquifer below Staten Island. This CO_2-brine solution would be pumped back into the aquifer at a depth of 500m, where the partial pressure of CO_2 and temperature were calculated to be 50 bars and 40°C, respectively. Assuming a grain size of 7mm, this study estimated a carbonation rate of 1.04×10^9 tons CO_2 yr^{-1} for the formation as a whole. This was shown to greatly surpass the CO_2 emissions from the landfill gas processing plant. This study also measured prior natural carbonation, which is an important factor in storage capacity. Carbon was found to comprise <0.1% of the serpentinite matrix, and ~10% of carbonate veins, indicating that veins are nearly pure carbonate and that the rest of the rock contains ample metal cations available for carbonation reactions. The natural rate of mineral carbonation was gauged by radiocarbon dating carbonate veins within the serpentine matrix. Two younger ages of 1,270±25 yr and 3,420±25 yr indicate that the Serpentinite is actively carbonating, while three older ages of ~40,000 years are near the upper end of ^{14}C age resolution, and could be the result of minor amounts of solution and redeposition rather than the age of initial formation of carbonate. If a CCS project proves feasible, it may serve as a model for innovative urban CCS and waste management initiatives.

AUTHOR INDEX

AUTHOR INDEX